AF664467

ENVIRONMENTAL AWARENESS

ENVIRONMENTAL AWARENESS

Arvind Kumar

2024

Daya Publishing House®

A Division of

Astral International Pvt. Ltd.

New Delhi – 110 002

First Published, 2008

Reprinted, 2024

ISBN: 978-81-7035-540-3 (HB)

Published by : **Daya Publishing House®**
A Division of
Astral International Pvt. Ltd.
– ISO 9001:2015 Certified Company –
4736/23, Ansari Road, Darya Ganj,
New Delhi-110 002
Ph. 011-43549197, 23278134
E-mail: info@astralint.com
Website: www.astralint.com

Printed at : **Replika Press Pvt. Ltd.**

Preface

Earth Day in 1970 brought forth a new era in environmental consciousness. Over the last several decades, humans have made significant improvements in restoring some of the Earth's fragile environment. In the United States, the return of pollution-sensitive fish to many freshwater ecosystems, the cleaner air over some urban areas, and the rise from near extinction of the California gray whale and bald eagle are several examples of environmental "progress" since the first Earth Day. However, the world's environmental record leaves ample room for improvement. Approximately half of all species are likely to become extinct by the end of the twenty-first century. The principal underlying causes of extinction are the expansion of the human population, trophy hunting economic harvesting, deforestation, wetland drainage, urbanization, agricultural clearing, and pollutants. Humans are changing the environment and destroying natural habitats too rapidly for most species to adapt.

In addition, environmental disasters, such as the Bhopal gas leak in India (1984), the discovery of a growing hole in the stratospheric ozone layer (1985), the explosion of the Chernobyl nuclear reactor in the former Soviet Union (1986) and the predicted global warming during twenty-first century have forced people to recognize that human actions have destructive influences on the environment. If this trend of environmental disaster continues, our coming generation will not forgive us. Therefore, proper environmental management and mass awakening right from the grass root level to top ecoplanners are the crying need of time, only then our long cherished dream of establishing eco-socialism on this watery planet cold come true.

This book entitled "*Environmental Awareness*" is the unique compilation of the most recent research articles of internationally acknowledged experts in the field of Environmental sciences with the intention of providing a sufficient depth of the subject to satisfy the needs at a level which will be comprehensive and interesting. This book will be useful to the students, research scholars, scientists in the field of Environmental management and ecoplanners, politicians and other people with similar interest.

I consider myself extremely fortunate in having got the invaluable support and erudite suggestions of eminent persons like Prof. (Dr.) Christian Ulrichs of Germany, Dr. M. A. Kabir Chowdhury of Malaysia, Prof. Dr. Ahmed H. Al-Harbi of Saudi Arabia, Prof. Tej Kumar Shrestha of Nepal, Prof. P. V. Dehadrai, Former Director, CIFRI, Dr. Dilip Kumar, Director, CIFE, Mumbai, Prof. K. C. Pandey, Former V. C. (Lucknow), Prof. A. R. Yousuf of University of Kashmir, Srinagar (J & K), Professor N. C. Datta of Calcutta University, Professor S. K. Konar of Kalyani University, Professor D. K. Belsare of Bhopal University, Professor U.C. Goswami of Gauhati University, Professor P. C. Mishra of Sambalpur University, Professor P. Natarajan of Kerala University, Professor P. S. Murthy of Bangalore University, Professor Ajit Varma of JNU, Professor A. L. Bhatia of Jaipur University, Professor A. K. Mittal of BHU, Professor S. P. Hosmani of Mysore University, Professor K. Kapoor of Udaipur University, Professor K. B. Reddy of Nagarjuna University, Professor M. Vikram Reddy of Pondicherry University, Professor B. K. Tiwari of NEHU, Professor G. Tripathi of Jodhpur University, Professor R. Ramalingam of Annamalai University, Professor Sharif U. Ahmad of Nagaland University, Professor G. K. Kulkarni of Aurangabad University, Professor S. U. Mehram of Nagpur University, Professor S. K. Battish of PAU, Professor B. M. Sharma of Manipur University, Professor B. D. Joshi of Hardwar University, Professor G. C. Pandey of Faizabad University, Professor K. C. Sharma of Ajmer University, Professor M. Raziuddin of Hazaribag University, Professor U. S. Bagde of Mumbai University, Professor Gurdeep Singh of I. S. M., Dhanbad, Dr. P. K. Goel of Karad, Professor S. P. Roy and Shri Tribhuwan Poddar of Bhagalpur University, Bhagalpur for encouragements.

I also express my deep sense of gratitude to my parents whose blessings have always prompted me to pursue academic activities deeply. I am also thankful to my wife, *Kumari Bimla* and my two lovely sons, *Kumar Pallav Shivshankaran* and *Kumar Prasun Ramakrishnan* whose natural smiles extended to me relief all through this tiresome endeavour.

Last but not the least, I am also thankful to Mr. Anil Mittal, Proprietor, Daya Publishing House, New Delhi for taking keen interest in bringing out of this book. Finally, I will always remain a debtor to all my well-wishers for their blessings, without which this book would not have come into existence.

Professor A. Kumar

Contents

Preface *v*

1. **Awareness of Environmental Impact Assessment of Power Sector Reform in India** 1
   P.C. Mishra and Arvind Kumar

2. **Assessment of Knowledge of Students at Primary Level Towards Environmental Concepts and Environmental Problems** 17
   R. Vijaya and Anandhavalli Mahadevan

3. **Vehicular Pollution Abatement and Control Measures Implemented in Delhi** 24
   D.S. Kharat, Paritosh Kumar and P.M. Ansari

4. **Review of Market Potential for Two-wheelers Using Time Series Analysis** 39
   V. Joseph Paul Raj

5. **Positional Survey of the Groundwater Quality of the Bharatpur Area during the Monsoon Season 2006** 46
   Deepshika Garg, Sunita Goyal, Samiksha Chaturvedi and R.V. Singh

6. **Modelling Tree-Grass Interactions in Semi-Arid and Temperate Silvopastoral Systems** 53
   T.A. Khan, W.R. Eason, G. Suresh and R.K. Bhatt

7. **Phytoplankton Productivity Assessment of Kadinamkulam Lake with Special Reference to the Retting Activity, Thiruvananthapuram, Southern Kerala** 62
   V. Sobha, Anuradha Rammohan, S. Santhosh and J. Chandra Thara

8. **The Impact of Industrial Effluents and Sewage on the Water Quality of River Godavari at Bhadrachalam: A Temple Town in South India** **70**
B. Sesha Sreenivas, V.V. Ramana, K. Ramesh and M.A. Singara Charya

9. **Influence of Yeast Extract on Aerobic Biodegradation of 2-Chlorophenol and 4-Chlorophenol** **82**
J.B. Chawla and A.O. Ingle

10. **Biodiversity of Blue-Green Algae from Paddy Sugar Cane and Banana Fields of Karnataka State, India** **97**
B.S. Giriyappanavar

11. **Composition of Vermicast of *Eisenia fetida* Savigny (1826) (Lumbricidae), *Metaphire posthuma* Vaillant (1868), *Lampito mauritii* Kinberg (1866) (Megascolecidae) and *Eutyphoeus waltoni* Michaelsen (1907) (Octochaetidae): A Comparative Study** **104**
Neena Dhiman

12. **Larvicidal and Ovicidal Activity of the Leaf Extract of Two Cucurbitacious Plants Against Filarial Vector, *Culex quinquefasciatus* Say** **111**
K. Mullai and A. Jebanesan

13. **Kinetics Study of Auto-Gasification of Bagasse for Energy Augmentation** **117**
V. Sivaramakrishnan, V. Kirubakaran, M. Premalatha and P. Subramanian

14. **Influence of Diet Composition and Starvation on the Growth of the Fingerlings of the Grass Carp, *Ctenopharyngodon idella*** **128**
R. Sivakami, R. Ramya, M. Suganya, M. Marieeswari and G. Premkishore

15. **Comparison of Energy Interaction Parameters for the Complexation of Pr(III) with 1,10-Phenanthroline and 2,2′-Bipyridyl in Different Organic Solvents using 4f-4f Transition Spectra as Probe: An Absorption Spectral Study** **134**
N. Yaiphabaa, Th. David Singh, Ch. Sumitra, M. Indira Devi, N. Rajmuhon Singh

16. **Sequential Procedures for the Class of Exponential Family of Distributions** **145**
C.P.S. Hungund

17. **A Survey of Readymade Chilli Powders for Aflatoxin B_1 Contamination** **152**
K. Aruna Lakshmi, M.V.V. Chandana Lakshmi and V. Sridevi

18. **Effect of Weed Control Treatments on Nutrient Removal by Weeds in Direct Sown Wet Seeded Rice**
B. Rajendra Kumar and A. Christophrer Lourduraj

19. **Qualitative Evaluation of Antimicrobial Efficacy of Novel N-Heterocyclic Compounds** **160**
Megha S. Kulkarni, Arti S. Shanware and Jyotsna S. Meshram

20. **Impact of Tsunami on Physico-chemical Characteristics of Coastal Water in Kulachel Area of Kanyakumari District** **171**
V. Ramasubramanian, R. Jeyaprakash and S. Jeyakumar

21. A Brassinosteroid as an Antihyperglycaemic in Alloxan Induced Diabetic Rats 176
P. Muthuraman and K. Srikumar

22. Production of Calcium Gluconate by Fermentation 184
Sanjay Timande and Anshu Chopda

23. Neem Derivatives as a Larvicide for the Management of Country Bean Pest *Pericallia ricini* F. (Lepidoptera : Arctidae) 195
N. Revathi and S. Kingsley

24. Impact of Aqueous Litter Leachate of *Eucalyptus* on Germination and Seedling Growth of Red Gram Seed 199
P.K. Gantayat, B. Padhy, M. Behera and S.K. Padhy

25. Weed Flora, Relative Density and Summed Dominance Ratio of Weeds, as Influenced by Weed Management Practices in Direct Sown Wet Seeded Rice 207
B. Rajendra Kumar and A. Christophrer Lourduraj

26. Effect of Sugar Mill Effluent on Biochemical Changes in the Liver of Freshwater Fish *Cirrinus mrigala* 211
K. Shanthi and V. Dhanalakshmi

27. Studies on Physico-chemical and Heavy Metal Accumulation in Selected Sites along the Coastal Belt of Kerala 216
K. Murugan

28. *In vitro* Plant Regeneration from Different Explants of *Centella asiatica* (L.) 221
Surjeet S. Bisht, Snehlata Bhandari and N.S. Bisht

29. Influence of Weather Factors on Dynamics of Bud Fly Incidence on Three Common Varieties of Linseed 228
Rabindra Prasad, Uday Kumar Prasad and Devendra Prasad

30. Integrated Ecofriendly Management of Solid Wastes of Pulp Industry: Review 232
P. Jothimani, A. Bhaskaran and S. Vidhya

31. Mycoflora of the Soil Sediment of Two Contrasting Lentic Systems, Tiruchirappalli, Tamil Nadu 241
R. Sivakami, R. Ramya, V. Guru, M. Suganya, and G. Premkishore

32. Estimation of Economic Feasibility of Integrated Management of Sucking Insect Pests of Brinjal Through Inter-cropping and Insecticides 246
Uday Kumar Prasad and Rabindra Prasad

33. Succession and Status of Insect Pests Infesting Linseed 252
Rabindra Prasad, Uday Kumar Prasad and Devendra Prasad

34. Water Quality Monitoring in Respect to Physico-chemical Characteristics of a Tropical Lake of Udaipur City of Rajasthan 256
Rajesh Kumar and Kamal Kapoor

35. Variations in Different Biochemical Parameters of Commercial Pearl Millet [*Pennisetum glaucum* (L.) R.Br.] Hybrids Under Irrigated and Terminal Water Stress Conditions 264

Harshal E. Patil

36. Life-Form Analysis and Biological Spectrum of the Macrophytes of the Laisoipat Lake, Manipur 270

Laishram Kamala Devi and B. Manihar Sharma

37. Integrated Nutrient Management System for Productivity Potential for Sorghum-Wheat Sequence Under Irrigated Conditions 278

H.M. Patil and A.G. Wani

38. Comparison of Energy Interaction Parameters for the Complexation of Pr(III) with Guanosine in Absence and Presence of Zn(II) in Aqueous and Aquated Organic Solvents Using 4f-4f Transition Spectra as PROBE 289

T. Kriyananda Singh, Th. David Singh, N. Yaiphaba, Ch. Sumitra and N. Rajmuhon Singh

39. Mangroves: An Ecosystem in Peril 302

A. Saleem Khan and M. Sheik Ali

40. Plant Regeneration and In vitro Flowering of *Solanum trilobatum* L.: An Important Medicinal Plant 306

S.M. Nagarajan, T.M. Sathees Kannan, S. Rajasekaran and S. Kulothungan

41. Postharvest Soil Fertility Status and Nitrogen Use Efficiency of Hybrid Rice as Influenced by Nitrogen Management 311

S. Ravi, S. Ramesh, S.D. Sundar Singh and B. Chandrasekaran

42. Structure and Optical Properties of DC Reactive Magnetron Sputtered Zinc Oxide Films 320

M. Lavanya, Sa K. Narayandass and D. Mangalaraj

43. Antibacterial and Antifungal Activity of Marine Bacteria from Gulf of Mannar Resources 326

N.K. Asha Devi, R. Gopal, K. Balakrishnan and S. Padmavathy

44. Estimation of Understorey Grass Production in *Acacia tortilis–Cenchrus ciliaris* Silvopastoral System through Principal Component Regression Analysis 331

T.A. Khan, R.K. Bhatt and G. Suresh

45. Enumeration and Analysis of Weed Flora of Agricultural Fields in Warangal District, Andhra Pradesh 338

M. Dharmender Rao

46. Evaluation of Biopesticides Against *Spodoptera litura*, Fab. Infesting *Kharif* Groundnut 348

M.M Sanap, S.T. Aghav, P.K. Dharne, D.A. Shambharkar, A.J. Patil and R.B. Jadhav

47. Euglenophyceae of Hatnur Dam, Jalgaon District, Maharashtra **352**
P.N. Narkhede

48. Larvicidal Activity of Plant Oils Against the Mosquito, *Culex quinquefasciatus* (Say) **357**
K. Manimegalai, S. Agalya, S. Sajani and R. Uthayakumari

49. Shigometric Studies in Regard to Heartwood Formation **360**
S.V. Deore and P.V. Ramaiah

50. Impact of Health Education on Knowledge Regarding Hepatitis B Disease Vaccine and Acceptance of Hepatitis B Vaccine Among Non-immunized Nurses in Civil Hospital **364**
Bharat H Patel, Deepak Saxena, Deepika Singhal, Amul B. Patel, V.K. Sharma, and M.M. Prakash

51. Analytical Study of Molecular Interactions in Binary and Ternary Liquid Mixtures **369**
D.K. Gupta and G. Meenakshi

52. Silver Toxicity to *Anabaena doliolum* Bharadwaja **375**
Praveen K. Chandel, Dilip K. Rathore and Krishna Chandra Sonie

53. Water Quality Studies on Freshwater Lakes of Kolar Taluk, Kolar, Karnataka **382**
H. Krishna Ram, M. Ramachandra Mohan, Chithanagiri and N. Latha

54. Effect of Mulches on Soil Properties and Yield of Maize in Mizoram **391**
U.K. Sahoo and Nigel J.S. Kharmalki

55. Biological Treatment of Azodyes **399**
P. Dhevagi and K. Sujatha

56. Cyanobacterial Diversity in Polluted Lakes of Jalgaon District of North Maharashtra **416**
S.N. Nandan and S.R. Mahajan

57. Defoliation/Detasselling Studies in Maize (*Zea mays* L.): A Review **451**
Sumit Mishra and D.S. Kler

58. Wastewater Quality of Major Drains of Delhi Draining Wastewater to River Yamuna and Assessment of Water Quality of River Yamuna at Delhi Stretch **468**
P.K. Behera, R.C. Trivedy and P.C. Mishra

59. Limnology of Lentic Freshwater Systems in North Cachar Hills, Assam, India **546**
T. N. Majumdar, A. Gupta and M. Daolagupu

60. Studies on Selected Major Elements and Nutrients in Rushikulya Estuary (East Coast of India) **559**
Tapan Rani Mahapatro

Index **577**

Chapter 1

Awareness of Environmental Impact Assessment of Power Sector Reform in India

P.C. Mishra[1] *and Arvind Kumar*[2]

[1]*Department of Environmental Sciences, Sambalpur University, Jyoti Vihar – 768 019, Orissa, India*

[2]*Pro-Vice Chancellor, S.K.M. University, Dumka – 814101, Jharkhand, India*

ABSTRACT

The current study on environment impact assessment of power sector reform in India looks at macro level issues, such as: (*i*) compliance status of the thermal power plants with respect to emission of pollutants; (*ii*) the implications of fuel mix/captive plant capacities, generation efficiency etc.; (*iii*) disposal and utilization of fly ash generated; (*iv*) optimization of resource consumption through loss reduction, better planning and management and appropriate technology; (*v*) implications for choice of or access to clean energy; (*vi*) environmental impacts on account of sectoral policies, pricing distortions and incentives; (*vii*) groundwater utilization and degradation; (*viii*) environment governance with respect to tariff structure and institutional framework such as policies, statutes, generation and distribution norms, standards and procedures; and (*ix*) consideration of environmental management and quality systems in improving operational efficiency and financial viability.

Keywords: Environmental impact, Power sector, Reform.

Introduction

Power Sector Reforms in India brought significant changes in the power supply system of the State, initially restructuring it by the unbundling of generation, transmission 'and distributions segments, followed by the corporatization and privatization of the utilities and subsequently by the establishment of an independent regulatory authority. In spite of initial setbacks, the reform programme has improved the availability and reliability in power supply in certain states, specially in Orissa. In recent times there are signs of a turn around in the commercial performance of the distribution and transmission segments of the sector with emerging indications of sustenance of the reform process.

Since the introduction of power sector reforms, eight years have passed. The major thrust areas so far have been to bring changes in organization structure, revamping of distribution system and independent regulation of the commercial and financial aspects. With the growing awareness on environment-related issues in the country, besides socio-economic aspects, both direct and indirect environment impacts of power sector reforms deserve attention to bring further sustainability to the restructuring process and its achievements (OERC, 1995; Mishra and Naik, 1999).

Methodology

The data have been collected mostly through secondary sources from government records, research papers and the response of people has been obtained through a structural questionnaire (OERC, 1995; Mishra and Naik, 1999).

Results

The likely environmental impacts due to power sector reform in the generating segment of the power sector are as follows:

1. When there is increased demand for electricity on account of the improved supply situation occasioned by the reform programme, it would induce additional power generation, which, in the case of thermal power, would require more coal. Thus, there would be (*a*) increased mining activities with more environmental damage in mining areas and (*b*) more emissions of SO_2, NOx, SPM and CO_2 in power plants. However, reform also encourages operational efficiency in power plants and this would be limiting the negative environmental impact caused by plant emissions.
2. If power becomes costlier than before due to the tariff rationalization process under reform, the demand would naturally decrease. In rural areas, when electricity becomes costly, people tend to use it in unauthorized ways or switch over to wood and cow dung, thus putting a pressure on the forest.

Thermal Power Generation

Environmental degradation occurs in the power sector largely due to thermal power generation because of the coal mining that is induced as well as the use of coal as fuel in the thermal power plants. The physical and biological environment get affected in numerous and complex ways through loss of vegetation cover and biodiversity, human displacement, water and air pollution etc. Thermal power plants in India, where poor quality of coal is used, add to environmental degradation problems through gaseous emissions, particulate matter and ash. Emissions from the thermal power plants like SO_2, NOx, SPM, CO_2 also affect the environment in local, regional and global scale. When there is greater demand of power, obviously it would result in more mining activities leading to greater damage to mining environment unless mining activities are made sustainable. Disposal of flyash generated from the power plants have become a difficult task because of huge land requirement. Disposal of ash through back-filling in abandoned mines is an area which can take care of a large amount of ash but this needs a concerted effort and coordination between the power generators and coal suppliers.

Generation of power by thermal power plants in Orissa has increased in recent times. NTPC Kaniha's capacity has increased with addition of new units. Generation at the captive power plant of NALCO, Angul has increased with addition of new units to meet the demand of its smelter after expansion. Captive generations of Indal, Hirakud, TSP, Rourkela and ICCL, Choudwar have increased marginally to meet the internal demand.

Thermal power generation is going to increase further in the State with expansion of NTPC Kaniha plant from 1000 MW to 1500 MW to meet the demand of Eastern region states. NALCO's capacity has already increased from 720 MW to 960 MW and would increase soon to 1200 MW to meet the second phase expansion of its smelter. Indal, Hirakud has plans for expansion of generation capacity and a new thermal power plant based on Ib valley coal is likely to come up. Captive power generation of proposed and existing industries of mini steel plants and sponge iron plants are being planned.

As per 2002–03 figures, GRIDCO procured 8600 MU power from thermal plants in the State, which is 72 per cent of the total power it purchased in the same year. While no further hydel projects are in prospect, the thermal power generation would increase in future to meet the growing demand. With the availability of plenty of coal, thermal power generation in the state is expected to double in the coming decade.

Table 1.1: Thermal Power Generation in Orissa

	2000–01	*2001–02*	*2002–03*	*Remarks*
NTPC, Kaniha	4,845	5,753	5,964	Surplus power sold to eastern region grid (central) which is increasing recently
NTPC, Talcher	2,488	2,465	2,233	After auxiliary consumption rest is sold to GRIDCO
ITPC, Banharpali	3,382	2,599	2,621	–do–
RSP, Rourkela	325	445	–	Own consumption mostly
INDAL, Hirakud	524	543	–	–do–
NALCO, Angul	4,283	4,419	4,777	After auxiliary consumption and meeting own demand, surplus is sold to GRIDCO
ICCL, Chowdar	770	778	842	–do–

Figures in MU.

Coal Mining

The thermal power plants are the major consumers of coal within the State. Proposed power plants, addition of new units at CPP, NALCO and NTPC, Kaniha, development of captive coalmines by the new steel plants with CPPs, would increase the coal production and also consumption by thermal power plants. While dealing with the environmental impact of power sector reform, the expected increase in production from the coalmines and related environmental issues also deserve consideration. All the coalmines in Orissa are operated by Mahanadi Coalfields Limited (MCL) and confined within two coalfields *i.e.* Ib Valley and Talcher. In the past 7 years, from 1996–97 to 2002–03, increasing capacity of existing mines and the development of new mines (mostly opencast operations) have led to consistently increasing annual coal production by MCL (Table 1.2).

All the mines are presently operating with approved mining plans and incorporating adequate environment management systems (EMS) that have been cleared by appropriate statutory authorities and received the consent of the Orissa State Pollution Control Board (OSPCB). However, there are still some important environmental issues, which require attention. The dust pollution problem particularly continues to be serious in the coalfield areas and this is brought out by the figures given in Tables 1.3–1.5.

Table 1.2: Coal Production by MCL from 1996–97 to 2002–03

Year	*Coal Production in Million Tonnes*
1996–97	37.37
1997–98	42.17
1998–99	43.51
1999–00	43.55
2000–01	44.80
2001–02	47.81
2002–03	52.23
2006–07 (Expected)	68.00
2011–12 (Expected)	120.00

Table 1.3: SPM and RPM Concentration in Coalfields

Year	*Annual Average of SPM (ke/m³)*	*Annual Average of RPM (ke/m³)*
1999–00	246.3	120.4
2000–01	189.3	99.2
2001–02	173.9	87.5
2002–03	162.9	78.5

Table 1.4: Ambient Air Quality in Ib Valley and Talcher Coalfields During 2002–03

OCP	*Concentration (μg/m³)*			
	RPM	*SPM*	*SO_2*	*Nox*
Samaleswari OCP (Gopivihar colony)	31.6–162.3	72.3–297.3	< 10.0	<10.0–31.8
Lajkura OCP (Madhuban Nagar)	32.1–118.4	61.8–178.1	<10.0	<10.0–32.8
Ananta OCP (Ananta vihar colony)	7–519	64–881	<10.0–27	<10.0–117
Hingula OCP (Project office)	21–111	42–206	<10–65	14–124

Coal mining also involves significant loss of forest area and biodiversity (Table 1.6). However, the loss is largely compensated through land reclamation and plantation activities undertaken by MCL in the mined areas (Table 1.7). Coal fires in mines and stockpiles during summer cause carbon emissions and air pollutions in the region. Loss and wastage of coal within the pits and coal fire, besides loss during handling can be minimized. This is one area where no statutory agency is enforcing control over the coalmines.

Table 1.5: Environmental Standards for Ib Valley Coalfield of MCL [as per G.S.R. 742(E) dated 25th September, 2000]

Pollutant	*Time weighted Average*	*Concentration in Ambient Air*	*Method of Measurement*
Suspended Particulate Matter (SPM)	Annual average* 24 hours**	430 μg/m³ 600 μg/m³	High volume sampling (Average flow rate not less than 1.1 m³/minute)
Respirable Particulate Matter (size less than 10 mm) (RPM)	Annual average* 24 hours**	215 μg/m³ 300 μg/m³	Respirable Particulate Matter sampling and Analysis
Sulphur Dioxide (SO_2)	Annual average* 24 hours**	80 μg/m³ 120 μg/m³	1. Improved West and Gaeke method 2. Ultraviolet fluorescence
Oxides of Nitrogen As (NO_2)	Annual average* 24 hours**	80 μg/m³ 120 μg/m³	1. Jacob and Hochherse Modified (Na Arsenite Method) 2. Gas Phase chemolumilnesce science

*: Annual Arithmetic mean for the measurements taken in a year, following the guidelines for frequency of sampling as laid down.

**: 24 hours/8 hourly values shall be met 92 per cent of the time in a year. However, 8 per cent of the time it may exceed but not on two consecutive days.

Table 1.6: Status of Forest Area Loss in Angul–Talcher Mining Complex

OCP	*Total Forest Land (ha) Including Reserve Forest*	*Total Forest Area Lost for Mining Activities (ha)*	*Total Forest Area to be Lost for Mining (ha)*
Bharatpur	198.171	65.01 (32.80 per cent)	133.161 (67.19 per cent)
Bolanda	1045.75	245.0 (23.42 per cent)	315.0 (30.12 per cent)
Kalinga	85.201	–	60.523 (71.03 per cent)
Lingaraj	240.804	6.71 (2.78 per cent)	109.687(45.55 per cent)
Ananta	146.0	74.0 (50.68 per cent)	82.0 (56.16 per cent)
J agannath	106.16	106.16 (100.0 per cent)	–
Total	1822.086	496.88 (27.26 per cent)	700.371 (38.43 per cent)

Table 1.7: Status of Land Reclamation and Plantation in Mahanadi Coalfields

Sl.No.	*Item*	*Status as on 31-3-2003*
1.	Area excavated (ha)	2160
2.	Area Technically reclaimed (ha)	1213
3.	Area Biologically reclaimed (ha)	983
4.	Total plantation (over burden, vacant, avenue)	No. 34.0 1akhs

Groundwater Utilization and Degradation

The thermal power plants consume substantial quantities of water, and considering the apprehensions relating to severe water shortage in the future, the impact on water resources need to be a priority concern. Unfortunately, the impact on groundwater table and pollution of surface and groundwater in regional scale has not been critically assessed till date. The data generated by Central Groundwater Board (Table 1.8) reveal no significant depletion in groundwater level although water scarcity is reported in coalfield areas during summer.

Table 1.8: Groundwater Level (Meters Below Ground Level) in Talcher and Ib Valley Coalfields as Reported by Central Groundwater Board

Year	*January*	*April*	*August*	*November*
		(1) Talcher		
1998	6.24	8.05	3.40	4.8
1999	6.68	8.80	3.10	–
2000	6.35	8.85	2.69	6.57
2001	7.69	9.25	1.62	4.40
2002	6.46	8.96	–	–
		(2) Balanda		
1998	4.09	4.42	3.58	4.06
1999	4.43	4.9	2.38	–
2000	4.38	4.88	2.11	4.31
2001	4.63	5.28	1.96	3.89
2002	4.4	4.41	–	–
		Ib Valley Coalfields		
		(1) Lakhanpur		
1996	4.45	6.35	2.06	3.96
1997	5.29	6.25	1.73	4.55
1998	4.20	6.46	3.02	4.1
1999	5.05	6.35	1.96	–
2000	4.96	6.28	–	–
		(2) Brajrajnagar		
1996	5.19	7.93	1.87	4.87
1997	5.11	5.76	1.97	5.0
1998	5.03	5.45	4.21	4.28
1999	5.35	5.88	2.39	–
2000	5.43	8.37	–	–

Plant Emissions and Acid Rain Potential

While the recent plants are having zero discharge to take care of water pollution, the air pollution, CO_2 emissions by the thermal plants and their acid rain potential deserve immediate attention (Tables 1.9 and 1.10). Further, the impact of heat generated by the plants and the effect of green house gases produced may disturb the ecological balance.

Table 1.9: Coal Consumption and Generation of Pollutants from Thermal Power Plants of Orissa (S, N and C per cent in Coal are 0.4, 0.7 and 32 respectively) (Dash *et al.*, 2003)

Power Plant	*Coal Consumption TDP*	*Ash Generation TDP*	*SO_2 Generation TDP*	*NOx TDP*	*CO_2 TDP*
NTPC, Kaniha	15550	6510	124	356.0	18154.0
NTPC, Talcher	5000	2100	40	115.0	59.0
NALCO, Angul	12000	5040	96	276.0	14055.0
ICCL, Choudwar	1500	630	12	34.0	1757.0
Ib Thermal	4500	1890	36	103	527.0
Hirakud Power, INDAL	1530	643.0	12.0	35.0	1792.0
Rourkela Steel Plant	2500	1050	20.0	57.0	2928.0
Total	42530	17863.0	340.0	976.0	36811.0

Table 1.10: Current and Future Acid Rain Producing Potential in Dome Parts of Orissa (The values are based on Stochiometric equation) (Dash *et al.*, 2003)

Electricity Generation (Current generation 3020 mw)	*H_2SO_4 TDP*	*HNO_3 TDP*	*H_2CO_3 TDP*
Angul-Talcher Area 2288 mw	416.43	1096.19	377790.79
Ib Valley 487.5 mw	73.85	189.52	9950.82
Rourkela Steel Plant 245 mw	30.62	78.58	4125.55
Future Projection 12300 mw (Additional)	2239.46	5746.19	201818.40

Disposal and Utilization of Flyash Generated

The environmental impact of flyash generated by the thermal power plants is a matter of serious concern today. Currently, these plants consume about 14 million tonnes of coal (Table 1.11) and generate about 6 million tonnes of ash annually (Table 1.12). Inspite of pressure from Ministry of Environment and Forest, Government of India and the State government, the percentage utilization of ash is very low (Table 1.13) for various practical reasons. Void filling by 'Fly Ash' from thermal power plants have not been taken up so far. However, recently there has been an MOD between NALCO, Angul and MCL for backfilling of Flyash in mined void. This very serious as the surrounding areas are highly populated and there is shortage ofland for agriculture and other purposes. Filling of voids in adjacent coalmines, use in road buildings, wasteland agricultural and industrial applications/uses can only solve the stupendous problem the state is going to face.

Table 1.11: Coal Consumption by Thermal Power Plants

Power Plant	*2000–01 (Million Tonnes)*	*2001–02 (Million Tonnes)*	*2002–03 (Million Tonnes)*	*Consumption (Kg per unit Generation)*
NTPC, Kaniha	3.57	4.20	4.54	0.76
NTPC, Talcher	2.17	2.19	1.92	0.86
ITPC, Banharpali	2.64	2.20	2.22	0.85
RSP, Rourkela	0.36	0.43	0.50	0.96
INDAL, Hirakud	0.43	0.43	0.40	0.79
NALCO, Angul	3.35	3.43	3.70	0.77
ICCL, Chowdar	0.65	0.66	0.69	0.82

Table 1.12: Fly Ash Generated by Thermal Power Plants

Power Plant	*2000–01 (Million Tonnes)*	*2001–02 (Million Tonnes)*	*2002–03 (Million Tonnes)*	*Consumption (Kg per unit Generation)*
NTPC, Kaniha	1.43	1.55	1.62	36 per cent
NTPC, Talcher	0.87	0.90	0.77	40 per cent
ITPC, Banharpali	1.00	0.85	0.90	40.5 per cent
RSP, Rourkela	0.07	0.10	0.15	–
INDAL, Hirakud	0.16	0.16	0.14	37 per cent
NALCO, Angul	1.20	1.31	1.42	38 per cent
ICCL, Chowdar	0.30	0.30	0.30	43 per cent

These figures of fly ash generated are not realistic; the reported figures for NTPC, Kaniha, INDAL and NALCO may be after excluding Fly Ash used in various ways.

Environmental Management and Monitoring

The present status with respect to environmental management by the thermal plants and the control/monitoring by statutory authorities have been analyzed and the major findings are given below:

1. As per records all the thermal power plants are mostly meeting the pollution/environment standards, except periodical failures in meeting SPM level and reported water pollutions by ash disposals.
2. Coal consumption in these plants has been proportionally varying with the generations indicated at Table 1.11. The specific consumptions (norms) of coal have marginal variations depending upon quality used, size and type of plant with technology. However, the best operation for this reason is of NTPC Kaniha and NALCO, Augul. The norms for these coals are accepted to be between 0.7 to 0.8 kg/Unit.
3. Fuel oil consumption is accepted to be around 0.61 ml/unit. All most all plants are meeting the norms, while some are marginally higher which is mostly due to plant and grid failures.

Table 1.13: Status of Fly-ash Utilisation by Thermal Power Plants of Orissa (2002–03)

Sl. No.	Name of the Plant, Address and Capacity	Coal Consumption During 2002–03	Power Generated During 2002–03 MU (Million Units)	Qty. of Ash Generated During 2002–03 (MT)	Disposal Methods	Brick Making (MT)	Supplied to Industry Other Agency (MT)	Land Filling (MT)	Embankment Dyke	Other Purposes	Total Ash Utilized	Per cent of Utilisation
1.	NALCO (CPP), Dist. Angul, 7 × 120 = 840 M	3702990		1423400	Hydro sluicing ash disposal system	Nil	Nil	5000	20000	600 Agriculture and Research	205600	14.4
2.	Ib Thermal Power Plant, Jharsuguda, 2 × 210 = 420 MW	2228453	2621.13	890053	Wet disposal and dry disposal (small scale)	233	Nil	Nil	18907	Nil	19140	2.15
3.	Indian Aluminium Company Ltd., Hirakud, Sambalpur, 67.5 MW	404137		143893	Dry ash disposal	1008	75830 Cement industry	–	–	–	76838	53.4
4.	Indian Charge Chrome Ltd., Choudwar, Cuttack, 108 MW	675809	842677	314887	By dumpers (dry disposal)	110	Nil	289182 (Waste and degraded land)	Nil	–	289292	91.9
5.	NTPC Kaniha Deepsikha, Angul, 2 × 500 = 1000 MW	454000		1621173	Wet disposal	16085	24 Cement industry	63579	78190	14222 Ash product	1712109	10.61
6.	Rourkela Steel Plant (CPP-1), Rourkela, 3 × 25 = 75 MW	502000	150000	Wet disposal	–	–	9200 Cement industry	15600	34300	–	59100	39.4
7.	NTPC–SAIL Power Company RSP (CPPII), Rourkela, 128 MW	–	–	245100	Wet disposal	–	–	–	150960	–	150960	61.6

Contd...

Table 1.13–Contd...

Sl. No.	Name of the Plant, Address and Capacity	Coal Consumption During 2002–03	Power Generated During 2002–03 MU (Million Units)	Qty. of Ash Generated During 2002–03 (MT)	Disposal Methods	Brick Making (MT)	Supplied to Industry Other Agency (MT)	Land Filling (MT)	Embankment Dyke	Other Purposes	Total Ash Utilized	Per cent of Utilisation
8.	CPP Birla Tyres 25MW	142494	73107	30794	Dry disposal	6480	–	24314	–	–	30794	100
9.	Talcher Thermal Power Station Angul 4 × 60, 2 × 110 = 460MW	1925000		770116	Wet disposal	7825	98 Cement Industry	–	166000	1889 Ash Based Products	175812	26.02
10.	Bilt Unit Sewa Gaganpur, Orrisa	139726	64.73	49568	Dry	4000 Disposal	–	45568	–	–	49568	100
11.	M/s JK Paper Mills, Jakaypur, Rayagada	174000	132812	73000	Dry disposal	14000	–	59000	–	–	73000	100
12.	M/s Alumina Refinery NALCO Damanjodi, 3 × 18.5 = 55.5 MW	904563		381183	Wet disposal	–	–	–	–	–	–	0
13.	M/s Ballarpur Industries Ltd., Choudwar	Closed	Closed	Closed	Dry and wet disposal	Closed	Closed	Closed	Closed	Closed	Closed	Closed
14.	M/s Enami Paper Mill, Balgopalpur Balasore, 5 MW	48184	–	16864	Dry and wet disposal	3375	–	13489	–	–	16864	100

4. Auxiliary consumption norm of 9 to 10 5 of generation is mostly met by the plants if additional consumptions by peripherals for local considerations are excluded.
5. Emission norms CO_2 *i.e.* 806.5 to 865 grams/unit has been confirmed by Nalco and the figures of other plants are not available. Detailed study and assessment is required for future.
6. It is understood that power sector reforms have not considered the impact on environment management of thermal power plants, which happen to be internal issues only. On the other hand improvement in grid performance has reduced frequency of tripping, resulting in marginal reduction in fuel oil consumption and efficient operation of boilers. This would obviously be a step towards energy efficiency operation and reduction in carbon emission in auxiliary systems. It could not be confirmed, as plants do not keep records to assess these parameters.

The norms/standards to evaluate energy efficient operation and monitoring of air pollution directly from stack deserve to be reviewed and implemented. Technology improvements and attitudinal changes for fly ash utilization is essentially required. Better systems of handling, storage, disposal and use have to be there to meet the future need.

Quality of Life Index in Mining Areas

The present study involved a sample survey of 367 households from different villages located in the three proposed mining areas *viz.*, Gopalprasad, near Talcher Mining area and Talabira near Brajarajnagar. The sample survey is conducted covering about 367 households from different villages in the above mentioned regions belonging to different caste groups, occupation groups and land size groups to make it approximate to a stratified sampling method. The importance of conducting such a survey lies in assessing the level of social infrastructure as well as individual well being in the region which would indicate the role of industrialization and mining in the upliftment of standard of living of the people. It may, however, be mentioned that since base line data (prior to onset of reform process) do not exist for assessing the specific impact of power sector reforms on the quality of life, the results of quality of life study would only help in providing a direction towards the possible areas where the reform process must intervene to check further deterioration of quality of life of the people.

Quality of Life Index[1] for the sample households is very poor, with higher index for service class, general caste group and top land size group. Since the quality of life index for people of the region is very low, which is indicative of both low level of social infrastructure in the region as well as low level of individual prosperity, the reform process must supplement in the creation of conditions for the

---

[1] The quality of life index (QOL) has been computed for these sample households with the methods adopted in a study "*Quality of Life Index of the Mining Area*" by Saxena, Pal, Prusti and P. Kumar, (1998) of Centre of Mining Environment, Indian School of Mines, Dhanbad, published in special issue on Environment of the Indian Mining and Engineering Journal.

The following parameters are included for the computation of the Quality of Life index: Houses, Water availability, Sanitary facilities, Food and nutrition status, Health and safety status, Education status, Medical facilities, Public transport and communication facilities, Fuel and energy availability, Assets, Own transportation means, Per capita income and Recreation facilities. The minimum desired level for the above parameters for a fair living condition is defined with a value of 0.5 on a scale of 0 to 1.0. All the parameters are given an equal weightage and the total score of QOL is 13. The grading on the basis of total score is: <3 Very poor; >3–5 Poor; >5–7 Fair; >7–10 Good; >10–13 Very good.

improvement of social infrastructure in the region like improved road conditions, provision of transport facilities and medical facilities and generation of employment opportunities for the people to improve their conditions of living. The management structure in the power sector changing from state ownership to privatization in the reform process may lead to increased extraction of coal resulting in increased deforestation in the region, loss of biodiversity, pressure on water sources and increased air and water pollution. These processes are likely to contribute in the further deterioration of individual livelihood conditions if proper management of environmental aspects are not looked into.

Response of People Around Thermal Power Plants

In the present study, responses of sample respondents selected from among residents around power plants in Talcher and Ib valley have been analyzed to gain insights about the perception of the people in the Thermal Power Plant belt regarding the operation of power sector in general and reform process in particular. The sample study has coverage of 150 sample respondents randomly selected from the towns where the power plants are situated and adjacent villages.

Majority (60 per cent) of the respondents related all types of environmental problems such as air and water pollution, climate change, water scarcity and forest and wild life loss in the region to coal mining and power plant. Regarding the existence of changes in the power sector during past 10 years (after reforms), about 71 per cent of the respondents visualized no change. About 60 per cent of the respondents viewed air and water pollution as the major negative consequence of the use of coal by thermal plants. 11.33 per cent of the respondents indicated adverse effect on health as the major negative consequence. The remaining 29 per cent indicated in terms of combinations of forest loss, pollution, heat generation etc.

Three-fourths of the respondents are aware of increased coal extraction during the past 10 years. Similarly 51 per cent of the respondents are found to have expressed increase in transport activities of coal during last 10 years. On a specific question on forest loss it is found that 90 per cent of the respondents indicated that there has been loss of forest during past 10 years.

Similarly on the consequences of water scarcity it is found that 66 per cent of the respondents expressed decline in water table in dug wells during past 10 years while 32.66 per cent expressed lack of knowledge about this phenomenon. However, 89.33 per cent of the respondents indicated that scarcity of water has increased during past 10 years. The major reason is expressed to be decrease in rainfall by most of these respondents. On a specific question on air and water pollution, 94 per cent of the respondents indicated increase in pollution during past 10 years and they have attributed it to the power plants. Regarding problems of fly-ash, one finds that 35.33 per cent did not respond to the question while 28 per cent indicated adverse effect on agriculture, 14.66 per cent on generation of air pollution and remaining 22 per cent on various combinations of pollution, diseases and agricultural loss.

Similarly on a specific question regarding what way operation of thermal power plants bring inconvenience to people, around 80.67 per cent expressed combinations of various inconveniences such as contamination of ponds, loss of life, loss of forest, loss of agriculture, health disturbances and increase in temperature etc. Regarding number of members of the families suffering from attack of the diseases such as respiratory, gastro-intestinal and skin one finds that 32.66 per cent reported no member in the family is suffering from any disease regularly and almost the same number indicated that the frequency of attack has not increased during past 10 years. This indicates that a sizeable majority of the respondents are of the opinion that not only members are suffering from these diseases but also the frequency of their occurrence has increased during the past 10 years.

The importance of the response study lies in documenting. It may be mentioned that most of the findings of the response study are applicable to operation of power sector irrespective of the ownership structure. Since majority of the respondents viewed environmental deterioration characteristic of power production in Orissa through thermal power plants and people around have been suffering from various ailments, which are resultant of the power production, the reform process must take cognizance of the environmental problem and create safeguards so that health conditions of the people around are not adversely affected. Environmental safeguards and medical facilities being at a low level before the onset of reform process, as indicated by the quality of life index study, initiatives must be taken for their improvement during the reform process. The perception of the people being deterioration of health conditions, loss of bio-diversity, increased pollution during the past 10 years, specific measures are to be taken up through afforestation, proper disposal of fly ash, measures to reduce water scarcity and water contamination. Specific development package must be worked out especially for people engaged in agricultural and allied activities to improve their conditions of living.

Hydel Power Generation

Environmental damage in hydel power generation occurs mainly during the construction of large dams. Apart from the submergence of large tracts of land area that often leads to significant biodiversity loss, there is the problem of human displacement involved in the construction of large dams.

Gridco would obviously prefer hydel power for its low cost and as such Gridco purchases the total hydel power generated. However, generation from this source depends upon monsoon and hence is not reliable and perennial. The environmental issues for this sector apparently looks to be only socioeconomic related to land acquisitions and deforestation over extensive areas, which are taken care of during the initial project stage itself. Impact of this however continues with the affected people for longer period in terms of the impact on hydrology in downstream areas, silting, changes in landscape, land use and agricultural outputs, which are invariably not examined.

Environmental Impact of Reform in Transmission and Distribution

As the reform aims at controlling losses in the transmission and distribution of power through greater operational efficiency, infrastructure upgradation and better system maintenance, there would be certain environmental improvements on account of:

1. Efficient use of energy,
2. Long life of equipment in use,
3. Less tripping and breakdown of equipment in use.
4. It will also result in reducing pollution of SF_6 used for insulation, current intensive equipment used, in transmission and distribution systems, circuit breakers, leakages from storage tanks and electrical equipment etc.

Tariff reforms bringing rationalization in use of electricity would discourage wastage of water and over-drawal in both agriculture and also domestic sectors. It will help in conservation of water and energy both. It may help in sanitation and hygiene by avoiding water logging and also reducing the impact on groundwater level.

Environmental Impacts without Reform and Due to Reform

When we try to discuss the impacts that would have been observed even without reforms, the answer is "No difference due to reform". Rather, it is because of increased coal mining and thermal power generation, which would have occurred even without reform, that the negative impact on environment has definitely increased.

Checks, Balances and Incentives

While penalizing the generators whose compliance does not match with the norms, it may be desirable to incorporate the provision of incentives to private generators who comply with environmental norms in the following ways:

1. Power Generators using coal from the coal companies certified with ISO9000 and 14000.
2. Power Generators with ISO9000 and 14000 certification.

The standard prescribed by Central Pollution Control Board for stack emission in Thermal Power Plants is only for SPM. Other emissions like SO_2, NOx and CO_2 are not taken in to account. Report by Dash *et al.*, 2003 reveals that power grade coal in Orissa contains about 0.4 per cent sulphur, 0.7 per cent nitrogen, 32 per cent carbon and 42 per cent ash. They have estimated that the existing power plants currently generate 272,781, 26821 tpd SO_2, NOx and CO_2 in Angul-Talcher area and 48.24, 138.44 and 7062.33 tpd in Ib valley area, respectively. Thus, these gases contribute to acid rain. Therefore, it may be debated if there is a necessity to set standards for these emissions for unit power produced looking to the economic aspects of it. Stack emission limits for large Thermal Power Plants in respect of SO_2, NOx in few countries are available (Table 1.14). At least, there could be a beginning for large Thermal Power Stations generating more than 1000 MW.

Table 1.14: Stack Emission Standards from Thermal Power Plants of Selected Countries (CPCB, 2003)

Country	*Particulate (mg/Nm³)*	*SO_2 (mg/Nm³)*	*NOx (mg/Nm³)*
Austria	50	400	200–300
Canada	125	715	740
Germany	50–125	400–2000	200
Japan	100	170–860	410
UK	50–140	400	650
USA	40–125	740–1480	555–860
India	150 (of capacity > 200 MW) 350 (of capacity < 200 MW)		

Clean Technology and Energy Conservation

Adoption of Clean Technology definitely leads to better environmental management. Use of Beneficiated Coal, as advised by Ministry of Environment and Forests and CPCB has not been implemented in Orissa. The CFBC (Circulation Fluidised Bed Combustion), FGD (Flue gas desulphurisation) are some of the clean technologies in power sector. In Orissa most of the power plants have not adopted such technologies except at INDAL, Hirakud in respect to CFBC (with inbuilt SO_2 capturing system) and NALCO at Angul and NTPC at Kaniha in respect to FGC. Thus the provision of incentives should also be linked to the adoption of Clean Technology in Power Sector.

Linkage Between OERC and the State Pollution Control Board

OERC can play a positive role in ensuring environmental compliance by the power generators provided a provision is made in the Orissa Electricity Reform Act through amendments to make it mandatory for the Generators for ISO Certifications (9000 and 14,000) in order to be eligible to supply sale power and necessity of Orissa Pollution Control Board certificate in respect to the standards prescribed by the Board for emissions. The provision of incentives should also be there for the generators using clean technologies as well as proper and large scale disposal of fly ash.

Environmental Awareness Review

The awareness survey data collected from personnel associated in power sector operation, administration, supply and distribution reveal that all are well aware of environmental issues involved. Air and water pollution are the major concern. Mining, thermal power, and hydropower come in decreasing order of importance as the major environmental degrading activities in the power sector. Specific reference has been made about ash disposal of thermal power plants. Majority of the respondents highlighted deforestation as a major environmental problem in the coal-mining belt of the State. Loss of biodiversity and problems of human settlement are not considered as critical issues, while a few have viewed that land use pattern and transport sectors are to be properly controlled. Almost all have expressed their inability to influence the reform process in taking care of the environmental problems, though sometimes they are involved in some related activities.

Majority feels that:

1. There is more scope for improvement in environmental issues by proper policy decisions and better management.
2. A system for more information on the trade offs between environmental impacts and economic and social costs of power sector activities need to be worked out.
3. Technical solution and better governance are the main answers to environmental problems in power sector and significant institutional reform may not be required. The major ones in order of decreasing preference are:
   (*a*) Technical solution–ESP, low NOx, burners etc.
   (*b*) Energy efficiency operations
   (*c*) Inter fuel substitution, renewable gas etc.
   (*d*) Policy instruments, environmental issues, energy efficient campaigns etc.
   (*e*) Plant sitting and interstate trade
4. The environmental problems of power sector are not solved due to:
   (*a*) Poor understanding of environmental impacts.
   (*b*) Regulations are not adequately enforced.
   (*c*) Lack of resources in the agencies to monitor pollution.
   (*d*) Inadequate investment funds for pollution abatement.
5. Majority feels that low residential and agricultural electricity prices still continue to be serious problems.
6. Raising public awareness about environmental issues and the power sector would be helpful. Global warming is an important problem for Orissa. Large-scale plantations from the very beginning of any project plan should be ensured.

7. Competition in electricity supply will benefit the environment by encouraging operational efficiency in generation, transmission and distribution of electricity.

Recommendations

1. Reform should consider environmental issues related to power sector.
2. Sale value of power from polluting thermal power plants should be taxed and plants be penalized. Procurement can also be limited or reduced from these units, if possible.
3. Norms for consumption of raw materials, losses and wastes are to be monitored to ensure optimum utilization of resources and efficiency in management. Excellence in achievements should be rewarded.
4. SPM, NOx and SO_2 to be monitored at stack on regular basis. CO_2 equivalent emissions to be calculated and reported.
5. Norms and standards of pollutants are to be reviewed for bringing improvements in environment management.
6. Monitoring system has to be improved, strengthened and also reviewed periodically.
7. Regional impact studies need to be undertaken for future planning purpose while sitting of plants should be based after carrying capacity and Zoning Atlas Project of OSPC Board recommendations.

Acknowledgements

The authors are thankful to Prof. M.C. Dash, former Chairman, Pollution Control Board, Orissa and former Vice-Chancellor and Professor, Sambalpur University and Dr. R.C. Mohanty, former Executive Director, National Aluminium Company Ltd. for expert advice and critical comments. Thanks are due to CMDR, Dharward for associating the author in the study.

References

Central Pollution Control Board, 2003. Standards for Water and Air Pollution Control in Some Countries. *Information Manual on Pollution Abatement and Cleaner Technologies Series*: impacts/8/20022003.

Dash, M.C., Panda, R. and Pati, S.S., 2003. Potential generation of acid rain from thermal power plants, Orissa. *SGAT Bulletin*, 4(1): 36–40.

Electricity Act, 2003 (Salient Features and Requirements), Price Water House Coopers.

Mishra, P.C. and Naik, A. (Ed.), 1999. *Environmental Management in Coal Mining and Thermal Power Plants*. Techno Science Publ., Jaipur.

Orissa Electricity Regulatory Commission, 1995. Orissa Electricity Reform Act.

Chapter 2

Assessment of Knowledge of Students at Primary Level Towards Environmental Concepts and Environmental Problems

R. Vijaya and Anandhavalli Mahadevan

Department of Futures Studies, Madurai Kamaraj University, Madurai – 625 021, Tamil Nadu, India

ABSTRACT

Inclusion of Environmental Education (EE) contents in the syllabi to impart EE to the students at primary level was done by NCERT with care. Paper tries to assess the general awareness among the primary level students towards environmental concepts, environmental problems and management aspects. Students get environmental awareness through textbooks and from other sources. The sources of knowledge for EE contents were identified through interview schedule. Knowledge about environmental problems is also essential. So, this study tries to identify whether the students understood the environmental problems. This study focuses on the need for creating environmental awareness among the primary level students.

***Keywords**: Environmental Education (EE).*

Introduction

Educational system should prepare the future citizens because each citizen needs to develop an environmental ethic and a priority to accept and internalize the social costs of his actions. There are many ways to prepare the children as the future citizens. One way is to develop awareness, attitudes, values and skills among students in an interdisciplinary Environmental Education process where students and teachers work together to learn how to live harmoniously with their environment.

Environmental education (EE) will help children to develop the relationships with their surroundings and nature. If children are well informed about the environment and its problems, they may not take part in spoiling nature in their day-to-day life. So EE is necessary in school days especially in primary level because it is very essential to create general awareness about environment, nature, and importance of cleanliness, protection of trees and other environmental concepts at this stage. Education is the important category of the developmental plans, which change the behaviour of the children. Primary education is very essential for children because it paves the base for a good citizenship. Since the concern over environment and its problems is gaining more attention, Environmental Education is given an essential status. To impart EE at school level, environmental concepts, environmental problems and environmental management aspects are included in the syllabi of the textbooks at various levels of school education.

When compared to the educational achievements of children of developed countries, our Indian children are doing very poor. The quality of the educational system in India today compares unfavorably with that in the developed countries. The International Association for Evaluation of Educational Achievement (conducted evaluation in science and reading) revealed that the mean scores of Indian children in an international comparison is staggeringly low (Rohatji, 1982).

Lakshmi Perindevi (1996) says, "In the contest of the present day increase in environmental problems, Environmental Education for children need not be over emphasized. Children do learn certain concepts about environment and environmental problems in their classrooms, without proper teaching aids, no lasting impression are made in the minds of children". She also tested the knowledge related to the environmental concepts through a questionnaire and came to a conclusion that play way methods can create more awareness among students and she had developed a kit for that purpose.

Objectives of the Study

1. To assess the level of general awareness towards environmental concepts, environmental problems and environmental management.
2. To identify the sources of environment related knowledge.
3. To identify whether the students understand how they are exposed to environmental problems.

Material and Methods

The sample for the study constituted 100 students at the primary level. A questionnaire was designed to assess the knowledge of children related to four specific categories namely: general awareness towards environment, knowledge of environmental problems and environmental management, source of environmental knowledge and the knowledge of exposure to environmental problems. The number of questions was 25. The investigator collected the primary data by interviewing the children using the questionnaire. Rating Scale (Goode *et al.*, 1981) and Karl Pearson's Product Moment correlation methods (Gupta, 1990) were used. The scores were converted into scores. Correlation between the average mark of the academic achievement and the Environmental Awareness tests was calculated to know whether the goals of NCERT are achieved through the syllabus prescribed by them for the primary level.

For this study, a questionnaire was framed on the basis of NCERT guidelines for the primary level education. Questions were prepared on the basis of Environmental Concepts, Environmental Problems and Environmental Management aspects in which the students are expected to gain knowledge and awareness. Questions were explained to the children and the answers were scrutinized through Rating Scale method. 5 marks were given for the correct answer; 3 marks were given if they gave one or two points and 1 mark was given if they gave vague answer.

Results and Discussion

Mean and standard deviation for these tests were given in the Table 2.1. Standard deviation for the Awareness test was ±9.38. According to this study, 9 students got above 36.4 marks, which was the highest score and 3 students got 15.6, which was the lowest score. Standard error was 0.19143. Standard scores (Z) for the academic achievement and the environmental awareness tests were calculated and the correlation coefficient (r) between the scores was 0.1162, which is below 0.2 ($r < 0.2$). Since the correlation coefficient is below 0.2, we can conclude the awareness test reveals that there is almost no correlation between the two tests. This shows that the environmental concepts, problems and management aspects in the lessons did not make any impact on the students and the overall performance of the students wag poor. 76 per cent of the students were average; 6 per cent of them were high achievers and 18 per cent were low achievers.

Table 2.1: Marks of Students

Tests	*Mean*	*SD*	*Correlation (r)*
Academic Achievement Test	56.6	17.8±	0.1162
Environmental Awareness Test	27.0	9.4±	

Favourite Subjects

According to the present study, 46 per cent of the students like Tamil as their optional subject. 28 per cent of them gave priority to Science; 10 per cent gave priority to English. 10 per cent gave preference to Social Studies and 6 per cent to Mathematics. So incorporation of more environmental concepts in Tamil and Science can create more interest among students (Figure 2.1).

General Awareness

General awareness was less among the students. 24 per cent of them answered clearly. 36 per cent answered the questions to some extent. Rest of the 40 per cent students did not aware of the Environmental Concepts (Figure 2.2).

Environmental Problems

17 per cent of the students knew the Environmental Problems in Tamil Nadu mentioned in the questionnaire. They answered well. 39 per cent of the students knew some Environmental Problems. They identified at least one or two. 44 per cent of the students did not know the Environmental Problems in Tamil Nadu. They answered with difficulty even after explanation (Figure 2.3). This shows that the regional environmental problems should be given more importance during the preparation of syllabi for EE portion. Teachers must be informed well about the Environmental Problems, so that they can impart that knowledge to the students.

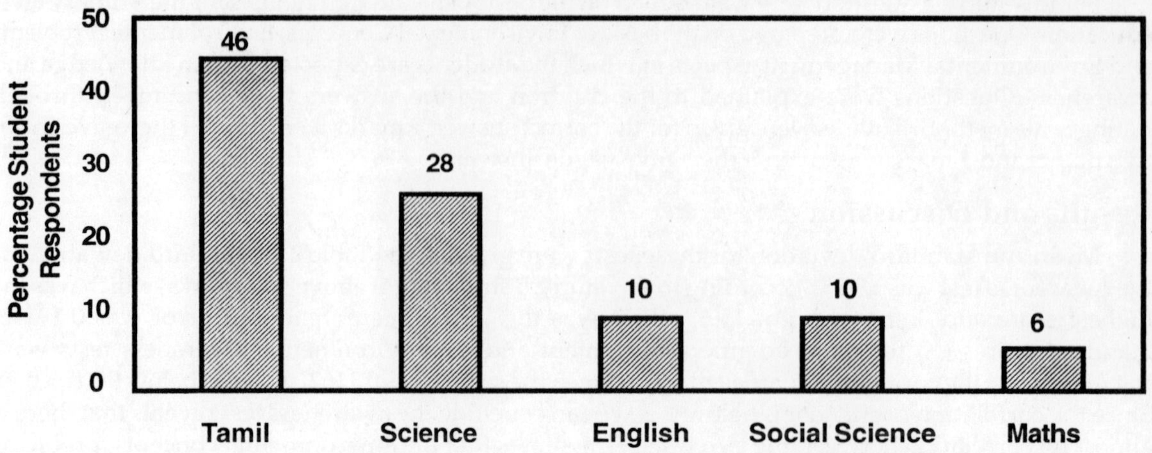

Figure 2.1: Favourite Subjects of Students

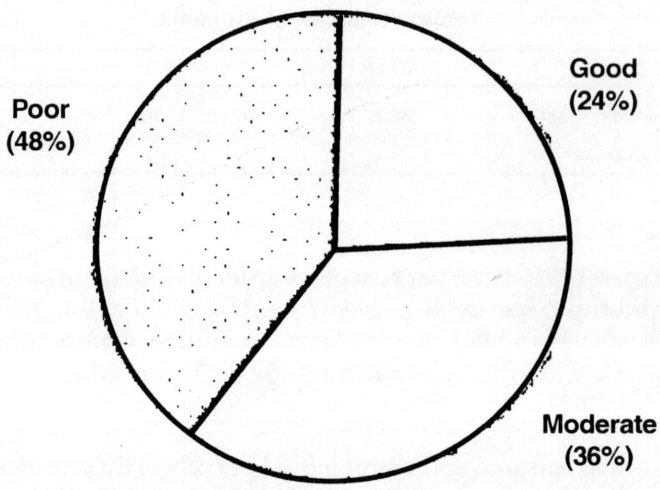

Figure 2.2: General Awareness of Environmental Concepts

Environmental Management

Environmental management was not clearly known to the students due to poor inclusion of the management portion. Only 17 per cent of the students answered the questions. 24 per cent answered to some extent and 59 per cent did not answer well. This reveals that the portion related to environmental management should be included more than the present level (Figure 2.4).

Sources of Knowledge

Students get environmental related knowledge from textbooks, teachers, parents, relatives and friends. When tested for the extent of help from all the sources, 18.6 per cent of the students told that

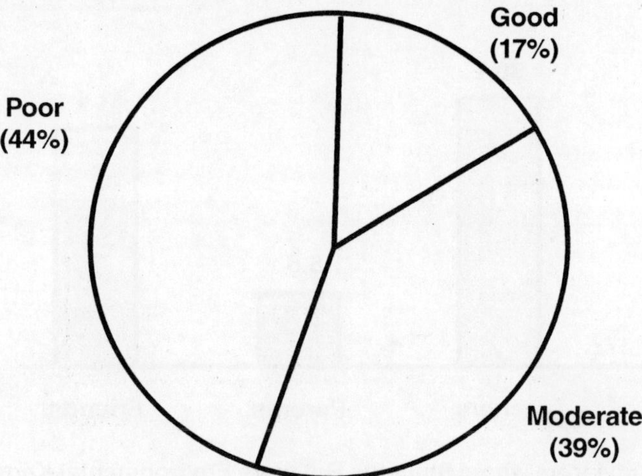

Figure 2.3: Knowledge of Environmental Problems

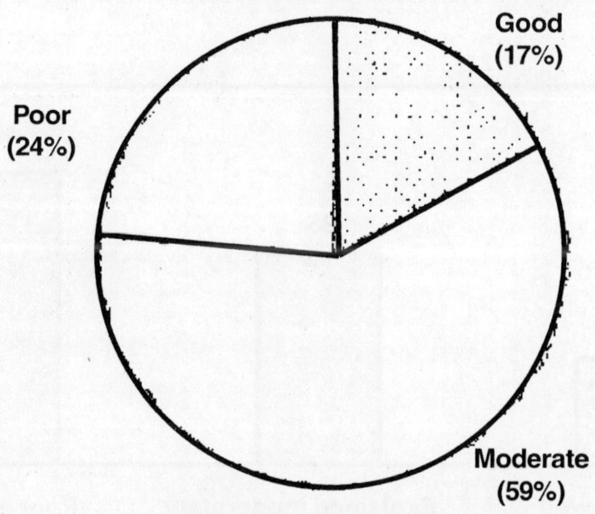

Figure 2.4: Environmental Management

they gained more knowledge from the textbooks, 38.1 per cent of the students told that they got more from teachers; 8.5 per cent got knowledge from parents and 31.4 per cent of the students got help from friends to understand the environmental concepts. Only 3.4 per cent of the students got help from their relatives (Figure 2.5). This shows that the participation of these sources is not adequate.

Exposure to Environmental Problems

For the question related to the knowledge about the environmental problems which affected the students) 16.5 per cent answered well. They explained how they were affected by water pollution, air

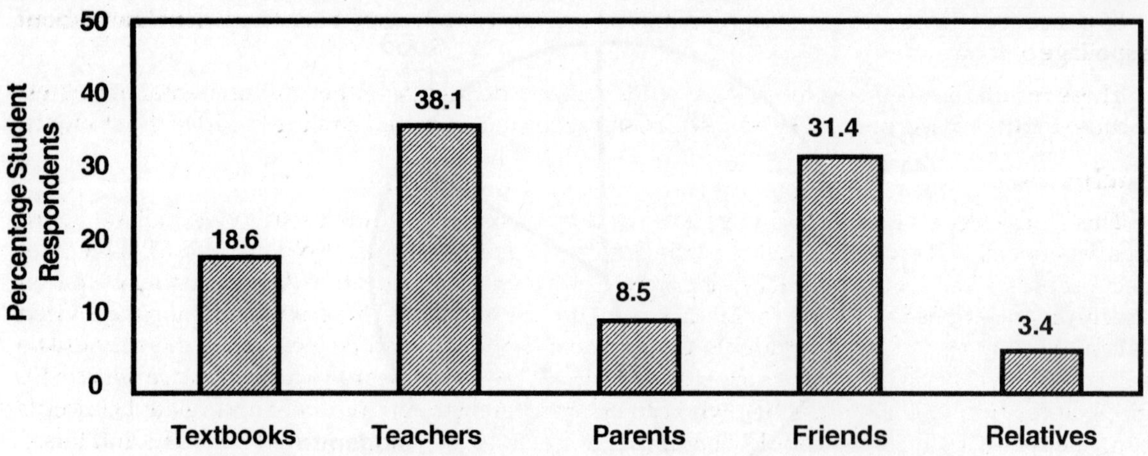

Figure 2.5: Source from whom Students Got more Environmental Knowledge

pollution and unclean surroundings. 35 per cent explained to some extent and 48.5 per cent did not explain well (Figure 2.6).

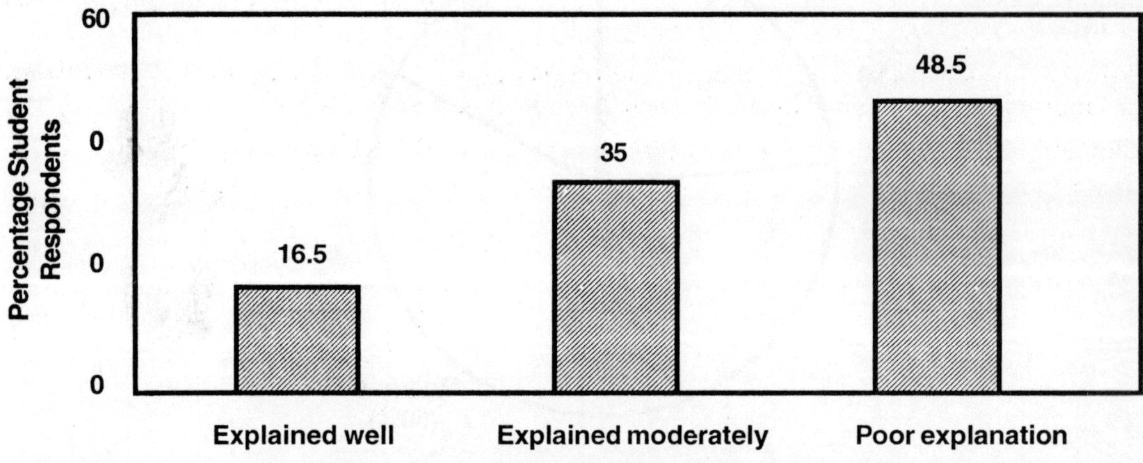

Figure 2.6: Knowledge of Students about the Exposure to Diseases Induced by Environmental Pollution

When they were asked to specify the diseases they were prone to, 38 per cent told the diseases. But they were unable to differentiate between water borne, air borne and other diseases. Others did not specify any disease.

For the question related to the effects of polluted surroundings, 2 per cent answered that they were affected seriously and they were able to explain the causes of polluted surroundings. 56 per cent explained moderately and 42 per cent, were unable to explain.

86 per cent of the students did not explain the causes of spoilage of food. 16 per cent knew about the spoilage of food.

These results showed that most of the students were not aware of the environmental problems. The cause for the environment related diseases should be incorporated and explained to the students.

Conclusion

This study reveals that the primary level students, especially students studying in State Board Primary Schools, were not well aware of the Environmental Concepts, Problems and Management aspects as prescribed by the NCERT guide lines. Environmental Education Aids are being prepared by many organizations like Centre for Environment Education (CEE). These aids and other improvised aids can be used to motivate the students. Government should take proper action. Teachers should be assessed for their environmental awareness and should be given in-service as well as pre-service EE training accordingly. It is the teacher, who can create interest in environment and related concepts among the students. Teachers should definitely use teaching aids, other related materials and kits.

References

Best, John W., 1996. *Research in Education.* Prentice-Hall of India Private Ltd., New Delhi, p. 293–308.

Goode, Willimn, J. and Hatt, Paul, K., 1981. *Methods in Social Research.* McGraw-Hill, New Delhi, p. 255–260.

Gupta, S.C., 1990. *Fundamentals of Statistics.* Himalaya Publishing House. New Delhi, p. 380, 519–521, 1043.

Lakshmi Perin Devi, T., 1996. *Proceedings of Fourth Seminar-cum-Exhibition on Students' Project Scheme.* Tamil Nadu State Council for Science and Technology, Chennai, p. 47–49.

NCERT, 1993. *Minimum Levels of Learning (Primary Education).* NCERT, Chennai, p. 59–77.

Rohatgi, K., 1982. *Technological Forecasting.* Tata McGraw Hill, New Delhi, p. 137–138.

Chapter 3

Vehicular Pollution Abatement and Control Measures Implemented in Delhi

D.S. Kharat, Paritosh Kumar and P.M. Ansari*
Central Pollution Control Board, Ministry of Environment and Forests, East Atjun Nagar, Shahdara, Delhi – 110 032, India

ABSTRACT

Vehicular emissions are the main source of air pollution in Delhi. In an effort to address the air pollution problems caused by vehicles, series of steps have been taken up in recent past at the instance of the Hon'ble Supreme Court of India. During implementation of various pollution control measures, main emphasis was given to introduction of natural gas as an automotive fuel on large scale and necessary infrastructure facilities were strengthened. On account of various measures for abatement and control of pollution, the air quality in Delhi has indicated significant improvements. The present paper gives the details of the activities undertaken for implementation of the abatement and control measures between 1998 and 2005. The aspects covered include approach, action plan, different control measures, constrains, effect on ambient air quality and future plans.

Keywords: *CNG, Control measures, Delhi, Vehicular emission.*

Introduction

Besides being capital, Delhi is a major centre of commerce, industry and education. Rise in population and growth in economic activity have led to increase in air pollution in the city. Main

* *Corresponding Author*. E-mail: kharatds@hotmail.com.

sources of air pollution in Delhi are emissions from vehicles, coal based thermal power plants, industrial units and domestic activities. As per the published data, vehicular pollution contributes about 67 per cent of total air pollution of Delhi. With their contribution of 13 per cent, power plants are the second most polluting source in the city. The industries and domestic activities generate 12 per cent and 8 per cent of pollution. The air quality in the city is being monitored in terms of major pollutants since 1978. In the late 1980s, Delhi was reported to be one of the most polluted cities in the world. This assessment was mainly based on suspended particulates in the ambient air.

The data published by Delhi Transport Department indicate that vehicular population in 1991 was 1.9 million, which has increased to 4.1 million as on 1-4-2004. The average daily addition of new vehicle is 1679 (328 commercial and 1351 private). It is to be noted that Delhi's vehicular population is more than the number of vehicles of Mumbai, Kolkata and Chennai put together. Considering the severity of air pollution in Delhi, which is predominantly contributed by vehicular emissions, series of measurers have been taken up since 1998 at the instance of the Hon'ble Supreme Court of India. On account of various measures for abatement and control of pollution, the air quality in Delhi has indicated significant improvements.

Approaches and Action Plan

While framing the action plan for pollution control, more focused attention was given to tackle the emissions of commercial vehicles, as these vehicles are poorly maintained and heavily used as compared to private vehicles. Natural gas, which is inherently less polluting than diesel and petrol, was recommended for use as automotive fuel on large scale. While buses were directed to convert to compressed natural gas (CNG) mode, use of CNG or clean fuel was recommended for three-wheeler autorikshaws and taxis. It was also decided to phase out the old commercial vehicles in a phased manner. For the non-commercial private vehicles, improved engine technology was made mandatory. In case of two-wheelers, no decision was taken. However, it was felt that use of LPG could be a long term solution to two-wheelers of the city. The approach for talking vehicular pollution is illustrated in Figure 3.1.

The abetment and control measures implemented in Delhi include improvement in vehicle technology and fuel quality, and instruction of CNG vehicles, which directly helps to cut down vehicular emissions. In addition, indirect control measures such as inspection and maintenance centers, pre-mixed 2T oil dispensers, interstate bus terminals at periphery of city etc. were also considered. The action plan with priority measures is given in Table 3.1.

While hearing the matter of writ petition no. 13029 of 1985; *M.C. Mehta* versus *Union of India and others* on July 28, 1998, the Hon'ble Supreme Court agreed to these priority measures. Most of the works were not completed as per the schedule. During the course of their implementation, the Hon'ble Supreme Court extended target dates on appeals from the concerned implementing agencies. From time to time many more directions were issued by the Hon'ble Court while hearing the matter and related interlocutory applications filed by concerned agencies. Other issues witch were examined by the Hon'ble Court for issuance of the direction include engine technology, fuel quality, non-destined transit trucks and express bye-pass roads.

Implementing Agencies

The concerted efforts for abatement and control of vehicular pollution were initiated after the Ministry of Environment and Forests, Government of India published the White Paper on Pollution With Action Plan in 1987. Subsequently, a committee called Environment Pollution (Prevention and

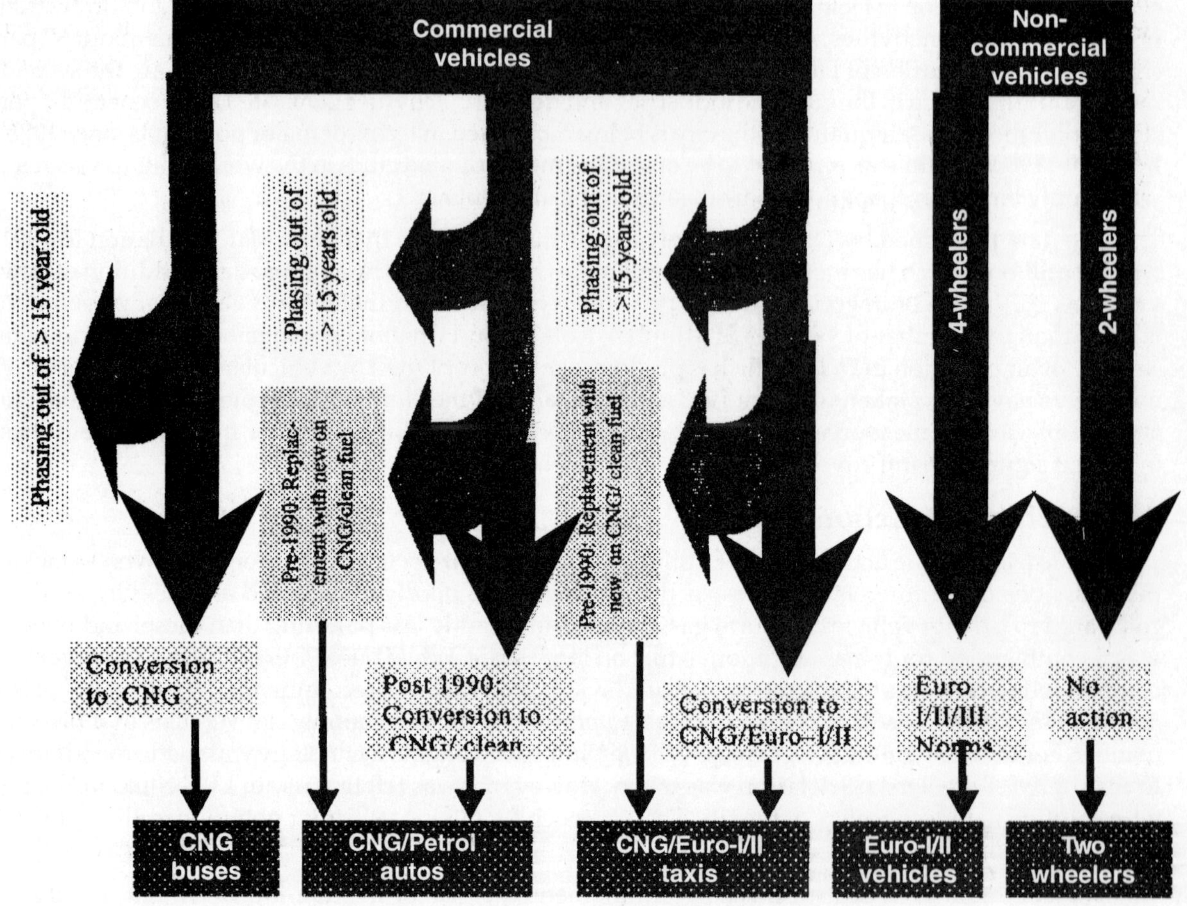

Figure 3.1: Approach for Containment of Vehicular Emissions

Control) Authority for the National Capital Region (EPCA) was constituted on January 29, 1998 to monitor the implementation of the action plan at the instance of the Ministry of Environment and Forests and the Hon'ble Supreme Court of India. Central Pollution Control Board (CPCB) provided necessary technical and secretarial support to the EPCA. Main stakeholders in implementing the action plan include Indraprashtha Gas Ltd., Delhi Transport Department, oil companies and the automobile manufacturers.

Abatement and Control Measures

CNG Supply Network

Initially, there were 9 CNG refilling stations in Delhi in 1998. These stations were mostly located in South Delhi. With the decision to convert city buses, autorikashaws and taxies on CNG mode, it was required to increase the number of CNG stations and also to spread the same in the entire Delhi. Accordingly, in 1998, it was planned to increase the number of CNG stations to 80 based on a survey conducted by the Central Road Research Institute, New Delhi. An independent agency called

Indraprashtha Gas Limited (IGL) was incorporated on December 23, 1998 to carry out work for installation of additional CNG refilling stations and other necessary infrastructure. The IGL is a joint venture of Gas Authority of India Limited, Bharat Petroleum Corporation and Government of National Capital Territory (NCT) of Delhi. In the initial stage, IGL succeeded in installation and commissioning of 44 CNG refilling stations by March 31, 2000. The target of 80 CNG station was achieved by IGL by the end of year 2001.

Table 3.1: Action Plan for Abatement and Control of Vehicular Pollution

Sl.No.	*Abatement and Control Measures*	*Target*
1.	Augmentation of public transport (stage carriage) to 10,000 buses from existing 6,600 buses	01-04-2001
2.	Elimination of leaded petrol from Delhi	01-09-1998
3.	Installation of pre-mix dispensers for supply of only pre-mix petrol in all petrol stations to two stroke engines	31-12-1998
4.	Replacement of all pre-1990 autos and taxis with new vehicles using clean fuel	31-03-1998
5.	Replacement with financial incentives of post 1990 autos and taxis with new vehicles on clean fuel	31-03-2001
6.	Ban on plying of buses more than 8 years old except on clean fuel	01-04-2000
7.	Entire city bus fleet (DTC and Private) to be steadily converted to single fuel mode on CNG	31-03-2001
8.	New ISBT to be built at North and South-West borders of NCT Delhi to avoid pollution due to entry of inter-state buses	31-03-2001
9.	Gas Authority of India Ltd., to ensure availability of CNG by increase of CNG supply outlets in the cities from 9 to 80	31-03-2001
10	Tow autonomous fuel testing laboratories to be established for monitoring fuel quality specifications and adulteration	01-06-1999
11.	Automated inspection and maintenance facilities to be set up for commercial vehicles in coordination with private sector, for which proposals received, *e.g.* Society of India Automobile Manufactures (SIAM)	Immediate

Till March 31, 2002, a total of 87 stations were installed and made operation. At this stage, it was realized that 87 CNG refilling stations were not adequate to meet the demands of CNG vehicles. Long queues of vehicles waiting for their turn could be seen at all the stations during busy hours primarily due to lack of CNG stations in all parts of the city. As such, the work of installation of new stations had to be continued.

In 2002, the gas transmission pipeline was available in South, North and East Delhi only. There was no facility available in western region. The entire bus fleet of western region was required to come to South or North Delhi for CNG filling leading to overloading of CNG stations in these areas. This also caused loss of mileage in traversing for the CNG filling. The demand of CNG in western Delhi was met by laying of gas pipe line connecting Dhaula Kuan to G.T. Karnal by-pass through West Delhi. By March 31, 2003, 106 numbers of CNG stations were installed which increased to 121 in March 2004. These stations are able to meet the CNG requirements of vehicles in the city. Figure 3.2 shows growth of CNG refilling station over the last six years.

The IGL has planned and installed the above CNG stations under three categories namely mother station, on-line station and daughter station. Mother stations and on-line stations have compression

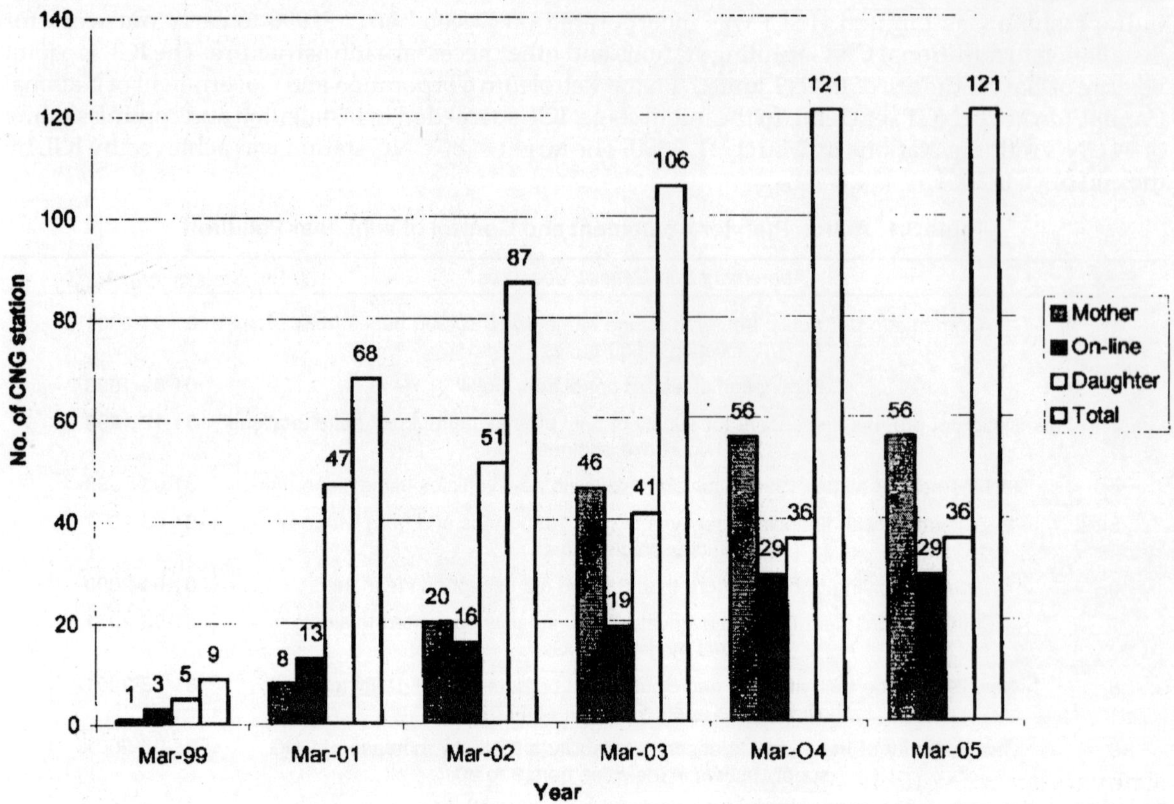

Figure 3.2: Growth of CNG Refilling Stations

facilities, where natural gas is compressed to desired pressure of 200 to 250 bars. This pressure is sufficient for refilling CNG in the vehicles. The daughter stations do not have their own compression facilities but get the CNG cascades though trucks from the mother or on-line stations. These cascades are replaced once the CNG is exhausted. The mother stations and on-line stations can supply CNG to both heavy vehicles and light vehicles. The daughter stations cater to the needs of only light vehicles including taxis, cars and three-wheeler autorikshaws.

While providing the facility of CNG refilling stations, proper planning is needed so that the facility covers all the region and parts of Delhi and hassle-free supply of CNG to all CNG vehicles can be assured. As the city buses are the largest consumer of CNG, attempts were made to locate adequate number of CNG refilling station as close as possible to the bus-depots and wherever possible stations were installed at the bus-depot itself. Out of the existing 121 CNG stations, 56 are mother stations (19 for DTC at its depots), 29 are on-line stations and 36 are daughter stations. Among the 56 mother stations, 19 are installed in the bus-depots.

The combined CNG dispensing capacity of 121 CNG refilling stations works out to be 16.30 lacs kg. per day against the initial target of 16.11 lakh kg per day. With increase in number of CNG refilling stations, sales of the CNG has proportionately increased. Actual sale of CNG in March 2004 was 7.96 lakh kg per day. The installed capacity and sale of CNG are given in Figure 3.3.

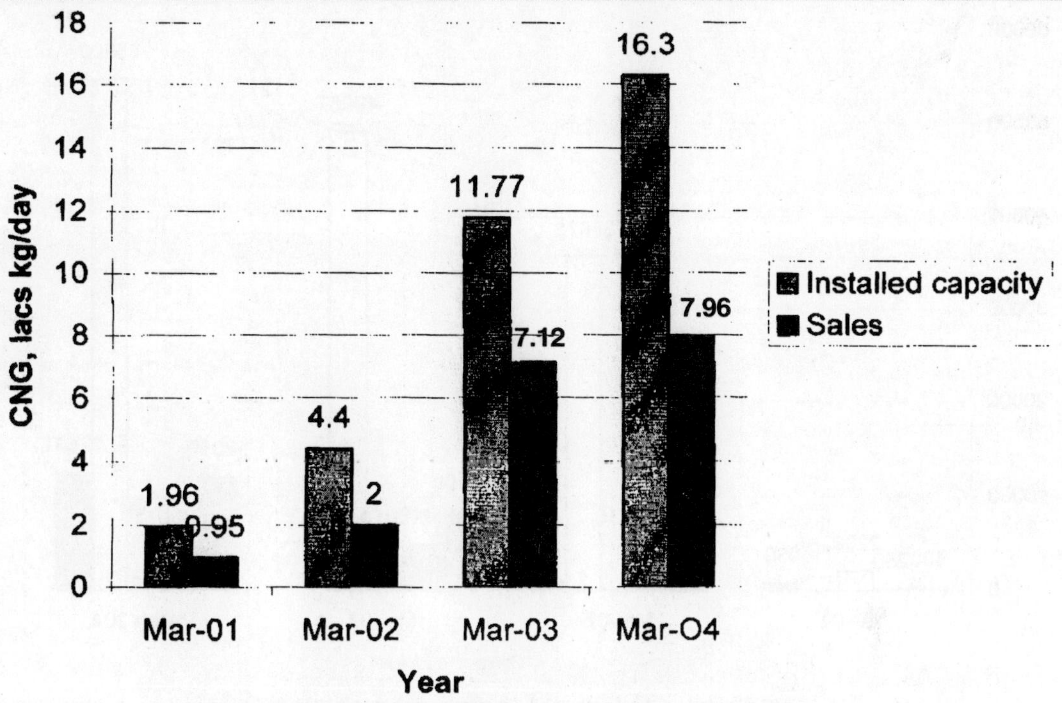

Figure 3.3: CNG Dispensing Capacity and Sales

Conversion of Vehicle Fleet to CNG Mode

Under the conversion programme, three approaches were adapted: (*a*) purchase of new dedicated CNG vehicles; (*b*) conversion of existing engine to operate on CNG mode by incorporating necessary modification; and (*c*) by retrofitting of new CNG engine in existing vehicles. In case of three-wheeler aoutorikshaws, which use two-stroke engine, conversion of the old petrol engine to CNG was not feasible. Introduction of CNG operated autorikshaws were possible when leading manufactures like Bajaj Aoutos and Scooter India started production of duel fuel autorikshaws, which operate on petrol as well as on CNG mode. Use of clean fuel, *i.e.* lead-free petrol with low sulphur and benzene contents was allowed for autorikshaws.

Large commercial vehicle fleet was converted to CNG by using all the three approaches. Two principle bus suppliers TELCO and Ashok-Leyland strengthened their infrastructure to manufacture dedicated CNG buses. They also undertaken work of retrofitment of new CNG engines in their old buses. Few other leading workshops also modified old diesel engines to operate on CNG mode. Conversion of old buses to CNG was a cheaper proposition as compared to purchasing of new CNG buses. CNG being cheaper as compared to petrol and diesel, can partly off set cost involved in purchase or conversion operation. As in March 1999, only one bus and 1451 cars were playing on CNG mode in Delhi. There was no autorikshaw at this point of time. The growth of different CNG vehicles in Delhi is presented in Figure 3.4.

In conversion of vehicle fleet to CNG, there were some teething problems initially due to lack of technical know-how and defined regulatory mechanism to deal with CNG vehicle. While the conversion of a petrol engine to CNG required only minor modification and no elaborate procedure to certify its

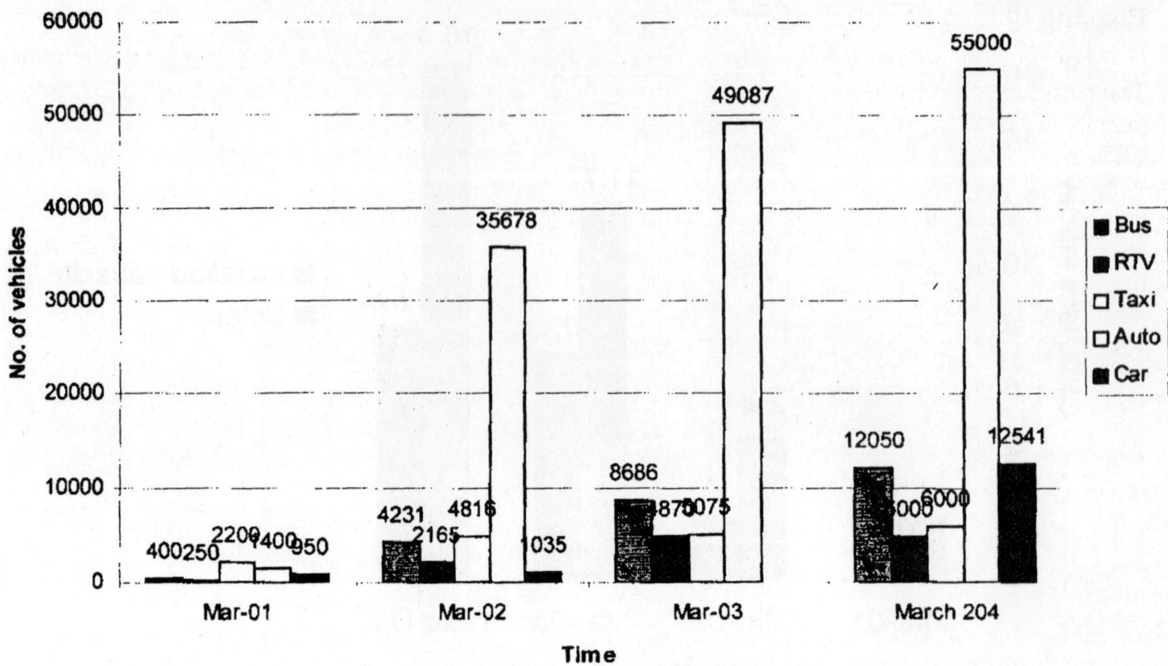

Figure 4.4: Population of Different CNG Vehicles

safety and road worthiness were required, a diesel engine had to undergo serious modifications and therefore had to be placed on a different footing for which knowledge on diesel conversion and kits were limited. The Ministry of Road Transport and Highways in association with the statutory recognized vehicle testing agencies, *i.e.* Indian Institute of Petroleum, Dehradun and Automotive Research Association of India, Pune evolved and notified procedure for the type approval, safety and emission norms for CNG vehicles which helped concerned agencies to fulfil the task of conversion of vehicle fleet on CNG mode.

In beginning of transition, some CNG vehicles, particularly buses were reported to have caused problems like leakages of gas, miss-firing, seizure of engines, starting problem, inability to negotiate flyovers, heating up of engine etc. There have been few cases of fire due gas leakages. The defects were rectified by the vehicle manufacturers to ensure trouble-free operation of the vehicles.

Despite the enormous efforts by the concerned implementing agencies, the target set for the conversion of the entire bus fleet to CNG mode by the March 2001 was not met initially due to shortage of CNG buses and conversion facilities or delay in procurement or placement of orders for CNG buses. As such, diesel buses were allowed to ply after March 2001, if their owners filed affidavits that they had placed a firm order for either for same number of new CNG buses or for the conversion/retrofitment of their old buses. These diesel buses were directed to carry a sticker from the Delhi Transport Authority stating that this bus has been allowed to ply against a firm order for CNG bus that will replace it from September 30, 2001. The diesel buses which were left after September 30, 2001 were further allowed to ply subject payment of fine on daily basis. These relaxation were considered in the interest of commuters and to ensure that transition takes place as smoothly as possible without creating unnecessary disruption in to the public transport.

Phasing Out of Old Vehicles

For phasing out of old commercial vehicles, the schedule as indicated in Table 3.2 was directed. The commercial vehicles were phased out in stages. In the first stage, vehicles of age more than 20 years were phased out by October 2, 1998. The vehicles of age between 17 to 19 years were phased out till November 15, 1998. All the vehicles of 15 to 18 years were phased out by December 12, 1998. For vehicles below 15 years of age, conditional phasing out was implemented. Diesel buses of more than 8 years age were directed to convert to CNG by April 1, 2000. The date was subsequently extended. In case of taxis and autos different actions were taken for phasing out depending on their date of manufacture. Pre-1990 autos and taxis were directed for replacement with new vehicles using CNG or clean fuel. In case of post-1990 vehicles, replacement with financial incentives fro m the Government was effected by March 31, 2001.

Table 3.2: Schedule for Phasing Out of Old Vehicle

Sl.No.	*Age of vehicle*	*Target*
1.	More than 20 years old	2nd October 1998
2.	17–19 years	15th November 1998
3.	15–18 years	31st December 1998

The retrofitment of CNG kits in the post-1990 autos and taxis was allowed. However, retrofitment of CNG kits in pre-1990 autos and taxis, was not allowed because these vehicles were manufactured before the imposition of the emission standards of 1991, and were gross emitters.

The Delhi Government has to initiate stringent action including cancellation of permits and impounding of vehicles. Till 31st December 1998, 7696 numbers of three-wheeler autorikshaws were deposited with the Delhi Government. Out of these, 4403 were scrapped. The impounded vehicles were released only after the vehicle owners had given an undertaking to sale the vehicle outside Delhi or to scrape. Delhi Government also launched a scheme for purchase of new vehicles for the owners of phased out vehicles for post-1990 categories under the scheme, exemption of sale tax @ 8 per cent and subsidy of 4 per cent on interest rates on loan were given.

The transition of buses to CNG has not been smooth. The Action Committee of private schools and other institutions filed an interlocutory application to the Hon'ble Supreme Court for allowing use of their school buses more than 8 year old for transporting students. They submitted to the Hon'ble Court that these buses cater to the need of students only within a certain distance of the location of school and ply only for a shorter distance and therefore, do not strictly speaking, deserve to be treated as commercial buses. After considering the an interlocutory application, the Hon'ble Court vide its order dated April 24, 2000, directed that the buses be steadily converted to single fuel mode on CNG on or before March 31, 2001. The Hon'ble Supreme Court further directed that educational institution buses shall be got checked for pollution control to avail the benefit of the order. In the event it is found that any bus does not have pollution under control certificate, that bus shall not ply under the force of the order.

Improvement in Vehicle Technology

With increasing concern for controlling air pollution caused by automobiles, the emission norms for various categories of automobiles were progressively tightened. It was made mandatory the registration of Bharat Stage I (Euro I) compliant light vehicles from June 1, 1999. Subsequently, registration of only Bharat Stage II (Euro II) compliant light vehicles was started from April 1, 2000. In case of heavy vehicles, Bharat Stage II norms were enforced from October 23, 2001. Before the intervention of the Hon'ble Supreme Court, Bharat Stage I and Bharat Stage II norms were planed to be implemented from April 1, 2000 and April 1, 2005 respectively. In case of two-wheelers and three-wheelers, Barat Stage II norms are implemented from April 1, 2005. For the other vehicles, Bharat Stage III (Euro III) norms have been enforced. At present only Bharat Stage III compliant vehicles (except two-wheelers and thee-wheelers which are Bharat Stage II) are being registered in Delhi. Further improvements in emission norms are expected in days to come. For compliance of the emission norms, the automobile manufacturers need to improve the engine technologies.

Improvement in Fuel Quality

Fuel quality is one of the predominant factors affecting the emissions and the ambient air quality. From time to time, the quality specifications have been laid down by the Bureau of Indian Standards and the Ministry of Road Transport and Highways. For the first time, a major initiative towards fuel quality improvement was taken when the emission related parameters were notified under the Environment Protection Act in 1996. Based on the local ambient quality and health effect of auto exhaust pollutants, critical parameters including lead, sulphur and benzene is considered for fuel quality enhancement, which was implemented as per the schedule given below.

In 1997, the sulphur content in diesel was 0.25 per cent, which was brought down to 0.05 per cent with effect from April 1, 2000 and made available at selected outlets in NCT of Delhi. Supply of this diesel was extended to the entire NCT of Delhi from March 31, 2001.

Petrol quality was also improved in respect of benzene, sulphur and lead content. Unleaded petrol was introduced in April 1995 for new four-wheeled vehicles in Delhi. Introduction of unleaded petrol enabled use the catalytic converters in the vehicles. Leaded petrol was completely phased out from National Capital Region (NCR) from September 1, 1998 and throughout country from January 1, 2000. Petrol supply with 0.05 per cent sulphur content from April 1, 2000 and 1 per cent benzene content from November 1, 2000 was made available in NCT of Delhi. In 1997, these parameters were 0.1 per cent and 5.0 per cent respectively.

Prevention of Fuel Adulteration

Adulteration of fuels is one of major causes of excess emission of pollutants. The Ministry of Petroleum and Natural Gas (MoP&NG) was asked to take effective measures to prevent fuel adulteration. The MoP&NG has taken preventive measures, which include issuance of the Solvent, Raffinate and Slop (Acquisition, Sale, Storage and Prevention of Use in Automobiles) Order, 2000; and the Naphtha (Acquisition, Sale, Storage and Prevention of Use in Automobiles) Order, 2000. Naptha and Solvent orders are in various stages of implementation. An Anti-adulteration Wing has also been set up under the aegis of the MoP&NG and Department of Food and Civil Supply, Delhi Government.

Besides the above regulatory mechanism, one independent fuel-testing laboratory has been commissioned at NOIDA in 2000 to check adulteration of fuels. The laboratory is maintained by Society for Petroleum Laboratory with members from Indian Oil, Oil Coordination Committee, Central Pollution Control Board, Indian Institute of Petroleum, Delhi Government, Bharat Peroleum, Hindustan Petroleum, IBP, SIAM and Tata Energy Research Institute. The capital cost of Rs. 13 crores for the laboratory was borne by the MoP&NG. The laboratory was made operational in 2000, but the issue of mobilizing funds for the recurring cost is yet to be resolved completely. The operation and maintenance cost of the laboratory is being presently shared by SIAM, Ministry of Road Transport and Highways, Ministry of Heavy Industries, Ministry of Petroleum and Natural Gas, Government of NCT of Delhi and EPCA (Ministry of Environment and Forests). Since its commissioning in 2000, the laboratory has tested samples which have been mostly collected by the Anti-adulteration Wing. In the year 2003, an independent agency "Centre for Science and Environment, New Delhi" was also given the task to collect samples of fuels and perform testing to know the status of adulteration. It was found that 8.6 per cent of samples failed to confirm to the prescribed standards.

Premixed 2T Oil Dispensers

Before the pollution control programme was initiated, the two-stroke engined vehicles such as two-wheelers and three-wheeler autorikshaws were using manually mixed 2T engine oil with petrol while fuelling their vehicles. Manual mixing of 2T oil with the petrol does not ensure the required proportioning of fuel and the oil. For proper lubrication of two-stroke engine, about 2 per cent of 2T oil is sufficient. It was observed that the owners of two-stroke engined vehicles mostly use excessive quantity of 2T oil amounting to more than 5 per cent. This is one of causes for the higher emissions of pollutants. To check excessive use of 2T oil, it was decided to install pre-mixed oil dispensers at the petrol pump. It was also decided to ban the sale of loose 2T oil at petrol pumps, service garages etc. These decisions were implemented by the Ministry of Petroleum and Natural Gas and Oil Companies within a span of tow years. By the end of year 2000, 646 numbers of premixed 2T oil dispenses were installed at all the 294 petrol pumps located in Delhi.

Emission Warranty

The issue of emission warranty was taken up with Society of Indian Automobile Association. The SIAM agreed to give emission warranty for all the categories of vehicles including passenger cars, multi-utility vehicles (MUV), commercial vehicles and two-wheelers and three-wheelers. The warranty is valid for all passenger cars, multi-utility vehicles and two-wheelers and three-wheelers soled after July 1, 2001, and for commercial vehicles from the date of implementation of Bharat Stage II emission norms. To begin with, the warranty was started in Delhi, Mumbai, Kolkata and Chennai where Bharat Stage II emission norms were implemented. The warranty period of each vehicle category and criteria are given in Table 3.3. To avail the benefits of the emission warranty, the vehicle owner is required to sign an agreement with the manufacturers.

Table 1.3: Criteria for Emission Warranty

Sl.No.	Vehicle Category	Criteria
1.	2 wheelers	30,000 km or 3 years whichever occurs earlier
2.	3 wheelers	30,000 km or 1 year whichever occurs earlier
3.	Passenger cars	80,000 km or 3 years whichever occurs earlier
4.	Multi-utility Vehicles	80,000 km or 3 years whichever occurs earlier
5.	Commercial Vehicles	80,000 km or 1 year whichever occurs earlier

Provision of New ISBTs

In order to ensure that the interstate polluting diesel buses do not enter the city, it was proposed to have two interstate bus terminals (ISBTs) at the periphery of the city. At present, there are two ISBTs in addition bus terminal at Kashmere Gate. These tow new bus terminals are at Anand Vihar and Sarai Kale Khan. The Anand Vihar Bus Terminal caters to the buses coming from the eastern side of the city The buses arriving towards the south Delhi side is terminated at Sarai Kale Khan Bus Terminal.

For the buses arriving from North and South-West side of Delhi, two interstate bus terminals have been planned at Narela and Dwarka.

For the Dwarks ISBT payment of land charges amounting to Rs. 8.0 crores has been made by Transport Department to the Delhi Development Authority. In case of ISBT at North border, at Narela the land allotment is yet to be settled the by DDA and Transport Department.

Inspection and Maintenance Centre

Setting up of Inspection and Maintenance Centre for certification of fitness and emission is necessary to reduce emissions from in-use vehicles. Looking into the number of vehicles many Inspection and Maintenance Centres are required, which may not be possible for the government to create, operate and maintain the same. As such, it was desired to give permission to private independent bodies. One such new centre is planned to be set up at Burari, for which Delhi Government has tied up with ARAI, Pune for consultation.

Other Measures

There is ban on playing of good vehicles in day time, which helps to reduce traffic congestions as well as air pollution problems in pick hours. This direction is being implemented by Delhi Traffic Police. It is also to mention that over loading of goods vehicles has been strictly banned.

The pollution under certificate (PUC) programme is also being enforced by the Delhi Government, for which computerized emission checking facilities have been set up at petrol pups and workshops in deferent parts of the city. The centres issue PUC certificate to the vehicles meeting emission standards. In case the vehicle is found polluting beyond prescribed emission norms, necessary repairs or tuning in the vehicle is required. At present, there are 280 centres for petrol and diesel vehicles and 73 centres for the diesel vehicles in Delhi.

Constraints

The work for setting up of 80 CNG stations was not accomplished within target date. Cause for the delay was mainly attributed to procedural delays in allotment and acquisition of land, and getting clearances for setting up of CNG stations and pipelines etc. In case of conversion of vehicle fleet on CNG also, the target date had to be extended. Initially, two independent fuel testing laboratory were planned. Out of these, one has been commissioned at NOIDA. The decision to set up other fuel testing laboratory has been kept under abeyance till the issue of mobilizing funds to meet the operating cost of existing laboratory at NOIDA is completely resolved. The action on setting up of inspection and maintenance centres has been delayed. Delhi Government has expressed its inability to initiate action due to stay order granted by the Calcutta High Court in the matter of All India Federation of Motor Vehicles Department Technical Executives Officer's Association and others versus Union of India and others [case no. 19138 (W) of 1992] against Section 56(2) of the Motor Vehicles Act regarding authorization of private testing station. The stay order has been vacated now.

In an effort to promote cleaner fuels other than natural gas, a project 'Introduction of Propane as an Automotive for Three-wheeler Autorikshaws' was commissioned by M/s G&T Resources Worldwide in 1998. Under the project, G&T Resources Worldwide was allowed to carry out experimentation on 25 new and 25 old autorikshaws using propane as fuel. The project did not reached to the conclusion.

Effect of Abetment and Control Measures on Ambient Air

The CPCB has been regularly monitoring ambient quality in Delhi for parameters including sulphur dioxide (SO_2), nitrogen dioxide (NO_2), suspended particulate matters (SPM), carbon monoxide (CO_2) and lead (Pb). The annual average values of these parameters during and after implementation of control measures *i.e.* 1998 to 2005 are presented in Figures 3.5(a) to (e). It can be seen that the concentrations of SO_2, CO_2 and Pb in the ambient air have shown decreasing trend since 1998. As regards the NO_2 and SPM are concerned, their concentrations in the ambient air have slightly increased during the period.

The reduction in SO_2 level in ambient air was observed to be 64 per cent on account of various measures for abatement and control of pollution. Similarly, CO level in ambient air has decreased by 53 per cent. As much as 79 per cent reduction in Pb level was also observed. The reduction of sulphur and lead content in the ambient air is primarily due to improvement in fuel quality as discussed earlier. Improved vehicle technology and three way catalytic converters contributed to reduction of CO emissions. As per the monitoring results, an increase of 31 per cent in NO_2 level and 20 per cent increase in SPM level in air were observed after implementation of control measures. Higher ignition point of CNG may be the reason for more NO_2 emissions in auto exhaust. Marginal increase in SPM level in ambient air indicates presence of other sources of SPM emissions. It needs to be noted that the improvement in air quality in terms of SO_2, CO_2 and Pb as observed during monitoring is despite the addition of new vehicles in Delhi, which is average 5,03,963 vehicles per year.

Future Plans of Implementing Agencies

Introduction of CNG as automotive fuel has been one of major step for containment of vehicular pollution in Delhi. Under this programme, emphasis has been given to conversion of commercial vehicles such as buses, three-wheeler autorikshaws and taxis to CNG mode. In near future emphasis will be given to conversion of light goods vehicles playing in Delhi for which time schedule has already been worked out. At present, CNG facility is not available in neighbouring towns of Delhi. For the supply of CNG in these areas, feasibility study is underway in towns of Noida, Greater NOIDA, Faridabad and Gurgaon in the first phase. IGL has proposed to cover towns of Ghaziabad and Bahadurgarh in the second phase.

In the future, high capacity bus system is to be introduced at selected routes. One such bus is presently being operated on experimental basis at Mulchand Hospital–Dr. Ambedkar Nagar route. There is also plan to operates electric trolley buses at least on two routes. One such route is Badarpur–Pragati Maidan. Mass rapid transport system is being implemented by Delhi Government with a view to place non-polluting and efficient rail based system, which will be integrated with other modes of transport.

At present, there is one inspection and maintenance system for commercial vehicles, which is inadequate. Additional inspection and maintenance centre has been planned at Burari. In order to avoid entry of non-destined vehicles, a plan for the construction of express by-pass road around Delhi has been approved. The bye-pass road will be 240 km long.

Conclusion

Most of the control measures discussed in the chapter have been tried and implemented for the first time in the country. Supply of lead free petrol was started in Delhi with few outlets and spread to the entire city and then in the entire Nation Capital Region. Similarly, the fuel with low sulhpur and benzene content was introduced initially in Delhi. Registration of only Euro I (Bharat Stage I) and

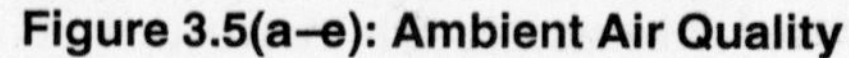
Figure 3.5(a–e): Ambient Air Quality

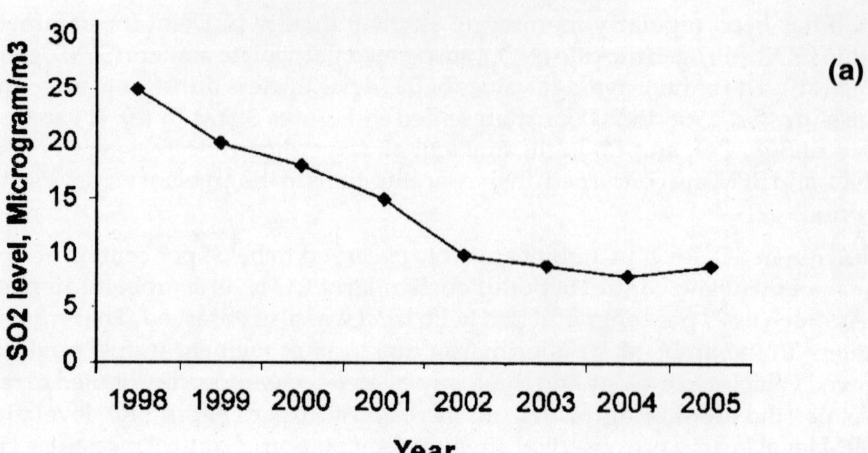

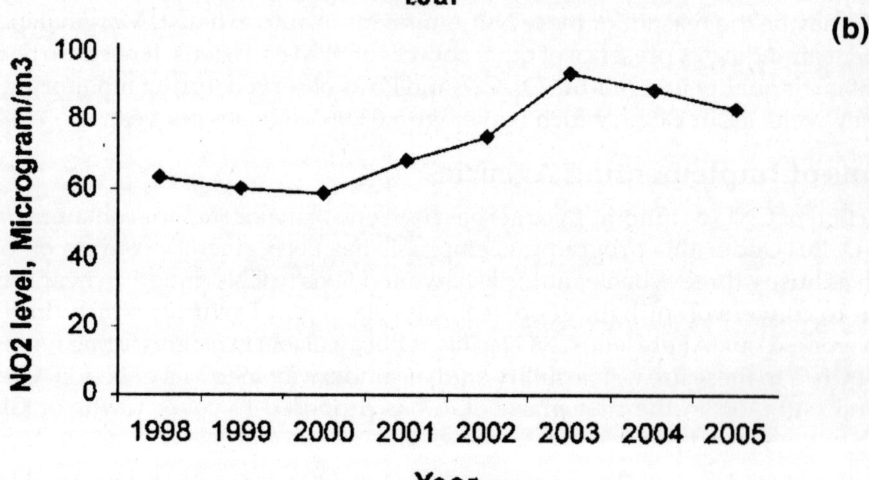

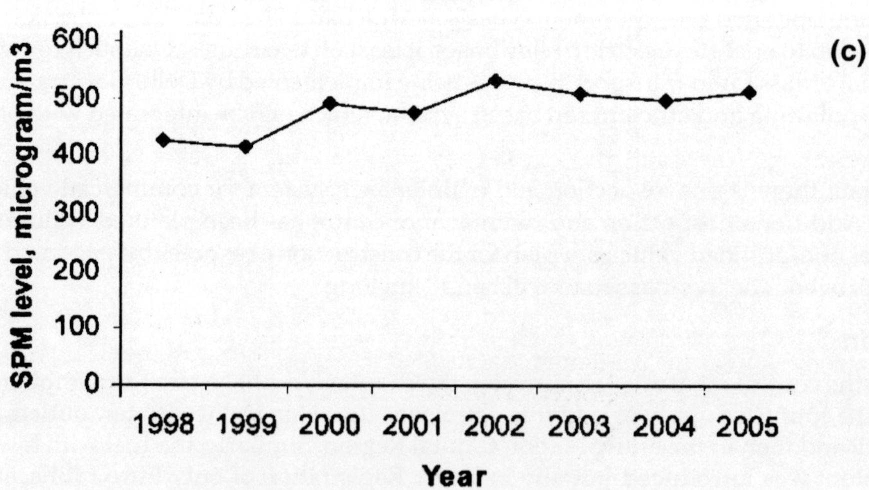

Contd...

Figure 3.5(a–e)–Contd...

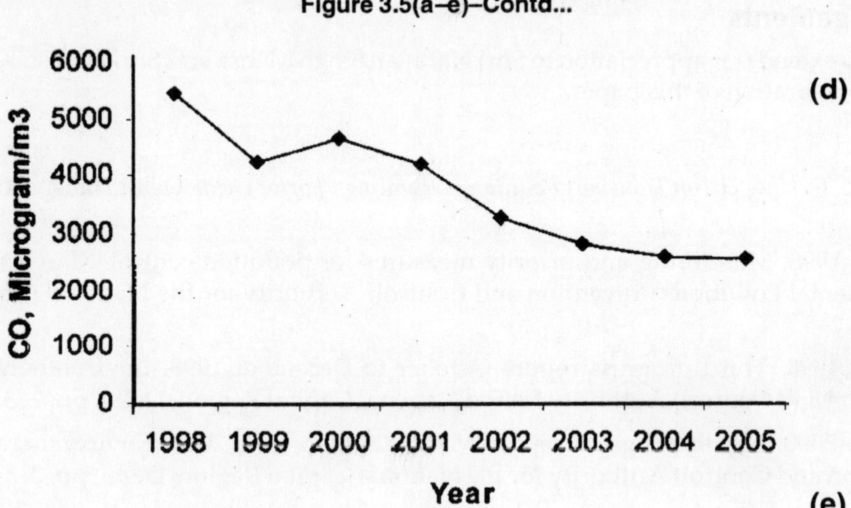

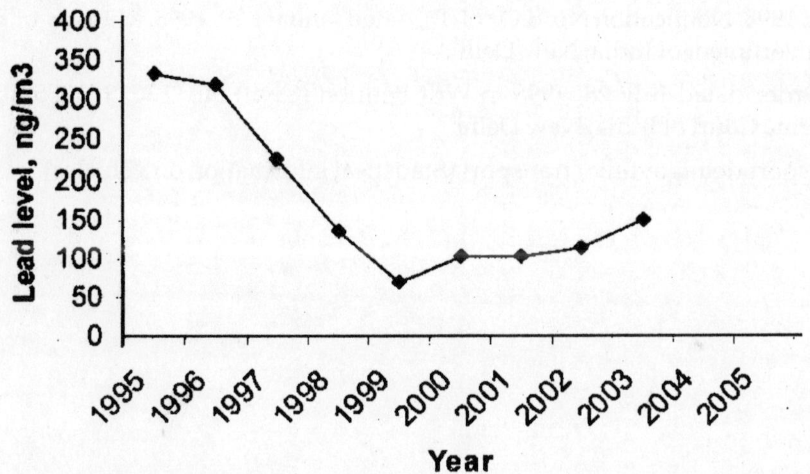

Euro II (Bharat Stage II) compliant vehicles have been started in Delhi. In case of CNG vehicles, experiment with few buses with modified engines to operate on CNG was initiated in Mumbai. However, introduction of CNG as an automotive fuel on mass scale has been taken place in Delhi. With more than 90,000 vehicles playing on CNG, Delhi is now one of the cities in the world to have large number of vehicles operating on CNG.

The GAIL India Limited is in the process of expanding CNG supply network in 22 different cities. Experience gained though the implementation of vehicular pollution control measures in Delhi will be useful in future. The practice followed in Delhi for the containment of vehicular pollution so far can serve as role model for the development of pollution control programme of other cities in India.

Acknowledgements

We wish to extend our appreciation to Shri Narayan Singh Mehra and Shri Lokesh Kumar for the assistance in preparation of this paper.

References

DTD Plan, 2002. *Tackling Urban Transport Pollution: Operating Plan for Delhi*. Delhi Transport Department, Delhi.

EPCA Report, 1998. Monitoring and priority measures for pollution control–March to June 1998. Environmental Pollution (Prevention and Control) Authority for the National Capital Region, Delhi, pp. 2–3.

EPCA Report, 1998. Third progress report–October to December 1998. Environmental Pollution (Prevention and Control) Authority for the National Capital Region, Delhi, pp. 2–3.

EPCA Report, 2000. Eighth progress report–March 2000 to June 2000. Environmental Pollution (Prevention and Control) Authority for the National Capital Region, Delhi, pp. 2–5.

MoEF Report, 1997. White paper on pollution with an action plan. Ministry of Environment and Forests, Government of India, New Delhi, pp. 15–17.

MoEF Notification, 1998. Notification No. S.O. 93 (E), dated January 29, 1998. Ministry of Environment and Forest, Government of India, New Delhi.

SC Order, 1998. Order dated July 28, 1998 in Writ Petition (Civil) No. 13029/1985, issued by the Hon'ble Supreme Court of India, New Delhi.

http://www. transport.delhigovt.nic/transport (Statistical information on vehicles).

Chapter 4

Review of Market Potential for Two-wheelers Using Time Series Analysis

*V. Joseph Paul Raj**

Department of Management Studies, St. Joseph's College of Engineering, Chennai – 119

ABSTRACT

Market potential is the maximum possible sales of all sellers of a given product in a defined market during specified time period (William, 1977). Maximum sales opportunities for a product of an individual company is referred to as sales potential, which is the maximum share of market potential. In this paper an attempt has been made to analyze market potential of two-wheelers in terms of seasonal, cyclical and erratic (demand) sales of during the period from 1997–98 to 2004–05. The analysis of the present paper has been classified under the heads namely time series approach to sales forecast and variability in sales with respect to time series components.

Introduction

The two-wheeler market has been more than a century old, since the time when just a few two-wheelers hit the Indian roads. The concept of manufacturer, brand name, etc., although present, was not really the deciding factors when it came to the buying of a two-wheeler. What mattered to the consumer then was the fact that this product merely provided personal transport. It was deemed to be luxury to own a two-wheeler and the owner of the two-wheeler was considered to be one among the privileged class.

* E-mail: vjpraj1971@yahoo.com

Over the decades a two-wheeler product category of the automobile industry has come a long way to its current state of being a basic necessity which was fostered mainly by the revolution in the two-wheeler industry, and changing faces of need and want of the two-wheeler being a product of the luxury class meant merely for premier mode of transports has become today more of basic need or want of a family or an individual in his every day life.

The critical factor for over all marketing developments in the coming decade was the company's attitude towards marketing and marketing implications (Avlonitis and Gouremenous, 1995) and the battle for leadership will be between production volumes and product image, between utility and style and between price and image (Ramachandran and Sehrawat, 1996). During the early days, the power driven two-wheelers were preferred in the place of bicycle (Valsan, 2002), later the vehicle having high mileage, post sales service, flexibility and feasibility for transportation were the reasons for choosing the two-wheeler (Xavier, 2004).

Database

Both primary and secondary data have been used for the present study. A survey was made with selected dealers to get acquainted with marketing and marketing strategy followed actually to promote the sale of two-wheelers.

Secondary sources of data relating to the month-wise sale of the selected brand of bikes for 8 years from 1997–98 to 2004–05 were collected from the records of Regional Transport Office (RTO), Madurai.

Model Description

In order to analyse the trend, the data on sales of two-wheelers were subjected to 12 months moving average so as to remove seasonal effect from the time series. Further to rest the significance of the trend over time, the following linear regression equation has been fitted (Grewal, 1987).

$$Y = a + bt$$

where,

Y: Indicates the sales of two-wheelers

t: Indicates the time (years)

Exponential growth rates were worked out by using the following formula:

$$Y = ab^t$$

where,

Y: Indicates the sales of two-wheelers indicates the time (years)

t: Indicate the time (years)

Compound growth rate is (Antilog b-1) × 100

Cyclical and irregular variations are computed with the help of residual Method.

$$C = TC/T$$

$$I = SI/S$$

In order to measure the variability in sales due to trend, cyclical, seasonal and irregular fluctuations, the sales of two-wheelers, and the co-efficient or variations are worked out separately for the brands of two-wheelers.

Time Series Approach to Sales Forecasting

A study on time series approach in sales of two-wheelers would be useful in forecasting the market potential in future. This would in turn help the producers to make decisions on production and marketing of two-wheelers.

The details of seasonal indices of sales of two-wheelers in Madurai city during the period 1997–98 to 2004–2005 are furnished in Table 4.1.

Table 4.1: Seasonal Indices of Sales of Two-wheelers in Madurai City from 1997–98 to 2004–2005

Month	*Seasonal Indices*				
	Hero Honda	*Bajaj*	*TVS*	*Yamaha*	*Enfield*
January	120.44	96.96	92.48	9.63	134.33
February	96.47	108.22	92.48	84.54	116.42
March	97.72	106.49	96.58	98.63	107.46
April	99.67	86.58	86.63	96.28	80.60
May	99.57	83.12	92.48	98.63	98.51
June	110.79	91.78	108.87	100.98	107.46
July	110.18	129.0	119.99	110.07	I 07.46
August	114.84	115.15	105.94	122.11	116.92
September	104.57	121.21	110.04	126.81	98.51
October	95.85	96.10	98.91	105.08	89.55
November	103.01	80.58	106.53	98.63	80.60
December	68.47	96.10	88.97	58.71	62.69

The seasonal changes in the sales of two-wheelers namely Hero Honda, Bajaj, TVS, Yamaha and Enfield are mainly caused by the seasonal nature of demand. Apart from this, the other reasons are fuel consumption and the resale value of the brand of two-wheelers. Due to these reasons, the number of two-wheeler in the market during a certain period creates a glut leading to fall in the sales of two-wheelers while at other times, scarce supply results in higher demand. The seasonal indices of Bajaj sales were observed in the range of 83.12 per cent in May to 129.00 per cent in July. The peak period of sale of Hero Honda range form June to September. These seasons fall on school/college reopening, festival as well as bonus paid periods. The lean period of Bajaj varied from 83.12 per cent in May to 96.10 per cent in October. It is inferred from the analysis that seasonal variability in Hero Honda was found to be less than that of Bajaj indicating the steady sale of Hero Honda even during seasonal variations.

The seasonal indices of Enfield observed in the range of 62.69 per cent in December to 134.33 per cent in January.

In striking similarly during the lean period of TVS and Yamaha, the Enfield varied from 134.33 per cent in January to 98.51 per cent in May. It is inferred from the analysis that seasonal variability in TVS and Yamaha was found to be less than that of Enfield.

Irregular Variations

Irregular variations are the behaviour of sales due to random factors and the computed data on irregular indices of sales of two-wheelers namely Hero Honda, Bajaj. TVS. Yamaha and Enfield in Madurai city for the period from 1997–98 to 2004–2005 are presented in Table 4.2.

Table 4.2: Irregular Indices of Sales of Two-wheelers in Madurai City from 1997–98 to 2004–2005

Month	*Irregular Indices of*				
	Hero Honda	*Bajaj*	*TVS*	*Yamaha*	*Enfield*
January	250.75	55.69	131.92	46.63	12.66
February	305.77	97.95	164.36	43.77	10.31
March	271.18	112.69	194.66	36.50	9.31
April	347.68	124.74	193.93	40.51	9.93
May	298.28	107.07	168.69	35.49	11.17
June	252.73	92.61	202.08	38.62	12.10
July	285.90	82.17	156.68	39.07	9.36
August	289.74	94.66	169.91	32.75	10.45
September	298.36	89.93	173.57	41.79	10.15
October	251.43	83.25	111.21	37.85	12.28
November	281.0	85.47	112.64	44.61	11.17
December	322.77	98.86	125.89	37.47	12.76

It is observed from Table 4.2 that the irregular indices of Hero Honda varied from 250.75 per cent in January to 347.68 per cent in April. Similarly the irregular fluctuations in respect of sales of Bajaj ranged from 55.69 per cent during January to 124.74 per cent during April. It can be inferred from the analysis that the irregular fluctuations found in sales of Bajaj were much less than those of Hero Honda over a period of time.

The irregular indices of sale of TVS varied from 111.21 per cent in October to 202.08 per cent in June. In the case of Yamaha irregular fluctuations in sale varied from 32.75 per cent in August to 46.63 per cent in January. It is understood that the variation in the sale of Yamaha was not found high due to erratic change.

Similarly, the irregular fluctuations in sale of Enfield ranged from 9.31 per cent during March to 12.70 per cent during December. It can be inferred from the analysis that the irregular fluctuations in the sale of Enfield were much less than those of Hero Honda, Bajaj, TVS and Yamaha over a period of time.

Thus, it may be concluded from the analysis that the sale of two-wheelers Hero Honda, Bajaj and TVS has been much affected due to erratic change over a period under study. In the case of Yamaha and Enfield, the impact was not so conspicuous due to erratic change.

Cyclical Variations

Cyclical variations are the movement with periods longer than a year and all cycles are not of the same length.

Cyclical variations are computed by using Residual method. The computed indices of the cyclical variations in the sale of two-wheelers namely Hero Honda, Bajaj, TYS, Yamaha and Enfield in Madurai city during 1997–98 to 2004–05 are provided in Table 4.3.

Table 4.3: Cyclical Indices of Sales of Two-wheelers in Madurai City from 1997 to 2005

Year	Cyclical Indices of				
	Hero Honda	Bajaj	TVS	Yamaha	Enfield
1997	117.41	323.45	354.01	83.05	109.86
1998	108.21	124.77	331.77	128.99	117.49
1000	90.21	60.95	183.81	119.14	73.95
2000	86.59	73.57	82.02	93.01	97.50
2001	89.11	79.45	49.20	81.30	88.58
2002	105.61	82.52	87.47	97.51	80.25
2003	100.96	103.04	116.29	80.39	99.97
2004	107.51	133.33	96.54	109.18	125.19
2005	98.44	131.23	61.42	109.76	122.24

It is observed from Table 4.3 the cyclical variations ranged from 86.59 per cent in 2000 to 117.41 per cent in 1997 in the case of Hero Honda. The cyclical indices of Bajaj was ranging from 60.95 per cent in 1999 to 323.45 per cent in 1997. Again it supports the earlier views of greater variations in sale of Bajaj when compared to Hero Honda.

It is also observed from the analysis that the cyclical variations ranged from 49.20 per cent in 2001 to 354.01 per cent in 1997 in the case of TVS. 80.39 per cent in 2003 to 128.99 per cent in 1998 in the case of Yamaha. The cyclical indices of Enfield is ranging from 73.95 in 1999 to 125.19 per cent in 2004. Again it supports the earlier views of greater variations in Enfield when compared to TVS and Yamaha.

The peak periods of sale of Hero Honda as observed from Table 4.3 are 1997–1998 and 2004 and then lean periods of sale of Hero Honda are 1999, 2000, 2004 and 2005. The peak period of sale of Yamaha are 1998, 1999, 2004 and 2005 and then lean periods of Yamaha are 1997, 2000, 2001, 2002 and 2003.

In the case of Enfield the lowest sale was found during the period from 1999 to 2003 and highest Enfield sale was found in 2004 and 2005.

Secular Trend and Growth Rate

Trend is the general tendency of sale of the two-wheelers to increase or decrease over a long period. In order to understand the secular trend of sale of two-wheelers namely Hero Honda, Bajaj, TVS, Yamaha and Enfield, the equations are estimated by the Method of Least squares. The growth rate is also calculated. The estimated values of secular trend and growth rate of sale of Hero Honda, TVS, Yamaha, Bajaj, and Enfield are presented in Table 4.4.

The results of Table 4.4 indicate that the trend co-efficient of sale of Hero Honda, Bajaj, TVS and Yamaha are positive and statistically significant at 5 per cent level. The growth rates are 11.45 per cent 21.03 per cent. 15.98 per cent and 13.61 per cent for Hero Honda, Bajaj, TVS and Yamaha respectively. It reveals that the sale of Bajaj had increased to the tune of 21.03 per cent per annum compared to the other two-wheelers namely Hero Honda, TVS and Yamaha which are 11.45 per cent, 15.98 per cent and 13.61 per cent respectively per annum.

Table 4.4: Secular Trend and Growth Rate of Sales of Two-wheelers in Madurai City from 1996–97 to 2003–04

Name of Brand	Regression Co-efficient		$\bar{R}^2$	Growth Rate in Percentage
	a	B		
Hero Honda	7.67	0.108* (3.011)	93.50	11.45
Bajaj	6.12	0.053* (3.191)	78.33	21.03
TVS	6.76	0.148* (2.053)	53.07	15.98
Yamaha	15.53	0.128* (4.024)	79.55	13.61
Enfield	5.50	0.135 (0.019)	87.46	–12.65

Figures in parentheses are the t–values.

* Indicates the trend co-efficient are statistically significant at 5 per cent level.

In the case of Enfield, trend coefficient of sale is not statistically significant at 5 per cent level and a declined growth rate (–12.65 per cent) is observed during the period under study.

Variability in Sales with Respect to Time Series Components

A perusal of Table 4.5 indicates the variability in sales due to trend, cyclical, seasonal and irregular fluctuations.

Table 4.5: Variability in Sales Due to Trend, Cyclical, Seasonal and Irregular Fluctuations

Components	Coefficient of Variations				
	Hero Honda	Bajaj	TVS	Yamaha	Enfield
Trend	14.35	38.79	26.35	19.71	16.14
Cyclical	9.70	71.08	71.79	16.44	17.07
Seasonal	12.29	13.82	9.75	16.60	18.61
Irregular	9.83	17.74	19.12	9.78	11.48
Total	46.17	141.43	127.01	62.53	63.30

It is apparent from Table 4.5 that the total coefficients of variations for Hero Honda, Bajaj, TVS, Yamaha and Enfield were 46.17 per cent, 141.43 per cent, 127.01 per cent, 62.53 per cent and 63.30 per cent respectively. It indicates that less variation was found in Hero Honda followed by Yamaha and Enfield. Higher variations could be observed both in Bajaj and TVS due to cyclical component.

The variations due to trend component ranked second for both Bajaj and TVS followed by irregular component with variations of 17.74 per cent and 19.12 per cent for Bajaj and TVS respectively. The coefficients of variations due to seasonal fluctuations were minimum, for the sale of all brands among all the components under study.

Conclusions

It has been observed that the sales of two-wheelers were found to be peak during the month of June to September. The seasonal variability in the case of Hero Honda was found to be less than that of Bajaj indicating the steady sales during seasonal variations. Further sales of two-wheelers such as Hero Honda, Bajaj, and TVS have been much affected due to erratic change over a period under study. The irregular fluctuations in sales of Hero Honda, Bajaj and TVS were higher than that of Yamaha and Enfield. Less cyclical variations were observed in the sale of Hero Honda, Yamaha and Enfield brands, while in the case of Bajaj and TVS it has been found to be high compared to other brands. The effect of trend component in sales of Hero Honda is comparatively less than that of other brands in the market.

References

Avlonitis, G. and Gouremenous, A., 1995. The changing consumer in Greece. *International Journal of Research in Marketing*, 12: 435–448.

Grewal, P.S., 1987. *Numerical Methods of Statistical Analysis.* Sterling Publishers Private Limited, New Delhi, p. 555.

Ramachandran, J. and Sanjeev Sehrawat, 1996 The Indian two-wheeler industry: Riding in the fast lane. *South Asian Journal of Management*, 5(1 and 2): 1–22.

Valsan, Anil, 2001. *Automobile Industry: Characteristics and Organization.* Indian Industry a Geographical Perspective, pp. 126–127.

William, E. Cox Jr. and George N. Havens, 1977. Determination of sales potential and performance for an industrial goods manufacturers. *Journal of Marketing Research*, 14: 574.

Xavier, Francis, 2004. The year of dawn, splendor and passion. *The Hindu, Business Line*, July 29, p. 3.

Chapter 5

Positional Survey of the Groundwater Quality of the Bharatpur Area during the Monsoon Season 2006

Deepshika Garg[1], Sunita Goyal[2], Samiksha Chaturvedi[3] and R.V. Singh[1*]

[1]Department of Chemistry, University of Rajasthan, Jaipur – 302 004, India

[2]Lecturer, Department of Chemistry, J.E.C., Kukas, Jaipur, India

[3]Lecturer, Department of Chemistry, Arya College of Engineering and I.T., Kukas, Jaipur, India

ABSTRACT

Bharatpur is a well known place because of world famous 'Keoladeo Ghana National Park'. This paper presents the important results of the physico-chemical analysis of the groundwater samples of bore wells, hand pumps and open wells of this district during the monsoon season, 2006. During these investigations some of the important parameters which determine the status of drinking water quality were studied. Such parameters are pH, DO, nitrate, fluoride, total alkalinity, colour, total dissolved solids, total hardness, chloride, sulphate, calcium hardness, magnesium hardness, carbonate hardness and noncarbonate hardness. It was found that the water of almost all the sampling points is highly contaminated with nitrate and total dissolved solids.

Keywords: *Bharatpur, Keoladeo, Monsoon season 2006, Physico-chemical parameters.*

* *E-mail: rvsjpr@hotmail.com

Introduction

Bharatpur is the world famous place because of 'Keoladeo National Park'. Keoladeo National Park is popularly known as 'Bird Paradise'. Over 370 bird species have been recorded in the Park. It is the only regular wintering area in India for the rare and highly endangered Siberian Crane (Pain *et al.*, 2004). The mammalian fauna of KNP is equally rich with 27 identified species. Blue bull, feral cattle, and spotted deer are common while sambhar are few. The geographical location is ideal as it is on the main north-south avian route of India. About 11 sq.km area of the park is covered with water. Annual rain fall in Keoladeo Ghana National Park is approximately 690 mm and climate in summer with maximum 42°C and minimum 29°C and in winter with 25°C and minimum 8°C. "Ajan Bandh" is the main water source to fill the various lakes and ponds of the park. Today water pollution is the biggest problem for human beings, because of various pollution conditions water gets unfit for drinking purpose. Water pollution is a phenomenon that is characterized by the deterioration of the water quality as a result of various human activities (Kulshrestha *et al.*, 2004). The chief source of water pollution is the increasing use of chemical fertilizers and pesticides letting in of untreated sewage and industrial effluents into rivers and streams running close to the cities and other lowland is yet another cause. Many dangerous diseases are caused by the use of the polluted water. Keeping these facts in view and the importance of this place due to a large number of tourists used to visit this place, it was thought worthwhile to study the water quality of Bharatpur city and Keoladeo Ghana National Park. The findings of these studies have been discussed in this article.

Experimental

The water samples were collected from the twelve different sampling points which are situated at the distance of approximately 5 km. The samples were collected during the month of August and September 2006. Samples for analysis were collected in sterilized bottles using the standard procedure for grab or catch samples in accordance with the standard methods (APHA, 1995 and Kulshrestha *et al.*, 2002). Areas which were selected for these studies include two different sites at Ghana, two different areas in Jawahar Nagar, Tilak Nagar, Ranjeet Nagar, Namak Katra, Pai Bag, Krishna Nagar, Adarsh Colony, Ganga Mandir and Chomukha Mahadev. All the samples were properly labelled indicating the source of the sample, the time and the date of the collection. The analysis of various physico-chemical parameters were carried out with the help of standard methods. The parameters, methods, standard values and units selected for the said examination are detailed in Table 5.1.

Results and Discussion

The physico-chemical characteristics of the various hand pumps, open wells and bore wells have been shown in Table 5.2.

Colour

All samples are colourless.

Odour

All samples are odourless.

Temperature

The temperature of water samples ranged between 32.0°C–36.5°C. The temperature is the important factor of various chemical and biological reactions in water. Also the solubility of most of the substances is largely depends on the temperature.

Table 5.1: Parameters, Methods, Standard Values and Unit Employed in Physico-chemical Examination of Samples

Parameters of Water Analysis	*Methods*	*Standard Values as Guided by ICMR*		*Unit*
		Desirable Concentration	*Maximum Permissible*	
Colour	By sight	–	–	–
Odour	Semelling	–	–	–
Temperature	Thermometric	–	–	°C
pH	pH Meter	7.0–8.5	6.5–9.2	–
Dissolved Oxygen (DO)	Azide modification	7 mg/L at 35°C	–	mg/L
Total Alkalinity	Titrimetric	200	600	mg/L
Total Hardness	Titrimetric	300	600	mg/L
Calcium Hardness	Titrimetric	75	200	mg/L
Magnesium Hardness	Titrimetric	50	150	mg/L
Calcium	Titrimetric	–	–	mg/L
Magnesium	Titrimetric	–	–	mg/L
Carbonate Hardness	Titrimetric	300	600	mg/L
Non-carbonate Hardness	Titrimetric	300	600	mg/L
Chloride	Argentometric	200	1000	mg/L
Sulphate	Turbidity meter	200	400	mg/L
Nitrate	Ion meter	20	50	mg/L
Total Dissolved Solids	Conductivity meter	500	1500	mg/L
Fluoride	Ion selective electrode	1.0	1.5	mg/L
Electrical Conductivity	Conductivity meter	–	–	mg/L

pH

The hydrogen ion concentration affects the taste of water..The pH value of natural water changes due to the biological activity, temperature and disposal of industrial wastes as well as due to the acid mines drainage wastes. The pH of the various sampling area varied from 7.8 to 8.2 which shows slight alkalinity of water.

Dissolved Oxygen (DO)

The presence of oxygen is essential for the survival of aquatic life in water. The DO level in natural water depends on physical, chemical and biological activities in water body. Thus, it is one of the important parameters for assessing the quality of water bodies. As DO level falls; undesirable odours, tastes and colours reduce the acceptability of water. In the present case the DO ranged between 3.1 to 7.5 mg/L. The desirable DO concentration is 7 mg/L at 35°C according to ICMR. Thus DO in the water of all sampling stations except point-1 is less than the desirable limit.

Total Alkalinity

The values of total alkalinity of the water samples ranged between 190 to 850 mg/L. The maximum permissible limit of total alkalinity is 600 mg/L (ICMR). Points 7, 8 and 11 contain high total alkalinity

Table 5.2: Physico-chemical Parameters of Water of Bharatpur City During Monsoon Season, 2006

Parameters	*Point 1 Ghana O/W*	*Point 2 Ghana H/P*	*Point 3 J.N. O/W*	*Point 4 J.N. O/W*	*Point 5 T.N. H/P*	*Point 6 R.N. O/W*	*Point 7 N Katra H/P*	*Point 8 Pibag H/P*	*Point 9 K.N. H/P*	*Point 10 A.C. H/P*	*Point 11 G.M. H/P*	*Point 12 C.M. B/W*
Colour	Colour-less	Colour-less	Colour-less	Colour-less	Colour-less	Colour-less	Colour-less	Colour-less	Colour-less	Colour-less	Colour-less	Colour-less
Odour	Odour-less	Odour-less	Odour-less	Odour-less	Odour-less	Odour-less	Odour-less	Odour-less	Odour-less	Odour-less	Odour-less	Odour-less
Temperature	35.0°C	34.5°C	35.0°C	35.5°C	34.5°C	33.0°C	33.0°C	32.0°C	35.0°C	36.5°C	34.0°C	35.5°C
pH	7.8	7.9	8.0	8.0	7.8	7.9	8.1	8.2	8.0	7.9	8.0	8.1
Dissolved Oxygen (DO)	7.5	3.1	3.1	5.6	4.2	3.9	3.6	4.0	4.5	3.4	5.8	5.8
Total Alkalinity (as$CaCO_3$)	380	190	330	500	340	200	620	850	340	600	800	450
Total Hardness (as $CaCO_3$)	130	420	140	410	220	440	300	880	920	500	600	310
Calcium Hardness (as $CaCO_3$)	50	160	60	170	90	180	120	350	400	200	200	140
Magnesium Hardness (as $CaCO_3$)	80	260	80	240	130	260	180	530	520	300	400	170
Calcium	20	64	24	68	36	72	48	140	160	80	80	56
Magnesium	19.2	62.4	19.2	57.6	31.2	62.4	43.2	127.2	124.8	72	96	40.8
Carbonate Hardness (as $CaCO_3$)	130	190	140	410	220	200	300	850	340	500	600	310
Non-Carbonate Hardness (as $CaCO_3$)	Nil	230	Nil	Nil	Nil	Nil	Nil	30	580	Nil	Nil	Nil
Chloride	200	240	690	1090	590	300	760	260	900	670	425	690
Sulphate	48	55	65	178	82	90	110	140	102	100	105	120
Nitrate	45	48	96	80	82	70	110	150	108	110	112	114
Total Dissolved Solids	1920	2080	2560	3200	2720	2840	3840	4560	3280	3600	3360	2800
Fluoride	1.2	1.1	0.5	1.5	1.0	12	0.3	0.8	1.0	1.2	0.9	0.6
Electrical Conductivity	2.7×10^3	2.9×10^3	3.6×10^3	4.5×10^3	3.8×10^3	4.0×10^3	5.4×10^3	6.5×10^3	4.6×10^3	5.1×10^3	4.8×10^3	4.0×10^3

O/W: Open well; H/P: Hand Pump; B/W: Bore Well.

than the permissible limit. Alkalinity in itself is not harmful to the human being, still the water supplies with less than 100 mg/L are desirable for domestic use. However, in large quantities, alkalinity imparts bitter taste of water.

Total Hardness

The hardness of water, caused by multivalent metallic cations varies considerably from place to place. The hardness of water reflects the nature of geological formation with which it has been in contact. In the present case total hardness varied between 130 to 920 mg/L. The maximum permissible value to total hardness is 600 mg/L (ICMR). Thus, the waters of points 8, 9 and 11 are very hard in character.

Calcium Hardness

The quantities of calcium in natural water depend on the type of rocks. Being presence of calcium in high quantities in the rocks, it is leached and contaminates the water. Calcium was found in these samples ranging from 50 mg/L to 400 mg/L. The maximum permissible limit is 200 mg/L according to ICMR.

Magnesium Hardness

The maximum allowed value of magnesium according to ICMR is 150 mg/L. In the present study, magnesium hardness ranged between 80 to 530 mg/L. Magnesium has been considered as non-toxic to human at the concentration expected in water.

Carbonate and Non-carbonate Hardness

The maximum permissible limit in the absence of alternate source is 600 mg/L. The carbonate hardness values varied from 130 to 850 mg/L. The non-carbonate hardness values varied from Nil to 580 mg/L.

Chloride

The higher concentration of chloride above the 250 mg/L makes the water salty in taste. The maximum permissible limit is 1000 mg/L according to ICMR. Chloride values varied from 200 mg/L to 1090 mg/L. Except point 1 the chloride content is higher for all water samples.

Sulphate

Sulphate does not affect the taste of water. The sulphate in water samples of study area ranged between 48 to 178 mg/L.

Nitrate

Nitrate is the highest oxidisable form of nitrogen and occurs in trace quantities in surface waters but may attain high level in some groundwater. High concentration of nitrate in groundwater is caused because of runoff of the nitrate rich fertilizers and animal manure into the water supply. The maximum permissible limit is 50 mg/L (ICMR). In the study area nitrate concentration was varied from 45 to 150 mg/L. Thus except points-1 and 2 all water samples contain high concentration of nitrate. In rainy season there is the greater possibility of runoff the nitrate rich fertilizers and animal manure into the water bodies, that's why the nitrate was found in higher concentration. Nitrate in water supply in concentration over 100 mg/L causes methamoglobinemia, particularly in infants upto six months of age, whose main liquid intake is powdered milk formula made up with tap water containing high concentration of nitrates. Nitrate forms nitrosoamine in stomach and causes gastric cancer (Majumdar and Gupta, 2000).

Total Dissolved Solids (TDS)

The maximum permissible limit of TDS is 1500 mg/L (ICMR). Gastrointestinal irritation is caused due to higher concentration of TDS. The TDS values varied from 1920 to 4560 mg/L in the study area.

Fluoride

The guideline value of fluoride is 1.5 mg/L in drinking water. Fluoride is more commonly found in groundwater than the surface water. It is released to the groundwater through weathering of primary silicates and associated accessory minerals (Thakare *et al.*, 2005). The excessive amount of fluorides causes disfigurement of teeth known as mottled enamal (Kulshreshtha *et al.*, 2004). However, presence of less than 0.8 mg/L fluoride in water causes dental caries in children. In the study area fluoride varied from 0.3 mg/L to 1.5 mg/L. Thus it is completely in the range.

Electrical Conductivity

Conductivity is a measure of the current carrying capacity. Thus, as the concentration of dissolved salts increases, conductivity also increases. As in sampling areas TDS was found in higher concentration, thus electrical conductivity is also high and pollution level increases. Electrical conductivity in the study area ranged between 2.7×10^3 to 6.5×10^3 μmhos/cm.

Conclusions and Recommendations

The present investigations show that there is high nitrate content in almost all the water samples. This may be caused because of contamination of groundwater by sewage, septic tank effluents and other wastage rich nitrate. The waters of study area is also hard and contaminated with high concentration of total dissolved solids. That's why many people of Bharatpur area suffered from gastro-intestinal troubles. DO level of almost all water samples is also found in low level, so there is need to improve the water quality as to make it potable.

Acknolwedgements

The authors are thankful to Dr. S.S. Dhindsa, Chief Chemist and Shri H.S. Devenda, Senior Chemist, P.H.E.D., Gandhi Nagar, Jaipur for valuable suggestions. Drs. Samiksha Chaturvedi and Sunita Goyal are also thankful to their respective College Administration for their kind cooperation.

References

APHA (American Public Health Association), 1995. American Water Works Association and Water Pollution Control Federation, *Standard Methods of Examination of Water and Waste Water*, 19th Edition, New York, U.S.A.

Kulshreshtha, S., Chaturvedi, S., Dave, S., Dhindsa, S.S. and Singh, R.V., 2004. Important studies on the assessment and impact of industrial effluents of Sanganer town of Jaipur city on the quality of soil and water. *Indian J. Environmental Ecoplan.*, 8(2): 485.

Kulshrestha, S., Dhindsa, S.S. and Singh, R.V., 2002. Physico-chemical characteristics of undergroundwater and effluent water in Sanganer town of Jaipur city during pre-monsoon season. *Nat. Environ. Poll. Tech.*, 1(4): 453.

Kulshreshtha, S., Sharma, S. and Singh, R.V., 2004. *Jnt. J. Chem. Sci.*, 2(1): 27.

Majumdar, D. and Gupta, N., 2000. Nitrate pollution of groundwater and associated human health disorders. *Indian J. Env. Hlth.*, 42(2): 168.

Pain, D.J., Gargi, R., Cunningham, A.A., Jones, A. and Prakash, V., 2004. Mortality of globally threatened Sarus Cranes Grus Antigon from monocrotophos poisoning in India. *Science of the Total Environment*, 326: 55.

Thakare, S.B., Parwate, A.V. and Rao, M., 2005. Analysis of fluoride in the groundwater of Akola district: A case study. *Indian J. Environ. and Ecoplan.*, 10(3): 657.

Chapter 6

Modelling Tree-Grass Interactions in Semi-Arid and Temperate Silvopastoral Systems

T.A. Khan[1], W.R. Eason[2], G. Suresh[1] and R.K. Bhatt[1]*
[1]*Indian Grassland and Fodder Research Institute, (IGFRI), Jhansi, India*
[2]*Institute of Grassland and Environmental Research, (IGER), Aberystwyth, U.K.*

ABSTRACT

Two silvopastoral systems *Acacia tortilis* + *Cenchrus ciliaris* (tropical semi-arid site, Jhansi, India) established in 1982 and *Fraxinus excelsior* trees + *Lolium perene-Trifolium repens* ground vegetation (temperate, at Aberystwyth, Wales, UK) established in 1988 in two contrasting situations. Mathematical models for tree/grass growth and production of trees and understorey grass production with respect to microclimate; soil moisture, soil temperature and tree roots were evolved for both the situations. These relationships were integrated over a spreadsheet as per an assumed flowchart and predicted outputs in the form of growth and production of trees and understorey vegetation was compared.

The regression coefficients of tree growth/prediction models (growth diameter, height-diameter, crown width and production–above ground tree biomass) and understorey ground vegetation production evolved for both the situations were found significant ($P < 0.01$). Among various micro climatic parameters and below-ground interactions studied, soil moisture in the surface layers and tree root parameters had a significant influence on understorey pasture yield at both sites.

* *Corresponding Author*: E-mail tanweer_khan234@rediffmail.com; Telephone (O): 091-510-2730666; Fax: 091-510-2730833

This study has demonstrated that through an integrated approach various growth and production models could be developed under different climatic situations. These models can be inter-locked on any spreadsheet to facilitate the comparison. Growth and production models in both the situations suggested the need for their validation in similar climatic conditions.

Keywords: *Acacia tortilis, Current annual increment, Fraxinus excelsior, Mean annual increment, Tree-grass interaction, Spreadsheet model.*

Introduction

Agroforestry systems in which trees are planted with agricultural crops offer significant potential for more efficient use of resources. These systems are found in many socio-economic and climatic regions of the world (Lundergren and Raintree, 1982).

In India, silvopastoral systems are being adopted, for meeting the growing needs of fodder and firewood of rural population and conservation of natural resources (Paroda, 1990). Several authors reported silvopastoral systems as a low input technology (Deb Roy and Pathak, 1974, 1983). Within the European community, where there is pressure to find forms of land use which address production and environmental concerns. Computer models developed suggest that silvopastoral land use systems, in which animal and forest production are combined on the same land area may be a viable alternative land use option for temperate European countries (Tabbush *et al.*, 1985 and Doyle *et al.*, 1986). In an agroforestry system, competition for resource capture (light, water and nutrients) between the tree and ground crop will be critical in determining the efficiency of such systems (Monteith *et al.*, 1991).

In conventional forestry systems growth and yield models are needed to describe the interaction of stand management and environment. Earlier, yield tables served as a valuable tool to calculate allowable cuts and to regulate the forest production (New Bould, 1967). Trenbrath (1999) described case studies using mathematical models based largely on Indian data to illustrate productivity benefits of growing two species together rather than only a single best species. Spreadsheet based agroforestry system models for temperate silvopastoral systems have been developed by Thomas and Willis (1995). Spreadsheet models are powerful and flexible tools. However, in this context it is seen as a framework within which preliminary analysis can be undertaken, continually refined as improved functions are incorporated as a result of further research (Thomas, 1991). A number of approaches have been used in simulation of decline in agricultural production over time in agroforestry systems. In New Zealand, it has been observed that the decrease in annual pasture dry matter production correlated well with crown area and the number of crowns per hectare (Percival *et al.*, 1984). In a silvopastoral study conducted in Greece, it was found that the combined growth of trees and herbaceous plants requires a continuous and well-balanced access to solar energy, soil moisture and nutrients (StergioGakis *et al.*, 2004).

The results reported in the study are the joint effort made by Indian Grassland and Fodder Research Institute (IGFRI), Jhansi, India and Institute of Grassland and Environmental Research (IGER), Aberystwyth, UK during 1995 in a collaborative project. The aim of the present study was to quantify and compare the above and below ground factors influencing understorey pasture production of established silvopastoral systems through spreadsheet modelling in two situations (tropical and temperate) (Figure 6.1). Two contrasting systems offered the opportunity to compare the relative importance of different factors under different conditions.

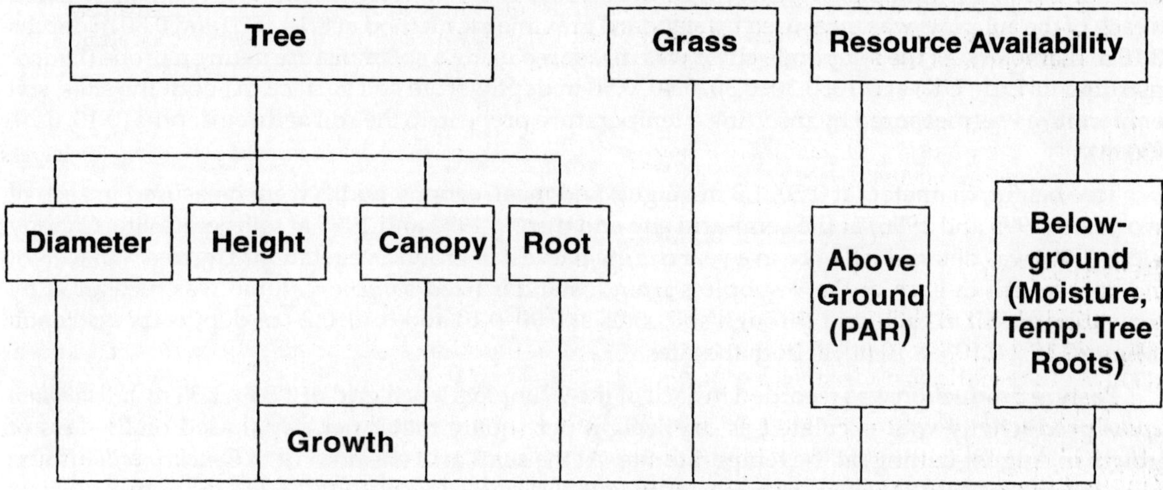

Figure 6.1: Component Interactions in Silvopastoral Systems

Materials and Methods

Site Characteristics

The comparative site characteristics and species composition of both study sites are described in Table 6.1. At each site three four-tree plots were divided into a 3 × 3 grid consisting of 9 subplots. Within each subplot light, soil moisture and tree root density were recorded. The sampling position within each grid was the centre.

Table 6.1: Main Site Characteristics and Treatment Components at Two Contrasting Silvopastoral Systems in India and the UK

Site	*Semi-arid*	*Temperate*
Tree species	*Acacia tortilis*	*Fraxinus excelsior*
Ground vegetation	*Cenchrus ciliaris*	*Lolium perenne-Trifolium repens*
Tree spacing	4 × 4 m-regular grid	4 × 4 m-regular grid
Tree established	1982	1988
Site position	25°27' N and 78°37'E 275 m above mean sea level	52°35'N and 4°15'W 25 m above mean sea level
Rainfall	900 mm	1471 mm
Mean temperature range	32.5°C to 17.7°C	15.5°C to 5.0°C
Soil type	Shallow alfisols with low fertility pH: 6.89–7.69	Permeable medium loam over gravel pH: 5.07–5.76

Data Measurement

At semi-arid site, Quantum sensors (Li-cor Instruments) determined PAR at 10.00 hrs (local time). At the UK site mean PAR was recorded through sensor network wired to a data logger (CR 10, Campbell

Scientific Ltd.) via a multiplexer (AM416, Campbell Scientific Ltd.). At the semi-arid site, soil moisture in each of the subplots was measured by standard gravimetric method at 0.10, 0.20 and 0.30 m depths (0.10 m diameter). At the temperate site it was measured using a soil moisture neutron probe (Didcot Instruments Ltd., UK) at 0.10, 0.20, 0.30, 0.40. 0.50 m depths from soil surface. At both the sites, soil temperature was measured by inserting a temperature probe in to the soil at three depths (0.10, 0.20, 0.30 m).

Tree height, diameter (at 0.20, 1.3 m height) and mean canopy width were measured in each of two years (1995 and 1996) at the semi-arid site and during 1994 and 1995 at temperate site. Canopy width (CW) was determined once in a year during harvest of ground vegetation. The tree root weight and root length in each of the 9 subplots from a standardized sample volume was measured by excavation of soil in each grid through soil cores at 0.00–0.10 and 0.10–0.20 m depths (per core soil volume $0.10 \times 0.10 \times 0.10$ m^3) at both the sites.

Pasture production was recorded in each of the 9 subplots (each grid of 1.33×1.33 m^2). *Trifoloium repens* productivity was calculated as dry weight per square meter per day (based on 21-days of growth in August-cutting) at the temperate site. At the semi-arid site one cut of *Cenchrus ciliaris* was taken in September following whole season growth.

Destructive sampling of 59 *Acacia tortilis* and 140 *Fraxinus excelsior* were done for above ground biomass estimation as per the procedures outlined by New Bould (1967).

The following tree growth/biomass prediction equations (Khan *et al.*, 1994) were evolved:

Trunk Diameter Growth Model

The model taken as:

$$dy = a\, y^{b}\, c^{y} \quad (1)$$

where,

y is the Diameter (D) and dy is the change in the diameter. The a, b and c are constants determined through regression analysis using initial growth data on trees from planting to 13-year and 8-year respectively (Draper and Smith, 1981).

Height-Diameter Model

$$H(t) = 20\, e^{[-a'\, e(-b'\, (CD)c')]} \quad (2)$$

where,

a′, b′ and c′ are constants.

Above Ground Tree Biomass

It was assumed that tree biomass (B) = Function (height H, Diameter D), predicted

$$B\,(t) = a_2 + b_2 Y_1 + C_2 \int dy_2{}' \quad (3)$$

where,

Y_1 = Height and dy_2 = diameter (initial diameter + change in D) derived from (2) and (1) respectively and $\int$ is the summation over the period depending upon the growth of tree species. a_2, b_2 and c_2 are constants.

Crown Width Model

Crown width of trees for the tropical situation was evolved as:

Crown spread (m^2) = a $(H)^b$ $(D)^c$ $(dbh)^d$,

where,

a, b, c and d are constants determined through regression analysis. [H= height of tree (m), D = diameter at 20 cm height (cm) and dbh = diameter at breast height (cm, at 10^3 m height of the tree)] as predictors.

The crown width model suggested by White (1981) was used for temperate situation.

Canopy width (m^2) = α_o $(D)^a$

where,

α and α_o are constants.

Prediction equations were decided on the basis of coefficient of determination (R^2) and analysis of residuals (Draper and Smith, 1981) using stepwise procedure of regression analysis module of SPSS (Ver. 5.0).

Crown Width Vs. Pasture Production

Stochastic modelling was used for the prediction of understorey grass production, assuming that the production of pasture is effected by crown width of tree. Pasture yield was determined taking random variation at 5 per cent deviation from the predicted yield. Potential yield of crop is predicted with respect to above and below ground factors. It was assumed that when crown width is less than 5 m^2, the reduction in the pasture yield is zeroed. When crown width is between 5–15 m^2, the reduction follow the trend based on relationship of crown width with the actual pasture production observed in the plots, and when crown area is more than 15 m^2 reduction was estimated to be 0.3 t/ha (Khan *et al.*, 1994).

Spreadsheet Modelling

Relationships amongst the system components were integrated on a spreadsheet (Microsoft Excel version 5.0, Microsoft Corporation USA) for predicted performance of the systems at a given specific site in semi-arid and temperate situations according to assumed flow chart (Figure 6.1) and as per guidelines suggested by Thomas (1991).

Results and Discussion

Tree Growth and Production Models

The regression coefficients of various growth models (Table 6.2) were found to be significant ($P < 0.01$) in both the situations. The tree diameter, height and crown spread growth models established in the semi-arid condition gave a better fit than in temperate conditions except the tree biomass production model. Initially the biomass accumulation was slow in *Acacia tortilis* compared to *Fraxinus excelsior* but at the end of 6th year *Acacia tortilis* overtook *Fraxinus excelsior.* The above ground biomass prediction at the end of 25th year was more (46.35 t/ha) in *Acacia tortilis* compared to *Fraxinus excelsior* (14.8 t/ha) (Figure 6.2).

Pasture Yield

At the semi-arid site, the understorey grass production was significantly affected by PAR, tree root weight (at 0.10–0.20 m depth) and crown width. At the temperate site, PAR, soil moisture (0.30 m

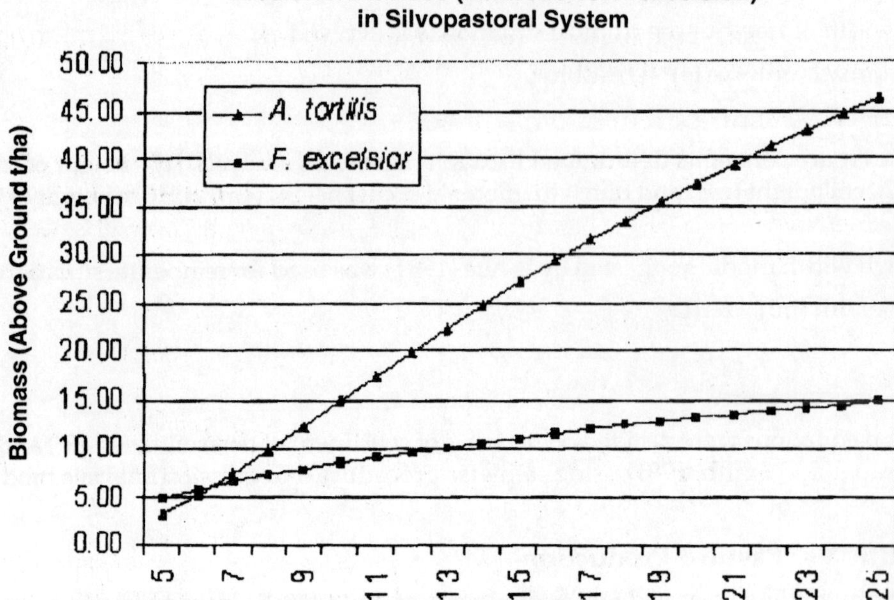

Figure 6.2: Predicted Above Ground Biomass (Temperate and Tropical) of Trees in Silvopastoral Systems

depth) and tree root length influenced the pasture production. The prediction equations evolved are as under:

1. *Acacia tortilis* Based System (Semi-arid)

Pasture (g/m^2)= 21.0058 EXP (0.6694** PAR – 0.0326** Root weight (0.10–0.20 m) –0.013** crown width), $R^2 = 0.53$

Table 6.2: Tree Growth Prediction Models for Growth and Biomass Production at the Contrasting Situations

Indian Semi-arid Situation (IGFRI, India)		*Temperate Conditions (IGER, UK)*	
Description of the Model	R^2	*Description of the Model*	R^2
Diameter growth Dy = 0.7078 $(D)^{1.1016}$ $(0.7402)^D$	Diameter growth 0.744**	dy = 1.091 $(D)^{1.406}$ $(0.659)^D$	0.620**
Height-diameter model 20 EXP (–4.1911 EXP (–0.3426 $(D)^{0.758}$))	0.983**	Height-diameter model 20 EXP (–3.621 EXP(–0.337$(D)^{0.502}$))	0.590**
Tree above ground biomass 0.059911 $(H)^{0.7588}(D)^{1.9435}$	0.867**	Tree above ground Biomass 0.0259 $(D)^{1.7590}$ $(H)^{0.7620}$	0.941**
Crown Spread EXP (0.613 + 0.312 D – 0.067 dbh – 0.150 Ht)	0.810**	Crown Spread –1 + EXP (–0.161+0.147D)	0.460**

H (m): Height of tree; D (cm): Diameter of tree; dbh: $(D)^{0.976}$; Biomass: Kg/tree; Canopy spread: m^2; **: Significant ($P < 0.01$)

2. *Fraxinus excelsior* Based System (Temperate)

Pasture (g/m²) = 0.5665 EXP (0.0025** soil moisture (0.30 m) – 0.0025** PAR + 0.0003** Root length (0.0–0.15 m) – 0.0005** Root length (0.15–0.30 m), $R^2 = 0.97$,

Crown Width Vs. Pasture Production

The predicted understorey pasture yield under *Acacia tortilis* varied between 2.7–1.5 t/ha with the rise in crown width from 2868 to 24447 m²/ha between 5–25 years. In *Fraxinus excelsior* stand pasture yield varied between 2.7–1.7 t/ha with the increase in crown width from 212 to 3268 m²/ha in the same growth period (Figure 6.3).

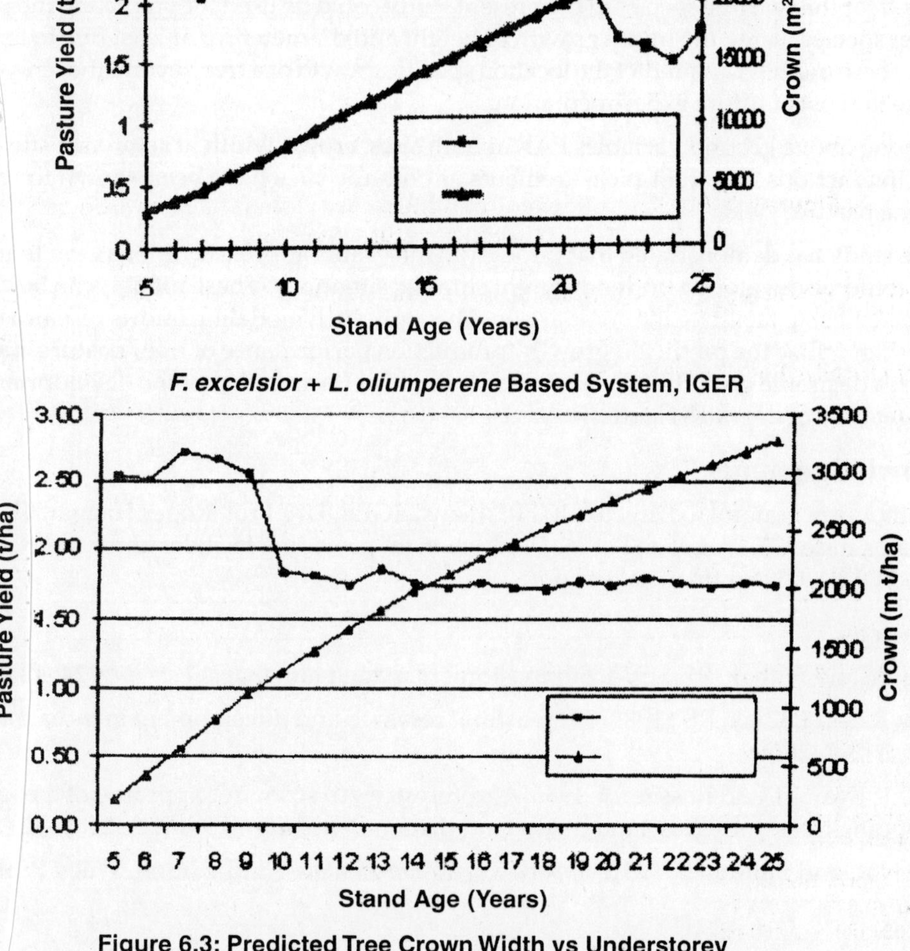

Figure 6.3: Predicted Tree Crown Width vs Understorey Pasture Production (Temperate and Tropical) Silvopastoral Systems

The prediction equations evolved for growth and production indicated significance of interactions at both the semi-arid and temperate sites. The diameter growth model gave a better fit for the semi-arid site. An overview of prediction revealed that the biomass in the *Fraxinus excelsior* had slow quantum of growth compared to growth in *Acacia tortilis*. This might be because of tropical Indian climate, abundant sunshine; it is presumed that the growth and biomass accumulation are rapid and fast under semi-arid conditions.

Despite contrasting climates, different species composition, there was some commonality between the influences of below ground factors (tree root weight/length). Further, understanding the inherent conflict between facilitative, complementarily and competitive effects of root systems in silvopastoral agroforestry is a complex task, which requires a detailed assessment of root-soil, and root-root interactions (Scroth, 1999; Ong *et al.*, 1991).

At semi-arid site, the understorey grass production was significantly influenced by both PAR and crown width and by PAR at the temperate site. The results obtained are in line with our earlier findings in the temperate regions as reported by Eason (1994) and Khan and Pathak (1996) in *Acacia tortilis* based silvopastoral system of semi-arid regions. The tree growth (diameter and height) models as evolved for the two tree species in the present study could be used for predicting the performance of any other species, using the initial growth of height and diameter for at least five to tenyears' actual growth. These models can predict the location specific growth of a tree beyond the ten-year period and may save the cost of future experimentation.

Among above ground variables PAR at both sites, crown width at semi-arid site;among below ground interactions, tree-root measurements at both the sites have been shown to be important in predicting pasture yield.

This study has demonstrated that through an integrated approach various growth and production models could be developed under different climatic situations. These models can be inter-locked on any spreadsheet to facilitate the comparison. Through, such modeling efforts research managers and planners can utilize the predicted growth/production performance of tree/pasture in a silvopastoral system at a desirable growth stage and may facilitate in formulation of the developmental programs for revegetation of degraded lands.

Acknowledgements

Authors are grateful to Director, IGFRI, Jhansi, IGER, UK, Prof. Roger Haggar, UK Technical Co-coordinator, Indo-UK Project on Forage Production for providing facilities and support, for conducting trials and DFID, UK for financial support.

References

Deb Roy, R. and Pathak, P.S., 1974. Silvipastoral management. *Indian Farming*, 24: 41–42.

Deb Roy, R. and Pathak, P.S., 1983. Silvipastoral research and development in India. *Indian Rev. of Life Sci.*, 3: 247–264.

Doyle, C.J., Evans, J. and Rossiter, J., 1986. Agroforestry: An economic appraisal of the benefits of inter-cropping trees with grassland in lowland Britain. *Agricultural Systems*, 21: 1–32.

Draper, N.R. and Smith, H., 1981. *Applied Regression Analysis*, 2nd edition. Wiley Publication pp. 1–709.

Eason, W.R., 1994. Silvopastoral research in temperate upland and lowland grassland in the United Kingdom. In: *Agroforestry Systems for Sustainable Land Use,* (Eds.) Singh, P., Pathak, P.S. and M.M. Roy. Oxford and IBH Publishing Co. Pvt. Ltd., New Delhi, pp. 133–153.

Khan, T.A., Pathak, P.S. and Gupta, S.K., 1994. Modelling *Acacia tortilis* based silvipastoral systems using system analysis methodologies. *Range Mangt. and Agroforestry,* 15(2): 203–210.

Lundgren, B.O. and Raintree, J.B., 1982. Sustained agroforestry. In: *Agricultural Research for Development: Potentials and Challenges in Asia,* (Ed.) B. Nestel. ISNAR, The Hague, The Netherlands, pp. 37–49.

Monteith, J.L. Ong, C.K. and Corlett, J.E., 1991. Microclimatic systems interactions in agroforestry. *For. Ecol. Manage.,* 45: 31–44.

New Bould, P.J., 1967. *Methods for Estimating the Primary Production of Forests.* IBP Handbook 2. Blackwell Scientific Publication, Oxford, pp. 62.

Ong, C.K. Corlett, J.E., Singh, R.P. and C.R. Black, 1991. Above and below ground interactions agroforestry systems. *For. Ecol. Manag.,* 45: 45–57.

Paroda, R.S., 1990. Silvopastoral approach for wasteland development: Present options and future strategy-inaugural address. In: *Silvopastoral Systems in India,* (Eds.) Pathak, P.S. and P. Singh. Management Society of India. Indian Grassland and Fodder Research Institute, Jhansi, p. 1–3.

Percival, N.S., Bond, D.I., Hawke, M.F., Cranshaw, C.J., Andrew, B.L. and Knwles, R.L., 1984. Effects of radiata pine on pasture yields, botanical compositions, weed populations and production of a range of grasses. In: *Proceedings of a Technical Workshop in Agroforestry,* Dunedin, New Zealand, p. 25–32.

Schroth, G., 1999. A review of below ground interactions in a agroforestry, focusing on mechanisms and management options. *Agroforestry Systems,* 43: 5–34.

StergiosGakis, Konstantinos Mantzanas, Dimitrios Alifragis, Vasilios, P. Papanastasis, Athanasios Papaioannou, Dimosthenis Seilopoulos and Panagiotis Plates, 2004. Effect of understorey vegetation and tree establishment and growth in a silvopastoral system in Northern Greece. *Agroforestry Systems,* 60: 149–157.

Tabbush, P.M., White, L.M.S., Maxwell, T.J. and Sibbald, A.R., 1985. *Tree Planting on Upland Farms*–Study Team Report, UK, Forestry Commission and Hill Farming Research Organisation.

Thomas, T.H. and Willis, R., 1995. The Bio-economic modelling of agroforestry systems: A compendium of *technical overviews for the BEAM spreadsheet models.* The BEAM Project, School of Agricultural and Forest Sciences, University of Wales, Bangor, Gwynedd, LL57 2 UW, United Kingdom, p. 1–94.

Thomas, T.H., 1991. A spreadsheet approach to economic modelling of agroforestry systems. *For. Ecol. and Manag.,* 45: 207–235.

Trenbath, B.R., 1999: Multispecies cropping systems in India: Predictions of their productivity, stability, resilience and ecological sustainability. *Agroforestry Systems,* 45: 81–107.

White, J., 1981. The allometric interpretation of self–thinning rule. *Journal of Theoretical Biology,* 89: 475–500.

Chapter 7

Phytoplankton Productivity Assessment of Kadinamkulam Lake with Special Reference to the Retting Activity, Thiruvananthapuram, Southern Kerala

V. Sobha, Anuradha Rammohan, S. Santhosh and J. Chandra Thara
Department of Environmental Sciences, Kariavattom Campus,
University of Kerala, Thiruvananthapuram – 695 581, India

ABSTRACT

The intensive coconut husk retting in Kadinamkulam lake cause much environmental problem and affect the phytoplankton productivity and chlorophyll 'a' content. The present study is to assess the relation between the water nutrients and phytoplankton productivity. Low rate of primary productivity and chlorophyll 'a' content were recorded at husk retting area (Station I and II), when compare to non-polluting station I.

Keywords: *Chlorophyll 'a', Gross primary productivity, Net primary productivity, Nutrients.*

Introduction

The Kadinamkulam backwater (Latitude 8°35′N; Longitude 76°45′E–76°52′E), the chosen site for the present study, is the largest among the backwaters of Trivandrum district. It lies parallel to the adjoining Arabian Sea at the North-western region. The Vamanapuram river originating from the Western Ghats, after traversing for about 80 km and carrying a drainage area of 540 km^2, empties into the backwater through the Anchuthengu lake in the vicinity or the temporary bar mouth. The formation and closure of the sand bar at the mouth leads to pronounced changes in the hydrobiological conditions of the backwater. The water levels increases engulfing the coconut husk retting pits situated along the banks as semi-enclosures, resulting in the admixture of heavily polluted rets liquor with the water mass. When the rising water level threatens to inundate low-lying farm areas in the vicinity, the local inhabitants cut open the sand bar.

The backwater is shallow and a mixture of sand, clay, detritus and even coconut pith carpets the basin. Laterite deposits are also common in some locations. Laterite and granite blocks protect the banks as well as the reclaimed portions or the back waters. Climate of the areas is typically tropical and the influence of both the South-West and North-East monsoon are felt here, the former from June–September and the latter which is normally less intense during October–November. Three stations namely Murukkumpuzha (Station I, region of extensive retting); Channakkara (Station II, moderate retting) and Thalamukku (station III, Unpolluted region), where selected from the backwater system for the regular observation of the present study. The present study includes the phytoplankton productivity and chlorophyll 'a'. Organic matter available in the estuary is primarily synthesis of the primary producer and the products are transferred to the consumers through different trophic levels. The amount of primary production is the most important factor that determines whether or not a particular level of water is important from the fisheries angle (Nair and Ganapathy, 1983). Despite its growing importance, studies on the primary productivity of the estuarine waters of India are few and far between. Nutrient levels in relation to phytoplankton productivity and chlorophyll 'a' in estuaries and lakes had been studied by Ketchum (1967), Williams (1972), Santhosh (2002) and McCauley *et al.* (1989). While influence of salinity in diatoms (Desikachary and Rao, 1972) and in the rate of photosynthesis and abundance of phytoplanktons (Qasim *et al.*, 1972) have also been reported.

Kerala is endowed with an extensive tract of perennial and temporary estuaries lying almost parallel to the Arabian Sea. The present invegetation was undertaken in the light of lacuna that exists in our scientific understandings the nature of productivity of the large number of these estuaries in Kerala with a view to preparing detailed field plans to establish aquaculture farms along the back water tracts of Kerala. The study is particularly relevant in the context of the ever increasing threat to the estuaries of Kerala from various source of pollution such as husk retting, sewage effluents, factory effluents etc. It is intended as a baseline study, which would be helpful to plan out programmes of fish culture in the state. On this background, the present study is an attempt to study the relation between the productivity and different hydrographical parameters of Kadinamkulam backwater.

Materials and Methods

The present study was undertaken during October 1991 to September 1992 in the surface and bottom waters at three stations in the Kadinamkulam Lake. The gross and net primary productivity was estimated by following the Method of Light and Dark bottle technique and Chlorophyll 'a' by following the standard method (APHA, 1984; Strickland and Parson, 1968; Vollenweider, 1969). Analysis of variance technique as given by Snendcor and Cochran (1967) was used for various statistical tests.

Results

The gross primary productivity of the surface water ranged from 34.77 to 27.25 mg $C.m^{-3}.hr^{-1}$ at station I, from 28.95 to 284.06 mg C $m^{-3}.hr^{-1}$ at Station II and from 46.34 to 397.69 mg $C.m^{-3}.hr^{-3}$ t

Station III. In all the three cases, the lowest value was recorded in January and the highest in September. On the other hand, in the bottom water values ranged from undetectable levels to 69.58 mg $C.m^{-3}.hr^{-1}$ (December) at Station I and to 167.28 mg $C.m^{-3}.hr^{-1}$ (July and August) at Station II. However, GPP at Station III ranged from 28.41 mg $C.m^{-3}.hr^{-1}$ (May) to 124.99 mg $C.m^{-3}.hr^{-1}$ (September). Gross productivity was undetectable in the bottom water from April to July and also in September at Station I. In Station II, such a state prevailed in June. Mean seasonal values showed that in the surface water, low values of GPP were recorded during the postmonsoon and high values during monsoon in the three stations (Table 7.1). The average annual GPP was lower at station I and highest at Station III, in the surface water while in the in the bottom water, the minimum value was at Station I and the maximum at station III. Statistical analysis revealed that the variations in GPP were significant at 1 per cent level between seasons and between months in the season in surface water. On the other hand, in bottom water variations were found to be significant at 1 per cent level only between the three stations (Table 7.2).

The net primary productivity of surface water was lowest in January and highest in September, similar to the trend observed in the monthly variation of GPP. At Station I, the surface water NPP ranged between 11.59 at 214.63 mg $C.m^{-3}.hr^{-1}$, 11.59 at 233.56 mg $C.m^{-3}.hr^{-1}$ a Station II at 16.91 at 280.28 mg $C.m^{-3}.hr^{-1}$ at Station III. However, in the bottom water, undetectable levels to 46.31 mg $C.m^{-3}.hr^{-1}$ (December) at 91.48 mg $C.m^{-3}.hr^{-1}$ (July) were recorded at Station I and II respectively. Whereas at Station III the NPP ranged from 1 1.58 mg $C.m^{-3}.hr^{-1}$ (January) to 69.44 mg $C.m^{-3}.hr^{-1}$ (July and August). The mean seasonal value revealed that minimum values were recorded during post-monsoon in the surface water in all the three stations (Table 7.1). The annual average of NPP was a low at Station III at a high at Station II in surface water while in the bottom water a low at Station I, at a high at Station II was recorded.

Analysis of variance revealed that the variation in the surface water NPP between the three seasons and between months with all seasons were significant at 1 per cent level. While in the bottom water, variations, between the three stations were significant at 1 per cent level and between months with the season were significant at 5 per cent level (Table 7.2).

The Chlorophyll 'a' value showed wide fluctuation and it ranged from 1.60 to 52.07 mg $C.m^{-3}.hr^{-1}$. At Station I the values ranged from 1.60 mg $C.m^{-3}.hr^{-1}$ (September) to 36.94 mg $C.m^{-3}.hr^{-1}$ (June), from 2.14 mg $C.m^{-3}.hr^{-1}$ (October) to 52.07 mg $C.m^{-3}.hr^{-1}$ (May) at Station II and from 1.74 mg $C.m^{-3}.hr^{-1}$ (October) to 45.83 mg $C.m^{-3}.hr^{-1}$ (May) at Station III. The chlorophyll 'a' values showed an increase during the premonsoon, with the highest values being recorded in the May/June. By the onset of monsoon, a steady decline was observed which reached its lowest during post monsoon, more so in October. The mean seasonal values of Chlorophyll 'a' are shown in Table 7.1. During post monsoon the chlorophyll a concentration was the lowest and during premonsoon, it was higher in all the three stations. The annual mean of chlorophyll 'a' was lowest at Station I and highest at Station III. Analysis of variahce revealed that the variation between seasonal and between months in the seasons were significant at 1 per cent levels (Table 7.2). Correlation coefficient shows a significant relation between the hydrographical parameters and biological parameters in the Kadinamkulam backwater system (Table 7.3).

Discussion

The estuaries in Kerala have been found to be highly productive (Abdul Azis, 1985; Nair *et al.*, 1984; Pillai *et al.*, 1975). Similarly the average rate of primary production in the Kadinamkulam lake has also been high, with a large range. Nair and Pillai (1983) have mentioned that the average rates of

Table 7.1: Seasonal and Annual Mean of Gross and Net Primary Productivity, Chlorophyll 'a' of Kadinamkulam Backwater During 1991–92

	Station	Surface Water			Station	Bottom Water			Annual Mean
		Seasonal Mean				Seasonal Mean			
		Premonsoon	Monsoon	Postmonsoon		Premonsoon	Monsoon	Postmonsoon	
GPP	I	131.78±43.09	138.09±32.94	65.69±11.01	I	31.74±18.36	7.87±7.89	40.09± 12.0l	26.57±7.80
	II	134.05±31.46	174.38±38.66	62.96±12.53	II	61.72±14.94	123.09±41.09	58.31± 12.20	81.04±15.65
	III	91.02± 16.18	247.77±63.58	77.39±19.33	III	53.67±9.21	106.21±8.73	56.86± 18.33	72.24±9.47
NPP	I	110.68±39.42	116.78±35.08	40.16± 10.77	I	8.42±5.72	7.10±7.10	19.36±9.02	11.63±4.03
	II	110.32±33.59	142.03±34.16	31.86±8.69	II	25.46±5.51	55.81±19.61	23.34±9.68	34.87±7.79
	III	56.69± 16.48	132.36±49.52	34.10±5.75	III	24.91 ±4.66	46.24±13.72	22.25±4.20	31.13±5.35
CHL 'a'	I	15.45±3.37	12.92±8.09	2.93±0.36					
	II	21.59± 10.17	12.41±6.51	3.14±0.36					
	III	22.62±7.95	15.37±6.71	2.93±0.77					

primary productivity on and magnitude of variations in tropical brackish water bodies are high. Probably this is due to high fertility, with regards to the availability of nutrients and large standing crop of mixed phytoplankton derived from marine, estuarine and fresh water masses suspended in a widely fluctuating ambient medium. When the three stations were compared it was found that Station I revealed the lowest production and Station II recorded the highest net primary productivity and station III recorded highest gross primary productivity. Low production at station I was due to the highly polluted nature of the lake in this region, which lead to low phytoplankton density, and low light intensity was because of the turbid nature of water in this region due to the presence of considerable amount of ret liquor. Although Station II was also polluted, the net primary productivity was rather high due to the relatively stable environment with higher availability of nutrients especially nitrite and nitrates, while high GPP at Station III may be because of high light intensity, rich phosphate content, high chlorophyll 'a' concentration and also maximum phytoplankton density prevalent in this station.

Table 7.2: Analysis of Variance of Chlorophyll 'a' and Surface and Bottom Water Primary Productivity Rates

Source	*df*	*F Ratio*				
		Chlorophyll 'a'	*GPP*		*NPP*	
			SW	*BW*	*SW*	*BW*
Stations	2	1.4537	0.7692	6.5467**	0.9521	6.3249**
Seasons	2	40.2765**	14.8960**	2.1115	19.6627**	3.4041
Months	9	18.2899**	3.4614**	1.1242	5.7382**	2.5115*

Table 7.3: Correlation Coefficients Between Rates of Primary Production and Physico-chemical Parameters

	Temp	*pH*	*Salinity*	*DO*	*Phosphate*	*Silicate*	*Nitrite*	*Nitrate*
				Station I				
GPP	0.3390	0.1117	–0.2536	0.4209	–0.1064	0.6154*	–0.2991	–0.2191
NPP	0.3349	0.0329	–0.3023	0.3842	–0.1293	0.6426*	–0.2959	–0.2293
				Station II				
GPP	0.2384	–0.3122	–0.2753	0.6131*	–0.3593	0.3870	–0.3953	–0.0186
NPP	–0.3060	–0.42	–0.3060	0.5192	–0.3115	0.4475	–0.3813	0.1307
				Station III				
GPP	0.1418	–0.5178	–0.61324*	0.8014**	0.4674	–0.2384	–0.3327	–0.3777
NPP	0.1630	–0.5717*	–0.5337	0.5455	0.3857	–0.0187	–0.3816	–0.3001

Surface water production was observed to be higher than that of bottom water. The observed higher production in the surface water could be on account of the prevalent high light intensity and large phytoplankton population. Moreover, the nutrients were also more in surface water than the bottom water. Earlier, Nair *et al.* (1984) have made similar observations in the same lake, while Devassy and Ooes (1989) observed similar variations in the surface and bottom water production in the Mandovi and Zuari estuarine system. The primary production in the present study showed distinct seasonal variation. Statistical analysis, has also revealed significant variations in primary production between the three stations. In general, the maximum production in the lake was during the monsoon season and the minimum during postmonsoon. This is in conformity with the observation made by Nair *et al.* (1984) in the same lake. Maximum production during monsoon has also been observed in the near shore waters off Thal (Varshney *et al.*, 1983), and according to them the high production during monsoon have been attributed to the relatively low salinity when the photosynthesis would be maximum and also due to the euryhaline nature of the phytoplankton crop prevailing at that time. Fogg (1982) has suggested that nutrient concentration and availability of light are the major parameters, which govern phytoplankton primary production. In the present study, a relationship between light and production was not discernible, *i.e.*, the high and low values in gross and net primary productivity did not synchronize with he high and low illuminations (Transparency) revealed in the area. Similar observations have been reported from the Ashtamudi lake (Nair *et al.*, 1984) and Paravur lake (Shibu, 1991).

Seasonal variations in pH and salinity seemed to have an inverse effect on the productivity in Kadinamkulam lake during the present study. This was more pronounced at Station III where statistical analysis has also shown a distinct negative correlation between NPP and pH, GPP and salinity. Ragothaman and Reddy (1982) have observed that less pH and salinity coincided with less productivity in the Tapi estuary. In the present study primary productivity showed high significant correlation with dissolved oxygen at Station II and III. The maximum dissolved oxygen values during monsoon coincided with high productivity during the same seasons, but low values of dissolved oxygen during premonsoon did not coincide with the least values of primary productivity on, which was during post monsoon, although low values were observed during premonsoon. Nair *et al.* (1983) did not observe any relation between dissolved oxygen and productivity in an earlier work conducted in the same lake. But in Ashtamudy estuary Nair *et al.* (1984) and Nair and Abdul Azis (1987) have observed that productivity increased and decreased along with fluctuations in the dissolved oxygen concentrations. In the present study it was observed that variation in the nutrient concentration had an inverse effect on primary production and apart from silicate, which was highly significant (5 per cent level), the other inorganic nutrients did not reveal any significant correlation with the rates of production. The nutrients were low during monsoon when the primary production was at its peak. This may be due to the depletion in the nutrient in monsoon at the time of increase of phytoplankton biomass. Which in its turn would have increased primary production during the seasons. Similar observation have been made by Harvey (1955), who has opined that in temperate regions a total depletion of the nutrients during spring diatom outburst acts as a limiting factor for primary production. Gopinathan and Rodrigo (1991) clearly confirmed and supported the findings of the present study.

The chlorophyll 'a' was minimum during post monsoon and highest during the premonsoon. Blooms of *Anabaena* spp. and large number of *Microcystis* sp., during the premonsoon months would have caused the peak in chlorophyll 'a' during the premonsoon. It was however not possible to estimate how much pigment actually came from the living organism of the backwater According to Sheela *et al.* (1995) high chlorophyll a may also be due to detrited chlorophyll associated with stirred up sediments. Balachandran *et al.* (1989) have observed a similar fluctuation of chlorophyll 'a' in the waters of Cochin as that observed during the present study. Although the chlorophyll a concentration was low during low productivity periods, *i.e.* during premonsoon a similar trend was not observed in the peak values, because chlorophyll 'a' was at its peak during premonsoon while productivity was its maximum during monsoon. In the present study temperature seems to be a limiting factor in the synthesis of chlorophyll 'a' pigment. However, other parameters like pH, salinity and dissolved oxygen did not have any distinct relationship with the fluctuation in chlorophyll 'a' concentration, phosphate and nitrite seemed to have an inverse relationship with fluctuations in chlorophyll concentration, But silicate and nitrate seem to have a similar seasonal fluctuation as that of chlorophyll "a', Desai *et al.* (1984) have also observed high nitrate concentration associated with high pigment value. These in general nutrients dissolved oxygen seem to be a limiting factor in the synthesis and production of chlorophyll a pigment in Kadinamkulam Lake.

Conclusion

It was concluded that low primary productivity, chlorophyll 'a' were reported the study area due to the retting activity. The effect of retting activity is maximum at station I and II and minimum at station III, which is considered as non-polluted area. This is a serious environmental concern, which call for an effective management action plan.

References

Abdul Azis, P.K., 1985. *Fisheries Research Report*. Regional Agricultural Research Stations, Kurnarakom, 26 pp.

APHA, 1984. *Standard Methods for the Examination of Water and Wastewater*, 15th ed. APHA, AWWA, WEF, Washington.

Balachandran, V.K., M.S. Rajagopalan and Pillai, V.K., 1989. Chlorophyll 'a' and phaeo pigment as indices of biological productivity in the inshore waters of Cochin. *Indian J. Fish.*, 36(3): 227–237.

Desai, B.N., M. Jiyalal Ram, S.A.H. Abidi and V.R. Nair, 1984. Distribution of phytoplankton pigments in Auranga, Ambika, Puma and Mindola estuaries of Gujarat. *Mahasagar Bull. Nat. Inst. Oceanogr.*, 17(2): 79–87.

Desikachary, T.V. and Rao, Y.N.R., 1972. Salinity and diatoms. *J. Mar. Biol. Ass. India*, 14(2): 524–538

Devassy, V.P and Ooes, J.I., 1989. Seasonal patterns of phytoplankton biomass and productivity in a tropical estuarine complex (West Coast of India). *Proc. Indian Acad. Sci. (Plant Sci.)*., 99(5): 485–501.

Fogg, G.E., 1982. Phytoplankton primary production. In: *Fundamentals of Aquatic Ecosystem*, Blackwell Scientific Publications, Oxford.

Gopinathan, C.P. and Rodrigo, J.X., 1991. Investigations on primary production and related parameters in the inshore waters of Tuticorin. *J. Mar. Biol. Ass., India*, 33(1 and 2): 33–39.

Harvey, H.W., 1955. *The Chemistry and Fertility of Seawaters*. Cambridge University Press, London, 240 pp.

Ketchum, B.H., 1967. Phytoplankton nutrients in estuaries. In: *Estuaries*, (Ed.) G.H. Lauff. No. 83, American Association for Advancement of Science, Washington D.C., p. 329–335.

Mc Cauley, E., J.A. Dowing and Watson, S., 1989. Sigmoid relationships between nutrients and chlorophyll among lakes. *Can. J. Fish. Aquat. Sci.*, 46(7): 1171–1175.

Nair, K.V.K. and Ganapathy, S., 1983. Baseline ecology of edaiyur sadras estuarine system at Kalpakkam. Part I: General hydrographic and chemical feature. *Mahasagar*, 16(2): 143–151.

Nair, N.B., P.K. Abdul Aziz, K. Krishnakumar, K. Dharmaraj and Arunachalam, M., 1984. Ecology of Indian estuaries. Part VI: Physico-chemical conditions in Kadinamkulam backwater, SW coast of India. *J. Mar. Sci.*, 13(2): 69–74.

Nair, N.B., P.K., Abdul Aziz, K. Dharmaraj, M. Arunachalam, K. Krishnakumar and Balasubramanian, N.K., 1984. Ecology of Indian estuaries. Part V: Primary productivity of the Ashtamudi estuary south-west coast of India. *Proc. Indian Acad. Sci* (*Anim. Sci.*), 93: 9–23.

Nair, P.V.R and Pillai, V.K., 1983. Productivity of the Indian seas. *J. Mar. Biol. Ass., India*, 25 (1–2): 41–50.

Nair, N.B. and Abdul Aziz, P.K., 1987. The estuarine scenario of Kerala with reference to the status of aquaculture development. *Proc. Nat. Sem. Estuar. Management*, p. 532–543.

Pillai, V.K., K.J. Joseph and Kesavan Nair, A.K., 1975. The plankton production in the Vembanad lake and adjacent water in relation to the environmental parameters. *Bull. Dept. Mar. Sci. Univ., Cochin*, 7(1): 137–150.

Qasim, S.Z., P.M.A. Bhattathiri and Devassy, V.P., 1972. The influence of salinity on the rate of photosynthesis and abundance of some tropical phytoplankton. *Mar. Biol.*, 12(3): 200–206.

Ragothaman, G. and Reddy, Y.R., 1982. Hydrobiology of Tapi estuary, Surat. *Indian J. Mar. Sci.*, 11(3): 256–258.

Santhosh, S., 2002. Hydrogeochemistry of Paravur-Kappil backwater with special reference to plankton. *Ph.D. Thesis*, University of Kerala, Trivandrum.

Sheela, P., K. Sarala Devi, T. Balasubramanian and Sankaranarayanan, V.N., 1995. Chlorophyll a, particulate organic carbon and suspended load from the mangrove areas of Cochin Backwaters. *Proc. Seventh Kerala Sci. Congr.*, Palakkad, p. 87–89.

Shibu, S., 1991. Ecology of the Paravur lake. *Ph.D. Thesis*, University of Kerala, Trivandrum.

Snedecor, G.W. and Cochran, W.G., 1967. *Statistical Methods*, 6th edn. Oxford & IBH Publishing Co., New Delhi, 593 pp.

Strickland, J.D.H. and Parsons, T.R., 1968. A practical handbook of seawater analysis. *Bull. Fish. Res. Biol.* Canada, 167: 311.

Varshney, P.K., V.R. Nair and Abidi, S.A.H., 1983. Primary productivity in near shore waters of Thai, Maharashtra Coast. *Indian J. Mar. Sci.*, 12(1): 27–30.

Vollenweider, R.A., 1969. *A Manual of Methods for Measuring Primary Production in Aquatic Environment.* Blackwell Scientific Publications, Oxford IB Handbook, 12: 213.

Williams, R.B., 1972. Nutrient levels and phytoplankton productivity in the estuary. In: *Proceeding of the Coastal Marsh and Estuary Symposium*, Louisiana State University, Louisiana, (Ed.) R.H. Chabreck, p. 59–89.

Chapter 8

The Impact of Industrial Effluents and Sewage on the Water Quality of River Godavari at Bhadrachalam: A Temple Town in South India

B. Sesha Sreenivas[1], V.V. Ramana[2], K. Ramesh[3] and M.A. Singara Charya[4]

[1]*Department of Civil Engineering, University College of Engineering (K.U.), Kothagudem, A.P.*
[2]*Department of Botany, S.C.W.D. College, Kothagudem, A.P.*
[3]*Execitive Engineer, Regional Office, APPCB, Kothagudem, A.P.*
[4]*Department of Microbiology, Kakatiya University, Warangal, A.P.*

ABSTRACT

Physico-chemical parameters of Godavari waters at Bhadrachalam was monitored during April, 2005 to March, 2006 to study the impact of sewage and industrial effluents on the water quality. Five sampling sites were selected over a distance of about 45 km to study the parameters, pH, Electric Conductivity, Dissolved Oxygen, Chemical Oxygen Demand, Biological Oxygen Demand, Chlorides, Total Coliforms and Temperature. The maximum pH recorded was 9.0 and highest water temperature was 31. 8°C. The discharge of industrial effluents and sewage drastically reduce the DO (3.2 mg/L). The maximum BOD and COD levels were 8.0 and 28 mg/L respectively in river waters. High densities of coliforms were recorded in 84 and 85 sites (>2400).

***Keywords**: Godavari river, Industrial effluents, Physico-chemical parameters, Sewage and Water quality.*

Introduction

Godavari, a river of Central India flowing about 1448 km from the Western Ghats, southeast across the Deccan plateau to the Bay of Bengal. It rises at Trimbakeshwar in Nasik in Maharastra, flows southeast across South Central India, enters into Andhra Pradesh at Basar in Nizamabad district, flows through Vemulawada and enters into Bhadrachalam town.

Bhadrachalam town is located at 179°45′ latitude and 80°58′ longitude. The average rainfall is 1000 mm. The temperature during summer months varies from 40° to 46°C. Bhadrachalam town is an important pilgrim centre in India in Khammam district of Andhra Pradesh. The historic shrine of Lord "Sita Rama Chandra Swami" is situated on the left bank of river Godavari. Normal floating population to visit the shrine will be in the order of few thousands in addition to the local population.

The town is provided with protected water supply scheme executed by the Public Health Engineering Department. The town does not have any underground drainage system. Many a places people defecate on the road margins and open spaces available on the river banks and pollute the surroundings. The problem is very much compounded when huge floating population congregate during the celebrations of Sri Rama Navami, Mukkoti, Godavari Pushkaralu and other Hindu festivals.

The prominent industries that are located on the banks of river Godavari on either side of the off-take of existing water supply scheme are as follows:

1. Manuguru (Aswapuram village)–Heavy Water Project–about 40 km upstream of Bhadrachalam. The total effluent discharge from Manuguru is 143 mld.
2. Kamalapuram Rayon factory, 100 km upstream of the Bhadrachalam water supply head works. The total effluent discharg is 23 mld.
3. Bhadrachalam Paper Boards (Indian Tobacco Company) at Sarapaka 2.5 km from Bhadrachalam (downstream of water supply off-takes). The total PSPD (Paper and Speciality Paper Division) effluent discharge is 55 mld.
4. Kothagudem Thermal Power Station (KTPS) of AP GENCO at Paloncha, which lets out the accumulated ash into Kinnerasani river a tributary of river Godavari wich joins at 10 km downstream of Bhadrachalam water supply head works. The total effluent discharge of KTPS is 99000 kld. All these industries have installed their own treatment plants.

The intensity, dispersion, concentration and effects on river water at the points where the effluents join the flow were studied. Water samples were collected at all five points regularly every month and the samples were analysed for various physico-chemical parameters. The multiple points of pollution of river Godavari due to various pollutants have been studied.

Sl.No.	*Sampling Site*
1.	S1: Kinnerasani after confluence of KTPS ash pond effluent, Khammam district.
2.	S2: Sabari at Kunavaram, Khammam district.
3.	S3: Bhadrachalam U/S before confluence of sewage from town, Khammam district.
4.	S4: Bhadrachalam D/S after confluence of sewage from town, Khammam district.
5.	S5: Burgamphad after confluence of ITC Ltd., PSPD effluents, Khammam district.

Material and Methods

The present study area includes five prominent sites on the banks of river Godavari where sewage and industrial effluents join the flow (Figure 8.1).

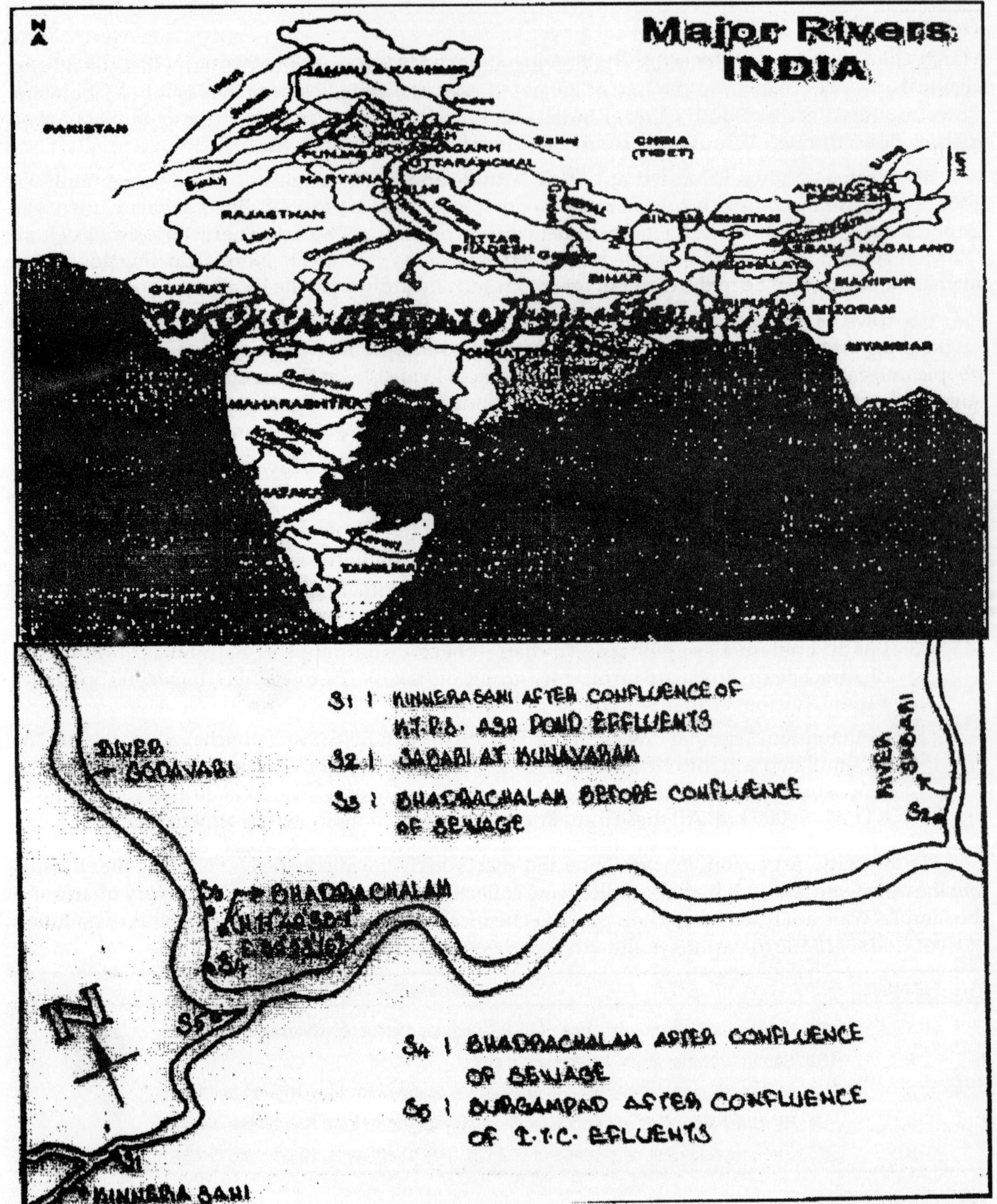

Figure 8.1: Details of Site Locations

Water samples were collected at all these five points regularly every month during April, 2005 to March, 2006 and the samples were analysed for various physico-chemical parameters like pH, Temperature, Electrical Conductivity (EC), DO, BOD, COD, Chlorides and Total Coliforms.

The standard methods for examination of water and waste water (APHA, 1989) were followed for the analysis of physico-chemical parameters. Water samples were collected in sterile bottles and were immediately transported to the laboratory for analysis. pH and Temperature were measured at the site.

Results and Discussion

pH

The pH values measured at all the five sites were presented in Table 8.1. The results indicate that the waters of river Godavari at Bhadrachalam are slightly alkaline. The values varied between 7.0 to 9.0. The maximum value of 9.0 was recorded at 84 (after the confluence of sewage). The pH values of natural waters changes due to biological activity, temperature and disposal of industrial waste. The impact of seasonality is not conspicuous on the pH of waters.

Table 8.1: pH in River Godavari at Five Sampling Sites During 2005–06

Month and Year	*Sampling Site–1*	*Sampling Site–2*	*Sampling Site–3*	*Sampling Site–4*	*Sampling Site–5*
Apr-05	8.8	8.5	8.8	9.0	8.2
May-05	8.2	8.0	8.5	8.4	8.6
Jun-05	8.1	8.3	8.4	8.6	8.2
Jul-05	8.7	8.0	8.4	8.5	8.3
Aug-05	8.1	7.8	8.1	7.8	7.9
Sep-05	8.1	8.2	8.4	8.0	8.2
Oct-05	8.2	8.2	8.5	8.2	8.4
Nov-05	8.4	8.6	8.9	8.9	8.6
Dec-05	7.0	7.3	7.6	7.9	7.9
Jan-06	7.9	7.7	7.9	8.0	8.2
Feb-06	8.4	7.8	7.7	7.8	8.1
Mar-06	8.2	7.9	7.8	7.8	8.0

Temperature

The temperature levels of surface water in all the sampling points observed were presented in Table 8.2. The values ranged from 26.7°C (January 2006–S4) and 31.8°C (March, 2006–S2, S3, S5). The impact of effluents or sewage did not show any influence on the temperature levels of river water.

Electrical Conductivity (EC)

It is a measure of capacity of a solution to carry an electrical current. It is a total parameter for dissolved and dissociated substances and indicates the concentration of dissolved electrolytes. The permissible value of conductivity for drinking water is 300 micro mhos/cm (USPHS).

Table 8.2: Temperature (°C) in River Godavari at Five Samplinag Sites During 2005–06

Month and Year	*Sampling Site–1*	*Sampling Site–2*	*Sampling Site–3*	*Sampling Site–4*	*Sampling Site–5*
Apr-05	31.0	31.0	31.0	31.0	31.0
May-05	30.3	30.5	30.5	30.4	30.6
Jun-05	29.5	30.1	29.9	29.9	29.9
Jul-05	28.8	28.6	28.9	28.9	28.8
Aug-05	28.4	28.6	28.6	28.6	28.6
Sep-05	28.4	28.5	28.4	28.5	28.5
Oct-05	28.8	28.9	28.8	28.7	28.9
Nov-05	27.1	27.1	27.2	27.2	27.2
Dec-05	27.8	27.8	27.9	27.8	27.9
Jan-06	26.7	26.7	26.8	26.7	26.8
Feb-06	30.6	30.6	30.7	30.6	30.6
Mar-06	31.7	31.8	31.8	31.7	31.8

The EC values at five sampling points were presented in Table 8.3. The EC levels were maximum at 85 (650 μmho/cm) in the month of August, 2005 and the minimum was recorded at 82 (85 μmho/cm) in July, 2005. In general, the sampling points S4 and S5 which were loaded with sewage water and effluents from ITC Bhadrachalam respectively showed maximum levels of conductance (333–650 μmho/cm). The sampling point S1, located in Kinnerasani after the confluence of KTPS ash pond effluent was with moderate to high levels of conductivity and the range varied from 364 μmho/cm to 475 μmho/cm. The sampling point S2, in the river Sabari at Kunavaram was with low levels of conductance throughout the year and ranged from 92 to 200 μmho/cm.

Table 8.3: Electrical Conductivity (μmho/cm) in River Godavari at Five Sampling Sites During 2005–06

Month and Year	*Sampling Site–1*	*Sampling Site–2*	*Sampling Site–3*	*Sampling Site–4*	*Sampling Site–5*
Apr-05	382.0	100.0	327.0	350.0	382.0
May-05	396.0	92.0	320.0	358.0	379.0
Jun-05	411.0	98.0	335.0	380.0	367.0
Jul-05	475.0	85.0	124.0	236.0	252.0
Aug-05	420.0	98.0	205.0	600.0	650.0
Sep-05	364.0	192.0	228.0	325.0	389.0
Oct-05	387.0	184.0	236.0	340.0	381.0
Nov-05	392.0	188.0	231.0	343.0	386.0
Dec-05	379.0	102.0	200.0	333.0	353.0
Jan-06	387.0	180.0	280.0	398.0	457.0
Feb-06	388.0	195.0	300.0	391.0	420.0
Mar-06	366.0	200.0	325.0	403.0	412.0

DO

DO is an important parameter to determine the water quality for various purposes. DO concentration in a water body indicates its ability to support aquatic life. In natural water, DO levels depend on the physical, chemical and biological activities of the water body. Clean surface water is normally saturated with DO (7.6 mg/L at 30°C). Oxygen can be rapidly removed from water by discharge of the oxygen demanding wastes.

The quantity of DO in five sampling points were analysed with monthly intervals and presented in Table 8.4. From the table it was clear that S2 is with very less amount of organic pollution, where the DO levels reached 6.2 mg/L (May, 2005). The sampling site S3 which is upstream of Bhadrachalam before the confluence of sewage was with high levels of DO and ranged from 5.0 to 6.2 mg/L, while the sample collected after the confluence (S4) recorded rapid decline in the DO levels due to addition of sewage. The DO levels were comparatively low and ranged from 3.2 to 5.4 mg/L in the river waters of Godavari at Burgamphad with the discharge of effluents from ITC, Bhadrachalam (S5). Inspite of contamination of Kinnerasani with the discharge of KTPS ash pond effluents no remarkable decline in the DO levels were noticed at S1.

Table 8.4: DO (mg/L) in River Godavari at Five Sampling Sites During 2005–06

Month and Year	*Sampling Site–1*	*Sampling Site–2*	*Sampling Site–3*	*Sampling Site–4*	*Sampling Site–5*
Apr-05	2.8	2.2	5.0	3.0	3.2
May-05	5.6	6.2	6.0	5.0	4.8
Jun-05	5.8	5.4	5.6	4.8	4.6
Jul-05	5.2	5.8	5.0	4.6	4.2
Aug-05	5.6	5.8	6.2	4.2	5.4
Sep-05	6.2	5.6	5.4	5.4	4.8
Oct-05	5.2	5.8	5.0	4.6	4.2
Nov-05	5.6	5.8	6.2	4.2	5.4
Dec-05	6.2	5.6	5.4	5.4	4.8
Jan-06	5.8	4.6	5.3	4.5	4.1
Feb-06	6.0	5.4	5.6	5.0	5.2
Mar-06	6.4	5.6	6.2	4.8	5.0

Chlorides

Chlorides in natural waters can be attributed to discharge of effluents or sewage disposal which may result in local contamination of river water. Domestic sewage contains more chlorides. Chlorides often serve as a chemical pollution indicator of sewage contamination. The permissible value of chloride content in potable water is 600 ppm (ISI).

The chloride contents in the five sampling sites analysed during April, 2005 to March, 2006 are presented in Table 8.5. The maximum chloride contents were recorded at S4 and S5; which are contaminated with sewage and ITC effluents respectively. In general, the months July, 2005 to October, 2005 have recorded meager amounts of chlorides in the river waters. The moderate levels of chlorides

(70–90 mg/L) were recorded at S1 and S3. The site S2 which is not contaminated by any discharges recorded very less amounts of chlorides, which ranged from 1.0 to 74 mg/L.

Table 8.5: Chloride (mg/L) in River Godavari at Five Sampling Sites During 2005–06

Month and Year	*Sampling Site–1*	*Sampling Site–2*	*Sampling Site–3*	*Sampling Site–4*	*Sampling Site–5*
Apr-05	14.0	6.0	12.0	12.0	14.0
May-05	36.0	8.0	42.0	36.0	48.0
Jun-05	42.0	10.0	38.0	40.0	42.0
Jul-05	1.0	BDL	1.5	2.5	4.0
Aug-05	8.0	1.0	2.0	4.0	2.0
Sep-05	3.0	BDL	BDL	2.0	3.0
Oct-05	36.0	12.0	18.0	15.0	42.0
Nov-05	44.0	12.0	18.0	22.0	34.0
Dec-05	37.0	19.0	20.0	22.0	45.0
Jan-06	40.0	70.0	90.0	130.0	140.0
Feb-06	30.0	64.0	55.0	90.0	120.0
Mar-06	39.0	74.0	49.0	84.0	115.0

BDL: Below Detectable Limits.

Biochemical Oxygen Demand

BOD and COD are the most important parameters used to determine the degree of pollution in river water. The biochemical oxygen demand (BOD) is a measure of the oxygen utilized by micro organisms during biological oxidation of organic matter contained in liquid water under a specified experimental conditions. BOD levels were analysed at five sampling sites and the obtained data was recorded in Table 8.6. The data clearly indicated that S4 and S5 were with maximum BOD (8.0 mg/L), while S1 and S3 showed moderate levels of BOD (2.0–4.0 mg/L). Among all the sites S2 was with least BOD levels (0.2–1.5 mg/L).

Chemical Oxygen Demand

COD is the oxygen required for the chemical oxidation of organic matter. The test indicates the total oxidizable organic material present in the sample. The data of COD in five sampling sites of river Godavari were recorded in Table 8.7. From the table it was clear that S1, S4 and S5 which were contaminated with effluents from KTPS, sewage and PSPD effluents from ITC Bhadrachalam respectively showed maximum levels of COD (24 mg/L). Moderate COD levels were noticed at S2 and S3, where it ranged between 2.0 and 9.0 mg/L.

Total Coliforms

Bacteriological tests have been designed which are extremely sensitive and specific in revealing evidence of pollution. Total coliform counts were recorded during April, 2005 to March, 2006 and the data obtained is presented in Table 8.8.

Table 8.6: BOD (mg/L) in River Godavari at Five Sampling Sites During 2005–06

Month and Year	*Sampling Site–1*	*Sampling Site–2*	*Sampling Site–3*	*Sampling Site–4*	*Sampling Site–5*
Apr-05	2.0	0.8	4.0	5.0	6.0
May-05	2.5	0.4	2.5	3.0	3.5
Jun-05	2.4	0.2	2.0	4.0	4.0
Jul-05	2.4	0.2	3.0	5.0	6.0
Aug-05	2.5	0.2	2.2	4.0	4.8
Sep-05	3.0	0.8	2.0	2.8	3.5
Oct-05	3.0	0.8	2.5	4.0	5.0
Nov-05	2.0	0.4	2.8	5.0	6.0
Dec-05	2.0	0.8	3.0	4.0	4.5
Jan-06	2.4	0.8	4.0	4.5	5.0
Feb-06	2.2	0.8	4.0	5.0	6.5
Mar-06	2.0	1.5	3.8	6.0	8.0

Table 8.7: COD (mg/L) in River Godavari at Five Sampling Sites During 2005–06

Month and Year	*Sampling Site–1*	*Sampling Site–2*	*Sampling Site–3*	*Sampling Site–4*	*Sampling Site–5*
Apr-05	5.2	4.0	6.0	10.0	11.0
May-05	5.0	4.0	5.0	6.0	6.5
Jun-05	4.5	3.5	5.5	7.0	9.0
Jul-05	5.0	4.0	9.0	24.0	23.0
Aug-05	20.0	7.0	8.0	16.0	18.0
Sep-05	17.0	4.0	5.0	11.0	14.0
Oct-05	28.0	5.0	6.0	12.0	16.0
Nov-05	11.0	3.0	7.0	17.0	25.0
Dec-05	11.0	2.0	8.0	13.0	16.0
Jan-06	8.0	4.0	6.0	11.0	14.0
Feb-06	3.0	4.5	7.0	10.0	12.0
Mar-06	5.0	4.0	5.0	15.0	16.0

From the Table 8.8 it is observed that the total coliform densities were very high and exceeded 2400 in all the months under study at S4 and S5. Moderate to high densities of bacterial population was noticed in sampling points S1 and S3 and MPN ranged from 120–2400 and at S2, the coliform population was less (93–1100).

The correlation coefficient studies among the above eight parameters at all five sites were analysed and recorded in Tables 8.9 to 8.13. The correlation was positive and significant between BOD and COD (r 0.66217) ($P < 0.01$) at S1. Maximum correlation was drawn between total coliforms and COD (r 0.98864) ($P < 0.001$) at S2. The correlation between conductivity and BOD was significant at 1 per cent level (r 0.631681) and while it was (r 0.606593) ($P< 0.01$) in between conductivity and total coliforms at S2. The correlation was negative between conductivity and COD (r –0.77828) ($P<0.001$) at S3. The correlation was significant at 5 per cent between COD and BOD (r 0.467558) ($P< 0.05$) at S4. The BOD was correlated significantly with COD (r 0.405955) ($P< 0.05$) and chlorides (r 0.416973) ($P< 0.05$) at S5.

Table 8.8: Total Coliforms Most Probable Number (MPN) in River Godavari at Five Sampling Sites During 2005–06

Month and Year	Sampling Site–1	Sampling Site–2	Sampling Site–3	Sampling Site–4	Sampling Site–5
Apr-05	1100.0	480.0	1200.0	>2400.0	>2400.0
May-05	800.0	240.0	1200.0	>2400.0	>2400.0
Jun-05	210.0	120.0	1100.0	>2400.0	>2400.0
Jul-05	240.0	93.0	460.0	>2400.0	>2400.0
Aug-05	380.0	120.0	1300.0	>2400.0	>2400.0
Sep-05	240.0	480.0	1100.0	>2400.0	>2400.0
Oct-05	120.0	460.0	240.0	1100.0	>2400.0
Nov-05	1100.0	240.0	480.0	>2400.0	>2400.0
Dec-05	480.0	480.0	1600.0	>2400.0	>2400.0
Jan-06	240.0	480.0	1800.0	>2400.0	>2400.0
Feb-06	480.0	480.0	1200.0	>2400.0	>2400.0
Mar-06	1200.0	1100.0	480.0	>2400.0	>2400.0

Table 8.9: Correlation Coefficient (r) among the Physico-chemical Characteristics at Site 1 in River Godavari During 2005–06

	pH	EC	DO	BOD	COD	Cl^-	TCF	Temp
pH	1	0.34578	–0.57443	0.02252	–0.21548	–0.37038	0.26993	0.42125
EC		1	–0.1452	0.04482	–0.13971	–0.40765	–0.34682	–0.1458
DO			1	0.15515	0.06167	0.30877	–0.27202	–0.24828
BOD				1	0.66217	–0.32696	–0.70406	–0.21704
COD					1	–0.15468	–0.40938	–0.41541
Cl^-						1	0.24290	–0.05453
TCF							1	0.45475
Temp								1

Similar to the present observations, Rai (1964), Saxena *et al.* (1966), Manivasakam (1980), Bilgrami and Munshi (1985), Rajani *et al.* (1987), Ghosh and Sharma (1988), analysed the impact of industries and municipal sewage on the physico-chemical and bacteriological characters of different rivers in India. Somasekhar (1984), studied the water pollution of river Cauvery with special emphasis on physico-chemical characteristics altered by sewage contamination. Azmal and Razi-Ud-Din (1988), investigated the pollution levels of Hindon and Kalinadi rivers and reported the relations between organic contaminants and coliform densities. Mittal and Sengar (1990), evaluated pollution status of

Karwar river and noticed the abnormal levels of BOD and COD. Sharma and Chakrabarthi (1995), studied the influence of seasonal changes and industrial effluents on the pollution levels of river Yamuna and suggested conservation practices. The heavy metal distribution and other pollutants in the upper reaches of river Beas was studied and their impact on the flora and fauna was evaluated by Singh *et al.* (1996). Kisku and Bhargava (1998), assessed the water quality in river Damodar at Durgapur industrial belt and noticed abnormal levels of BOD and COD. Madhyastha *et al.* (1999), determined the impact of anthropogenic activities on the physico-chemical and microbial characteristics of riverine stations located in the vicinity of the major temples of Dakshina Kannada District. Mohanty *et al.* (2003) evaluated the relationship between BOD and COD for river Nagavali and Kolab and noticed their impact on the proliferation of bacteria in the river waters. Singh and Rai (2003) studied the impact of the industrial effluents and domestic sewage on river Ganga at Allahabad and reported that all the pollution parameters are beyond the permissible limits and unfit for human consumption. Tiwari (2004), studied the pollution potential of river Pandu contaminated heavily by the discharges of various industrial effluents. Rajyalakshmi and Sreelatha (2005) studied the physico-chemical parameters of river Goutami-Godavari flowing across Yanam reported the negative impact of agro-chemicals on the quality of water. Gupta and Pankaj (2006) studied the eutrofication and heavy metal pollution in river Ganga and Gomti with reference to human activity. Kulshresta and Sarma(2006), Chetna *et al.* (2006) observed the impact of mass bathing during festive days and analysed the bacteriological water quality in different rivers in India.

Table 8.10: Correlation Coefficient (r) among the Physico-chemical Characteristics at Site 2 in River Godavari During 2005–06

	pH	*EC*	*DO*	*BOD*	*COD*	*Cl⁻*	*TCF*	*Temp*
pH	1	0.10418	–0.28967	–0.213	0.026507	–0.4043	–0.20254	0.160208
EC		1	0.127679	0.631681	–0.01292	0.596079	0.606593	–0.08874
DO			1	–0.22222	0.064418	–0.05199	–0.17405	–0.21913
BOD				1	–0.19025	0.67744	0.988644	0.340764
COD					1	–0.08334	–0.16733	0.146619
Cl⁻						1	–0.16733	0.170025
TCF							1	0.395687
Temp								1

Table 8.11: Correlation Coefficient (r) among the Physico-chemical Characteristics at Site 3 in River Godavari During 2005–06

	pH	*EC*	*DO*	*BOD*	*COD*	*Cl⁻*	*TCF*	*Temp*
pH	1	0.006668	–0.10573	–0.34482	–0.17156	–0.45067	–0.44083	–0.04968
EC		1	0.234658	0.294644	–0.77828	0.560838	0.209858	0.596571
DO			1	–0.17624	–0.15182	0.086882	–0.03126	0.122158
BOD				1	0.042397	0.484566	0.172309	0.258231
COD					1	–0.54875	–0.00601	–0.37587
Cl⁻						1	0.43228	–0.00559
TCF							1	–0.21483
Temp								1

Table 8.12: Correlation Coefficient (r) among the Physico-chemical Characteristics at Site 4 in River Godavari During 2005–06

	pH	*EC*	*DO*	*BOD*	*COD*	*Cl⁻*	*Temp*
pH	1	–0.43125	–0.6124	0.083335	0.002885	–0.37984	0.030606
EC		1	–0.17648	–0.02366	–0.16223	0.156623	0.036881
DO			1	–0.42061	–0.11889	0.137938	–0.14297
BOD				1	0.467558	0.378023	0.283297
COD					1	–0.28444	–0.28174
Cl⁻						1	0.045583
Temp							1

Table 8.13: Correlation Coefficient (r) among the Physico-chemical Characteristics at Site 5 in River Godavari During 2005–06

	pH	*EC*	*DO*	*BOD*	*COD*	*Cl⁻*	*Temp*
pH	1	–0.41787	–0.10355	–0.2019	0.03608	0.085779	–0.10805
EC		1	0.39312	–0.04921	–0.05332	0.085779	–0.0956
DO			1	0.01331	0.263391	0.096971	–0.12523
BOD				1	0.405955	0.416973	0.361906
COD'					1	–0.23557	–0.5285
Cl⁻						1	0.118854
Temp							1

Based on the present study, it is felt that the multiple points of pollution of river Godavari due to various pollutants have adverse effect on the quality of river waters. Hence, to preserve its purity, river ecology and environment the pollutants must be treated properly, to get the BOD, DO and other pollution parameters to the desired quality, before they are discharged into the river.

References

Ajmal, M. and Razi-Ud-Din, 1988. Studies on pollution of Hindon river and Kalinadi (India). In: *Ecology and Pollution of Indian Rivers* (Ed.) R.K. Trivedy. Asian Publishing House, New Delhi, p. 87–111.

APHA, (American Public Health Association), 1989. *Standard Methods for Examination of Water and Wastewater.* 17th edn., Washington DC.

Bilgrami, K.S. and Munshi, J.S.D., 1985. Ecology of River Ganga: Impact of human activities and conservation of aquatic biota (Patna to Farakka). *Final Technical Report* (*D.O.E*), Allied Press, Bhagalpur, pp. 97.

Chetna Anand, Pratima Akolkar and Rina Chakrabarthi, 2006. Bacteriological water quality status of River Yamuna in Delhi, *J. of Environmental Biology*, 27: 97–101.

Ghosh, N.C. and Sharma, C.B., 1988. Effect of drain water on the physicochemical and bacteriological characters of the river Ganga at Patna, Bihar. In: *Ecology and Pollution of Indian Rivers*, (Ed.) R.K. Trivedy. Asian Publishing House, New Delhi, p. 225–269.

Gupta, A.K. and Pankaj, P.K., 2006. Comparative study of eutrophication and heavy metal pollution in rivers Ganga and Gomti with reference to human activities. *Nature Environment and Pollution Techno.*, 5: 229–232.

Kisku, G.C. and Bhargava, S.K., 1998. Water quality around Durgapur industrial belt and its impact of environment. *Indian J. Env. Proto.*, 18: 898–903.

Kulshresta, H. and Sharma, S., 2006. Impact of mass bathing during Arth Kumbh on water quality status of river Ganga. *J. of Environmental Biology*, 27: 437–440.

Madhyastha, M.N., Rekha, P.O. and Shashikumar, K.C., 1999. Impact of anthropogenic activities on the river water quality along pilgrimage centres at Dakshina Kannada. *J. Hum. Ecol.*, 10: 29–34.

Manivasakam, N., 1980. *Physico-chemical Examination of Water, Sewage and Industrial Effluents*. Pragati Prakashan Publ., Meerut, India.

Mittal, S. and Sengar, R.M.S., 1990. Studies on the assessment of water pollution in Karwar river. *Poll. Res.*, 9: 91–94.

Mohanty, S. Kumar, Patnaik, D. and Rout, S.P., 2003. Evaluation of relationship between BOD and COD for river Nagavali and river Kolab in Koraput district of Orissa. *Indian J. Env. and Ecoplaning.*, 7: 549–552.

Rai, H., 1964. The bacteriological studies on the Yamuna river (Delhi, India), Scientific Paper from Institute of Chemical Technology, Prague. *Technology of Water*, 8: 319–335.

Rajani, K., Behera, R.B., Panda, B.K., Sahu, B.K. Sinha and Nayak, A., 1987. Assessment of Brahamani river Part–1: Water quality U/S and D/S of Rourkela industrial complex. *Poll. Res.*, 8: 195–200.

Rajyalakshmi, S. and Sreelatha, K., 2005. Physico-chemical parameters of river Goutami–Godavari, Yanam (Union Territory of Pondicherry). *J. Aqu. Biol.*, 20: 110–112.

Saxena, K.L., Chakrabarthi, R.N., Khan, A.A., Chattopadhya, S.N. and Chandra, H., 1966. Pollution studies of river Ganges near Kanpur. *Environ. Health*, 8: 270–285.

Sharma, J.G. and Chakrabarthi, 1995. Seasonal Changes in the physico-chemical and biological parameters of an Ox-bow lake of river Yamuna. *Proc. Indian Sci. Congr.*, p. 100.

Singh, J.B., Gupta, D. and Sharma, A., 1996. Heavy metals distribution and other pollutants in the upper reaches of river Beas in Himachal Pradesh. *Indian J. of Env. Protection*, 17: 43–46.

Singh, S.K. and Rai, J.P.N., 2003. Pollution studies on river Ganga,in Allahabad district. *Poll. Res.*, 22: 469–472.

Somashekhar, R.K., 1984. Studies of water pollution of the river Cauvery physico-chemical characteristics. *Intern. J. Env. Studies*, 23: 209–216.

Tiwari, D., 2004. Pollution potential of the wastes polluting river Pandu. *Nature Env. Poll. Tech.*, 3: 219–221.

Chapter 9

Influence of Yeast Extract on Aerobic Biodegradation of 2-Chlorophenol and 4-Chlorophenol

J.B. Chawla and A.O. Ingle*

Department of Microbiology, LIT Premises, RTM Nagpur University, Nagpur, Maharashtra

ABSTRACT

The influence of yeast extract on the aerobic biodegradation of 2-chlorophenol (2-CP) by *Alcaligenes* sp. strain J-3 and 4-chlorophonol (4-CP) by *Acinetobacter* sp. strain A-2 was studied. The strains were isolated from soil by enrichment culture technique. Addition of yeast extract in the range 50 to 200 mg/L to the basal medium increased the degradation rate of 2-CP up to 20 per cent, while yeast extract in the range from 100 to 400 mg/L increased the degradation rate of 4-CP up to 203 per cent. Presence of yeast extract always increased the biomass yield and mostly resulted in an increased simultaneous utilization of target substrate. Increase in biomass yield and increased simultaneous utilization of 4-CP by strain A-2 were also noticed with the inclusion of meat extract and peptone in the medium. However, yeast extract was found to be superior secondary substrate than meat extract and peptone for biodegradation of 4-CP. Outcome of addition of yeast extract in the medium is not always uniform and differed depending on the type of strain involved, nature and concentration of target substrate and concentration of yeast extract itself. Influence of yeast extract was more pronounced on the degradation of 4-CP by strain A-2 as compared to degradation of 2-CP by strain J-3.

Keywords: *2-chlorophenol, 4-chlorophenol, Alcaligenes, Acinetobacter, Biodegradation.*

* *Corresponding Author.* Phone (M) 9422834758

Chlorophenols are widely used in manufacture of various insecticides, fungicides and wood preservatives. The recalcitrant and carcinogenic nature of chlorophenols coupled with bioaccumulation potential is a matter of great concern (Nesnow *et al.*, 1986; Frank *et al.*, 1989). Both aerobic and anaerobic degradation of polychlorophenols may lead to formation of monochlorophenols (Neilson, 1990; Armenante, 1992). It required 120–220 days for the degradation of 0.1 mM 2-, 3- and 4-chlorophenols under sulphate reducing conditions (Haggblom *et al.*, 1990). Under aerobic conditions the logarithmic biomass growth of *Pseudomonas putida* on pure monochlorophenols did not occur eyen after 2 weeks of incubation (Dapaah *et al.*, 1992).

It has been reported in several studies that addition of easily metabolised organic substrates in the medium enhances the degradation of various xenobiotics. The presence, nature and concentration of supplemental substrates such as glucose and phenol are important factors influencing biodegradation rates of toxic compounds, like 2-CP when being degraded by mixed population of activated sludge (Basu, 1995). Auxiliary readily utilizable carbon sources such as propionate enhanced the rate of 4-CP degradation (Haggblom *et al.*, 1993). The logarithmic growth phases of *Pseudomonas putida* were only observed when monochlorophenols were cometabolised with phenol (Dapaah, 1992). The yeast extract (YE) and to lesser extent vitamins and trace elements can significantly enhance the aerobic degradation and dechlorination of some chlorobenzoic acid isomers (Fava *et al.*, 1995; Armenante *et al.*, 1992). However on the contrary the presence of glucose, pyruvate and lactose in the medium reduced the capability of *Pseudomonas putida* 10.2 to degrade 3-chlorobenzoate (Chobchnohum *et al.*, 1996). Similarly the addition of glucose, adenine or propionate inhibited the metabolism of p-nitrophenol in lake water (Rubin, 1983).

All these findings are of much importance in designing the biological treatment processes. Uptil now comparatively less attention has been paid to the optimization of nutritional composition of culture medium with respect to biodegradation of recalcitrant xenobiotics. In the present study the effect of YE on the biodegradation of 2-CP and 4-CP under aerobic conditions has been studied. A comparative study with meat extract (ME) and peptone (PEP) under similar conditions on the biodegradation of 4-CP has also been made.

Materials and Methods

Chemicals

Chemicals 2-CP and 4-CP were obtained from Aldrich Chemical Co. Inc. All other chemicals were of analytical grade. YE, ME and PEP obtained from HI Media, Mumbai.

Organisms

Acinetobacter sp. strain A-2 and *Alcaligenes* sp. strain J-3 were isolated from the soil by enrichment culture technique and used in this study. Strain A-2 can utilize 4-CP while strain J-3 can utilize 2-CP as their sole sources of carbon.

Basal Medium

It consisted of following components per litre of distilled water: Potassium dihydrogen phosphate 1 g; Dipotassium hydrogen phosphate 5 g; Magnesium sulphate 100 mg; Ammonium sulphate 1 g; Ferric ammonium citrate 5 mg and Calcium nitrate 50 mg.

The stock solutions of various ingredients were prepared, sterilized and mixed in appropriate amounts to give required concentration of each ingredient. In addition filter sterilized 2-CP/4-CP was added in required amount to obtain desired initial substrate concentration in the medium. The 50 mg/L concentration of YE in the medium was obtained by adding 0.1 mL of filter sterilized stock solution of YE (50 mg/mL) to 100 mL of basal medium. In place of YE solution; ME or PEP stock solution was used in comparative studies.

Preparation of Inoculum

The organisms were grown overnight at 33°C on nutrient agar slant at pH 7.8, supplemented with respective isomer of monochlorophenol at 0.2 mM concentration. The cells from the slant were collected and washed twice with the mineral medium and resuspended in the mineral medium to obtain extinction of 0.02 at 540 nm. 2 mL aliquot of this cell suspension was used to inoculate 100 mL of medium.

Experimental Procedure

Experiments were conducted in 250 mL Erlenmayer flasks containing 100 mL of basal medium with and without YE. All experiments were performed in duplicate. After inoculation of medium with appropriate culture, the flasks were incubated at 33°C. Samples were withdrawn at regular intervals and after centrifugation at 7000 rpm for 20 mins, cell free supernatants were analysed for chloride ions and residual chlorophenols.

A control flask inoculated with *E. coli* (a non-degrader of monochlorophenols) was also simultaneously incubated along with every batch of experimental flasks to check losses of chlorophenols from the medium due to auto-oxidation, adsorption on the cells and other unknown reasons. Losses observed in control flask were subtracted from the observed amount of disappeared substrate in experimental flasks. The purity of culture in each flask was checked at the end of the experiment by streaking the culture on nutrient agar plates.

Analytical Procedure

4-Aminoantipyrine method was used for estimation of phenols (Clesceri *et al.*, 1989). The chloride ion concentration in the culture supernatant was determined by the method reported by Higson and Focht (1990). Extinction of culture was measured at 540 nm by Systronics spectrophometer model 106.

Results and Discussion

Effect of Yeast Extract on 4-CP Degradation

The effect of various concentrations of YE, on the degradation of 4-CP has been illustrated in Figures 9.1–9.3. Strain A-2 degraded 250 μM 4-CP in 50 hrs (Figure 9.1a). The inclusion of YE in the medium at 100, 200, 400 and 600 mg/L concentrations resulted in degradation of 4-CP in 42, 30, 26 and 24 hrs respectively. 19 to 108 per cent increase in degradation rate (Table 9.1) and 4 to 8-fold increase in biomass concentrations were observed with increase in YE concentration in the medium (Figure 9.1c). In general the degradation rate of 4-CP increased with increase in YE concentration. However at 400 mg/L and 600 mg/L YE concentrations, the degradation rates of 4-CP initially slowed down but picked up rapidly in next few hours and ultimately resulted in overall higher degradation rates (Figure 9.1a and Table 9.1). At 500 μM concentration of 4-CP in the medium, strain A-2 could not bring about complete degradation up till 80 hrs and residual concentration of 4-CP was 225 μM at this juncture. The addition of YE at 100 mg/L concentration increased the degradation rate of 4-CP and the residual concentration of 4-CP was only 40 μM after 80 hrs of incubation. Similarly when YE was added in the medium at concentrations 200, 400 and 600 mg/L, the degradation of 4-CP occurred at faster rates and the degradation of 4-CP was over in 60, 48 and 50 hrs respectively (Figure 9.2a). 67 to 203 per cent increase in degradation rate (Table 9.1) and 5 to 9-fold increase in biomass concentrations were observed with respect to increase in YE concentrations (Figure 9.2c). As in the previous case (*i.e.* at 250 μM 4-CP) here also at 400 mg/L YE concentration in the medium, the degradation rate initially slowed down but picked up later. However, slightly lower degradation rate was noticed at 600 mg/L as compared to 4(1) mg/L YE concentration (Figure 9.2a).

Table 9.1: Effect of Supplementary Substrates (YE/ME/PEP) on the Degradation Rate of 4-CP with Strain A-2

Initial Amount of 4-CP in the Medium (μM)	*Initial Amount of YE in the Medium (mg/L)*	*Amount of 4-CP Degraded (μM)*	*Incubation Time (h)*	*Degradation Rate μM h^{-1}*	*% Increase in Degradation Rate*[Δ]
250	00	250	50	5.00	–
250	100	250	42	5.95	19.0
250	200	250	30	8.33	66.6
250	400	250	26	9.61	92.2
250	600	250	24	10.41	108.2
500	00	275	80	3.43	–
500	100	460	80	5.75	67
500	200	500	60	8.33	142
500	400	500	48	10.41	203.4
500	600	500	50	10.00	191.5
1000	00	00	100	0.0	–
1000	100	15	100	0.15	Ø
1000	200	20	100	0.20	Ø
1000	400	30	100	0.30	Ø
1000	600	40	100	0040	Ø
500+	00	500	108	4.62	–
500+	400*	500	70	7.14	54.5
500+	400**	500	55	9.09	96.7
500+	400	500	48	10.41	125.3

* In place of YE, PEP was included in the medium; ** In place of YE, ME was included in the medium; Δ With respect to degradation ,ate of substrate in the absence of YE at corresponding concentration and conditions; Ø Could not be calculated; + Refer Figure 9.7.

In the absence of YE the highest degradation rate of 4-CP was observed at 250 μM concentration. However out of all the tested concentrations in presence or absence of YE in the medium, the highest degradation rate was observed at 250 μM 4-CP with 600 mg/L YE as well as at 500 μM 4-CP with 400 mg/L YE in the medium (Table 9.1).

Rapid increase in biomass from the beginning and simultaneous increase in 4CP utilization suggest that both 4-CP and YE are being utilized simultaneously (Figures 9.1 and 9.2). This is in line with the findings of Fava *et al.* (1995) who reported that an addition of 50 mg/L YE to the cultures

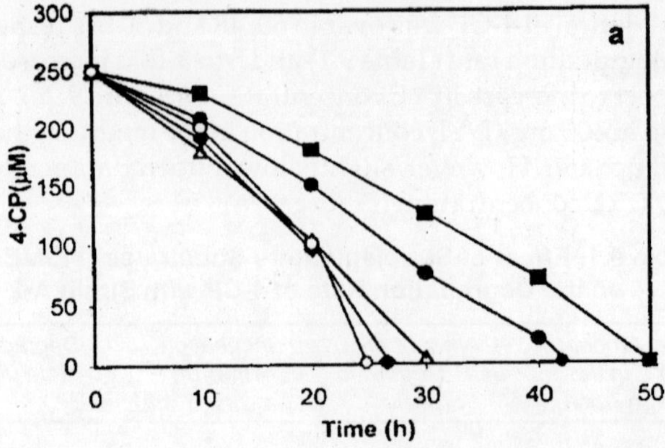

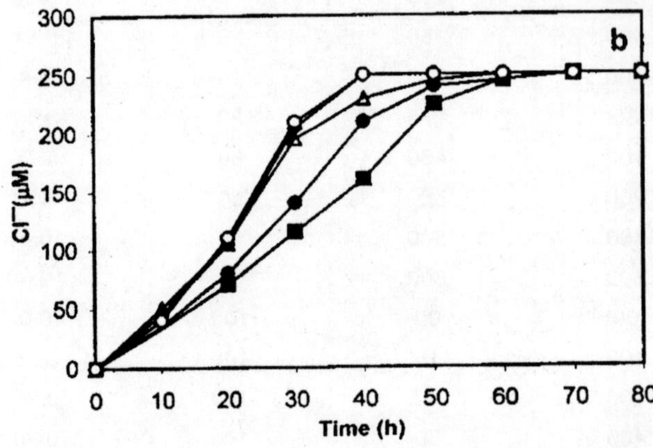

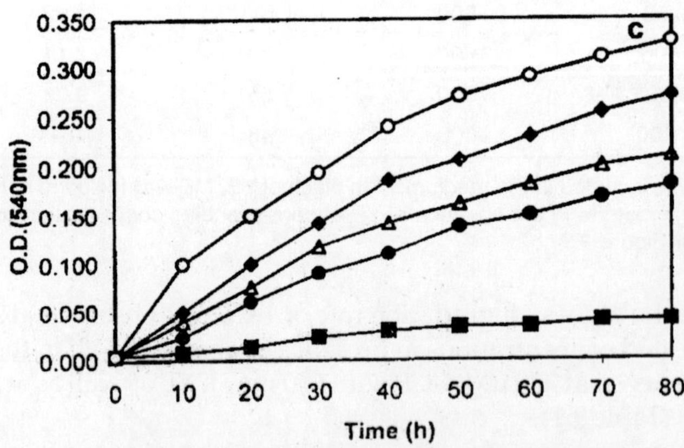

Figure 9.1: Effect of Yeast Extract on the Degradation of 250 μM 4-CP by Strain A-2
(*a*) Degradation of 4-CP; (b) Chloride ion concentration; (c) Cell density
Symbols: ■: 4-CP; ●: 4-CP+YE 100 mg/L; Δ: 4-CP+YE 200 mg/L;
♦: 4-CP+YE 400 mg/L; O: 4-CP+YE 600 mg/L

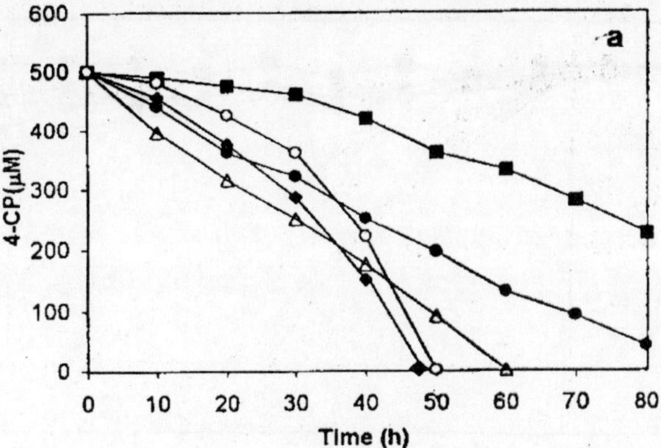

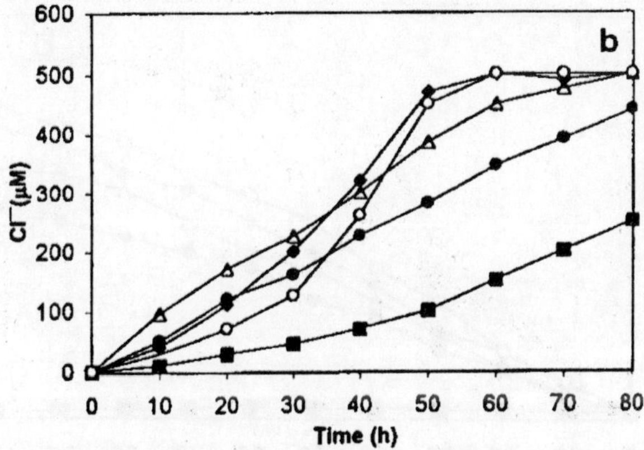

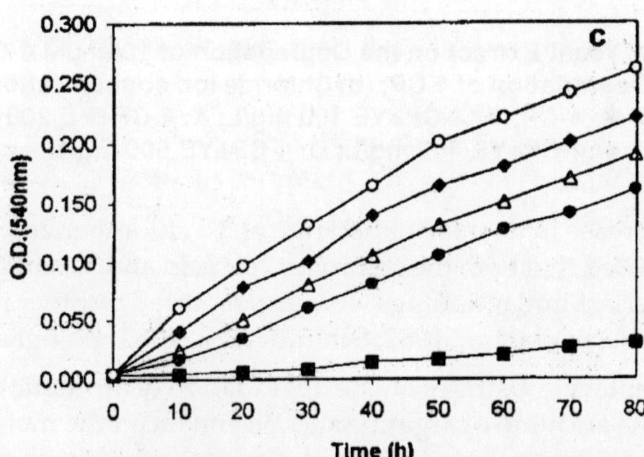

Figure 9.2: Effect of Yeast Extract on the Degradation of 500 μM 4-CP by Strain A-2
(*a*) Degradation of 4-CP; (b) Chloride ion concentration; (c) Cell density
Symbols: ■: 4-CP; ●: 4-CP+YE 100 mg/L; Δ: 4-CP+YE 200 mg/L;
♦: 4-CP+YE 400 mg/L; O: 4-CP+YE 600 mg/L

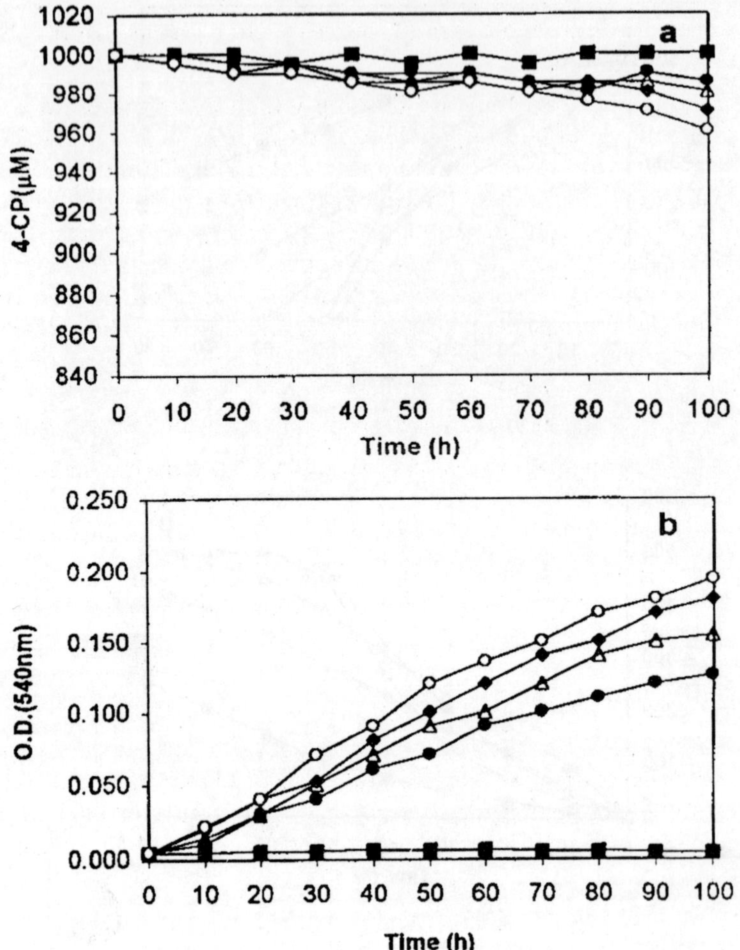

Figure 9.3: Effect of Yeast Extract on the Degradation of 1000 µM 4-CP by Strain A-2 (*a*) Degradation of 4-CP; (b) Chloride ion concentration Symbols: ■: 4-CP; ●: 4-CP+YE 100 mg/L; Δ: 4-CP+YE 200 mg/L; ♦: 4-CP+YE 400 mg/L; O: 4-CP+YE 600 mg/L

resulted in a 70 per cent increase in dechlorination rate of 2,5 chlorobenzoic acid by *Pseudomonas* sp. CPE2 strain and also concluded that both the chlorobenzoic acid and YE carbon sources were utilized simultaneously by the bacteria. Similar findings were also reported by other investigators (Haggblom, 1993; Reber *et al.*, 1981; Panastasion *et al.*, 1982; Schmidt *et al.*, 1987; Wijngaard *et al.*, 1993).

The effect of supplementary substrates on the degradation of aromatic compounds could not be predicted as on one hand the addition of arginine and YE enhanced the metabolism of phenol and p-nitrophenol in lake water whereas the addition of glucose, adenine or propionate inhibited the metabolism of the pollutants (Rubin, 1983). A delay in transformation of phenolic compounds was also observed by the inclusion of 0.5 per cent (w/v) of proteose peptone in the medium (Bisaillon *et al.*, 1993). The capability of *Pseudomonas putida* 10.2 to degrade 3-chlorobenzoate was greatly reduced by the presence of 1 per cent easily degradable carbon sources such as glucose, pyruvate and lactose in the medium (Chobchuenohom *et al.*, 1996).

In our study the results obtained indicate that YE-which is a complex mixture of several aminoacids, peptides, proteins, carbohydrates, vitamins and trace elements acted as alternative source of nutrients and resulted into rapid increase in biomass yield. This ultimately resulted into rapid chlorophenol utilization in most of the cases. Although a drop in degradation rate of 4-CP was noticed particularly in the beginning, when concentration of YE was increased to 400 mg/L in the medium; the degradation rate of 4-CP picked up rapidly and ultimately resulted in a higher degradation rate (Figure 9.1a). A further increase in YE concentration did not lead to much enhancement of 4-CP utilization or in fact resulted in decreased substrate utilization as seen in case of 500 µM 4-CP when YE concentration was increased from 400 mg/L to 600 mg/L (Figure 9.2a and Table 9.1). All these findings indicate that strain A-2 prefers YE as compared to 4-CP and as biomass increases there is a simultaneous increase in 4-CP degradation rate within certain range of YE concentration.

The concentration of 4-CP in the medium has pronounced effect on the biomass yield. The biomass yield decreased with increase in 4-CP concentration in the medium. This effect has been observed in absence as well as in the presence ofYE at corresponding concentrations, indicating inhibitory effect of 4-CP on strain A-2 (Figures 9.1c, 9.2c and 9.3b).

When the degradation rate of 4-CP at 500 µM concentration in the absence of YE by strain A-2 is calculated out at 80 hrs and at 108 hrs of incubation (Table 9.1), it was found that degradation rate was comparatively higher at the end of 108 hrs. This indicates that increasing biomass continuously increased the degradation rate as time passed on (Figures 9.2a and 9.7a).

At 1000 µM concentration of 4-CP in the medium strain A-2 failed to degrade the substrate (Figure 9.3a). Chloride ion concentration was below the detectable limit of analytical method used. In previous cases higher biomass resulted in higher degradation rates. Although at 1000 µM 4-CP, the addition of YE resulted in significant biomass yield but unexpectedly no or very little 4-CP utilization was noticed (Figure 9.3a).

Effect of Yeast Extract on 2-CP Degradation

The effect of YE on the degradation of 2-CP by strain J-3 has been illustrated in Figures 9.4 to 9.6. Strain J-3 degraded 500 µM 2-CP in 36 hrs. Supplementation of in the medium with 50, 100 and 200 mg/L YE, the degradation of 500 µM 2-CP occurred in 33, 32 and 30 respectively (Figure 9.4a). 9 to 20 per cent increase in degradation rate (Table 9.2) and 1.5 to 4 fold increase in biomass yield was noticed with increase in YE concentration in the medium (Figure 9.4c).

Strain J-3 degraded 1000 µM 2-CP in 80 hrs. While in the presence of 50, 100 and 200 mg/L YE, the degradation of 2-CP occurred in 78, 76 and 74 hrs respectively (Figure 9.5a). Thus up to 8 per cent increase in degradation rate (Table 9.2) and 1.5 to 3.5 fold increase in biomass yield was observed due to inclusion of YE in the medium (Figure 9.5c).

At 2000 µM 2-CP concentration in the medium the strain J-3 degraded 2-CP at a slower rate and at 180 hrs the residual concentration of 2-CP was around 85 µM. However in the presence of YE at concentrations of 50.100 and 200 mg/L, the degradation of 2-CP occurred in 180. 174 and 170 hrs respectively (Figure 9.6a). Up to 10°C increase in degradation rate and (Table 9.2) 1.5 to 3 fold increase in biomass yield were observed in response to YE included in the medium (Figure 9.6c).

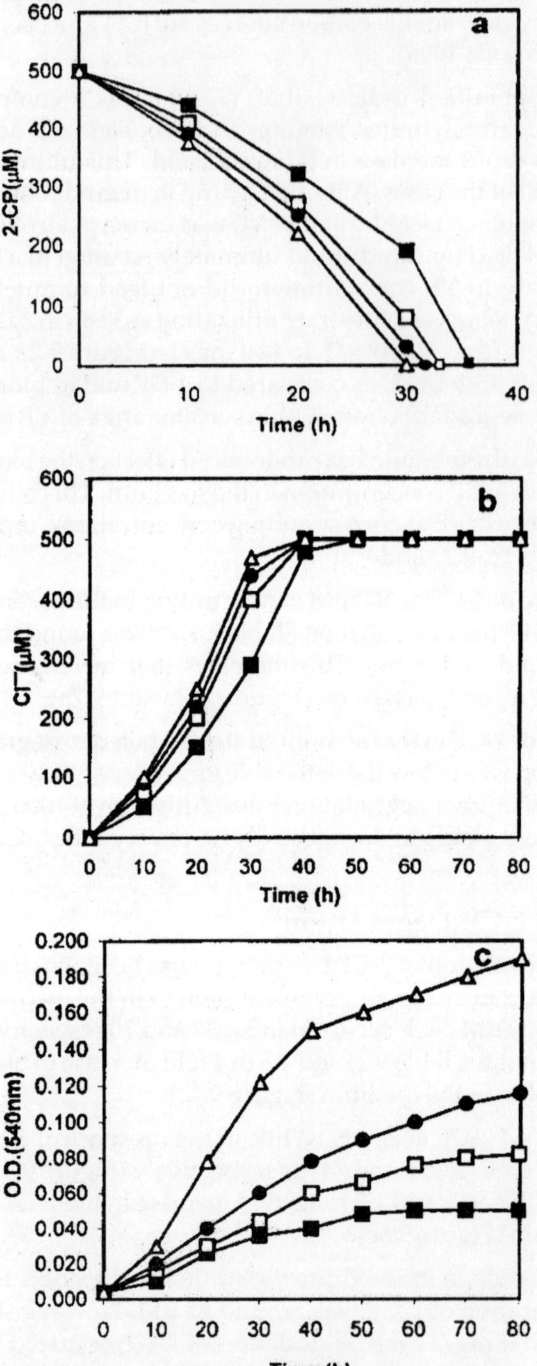

Figure 9.4: Effect of Yeast Extract on the Degradation of 500 µM 2-CP by Strain J-3
(*a*) Degradation of 4-CP; (b) Chloride ion concentration; (c) Cell density
Symbols: ■: 2-CP; □: 2-CP+YE 50 mg/L; ♦: 2-CP+YE 100 mg/L; Δ: 2-CP+YE 200 mg/L

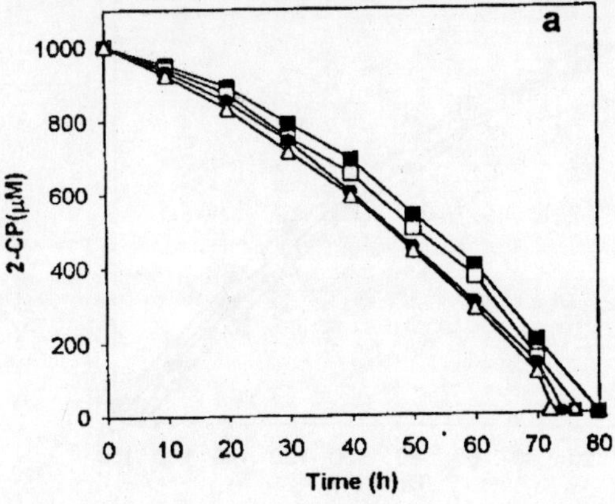

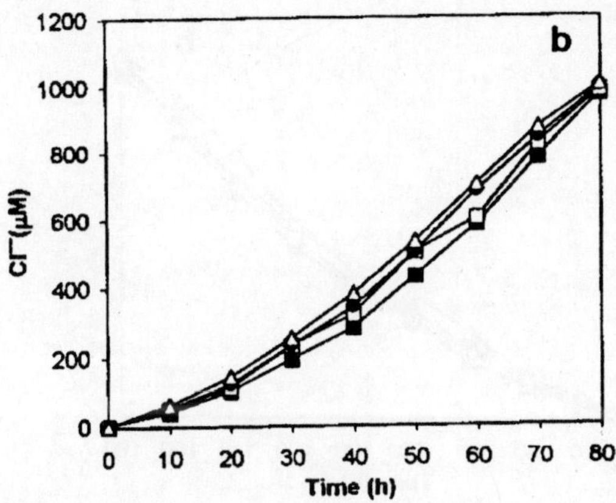

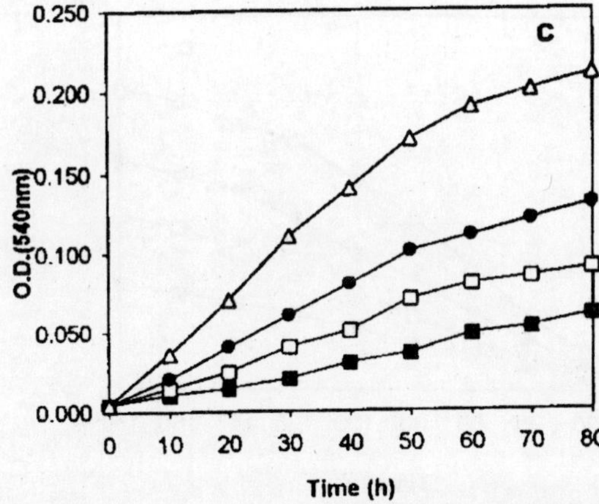

Figure 9.5: Effect of Yeast Extract on the Degradation of 1000 μM 2-CP by Strain J-3
(*a*) Degradation of 4-CP; (b) Chloride ion concentration; (c) Cell density
Symbols: ■: 2-CP; □: 2-CP+YE 50 mg/L; ♦: 2-CP+YE 100 mg/L; Δ: 2-CP+YE 200 mg/L

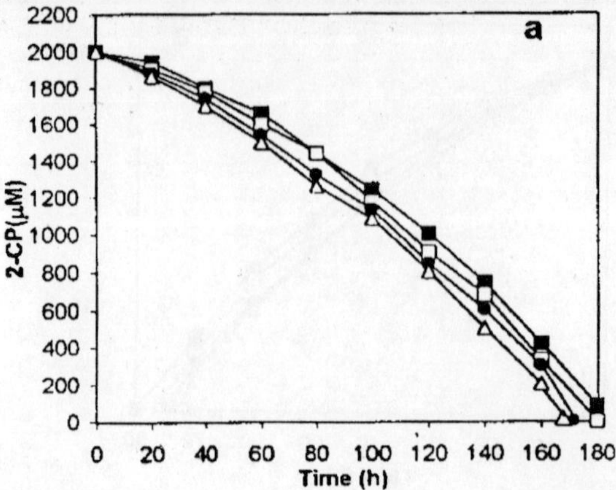

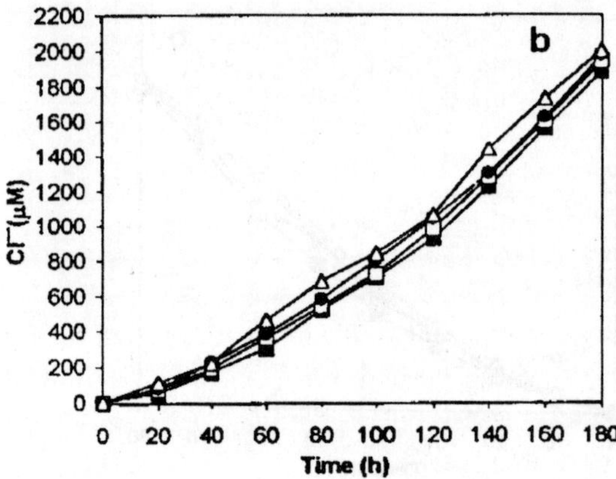

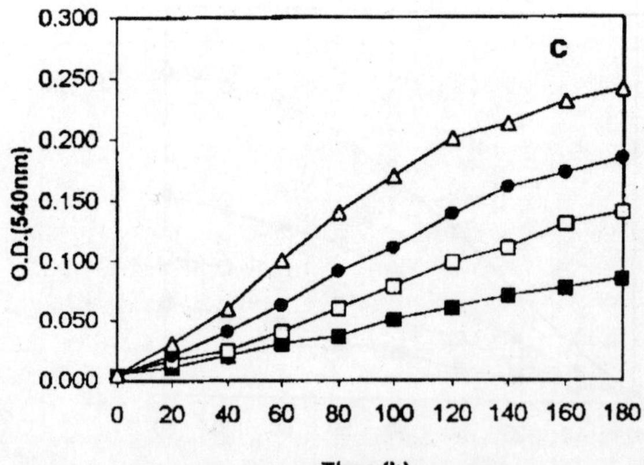

Figure 9.6: Effect of Yeast Extract on the Degradation of 2000 μM 2-CP by Strain J-3
(*a*) Degradation of 4-CP; (b) Chloride ion concentration; (c) Cell density
Symbols: ■: 2-CP; □: 2-CP+YE 50 mg/L; ♦: 2-CP+YE 100 mg/L; Δ: 2-CP+YE 200 mg/L

Table 9.2: Effect of YE on the Degradation Rate of 2-CP with Strain J-3

Initial Amount of 4-CP in the Medium (μM)	*Initial Amount of YE in the Medium (mg/L)*	*Amount of 4-CP Degraded (μM)*	*Incubation Time (h)*	*Degradation Rate μM h^{-1}*	*% Increase in Degradation Rate*[Δ]
500	00	500	36	13.88	–
500	50	500	33	15.15	9
500	100	500	32	15.62	12
500	200	500	30	16.66	20
1000	00	1000	80	12.50	–
1000	50	1000	78	12.82	2
1000	100	1000	76	13.15	5
1000	200	1000	74	13.51	8
2000	00	1915	180	10.63	–
2000	50	2000	180	11.11	4
2000	100	2000	174	11.49	8
2000	200	2000	170	11.76	10

Δ With respect to the degradation rate of substrate in the absence of YE at corresponding concentration and conditions.

In the absence of YE in the medium highest degradation rate was observed at 500 μM 2-CP. While at various concentrations of YE in the medium the highest degradation rate was observed at 500 μM 2CP with 200 mg/L YE (Table 9.2).

Similar to the inhibitory effect of 4-CP seen on the biomass yield of strain A-2, the biomass yield of strain 1-3 also decreased with increasing concentration of 2-CP particularly when concentration increased from 1000 μM to 2000 μM (Figures 9.5c and 9.6c). However extent of inhibition was less in 2-CP as compared to 4-CP; with corresponding strains (Figures 9.1c to 9.6c). As evident from Figures 9.4, 10.5 and 10.6 the rate of degradation of 2-CP by strain J-3 increased with increase in YE concentration and resulted into around 1.5 to 4 fold increase in biomass yield. However YE had higher growth stimulating effect on strain A-2 as compared to strain J-3. This can be seen from the fact that in the presence of 200 mg/L YE with 500 μM of corresponding monochlorophenol; around 6 fold increase in biomass yield of strain A-2 while around 4 fold increase in case of strain J-3 was noticed at the end of 80 hrs incubation time (Figures 9.2c and 10.4c). This was also reflected during the degradation of monochlorophenols at abovementioned concentration in the presence of 200 mg/L YE; as this resulted into an increase of 142 per cent in degradation rate of 4-CP as compared to 20 per cent in 2-CP (Tables 9.1 and 9.2). Similarly in the presence of 100 mg/L YE, 5 fold increase in biomass yield of strain A-2 resulted in to 67 per cent increase in the degradation rate of 500 μM 4-CP as compared to 2 fold increase in biomass yield of strain J-3 which resulted to 12 per cent increase in degradation rate of 500 μM 2-CP (Figures 9.2c, 9.4c, Tables 9.1 and 10.2). The above findings indicate that the stimulatory effect of YE varies with respect to strain involved and the recalcitrant substrate to be biodegraded.

Effect of Meat Extract and Peptone on 4-CP Degradation

A comparative study of 4-CP degradation in presence of YE, ME and PEP was undertaken and results are depicted in Figure 9.7. The degradation of 500 μM 4-CP in presence of 400 mg/L concentration of YE, ME as well as PEP was studied up to 110 hrs of incubation time.

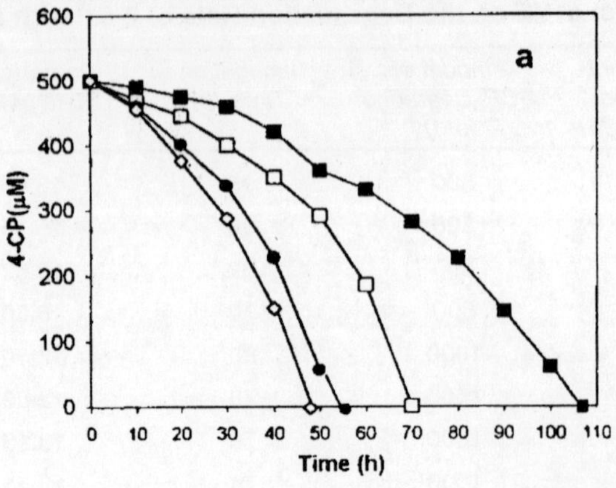

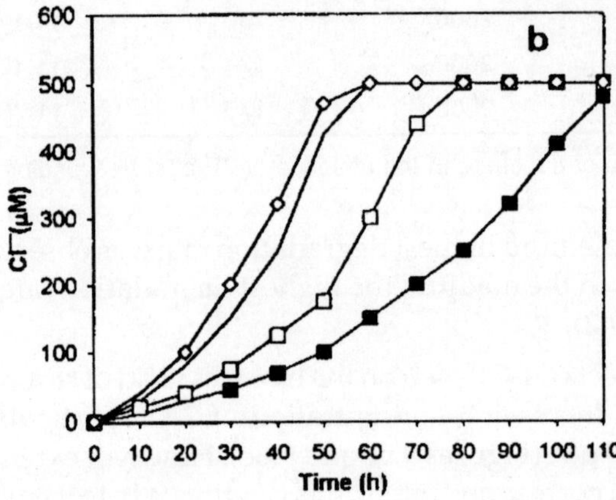

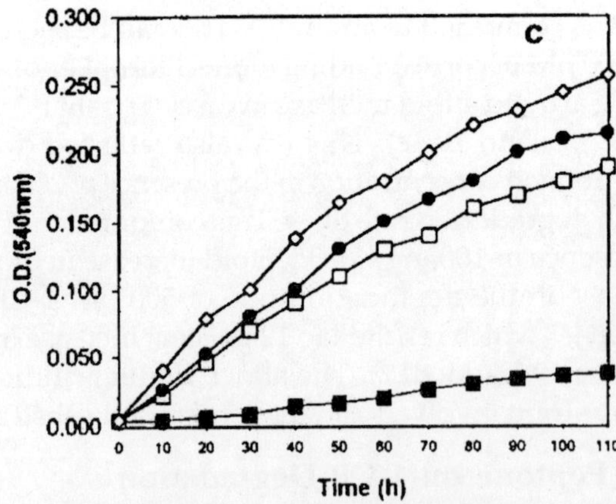

Figure 9.7: Effect of Yeast Extract, Meat Extract and Peptone on the Degradation of 500 μM 4-CP by Strain A-2

(*a*) Degradation of 4-CP; (b) Chloride ion concentration; (c) Cell density

Symbols: ■: 4-CP; □: 4-CP+YE 40 mg/L; •: 4-CP+YE 400 mg/L; ◊: 4-CP+YE 400 mg/L

The effect of supplemented substrate on the degradation of 4-CP occurred in the following order: YE > ME > PEP. Thus YE was found to be the most preferred secondary substrate to accelerate the degradation of 4-CP. 54 per cent faster degradation was noticed when PEP was included in the medium while 97 per cent and 125 per cent faster degradation of 4-CP was observed in the presence of ME and YE respectively (Figure 9.7a and Table 9.1).

Chloride Release

Stoichiometric release of chloride was noticed in all the degradation experiments, both with 4-CP as well as 2-CP as the substrates (Figures 9.1b, 9.2b and 9.4b to 9.7b). At 1000 μM 4-CP concentration the chlorides could not be estimated because of the detectable limit of analytical method used (Higson *et al.*, 1990).

Conclusion

From the above findings it could be concluded that in general YE enhances the biomass yield and also enhances the degradation rate of the substrate; however its effect may vary depending on the type of strain involved, nature and concentration of target substrate and the concentration of YE itself. The increase in biomass yield with simultaneous increase in 4-CP utilization was also noticed due to inclusion of ME or PEP in the medium with strain A-2. However as evident from fig. 7, YE seems to be a better secondary substrate as compared to ME and PEP for biodegradation of 4-CP. The findings of the study suggest that availability of the growth supplements such as YE, ME and PEP can play a significant role in the degradation of recalcitrant compounds and may be of great significance in improving the efficiency of biological treatment processes.

References

Armenante, P.M., Fawa, F. and Kafkewitz, D., 1995. Effect of yeast extract on growth kinetics during aerobic biodegradation of chlorobenzoic acids. *Biotechnol. Bioeng.*, 47: 227–233.

Armenante, P.M., Kafkewitz, Lewandowski, G. and Kung, C.M., 1992. Integrated anaerobic-aerobic process for the biodegradation of chlorinated aromatic compounds. *Environ. Prog.*, 11: 113–122.

Basu, S.K. and Oleszkiewicz, J.A., 1995. Factors affecting aerobic degradation of 2-chlorophenol in sequencing batch reactors. *Environ. Technol.*, 16(12): 1135–1143.

Bisaillon, J.G., Lepine, F., Beaudet, R. and Sylvestre, M., 1993. Potential for carboxylation-dehydroxylation of phenolic compounds by a methanogenic consortium. *Appl. Environ. Microbiol.*, 59(4): 1162–1167.

Chobchuenohom, Mong, Kolsuk, S. and Bhumiratana, A., 1996. Biodegradation of 3-chlorobenzoate by *Pseudomonas putida* 10.2. *World J. of Microbio. and Biotech.*, 12: 607–614.

Clesceri, L.S., Greenber, A.E. and Trussell, R.R. (ed.), 1989. *Standard Methods for the Examination of Water and Wastewater,* 17th edn. American Public Health Association, Washington D.C., p. 548–554.

Dapaah, S.Y. and Hill, G.A., 1992. Biodegradation of chlorophenol mixtures by *Pseudomonas putida. Biotechnology and Bioengineering,* 40: 1353–1358.

Fava, F., Armenante, P.M., Kafkewitz, D. and Marchetti, L., 1995. Influence of organic and inorganic growth supplements on the aerobic biodegradation of chlorobellzoic acids. *Appl. Microbial Biotechnol.*, 43: 171–177.

Frank, R. and Braun, H.E., 1989. PCB and DDE residues in milk supplies of Ontario, Canada 1985–1986. *Bull. Environ. Contam. Toxicol.*, 42: 666–669.

Haggblom, M.M. and Young, L.Y., 1990. Chlorophenol degradation coupled to sulphate reduction. *Appl. Environ. Microbiol.*, 56: 3255–3260.

Haggblom, M.M., Rivera, M.D. and Young, L.Y., 1993. Effects of auxiliary carbon sources and electron acceptors on methanogenic degradation of chlorinated phenols. *Environmental Toxicology and Chemistry*, 12(8): 1395–1403.

Higson, F.K. and Focht, D.D., 1990. Bacterial degradation of ring chlorinated acetophenones. *Appl. Environ. Microbiol.*, 56: 3678–3685.

Neilson, A.H., 1990. The biodegradation of halogenated organic compounds. *J. Appl. Bacteriol.*, 69: 445–170.

Nesnow, S., Argus, M., Bergman, H., Cnu, K., Frith, C., Helmes, T., McGaughy, R., Ray, V., Slaga, T.J., Tennant, R. and Weisburger, E., 1986. Chemical carcinogens: A review and analysis of the literature of selected chemicals and establishment of the Gene Tox Carcinogen Data Base. *Mutation Res.*, 185: 1–195.

Papanastasion, A.C. and Maier, W.J., 1982. Kinetics of biodegradation of 2,4-dichlorophenoxy-acetate in the presence of glucose. *Biotechnol. Bioeng.*, 24: 2001–2011.

Reber, H.H. and Kaiser, P., 1981. Regulation of the utilization of glucose and aromatic substrates in four strains of *Pseudomonas putida*. *Arch. Microbiol.*, 130: 243–247.

Rubin, H.E. and Alexander, M., 1983. Effect of nutrients on rates of mineralization of trace concentrations of phenol and p-nitrophenol. *Environ. Sci. Technol.*, 17: 104–107.

Schmidt, S.K., Scow, K.M. and Alexander, M., 1987. Kinetics of p-nitrophenol mineralization by a *Pseudomonas* sp. rffect of secondary substrates. *Appl. Environ. Microbiol.*, 53: 2617–2623.

Wijngaard, A.V., Van Der Kleij, R.G., Doomweerd, R.C. and Jansen, D.B., 1993. Influence of organic nutrients and co cultures on the competitive behaviour of 1,2-dichloroethane degrading bacteria. *Appl. Environ. Microbiol.*, 59: 3400–3405.

Chapter 10

Biodiversity of Blue-Green Algae from Paddy Sugar Cane and Banana Fields of Karnataka State, India

B.S. Giriyappanavar

Algal Laboratory, Department of Botany, Karnatak Science College, Dharwad, Karnataka, India

ABSTRACT

A total of 19 genera including 8 heterocystous, 7 unicellular and 4 non-heterocystous forms were recorded from 75 soil samples. Paddy field soils had maximum numbers of algal genera followed by sugarcane field soils and banana field soils of these *Chroccoccus* was common to the all the three crop soils under study. While among heterocystous forms *Nostoc* and *Calothrix* were observed from all soils, *Oscillotoria* was more frequent among non-heterocystous forms.

Out of 106 algal species recorded, 63 belong to heterocystous, 30 to non-heterocystous and 13 to unicellular.

Paddy field soils harboured maximum number of taxa belonging to heterocystous, compared to non-heterocystous and unicellular among all the three soil groups. Heterocystous, non-heterocystous and unicellular algae were generally very poorly represented in banana field soil than sugarcane field soil.

Introduction

To build up and maintain the fertility status of the soil so on to increase the crop production, it is important to appreciate that soil, besides being composed of physical and chemical constituents, is a

* *Corresponding Address*: Dr. B. S. Giriyappanavar "Basav Sadana", Gulaganji-koppa, Dharwad – 580 008 Karnataka

biomass, teeming with mutualistic, antagonistic and sometime, pathogenic activities of countless microorganisms, inhabiting it. In recent years, blue-green algae, an ubiquitous group of soil microorganisms, have been considered a potential biological system for increasing various crop yield at comparatively low cost.

Studies on soil alga (BGA) and importance of soil fertility have attracted the attention of scientists for a few decades, although, there are many reports on soil algae from different regions Lund, 1947; Shilelds and Drouet, 1962; De and Mandal, 1956; Datta and Venkataraman, 1958; Marathe, 1967; Ananthani and Marathe, 1972; Bongale and Bharati, 1980; Bongale, 1981 and Giriyappanavar, 1988.

The present investigation deals with relation between different bluegreen algae and crop soils from four places of Dharwad district, 11 from Belgaum district where paddy, sugarcane and banana were cultivated.

Materials and Methods

Dry soil samples from 3–4 spots in a field were mixed and taken as representative of the respective field and were studied in the laboratory for blue-green algal flora by enrichment cultures using De's modified Seneck medium (Singh, 1961) in most and biphasic cultures. Blue-green algae were collected and identified by camaralusida drawings, every fortnight, for three months. A total for 91 soil Samples from paddy, sugarcane and banana fields of Dharwad and Belgaum districts have been made in the present investigation.

A total of 70 soil samples from 2 districts (Dharwad and Belgaum), were collected; of which 39 were from paddy field soils, 24 were from sugarcane field and 7 were from banana field soils.

In the present investigation a comparative study of the composition of blue-green algal flora of different habitats namely paddy field, sugarcane field and banana field soil has been given. Out of 106 taxa recorded in the present study from different habitats, 63 taxa belong to heterocystous, 30 to non-heterocystous and 13 to unicellular were recorded from 91 soils under study. However, paddy field soils and sugarcane field soils have larger number of taxa than banana field soils.

The composition of the blue-green algal flora indicates that as many as 7 taxa are common to the all three different habitats of paddy, sugarcane and banana field soils. They are *Chroococcus turgidus, Oscillatoria limnetica, Nostoc carneum, N. epiphvtica, N. hatei, N. sphaeicum* and *Calothirx epiphvtica*. In addition to 32 taxa are common between paddy field soils and sugarcane field soils.

By Singh, 1961; Gonzalves and Gangla, 1949; Mitra, 1951; Datta and Venkataraman, 1958 have reported the preponderance of Cyanophyceae over other groups in the cultivated soils of India, and attributed it to the tropical and subtropical conditions. Among the soil groups, Heterocystous represented the major constituents of soil algal flora varying from 62 per cent to 67 per cent of the total algae. Paddy soils recorded maximum, non-heterocystous species (31 per cent of total algae compared to 33 per cent in banana field soils). For the view point of nitrogen fixation in the tropical soils, the heterocystous species as percentage of total algae were 57 per cent, 62 per cent and 67 per cent in case comparatively Gonzalves and Gangla (1949) (42 per cent), Mitra (1951) (33 per cent), Pande (1962) (35 per cent) and Bongle (1984) (36 per cent) have also reported heterocystous algae. Banana soils, have poor representation of unicellular algae in the present study.

Blue-green Algal General Recorded

It is also evident that a total of 19 genera recorded in the present study include 8 heterocystous, 7 unicellular and 4 non-heterocystous representing about (Tables 10.1 and 10.2). Paddy field soils were maximum number of genera (18) while, sugarcane field soils and banana field soils harboured 14 and

10 genera respectively. However, a number of reports have appeared on the paddy field algae, indicating the preponderance of heterocystous algae (Singh, 1961). Not much information is available on the other crop soils groups. In the present study among various soil groups, the banana field soils appear to favour heterocystous forms in terms of species (67 per cent) and genera (50 per cent) sugarcane field soils, favoured more numbers of species of heterocystous (62 per cent) compared to non-heterocystous and unicellular blue-green algae respectively. Paddy field soils had maximum population of hetrocystous, expressed both as species (57 per cent) and genera (44 per cent). Similar observations were made by Bongale (1981).

Table 10.1: Distribution of Blue-Green Algae (70 Soil Samples)

Name of the Algae	*Paddy (39)*	*Sugarcane (24)*	*Banana (67)*
Heterocystous			
Nostoc Piscinale	+	+	
N. carnenum	–	+	+
N. ellipsosporum	+	+	+
N. rivalare	+	–	+
N. hatei	–	+	+
N. entopytum	+	+	–
N. commune	+	+	–
N. muscorum	+	+	–
N. sphaericum	+	+	+
N. paludosum	+	+	–
N. passrinianum	+	–	–
N. calcicola	+	+	–
N. linekia	+	+	–
N. qmplissimum	–	+	–
N. coeruleum	–	+	–
N. Punctiforme	+	–	+
Anbaena Sphaerica	+	–	+
A. variabilis	–	–	+
A. iyengari	–	+	–
A. torulosa	+	–	+
A. fertilissima	+	–	–
A. wisconsnense	+	–	–
A. oryzee	+	–	–
A. doliolum	–	–	+
Scytonema hofmanni	–	+	–
S. fritschii	–	+	–
S. fremyii	+	+	–
S. burmanicum	–	–	+
S. sps	–	–	+
S. simplex	+	–	–
S. coactile	+	–	–

Contd...

Table 10.1–Contd...

Name of the Algae	*Paddy (39)*	*Sugarcane (24)*	*Banana (67)*
Tolypothrix tenuis	+	–	+
T. nodesa	+	–	–
T. fragilis	+	–	–
T. bvssoidea	+	–	–
T. Conglutinata	+	+	–
Microchate uberrima	+	–	–
M. calothricbides	+	–	–
M. violacea	+	+	+
M. aequalis	–	–	–
M. loktakensis	–	+	–
M. goeppertiana	+	+	–
M. tenera	+	+	–
M. son	–	+	–
Calothrix membranacea	+	–	–
C. elenkinii	+	+	–
C. marchicalem	+	–	–
C. Marhica var crassa	–	–	+
C. scopulorum	–	+	–
C. con tarenii	–	+	–
C. braunii	+	+	–
C. castelli	+	–	–
C. epiphytica	+	+	+
C. fusca	–	–	+
C. marchiea	+	–	+
C. weberi	+	–	–
Hapolosiphon stuhlmannii	+	+	–
H. flagelliformis	+	–	–
H. sps	+	–	–
Cylindrospermum musicola	+	–	–
Non-Heterocystous	+	+	–
Oscillatoria Chalvbea	+	–	–
O. chlorina	–	+	–
O. subbrevis	–	+	–
O. curviceps	+	+	–
O. laete-virens	+	+	–
O. vixagapatensis	+	+	–
O. tere briformis	+	+	+
O. limnatica	+	+	–

Contd...

Table 10.1–Contd...

Name of the Algae	Paddy (39)	Sugarcane (24)	Banana (67)
Oscillatoria claricentrasa	+	–	+
O. satina	+	–	–
O. dimasa	+	–	–
O. anguina	+	–	–
O. psemdogeminata	+	+	–
O. ornate	+	+	–
O. martini	+	+	–
O. princeps	+	–	–
O. acuminata	+	+	–
O. sps	–	+	–
O. sps	–	–	+
Lyngbya confervides	+	–	–
L. martensiana	+	+	+
L. grasilis	–	–	+
L. shackletoni	–	–	+
L. dendrobia	+	–	–
L. spirilis	+	–	+
Phormidium mucosum	+	+	–
P. ladinianum	+	–	–
P. microtomum	+	+	–
P. purpurascens	+	–	
Plectonama nostoarum nicellular	+	–	–
Microcystis flosaquae	+	–	–
Chroococcus sctizodermations	+	+	+
C. turgidus	+	+	+
C. sps	+	+	+
Gloecapsa montana	+	+	–
G. knetzingiana	+	+	–
Aphanocpsa banaresentis	+	–	–
Gloeothese rupestris	+	+	–
G. samoentis	+	–	–
Aphanothece microscopica	+	–	–
A. stagnina	–	+	–
A. contagnei	–	–	–
Synechocystis apuatilis	+	–	–

+: Presence; –: Absence of algae.

Table 10.2: Abundance of Blue-Green Algal Taxa in Soils (Total Numbers of Taxa Recorded from Each Soil Group Pooled)

Field Soils		*Unicellular*		*Heterocystous*		*Non-heterocystous*		*Total Algal Taxa*
		No. of Taxa % of	*Total*	*No. of Taxa % of*	*Total*	*No. of Taxa % of*	*Total*	
I	Species							
	Paddy	9	12	44	57	24	31	77
	Sugarcane	5	10	31	62	14	28	50
	Banana	3	10	20	67	7	23	30
II	Genara							
	Paddy	6	33	8	44	4	22	18
	Sugarcane	4	29	7	50	3	21	14
	Banana	3	30	5	50	2	20	10

Table 10.3: Distribution of Soil Blue-Green Algal General

Algal Genera	*Number of Samples Recorded from*	*Paddy Field*	*Sugarcane Field*	*Banana Field*
Microcystis	1	+	–	–
Chroococcus	8	+	+	–
Gloeocapsa	5	+	+	–
Aphanacapsa	1	+	–	–
Gloeothece	2	–	+	+
Aphanothece	9	+	+	–
Synechocystia	1	+	–	–
Oscillatoria	29	+	+	+
Lyngbya	11	+	+	+
Phormidium	11	+	+	–
Flectonema	2	+	–	–
Nostoc	41	+	+	+
Anabaena	21	+	+	+
Scytonema	16	+	+	+
Tolypothrix	9	+	+	+
Microchaete	13	+	+	+
Calothrix	28	+	+	+
Hapalosiphon	8	+	+	–
Cylindrospermum	5	+	–	–

+: Presence; –: Absence of algae.

In the present investigation, non-heterocystous species were more in paddy field soils compared to other soil groups of sugarcane soils. The genera recorded were *Oscillatioia* (29 samples), *Lyngbya* (11 samples), *Nostoc* (41 samples), *Anebaena* (21 samples), *Scvtonema* (16 samples), *Tolvpothrix* (9 samples), *Microchaete* (13 samples) and *Calothrix* (28 samples) from 70 soil samples studied and were common to all the three soil groups. *Microcystis* (1 sample), *Pletonema* (2 samples) and *Cylindrospermun* (5 samples) were represented only in paddy field soils. The general Nostoc is the most common soil algae recorded.

It is evident from the above discussions that paddy field soils favoured maximum numbers of algal genera compared to other crop soil groups studied. Heterocystous species predominated over non-heterocystous and unicellular algae in all the soil groups. Banana soils were very poorly represented the blue-green algae.

Acknowledgements

I am thankful to the Principal, Karnatak Science College, Dharwad and Dr. S.P. Hosmani, Professor (Retired) P.G. Studies in Botany, Mysore University for encouragement.

References

Bharati, S.G. and Bongale, U.D., 1975. Studies on soils and soil algae of Karnataka State, Part-I: soils and soil algae of Dharwad, Karnataka. *J. Sci.*, 20: 142–150.

Bongale, U.D and Bharati, S.G., 1980. On the algal flora of cultivated soils of Karnataka State, India. *Phykos*, 19(1): 95–109.

Bongale, U.D., 1981. On the soil algae from paddy fields of Panjion (Goa and Chikkamachati, Raichur Districts, Karnataka. *India J. Indian Bot. Soc.*, 60: 326–329.

Bongale, U.D., 1984. Soil algae in relation to crop soil groups in certain soils of Karnataka State, India. *Compo. Physiol. Ecol.*, 9: 384–393.

Datta, N. and Venkataraman, G.S., 1958. An exploratory study of the algae of some cultivated and uncultivated, soils. *Indian J. Agronamy*, 3: 109–115.

Desikachary, T.V., 1959. *Cyanophyta*, ICAR–Monograph, New Delhi.

Giriyappanavar, B.S., 1988. Studies on the nlue-green algae and the effect of certain chemicals on them. *Ph.D. Thesis*, Karnatak University, Dharwad.

Gonzalves, E.A. and Gangla, G.S., 1949. Observations on the algae of paddy field soils. *J. Uni. Bombay*, 18, 51.

Mitra, A.K., 1951. The algal flora of certain Indian soils. *Indian J. Agri. Sci.*, 21(4): 357–373.

Prescott, G.W., 1951. *Algae of the Western Great Lake Area*. Crabrook Inst. of Sci. Publ.

Srieds, L.M. and Durrell, L.W., 1964. Algae in relation to soil fertility. *The Bot. Rev.*, p. 92–128.

Singh, R.N., 1961. *Role of Blue-green Algae in Nitrogen Economy of Indian Agriculture*. ICAR Monograph, New Delhi.

Chapter 11

Composition of Vermicast of *Eisenia fetida* Savigny (1826) (Lumbricidae), *Metaphire posthuma* Vaillant (1868), *Lampito mauritii* Kinberg (1866) (Megascolecidae) and *Eutyphoeus waltoni* Michaelsen (1907) (Octochaetidae): A Comparative Study

*Neena Dhiman**

Department of Zoology and Fisheries, Punjab Agricultural University, Ludhiana – 141 001, Punjab

ABSTRACT

The paper reports the comparative account of assessment of vermicast of *Eisenia fetida* Savigny (1826) (Lumbricidae), *Metaphire posthuma* Vaillant (1868), *Lampito mauritii* Kinberg (1866) (Megascolecidae) and *Eutyphoeus waltoni* Michaelsen (1907) (Octochaetidae) as a bio-manure towards the sustainable management of agro-eco-regimes in terms of physical, chemical and biochemical parameters. Vermicasts were analyzed for electrical conductance, percentage moisture and pH values. Vermicasts were assessed for nutrient status in terms of essential macro elements, Phosphorus (P) and Potassium (K); essential and non essential micro elements, Boron (B), Calcium (Ca), Copper (Cu), Iron (Fe), Magnesium (Mg), Manganese (Mn), Sulphur (S)

* Ex-Ph.D. Student, Department of Zoology and Fisheries, Punjab Agricultural University, Ludhiana – 141 001, Punjab
Lecturer in Zoology, GNG College, Model Town, Ludhiana – 141 002, Punjab

and Zinc (Zn) and potentially toxic elements for plants, Aluminium (Al), Arsenic (As), Cadmium (Cd), Cobalt (Co), Chromium (Cr), Mercury (Hg), Molybdenum (Mo), Sodium (Na), Nickle (Ni) and Lead (Pb) with ICAP (Inductively Coupled Argon Plasma Photo Emission Spectroscopy) technique. Nitrogen (N) was estimated with standard Kjeldahl's method. Analysis was made towards the activity of urease, phosphatase (acidic and alkaline) and dehydrogenase enzyme complex. Results revealed NPK status and mean values for concentration of essential and non-essential elements (μg g^{-1} soil) of vermicast of *Eisenia fetida* to be higher than the other three casts. High concentration of potentially toxic elements was found to be lesser than the prescribed toxicity levels. Activity of urease (μg urea/g dry soil/min.), acidic and alkaline phosphatases (μg PNP/g soil/hour) and dehydrogenase enzyme complex (μg TPFI g soil/hour) were compared in four casts. Results were compared by two way analysis of variance (ANOVA-CRD, $CPCS_1$ Program) for enzyme analysis. Activity of alkaline phosphatase and dehydrogenase enzyme complex were seen maximum in casts of *E. fetida* followed by casts of *M. posthuma, L. mauritii* and *E. waltoni.* The values for activity of acidic phosphatase and urease followed a reverse trend for the four types of casts. These values conform to the pH-values in the same order. Vermicast of *E. fetida* showing maximum alkalinity revealed its usefulness in vermiculture practices to make soils alkaline and the most acidic casts of *E. waltoni* ensured its usefulness for highly alkaline to neutral soils. Casts of *M. posthuma* and *L. mauritii* show moderate values towards pH.

Keywords: *Earthworm, Eisenia fetida, Eutyphoeus waltoni, Lampito mauritii, Metaphire posthuma, Vermicast.*

Introduction

The pioneering work of Darwin (1881) established the role of earthworms in improving the fertility of the soil. The major influence of earthworms towards increased soil physical properties and nutrient status has been the major concern since that time (Russell, 1973). In particular, the enrichment of earthworm casts in essential plant nutrients such as N, P, K, Ca and Mg (Lunt and Jacobson, 1944; Parle, 1963; Graff, 1970) has been demonstrated. Paucity of information on the complete chemical and biochemical parameters of worm casts poses an obstacle in using the vermiculture practices for production potentialities of the soils. Such a detailed account substantiate determining the potential of exploring possible earthworm species for vermiculture so as to get unprecedented gains in terms of 4R's of management, reduction at source, reuse, recycling and recovery of energy and also identifying the constraints for successful vermiculture programme besides imparting detailed information on vermicast soil science for increasing crop production. The concept of vermiculture not only gives hope for healthy ecology by providing most useful organic manure having more N, P, K and other micro and macronutrients than normal manure (Srivastava and Beohar, 2004), but also aids in minimizing the environmental pollution and health hazards (Bhawalkar, 1992) caused by indiscriminate and injudicious use of chemical fertilizers. All the xenobiotics on one hand have shown deleterious and detrimental impacts at the reproductive and behavioural levels and on the other hand having access to genetic levels pose serious threat to the survival of mankind. The present knowledge of detailed analysis of vermicast of four species is very essential both for elevating the soil productivity and to check their potential as bio manure towards the sustainable management of agro-eco-regimes.

Materials and Methods

Vermicast of *M. posthuma, L. mauritii* and *E. waltoni* were collected from natural habitats within the university campus and the cast of *E. fetida* were harvested from the cultures maintained. Each type of cast is having its own characteristic features enlisted in Table 11.1. The worm castings were analyzed for chemical and biochemical parameters. Five samples were taken for each replication. For electrical conductance and pH values, 20 g of each of powdered sample was stirred with 40 ml of distilled water and allowed to stand for 2 hours followed by recording of pH with pH-meter electrode. Same samples were used for recording electrical conductance (in decisiemens per meter) by conductivity meter.

Table 11.1: Four Different Types of Vermicasts

Cast Type	*Name of Species*	*Place of Collection of Cast*	*Features*
S-casts	*M. posthuma* Vaillant	Small Animal Colony, PAU, Ludhiana	Pallet like
D-casts	*L. mauritii* Kinberg	Dairy Farm, PAU, Ludhiana	Granular and of moderate size
T-casts	*E. waltoni* Michaelsen	Department of Agronomy, PAU, Ludhiana	Tower like with a holed channel in centre of each tower
V-casts	*E. fetida* Savigny	Harvested from cultures	Loose, fine, light in weight and granular

For percentage moisture, 50 g of each soil sample was kept at 60°C in oven for 15 days till a constant weight is obtained. Weight loss was expressed w.r.t. initial weights in terms of percentage moisture by the formula as per cent M = [(b–a)/a] x 100 *where*, (b–a) = x is the loss in weight after 15 days interval, a = initial weight in grams and b = final weight in grams (Tables 11.2 and 11.3).

Table 11.2: Analysis of Physical Parameters: Four-Cast Comparison

Parameter	*S-Cast (X_S±S.E.)*	*D-Cast (X_D±S.E.)*	*T-Cast (X_T±S.E.)*	*V-Cast (X_V±S.E.)*	*Trend*
Electrical Conductance (EC) (ds/m)*	0.67±0.004	0.85±0.002	1.11±0.002	1.73±0.004	$X_V > X_T > X_D > X_S$
Percentage Moisture (per cent M)	3.20±0.02	2.69±0.024	4.86±0.012	11.2±0.007	$X_V > X_T > X_S > X_D$
pH-value	8.09±0.0 1	7.63±0.003	7.57±0.004	8.55±0.002	$X_V > X_S > X_D > X_T$

* ds/m: Decisiemens/meter.

Table 11.3: Analysis of Physical Parameters: Four Cast Comparison (in Terms of Number of Times)

Possible Combinations	*Parameters*		
	*Electrical Conductance (EC) (ds/m)**	*Percentage Moisture (% M)*	*pH-values*
$X_S V_S X_D$	$X_D > X_S$ = 0.26T	$X_D < X_S$ = 16.1T	$X_D < X_S$ = 0.06T
$X_S V_S X_T$	$X_T > X_S$ = 0.64T	$X_T > X_S$ = 51.9T	$X_T < X_S$ = 0.06T
$X_S V_S X_V$	$X_V > X_S$ = 1.57T	$X_V > X_S$ = 2.51T	$X_V < X_S$ =0.05 T
$X_D V_S X_T$	$X_T > X_D$ = 0.30T	$X_T > X_D$ =0.80T	$X_T < X_D$ = 0.01T
$X_D V_S X_V$	$X_V > X_D$ = 1.04T	$X_V > X_D$ =3.18T	$X_V < X_D$ = 0.12T
$X_T V_S X_V$	$X_V > X_T$ = 0.57T	$X_V > X_T$ =1.31T	$X_V > X_T$ = 0.13T

*ds/m: Decisiemens/meter.

Concentration (µg g^{-1} of soil) of twenty one elements was worked out and recorded under NPK status, essential element analysis (EEA) and potentially toxic element analysis (PT A) (Tables 11.4–11.6). Nitrogen was estimated by Standard Kjeldahl's method of N-estimation and remaining micronutrients were analyzed with Inductively Coupled Argon Plasma (ICAP) photo emission spectroscopy technique. For estimation with ICAP, 0.5 g of each of powdered and sieved sample was digested with 10 ml of double distilled nitric acid (HNO_3) till 2 ml remained. It was allowed to cool at room temperature followed by addition of 2 ml of hydrogen peroxide (H_2O_2) to each sample. Final volume was made to 10 ml with distilled water and these samples were loaded to ICAP instrument.

Table 11.4: NPK Status (Four Cast Comparison)

Elements (µg g^{-1} Soil)	*S-Cast*	*D-Cast*	*T-Cast*	*V-Cast*	*CD (P<0.05)*
N	1600	3175	3400	4250	217
P	770	949	692	2165	316
K	1757	1925	2046	3593	561

Table 11.5: Essential Element Analysis (Four Cast Comparison)

Elements (µg g^{-1} Soil)	*S-Cast*	*D-Cast*	*T-Cast*	*V-Cast*	*CD (P<0.05)*
B	155	99.1	198	143	31.9
Ca	29384	5850	12303	22090	2988
Cu	18.26	9.81	18.9	26.8	3.1
Fe	7259	6094	7567	8950	666
Mg	4842	2590	3175	4836	898
Mn	241	156	217	263	37.9
S	865	378	339	949	148
Zn	253	67.3	143	180	49.3

Table 11.6: Potentially Toxic Element Analysis (Four-Cast Comparison)

Elements (µg g^{-1} Soil)	*S-Cast*	*D-Cast*	*T-Cast*	*V-Cast*	*CD (P<0.05)*
Al	6107	5007	6502	5927	NS
As	17.5	11.2	16.3	17.3	NS
Cd	2.1	1.42	2.05	2.73	0.34
Co	21	12.4	15.8	18.1	2.4
Cr	83.5	39.2	62.3	162	11.9
Hg	17.9	16.9	23.5	26.8	4.51
Mo	4.38	1.64	4.3	4.32	0.98
Na	363	272	302	798	140
Ni	18.6	12.8	18.7	20.7	2.53
Pb	41.6	22.5	40.7	48.8	6.78

Standard methodology of soil enzyme analysis as suggested by Tabatabai (1994) was adopted for estimation of activity of urease, phosphatase (acidic and alkaline) and dehydrogenase enzyme complex (Table 11.7).

Table 11.7: Enzyme Activity: Four Cast Comparison

Enzyme Activity	*S-Cast*	*D-Cast*	*T-Cast*	*V-Cast**	*CD (P<0.05)*
Urease (μg urea/g dry soil/min)	4.18	4.69	4.68	5.25	0.13
Acidic Phosphatase (μg PNP/g dry soil/hour)	402	447	596	179	4.32
Alkaline Phosphatase (μg PNP/g dry soil/hour)	375	233	122	901	8.5
Dehydrogenase enzyme system (μg TPF/g soil/hour)	2.46	10.4	1.52	10.5	0.25

* Where V-cast was collected from culture of *Eisenia fetida* and S-, D- and T-casts were collected from natural habitats.

Results and Discussion

A comparison between the four types of vermicasts for the mean values of electrical conductance (decisiemens/ meter), percentage moisture and pH values is presented in Tables 11.2 and 11.3.

Whereas mean concentration values (μg g^{-1} of soil) for N, P, K (Table 11.4), eight essential elements needed for plant growth (Table 11.5) and potentially toxic elements for plants (Table 11.6) were found to be maximum in cast of *E. fetida* than the other three casts.

Activity of urease (μg urea/g dry soil/min), acidic and alkaline phosphatases (μg PNP/g dry soil/hour) and dehydrogenase enzyme complex (μg TPF/g soil/hour) were compared in four casts (Table 11.7). The data were subjected to statistical analysis and the results were compared by two-way analysis of variance (ANOVA-CRD, $CPCS_1$ Program). Critical differences were mentioned at 5 per cent level of significance ($p < 0.05$).

Maximum activity of alkaline phosphatase was seen in the cast of *E. fetida* followed by the cast of *M. posthuma, L. mauritii and E. waltoni.* These values conform to the pH values in the same order. The values for the activity of acidic phosphatase follow a reverse trend. Higher activity of alkaline phosphatase and lower activity of acidic phosphatase indicated against higher pH values reveals further validity of results.

Vermicast showing maximum alkalinity in terms of high pH values, high activity of alkaline phosphatase and low activity of acidic phosphatase ensured the usefulness of *E. fetida* in vermiculture and to make the soils alkaline. Casts belonging to *E. waltoni* are most acidic and can be employed for highly alkaline to neutral soils.

Half of the total nitrogen metabolized and voided was reported to be excreted mainly as ammonia and urea, with a small proportion as allantoin and uric acid. In actively feeding *Lumbricus terrestris* L., most of ammonia excreted was found to be voided in casts during experimental observations (Tillinghast, 1967) Stimulated activity of invertase, amylase and phosphatase was noted in a pot experiment on rye grass grown in a sub-soil in the presence of *Aporrectodea caliginosa* (Ross and Cairns, 1982).

Worm casts were found to contain higher NO_3^-, NH_4^+ moisture content and microbial respiratory quotients than the parent soils (Elliot *et al.*, 1991). Total nitrogen in worm-cast was reported to be higher than surrounding soils which had a very low level of NH_4^- N (Matsumoto, 1991).

The heavy metal content was assumed in compost of *E. fetida* which was characterized to possess greatest ability to accumulate Chromium (Filipeck *et al.*, 1999). Vermicast of *E. fetida* and *Eudrilus eugeniae* were described to contain high value of nitrogen, humic acid content and electrical conductance (Talashilkar *et al.*, 1999). Higher values of NPK in vermicompost of *Ipomoea* weed and *Parthenium* weed (Vasanthi and Kumaraswamy, 1999) were in agreement with the high values of these nutrients in vermicast from cultures of *E. fetida* in the present studies towards improved soil fertility.

Role of urease was reported in soil enrichment through degradation of organic nitrogen (Hasan, 2000)

Acknowledgements

Facilities provided by Dr. C.L. Arora and Dr. V. Beri, Department of Soil Science, Punjab Agricultural University, Ludhiana are duly acknowledged. The constructive suggestions by Dr. J.M. Julka, Zoological Survey of India, Solan in several phases of this work are greatly appreciated.

References

Bhawalkar, U.S., 1992. Chemical farming to natural farming: A quick change over without loss of yield. In: *Proc. Nat. Sem. Natural Fmg.*, (Eds.) Somani, L.L., Totawat, K.L. and B.L. Baser.

Darwin, C., 1881. *Vegetable Mould and Earthworms*. John Murray, London, pp. 326.

Elliot, P.W., Knight, D. and Anderson, J.M., 1991. Variables controlling denitrification from earthworm casts and soil in permanent pastures. *Biol. Fert. Soils*, 11: 24–29.

Filipeck, M.B., Mazur, K. and Gondek, K., 1999. Heavy metals accumulated from composted sewage sludges by *Eisenia fetida* earthworm. *Folia universitatis Agriculturae Stetinensis Agricultura*, 77: 99–104.

Graff, O., 1970. Phosphorus content of earthworm casts. *Landb Forschvolkenrode*, 20: 33–36.

Hasan, H., 2000. Ureolytic microorganisms and soil fertility: A review. *Communications in Soil Sci. and Plant Analysis*, 31: 2565–2589.

Lunt, H.A. and Jacobson, H.G.M., 1944. The chemical composition of earthworm casts. *Soil Sci.*, 58.

Matsumoto, S., 1991. A new method for evaluation of physiological properties of microorganisms in earthworm cast: Microorganism action and nitrogen components in earthworm casts. In: *Advances in Management and Conservation of Soil Fauna*, (Eds.) Veeresh, G.K., Raj Gopal, D. and C.A. Viraktanath. Oxford and IBH Publishing Co. Pvt. Ltd., pp. 591–597.

Parle, J.N., 1963. A microbiological study of earthworm casts. *J. Gen. Microbiol.*, 31: 13–22.

Ross, D.J. and Cairns, A., 1982. Effects of earthworms and ryegrass on respiratory and enzymatic activities of soil. *Soil Biol. Biochem.*, 14: 583–587.

Russell, E.W., 1973. *Soil Conditions and Plant Growth*. Longmans, London, pp. 196–205.

Srivastava, R.K. and Beohar, P.A., 2004. Vermicompost as an organic manure: A good substitute of fertilizers. *J. Curr. Sci.*, 5: 141–143.

Tabatabai, M.A., 1994. Soil enzymes. In: *Methods of Soil Analysis, Part II: Microbiological and Biochemical Properties*. SSSA Book Series No. 5, (Eds.) Weaver, R.W., Angle, J.S. and P.S. Bottomley. Soil Sci. Soc. of America Publishers, Madison, WI, pp. 775–833.

Talashilkar, S.C., Bhangarath, P.P. and Mehta, V.B., 1999. Changes in chemical properties during composting of organic residues as influenced by earthworm activity. *J. Ind. Soc. Soil Sci.*, 47: 50–53.

Tillinghast, E.K., 1967. Excretory pathways of ammonia and urea in the earthworm *Lumbricus terrestris* L. *J. Exp. Zool.*, 166: 295–300.

Vasanthi, D. and Kumaraswamy, K., 1999. Efficacy of vermicompost to improve soil fertility and rice yield. *J. Ind. Soc. Soil Sci.*, 47: 268–272.

Chapter 12

Larvicidal and Ovicidal Activity of the Leaf Extract of Two Cucurbitacious Plants Against Filarial Vector, *Culex quinquefasciatus* Say

K. Mullai and A. Jebanesan*

Division of Vector Biology, Department of Zoology, Annamalai University, Annamalainagar – 608 002, Tamil Nadu

ABSTRACT

Larvicidial efficacy of the leaf extract of *Cucumis pubescens* and *Luffa acutangula* with four different solvents namely benzene, petroleum ether, ethyl acetate and methanol were tested against late third instar larvae of *Culex quinquefasciatus.* The larval mortality was observed after 24 h exposure. The LC_{50} values of *C. pubescens* and *L.acutangula* against *Cx. quinquefasciatus* were 18.97, 58.82, 44.04, 49.07, 55.50, 59.49, 43.70 and 69.38 ppm respectively. The leaf extract of *C. pubescens* and *L. acutangula* was tested for ovicidal activity against the egg rafts of *Cx. quinquefasciatus.* The freshly collected egg rafts were exposed to different concentrations of leaf extract of *C. pubescens* ranging from 50 to 300 ppm and 100 to 600 ppm for *L. acutangula.* The mean per cent hatchability of the egg rafts were observed after 48 hrs of treatment. At 300 ppm dose of *C.pubescens* extracts killed all the eggs and showed 100 per cent mortality, whereas a

* E-mail: drjeban@rediffmail.com

dose of 500 ppm of *L. acutangula* extracts of benzene and methanol killed all the eggs and showed 100 per cent mortality, and the ethyl acetate and petroleum ether extracts of *L. acutangula* requires a concentration of 600 ppm for complete mortality.

Keywords: *Cucumis pubescens, Culex quinquefasciatus, Larvicidal, Luffa acutangula, Ovicidal.*

Introduction

The mosquito *Culex quinquefasciatus* acts as a vector for *Wuchereria bancrofti* responsible for filariasis in India. Control of mosquitoes has been principally provided by the use of synthetic insecticides. Adverse effects and regulatory status of these chemicals have highlighted the need for the development of selective mosquito control alternatives. The ideal control method is thus the systematic treatment of their breeding places through larvicides. Some of the plant species may possess substances with a wide range of activities, like antifeedant, antioviposition, repellent and growth-regulating activity (Schmutterrer, 1995; Prabakar and Jebanesan, 2004; Mullai and Jebanesan, 2006).

Both oviposition and ovicidal activities are employed for mosquito control in the early stages of its life cycle. Oviposition is one of the most important events in the life cycle of mosquito. If oviposition is prevented, the mosquito life cycle is disturbed and population growth is reduced, if oviposition is attractant the eggs are easily collected and destroyed (Prabakar, 2004). In ovicidal activity the plant extracts inhibit embryogenesis of the egg so as to prevent further stages of the mosquito (Su and Mulla, 1998; Rajkumar and Jebanesan 2004a&b). There has been little research on oviposition and ovicidal activity for mosquito control. The present study was an attempt to assess the larvicidal and ovicidal properties of leaf extracts of two *Cucurbitacious* plants *Cucumis pubescens* and *Luffa acutangula* with four different solvents against *Cx. quinquefasciatus.*

Materials and Methods

Plant Collection and Extraction

The plants of *C. pubescens* and *L. acutangula* were collected from Kothattai and Panruti, Cuddalore district, Tamil Nadu. Voucher specimens have been deposited in the Botanical Survey of India, Coimbatore, India. The plants were washed with tap water and were air dried under shade at room temperature and ground to mesh size 40–60. The powdered material of leaves were extracted separately with four different solvents namely benzene, ethyl acetate, petroleum ether and methanol using a soxhlet apparatus (Vogel, 1978). The solvent from the extract was removed using a vacuum evaporator. Standard stock solutions were prepared at 1.0 per cent dissolving the residues in the respective solvents. From this stock solution, different concentrations were prepared and these solutions were used for assaying mosquito larvicidal and ovicidal activity.

Mosquito Culture

The colonies of *Cx. quinquefasciatus* were cultured and maintained in the laboratory at 27±1°C and 85 per cent relative humidity. Larval forms were maintained in trays by providing dog biscuits and yeast powder in the 3 : 1 ratio. Adults were provided with 10 per cent sucrose solution and one week old chick for blood meal. Mosquitoes were held at 28 ± 2°C, 70 ± 5 per cent RH, and a photoregime of 16 : 8 (L : D) hr.

Bioassay

Five different test concentrations were prepared by adding different range of stock solution to 250 ml of water. Twenty five third instar larvae were exposed to the prepared 250 ml of test concentrations. Each experiment was replicated six times. The control experiments were also run parallel with each replicate (WHO, 1996).

The larval mortality was calculated after 24 hours of the exposure period. In the analysis, both dead and moribund larvae were considered as dead and they were removed from the containers. The corrected per cent mortality was calculated by applying Abbott's formula (Abbott, 1925). The data were subjected to Probit analysis (Finney, 1971).

For the ovicidal activity the method of Su and Mulla (1998) was slightly modified and followed. The leaf extract was diluted in the respective solvent to achieve the concentrations of 50 to 300 ppm for *C. pubescens* and 100 to 600 ppm for *L. acutangula*. Each egg raft containing 100 eggs of *Cx. quinquefasciatus* were exposed to each dose of leaf extract until they hatched or died. Eggs exposed to respective solvents in water served as control. After treatment the egg rafts from each concentration were individually transferred to distilled water cups for hatching assessment after counting the eggs in each raft under microscope. The test was replicated five times.

The hatch rate were assessed 48 h post-treatment by the following formula:

$$\frac{\text{Number of hatched larvae}}{\text{Total number of eggs in eggs rafts}} \times 100$$

Results and Discussion

The results of the experiments conducted for evaluating larvicidal efficacy of leaf extracts of *C. pubescens* and *L. acutangula* revealed that these extracts have larvicidal activity against *Cx. quinquefasciatus* (Table 12.1). The results revealed that *C. pubescens* exerted effective larvicidal properties against *Cx. quinquefasciatus* than *L. acutangula*. The maximum efficacy was observed in benzene extract of *C. pubescens* which caused 50 per cent mortality at 18.97 ppm and the petroleum ether extract was least effective with the LC_{50} value of 58.82 ppm. Ethyl acetate extract of *L. acutangula* was found to be more effective against *Cx. quinquefasciatus* with the LC_{50} value of 43.70 ppm followed by benzene, petroleum ether and methanolic extracts with the LC_{50} value of 55.50, 59.49 and 69.38 ppm respectively.

The percentage of egg hatchability for *C. pubescens* and *L. acutangula* were shown in Tables 12.2 and 12.3. The complete ovicidal activity (100 per cent mortality) was attained at 300 ppm for all the four extracts of *C. pubescens* against *Cx. quinquefasciatus* whereas *L. acutangula* extracts shows the 100 per cent mortality at 500 ppm for benzene and methanol extracts and the ethyl acetate and petroleum ether extracts shows the 100 per cent mortality at 600 ppm. *C. pubescens* leaf extract exhibits the maximum larvicidal and ovicidal properties against *Cx. quinquefasciatus* than *L. acutangula*.

The results of present study are comparable with earlier studies. The toxicity of the late third instar larvae of *Cx. quinquefasciatus* to methanolic leaf extract of *Memordica charantia, Trichosanthus anguina, L. acutangula, Benincasa cerifera* and *Citrullus vulgaris* was tested and the LC_{50} values were 465.85, 567.81, 839.81, 1189.30 and 1636.04 ppm respectively (Prabakar and Jebanesan, 2004). Mullai and Jebanesan (2006) reported the larvicidal efficacy of the leaf extract of *C. pubescens* with four different solvents against late third instar larvae of *Anopheles stephensi, Cx. quinquefasciatus* and *Aedes aegypti*.

Table 12.1: Larvicidal Activity of Leaf Extracts of Cucurbitacious Plants Against *Culex quinquefasciatus*

Plant Species	*Solvent*	*LC_{50} (ppm)*	*LC_{90} (ppm)*	*95% Confidence Limit (ppm)*		*Regression Equation*	*Chi-square Value (x^2)*
				LCL	*UCL*		
C. pubescens	Benzene	18.97±1.51	40.74±2.63	9.93±2.23	26.22±1.36	Y = 15.93 + 1.74 X	23.29*
	Petroleum ether	58.82±1.61	115.48±1.17	42.50±1.21	74.29±1.95	Y = 8.44 + 0.70 X	15.80*
	Ethyl acetate	44.04±0.41	88.30±0.80	29.48±1.10	57.23±0.77	Y = 10.44 + 0.89 X	18.62*
	Methanol	49.07±1.39	91.18±1.16	40.30±2.64	57.75±0.78	Y = 4.63 + 0.92 X	9.03*
L. acutangula	Benzene	55.50±0.54	106.88±0.66	45.61±1.94	64.92±0.59	Y = 8.00 + 0.74 X	7.34*
	Petroleum ether	59.49±1.09	113.93±0.31	47.07±1.40	71.57±1.39	Y = 6.72 + 0.72 X	10.64*
	Ethyl acetate	43.70±0.52	85.45±1.55	31.84±1.05	54.66±0.68	Y = 9.20 + 0.92 X	14.48*
	Methanol	69.38±1.43	137.59±1.15	53.08±0.60	84.83±1.18	Y = 9.03 + 0.59 X	11.43*

*: Significant at $P < 0.05$ level.

Each value (X±SD) represents mean of six values.

Table 12.2: Ovicidal Activity of *Cucumis pubescens* Leaf Extract Against Egg of *Culex quinquefasciatus*

Solvent	*Percentage of Egg Hatching*						
	Concentration (ppm)						
	Control	*50*	*100*	*150*	*200*	*250*	*300*
Ethyl acetate	1 00±0	82.3±0.94	57.3±0.71	43.2±1.22	28.3± 0.63	10.7± 0.99	NH
Benzene	100±0	84.7±0.81	61.3±0.63	47.6±0.97	23.6±0.90	9.3±0.83	NH
Petroleum ether	1 00±0	80.7±1.01	64.0±1.28	40.8±1.10	24.7±0.40	6.3±1.27	NH
Methanol	99.6±0.94	79.2±0.81	57.8±0.94	41.7± 1.01	19.8± 0.97	7.2±1.00	NH

NH: No hatchability (100 per cent mortality).

Each value (X±SD) represents mean of six values.

Table 12.3: Ovicidal Activity of *Luffa acutangula* Leaf Extract Against Egg of *Culex quinquefasciatus*

Solvent	*Percentage of Egg Hatching*						
	Concentration (ppm)						
	Control	*100*	*200*	*300*	*400*	*500*	*600*
Ethyl acetate	100±0	92.5±0.85	74.3± 1.22	47.5± 0.94	32.7±1.01	18.6±0.85	NH
Benzene	100±0	88.3±0.83	56.8±0.51	37.8±0.83	11.3±1.22	NH	NH
Petroleum ether	100±0	93.1±0.94	72.8±1.01	57.0± 0.97	34.1±0.37	21.3±0.94	NH
Methanol	100±0	91.6±0.97	65.1±1.35	44.3±0.55	17.4±0.31	NH	NH

NH: No hatchability (100 per cent mortality).

Each value (X±SD) represents mean of six values.

Very few reports are available on the egg hatchability of mosquitoes and the lethal values were ranging from 0.0025 mg/ml (Schmutterrer, 1995). The ovicidal activity of *Solanum trilobatum* leaf extract against *Cx. quinquefasciatus* and *Cx. tritaeniorhynchus* showed zero hatchability (100 per cent mortality) in eggs aged 0–3 and 3–6 h exposed to a ≥100 ppm concentration of the leaf extract. Eggs aged 9–12 and 15–18 h suffered incomplete ovicidal activity even at 200 ppm (Rajkumar and Jebanesan, 2004a). The bioactive compound Azadiarachtin (*Azadirachta indica*) showed complete ovicidal activity in the eggs of *Cx. tarsalis* and *Cx. quinquefasciatus* exposed to 10 ppm concentration (Su and Mulla, 1998). The ovicidal activity of *Moschosma polystachyum* leaf extract against the egg rafts of *Cx. Quinquefasciatus* showed 100 per cent mortality at 0–3 hand 3–6 h old egg rafts with the concentrations of 125, 150, 175 and 200 mg/l (Rajkumar and Jebanesan, 2004b).

The results of the present investigation indicates that the leaf extract of *C. pubescens* and *L. acutangula* exhibits both larvicidal and ovicidal activity against *Cx. quinquefasciatus*. Further analysis to isolate the active principle is underway in our laboratory, which will be reported in the near future.

Acknowledgement

The authors are greateful to Professor and Head, Department of Zoology, Annamalai University for the facilities provided and encouragements and University Grants Commission for financial assistance.

References

Abbott, W.S., 1925. A method of computing the effectiveness of an insecticide. *J. Econ. Entomol.*, 18: 265–267.

Finney, D.J., 1971. *Probit Analysis*. Cambridge University Press, London, p. 68–72.

Mullai, K. and Jebanesan, A., 2006. Bioefficacy of the leaf extract of *Cucumis pubescens* Willd (Cucurbitaceae) against larval mosquitoes. *Bull. Bio. Sci.*, 4(1): 35–37.

Prabakar, K., 2004. Development of mosquito repellents from certain plants of Tribal usage from western Ghats, Tamil Nadu, India. *Ph.D. Thesis*, Annamalai University.

Prabakar, K. and Jebanesan, A., 2004. Larvicidal efficacy of some Cucurbitaceous plant leaf extracts against *Culex quinquefasciatus* (Say). *Bioresource Technology*, 95: 113–114.

Rajkumar, S. and Jebanesan, A., 2004a. Ovicidal activity of *Solanum trilobatum* L. (Solanaceae) leaf extract against *Culex quinquefasciatus* Say and *Culex tritaeniorhynchus* Giles (Diptera : Culicidae). *International Journal of Tropical Insect Science*, 24(4): 340–342.

Rajkumar, S. and Jebanesan, A., 2004b. Ovicidal activity of *Moschosma polystachyum* Linn. (Lamiaceae) leaf extract against filarial vector *Culex quinquefasciatus* Say. *Tropical Biomedicine*, 21(1): 47–50.

Schmutterrer, H., 1995. *The Neem Tree (Azadirachta indica A. Juss.) and Other Meliaceous Plants*. VCH Publishers, Weinneim, Germany, pp. 696.

Su, T. and Mulla, M.S., 1998. Ovicidal activity of neem products (*Azadirachtin*) against *Culex tarsalis* and *Culex quinquefasciatus* (Diptera : Culicidae). *J. Am. Mosq. Cont. Assoc.*, 14: 204–209.

Vogel, A.I., 1978. *Textbook of Practical Organic Chemistry*. The English language Book Society and Longman, London, p. 1368.

WHO, 1996. *Report of the WHO Informal Consultation on the Evaluation and Testing Insecticides*. CTD/WHO PES/IC/96.1 p. 69.

Chapter 13

Kinetics Study of Auto-Gasification of Bagasse for Energy Augmentation

V. Sivaramakrishnan[1*], *V. Kirubakaran*[2], *M. Premalatha*[3] *and P. Subramanian*[4]

[1]*Research Scholar,* [3]*Assistant Professor,* [4]*Professor and Head (Retd)*

CEESAT, National Institute of Technology, Tiruchirapalli – 620 015, Tamil Nadu

[2]*Lecturer in Energy, Rural Technology Centre, Gandhigram Rural Institute, Dindugal, Tamil Nadu*

ABSTRACT

TGA studies were carried with bagasse, varying the particle size from less than 1 mm to more than 3 mm in static and purging nitrogen and static air environment at a heating rate, ranging from 5 to 50°C/min. Due to high porous nature of bagasse, continuous reaction model seems to apply. The heating rate of 5°C/min yielded complete thermal degradation of bagasse yielding ash at a temperature of 600°C, proving the feasibility of auto gasification. Using the TGA data the kinetics of auto gasification of bagasse was elucidated. Three zones corresponding to drying (30–120°C), devolatization (200–325°C) and gasification (325–600°C) were observed. The first order rate equation was used to evaluate the rate constant. The effect of temperature on rate constant was explained using Arrhenius equation. The calculated weight loss using kinetic parameters was found to match with the experimental weight loss satisfactorily, thus proving auto gasification.

***Keywords**: Auto gasification, Bagasse, Energy augmentation, Kinetics.*

---

* *Corresponding Address*: Professor of Mechanical Engineering, Saranathan College of Engineering, Tiruchirapalli – 620 012, Tamil Nadu; E-mail: vsmp1967@yahoo.com; Mobile: +91 98943 88495

Introduction

Ever increasing standard of living and subsequent energy demand lead to search for an alternate energy source. Extraction of energy out of these sources should be ecofriendly and economically viable. The continuous depletion of fossil fuel has also triggered the race of identifying alternate energy resources. Most of the resources have root on agricultural residues. These residues such as rice husk, coir pith, bagasse etc. are available both in quantity and quality in developing countries. The disposal of these bioresidues has created environmental issues. Identification of problems associated with disposal would pave way in extracting energy out of these residues and subsequently the energy demand.

The bio-residues are very bulky and occupy very large space. Burning them in open leads to pollute the environment. Transporting them to a central place for disposal would not be economical. Technology for then and there extraction of energy from them in an eco-friendly way or for then and there conversion of residues to solid and liquid fuels for easy transportation and storage should be developed. Bio-methanasion and thermal degradation of bio-residues are the two routes followed for extraction of energy. A reliable and viable technology could be developed, provided the composition of bioresidues, variables affecting the conversion, kinetics and mechanism of conversion are studied exhaustively.

Thermal degradation of bio-residues has been proved to be most suitable. Several attempts have been reported in the past. Antal, *et al.* (1980), Avni *et al.* (1985), Boateng, *et al.* (1990), Chakraverty (1985), Havens, *et al.* (1972), LipskaQuinn, *et al.* (1985), and Magin Lapuerta (2004), Ramar *et al.* (1981), Raveendran *et al.* (1996), Shafizadeh, (1968), Williams, *et al.* (1992).

Particle size, environment and heating rate appear to influence the thermal degradation of biomass. Upon heating, the biomass looses moisture up to 120°C, releases the volatile matter up to 400°C and then turns into ash above 600°C due to complete burning of char. The porosity of the bio-mass particle decides the type of reaction, namely Shrinking core reaction or continuous reaction that is taking place. The presence of oxygen during thermal degradation also controls the reaction, Levenspiel (1962). The heating rate is also responsible in the composition of producer gas.

The present investigation aims at studying the effect of these variables on the thermo-chemical conversion of bagasse into ash. Bagasse is a bulky bio-residue available from sugar industries. Every ton of sugarcane processed, leaves approximately 0.565 ton of bagasse as bio residue. Even though a major quantity is consumed for burning in boiler for cogeneration and in paper industries, still a sizable quantity is available.

The characteristics of devolatization of biomass can best be understood from the weight loss of biomass at different temperatures. The weight loss at different temperatures can be obtained from thermo-gravimetric analysis (TGA).

Materials and Methods

Bagasse from a neighbouring sugar industry was sieved into fractions having sizes less than 1 mm and more than 3 mm. The samples were sun dried and were stored in air-tight containers. Perkin Elmer Pyris-7 Thermo Gravimetric Analyzer was used to carry out TGA studies. The continuous measurement of sample weight as a function of temperature was provided by the instrument at different heating rates. The thermal degradation of bagasse has been studied to evaluate the effect of particle size, environment and heating rate.

Results and Discussion

Effect of Particle Size

Maa and Balilie (1973) have shown that the reaction controls for sizes less than 0.2 cm. Particle sizes ranging from less than 1.0 mm and more than 3mm was used in the present study. TGA studies were carried out with static air at a heating rate of 5°C/min. Figures 13.1 and 13.2 compare the per cent weight reduced at different temperatures for the different particle sizes. It is evident that the variation is negligible implying the absence of internal diffusion. Even though the particle size is varied by almost 3 times, no appreciable variation in weight reduction with size is observed. This is attributed to the characteristics of particle. Bagasse particle is highly porous allowing the passage of both reactant and product gases easily. Therefore the reaction assumes to be taking place throughout the particle continuously.

Effect of Environment

In order to understand the physico-chemical degradation of biomass, the effect of environment on the rate of reaction and on the product gases distributed should be assessed. Therefore the TGA studies were carried out with flowing nitrogen and with static air and nitrogen at the heating rate of 5°C/min.

Williams and Besler (1993) have carried out TGA studies on rice husk with nitrogen for purging. The flow of nitrogen removes the devolatilized gases as and when they form, thus preventing the secondary degradation of biomass and yielding char. This mode of operation has yielded gaseous and liquid fuels along with solid char. The liquid and solid fuel can be stored and transported economically while the gaseous fuel is consumed then and there. This overcomes the problems associated with biomass transportation and handling.

Figures 13.3–13.5 compare the TGA data obtained for all the three cases. With flowing nitrogen, char has resulted as a final product. Where as with static air and nitrogen, ash has resulted. Since flowing nitrogen prevented the secondary degradation, char formed as a product of devolatilization could not be gasified. This is also in accordance with observations made by Williams and Beseler (1993).

An interesting observation is that with static air and nitrogen, the biomass was completely converted into ash. In the absence of any carrier gas, the devolatilized gases further react with char to yield ash. Oxygen required for complete gasification of biomass could not be available in the balance section of TG Analyzer. Therefore it implies that the necessary oxygen should have come only from the biomass itself, proving the feasibility of auto gasification.

It is obvious that bagasse contains oxygen in significant quantity to sustain auto gasification without any external supply of oxygen. Further, Shafizadeh (1968) observed that the presence of impurity or trace metals even in low concentration would either catalyze or inhibit the degradation and would lower temperatures of thermal decomposition of bio mass. Therefore it is decided to carry out further experiments with static air.

Effect of Heating Rate

TGA data were collected at 5, 10, 20 and 30°C/min heating rates with static air. It is observed that at the heating rates above 20°C/min, char has resulted. At high heating rates the heat transfer through the particles controls the degradation. At these conditions, the exterior surface hardens, preventing

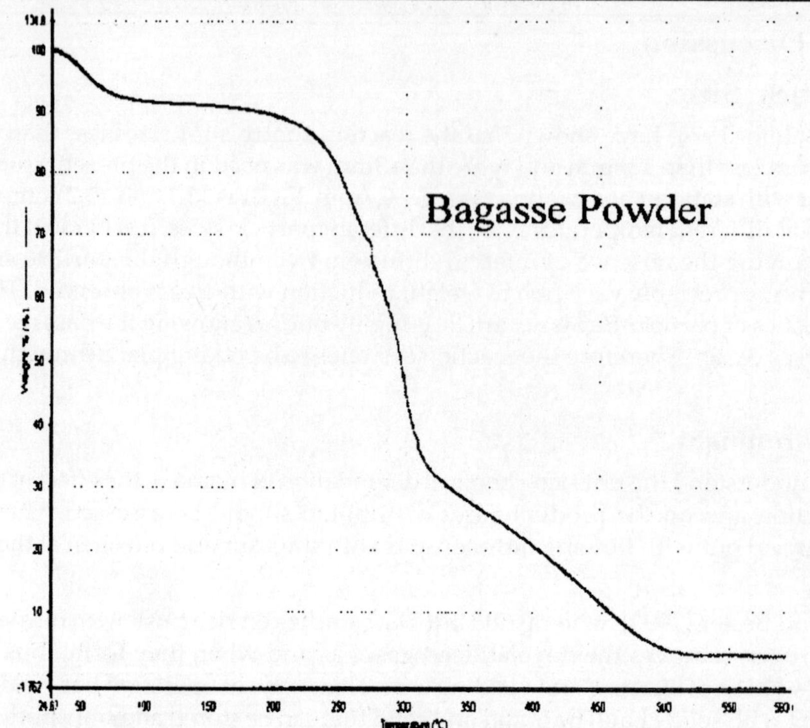

Figure 13.1: % Weight Reduced Vs Temp. for Particle Size (Bagasse Powder)

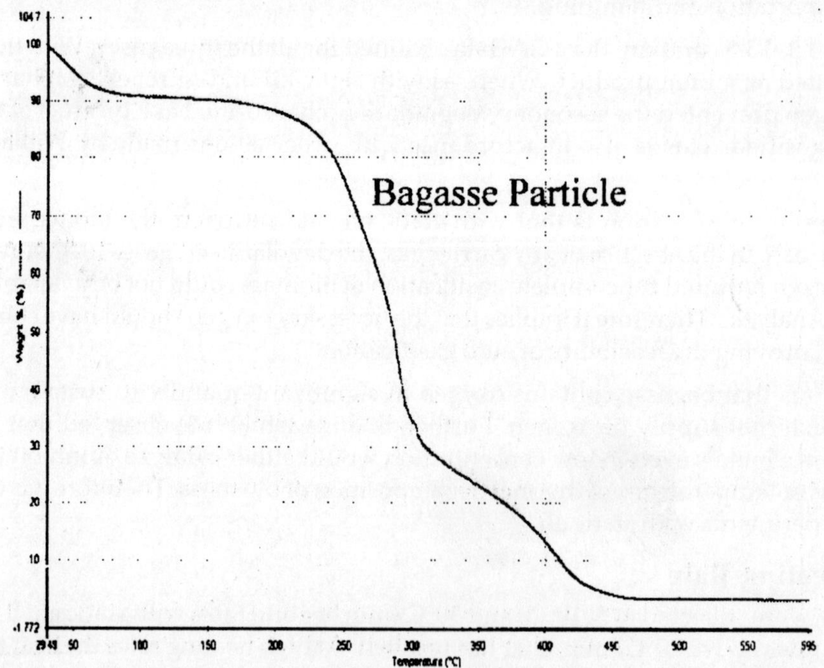

Figure 13.2: % Weight Reduced Vs Temp. for Particle Size (Bagasse Powder)

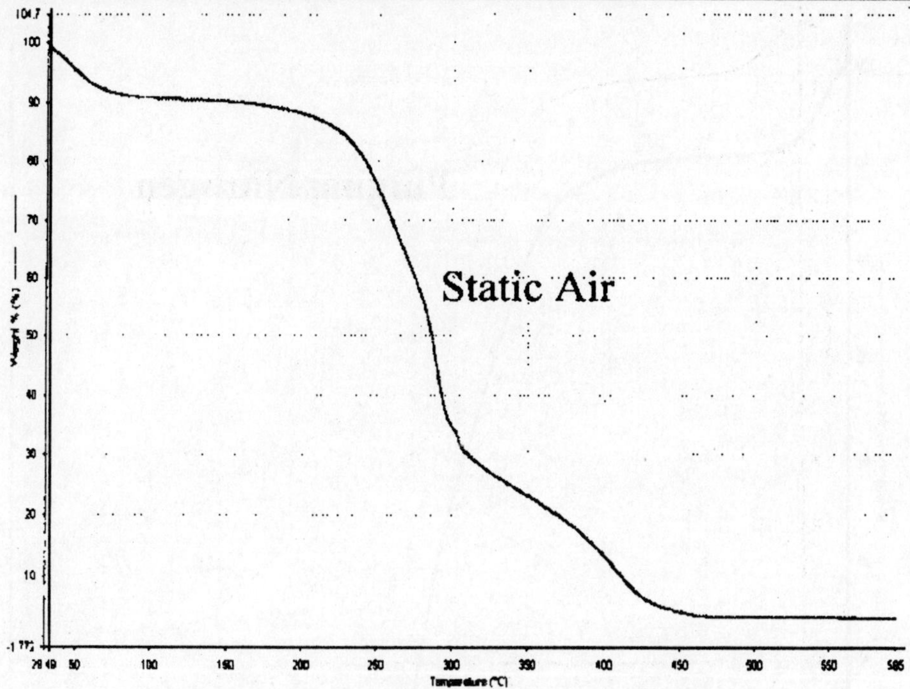

Figure 13.3: TGA Data for 3 Cases (Static Air)

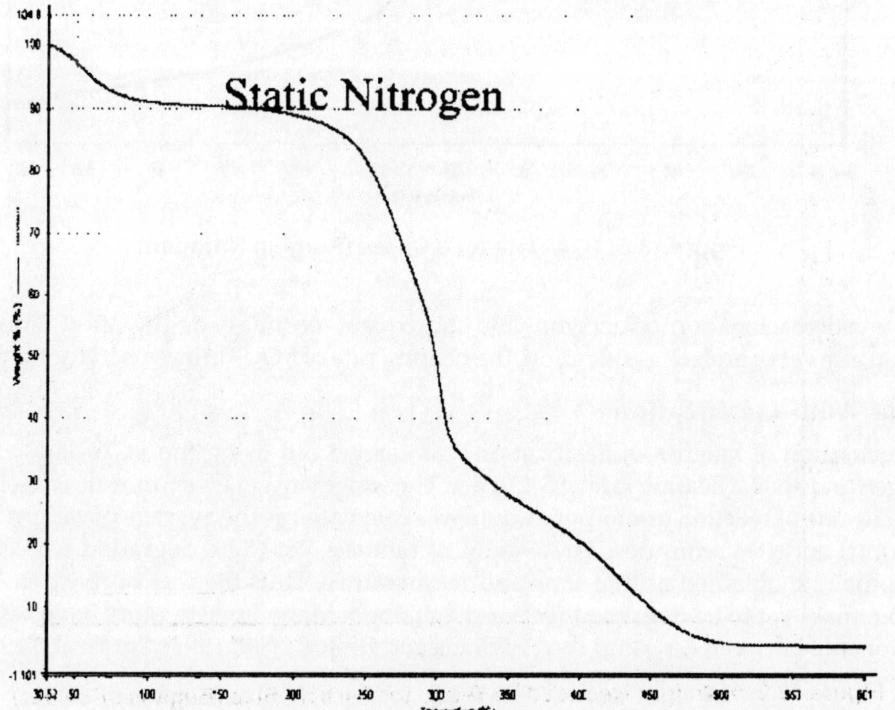

Figure 13.4: TGA Data for 3 Cases (Static Nitrogen)

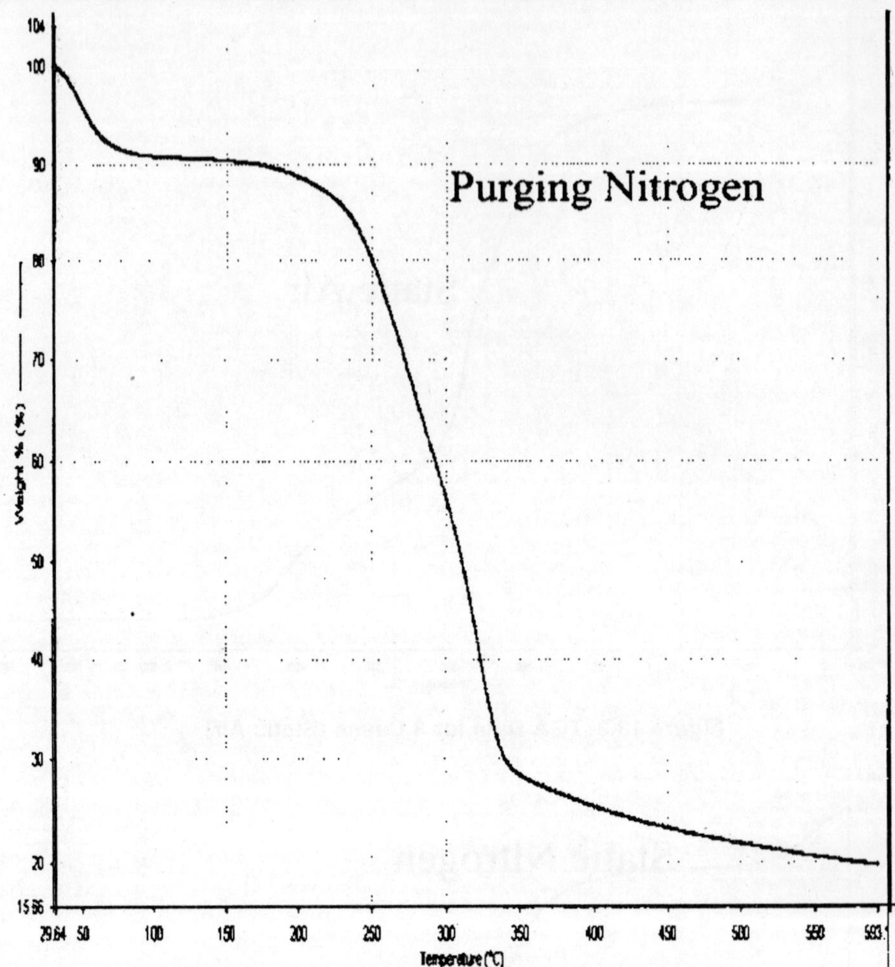

Figure 13.5: TGA Data for 3 Cases (Purging Nitrogen)

the penetration of reactant or product gases and subsequent secondary degradation. Since the present investigation aims at complete gasification, the heating rate of SOC/min is used for further studies.

Kinetics of Auto Gasification

The elucidation of kinetics of gasification was carried out using the TGA data obtained with static nitrogen and air at a heating rate of 5°C/min. The weight loss at every minute was obtained from TGA data. The rate of reaction grams per minute was calculated as the average of weight loss between (t-1) and (t), (t) and (t+1) minutes. The weight of biomass yet to be degraded was calculated by subtracting the rate obtained at that time and temperature. Thus the rate of reaction ($-r_d$) and the weight of bio mass yet to be degraded (w) were obtained. Many investigators have assumed a first order relationship between ($-r_d$) and (w) (Willams and Besler, 1993, 1996; Agrawal, 1998; Bhat *et al.*, 2001).

In the present investigation, the rate of reaction ($-r_d$) was assumed to be first order with respect to weight yet to be converted (w),

$$-r_d = k \times w \tag{1}$$

where,

k is first order rate constant. The rate constant was obtained for each temperature. The temperature dependency of k was established using Arrhenius equation.

$$k = k_0 \exp(-E/RT) \tag{2}$$

where,

k_0: Arrhenius frequency factor.
E: Activation energy.
R: Gas constant.
T: Temperature, Kelvin

Even though no reaction takes place during moisture removal, the rate of moisture removal was also modeled similar to reaction.

Figure 13.6 represents a plot of In (k) vs. (1/T). It is observed from Figure 13.6 that at least three zones exist. First zone represents moisture removal between 30°C to 120°C. The second zone of thermal degradation occurs between 200°C to 325°C, the third zone between 325°C and 600°C. A separate plot was prepared for each zone. Figure 13.7–13.9 represent these plots. From the slopes and intercepts the activation energy and frequency factor were determined for each zone and are given in Table 13.1.

Table 13.1: Activation Energy and Frequency Factor

Sl.No.	*Zone*	*Temperature Range °C*	*A*	*E Activation Energy*	*r^2*
1.	Moisture Removal	30–120	30454.83	8.00	0.7153
2.	Zone I	200–325	6.83×10^7	21.99	0.9748
3.	Zone II	325–600	18120.52	17.02	0.9643

Verification of the Model

Using activation energy and frequency factor, the weight of bio mass yet to be converted was calculated for every minute at the corresponding temperature. The calculated and experimented values are compared in Figure 13.10. It is evident that the kinetic parameters evaluated are representing the experimental data satisfactorily.

Conclusion

Gasification of bagasse was analyzed using the TGA data obtained with static air at a heating rate of 5°C/min in the temperature between 30°C and 500°C. These operative conditions were found to be conducive for auto-gasification, yielding complete gasification. The kinetics of gasification was elucidated using the first order equation. Four zones of gasification exist between 30°C and 500°C. A and E were evaluated for each zone. The weight yet to be converted was calculated using the kinetic parameters and was compared with experimental data satisfactorily.

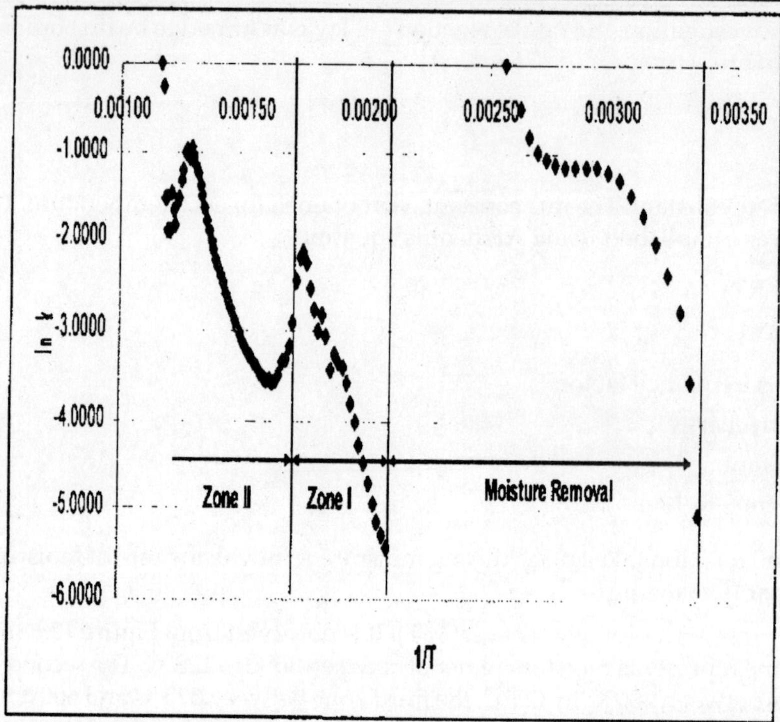

Figure 13.6: In (k) Vs. (1/T)

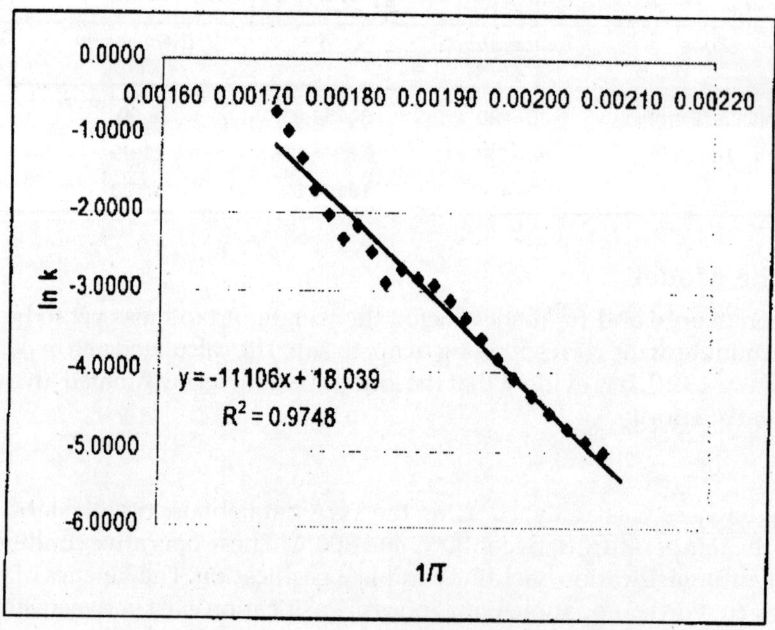

Figure 13.7: In (k) Vs. (1/T)–Zone I (200°C to 325°C)

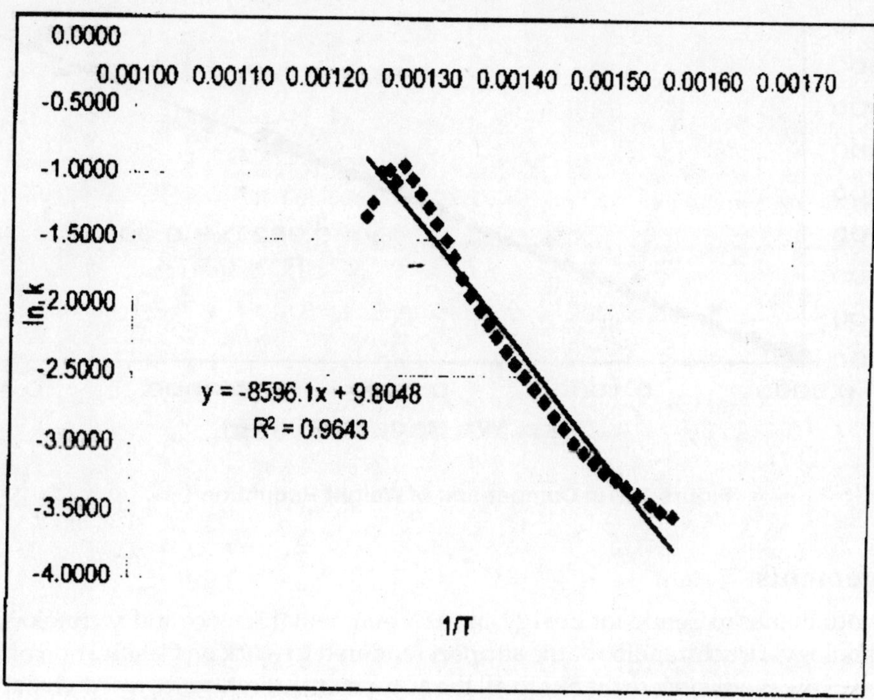

Figure 13.8: In (k) Vs. (1/T)–Zone II (325°C to 600°C)

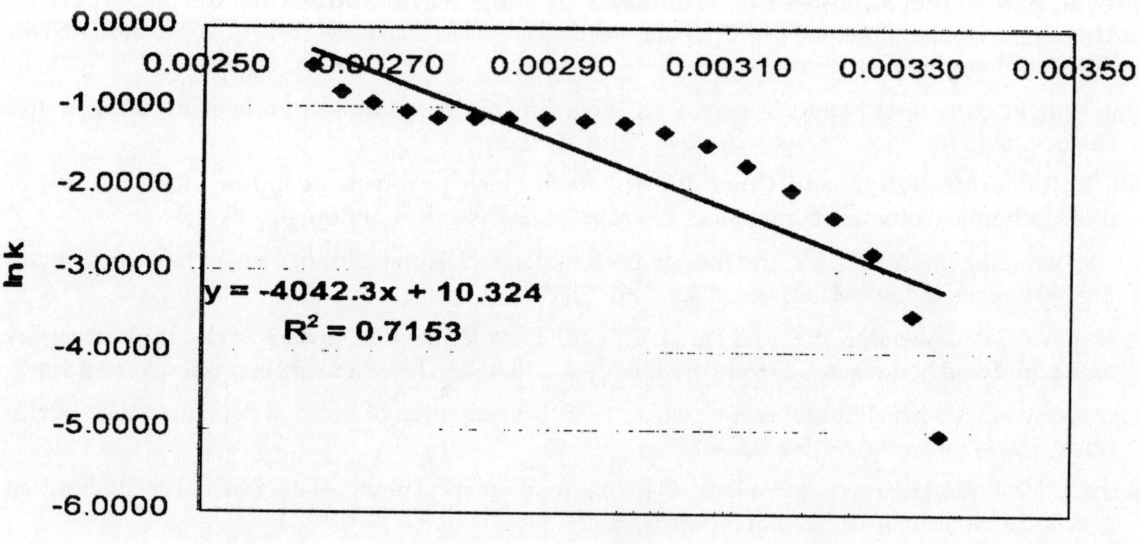

Figure 13.9: In (k) Vs. (1/T)–Moisture Removal Zone

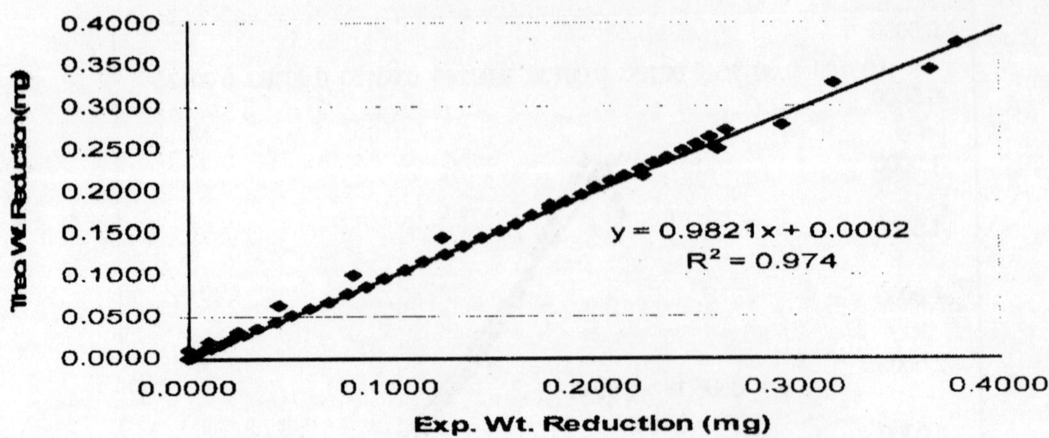

Figure 13.10: Comparison of Weight Reduction

Acknowledgements

We express our thanks to Centre for Energy and Environmental Science and Technology, National Institute of Technology, Tiruchirapalli for the support rendered to work on Gasification of Sugarcane Bagasse. We also express our sincere thanks to all the authors of our reference research papers which forms the source of information for this present study.

References

Agrawal, R.K., 1998. Compositional analysis of solid waste and refuse derived fuels by Thermogravimetery method D 997. In: *Annual Book of ASTM Standards*, (Ed.) C.M. Earnest. ASTM, Philadelphia, pp. 259.

Antal, M.J., Friedman, H.L. and Rogers, F.E., 1980. Kinetics of cellulose pyrolysis in nitrogen and steam. *Combustion Science and Technology*, 21: 141–152.

Avni, E., Davoudzadeh, F. and Coughlin, R., 1985. Flash pyrolysis of lignin fundamentals of thermochemical biomass conversion. *Elsevier Applied Science*, London, pp. 35–41.

Bhat, A., Ram Bheemarasetti, J.V. and Rajeswara Rao, T., 2001. Kinetics of rice husk char gasification. *Energy Conversion and Management*, 42: 2061–2069.

Boateng, A.A., Walawender, W.P. and Fan, L.T., 1990. Devolatilization studies of rice hulls in a TGA and a fluidized bed reactor. *Biomass for Energy and Industry*, Elsevier Applied Science, London.

Chakraverty, A., Mishra, P. and Banerjee, H.D., 1985. Investigation of thermal decomposition of rice husk. *Thermochimica Acta*, 94: 267–275.

Havens, J., Hashemi, H. Browl and Welker, R., 1972. A mathematical model of the thermal decomposition of wood. *Combustion Science and Technology*, 5: 91–98.

Levenspiel, 1962. *Chemical Reaction Engineering*. Wiley Eastern (P) Ltd., New Delhi, pp. 141.

Lipska-Quinn, A.E., Zeronian, S.H. and McGee, K.M., 1985. *Fundamentals of Thermochemical Biomass Conversion*. Elsevier Applied Science, London, pp. 453–471.

Maa, P.S. and Bailie, R.C., 1973. Influence of particle sizes and environmental conditions on high temperature pyrolysis of cellulose material. *Combustion Science Technology*, 7: 257–269.

Magin Lapuerta, Juan Jose Hernandez and Joaquin Rodriguez, 2004. Kinetics of devolatilization of forestry wastes from thermo gravimetric analysis. *Biomass and Bioenergy*, 27: 385–391.

Ramar, P., Walawender, W.P., Fan, L.T. and Howell, J.A., 1981. Thermo gravimetric analysis of biomass devolatilization studies on feedlot manure. *Ind. Eng. Che. Process Des. Dev.*, 20: 630–636.

Raveendran, K., Ganesh, A. and Khilar, K.C., 1996. Pyrolysis characteristics of biomass and biomass components. *Fuel*, 75: 987–998.

Shafizadeh, F., 1968. Pyrolysis and combustion of cellulosic material. *Adv. Carbohydrate Chem.*, 23: 419–474.

Shafizadeh, F. and McGinnis, G.D., 1971. Chemical composition and thermal analysis of cotton wood. *Carbohydrate Research Journal*, 16: 273–277.

Shafizadeh, F. and Bradbury, A.G.W., 1979. Thermal degradation of cellulose in air and nitrogen at low temperatures. *Journal of Appl. Polymer Science*, 23: 1431–1442.

Williams, P.T. and Besler, S., 1993. The pyrolysis of rice husks in a TGA and static atch reactor. *Fuel*, 72: 151–159.

Williams, P.T. and Besler, S., 1996. Influence of temperature and heating rate on the slow pyrolysis of biomass. *Renewable Energy*, 7: 233–250.

Williams, P.T., Shamsuddin, A.H. and Taylor, D.T., 1992. Pyrolysis of oil palm waste. *Biomass for Energy, Industry and Environment*. Elsevier Applied Science, London, pp. 757–761.

Chapter 14

Influence of Diet Composition and Starvation on the Growth of the Fingerlings of the Grass Carp, *Ctenopharyngodon idella*

R. Sivakami[1], R. Ramya[1], M. Suganya[1], M. Marieeswari[1] and G. Premkishore[2]

[1]*P.G. and Research Department of Microbiology, Kurinji College of Arts and Science, Trichirappalli – 620 002, Tamil Nadu*

[2]*Alan Feldman Public School and Junior College, Kazhzkuttam, Tirvandrum, Kerala*

ABSTRACT

With the exploding human population, all the third world countries especially India is faced with a precarious situation to feed its people with a protein rich diet. Aquaculture, specifically the production of fish, comes as a relief as fish is rich in protein. However, one of the serious factors that confront an aquaculturist is feed. Today, most of the feeds consist of fish meal. Hence, it was thought worthwhile to attempt a study on fish growth using a low cost commercial feed in the grass carp, *Ctenopharyngodon idella.* In addition, a study was also attempted to assess the ability of the fish to withstand starvation and hence growth during starvation was also studied. Results indicate that there was a significant difference ($p < 0.05$) in growth between feeding and starvation. Biochemical changes in the selected tissues of fish showed differences in their composition during starvation. Liver appeared to be the most affected tissue indicating that the energy stored in the liver is utilized first to overcome the stress caused during starvation.

Introduction

A clear understanding of the nutritional requirements of fish and feeding schedules is a must to develop a suitable cost effective balanced diet in fish culture (Naik *et al.*, 2000). The quantity and quality of feed given greatly influences the fish growth and thus the production. In order to assess the ecological success of a fish in a particular habitat, a knowledge of its efficiency to convert the food material into its body substance is of much importance (Kinne, 1960, 70, 71; Brett, 1970). For this, the growth and conversion in a fish has to be observed under different food items (Ritakumari and Nair, 1979).

Aquaculture production in India has been increasing rapidly and this has resulted in an increased demand for a variety of formulated feeds. A major share in expense in intensive fish culture is accounted for by supplementary feeds. The commonly used conventional feed in India is a mixture of rice bran and oil cake in equal propositions. However, it has been found that this conventional feed is nutritionally imbalanced to achieve fast growth of fish., Further, considerable increases in the cost of ingredients of conventional feed in these days have necessitated to look for alternate protein sources. Under these circumstances, it was thought worthwhile to attempt the use of a commercially available low cost animal feed commonly used for growing aquarium fishes in the present study. This food was tested in the fingerlings of the grass carp, *Ctenophayngodon idella*. In addition, starvation studies have also been attempted with a view to understand the ability of the fish to withstand unfavourable conditions.

Materials and Methods

Fingerlings of the grass carp, *Ctenopharyngodon idella* were collected from the Fish Farmers Development Agency, Trivandrum. In the laboratory, the fishes were stocked in glass aquarium tanks of 25 litre capacity. The fishes were acclimated to the lab conditions using well water for a fortnight before the start of the experiment. The fishes during this period were fed daily with the experimental diet.

For the experiment, a total of 20 fishes of almost similar size and weight were chosen. Only active and healthy fishes were taken. Prior to the experiment, the fishes were starved for a period of 72 hours in order to evacuate their gut contents completely (Windell, 1966). The fishes were then placed individually in glass containers containing the same well water with which they were acclimated. Before introducing the fish, the fish was blotted with a filter, paper and weighed. As the process of weighing was quick, the fish were not subjected to undue stress. Simultaneously, 3 fishes were weighed the same way and the muscle and liver dissected out and estimated for its protein, carbohydrate and lipid levels. This was used as the control.

The food was left in the experimental tank for 6 hours a day as the exposure of food for a duration of 6 to 8 hours a day was reported to be congenial for maximum consumption (Mercybai, 1970). The remaining of the food were collected after the exposure, blotted with a filter paper and weighed. The faeces released by the fish were also collected at regular intervals and weighed after blotting with a filter paper. In the present study, fed at the rate of 5 per cent of their initial body weight throughout the experiment.

After a period of 14 days feeding, the fish was starved for a period of 24 hours and then weighed individually. Five fishes were then sacrificed, its liver and muscle dissected out and estimated for its protein, carbohydrate and lipid levels.

The remaining fishes were then subjected to starvation to assess the effect of starvation on growth. At the end of the 14th day, the fishes were again weighed. In addition, specific tissues like the liver and muscle were dissected from 5 fishes, similar tissues pooled and estimated for its protein, carbohydrate and lipid contents. While protein content was analysed after Lowry *et al.* (1951), carbohydrate was estimated after Roe (1955) and lipids after Folch *et al.* (1975).

Some physico-chemical variables of the water which have an effect on the growth of fishes were also analysed. While water temperature was measured using a mercury thermometer, pH was measured using a pH meter (Elico). Dissolved oxygen, free carbondioxide, phenolphthalein and methyl orange alkalinities were analysed by following APHA (1989).

Results

The water quality variables of the media in which the experimental fishes were kept is presented in Table 14.1.The proximate composition of the commercial feed is provided in Table 14.2. The growth performance of the fingerlings of the grass carp while feeding and during starvation are provided in Table 14.3. As evident from this table, from an initial weight of 958.95 mg, the fish increased in weight to 991.35 mg thus recording a net gain of 32.40 mg during the period under feeding. The food conversion ratio at the end of the experiment was 4.45 with the survival percentage being 95 per cent. The growth of fingerlings after a period of 14 days starvation recorded a net decline in growth of 8.97 mg and the percentage survival during this period was 86.6 per cent.

Table 14.1: Physico-chemical Variables of Water

Sl.No.	*Variable*	*Range*
1.	Water Temperature (°C)	30.1–31.8
2.	pH	7.1–8.0
3.	Free carbon dioxide (mg/l)	0–10
4.	Dissolved oxygen (mg/l)	3.8–5.3
5.	Total alkalinity (mg/l)	200–250

Table 14.2: Proximate Composition of Experimental Feed

Sl.No.	*Variable*	*Content (per cent)*
1.	Protein	32
2.	Moisture	10
3.	Lipid	03
4.	Crude fibre	05

Table 14.3: Growth Performance of *C. idella* after Feeding and Starvation

Variable	*After Feeding*	*After Starvation*
Initial weight (mg)	958.95±2.7	990.26±2.18
Final weight (mg)	91.35±2.63	981.29±3.6
Weight gain (mg)	32.6±4.47	8.97±2.86
Relative growth rate (per cent)	3.61	–0.97
Survival (per cent)	95	86.8
Food Conversion Ratio	4.45±0.44	Nil

± indicates Standard deviation.

$$\text{Relative Growth Rate} = \frac{\text{Final weight} - \text{Initial weight}}{\text{Initial weight}} \times 100$$

$$\text{Food Conversion Ratio} = \frac{\text{Total food taken}}{\text{Total weight gain}}$$

The biochemical composition of the muscle and liver tissues after feeding and starvation are presented in Table 14.4. After feeding and starvation, the protein, carbohydrate and lipid contents showed changes in both the liver and muscle. While biochemical changes after starvation revealed no major changes in the muscle tissue, the liver tissue appeared to show more drastic changes, especially with regard to the carbohydrate and lipid levels than with the protein level.

Table 14.4: Biochemical Composition of Selected Tissues of the Grass Carp After Feeding and Starvation

Variable (per cent)	*Control*		*After Feeding*		*After Starvation*	
	Muscle	*Liver*	*Muscle*	*Liver*	*Muscle*	*Liver*
Protein	55.5	22.4	59.4	24.0	62.2	35.6
Carbohydrate	31.0	56.0	32.4	60.3	50.1	52.8
Lipid	13.5	21.6	8.2	15.7	7.7	11.6

Discussion

Information regarding the relationship between feed ratio and growth is important in ensuring minimal feed loss and economical production cost (Hashim *et al.*, 1999). In addition to the nutrient content in the artificial feed, feeding rate, water temperature and fish age are also considered to be the important factors affecting fish growth (Brett, 1979).

As far as the growth in terms of weight after feeding was concerned, the present study could record a growth of 32.4 mg. Comparing this with a related work done by Leena (2001) using an animal feed in the same fish, reveals almost comparable results. However, a comparison of the growth rate using a plant feed (Rani, 2001), revealed a lower growth rate in the present study. The differences in the growth rate obtained using different feeds in the same animal could be attributed to the composition of suitable organic and inorganic substances in different proportions in the various feeds.

When the fishes were exposed to starvation and then assessed for growth, it was found that after the starvation period, the fishes registered a loss in growth in terms of weight by 8.7mg. Comparing this with the growth obtained during feeding reveals a significant difference ($p < 0.05$).

Experiments with the present study indicated an FCR of 4.45. It is a well-known fact that a lower FCR value indicates abetter utilization of the feed while higher levels indicate poor utilization. Nevertheless, it is also common knowledge that FCR levels tend to be high during the initial days and the levels gradually decrease as the number of days increases. Hence, it is only natural that a high FCR level was obtained in the present study as it was only of a short duration and that there is every possibility for the FCR to decrease as days progressed. Nevertheless, an accurate measurement of the FCR can be made only after a prolonged study.

Biochemical studies were done to understand the changes in the protein, carbohydrate and lipid levels in the liver and muscle tissues after feeding. Results indicate that subtle changes have occurred in these tissues clearly indicating the role of feed in these changes. Biochemical changes after starvation revealed a decrease in the carbohydrate and lipid contents in both the muscle and liver tissues. However, among the two, liver appeared to be the most affected. The decrease in carbohydrates and lipid contents is due to their utilization under stress conditions. Carbohydrates represent the principal and immediate energy precursors of animals when exposed to stress conditions with the protein being the spare energy source during chronic conditions of stress. The reduction in carbohydrate content in the tissues may be due to the rapid degradation of stored glycogen or decreased rate of glycogenesis or glyconeogenesis or both (Sivakami *et al.*, 1995). Such a condition was also noticed by Jayachandran and Chockalingam (1986) in *Channa puntatus,* Premkishore and Chandran (1995) in freshwater turtle and Sivakami *et al.*, (1994) in *Mystus vittatus.* These observations are in line with the present study.

It does not, however, mean that all substances are simultaneously utilized. If the principal and immediate energy source is depleted, then the other sources of energy exhibit a proportional depletion as it is well known that the metabolism of these substances is inter linked through the metabolic pathway, the Tricarboxylic cycle.

References

APHA, 1989. *Standard Methods for the Examination of Water and Wastewater,* 17th edition. American Public Health Association, Washington, DC.

Brett, J.R., 1970. Temperature in animals: Fishes. In: *Marine Ecology, Vol. 1: Environmental Factors,* (Ed.) O. Kinne. Wiley Inter Science, London, 38–57 pp.

Brett, J.R., 1979. Environmental factors and growth. In: *Fish Physiology, Vol. 8: Bioenergetics and Growth,* (Eds.) Hoar, W.S., Randall, D.J and J.R. Brett. Academic Press, London, 599–675 pp.

Folch, J., Less, M. and Stoars-Stanely, A.H., 1975. A simple method for determination of total lipid from animal tissues. *J. Biol. Chem.,* 226: 497–508.

Hashim, R., Chong, A. and Ali, A., 1999. The effect of feeding rate on water quality and growth of red *Tilapia* fry cultured in portable canvas tanks. *J. Biosci.,* 10: 39–46.

Jayachandran, K.P.S. and Chockalingam, S., 1986. Effects of tannery effluent on the respiratory parameters and biochemical constituents in the air breathing fish, *Channa punctatus. Comp. Physiol. Ecol.,* 12: 197–200.

Kinne, O., 1960. Growth, food intake and food conversion in an europlastic fish exposed to different temperatures and salinities. *Physiol. Zool.,* 33: 288–317.

Kinne, O., 1970. Temperature in animals: Invertebrates. In: *Marine Ecology, Vol. 1: Environmental Factors.* Wiley Inter Science, London, 290–350 pp.

Kinne, O., 1971. *Marine Ecology, Vol. 2: Environmental Factors.* Wiley Inter Science, New York, 683–1244 pp.

Leena, S.N., 2001. A comparison of the growth of the finglerlings of the grass carp *Ctenopharyngodon idella* using an animal feed and a plant feed. *M.Sc. Dissertation,* University of Kerala, India.

Lowry, O.H., Rosebrough, N.J., Farr, A.L. and Randall, R.J., 1951. Protein measurement with folin phenol reagent. *J. Biol. Chem.,* 195: 265–275.

Mercybai, V., 1970. Studies on food requirements and rate of metaolism of a tropical fish. *Ph.D. Thesis,* Madurai Kamaraj University, Tamil Nadu, India.

Naik, M.K., Reddy, H.R.V. and Annappaswamy, T.S., 2000. Influence of diet composition, starvation and feeding frequency on gastric evacuation rates in catla, *Catla catla* (Hamilton) fingerlings. *Ind. J. Anim. Sci.,* 70: 1090–1093.

Premkishore, G. and Chandran, M.R., 1995. Effect of starvation on the tissue chemistry of the freshwater turtle, *M. trijuga trijuga* of Tamil Nadu, India. *Zoos Print,* 10: 44–45.

Rani, S.N., 2001. On the use of aquatic weeds (*Hydrilla* and *Pistia*) as feed for the growth of the grass carp, *Ctenopharyngodon idella. M.Sc. Dissertation,* University of Kerala, India.

Roe, J.H., 1955. The determination of sugar in blood and spinal fluid with anthrone reagent. *J. Biol. Chem.,* 212: 335–343.

Ritakumari, S.D. and Nair, B.N., 1979. Effect of feeding rate on the growth, conversion efficiency and chemical composition of *Noemacheilus triangularies* (Day) and *Lepidocephalus thermalis* (Cuv. and Val.). *J. Anim. Morphol. Physiol.,* 26: 296–306.

Sivakami, R., Premkishore, G. and Chandran, M.R., 1994. Sublethal effects of chromium on the feeding energetics and growth in the freshwater cat fish, *Mystus vittatus. J. Freshwater Biol.,* 6: 165–175.

Sivakami, R., Premkishore, G. and Chandran, M.R., 1995. Sublethal effects of mercury on the feeding energetics and body composition in the freshwater cat fish, *Mystus vittatus. J. Aqua. Trop.,* 10: 109–117.

Windell, J.T., 1966. Rate of digestion in the blue gill sunfish. Invest. *Indiana Lakes and Streams,* 7: 185–214.

Chapter 15

Comparison of Energy Interaction Parameters for the Complexation of Pr(III) with 1,10-Phenanthroline and 2,2'-Bipyridyl in Different Organic Solvents using 4f-4f Transition Spectra as Probe: An Absorption Spectral Study

*N. Yaiphabaa[1], Th. David Singh[2], Ch. Sumitra[2], M. Indira Devi[2], N. Rajmuhon Singh[1]**

[1]Department of Chemistry, Manipur University, Imphal – 795 003, India

[2]Department of Chemistry, Nagaland University, Mokokchung – 798 601, India

ABSTRACT

Lanthanide co-ordination chemistry in solution is of great importance with the increase use of lanthanide as probes in the exploration of the structural functions of biomolecular reactions. Comparative absorption and spectral analysis of 4f-4f transitions of the Pr(III) with 1,10 Phenanthroline (phen) and 2,2'-Bipyridyl (bpy) has been explored in aqueous and different aquated organic solvents, like DMF, CH_3CN, CH_3OH, Dioxane and their equimolar binary mixtures. The variation in energy parameters like Slator Condon (F_k) Racah (E^k) and Lande Spin orbit coupling constant (ξ_{4f}), Nephelauxetic ratio (β), bonding parameter ($b^{½}$) and per cent

* E-mail: rajmuhon@yahoo.co.in

covalency (δ) are discussed to explore the mode of interaction between Pr(III) and 1,10-Phenanthroline (phen) and 2,2′-Bipyridyl (bpy). The marginal decrease in the values of F_k, E^k, ξ_{4f} are correlated with the increase in the value of β. The difference in the energy parameters with respect to donor atoms and solvents, reveal that the chemical environment around the lanthanide ion has great impact on *f–f* transition and any change in the environment results in modifications of the spectra.

Keywords: *1,10-Phenanthroline, 2,2′-Bipyridyl, Absorption spectra, Nephelauxetic effect, Pseudohypersensitive transition.*

Introduction

The trivalent lanthanides Ln(III) constitute a series of hard Lewis acids with similar chemical behaviors and their complexes with anionic ligands containing oxygen as donar atoms have been the most studied (Sinha, 1966) and their co-ordination chemistry in solution is of great importance with the increase use of lanthanide as probes in the exploration of the structural functions of biomolecular reactions (Evans, 1990). The hard metal acceptors like lanthanide (III) prefer oxygen and nitrogen donor ligands for stable complexes with predominantly ionic character although the affinity for oxygen donar ligand is more. Shah and Shah (2001) studied comparative 4f–4f transition spectra of Pr(III) with lysozyme by using energy interaction parameters to explain the behaviour of binding between them. Studies on Ln(III) complexes derived from nitrogen donor ligands (Khan *et al.*, 2000) have shown that the Ln-N interaction is stronger than that previously believed. It has been found that the number of coordinated ligands to the lanthanides depends upon the ability of the anion to compete for the coordination sites and there relative size of the ligands.

The present work discusses the quantitative spectral analysis of energy interaction parameters of Pr(III) complexes with 1,10-phenanthroline and 2,2′-bipyridyl in MeOH, CH_3CN, DMF, Dioxane and their equimolar binary mixtures at 298 K. The ligand mediated pseudohypersensitive transitions of Pr(III), $^3H_4 \rightarrow {}^3P_0$, $^3H_4 \rightarrow {}^3P_1$, $^3H_4 \rightarrow {}^3P_2$ and $^3H_4 \rightarrow {}^1D_2$ which have been found to show substantial sensitivity with minor changes in the immediate coordination environment around it even in the presence of structurally related ligand have been utilised to calculate the magnitude and variation of energy interaction parameters *viz.*, Slater Condon (F_k), Lande spin orbit coupling (ξ_{4f}), Nephelauxetic ratio (β), bonding ($b^{1/2}$), Racah (E^k) and covalency (δ) parameters to discuss the interaction of Pr (III) with structurally related phen and bpy.

Materials and Methods

1,10-Phenanthroline and 2,2′-bipyridyl of analytical grade from Qualigens and Praseodymium(III) nitrate hexahydrate of 99.9 per cent purity from M/s Indian Rare Earths Ltd. are used for synthesis and spectralanalysis. The solvents used are MeOH, CH_3CN, DMF and Dioxane which are of A/R grade from E. Merk. All the spectra are recorded on a temperature controlled Perkin Elmer Lambda-35 UV-Visible Spectrophotometer in the range 350–910 nm in the concentration of 0.01 M. The temperature of all the observations is maintained by using water circulating HAAKE DC 10 thermostat.

The Nephelauxetic effect, a measure of covalency (Peacock, 1964; Jorgensen and Judd, 1964; Henrie *et al.*, 1976) has been interpreted in terms of Slater-Condon and Racah parameters (inter electronic repulsion parameters) as well as by the ratio of the free ion and complex ion (Sinha and Schmidtke, 1965; Sinha, 1966; Sinha *et al.*, 1972; Condon and Shortley, 1963).

$$\beta = \frac{F_K^c}{F_K^f} \text{ or } \frac{E_c^K}{E_f^K} \tag{1}$$

where,

F_k (K = 2,4,6) is the Slater-Condon parameter and E^k is the Racah parameters for complex and free ions respectively. The bonding parameter ($b^{1/2}$) is inter-related to Nephelauxetic effect as:

$$\beta^{1/2} = \left[\frac{1-\beta}{2}\right]^{1/2} \tag{2}$$

The electrostatic term E_0 is expressed in terms of the product of Slater radial integral known as Slater-Condon parameter, F_k and is given by

$$F_0 = \sum_{\kappa=0}^{\kappa=6} K^{\kappa} F_{\kappa} \tag{3}$$

The Slater-Condon parameters are also known as direct-integral and is decreasing function of K as given by the relation:

$$F_1^K = \int_0^{\infty}\int_0^{\infty} \frac{r^K}{r_>^{K+1}} R_i^2(r_i) R_j^2 R(r_j) r_i^2 r_j^2 dr_i dr_j \tag{3}$$

where,

R is the 4f-radial wave function; $r_<$ and $r_>$ are the radii of near and more distant electrons; i and j are the ith and j[th] electrons under consideration. Condon and Shortley (1963) redefined P integrals in terms of reduced integral F_k related to each other and the relation is

$$F_K = \frac{F^K}{D_K} \tag{5}$$

Combining relation (4) and (5), the reduced Slater-Condon integral can be written as:

$$F_{\kappa} = \frac{1}{D_{\kappa}} \int_0^{\infty}\int_0^{\infty} r_<^{\kappa} r_>^{\kappa+1} R_i^2(r_i) R_j^2 R(r_j) T_i^2 r_j^2 dr_i dr_j \tag{6}$$

Here D_k is the denominator and F_k are coefficient of linear combination and represent the angular part of the interaction F_k is the expectation value of scalar product $[C_1^{(K)} C_2^{(K)}]$.

The Racah energy interaction parameter E^{Ki}'s are linear combinations of F_K given by

$E^1 = (70 \times F_2 + 231\ F_4 + 20.02\ F_6)/9$

$E^2 = (F_2 - 3\ F_4 + 7\ F_6)/9$

$E_3 = (5\ F_2 + 6\ F_4 - 9\ F_6)/3$ (7)

The energy, E_{so} arising from the most important magnetic interactions, which are spin orbit interactions may be written as

$$E_{so} = A_{so}\,\xi_{4f} \tag{8}$$

where,

A_{so} is the angular part of spin orbit interaction and ξ_{4f} is the radial integral and is known Lande's parameter by first order approximation the energy E_j of the jth level is given by Wong (1961; 1963) as:

$$E_j(F_K, \xi_{4f}) = E_{oj}(F_K^0, \xi_{4f}) + \frac{\partial E_j}{\partial F_K}\Delta F_K + \frac{\partial E_j}{\partial \xi_{4f}}\Delta \xi_{4f} \tag{9}$$

where,

E_{oj} is the zero order energy of the jth level. The value of F_k and ξ_{4f} are given by

$$F_k = F_k^{\,0} + \Delta F_k$$

$$\xi_{4f} = \xi^0_{\;4f} + \Delta\xi_{4fR} \tag{10}$$

The difference between the observed E_j value and the zero order values, ΔE_j is evaluated by

$$\Delta E_j = \sum_{k=2,4.6} \frac{\partial E_j}{\partial F_K}\Delta F_K + \frac{\partial E_j}{\partial \xi_{4f}}\Delta \xi_{4f} \tag{11}$$

By using the zero order energy and partial derivatives of Pr(III) ion given by Wong (1961; 1963), the above equation can be solved by least square technique and the value of F_2 and can be found out. From these the value of F_2 and are found out by using relation (10). The estimated values of F_4 and F_6 are calculated by the relations,

$$\frac{F_4}{F_2} = 0.1380 \text{ and } \frac{F_4}{F_2} = 0.0150 \tag{12}$$

The per cent covalency parameter (δ) representing the Nephelauxetic effect was given by the relation

$$\delta = \left(\frac{1-\beta}{\beta}\right) \times 100 \tag{13}$$

Results and Discussion

The shape, energy and oscillator strength of hypersensitive or pseudohypersensitive transitions (Karraker, 1967) can be correlated with coordination number and diagnostic of immediate coordination environment around lanthanides ions. In Figure 15.1, red shift is observed when Phen/bpy is added to Pr(III). The variation of the magnitude of energy interaction parameters like Slater Condon (F_k), Lande factor (ξ_{4f}), Racah energy (E^K), Nephelauxetic ratio (β), bonding ($b^{1/2}$) and percentage covalency (δ) for Pr(III), Pr(III): bpy and Pr(III) : phen in DMF, CH_3CN, CH_3OH, Dioxane and their equimolar binary mixtures is shown in Table 15.1. The marginal decrease in the values of F_K, ξ_{4f} and E_K indicating complexation led to the increase in the values of Nephelauxetic ratio. Jorgensen and Ryan (1966) suggested that shortening in the metal ligand distance occurs with decrease in the co-ordination number. To interpret the correlation and analysis of the relationship between Nephelauxetic effect and geometry, energy parameters have been derived and evaluated as reported earlier (David *et al.*, 2003; 2004; 2005; 2006) for complex compound using the angular overlap model, the value of 'n' is proportional to the Nephelauxetic effect as

$$n = \left[\frac{1-\beta^{1/2}}{\beta^{1/2}}\right] \tag{14}$$

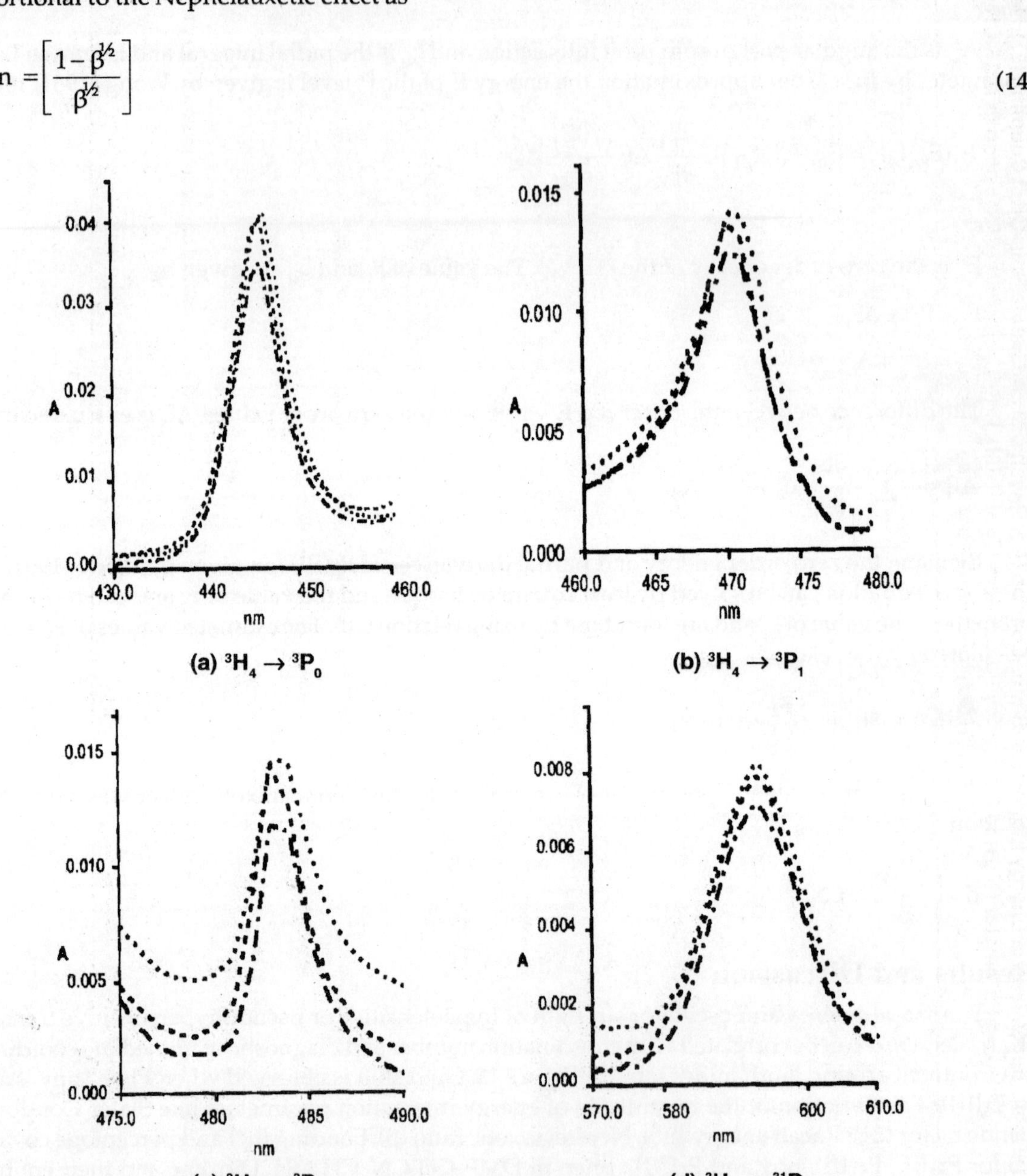

Figure 15.1: Comparative Absorption Spectra of Pr(III) –.–.–.–, Pr(III):bpy - - - - - - and Pr(III) : phen . . . . in DMF

Table 15.1: Computed Values of Energy Interaction Slator Condon (F_k), Racah (E^k) and Lande Spin Orbit Coupling (ξ_{4f}), Nephelauxetic Ratio (β) and Bonding Parameter ($b^{½}$) and Covalency Parameters (δ) of Pr(III), Pr(III) : bpy and Pr(III) : phen in DMF, CH_3CN, CHOH, Dioxane and their Equimolar Binary Mixtures

Sl.No.	System	F_2	F_4	F_6	ξ_{4f}	E^1	E^2	E^3	β	$b^{½}$	0
1.	Solvent–MeOH										
	Pr(III)	308.8467	42.6363	4.6636	719.6274	3506.8470	23.7314	586.0264	0.9444	0.1668	5.8904
	Pr(III) + Bpy	308.8586	42.6379	4.6638	719.8563	3506.9810	23.7323	586.0488	0.9445	0.1665	5.8709
	Pr(III) + Phen	308.8850	42.5638	4.6557	721.0336	3500.8850	23.6911	585.0302	0.9446	0.1665	5.8686
2.	Solvent–ACN										
	Pr(III)	308.3532	42.5682	4.6561	723.8837	3501.2430	23.6935	585.0900	0.9466	0.1635	5.6453
	Pr(III) + Bpy	307.2411	42.4146	4.6393	721.9964	3488.6150	23.6081	582.9798	0.9436	0.1679	5.9720
	Pr(III) + Phen	306.0091	42.2446	4.6207	722.3418	3474.6260	23.5134	580.6420	0.9421	0.1702	6. 1482
3.	Solvent–Dioxane										
	Pr(III)	308.10 13	42.5334	4.6523	722.6680	3498.3830	23.6742	584.6120	0.9454	0.1653	5.7794
	Pr(III) + Bpy	307.9974	42.5190	4.6508	722.0624	3497.2030	23.6662	584.4149	0.9448	0.1661	5.8429
	Pr(III) + Phen	306.9069	42.3685	4.6343	720.7680	3484.8210	23.5824	582.3456	0.9432	0.1698	6.1216
4.	Solvent–DMF										
	Pr(III)	308.1064	42.5341	4.6524	719.3425	3498.4400	23.6746	584.6216	0.9431	0.1687	6.0339
	Pr(III) + Bpy	308.2119	42.5487	4.6540	718.9176	3499.6380	23.6827	584.8218	0.9430	0.1689	6.0493
	Pr(III) + Phen	308.0735	42.5296	4.6519	719.2725	3498.0670	23.6720	584.5593	0.9430	0.1688	6.0447
5.	Solvent–MeOH + ACN										
	Pr(III)	308.2608	42.5554	4.6547	720.0623	3500.1940	23.6864	584.9146	0.9438	0.1676	5.9532
	Pr(III) + Bpy	308.2599	42.5553	4.6547	720.9904	3500.1840	23.6863	584.9129	0.9444	0.1667	5.8821
	Pr(III) + Phen	307.0358	42.5863	4.6362	721.2242	3486.2840	23.5923	582.5903	0.9428	0.1691	6.0652

Contd...

Table 15.1–Contd...

Sl.No.	System	F_2	F_4	F_6	ξ_{4f}	E^1	E^2	E^3	β	$b^{1/2}$	δ
6.	Solvent–MeOH + Dioxane										
	Pr(III)	308.0788	42.5303	4.650	721.4075	3498.1270	23.6724	584.5693	0.9445	0.1666	5.8798
	Pr(III) + Bpy	308.2658	42.5561	4.6548	721.3876	3500.2500	23.6868	584.9240	0.947	0.1662	5.8506
	Pr(III) + Phen	307.2837	42.4205	4.6400	721.3444	3489.0990	23.6113	583.0605	0.9433	0.1684	6.0152
7.	Solven–MeOH + DMF										
	Pr(III)	308.3668	42.5700	4.6563	719.6374	3501	23.6949	585.1157	0.9437	0.1678	5.9684
	Pr(III) + Bpy	308.3394	42.5663	4.6559	719.3452	3501.0860	23.6925	585.0637	0.9434	0.1682	5.9954
	Pr(III) + Phen	308.2668	42.5562	4.6548	719.0844	3500.2610	23.6869	584.9260	0.9432	0.1686	6.0274
8.	Solvent–DMF + CAN										
	Pr(III)	308.1657	42.5423	4.6533	719.5563	3499.1140	23.6791	584.7342	0.9433	0.1683	6.0077
	Pr(III) + Bpy	308.1514	42.5403	4.6531	719.3057	3498.9510	23.6780	584.7070	0.9431	0.1686	6.0294
	Pr(III) + Phen	307.8109	42.4933	4.6479	720.0342	3495.0850	23.6518	584.0609	0.9431	0.1686	6.0293
9.	Solvent–DMF + Dioxane										
	Pr(III)	308.2380	42.5523	4.6544	720.0548	3499.9350	23.6847	584.8713	0.9438	0.1677	5.9575
	Pr(III) + Bpy	308.2832	42.5585	4.6551	720.6835	35000.4480	23.6881	584.9572	0.9443	0.1669	5.9018
	Pr(III) + Phen	307.9737	42.5158	4.6504	721.0008	3496.9340	23.6644	584.3699	0.9440	0.1673	5.9282
10	Solvent–DX + CAN										
	Pr(III)	308.1142	42.5352	4.6525	723.3610	3498.5290	23.6752	584.6364	0.9459	0.1645	5.7243
	Pr(III) + Bpy	305.8562	42.2235	4.6184	719.5063	3472.8910	23.5017	580.3520	0.9399	0.1733	6.3926
	Pr(III) + Phen	305.8562	42.2235	4.6184	719.5063	3472.8910	23.5017	580.3520	0.9399	0.1733	6.3926

Table 15.2: Computed and Observed Values of Energies (cm^{-1}) and RMS Values for Pr(III), Pr(III) : bpy and Pr(III) : phen in DMF, CH_3CN, CH_3OH, Dioxane and their Equimolar Binary Mixtures

Sl.No.	System	$^3H_4$ Eobs	$^3P_2$ Ecal	$^3H_4$ Eobs	$^3P_1$ Ecal	$^3H_4$ Eobs	$^3P_0$ Ecal	$^3H_4$ Eobs	$^1D_2$ Ecal	R.M.S.
1.	Solvent–MeOH									
	Pr(III)	22490.10	22425.21	21311.51	21216.33	20737.42	20661.21	16893.89	17117.24	13 1.30
	Pr(III) + Bpy	22490.61	22427.16	21313.33	21218.07	20736.99	20662.48	16898.46	17118.45	129.47
	Pr(III) + Phen	22464.34	22396.81	21283.84	21185.13	20693.22	20627.05	16885.05	17097.19	126.18
2.	Solvent–ACN									
	Pr(III)	22519.99	22413.27	21311.96	21198.67	20681.67	20634.69	16890.75	17106.92	135.24
	Pr(III) + Bpy	22439.13	22328.65	21240.89	21113.24	20613.03	20553.06	16801.64	17050.31	136.25
	Pr(III) + Phen	22357.36	22247.14	21129.14	21028.28	20525.87	20467.26	16767.83	16994.68	138.94
3.	Solvent–Dioxane									
	Pr(III)	22508.33	22390.14	21316.51	21176.19	20669.70	20614.70	16837.00	17091.81	137.38
	Pr(III) + Bpy	22510.36	22380.07	21312.42	21166.50	20665.85	20606.26	16812.38	17085.27	135.51
	Pr(III) + Phen	22427.05	22299.89	21228.27	21084.94	20596.05	20527.27	16748.17	17031.38	136.89
4.	Solvent–DMF									
	Pr(III)	22432.09	22373.76	21255.34	21163.32	20686.80	20608.72	16862.27	17082.38	128.85
	Pr(III) + Bpy	22435.61	22378.75	21264.38	21169.03	20696.22	20615.32	16859.99	17085.99	130.75
	Pr(III) + Phen	22429.07	22370.97	21256.24	21160.54	20684.66	20606.12	16748.17	17031.38	130.96
5.	Solvent–MeOH + ACN									
	Pr(III)	22470.90	22387.81	21290.19	21177.01	20694.08	20620.93	16849.20	17091.57	137.75
	Pr(III) + Bpy	22473.43	22392.42	21297.44	21180.63	20688.94	20622.63	16860.85	17094.22	135.60
	Pr(III) + Phen	22416.50	22310.89	21244.50	21095.79	20601.57	20537.18	16770.64	17038.63	136.25

Contd...

Table 15.2–Contd...

Sl.No.	System	$^3H_4$ Eobs	$^3P_2$ Ecal	$^3H_4$ Eobs	$^3P_1$ Ecal	$^3H_4$ Eobs	$^3P_0$ Ecal	$^3H_4$ Eobs	$^1D_2$ Ecal	R.M.S.
6.	Solvent–MeOH + Dioxane									
	Pr(III)	22489.60	22382.28	21305.61	21169.60	20674.40	20610.72	16828.50	17087.11	138.86
	Pr(III) + Bpy	22492.13	22394.81	21296.99	21182.62	20688.09	20623.80	16857.72	17095.65	132.30
	Pr(III) + Phen	22454.25	22328.24	21257.60	21113.63	20619.41	20554.80	16773.74	17050.38	131.25
7.	Solven–MeOH + DMF									
	Pr(III)	22451.22	22392.83	21275.69	21182.74	20703.50	20627.55	16877.64	17095.21	127.63
	Pr(III) + Bpy	22450.22	22389.51	21275.69	21179.66	20703.08	20625.08	16868.81	17093.10	13 1.62
	Pr(III) + Phen	22440.14	22383.29	21268345	21173.54	'20699.22	20619.48	16865.11	17089.00	13 1.08
8.	Solvent–DMF + ACN									
	Pr(III)	22434.60	22378.84	21258.96	21168.33	20689.80	20613.29	16870.52	17085.72	125.99
	Pr(III) + Bpy	22435.11	22376.61	21256.70	21166.33	20690.23	20611.81	16864.83	17084.34	128.38
	Pr(III) + Phen	22420.52	22357.27	21243.15	21145.37	20662.44	20589.30	1685 1.19	17070.80	129.56
9.	Solvent–DMF + Dioxane									
	Pr(III)	22454.75	22386.23	21277.05	21175.38	20692.37	20619.31	16865.11	17090.50	133.39
	Pr(III) + Bpy	22458.29	22392.45	21277.50	21181.05	20692.37	20623.69	16881.06	17064.40	132.07
	Pr(III) + Phen	22444.67	22373.13	21266.64	21160.62	20668.84	20602.57	16861.13	17081.1 0	13 1.47
10.	Solvent–Dioxane + ACN									
	Pr(III)	22520.49	22394.49	21293.81	21179.84	20668.42	20616.92	16859.71	17094.41	135.14
	Pr(III) + Bpy	22360.53	22278.50	21148.68	21075.32	20575.13	20483.72	16673.92	16993.63	133.08
	Pr(III) + Phen	22354.36	22222.56	21125.57	21006.31	20530.09	20451.14	16695.61	16979.41	132.03

The Nephelauxetic effect increases with the co-ordination number decreases. The Ln - N distance shortens in spite of the addition nature of β and decrease in the number of the co-ordinating ligand. The variation in the value of E^k (K = 2, 4, 6); corresponds to that in the value of F_k, since they are inter-related. Misra *et al.* (1979; 1981) observed a general decrease in the values of F_k, E^k and ξ_{4f} parameters as compared to the corresponding parameters of the free ion. The values of nephelauxetic effect (β) in all the system is less then unity and the values of bonding parameters ($b^{1/2}$) are positive which indicates covalent bonding. The comparative absorption spectra of pseudohypersensitive transitions of Pr(Ill) ion, Pr(III): bpy (in 1: 1 molar ratio) and Pr(III): phen (in 1 : 1 molar ratio) in different organic solvents are given in Figure 15.1a to Figure 15.1d. The addition of bpy/phen in Pr(III) resulted in a marginal red shift of 4f-4f transition bands. The redshift is more in case of phen indicating a better ligand than bpy. The weak Ln-N interaction is enhanced by the presence of extra pi electrons as in the case of phen than in bpy, increasing the ligand character of the donor N-atom. The redshift is maximum in case when solvent is DMF or one of the solvent is DMF. Table 15.1 gives the computed values of Slator-Condon (F_k), Racah (E^k) and Lande (ξ_{4f}), Nephelauxetic parameter (β), bonding parameter ($b^{1/2}$) and covalency (δ) of Pr(III), Pr(III): bpy and Pr(III): phen in different organic solvents. Table 15.2 gives the computed and observed values of energies for the various transition bands. Further work on the evaluation of intensity parameters is going on.

References

Condon, S.U. and Shortley, G.H., 1963. *The Theory of Atomic Spectra.* University Press, Cambridge.

David Singh, Th., Singh, Mohondas N., Yaiphaba, N., Debecca Devi, H., Sumitra, Ch., Kriyananda, T., Indira Devi, M. and Rajmuhon Singh, N., 2003. *Chem. and Environ. Res.*, 12(3&4): 295.

David, Th., Sumitra, Ch., Bag, G.C., Indira Devi, M. and Rajmuhon Singh, N., 2006. *Spectrchimica Acta* (A), 63: 154.

David, Th., Sumitra, Ch., Yaiphaba, N., Debecca Devi, R., Indira Devi, M., Rajmuhon Singh, N., 2005. *Spectrchimica Acta,* (A), 61: 1219.

Debecca Devi, R., David Singh, Th., Yaiphaba, N., Sumitra, Ch., Indira Devi, M. and Rajmuhon Singh, N., 2004. *Asian Journal of Chemistry*, 16(1): 412.

Evan, C.H., 1990. *Biochemistry of Lanthanides.* Plenum Press, New York, Chapter 4.

Henrie, D.E., Fellow, R.L. and Choppin, G.R., 1976. *Coord. Chem. Rev.*, 18: 429.

Jorgensen, C.K. and Judd, B.R., 1964. *Molec. Phys.*, 8: 281.

Jorgensen, C.K. and Ryan, L., 1966. *J. Phys. Chem.*, 70: 2845.

Karrakar, D.G., 1967. *Inorg. Chem.*, 6: 1863.

Khan *et al.*, 2000. *Ind. Journal of Chemistry*, 39A: 1286.

Langhlin, R.D. and Conway, J.G., 1963. *J. Phys. Chem.*, 38: 1037.

Misra, S.N., Joshi, G.K. and Vaaishnav, P.P., 1979. *Indian J. Pure. App. Phys.*, 16: 553.

Misra, S.N., Megh Singh Joshi and G.K., 1981. *J. Inorg. Nucl. Chem.*, 43: 207.

Peacock, R.D., 1964. *Chem. Phys.*, 8: 281.

Shah, K.J. and Shah, M.K., 2001. *Bull. Pure Appl. Sci.*, C20(2): 81.

Sinha, S.P and Schmidtke, R.H., 1965. *Molec. Phys.*, 38: 2190.

Sinha, S.P. 1966. *Spectrochim. Acta.*, 22: 57.

Sinha, S.P., 1966. *Complexes of Rare Earths*. Pergamon Press, New York.

Sinha, S.P., Mehta, P.C. and Surana, S.S.L., 1972. *Molec. Phys.*, 23: 807.

Wong, E.Y., 1961, 1963. *J. Chem. Phys.*, 35: 544; 38: 976.

Chapter 16

Sequential Procedures for the Class of Exponential Family of Distributions

C.P.S. Hungund
Department of Statistics, Gulbarga University, Gulbarga

ABSTRACT

The present paper deals with the problem of statistical tests of hypothesis. Here we have obtained a method of obtaining an explicit formula for the Average Sample Number (ASN) for testing the variance of the class of exponential family of distributions. The result has been illustrated in detail for testing the variance of Poisson, Bionomial and Geometric distributions as they are particular cases of exponential family of distributions.

Keywords: *ASN function, Bionomial distribution, Geometric distribution, Poisson distribution.*

Introduction

The closed sequential decision procedures are obtained by Paulson (1964) for a number of decision problems including testing of hypotheses regarding mean, variance and ratio of variances of a normal distribution. The theory has been extended to a general class of distributions by Halakatti (1994). The authors have discussed the testing of hypotheses regarding the mean of a normal distribution and the procedure has been illustrated through some examples. In this paper, the method and formula for ASN function for the case of single parameter exponential family has been derived and also we have discussed the problem of testing the hypothesis regarding the variance of exponential family of distributions. Using the procedures suggested by Paulson (1964) the procedure of obtaining an ASN function for the class of exponential family of distributions has been discussed in the following section.

ASN Function for the Class of Exponential Family of Distribution

Let $X_1, X_2, \ldots$ be a sequence of independent random variables from common distribution whose p.d.f. is

$$f_x(\theta) = C(\theta) g(x) \exp[T(x) Q(\theta)],$$

$$X \in R$$

$$\theta \in \Theta \tag{2.1}$$

where,

$Q(\theta)$ is monotonically increasing in θ and $C(\theta) > 0$. Assume that $E[T(X)] = \theta$ and the closed sequential procedure to test $H_0 : \theta = \theta_0$ against $H_1 = \theta : \theta_0 / w$ is such that

P (rejected $H_0/\theta = \theta_0) \leq \alpha$ and

P (accepted $H_0/\theta = \theta_0/w) \leq \beta$

Using the procedure suggested by Paulson (1964) and Halakatti (1994) the procedure is as follows. Define

$$Z_N(1) = \ln\left[\frac{f_n\left(x; \frac{\theta_0}{w}\right)}{f_n\left(x; \frac{\theta_0 \lambda}{w}\right)}\right] \text{ and} \tag{2.2}$$

$$Z_N(2) = \ln\left[\frac{f_n\left(x; \frac{\theta_0}{\lambda}\right)}{f_n(x; \theta_0)}\right]$$

where,

$\lambda > 1$. Take observations one at a time and stop the experiment as soon as either

1. $Z_N(1) \leq \ln B$ (2.4)

or

2. $Z_N(2) \geq \ln A$ *happens, viz.,* (2.5)

1. Accept H_0 if

$$\sum_{i=1}^{n} T(X_i) \geq \frac{\left\{\ln\left[\frac{C\left(\frac{\theta_0}{w}\right)}{C\left(\frac{\theta_0 \lambda}{w}\right)}\right]\right\} - \ln B}{Q\left(\frac{\theta_0 \lambda}{w}\right) - Q\left(\frac{\theta_0}{w}\right)} \tag{2.6}$$

or

2. Reject H_0 if

$$\sum_{i=1}^{n} T(X_i) \le \frac{\left\{ \ln\left[\frac{C\left(\frac{\lambda_0}{\lambda}\right)}{C(\theta_0)}\right] \right\} - \ln A}{Q(\theta_0) - Q\left(\frac{\theta_0}{\lambda}\right)} \tag{2.7}$$

for each $\lambda > 1$, with $1 < \lambda < w$. We obtain a closed sequential procedure with obvious modifications the approximate formula for ASN function reduces to

$$E_\theta(N) = \frac{L(\theta)\ln B}{E_\theta[Z(1)]} + \frac{[(1-L(\theta)]\ln A}{E_\theta[Z(2)]} \tag{2.8}$$

where,

$$E_\theta[(Z(1)] = \ln\left[\frac{C\left(\frac{\theta_0}{w}\right)}{C\left(\frac{\theta_0\lambda}{w}\right)}\right] + \left[Q\left(\frac{\theta_0}{w}\right) - Q\left(\frac{\theta_0\lambda}{w}\right)\right]0 \tag{2.9}$$

$$E_\theta[(Z(2)] = \ln\left[\frac{C\left(\frac{\theta_0}{\lambda}\right)}{C(\theta_0)}\right] + \left[Q\left(\frac{\theta_0}{\lambda}\right) - Q(\theta_0)\right]\theta \tag{2.10}$$

On the similar lines one can obtain a closed sequential procedure for Poisson distribution,binomial distribution and negative binomial distribution as they are particular cases of (Eq. 2.1).

Particular Cases

Poisson Distribution

Let $X_1, X_2, \ldots, X_n$ be a sequence or independent Poisson random variables with mean θ variance θ having p.d.f $f_n(x)$ and the joint density

$$f_n(x;\theta) = \frac{\exp(-n\theta)\theta^{\sum_{i=1}^{n} x_i}}{\prod_{i=1}^{n} x_i!} \tag{3.1}$$

From the above *i.e.* ASN function for the class fo exponential family of distribution section we can obtain the closed sequential procedure to test $H_0: \theta = \theta_0$ against $H_1: \theta = \theta_0/w$ such that

P (rejected $H_0 | \theta = \theta_0) \leq \alpha$ and P (accepted $H_0 | \theta = \theta_0/w) \leq \beta$

By substituting $C(\theta) = e^{-\theta}$ and $Q(\theta) = \ln\theta$.

Hence from (Eq. 2.8) the approximate formula for ASN function for this procedure reduces to

$$E_\theta(N) = \frac{L(\theta)\ln B}{\left(\frac{n\theta_0(\lambda-1)}{w}\right) - (n\theta\ln\lambda)} + \frac{[1-L(\theta)]\ln A}{\left(\frac{n\theta_0(\lambda-1)}{\lambda}\right) - (n\theta\ln\lambda)} \tag{3.2}$$

In particular, when $\theta = \theta_0$

$$E_\theta(N) = \frac{(1-\alpha)\ln B}{\left(\frac{n\theta_0(\lambda-1)}{w}\right) - (n\theta\ln\lambda)} + \frac{\alpha\ln A}{\left(\frac{n\theta_0(\lambda-1)}{\lambda}\right) - (n\theta_0\ln\lambda)} \tag{3.3}$$

when $\theta = \theta_0/w$

$$E_{\theta_{\frac{0}{w}}}(N) = \frac{\beta\ln B}{\left(\frac{n\theta_0(\lambda-1)}{w}\right) - \left(\frac{n\theta_0\ln\lambda}{w}\right)} + \frac{(1-\beta)\ln A}{\left(\frac{n\theta_0(\lambda-1)}{\lambda}\right) - \left(\frac{n\theta_0\ln\lambda}{w}\right)} \tag{3.4}$$

when, $w = \lambda$ the procedure reduces to Wald's SPRT

$$E_{\theta_{\frac{0}{\lambda}}}(N) = \frac{\beta\ln B + (1-\beta)\ln A}{\left(\frac{n\theta_0(\lambda-1)}{\lambda}\right) - \left(\frac{n\theta_0\ln\lambda}{\lambda}\right)} \tag{3.5}$$

The ratio of E(N) is given by (Eq. 3.4) and (Eq. 3.5)

$$R_{E_{\theta_{\frac{0}{\lambda}}}}(N) = \frac{w\beta\ln B\left\{\left(\frac{\lambda-1}{\lambda}\right) - \left(\frac{\ln\lambda}{w}\right)\right\} + (1-\beta)\ln A\{[(\lambda-1) - \ln\lambda]\}}{\{\lambda[\beta\ln B + (1-\beta)\ln A]\}\left\{\left[\left(\frac{\lambda-1}{\lambda}\right) - \left(\frac{\ln\lambda}{w}\right)\right]\right\}} \tag{3.6}$$

and it can be easily seen that $R_E(N) \leq 1$ provided $\lambda < w$.

Binomial Distribution

Let $X_1, X_2, \ldots, X_n$ be a sequence of independent B $(n_1 . p)$ random variables having p.d.f. f(x) and the joint density

$$f_n(x) = \prod_{i=1}^{n} \binom{n}{\substack{1 \\ xi}} p^{\sum_{i=1}^{n} x_i} q^{nn_1 - \sum_{i=1}^{n} x_i} \tag{3.7}$$

from section ASN function above we can obtain the closed sequential procedure to test $H_0 : p = p_0$ against $H_1 : P = P_0/w$ such that.

P (rejected $H_0 | p = p_0) \le \alpha$ and P (accepted $H_0 | p = p_0/w) \le \beta$ by substituting $C(p) = e^{nn_1 \ln q}$ and Q(p) = ln(p/q). Hence from (Eq. 2.8) the approximate formula for ASN function for' this procedure reduces to

$$E_p(N) = \frac{L(p)\ln B}{n^2 p \ln\left\{\frac{(w-p_0\lambda)}{\lambda(w-p_0)}\right\} + nn_1 \ln\left\{\frac{(w-p_0)}{(w-p_0\lambda)}\right\}} + \frac{[(1-L(p)]\ln A}{n^2 p \ln\left\{\frac{(l-p_0)}{(\lambda-p_0)}\right\} + nn_1 \ln\left\{\frac{(\lambda-p_0)}{\lambda(1-p_0)}\right\}} \quad (3.8)$$

when p = p_0

$$E_p(N) = \frac{(l-\alpha)\ln B}{n^2 p \ln\left\{\frac{(w-p_0\lambda)}{\lambda(w-p_0)}\right\} + nn_1 \ln\left\{\frac{(w-p_0)}{(w-p_0\lambda)}\right\}} + \frac{\alpha \ln A}{n^2 p \ln\left\{\frac{(l-p_0)}{(\lambda-p_0)}\right\} + nn_1 \ln\left\{\frac{(\lambda-p_0)}{\lambda(1-p_0)}\right\}} \quad (3.9)$$

when, p = p_o/w

$$E_{\frac{p_0}{w}}(N) = \frac{\beta \ln B}{n^2 \frac{p_0}{w} \ln\left\{\frac{(w-p_0\lambda)}{\lambda(w-p_0)}\right\} + nn_1 \ln\left\{\frac{(w-p_0)}{(w-p_0\lambda)}\right\}} + \frac{(1-\beta)\ln A}{n^2 \frac{p_0}{w} \ln\left\{\frac{(l-p_0)}{(\lambda-p_0)}\right\} + nn_1 \ln\left\{\frac{(\lambda-p_0)}{\lambda(1-p_0)}\right\}} \quad (3.10)$$

when, W = λ

$$E_{\frac{p_0}{w}}(N) = \frac{\beta \ln B + (1-\beta)\ln A}{n^2 \frac{p_0}{\lambda} \ln\left\{\frac{(1-p_0)}{(\lambda-p_0)}\right\} + nn_1 \ln\left\{\frac{(\lambda-p_0)}{\lambda(1-p_0)}\right\}} \quad (3.11)$$

The ratio ofE(N) is given by (Eq. 3.10) and (Eq. 3.11)

$$R_E(N) = \frac{\left(\begin{array}{l} \beta \ln B \left(n^2 \frac{p_0}{w} \ln\left\{\frac{(1-p_0)}{(\lambda-p_0)}\right\} + nn_1 \ln\left\{\frac{(\lambda-p_0)}{\lambda(1-p_0)}\right\}\right) + \\ (1-\beta)\ln A \left(n^2 \frac{p_0}{w} \ln\left\{\frac{(w-p_0\lambda)}{\lambda(w-p_0)}\right\} + nn_1 \ln\left\{\frac{(w-p_0)}{(w-p_0\lambda)}\right\}\right) \times \\ \left(n^2 \frac{p_0}{\lambda} \ln\left\{\frac{(1-p_0)}{(\lambda-p_0)}\right\} + nn_1 \ln\left\{\frac{(\lambda-p_0)}{\lambda(1-p_0)}\right\}\right) \end{array}\right)}{\left(\begin{array}{l} n^2 \frac{p_0}{w} \ln\left\{\frac{(w-p_0\lambda)}{\lambda(w-p_0)}\right\} + nn_1 \ln\left\{\frac{(w-p_0)}{(w-p_0\lambda)}\right\} \times n^2 \frac{p_0}{w} \ln\left\{\frac{(1-p_0\lambda)}{(\lambda-p_0)}\right\} + \\ nn_1 \ln\left\{\frac{(\lambda-p_0)}{\lambda(1-p_0)}\right\} \times [\beta \ln B + (1-\beta)\ln A] \end{array}\right)} \quad (3.11)$$

and it can be easily seen that $R_E(N) \le 1$ provided $\lambda < W$.

Geometric Distribution

Let $x_1, x_1, \ldots x_n$ be a sequence of independent geometric random variables having, p.d.f f(x) and joint density.

$$f_n(x) = \frac{\prod_{i=1}^{n} \Gamma(x_i + n)(\theta)^{\sum_{i=1}^{n} x_i}}{\prod_{i=1}^{n} x_i!(n!)(1+\theta)^{n+\sum_{i=1}^{n} x_i}} \tag{3.13}$$

From section (2) we can obtain the closed sequential procedure to test $H_0 : \theta = \theta_0$ against $H_1 : \theta = \theta_0/w$ such that

P (rejected $H_0 | p = p_0) \leq \alpha$ and P (accepted $H_0 | p = p_0/w) \leq \beta$ by substituting $C(\theta) = e^{-n\frac{1}{2}n(1+\theta)}$ and $Q(\theta) = \ln(\theta/1+\theta)$. Hence from (Eq. 2.8) the approximate formula for ASN function for this procedure reduces to

$$E_\theta(N) = \frac{L(\theta)\ln B}{n^2 p \ln\left\{\frac{(w+\theta_0\lambda)}{\lambda(w+\theta_0)}\right\} + n\ln\left\{\frac{(w+\theta_0\lambda)}{(w+\theta_0)}\right\}} + \frac{[(1-L(\theta)]\ln A}{n\theta\ln\left\{\frac{(1+\theta_0)}{(\lambda+\theta_0)}\right\} + n\ln\left\{\frac{\lambda(1+\theta_0)}{(\lambda+\theta_0)}\right\}} \tag{3.14}$$

In particular when $\theta = \theta_0$

$$E_{\theta_0}(N) = \frac{(l-\alpha)\ln B}{n_{\theta_0}\ln\left\{\frac{(w+\theta_0\lambda)}{\lambda(w+\theta_0)}\right\} + n\ln\left\{\frac{(w+\theta_0\lambda)}{(w+\theta_0)}\right\}} + \frac{\alpha\ln A}{n_{\theta_0}\ln\left\{\frac{(1+\theta_0)}{(\lambda+\theta_0)}\right\} + n\ln\left\{\frac{\lambda(1+\theta_0)}{\lambda+\theta_0)}\right\}} \tag{3.15}$$

when $\theta = \theta_0/w$

$$E_{\theta\frac{0}{w}}(N) = \frac{\beta\ln B}{n_{\theta\frac{0}{w}}\ln\left\{\frac{(w+\theta_0\lambda)}{\lambda(w+\theta_0)}\right\} + n\ln\left\{\frac{(w+\theta_0\lambda)}{(w+\theta_0)}\right\}} + \frac{(1-\beta)\ln A}{n_{\theta\frac{0}{w}}\ln\left\{\frac{(1+\theta_0)}{(\lambda+\theta_0)}\right\} + n\ln\left\{\frac{\lambda(1+\theta_0)}{(\lambda+\theta_0)}\right\}} \tag{3.16}$$

for $w = \lambda$

$$E_{\theta\frac{0}{w}}(N) = \frac{\beta\ln B + (1-\beta)\ln A}{n_{\theta\frac{0}{w}}\ln\left\{\frac{(1+\theta_0)}{(\lambda+\theta_0)}\right\} + n\ln\left\{\frac{\lambda(1+\theta_0)}{(\lambda+\theta_0)}\right\}} \tag{3.17}$$

The ratio of E(N) given by (Eq. 3.16) and (Eq. 3.17)

$$R_{E_{\theta_{\frac{0}{w}}}}(N)=\frac{\left(\begin{array}{l}\beta\ln B\left(n_{\vartheta_{\frac{0}{w}}}\ln\left\{\dfrac{(1+\theta_0)}{(\lambda+\theta_0)}\right\}+n\ln\left\{\dfrac{\lambda(1+\theta_0)}{(\lambda+\theta_0)}\right\}\right)+\\(1-\beta)\ln A\left(n_{\vartheta_{\frac{0}{w}}}\ln\left\{\dfrac{(w+\theta_0\lambda)}{\lambda(w+\theta_0)}\right\}+n\ln\left\{\dfrac{(w+\theta_0\lambda)}{(w+\theta_0)}\right\}\right)\times\\\left(n_{\vartheta_{\frac{0}{w}}}\ln\left\{\dfrac{(1+\theta_0)}{(\lambda+\theta_0)}\right\}+n\ln\left\{\dfrac{(1+\theta_0)}{(\lambda+\theta_0)}\right\}\right)\end{array}\right)}{\left(\begin{array}{l}n_{\vartheta_{\frac{0}{w}}}\ln\left\{\dfrac{(w+\theta_0\lambda)}{\lambda(w+\theta_0)}\right\}+n\ln\left\{\dfrac{(w+\theta_0\lambda)}{(w+\theta_0}\right\}\times n_{\vartheta_{\frac{0}{w}}}\ln\left\{\dfrac{(1+\theta_0)}{(\lambda+\theta_0)}\right\}+\\n\ln\left\{\dfrac{(1+\theta_0)}{(\lambda+\theta_0)}\right\}\times[\beta\ln B+(1-\beta)\ln A]\end{array}\right)} \tag{3.18}$$

and it can be easily seen that $R_E(N)\leq 1$ provided $A\leq w$.

Conclusion

The formula derived for ASN function in this paper works for any distribution and the method of proof is well within the frame work of Waldo We have shown that they are elegant to apply for any distribution apart from normal mean (Paulson, 1964) situation. It can also be seen that the ratio in all the cases (Eqs. 3.6, 3.12, 3.18) $R_E(N)\leq 1$, which indicates that sometimes the procedure is more efficient than that of Walds SPRT.

References

Govindrajulu, Z., 1987. *The Sequential Statistical Anlaysis of Hypothesis Testing, Point and Interval Estimation and Decision Theory*. American Sciences Press, Inc, Clumbus, Ohio.

Muddapur, M.V. and Halkatti, C.P.P., 1994. Closed Sequential decision procedures and their efficiencies. *JISPS*, 1: 83–92.

Paulson, E., 1964. Sequential estimation and closed sequential decision procedures. *Ann. Math. Statist.*, 35: 1048–1059.

Wald, A., 1947. *Sequential Analysis*. Wiley, New York.

Chapter 17

A Survey of Readymade Chilli Powders for Aflatoxin B_1 Contamination

K. Aruna Lakshmi*, M.V.V. Chandana Lakshmi and V. Sridevi

Department of Biotechnology, College of Engineering, Gandhi Institute of Technology and Management, Rushikonda, Visakhapatnam – 530 045, A.P.

ABSTRACT

Mycoflora associated with eleven popular brands of chilli powders were enumerated. The mycoflora present were *Aspergillus flavus, Aspergillus niger, Aspergillus candida, Fusarium, Curvularia, Gliocladium, Rhizopus* and *Mucor.* Indirect competitive ELISA revealed that 45 per cent of the samples were contaminated with aflatoxin Bland the toxin ranged from 0–173.3 µg/kg.

***Keywords**: Aflatoxin B_1, Aspergillus flavus, Chilli powders.*

Introduction

Substantial quantities of food grains are being affected by mycotoxins (Bosch *et al.*, 1991; Busby *et al.*, 1979; Jelinke *et al.*, 1989; Pinstrup-Anderson *et al.*, 2000; Reddy *et al.*, 2001). Health hazards due to ingestion of aflatoxin contaminated food are much greater in developing countries, as they lie in tropics, where temperature and relative humidities are conducive for growth of aflatoxin producing fungi–*Aspergillus flavus* and *Aspergillus parasiticus*. Due to established carcinogenic effect of aflatoxins on various animals (Katiyar *et al.*, 2000), developed countries which import food and feed, have set

* *Corresponding Author.* Residential address: 8-1-71, Chinmaya Marg, Peda Waltair, Visakhapatnam – 530 017 Phone: 0891-2840246 (off), 0891-2532194 (R); E-mail: vellurusridevi@yahoo.co.in

aflatoxin contamination limits in food stuffs ranging from 0–10 µg/kg. This has resulted in import restrictions on many foods and feeds from India.

Foodstuffs are regularly surveyed for mycotoxin contamination in the first world countries. To discourage the marketing of high toxin contaminated foods it is essential to monitor various foods periodically. Occurrence of aflatoxins in stored chillies has been reported by Reddy *et al.* (2001), Ravi Kiran *et al.* (2005). No information about contamination levels of chilli powders is available. In the present study we examined eleven brands of commercially available chilli powders to determine the incidence of mold and aflatoxin B_1 content.

Materials and Methods

Eleven popular brands of chilli powders (500 g each) were collected from shops, super markets and households from various parts of Visakhapatnam, Andhra Pradesh. Aflatoxin B_1 in the powders was estimated using indirect competitive ELISA method of ICRISAT (Anjaiah *et al.*, 1989; Reddy *et al.*, 2002).

Methanolic extracts of the chilli powders (20 g of each sample) diluted 1 : 10 in PBS Tween were used for analysis. ELISA plates coated with AFB_1–BSA conjugate were washed and blocked with 0.2 per cent BSA. After thorough washing, 100 µl of diluted aflatoxin B_1 (100 ng to 10 pg) and 100 µl of methanolic extract of chilli powder intended for analysis were added to each well of the ELISA plate. Antiserum (50 µl) was added to each well. About 150 µl goat anti-rabbit IgG (1 : 1000) labelled with ALP was added to each well. The substrate used was P-nitrophenylphosphate. Absorbance was measured at 405 nm in an automatic ELISA reader.

Quantification of aflatoxin was done as follows:

$$AFB_1(\mu g/kg) = \frac{ADE}{G}$$

where,

A = AFB_1 concentration in diluted sample extract (ng/ml); D = Times dilution with buffer; E = Extraction solvent volume used (ml); G = Sample weight (g).

Results and Discussion

Different fungal species isolated from chilli powders are shown in Table 17.1. All the samples invariably harboured mycoflora. The total viable count ranged between 2×10^4 to 9×10^5/ml. Mold incidence was particularly high in some local brands (brands 3, 4, 7, 10). The predominant fungal species were *Aspergillus flavus* and *Aspergillus niger.* The incidence of these species was 100 per cent in samples 6, 8, 9 (Table 17.1). *Penicillium fusarium, Mucor Rhizopus,* and *Alternaria* were found in low percentage. *Fusarium* and *Alternaria* are known to be associated with stored chillies elaborating mycotoxins (Patel *et al.*, 1996; Hasmi and Thrane, 1997). *Cladosporium,* an aeroallergic fungi, and soil fungi *Trichoderma* and *Gliocladium* were found in high concentrations in chilli powders purchased from local mills suggesting cross contamination from other food grains like rice, wheat, and millet.

Our survey found widespread and considerable contamination of chilli powders stored in houses, supermarkets and marketplaces (Raghu Kishore *et al.*, 2006). Chillies stored in cold storage were often found to be contaminated with aflatoxigenic strains of *A. flavus* and *A. niger* (Ravi Kiran *et al.*, 2005). Chilli powders prepared from contaminated chillies would be containing aflatoxins (Ravi Kiran *et al.*, 2005; Raghu Kishore *et al.*, 2006). In India, the maximum permissible level of aflatoxin in foods is 30 µg/kg. The present study revealed that 45 per cent of the chilli powders are unfit for consumption as they had greater then 30 µg/kg of aflatoxin B_1. Only 55 per cent of the samples complied with the 30 µg/kg maximum level as set by Indian standards. Reddy *et al.* (2001) surveyed the occurrence of aflatoxins B_1 in 139 samples of chillies and showed that 65 per cent of chilli are contaminated by aflatoxin B_1 and 20 per cent had non-permissible levels (>30 µg/kg). According to Klieber (2000) the risk of aflatoxin contamination seems to increase for chilli powders compared to whole fruits probably due to the increased surface areas and associated increased moisture content. Aflatoxin B_1 as high as 969 µg/kg was found in some samples of grade three chilli powders (Reddy, 2001). In the present study 173 µg/kg aflatoxin B_1 has been found in one local brand. Being cheap in price it is mostly consumed by the weaker sections in Andhra Pradesh. All batches of powders should be tested to comply with relevant legal limits (30 µg/kg). Monitoring of foodstuffs periodically to discourage the marketing of high toxin contaminated foods is therefore inevitable in India.

Table 17.1: Mycoflora and Aflatoxin B_1 Associated with Various Brands of Chilli Powders

Brand	*Total Viable Count*	*Name of Fungi*	*Aflatoxin B_1 µg/kg*
1	3×10^4	*A. flavus* *Cladosporium* *Mucor*	7.3
2	2×10^4	*A. flavus* *A. niger* *Gliocladium* *Penicillium*	77
3	9.1×10^5	*A. flavus* *A. niger* *Rhizopus*	36
4	8×10^5	*A. candida* *A. flavus* *Fusarium* *Gliocladium*	5.1
5	2×10^4	*A. niger* *A. flavus* *Mucor* *Trichoderma*	0
6	2.9×10^5	*A. flavus* *A. niger*	38.0
7	3×10^6	*A. flavus* *A. niger* *Gliocladium* *Mucor* *Rhizopus*	60.0
8	5×10^4	*A. flavus* *A. niger*	0
9	2×10^5	*A. flavus*	173.3
10	3×10^6	*A. flavus* *Cladosporium*	4
11	3.1×10^5	*A. flavus* *A. niger* *Alternaria* *Fusarium* *Gliocladium* *Mucor*	8.1

References

Anjaiah, V., Mehan, V.K., Jayanthi, S., Reddy, D.V.R. and Donald, D.Mc., 1989. Enzyme linked immunosorbant assay (ELISA) for Aflatoxin B_1 Estimation. In: *Aflatoxin Contamination of Groundnuts*. ICRISAT, India, pp. 183–190.

Bosch, F.X. and Press, F., 1991. Aflatoxins: Data on human carcinogenic risk and relevance to human cancer of N–Nitroso compounds. In: *Tobacco and Mycotoxin*, (Eds.) O'Neill, J.K., Chen, J. and Barstch, H. International Agency for Research on Cancer, Lyon, France, pp. 48–53.

Busby, W.F. Jr. and Wogan, G.N., 1979. Food borne mycotoxins and alimentary mycotoxicoses. In: *Food-borne Infections and Intoxications*, (Eds.) Riemann, H. and Bryan, F.B. Academic Press, New York, pp. 519–610.

Hashmi, M.H. and Thrane, U., 1997. Mycotoxins and other secondary metabolites in species of Fusarium isolated from seeds of capsicum, coriander and fenugreek. *Pakistan J. Bot.*, 22: 106–116.

Jelinke, C.F., Pohland, A.E. and Wood, G., 1989. World wide occurrence of mycotoxins in foods and feeds. *AOAC*, pp. 223–230.

Katiyar, S., Thakur, V., Gupta, R.C., Sarin, S.K. and Das, B.C., 2000. P53 tumour suppressor gene mutations in hepatocellular carcinoma patients in India, pp. 1565–1573.

Klieber, A., 2000. *Chilli Spice Production in Australia*. RIDC Publication No. 20/33.

Patel, S., Hazel, C.M., Winterton, A.G.M. and Morbey, E., 1996. Survey of ethnic foods for contamination by mycotoxins. *Food. Add. Contam.*, 13: 833–841.

Pinstrup-Andersen, P., Pandya-Lorch, R.C. and Rosegrant, M.W., 2000. *Food Security: Problems, Prospects and Policies 2020 Vision*. Discussion Paper, Washington DC, International Food Policy Research Institute.

Raghu Kishore, G., Kanthi, P., Sujatha, S., Kalyan Chakravarthy, N. and Aruna Lakshmi, K., 2004. *AJMBES* (In Press).

Ravi Kiran, D., Narayana, K.J.P. and Vijayalakshini, M., 2005. Aflatoxin B_1 production in chillies (*Capsicum annum* L.) kept in cold stores. *African Journal of Biotechnology*, pp. 791–795.

Reddy, D.V.R., Thirumala Devi, B., Reddy, S.V., Waliyar, F., Mayo, M.A., Raffia Devi, K., Ortiz, R. and Lenne, J.M., 2002. Estimation of aflatoxin levels in selected foods and feeds in India. In: *Food Safety Management in Developing Countries*, (Eds.) Hanak, E., Boutrif, E., Fabre, P. and Pinerio, M. CIRAD–FAO, Montpellies, France, pp. 11–13.

Reddy, S.V., Kiran Mayi, D., Uma Reddy, M., Thirumala Devi, K. and Reddy, D.V.R., 2001. Aflatoxins B_1 in different grades of chillies (*Capsicum annum* L.) in India as determined by indirect competitive–ELISA. *Food Additives and Contaminants*, pp. 553– 558.

Chapter 18

Effect of Weed Control Treatments on Nutrient Removal by Weeds in Direct Sown Wet Seeded Rice

B. Rajendra Kumar[1]* and A. Christophrer Lourduraj[1]

[1]*Agricultural Officer, Agricultural Research Station, Amadalavalasa, ANGRAU, A.P.*

[2] *Associate Professor, Department of Agronomy, Tamil Nadu Agricultural University, Coimbatore, T.N.*

ABSTRACT

A field experiment was conducted at Tamil Nadu Agricultural University, Coimbatore during *rabi* season 2002–2003. The removal of N, P and K by weeds at 20, 35 and 50 DAS was lesser in Pretilachlor + Safener 0.45 Kg a.i.ha^{-1} at 4 DAS + 1 hand weeding at 35 DAS as compared to other weed control treatments favoured the crop to attain higher growth and yield attributes. Total weed population, weed DMP and N removal by weeds were negatively associated with grain yield, whereas rice DMP, N uptake by crop and filled grains panicle-1 had strong positive association with grain yield. The rate of decrease in N removal was proportional to the decrease in weed dry matter. The reduction in N removal was 56, 61 and 62 per cent in Pretilachlor + Safener, Butachlor and Butachlor + Propanil treatment when compared to hand weeding at 20 DAS. Better weed control efficiency as a result of lower weed dry matter has restricted the nutrient depletion by weeds.

Keywords: *Nutrient, Weed control, Weeds, Wet seeded rice.*

* *Corresponding Author.*

Introduction

Weeds compete with rice crop for nutrients, water, light and space and pose a serious problem to rice yields if not controlled timely. Weed competition is greater in wet seeded rice than in transplanted rice because the weeds and rice grow simultaneously making weed control by herbicides more critical and difficult. Weed control costs can be reduced by using competitive cultivars and enhancing its weed competitive ability through good crop management. Weeds deplete 107, 15.5 and 112.8 kg ha^{-1} of nitrogen, phosphorus and potassium, respectively which was 60.8, 6.85 and 46.6 Kg more than the total uptake of these nutrients by rice crop Rana and Angiras (1999).

Materials and Methods

A field experiment was conducted at Tamil Nadu Agricultural University, Coimbatore during *rabi* season 2002–2003. The experiment was laid out in split plot design with three varieties ASD 19, ADT 38 and CO 43 in the main plot and weed management practices (Pretilachlor + safener 0.45 kg a.i. ha^{-1} at 4 DAS + 1 hand weeding at 35 DAS; Butachlor 1 kg a.i. ha^{-1} at 8 DAS + 1 hand weeding at 35 DAS; Butachlor + Propanil 0.84 + 0.84 kg a.i. ha^{-1} at 10 DAS (tank mix) + 1 hand weeding at 35 DAS; two hand wee dings at 20 and 35 DAS; unweeded check) in the subplots with three replications.

The weed samples collected for recording dry matter accumulation were ground using willey mill and used for estimation of uptake of major nutrients at various stages of the crop. The N content was estimated using the micro kjeldahl method (Humphries, 1956) and P and K contents were estimated as per the method described by Jackson (1973). The total uptake was calculated by multiplying the nutrient content with DMP and expressed in kg ha^{-1}.

Results and Discussion

Nitrogen uptake by weeds was significantly influenced by varieties and weed control treatments. None of the varieties, exhibited significant difference in N removal by weeds at 20 DAS. However, significant difference was observed at 35 and 50 DAS. In both the stages, CO 43 (M_3) recorded maximum removal of N by weeds (25.35 and 17.88 kg ha^{-1}, at 35 and 50 DAS respectively), followed by ADT 38 (M_2). ASD 19 (M_3) recorded lowest N removal by weeds (20.11 and 14.16 kg ha^{-1}). Among the weed control methods, at 20 DAS Pretilachlor + safener 0.45 kg a.i. ha^{-1} + 1 hand weeding at 35 DAS (S_1), recorded lowest N removal by weeds (0.293 kg ha^{-1}), followed by Butachlor 1 kg a.i. ha^{-1} + 1 hand weeding at 35 DAS (S_2) which were on par with S_1. Two hand weedings at 20 and 35 DAS (S_4) and unweeded check (S_5) recorded highest N removal of weeds of (0.523 and 0.623 kg ha^{-1}, respectively). At both 35 and 50 DAS, S_1 recorded lowest removal of N by weeds (13.2 and 0.49 kg ha^{-1}, respectively), followed by S_4. S_1 and S_4 did not differ significantly with each other at both the stages. Highest removal of N by weeds was observed in S5 in all the stages of the crop growth. The interaction effect between varieties and weed control treatments was significant at all the stages. ASD 19 with Pretilachlor + safener (M_1S_1) recorded lowest N removal by weeds at 20 DAS while ADT 38 with Pretilachlor + safener (M_2S_1) and CO 43 with Pretilachlor + safener (M_3S_1) recorded lowest N removal by weeds at 35 and 50 DAS, respectively. Highest removal of N by weeds was recorded in CO 43 with unweeded check (M_3S_5) at all the stages.

The trend of P removal by weeds was similar to that of N. At 35 DAS, highest P removal by weeds (3.31 kg ha^{-1}) was recorded in CO 43 (M_3) and the lowest by ASD 19 (M_1) (2.62 kg ha^{-1}). At 50 DAS, ASD 19 (M_1) recorded highest P removal by weeds (2.60 kg ha^{-1}) and CO 43 (M_3), the lowest (2.05 kg ha^{-1}). Weed control treatments significantly influenced the removal of P by weeds. At 20 DAS S_1, recorded lowest removal of P by weeds (0.030 kg ha^{-1}), followed by (S_3) and (S_2). S_3 and S_2 did not significantly differ with each other. The highest removal of P was recorded in unweeded check (S_5). There was no significant difference among the varieties in potassium removal by weeds at 20 DAS. However, at 35 and 50 DAS significant difference was observed among varieties. At 35 DAS, ASD 19 (M_1) recorded lowest K removal by weeds (17.78 kg ha^{-1}) and CO 43 (M_3) recorded highest K removal by weeds (22.43 kg ha^{-1}). At 50 DAS, CO 43 (M_3) recorded lowest (12.46 kg ha^{-1}) K removal by weeds and ASD 19 (M_1), the highest (15.70 kg ha^{-1}). Among the weed control methods, S_1 recorded lowest K removal by weeds at 20, 35 and 50 DAS (0.16, 11.66 and 0.50 kg ha^{-1}, respectively). Two hand weeding at 20 and 35 DAS (S_4) was not significantly different from S_1 at 35 DAS and 50 DAS. The highest K removal by weeds was recorded in unweeded check (S_5) in all the three growth stages. At all the three stages of crop growth. The lowest removal of K by weeds was recorded in ASD 19 with Pretilachlor + safener (M_1S_1) (0.44 kg ha^{-1}) at 20 DAS, ADT 38 with Pretilachlor + safener (M_2S_1) (9.47 kg ha^{-1}) at 35 DAS, and M_1S_1 (0.44 kg ha^{-1}) at 50 DAS.

From the stand point of N removal by weeds, Pretilachlor + safener proved effective. Shetty and Gill (1974) observed that the rate of decrease in N removal was proportional to the decrease in weed dry matter. The reduction in N removal was 56, 61 and 62 per cent in Pretilachlor + safener, Butachlor and Butachlor + Propanil treatments, respectively when compared to hand weeding at 20 DAS. However at 50 DAS, the reduction in N removal under Pretilachlor + safener treatment was 17 per cent, and 22 and 15 per cent more in Butachlor and Butachlor + Propanil treatment when compared to hand weeding. This shows the importance of early and effective weed control through timely application of pre-emergence herbicides. This is in conformity with findings of Bayan (2000). The removal of P and K by weeds was also lower under Pretilachlor + safener, irrespective of the crop stages.

Table 18.1: Effect of Weed Control Treatments on Nitrogen, Phosphorus and Potassium Removal kg ha^{-1} by Weeds in Direct Sown Wet Seeded Rice

	N Removal kg ha^{-1} by Weeds			*P Removal kg ha^{-1} by Weeds*			*K Removal kg ha^{-1} by Weeds*		
	20 DAS	*35 DAS*	*50 DAS*	*20 DAS*	*35 DAS*	*50 DAS*	*20 DAS*	*35 DAS*	*50 DAS*
M_1	0.418	20.11	14.16	0.044	2.62	2.60	0.23	17.78	15.70
M_2	0.410	21.12	16.97	0.043	2.76	2.47	0.23	18.69	14.91
M_3	0.426	25.35	17.88	0.045	3.31	2.05	0.24	22.43	12.46
SEd	0.008	0.07	0.08	0.0012	0.03	0.014	0.009	0.05	0.24
CD (P = 0.05)	NS	0.20	0.24	NS	0.08	0.040	NS	0.15	0.68
S_1	0.293	13.20	0.49	0.030	1.72	0.074	0.16	11.66	0.50
S_2	0.323	18.56	0.72	0.034	2.42	0.077	0.18	16.41	0.72
S_3	0.326	18.02	0.68	0.034	2.35	0.101	0.18	15.94	0.67
S_4	0.523	15.82	0.59	0.055	2.06	0.089	0.29	13.99	0.60
S_5	0.623	45.38	79.20	0.066	5.93	11.548	0.35	40.16	69.30
SEd	0.010	0.10	0.18	0.0009	0.03	0.020	0.007	0.07	0.33
CD (P = 0.05)	0.200	0.20	0.37	0.0019	0.06	0.043	0.16	0.15	0.69

M_1: ASD 19; M_2: ADT 38; M_3: CO 43

S_1: Pretilachlor + safener 0.45 kg a.i. ha^{-1} + 1 hand weeding at 35 DAS; S_2: Butachlor 1 kg a.i. ha^{-1} + 1 hand weeding at 35 DAS; S_3: Butachlor + Propanil 0.84 + 0.84 kg a.i. ha^{-1} + 1 hand weeding at 35 DAS; S_4: Two hand weedings at 20 and 35 DAS; S_5: Un weeded check.

Better weed control efficiency as a result of lower weed dry matter has restricted the nutrient depletion by weeds as observed by Mathew and Alexander (1995). In general, wherever effective weed control was achieved, the nutrient loss due to weeds was minimum (Srinivasan and Choudhury, 1994; Umapathi, 1998). As a result of better weed control and lesser dry weight of weeds in Pretilachlor + safener, the N removal of weeds was lowered. Effective control of weeds, particularly grasses increased the weed control efficiency. This led to the least removal of N by weeds and ranked next to hand weeding treatment.

References

Bayan, H.C., 2000. Studies on integrated weed management under split application of nitrogen in wet seeded rice. *Ph.D. Thesis*. Tamil Nadu Agricultural University, Coimbatore.

Humphries, E.C., 1956. *Mineral Components and Ash Analysis*. Modern methods of plant analysis. Springer Verlag, Berten-l. pp. 468–502.

Jackson, M.L., 1973. *Soil Chemical Analysis*. Prentice-Hall of India Pvt. Ltd., New Delhi, pp. 1–495.

Mathew, Gracy and Alexander, D., 1995. Influence of inter cropped green manure crops on weed pressure and grain yield of semi dry rice. *Madras Agric. J.*, 82(1): 66–67.

Rana, S.S. and Angiras, N.N., 1999. Influence of integrated weed management practices on weed competition for nutrients in puddle sown rice. *Indian J. Weed Sci.*, 31(3 and 4): 161–163.

Shetty, S.V.R. and Gill, H.S., 1974. Critical period of crop weed competition in rice (*Oryza sativa* L.). *Indian J. Weed Sci.*, 6(2): 101–107.

Srinivasan, G. and Choudhury, S.K., 1994. Effect of management practices on nutrient removal by weeds in rice (*Oryza sativa*) based cropping system. *Indian J. Agron.*, 39(3): 363–367.

Chapter 19

Qualitative Evaluation of Antimicrobial Efficacy of Novel N-Heterocyclic Compounds

Megha S. Kulkarni[1*], *Arti S. Shanware*[2] *and Jyotsna S. Meshram*[3]

[1]*Department of Microbiology, Janata Mahavidyalaya, Chandrapur – 442 401, Maharashtra*

[2]*Rajiv Gandhi Biotechnology Centre, L.I.T. Campus, The Rashtrasant Tukdoji Maharaj Nagpur University, Nagpur – 440 033, Maharashtra*

[3]*PG Department of Chemistry, The Rashtrasant Tukdoji Maharaj Nagpur University, Nagpur – 440 033, Maharashtra*

ABSTRACT

Heterocyclic Chemistry is an important branch of organic chemistry amongst the heterocyclic compounds, condensed nitrogen containing heterocyclic compounds, *viz.*, Quinolones, Pyrazolones, Schiff bases, Pyrroles because of their effectiveness in many pharmacological areas have been largely investigated. Due to unique physical and chemical properties, their derivatives possess a great number of biological activities such as antifungal, antibacterial, anti-inflammatory, antiparasitic, antitumour, antiviral and anti hypertensive effects. Bacterial and fungal diseases have been of great concern in the field of agriculture and human physiology. For control of these diseases, various organic antimicrobial agents are being extensively used. A great deal of research is carried out to prepare new heterocyclic compounds having therapeutic value.

The antimicrobial activity of procured Novel N-heterocyclic compounds under study *viz.*, Quinolones, Pyrazolones, Schiff bases and Pyrroles was determined against wide micro flora

* E-mail: drmegha100@rediffmail.com

viz., Gram Positive bacteria, Gram negative bacteria and fungal isolates by agarcup method. Minimum Inhibitory Concentration values were recorded as the lowest concentration of the substances that had no visibility of zone of inhibition and found to be in the range of 100–200 µg/ 0.1ml. Quinolones, Schiff bases, Pyrroles were found to possess good antimicrobial activity, where as Pyrazolones were found to possess moderate to good antifungal activity.

Keywords: *Antimicrobial activity, Minimum inhibitory concentration values, Pyrazolones, Pyrroles, Quinolones, Schiff bases.*

Introduction

Amongst the heterocyclic compounds *viz.*, schiff bases (More *et al.*, 2002), quinolones (Powar *et al.*, 1994), pyrazolones (Mohamed *et al.*, 1995), pyrroles (Chavan *et al.*, 1999) possess diversified biological applications. These compounds are widely distributed in nature as several important compounds, ego antibiotics, amino acids, dyes, drugs, enzymes, DNA, Vitamin B complex, alkaloids contain N-Heterocyclic compounds. Ehrlich laid the foundations of chemotherapy and concluded that physiological action of drug depends upon its chemical nature. Infections caused by various pathogenic microorganisms can be treated with N-heterocyclic compounds having antimicrobial activity. Quinolones, due to good oral absorption activity against common Gram negative urinary pathogens and comparatively higher urinary concentrations, have been used clinically for treatment of urinary tract infections (Deigada *et al.*).

A number of researchers have reported antimicrobial activities of schiff bases. Biocidal activities of schiff bases have also been well established. These have been attributed to the toxophoric C=N linkge in them (Hodnett and Dunn, 1970).

Ludwig Knorr in 1884 (Modern Organic Chemistry), discovered 5 pyrazolone now known as antipyrine which is popularly used as analgesic. Phenyl butazone now being used for the releif of many forms of arthritis, reduces swelling and spasm by an antiinflammatory action. Pyrazoles and their derivatives have been investigated by the organic chemists due to their close association with biological activities (Penning *et al.*, 1997; Menozzi *et al.*, 1997; Naik *et al.*, 1998). In the present age of pharmacogenetics, chalcones due to wide spectrum of biological activities, have been given considerable interest owing to their antifungal, antihistaminic, antimatarial and anticancer properties; and agricultural chemicals (Karale *et al.*, 2002; Reddy *et al.*, 1993; Shingare *et al.*, 1999).

The rapid rise in bacterial resistance to the traditional antibiotics *viz.*, penicillins (Nathwani and Wood 1993) and tetracycline (Schanappinger and Hillen, 1996) has encouraged a continue search for new classes of compounds with novel modes of antibacterial activity. The pharmaceutical importance of N-Heterocycles as Nalidixic acid ciprofloxacin, phenyl butazone paved the way towards active research in this field. On this backdrop, Novel N-Heterocyclic compounds known from literature surveyed have been procured, characterised and screened for their antimicrobial activity.

Materials and Methods

Chemical compounds under study–Quinolone derivatives, schiff bases, pyrazolone derivatives, pyrrole derivatives

Solvent used: Ethanol, DMSO.

Media used: Muller Hinton agar, potato dextrose agar, nutrient broth

Bacterial and fungal isolates under study.

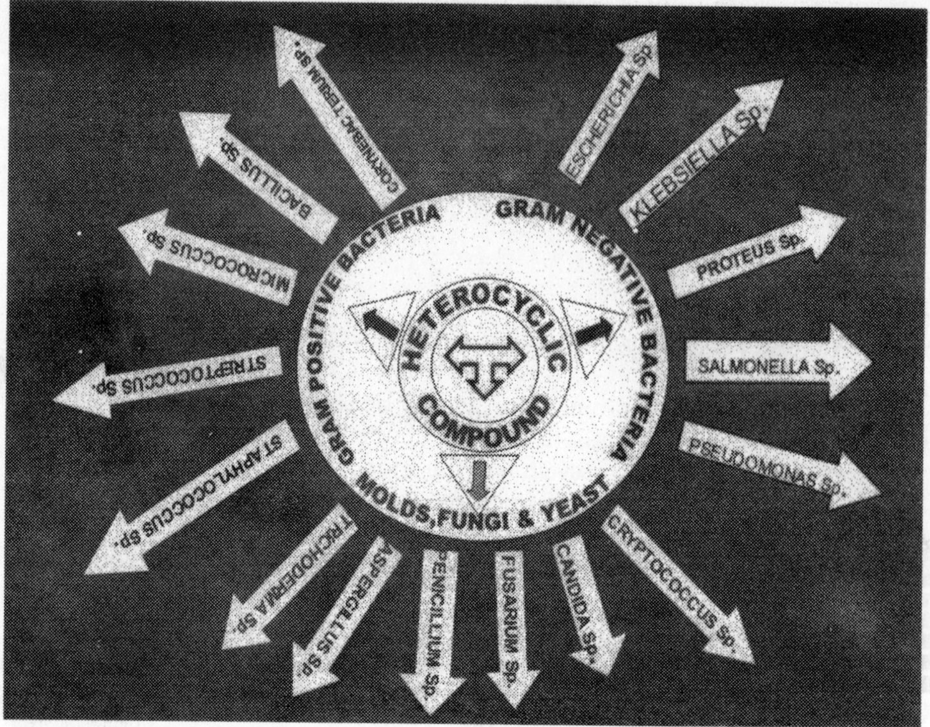

Figure 19.1

Experimental Section

Agar–Cup Method for Antibacterial and Antifungal Sctivity

The activity was determined using cup-plate agar diffusion method (Cruickshank and Duguid *et al.*, 1975) by measuring the inhibition zone in mm. All the compounds were screened *in vitro* for their antimicrobial activity against a variety of bacterial strains *viz.*, Gram positive, Gram negative and fungal isolates. Known antibiotics like streptomycin, ciprofloxacin and griseofulvin were used as standard.

The synthesized compounds were tested using solvent ethanol or DMSO. Sensitivity plates were seeded with bacterial inoculum and spore suspension of 1×10^6 CFU/ml and each cup (dia 10 mm) was loaded with 0.1 ml of test solution. The zone of inhibition was recorded after incubation for 24 hours for bacteria and after 6 days for fungi. Zone of inhibition measured in mm was compared with standard antibiotics

Serial Dilution Method for MIC *i.e.* Minimum Inhibitory Concentration

The compounds showing good antimicrobial activity were further screened for their Minimum Inhibitory Concentration values. The compounds under study and standard compounds were serially diluted, and applied for agar-cup method. Followed by incubation, zone of inhibition was recorded for each dilution and compared with norfloxacin used as standard.

Poisoned Food Technique

Novel compounds showing zone of inhibition about 10 mm or more for fungal isolates were further assessed by poisoned food technique (Flack, 1907; Nene, 1970) on potato dextrose agar containing 1 per cent test compound. Fungal disc of 10 mm diameter was kept in the centre of each petriplate containing poisoned food medium and also in control plate without poisoned food. The inoculated plates were incubated at 27°C for seven days and results were recorded in terms of diameter of fungal growth.

Percent inhibition was calculated using Vincent formula (Vincent, 1947)

$$I = C - T/C \times 100$$

where,

I: Per cent inhibition of growth

C: Growth of fungus in control

T: Growth of fungus in treatment

Results and Discussion

The attempts were made to screen bacterial and fungal isolates for their tolerance efficacy against novel N-Heterocyclic compounds under study. Test compounds were procured and screened for biological activities. Schill bases are important intermediates for the synthesis at some bioactive compounds such as β-lactams (Venturini *et al.*, 2002; Taggi *et al.*, 2002; Delpiccolo *et al.*, 2002). They are reported to show a variety of interesting biological actions including antibacterial (More *et al.*, 2001; Pandey *et al.*, 1999;) and antifungal (More *et al.*, 2001; Pandeya *et al.*, 1999). It has been suggested that azomethine linkages are responsible for biological activities displayed by schiff bases. Novel Schiff bases of furfuraldehyde with aminobenzoic acid were found to possess' moderate to good antimicrobial activity (Table 19.1). Novel quinolone derivatives containing β-napthol showed mark antibacterial and antifungal activity (Table 19.2, Figures 19.2 and 19.3) the results were compared to standard Ciprotloxacin (as shown in Figure 19.2).

The organism find difficulty in constructing N-N bonds, that limit the natural abundance of compounds having such bonds. Pyrazoles and their derivatives are a class of compounds containing N-N bonds which exhibit good antimicrobial and anti-inflammatory acitivity. Pyrazolones are associated with broad spectrum of biological activities including anti fungal, antibacterial, anti inflammatory acivities (Gan *et al.*, 2002; Turan-Zitouni *et al.*, 2001; Elguero *et al.*, 1984; Lung *et al.*, 2002; Kidwai *et al.*, 2002; Kidwai *et al.*, 2003) Novel pyrazolone derivatives *viz.*, Bis-pyrazolone chloride was found to possess good antimicrobial activity (Table 19.3).

In the past few decades, the chemistry of pyrrole and fused heterocyclic pyrrole derivatives has received considerable attention owing to their massive and effective biological importance (*G. baccolini* and *C. sandali*). Novel pyrrole derivatives synthesized from pyrrole, ethyl acetoacetate and carbon tetrachloride were found to possess good antimicrobial acitivity (Table 19.4 and Figures 19.4 and 19.5).

Minimum Inhibitory Concentration values of effective schiff bases were indicative of good antibacterial activity (Table 19.5 and Figure 19.6).

Table 19.1: Results for Antimicrobial Activity of Schiff Bases Solvent Used–DMSO Concentration–100 µg/0.1 ml; Method used: Agar-Cup method

Sl.No.	*Name of Organism*	*Zone of Inhibition in mm*						
		Z1	*Z2*	*Z3*	*Z4*	*Z5*	*Z6*	*Z7*
1.	*S. aureus*	25	6	8	15	20	–	8
2.	*Bacillus* sp.	10	8	10	6	12	–	–
3.	*Micrococcus* sp.	14	–	–	8	–	–	–
4.	*Steroptococcus* sp.	4	–	10	–	8	–	–
5.	*Corynebacterium* sp.	30	–	14	15	8	–	–
6.	*E. coli*	8	–	8	–	12	–	6
7.	*Salmonella*	6	–	10	–	8	6	–
8.	*Klebsiella* sp.	24	–	16	15	8	6	–
9.	*Pseudomonas* sp.	5	–	–	–	8	8	20
10.	*Candida* sp.	6	–	6	6	–	6	–
11.	*Penicillium* sp.	26	–	6	–	20	–	–
12.	*Trichoderma* sp.	–	–	6	–	–	–	–
13.	*Fusarium* sp.	08	–	6	6	–	6	20
14.	*Citrobacter* sp.	16	–	10	10	6	–	12
15.	*Aspergillus* sp.	20	–	6	–	20	–	16
16.	*Cryptococcus* sp.	14	–	6	–	–	–	12

Table 19.2: Result for Antimicrobial Activity of Quinolones Solvent Used: Ethanol; Concentration of Compound: 100 µg/0.1 ml

Name of Compound	*Zone of Inhibition (mm)*				
	Name of Organism				
	S. aureus	*E. coli*	*Pseudomonas sp.*	*Candida sp.*	*Aspergillus sp.*
Q1	10	12	13	–	20
Q2	20	40	23	20	30
Q3	20	15	15	16	–
Q4	11	12	12	–	–
Q5	10	10	10	10	20
Q6	10	13	13	–	–
Q7	–	–	–	–	–
Standard Ciprofloxacin	30	33	23	30	33
Norfloxacin				28 mm	
Griseofulvin				31 mm	

–: No zone of inhibition

Zone of Inhibition below 8 mm indicates inactive.

Zone of Inhibition between 8 mm to 13 mm indicates moderately active.

Zone of Inhibition above 13 mm indicates highly active.

Table 19.3: Results for Antimicrobial Activity of *Pyrazolones*
Concentration: 100 µg/0.1 ml; Method used: Agar cup method

Sl.No.	*Name of Organism*	*Zone of Inhibition (mm)*			
		Py–1	*Py–2*	*Py–3*	*PyA*
1.	*S. aureus*	–	14	–	8
2.	*E. coli*	–	10	–	8
3.	*Profeus* sp.	–	10	–	8
4.	*Klebsiella* sp.	–	12	–	–
5.	*Pseudemonas* sp.	–	8	–	–
6.	*Bacillus* sp.	–	15	–	–
7.	*Corynebacterium* sp.	–	13	–	–
8.	*Salmonella* sp.	–	4	–	–
9.	*Aspergillus* sp.	6	17	–	–
10.	*Penicillium* sp.	6	17	–	–
11.	*Candida* sp.	–	20	–	–
12.	*Trichoderma* sp.	–	18	–	–
13.	*Cryptococcus* sp.	–	–	4	–
	Standard Streptomycin	25			

–: No zone of inhibition

Zone of Inhibition below 8 mm indicates inactive.

Zone of Inhibition between 8 mm to 13 mm indicates moderately active.

Zone of Inhibition above 13 mm indicates highly active.

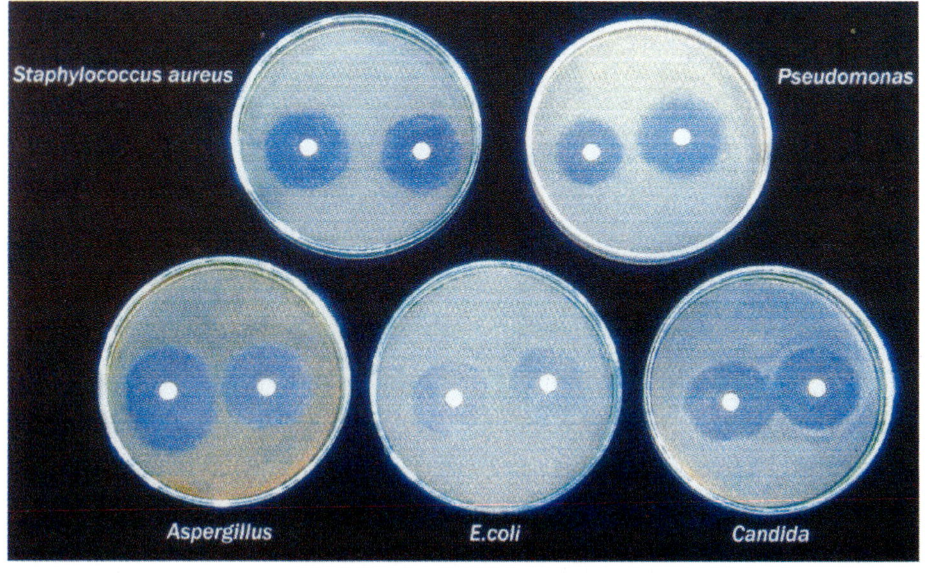

Figure 19.2: Ciprofloxacin

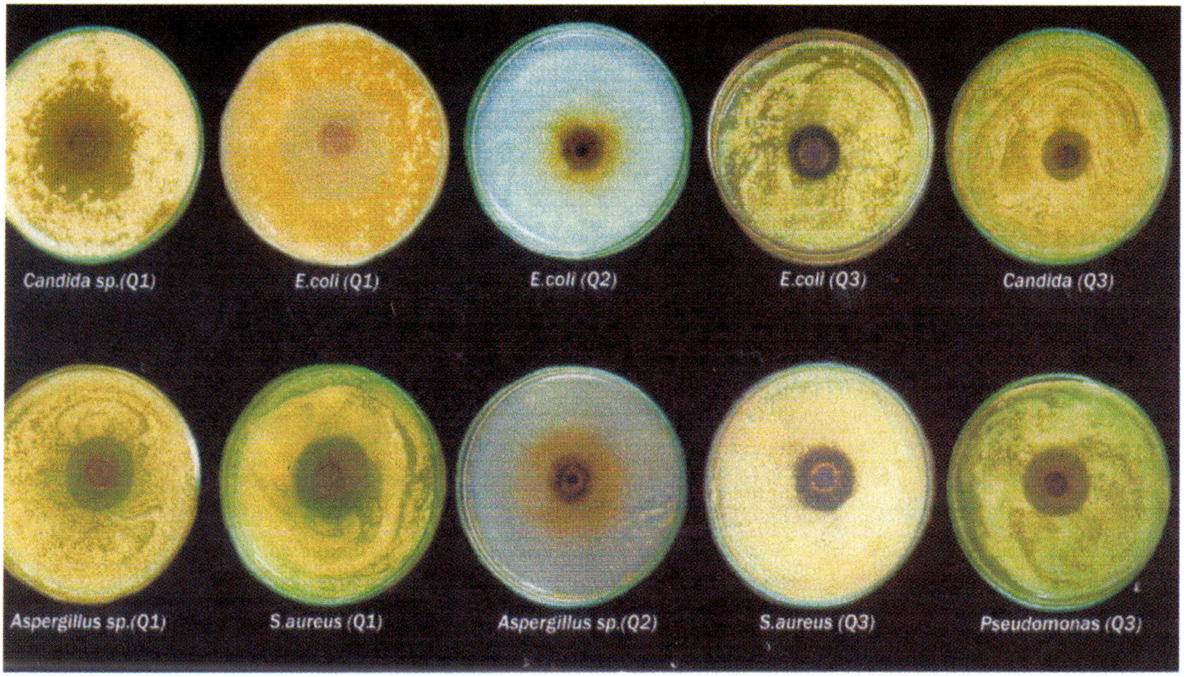

Figure 19.3: Antimicrobial Activity of Quinolone Derivatives

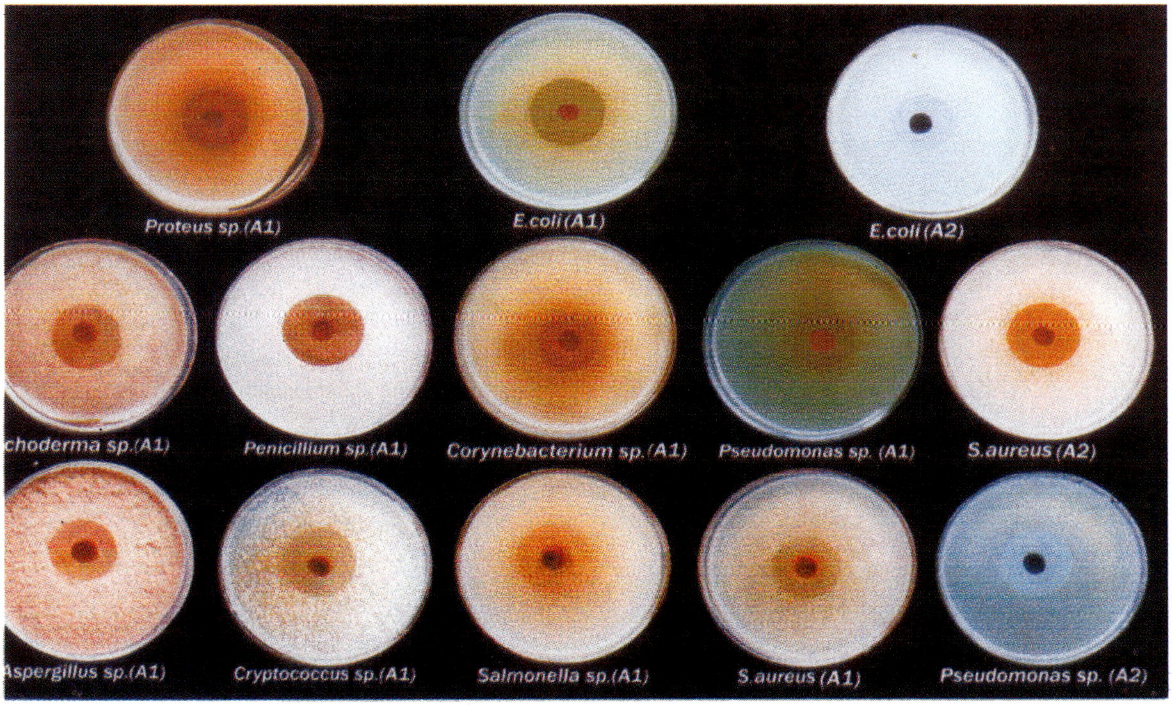

Figure 19.4: Antimicrobial Activity of Pyrroles (A1, A2)

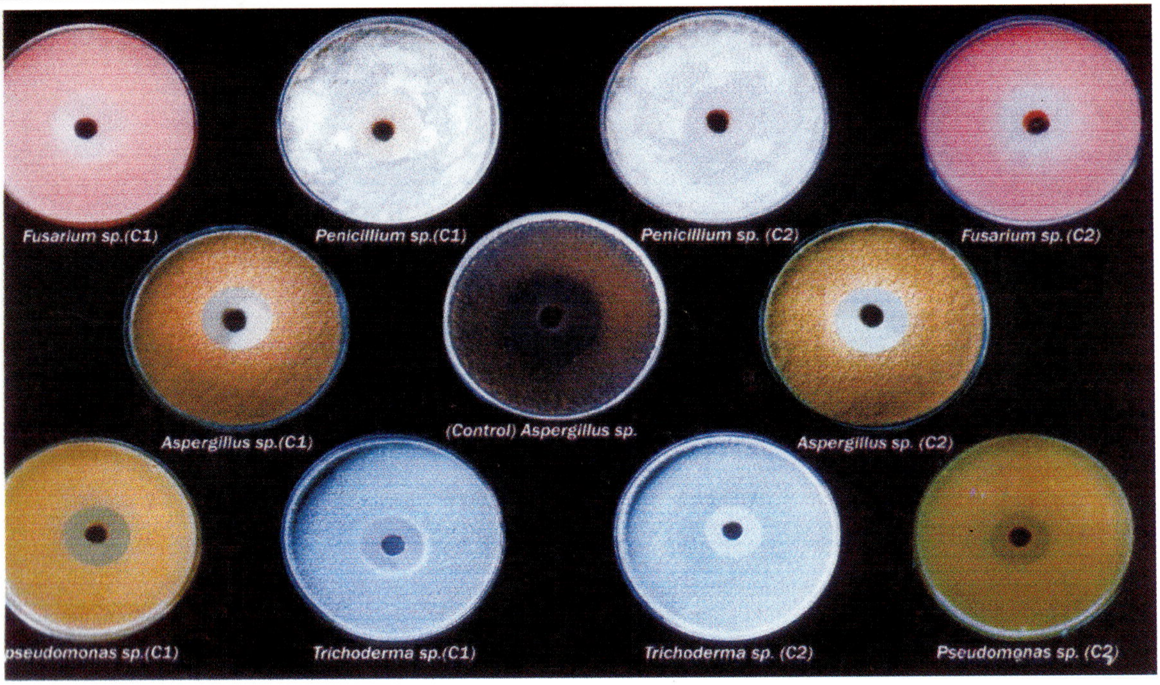

Figure 19.5: Antimicrobial Activity of Pyrroles (C1, C2)

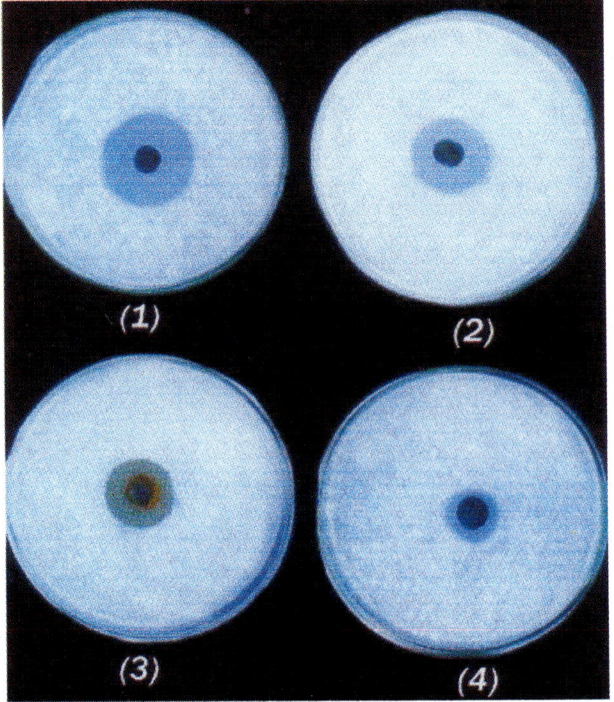

Figure 19.6: Minimum Inhibitory Concentration (MIC) of Schiff Base (Z1) for *S. aureus*

Table 19.4: Antimicrobial Activity of Pyrrole Derivatives
Solvent used: DMSO; Concentration: 100 μg/0.1 ml

Sl.No.	*Name of Organism*	*Zone of Inhibition in (mm)*			
		Al	*A2*	*C1*	*C2*
1.	*S. aueus*	12	11	10	–
2.	*E. coli*	12	12	–	–
3.	*Bacillus* sp.	8	–	6	–
4.	*Corynebacterium* sp.	8	–	–	–
5.	*Micrococcus* sp.	–	–	10	6
6.	*Pseudomonas* sp.	11	13	9	9
7.	*Klebsiella* sp.	–	–	–	–
8.	*Proteus* sp.	14	–	–	–
9.	*Salmonella* sp.	8	–	6	–
10.	*Aspergillus* sp.	23	–	14	10
11.	*Trichoderma* sp.	12	–	22	10
12.	*Penicillium* sp.	24	–	14	12
13.	*Candida* sp.	8	6	–	–
14.	*Cryptococus* sp.	16	6	–	–
15.	*Streptococcus* sp.	–	6	–	–
16.	*Fuzarium* sp.	8	6	12	–
	Standard Norfloxacin: 28 mm Griseofulvin: 31 mm Streptomycin: 24 mm				

–: No zone of inhibition

Zone of Inhibition below 8 mm indicates inactive.

Zone of Inhibition between 8 mm to 13 mm indicates moderately active.

Zone of Inhibition above 13 mm indicates highly active.

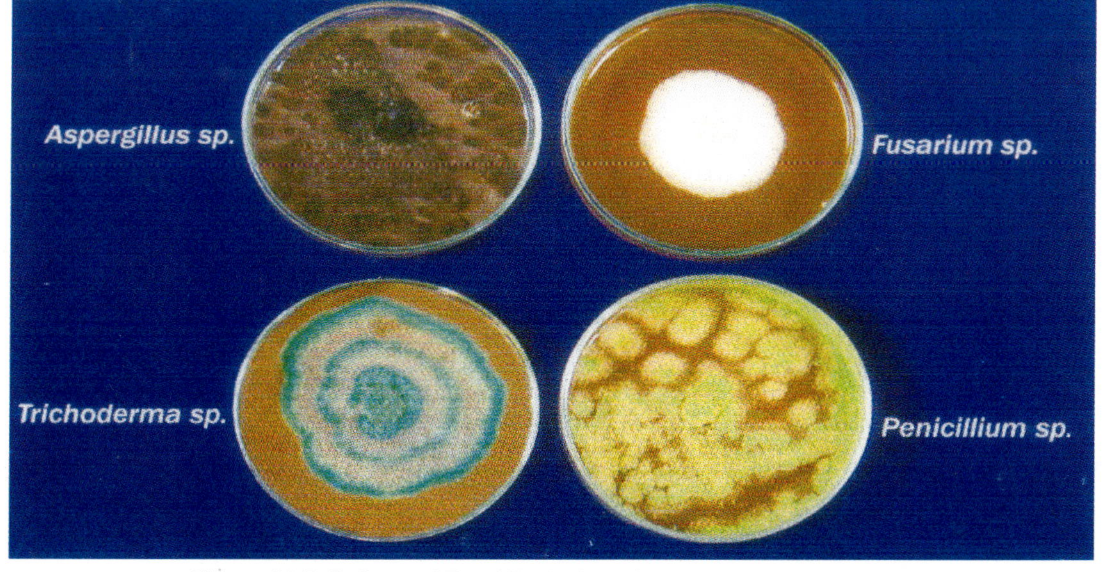

Figure 19.7: Poisoned Food Technique for Pyrazolone Derivative

Table 19.5: Minimum Inhibitory Concentration Values for Schiff Base

Name of Organism		*Compound*		
		Standard Streptomycin	*Z1*	*Z2*
	Concentration	*Zone of Inhibition in mm*		
S. aureus	20 mg/ml	25	18	12
	2 mg/ml	16	12	8
	0.2 mg/ml	8	10	4
	0.02 mg/ml	*4*	*4*	*No zone*

Poisoned food technique in the present investigation by pyrazolone-2 showed per cent inhibition of fungal isolates *viz., Trichoderma* sp. and *Fusarium* sp. by 50 per cent and 30 per cent (Table 19.6 and Figure 19.7).

Table 19.6: Percentage Inhibition of Fungi by the Pyrazolone (Py-2) Derivative by Poisoned Food Technique

Sl.No.	*Name of Organism*	*Diameter of Fungal Growth in mm on 7th Day*	*Percentage Inhibition*
1.	*Fusarium* sp.	19	50 per cent
2.	*Trichoderma* sp.	28	30 per cent
3.	*Aspergillus* sp.	–	–
4.	*Penicillium* sp.	–	–

Conclusion

Various parameters *i.e.* determination of antimicrobial activity by agar-cup method, MIC values, per cent inhibition by poisoned food technique led to the conclusion that Novel N-heterocyclic compounds are good antimicrobial agents. Increasing concentration of the above or combination of two or more would be significant for antimicrobial activity. Studies are being carried out to understand mechanism of action of these antimicrobial so that they could be extensively applied for medical purposes.

Acknowledgements

The authors are thankful to University Grants Commission for the financial assistance to carry out the present research work. The Heterocyclic compounds in the present study have been procured from P.G. Department of Chemistry, RTM Nagpur University, Nagpur.

References

Chavan, P.P., Mane, A.J., Bhawsar, S.B., Hangarge, R.V. and Shingare, M.S., 1999. Feb.–Nov., *Hindustan Antibiotic Bulletin*, 41(1–4): 32–1.

Cruickshank, R., *et al.*, 1975. *Medical Microbiology*, Vol. 2. Churchill Living Stone, London and New York, p. 190.

Deigada, J.M. and Ramers, W.A. *Text Book of Organic Medical and Pharmaceutical Chemistry*, 10th Edition, p. 197.

Delpiccolo, C.M.L. and Mata, E.G., 2002. *Tetrtahedron: Asymmetr*, 13: 905–910.

Elguero, J., 1984. In: *Comprehensive Heterocyclic Chemistry*, (Eds.) Katrizky, A.R. and Rees, C.W. Pergamon Press, Oxford, p. 1–5, 167–303.

Flack, R., 1907. Cronica Botnica Co. Pub. (USA), 1: 153–154.

Gan, Y., Lu, D., Liu, J., Tian, M., and Zhongguo Yaouri Zazhi, 2002. 11(2): 85, *Chem. Abstr.*, 2002, 136, 216696.

Hodnett, E.M. and Dunn, W.J., 1970. *J. Med. Chem.*, 13: 768 –770.

Jung, J.C., Watkins, E.B. and Avery, M.A., 2002. *Synth Commun.*, 32(24): 3767.

Karale, B.K. and Gill, C.R., 2002. *Indian Journal of Heterocyclic Chemistry*, 41B.

Kidwai, M., Rastogi, S., and Venkataramanan, R., 2003. *Bull. Chem. Soc.*, Japan, 76: 203.

Menozzi, G., *et al.*, 1997. *Journal of Heterocyclic Chemistry*, 3,4: 963.

Mohamed, E.A., Ismailm, M.M., Gabr, Y., Abass M., and Farraq, H.A., 1995. *Indian Journal of Chemistry*, 34B: 21–26.

More, P.G., Bhalvankar, R.B. and Pattar, S.C., 2001. *J. Indian Chem. Soc.*, 78: 474–475.

More, S.V., Dongarkhedkar, D.V., Chawan, R.N., Jadhao, W.M., Bhusare, S.R. and Powar, R.P., 2002. *Journal of Indian Chemical Society*, 79: 768–789.

Naik, R.D. and Desai, K.R., 1998. *Oriental Journal of Chemistry*, 14(1): 161.

Nathwani, D. and Wood, M.H., 1993. *Drugs*, 45: 866.

Nene, Y.L., 1970. Oxford and IBH Publ. Co., Bombay, p. 282–296.

Pandeya, S.N., Sriram, D., Nath, G., De Clercq, E. and II Farmaco, 1999. 54: 624–628.

Penning, T.D., *et al.*, 1997. *Journal of Medical Chemistry*, 40(4): 1347.

Powar, R.A. and Patil, A.A., 1994. *Indian Journal of Chemistry*, Section B. 33: 156.

Dincer, S., 1996. *Indian Journal of Chemistry*. Section B. 35: 1335.

Reddy, V.M., Rama, G.V.S. and Sarma, 1993. *Indian Journal of Heterocyclic Chemistry*, 3: 111.

Schanappinger, D. and Hillen, W., 1996. *Arch. Microbiology*, 165: 359.

Shingare, M.S. *et al.*, 1999. *Indian Journal of Heterocyclic Chemistry*, 9: 153.

Taggi, A.P., Hapz, A.N., Wack, H., Oung, B., Ferrari, S.D. and Lectke, 2002. *P.J. Am. Chem. Soc.*, 24: 56626–6635.

Turan-Zitouni, G., Sivaci, M., Kilic, F.S. and Erol, K., 2001. *Eur. J. Med. Chem.*, 326–685.

Venturini, A., and Ongalez, G., 2002. *J. Org. Chem.*, 67: 9089–9092.

Vincent, V.M., 1947. *Nature*, 96: 596.

Chapter 20

Impact of Tsunami on Physico-chemical Characteristics of Coastal Water in Kulachel Area of Kanyakumari District

V. Ramasubramanian, R. Jeyaprakash and S. Jeyakumar

Department of Plant Biology and Plant Biotechnology, Ayya Nadar Janaki Ammal College (Autonomous), Sivakasi – 626 124, Tamil Nadu, India

ABSTRACT

Comprising over 70 per cent of the earth's surface, water is undoubtly the most precious natural resource on our planet. The global and regional climatic changes, influence life process even at local level. The recent tsunami in the Indian Ocean has influenced the quality of water in coastal areas of Tamil Nadu. Thus, we programmed to assess the quality of water in tsunami severely hitted Kulachel area in Kanyakumari District.

Introduction

On 26th December 2004, an earthquake of *M* 9.0 shook the entire coast of India and Andaman and Nicobar Islands. Centred off the coast of Sumatra in the Indian Ocean, the earthquake caused extensive damage in the Andaman and Nicobar Island regions of India and was felt strongly all along the east coast of India from Srikakulam in Andhra Pradesh (AP) to Kanyakumari in Tamil Nadu (TN). The earthquake set off hugs tsunami, which travelled throughout the world oceans and severly affect the several countries in the Indian Ocean. The present study was made to assess the important water quality parameters due to unusual changes in the coastal region of Kulachel area in Kanyakumari District.

Material and Methods

The water samples were collected from the coastal region of Kulachel area in Kanyakumari District during 10th January 2005. The water samples were collected at four different parts of the area, *viz.*, bottom of the sea (20 m depth), estuary region, Poomparai Valkal area and middle of the sea (far away 1.5 km) from sea shore). Samples were bought to the laboratory and the water quality considering the parameters such as, pH, alkalinity, electrical conductivity, total dissolved solids, turbidity, total hardness, calcium, magnesium, nitrite, nitrate, chloride, fluoride, sulphate and phosphate contents were analysed. Sample taken in that area before the tsunami was maintained as control for comparison.

Results and Discussion

In the experimental samples, collected from different areas, turbidity, total dissolved solids and electrical conductivity were more compare to the control samples (Table 20.1). pH of the experimental samples ranged between 7.5–8.2 in all the stations, which is marginally higher than the control samples. The dissolved oxygen level in the water sample ranged from 1.5 to 6.1 mg/l during January 2005. While it is 6.8 mg/l control samples indicating a drastic reduction after the tsunami. The alkalinity in, experimental samples ranged between 98–188 ppt, but in control, the alkalinity was 94 ppt, the calcium, magnesium and nitrate concentration in experimental samples showed higher values than the control water samples. Sulphate and phosphate content of water samples were also increased after tsunami. The same result was also proposed by Chadha *et al.* (2005) and Reddy *et al.* (2005). Along the west coast of India, the presence of high salinity, alkalinity, low dissolved oxygen and nutrien trich water has been reported during south-west monsoon seasons, when subsurface water reaches the coastal zone due to upwelling (Sankaranarayanan and Jayaraman, 1972).

Table 20.1: Parameters of Water Samples Collected from Different Areas

Sl.No.	Parameters	Control	Inside the Sea	Estuary	Poomparai Valkal	Away from 1.5 km Seashore
1.	Appearance	C and C	C and C	C and C	C and C	Sl. Brownish
2.	pH	7.2	8.2	7.5	7.5	7.6
3.	Electrical Conductivity (mS/cm)	1451	34720	12885	28800	2100
4.	Turbidity	0.84	1	2	4	1.2
5.	Total dissolved solids (mg/l)	916	24304	9006	20160	1147
6.	Dissolved oxygen	6.8	5.80	3.7	6.1	1.5
7.	Alkalinity	94	116	120	188	98
8.	Hardness	1050	5500	2100	4650	1480
9.	Calcium	115	400	160	300	171
10.	Magnesium	46	1080	408	936	54
11.	Nitrite	0.01	0.07	1.38	0.58	0.05
12.	Nitrate	5	1	4	2	2
13.	Sulphate	146	2440	774	1845	184
14.	Phosphate	0.01	0.07	0.02	0.07	0.1

All the values are expressed in mg/l, except pH and Electrical conductivity

C and C: Clear and Colourless.

However, water samples during January 2005 from the coastal region of Kulachel area exhibited high alkalinity, hardness and low dissolved oxygen indicating the impact of tsunami on the characteristics of coastal water. This study reveals that, drastic change in the coastal water quality due to the tsunami has brought about a change in the coastal biodiversity.

References

Chadha, R.K., Latha, G., Harry Yeh, Curt Peterson and Toshitama Katada, 2005. The tsunami of the great Sumatra earthquake of *M.* 9.0 on 26th December 2004. Impact on the east coast of India. *Curr. Sci.*, 88(8): 1297–1301.

Reddy, H.R.V., Katti, R.J., Raveesha, K.P., Chandarasekara, Vikas, S.J., Nagendra Babu, R.S. and K.S. Shiv Kumar, 2005. Coastal water quality off Dakshina Kannada before and after tsunami. *Curr. Sci.*, 88(7): 1026–1027.

Sankaranarayanan, V.M. and N.R. Jayaraman, 1972. *Curr. Sci.*, 46: 204–205.

Chapter 21

A Brassinosteroid as an Antihyperglycaemic in Alloxan Induced Diabetic Rats

P. Muthuraman and K. Srikumar*

Department of Biochemistry and Molecular Biology, School of Life Sciences, Pondicherry University, Kalapet, Pondicherry – 605 014, India

ABSTRACT

Glucose transport into animal cells from circulating plasma is regulated by pancreatic insulin and specific glucose transporters. Investigations employing homobrassinolide (HB) a brassinosteroid isoform, on circulating glucose level in alloxan induced diabetic male albino wistar strain rats registered reduced (80 mg/dl) blood glucose level upto 144 hrs, and was observed to be very similar to a gliclazide test group. Increase in triglycerides (100 per cent), HDL (25 per cent), LDL (130 per cent), VLDL (100 per cent) and protein (100 per cent) within 6 hrs was also observed in the HB treated rats. The use of human plasma combined with HB yielded a protein bound HB fraction during gel filtration on Sephadex-G 50 suggestive of the existence of HB binding proteins/transporters in human plasma. HB's ability to reduce circulating glucose level eventhough affecting plasma insulin content suggests a role for HB as a natural antihyperglycemic compound available to animals and man in the diet.

Keywords: *Gliclazide, Glucose, Homobrassinolide, Plasma proteins, Regulation.*

* Phone: +91-413-265-4422; E-mail: kotteazeth_srikumar@excite.com

Introduction

Diabetes mellitus is characterized by hyperglycaemia due to insulin deficiency. A significant change in the form of fuel usage from carbohydrates to fats occurs in this condition (Andallu, 2000). Diabetic subjects exhibit either insulin insufficiency, or a progressive resistance to insulin action (Annapurna and Mahalakshmi, 2001). Diabetes also predisposed individuals to morbidity and premature mortality. Most complications encountered in diabetes are found to be irreversible. It is believed that despite suppression of insulin activity for ketogenesis, gluconeogenesis and glucose transport continued unabated in the cells of a diabetic tissue. Glucose remained the preferred energy source for cells for immediate use. A continuous source of glucose was therefore an essential requirement for man and animals. Diet, glycogenesis and glycogenolysis acts as the three primary sources of glucose for a cell. During glycogen depletion however denovo synthesis of glucose occurred through gluconeogenesis using ketogenic amino acids as a source of carbon.

Glucose availability to cells is further facilitated by glucose transporters (GLUT-1 to GLUT-7). A sodium dependent glucose transporter-1 (SDGT-1), is also found to be active in the kidney and the intestinal cells (Pantalcon and Kaye, 1998). Additionally, glucose level in the blood is maintained within a fairly narrow range (80-120 mg/dl) by regulatory hormones insulin, glucagon and epinephrine (Salt and Gabbriele, 1998). Plasma glucose is taken up by the cells when influenced by insulin. Insulin, therefore, decreased blood glucose level, and is counter regulated by glucagon, cortisol, epinephrine and growth hormones.

Following entry into cells, glucose is phosphorylated by hexokinase in most tissues or by glucokinase in the liver. Phosphorylated glucose can be channallized to one of the three metabolic pathways, such as for the formation of (1) pyruvate through glycolysis, (2) 6-phosphogluconate through hexose monophosphate (HMP) shunt (Henry and Wilson, 1997) catalyzed by glucose-6-phosphate dehydrogenase (G6-PD), while yielding pentoses, NADH and ATP, and (3) glucose-1-phosphate (G1-P) and uridine diphosphate-glucose (UDPG) through a third metabolic sequence in animal cells. The metabolic fate of G-6-P therefore remains governed by these pathways (Gidth *et al.*, 1993).

Clinical manifestations of deranged glucose utilization has been recognized in human as Insulin Dependent Diabetes Mellitus (IDDM/Type-I), as a consequence of insulin deficiency due to degenerating pancreatic β-cells and Non-Insulin Dependent Diabetes Mellitus (NIDDM/Type-II), due to deficiency in the availability and/or activity of hexokinase, or due to a deficiency in the glucose transporter activity (John, 1997). Whereas IDDM (Stoffel and Takeda, 1992) is a classical example of a genetic disorder, NIDDM is more often an age dependent manifestation of deranged glucose metabolism (Chakrabarti and Rajagopalan, 2002). Obesity, under activity and over eating contributed to the development of NIDDM characterized by diminished hepatic and muscle tissue sensitivity and progressively impaired (beta) cell function (Belmin, 2003).

The management of type II diabetes involved the use of oral agents and insulin therapy. Distinct classes of oral hypoglycemic agents are in current use such as sulphonylureas, meglinitides, biguanides, thiazolidinediones, α-glucosidase inhibitors, metformin, acarbose, pioglitazone and rosiglitazone. Many of the oral anti-hyperglycaemic agents while improving the dyslipidemic profile, reduced vascular inflammation and endothelial dysfunction (Robert, 2003). Search is, therefore, on to identity anti-diabetic drugs in folk and traditional systems of medicine like Ayurveda and Unani.

Studies were, therefore, initiated to investigate the effect of an isoform of the phytohormone brassinosteroid on glucose metabolism in male rats. Brassinosteroids in general influenced growth, seed germination, rhizogenesis, flowering and senescence in plants. The hormone is reportedly found at very low concentration in pollen, seeds and young vegetative tissues of plants (quantity per gram tissue weight unknown). Isomeric forms of this compound are known as epi and homo brassinolides. They exhibited a polyoxygenated steroid structure. They confered resistance to plants against various abiotic stresses (Marco and Burgos, 2003). A leucine-rich protein (BRL1), was identified as binding the polyoxygenated steroid in Arabidopsis thaliana and was considered a brassinolide receptor (Wang and Chory, 2006). Receptor binding by brassinolides caused activation of a receptor kinase domain and led to the phosphorylation of additional kinases within plant cells, similar to the cell signaling process found in animal cells. It was hypothesized by us therefore that either the epi or the homobrassinolide isoforms may potentially have receptors within animal cells, possibly mediating it's biological potency in animal cells while yielding recognizable circulatory responses. The study therefore investigated the effect of homobrassinolide in the reduction of blood sugar level in alloxan induced diabetic rats. Of the 44 and odd brassinosteroid varieties available, homobrassinolide [(6-Oxa-7-oxo-28-homobrassinolide) $C_{28}H_{48}O_6$] with molecular weight 480.69 was used for the study.

Materials and Methods

Male Albino wistar strain (8–10 week old) rats were used for carrying out the investigations. The animals were housed in propylene cages under controlled temperature and hygiene conditions, with 12 hours of light and dark cycle throughout the experimental period. The animals were provided free access to drinking water *ad libitum*. Groups of three rats were taken for the study. Blood was collected ftom tail vein by venous puncture and ftom the orbital sinus. Plasma was separated at low speed in a clinical centrifuge and used for biochemical studies. The equipments used included a refrigerated high speed centrifuge, eppendorf microfuge and Systronics-119 UV-visible spectrophotometer. Technical grade homobrassinolide was received courtesy of Dr. Vyas, Godrej Agrovet, Mumbai. Alloxan, Gliclazide, BSA, glucose and all other chemicals used were of analytical grade purchased from manufacturers in India. Glass distilled water was used for preparation of all reagents.

Experimental diabetes was induced in a group of 3 rats, each administered alloxan (150 mg/kg body weight), and allowed to grow for a further 15 days for the generation of diabetes. Controls were maintained normal. Following induction of diabetes, homobrassinolide (50μg) in 95 per cent ethanol, and gliclazide were administered intraperitoneal, intradermal and oral to different animals. Blood samples were collected from rats in each group at 2, 6, 12, 24 and 144 hrs. A comparative study was also carried out using heat treated (95°C, 5 min) homobrassinolide. Plasma and serum were separated for biochemical analysis. Mean of three investigations are reported here. No statistical analysis was carried out.

Blood glucose was estimated by the Asatoor and King method (Alan and Janet, 1954). The total cholesterol was estimated by the method of Zak *et al.* (1954). Triglycerides are enzymatically hydrolyzed to glycerol prior to estimation (Gruddy and Benjamin, 1999). The chylomicron, VLDL, and LDL are precipitated by addition of phosphotungstic acid and magnesium chloride, centrifuged and the supernatant fluid containing HDL fraction was assayed using cholesterol reagent (Bairaktari and Hatzidimou, 2000). Each sample was also estimated for total protein content by the method of Lowry *et al.* (1951). Measurement of plasma insulin level was carried out using RIA employing guinea pig anti-rat insulin primary antibodies.

The possible presence of a binding protein for HB in human plasma that possibly acted as a carrier of this phytohormone following absorption from the intestine was investigated by gel titration technique employing Sephadex G-SO at room temperature. Gel filtration column (14 cm × 0.5 scm) standardization (flow rate 60 ml/h) included (*a*) the establishment of elution volume Ve for free BSA (100 µg/ml), free HB and free rat blood plasma, (*b*) determination of the void volume of the column using a sample (5 mg/ml) of blue dextran. (*c*) Calibration of the column employing standard protein markers (1 mg/ml) in independent column runs, to establish the Ve of each marker. Following standardization, HB, BSA, plasma, HB+BSA, and HB+ plasma samples were applied to the column in separate runs to determine their respective Ve. The molecular weight was determined from a plot of Ve/Vo versus log molecular weight of standard proteins and blue dextran. The molecular weight of the complex was determined from such a plot.

Results and Discussion

The alloxan induced diabetic rat registered blood glucose level at 200 mg /dl. The glucose content of normal rat blood was 80 mg /dl. The blood glucose content of homobrassinolide treated rats at 2, 6, 12, 24 and 144 hrs post intraperitoneal, intradermal and oral treatment are as shown in Figure 21.1. In certain cases, the glucose content was observed to be below normal levels (Figure 21.1) suggestive of hypoglycemic state. The animals, however, did not become inactive or slip into coma. Estimate of insulin content of plasma in microunits is as given in Figure 21.2. Intradermal administration of heat treated homobrassinolide yielded result comparable to that of gliclazide at 6hr, and close to that of native homobrassinolide at 12 hr (Figure 12.1). At 24hr however, both compounds yielded similar responses. Native homobrassinolide maintained the lowest glucose level in the diabetic animals upto 144 hr, while gliclazide treated animals responded similarly only at 144 hr. The effect of heat treated homobrassinolide at 144 hr was found to be greater than that for native HB and gliclazide.

Determination of blood Cholesterol, Triglyceride, HDL, LDL, and VLDL content of each blood sample yielded data as shown in Figure 21.3. Blood cholesterol in homobrassinolide treated rat increased

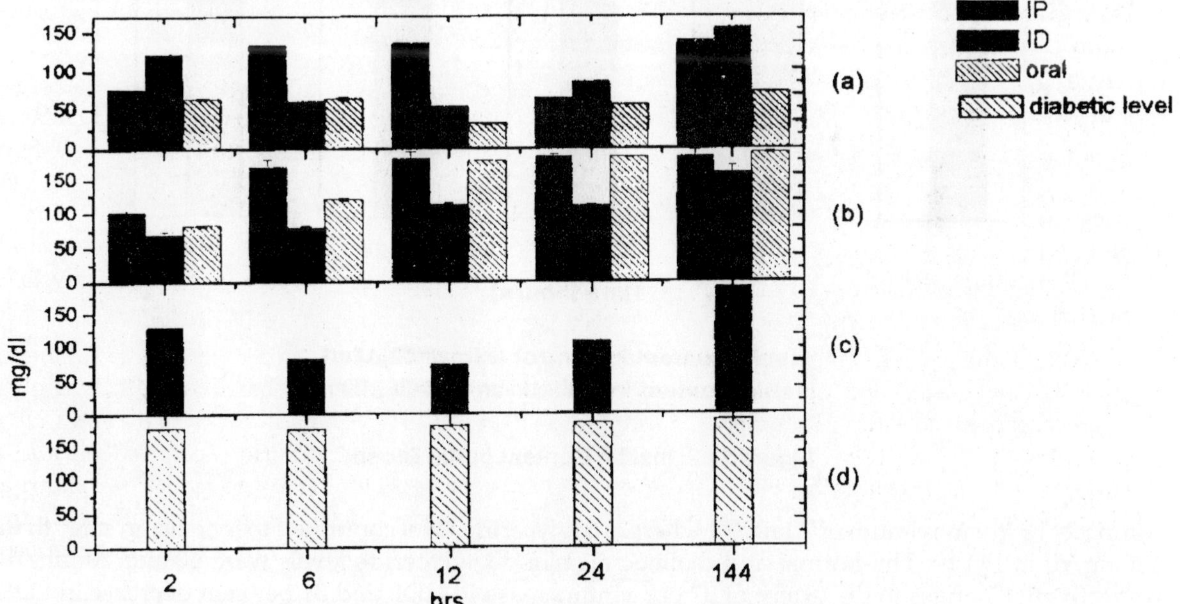

Figure 21.1: Glucose Content of Rat Blood
(*a*) HB treated; (*b*) Gliclazide treated; (*c*) Heat treated HB; (*d*) Diabetic Level (Alloxan Treated)
Glucose level in control animal = 80 mg/dl

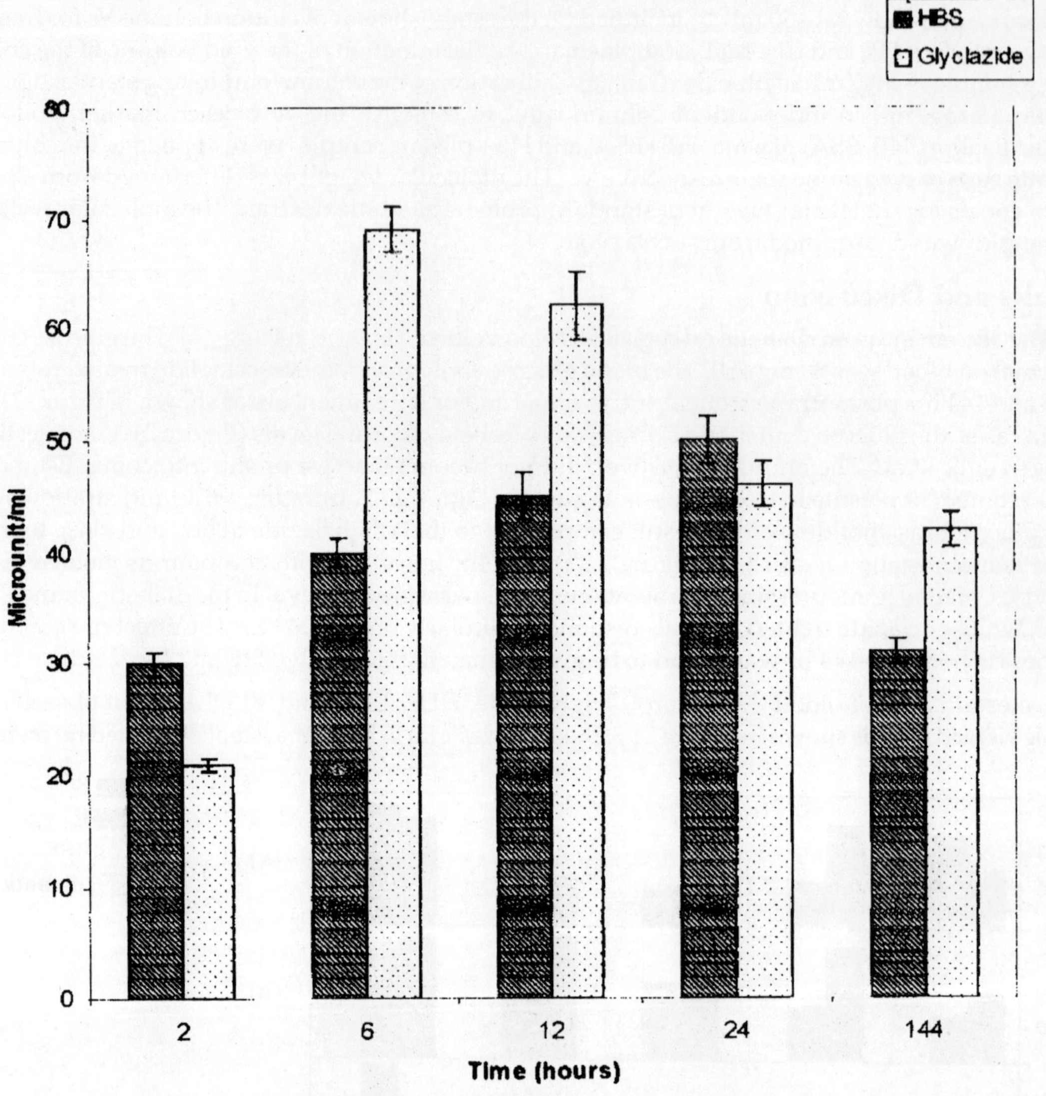

Figure 21.2: Insulin Content of Rat Blood

during 24 hr to a maximum of 70mg/dl whereas triglyceride level continued to increase greater than 120mg/dl in 144 hr. The normal and diabetic rat blood triglyceride levels were 62 and 70 mg/dl respectively. Changes to the extent of 37 per cent increase in HDL and 81 per cent decrease in LDL occurred in gliclazide treated rat, while 100 per cent increase in VLDL was observed in the same animals. Triglyceride increased nearly 90 per cent over the value in gliclazide treated rats and cholesterol content increased 32 per cent in these animals. Interestingly cholesterol increased nearly 81 per cent,

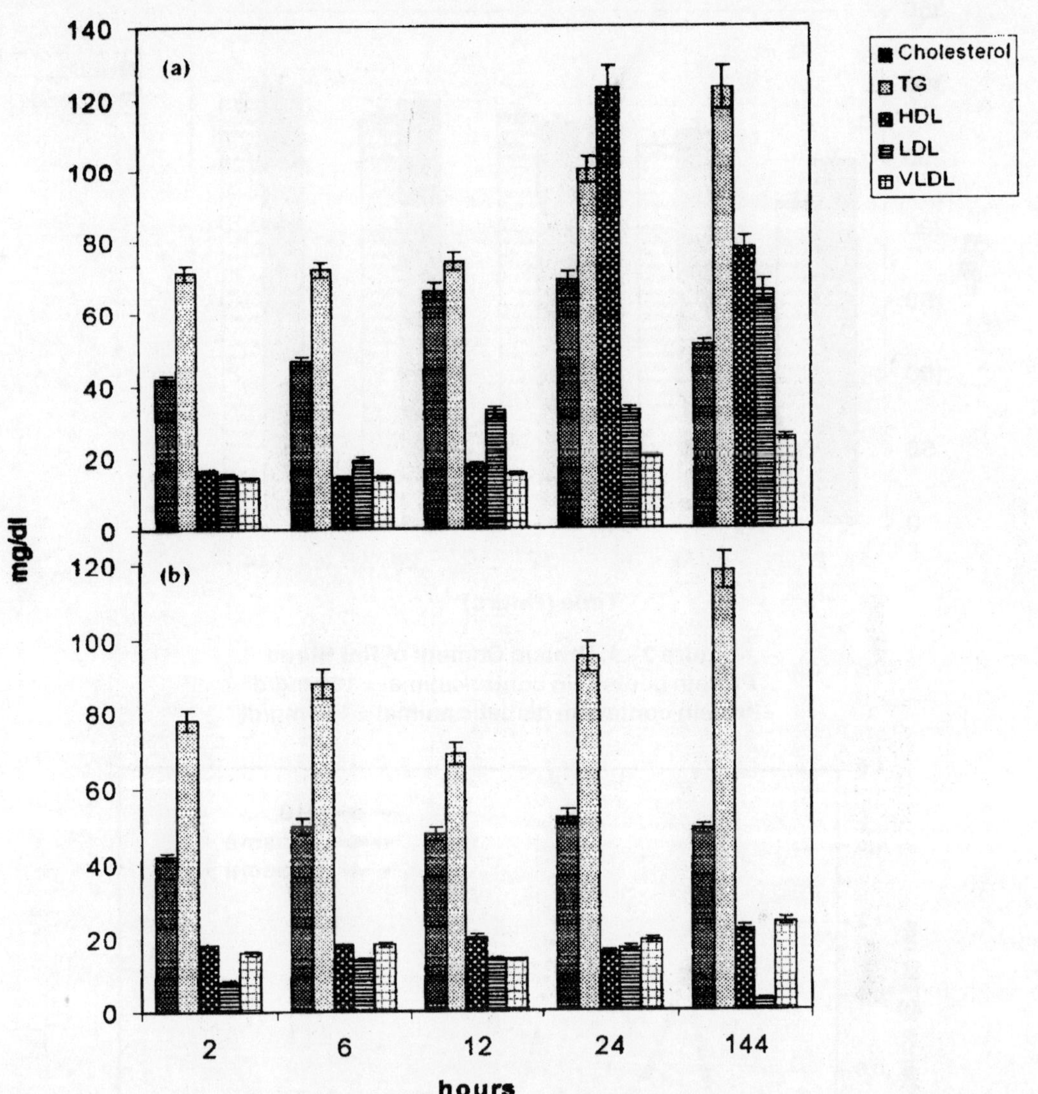

Figure 21.3: Lipid Content of Rat Blood
(*a*) HB treated; (*b*) Glyclazide treated

triglycerided 100 per cent, HDL 25 per cent, LDL 130 per cent and VLDL 100 per cent in the homobrassinolide treated rats. The total protein content of rat plasma (Figure 21.4) showed an increase of 110 per cent within 6 hr following homobrassinolide treatment compared to the total protein content of diabetic rat, and remained increased through a 24 hr period.

Gel permeation column chromatography using human plasma mixed with homobrassinolide yielded a distinct absorbance peak at 280 nm for the complex (Figure 21.5), that was independent of the absorbance peak either due to free plasma or free homobrassinolide. The use of bovine serum

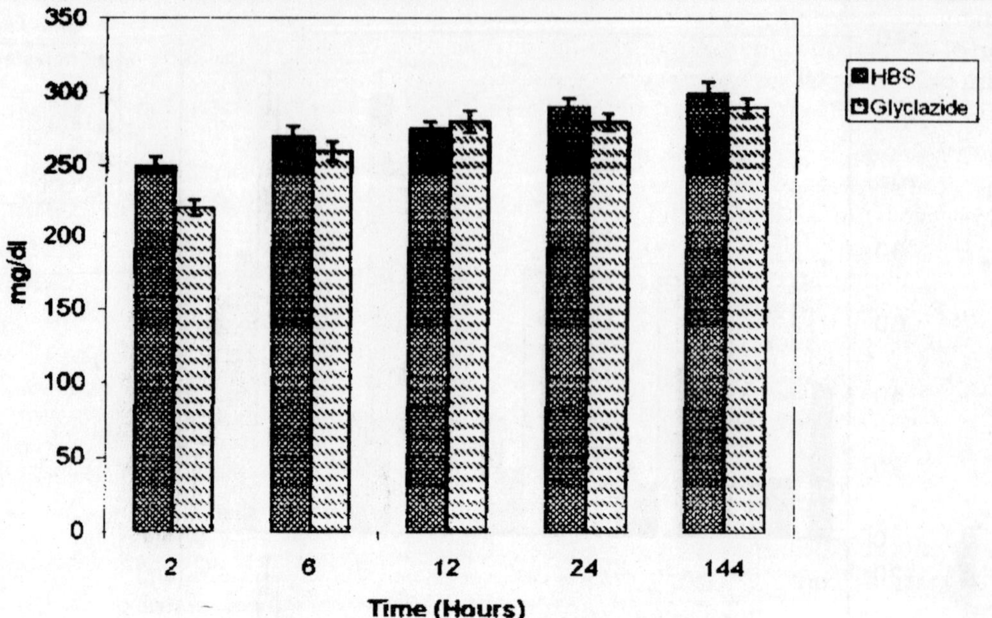

Figure 21.4: Protein Content of Rat Blood
Protein content in control animal = 130 mg/dl
Protein content in diabetic animal = 140 mg/dl

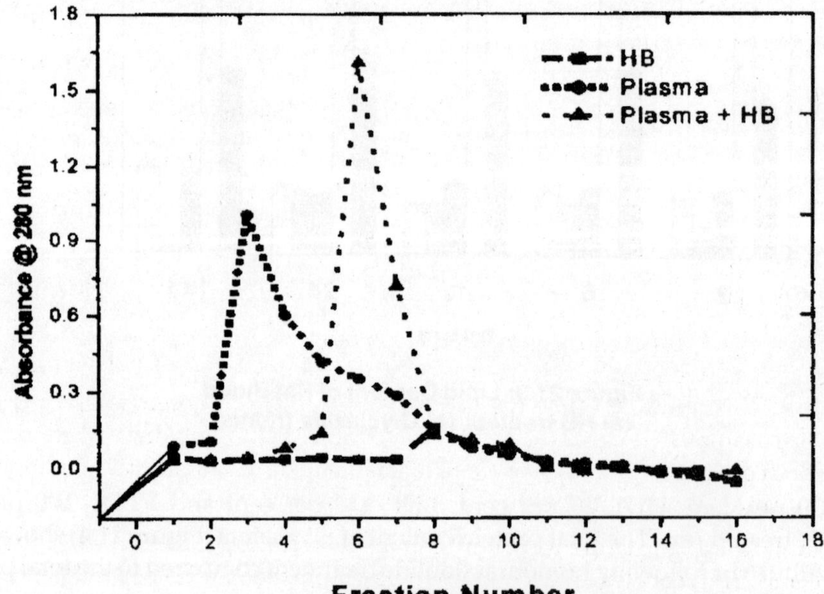

Figure 21.5: Gel Permeation Column Chromatography
Column Material: Sephadex G-50
Bed Volume: 20 ml
Fraction Volume: 2 ml
Flow rate: 60 ml/hr

albumin mixed with homobrassinolide also yielded a distinct absorbance peak at 280 nm during gel filtration of the mixture, independent of either the free BSA, or free homobrassinolide. The absorption spectrum of homobrassinolide generally yielded reportedly three consecutive peaks with Amax at 249, 255 and 26 nm, ahead of the absorbance peak due to plasma proteins at 280 nm.

The presence of homobrassinolide in the plasma and the BSA fractions was detected by spectroscopic analysis of each fraction at 249, 255 and 261 nm respectively in addition to taking absorbance measurements at 280 nm for the detection of protein in the fractions, based on the premise that any absorbance by the sample in the former three wavelengths wiil be suggestive of the presence of homobrassinolide in the plasma protein ftaction or the BSA fraction. The Gel filtered fractions containing human plasma yielded higher absorbance values at the lower uv wavelengths, suggestive of the presence of homobrassinolide associating with plasma protein in the fractions.

Steroids have come to be known to reduce basal glucose metabolism in isolated fat cells (Sutter-Dub *et al.*, 1988). Insulin stimulated glucose oxidation is known to be inhibited by some steroid antagonists (Sutter-Dub *et al.*, 1988). In castrated animals, estradiol increased basal insulin level and the quantum of insulin binding to fat, liver and muscle cell membranes. Progestrone reportedly increased basal insulin level and decreased insulin binding to fat cells. Thus, ovarian sex steroids exhibited significant effect on insulin binding to target cells (Cagnacci *et al.*, 1999). Estrogen augmented glucose metabolism whereas progestrin and androgen brought about deterioration in glucose metabolism (Ballejo *et al.*, 1983). Anabolic steroid diminished glucose tolerence while contributing to substantial increase in post glucose serum insulin concentration. It is reported that 72 hr alloxinated rat exhibited higher plasma cortisone levels and increased adrenal gland/body weight ratio (Bruno *et al.*, 1997). Changes in steroid feed back regulatory mechanism is implicated for this phenomenon that is further related to the hypothalamic–pituitary–adrenal cortical axis. Daily treatment of alloxan induced diabetic rat with an aquous extract of *Cissus sicyoides* (AECS) for seven days reportedly decreased blood glucose level in about 25 per cent of target animals. Measurement of AST and ALT enzyme activity levels in these treated rats were however reported to be within the normal range (Glauce *et al.*, 2004).

Differences in the level of glucose in the blood of control, diabetic and homobrassinolide treated diabetic rats indicated that the use of this compound resulted in the reduction of circulating glucose level, for nearly 144 hrs. Comparison with gliclazide treated animals in the animals indicated a sustained antihyperglycemic effect by homobrassinolide for upto atleast 24 hr. Neverthless a decline in the potency of the drug within 12 hr was noted when fed orally. The use of heat-treated HB, however, yielded a blood glucose profile similar to that for the normal compound, although marginal decrease in antihyperglycemic effect of the compound was observed. The animal's response to the compound was observed to be route dependent, such as whether administered intraperitonial, intradermal or oral. Orally fed HB exhibited extraordinary ability to maintain glucose level below 80 mg/dl upto 144 hrs. RIA estimates noted increase in plasma insulin content following gliclazide and HB treatment of the animal. The increase was however abrupt and maximal at 6 hr when gliclazide was administered. In contrast, the increase was gradual, reaching a maximum during the first 24 hr, and returning to normal level within 144 hr. when HB was employed.

Homobrassinolide and gliclazide were both capable of reducing the HDL, LDL and VLDL levels, while cholesterol and triglycerides levels remained undiminished in the normal and the diabetic rat blood. Differences in the gel elution profile of the complex compared to that of free plasma protein or of free homobrassinolide suggested the formation of a human plasma protein-homobrassinolide complex. The presence of homobrassinolide in the complex was confirmed by monitoring the absorbance @ λmax of HB in each column fraction. Absorbance values of the free plasma protein, free homobrassinolide, and the protein-homobrassinolide complex were then related to their respective elution volume (Ve). This study indicated an antihyperglycemic effect for HB as well as the existence of a plasma protein capable of complexing with homobrassinolide at room temperature and at pH 7. It is interesting to believe therefore that a carrier protein facilitated the vascular transport of homobrassinolide and functioned to release the hormone to the cells in a regulated manner to exert homeostatic control over glucose in circulation.

References

Alan, H. Gowenlock and Janet, R. McMurray, 1954. *Harold Varley*, 4: 56–58.

Andallu, B., 2000. Control of hyperglycemia and retardation of cataract by mulberry (*Morus indica* L). *Indian J. Exp. Biology*, 40: 791–795.

Annapurna, A. and Mahalakshmi, D., 2001. Antidiabetic activity of a poly herbal preparation (tincture punchapuma) in normal and diabetic rats. *Indian J. Exp. Biology*, 39: 500–501.

Bairaktari, E. and Hatzidimou, K., 2000. Estimation of LDL cholesterol based on Friedwald formula on apo-protein B levels. *Clin. Biochem.*, 33(7): 549–555.

Ballejo, G., Saleem, T.H., Khan-Dawood, F.S., Tsibris, I.C. and Spellacy, W.N., 1983. The effect of sex steroids on insulin binding by target tissues in the rat. *Contraception*, 28(5): 413–422.

Belmin, J. Valensip, 2003. Novel drug delivery system for insulin: Clinical potential for use in the elders. *Drugs and Aging*, 20(4): 303–312.

Bruno, L.D., Magda, F.S., Emiliano, O.B., Renato, S.B., Marco, A.M. and Patrica, M.R., 1997. Antigen-induced pleural eosinophilia is suppressed in diabetic rats: Role of corticosteroid hormones. *Mem Inst Oswaldo Cruz, Rio de Janeiro*, 92: 219–222.

Cagnacci, A., Paoletti, A.M., Arangino, S., Melis, G.B. and Volpe, A., 1999. Effect of ovarian suppression on glucose metabolism of young lean women with and without ovarian hyperandrogenism. *Hum. Reprod.*, 14(4): 893–897.

Chakrabarti, Rajan and Ramanujam Rajagopalan, 2002. Diabetes and Insulin resistance associated disorders: Disease and the therapy. *Current Science*, 83: 12–25.

Gidth, M. Jain and Takeda, J. Xulz, 1993. Glucokinase mutation associated with non-insulin dependent (type II) *Diabetes mellitus* have decrease enzymatic activity. Implications for structure/function relationships. *Proc. Natl. Acad. Sci.*, 90: 1932–1936

Glauce, S.B. Viana, Ana Carolina C. Medeiros, Ana Michelle R. Lacerda, Katyne A.M. Leal, Tigo G. Vale and Jose D. Abreu Matos, 2004. Hypoglycaemic and anti-lipemic effects of the aqueous extract from *Cissus sicyoide. BMC Pharmacology*, 4: 1–9.

Gruddy, S.M. and Benjamin, I.J., 1999. Diabetes and cardiovascular disease: A statement for health care professionals from the American heart association. *Circulation*, 100: 1134–1146.

Henry, J. Tsai and John, E. Wilson, 1997. Functional organisation of mammalian hexokinases: Characterization of rat Type III isozyme and its chimeric forms, constructed with the N and C terminal halves of the type I and II isozymes. *Archives of Biochem. and Biophysics*, 338: 183–192.

John, R. White, 1997. Combinational oral agent/insulin therapy in patient's with type II diabetes mellitus. *Clinical Diabetes*, 7: 34–35.

Lowry, O.H. and Rosebrough, N.J., 1951. Protein measurement with Folin-phenol reagent. *J. Biol. Chem.*, 193: 265–275.

Marco, Antonio Texeira Zollo and Mariangela de Burgos, 2003. Some notes on the terminology of Brassinosteroids. *Plant Growth Regulation*, 39: 1–11.

Maric, Pantalcon and Peter L. Kaye, 1998. Glucose transporters in preimplantation development. *Review of Reproduction*, 3: 77–81.

Robert, R. Henry, 2003. Insulin resistance from predisposing factor to therapeutic target in type II diabetes. *Current Therapeutics*, 25: B47–B63.

Salt, I. Ian and Gabbriele, 1998. AMP activated protein kinase is activated by low glucose in cell lines derived from pancreatic B-cells and may regulate insulin release. *Biochem. J.*, 335: 533–539.

Stoffel, M. and Takeda, J., 1992. Human glucokinase gene: Isolation, characterization and identification of two missense mutations linked to early onset non-insulin dependant (type II) diabetes mellitus. *Proc. Natl. Acad. Sci.*, 89: 7698–7702.

Sutter-Dub, M.T., Kaaya, A., Sfaxi, A., Sodoyez-Goffaux, F., Sodoyez, J.C. and Sutter, B.C., 1988. Progesterone and synthetic steroids produce insulin resistance at the post-receptor level in adipocytes of female rats. *Steroids*, 52(5–6): 583–608.

Wang, Xuelu and Joanne Chory, 2006. Brassinisteroids regulate dissociation of BKII, a negative regulator of BRII signaling, from plasma membrane. *Science*, 313: 1118–1121.

Zak, B. Dickenbaum and Whilte, R.L., 1954. *Amer. J. Clin. Path.*, 24: 1307–1311.

Chapter 22

Production of Calcium Gluconate by Fermentation

Sanjay Timande and Anshu Chopda
P.G. Department of Microbiology, Dhote Bandhu Science College, Gondia – 441 614, Maharashtra

ABSTRACT

A strain of *Aspergillus niger,* screened out from the soil collected from the vicinity of sugarcane industry, was found to produce notable level of calcium gluconate in fermentation. Fermentation was performed at 30°C for 3 days in medium containing KH_2PO_4: 0.1 per cent, $MgSO_4.7H_2O$: 0.05 per cent, Glucose: 10 per cent, Peptone: 0.2 per cent, and $CaCO_3$: 3.2 per cent w/v (Initial pH: 6.5). Other optimal cultural conditions were: volume of inoculum: 12 per cent v/v and volume of production medium: 25 ml. Of the factors tested, the sources of carbon and nitrogen in the medium had the greatest effect on calcium gluconate production. With carbon and nitrogen, at optimal level, *A. niger* accumulated 100.89 g/l calcium gluconate per liter. Of the carbon sources tested, glucose (10 per cent w/v) gave maximum production of calcium gluconate. Of the nitrogen sources tested, peptone was best for calcium gluconate production. A maximal level of calcium gluconate was produced, when 0.2 per cent w/v of peptone was used.

Keywords: *Aspergillus niger, Calcium gluconate production, Still fermentation.*

Introduction

Calcium gluconate is an imperative salt of mineral calcium. It is absorbed fairly well by the body. Calcium gluconate ($C_{12}H_{22}CaO_{14}H_2O$) occurs as a white, crystalline powder or granules. It is tasteless and odorless (Pharmacopoiea, U.S., 1990). It is extensively used in the treatment of calcium deficiency, both for oral and parenteral administration (www.nrdcindia.com). Calcium is also needed for the heart muscles and nervous system to work properly (www.healthtouch.com). It is also used as an oral antidote for fluorine and onalic acid poisoning (www.nrdcindia.com). Calcium is necessary for bone formation and maintenance. Calcium regulates heart rhythm, nerve transmission, and muscle contraction. Calcium can also lower blood pressure by up to 10 points, relieve insomnia, nervousness, and irritability, and relieve muscle cramping, especially in the lower extremities (www.health.ivillage.com). Calcium gluconate is also widely used in food, leather, pharmaceutical and chemical industries (Tripathi *et al.*, 1990). It is used as substitute for fruit acids in the manufacture of fruit jellies and dentifices and metal polishes (www.nrdcindia.com). Calcium gluconate can be commercially produced from glucose by fermentation (Shah and Kothari, 1993) as well as chemical (Kundu and Das, 1984) and electrolytic methods (www.nrdcindia.com). Calcium gluconate is produced by the electrolytic oxidation of glucose dissolved in sodium bromide solution used as electrolyte in a cell having graphite electrodes. Gluconic acid formed is neutralised by the suspended calcium carbonate present in the electrolyte. Calcium gluconate formed in the electrolyte is recovered by cooling (www.nrdcindia.com). In fermentation process, specific microorganisms are grown in medium containing glucose and other ingredient. The process involves an initial bioconversion by glucose oxidase. Increased commercial demand of calcium gluconate has stimulated various scientists to investigate efficient fermentation process for increased production.

This paper describes the selection of *Aspergillus niger* strain S9 as potential calcium gluconate producer via fermentation and the optimization of cultivation system and the medium composition in enhancing the production of calcium gluconate.

Materials and Methods

Chemicals

All bacteriological media were purchased from M/s Hi-media Pvt. Ltd., Mumbai. Chemicals of analytical grade were supplied by M/s SDS Laboratory Chemical Industries, Mumbai. All other media components used were of high purity grade available commercially.

Samples Collected

For isolation of fungi, soil samples were collected in polythene bags from the vicinity of sugar industry, tea stall, jaggery industry and juice centre and brought to laboratory.

Isolation of Molds

For the isolation of fungi, the standard pour plate isolation technique of Johnson *et al.* (1959) was used (Johnson *et al.*, 1959).

Maintenance of Cultures

Each fungus was cultivated on potato dextrose agar slant (pH: 5.2–5.8) for 7 days at room temperature. Isolated fungi were maintained on potato dextrose agar slants at 4°C in refrigerator with periodic sub-culturing on slants prior to use.

Selection and Screening

The calcium gluconate producing potential of fungi was determined by cultivating isolated fungi in liquid medium. Suspension of conidial spores washed out from the potato dextrose agar slants by scratching using 10 ml of sterile physiological sodium chloride solution. 1 ml conidial suspension was used to inoculate 100 ml sterilized mineral salt medium in 250 ml conical flask containing (g/l), glucose: 150, $(NH_4)_2SO_4$: 2.5, $MgSO_4.7H_2O$: 0.5, KH_2PO_4: 1.0, $CaCO_3$: 32 (Mariam and Nagra, 2002). The pH of medium was adjusted to 6.0 before sterilization. The medium was sterilized at 100°C for 10 minutes at pressure 10 lb before inoculation. The inoculated flasks were incubated for 10 days at room temperature in stationary condition. After incubation, culture broths were separated from mycelium by passing contents of flasks through Whatmann filter paper no. l. The culture filtrate were used for calcium gluconate estimation. Calcium gluconate produced in culture filtrates was determined for selection of mold strain efficient in production of calcium gluconate.

Estimation Methods

The contents of flasks after the completion of fermentation period were filtered through Whatmann filter paper no. 1 and the filtrates were used for estimation of glucose as well as for calcium gluconate.

Estimation of Glucose

DNS (dinitrosalicylic acid) method was used for glucose estimation (Miller, 1963). The transmittance was measured at 540 nm by spectrophotometer.

Estimation of Calcium Gluconate

The fermented broth was filtered and the filtrate was used for analysis of calcium gluconate by the method of pharmacopoiea U.S. (Pharmacopoiea U.S., 1990).

Mycelial Dry Weight Determination

Mycelial dry weight was determined according to Ali and Qadeer (2001).

Identification of Efficient Mold Strain

The isolated mold strain, which was found most efficient in calcium gluconate production, was identified by standard procedure (Elnaghy and Megalla, 1975).

Optimization of Fermentation Condition for Calcium Gluconate Production (Fermentation Studies)

Inoculum Preparation

A pure efficient strain of *A. niger* isolated in our laboratory was used throughout this study. For the preparation of inoculum, 1 ml of spore suspension, which was prepared by dispersing the mold spores in 10 ml of sterile distilled water from 7 days old PDA slant was used to inoculate 100 ml of sterile mineral salt medium (Mariam and Nagra, 2002) of composition (g/l), glucose: 150, $(NH_4)2SO_4$: 2.5, $MgSO_4.7H_2O$: 0.5, KH_2PO_4: 1.0, $CaCO_3$: 32. The pH of medium was adjusted to 6.0 before sterilization. The inoculated medium was incubated at room temperature for 7 days. After 7 days, spore suspension was used to inoculate sterile production medium (Mariam and Nagra, 2002).

Fermentation Conditions

The optimization of medium composition and cultural conditions was carried out based on the stepwise modification of governing parameters for calcium gluconate production. Each optimized parameter was employed in subsequent experiments. The effect of different physical parameters like incubation period (1 to 7 days), initial pH of the medium (3.5, 4.0, 4.5, 5.0, 5.5, 6.0, 6.5, and 7.0), inoculum densities (2, 4, 6, 8, 10, 12, 14, 16, 18 and 20 per cent v/v), temperature (20, 25, 30 and 35°C) was studied on production of calcium gluconate by selected strain of *A. niger* using sterile production medium (Mariam and Nagra, 2002). Most commonly used carbon sources like lactose, maltose, sucrose, mannitol and xylose are included in study. The respective carbon sources (15 per cent w/v) were included in production medium as a sole source of carbon by substituting glucose. The effect of glucose concentration (5, 10, 15 and 20 per cent w/v) on calcium gluconate production was also studied. The nitrogen sources, ammonium nitrate, ammonium chloride, peptone, ammonium dihydrogen phosphate, ammonium oxalate were used in our study. The respective nitrogen sources were added (0.25 per cent w/v) in production medium as a sole source of nitrogen by substituting ammonium sulphate. The effect of peptone concentration (0.15, 0.2, 0.25, 0.3, 0.35 per cent w/v) on calcium gluconate production was studied. The effect of $CaCO_3$ concentration (1.6, 3.2, 4.8 and 6.4 per cent w/v) on calcium gluconate production was studied. The effect of different volume of fermentation medium (25, 50, 75 and 100 ml) on calcium gluconate production was also tested.

Results and Discussion

It is evident that, calcium gluconate production strongly dependent on selection of strain as well as cultural conditions employed.

Potential of Mold Isolates to Produce Calcium Gluconate

The production of calcium gluconate by microorganisms is a great achievement in the field of fermentation technology. In our studies, 10 mold isolates from various soil samples were tested for the production of calcium gluconate and mold strain (S9) was found to be most efficient in calcium gluconate production. The selected *A. niger* strain produced 33.36 g/l of calcium gluconate after 10 days on conidial inoculation taking 150 g/l of glucose as initial concentration. The data in Table 22.1 show potential of mold isolates to produce calcium gluconate.

Table 22.1: Isolate Number and the Amount of Calcium Gluconate Produced

Mold Isolate Number	*Calcium Gluconate (g/l)*	*Calcium Gluconate (per cent)*
S_1	30.27	22.53
S_2	6.73	5.01
S_3	17.94	13.35
S_4	2.24	1.66
S_5	15.69	11.69
S_6	13.45	10.01
S_7	6.73	5.01
S_8	4.48	3.34
S_9	33.36	25.04
S_{10}	26.90	20.03

Identification of Selected Mold Strain

The S9 mold strain producing maximum calcium gluconate was identified by standard procedure and it was identified as *Aspergillus niger* on the basis of its following characteristics:

1. Black colored colonies on PDA after seven days incubation at 25°C.
2. Vegetative mycelium septate, branched.

3. Conidiophore unbranched.
4. Conidia are globose to subglobose in shape, rough surface texture, and produced in chain.

The above characteristics (3 and 4) were studied by observing *A. niger* stained with lactophenol cotton blue under 100 X magnification of bright field microscopy.

Optimization of Fermentation Condition for Calcium Gluconate Production

The recent studies developed the production of calcium gluconate by *A. niger*. For maximum bioconversion of glucose into calcium gluconate, optimization of cultural conditions is essential.

Impact of Fermentation Period

The optimum incubation time for maximum production was found to be 3 days after conidial inoculation. By increasing or decreasing, the duration of incubation, calcium gluconate production decreases (Figure 22.1). Prolong incubation did not increase the production which might be due to the over growth of mycelium.

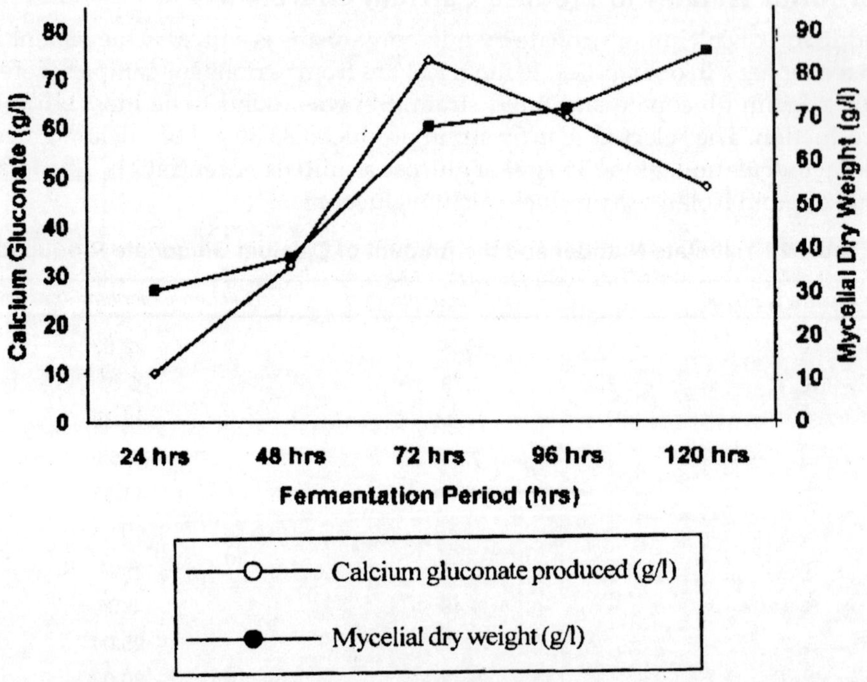

Figure 22.1: Impact of Incubation Period on Production of Calcium Gluconate

However, Panda *et al.* (1993) reported a period of 7 days for maximum yield (Panda and Subbarao, 1999). Hence, the present finding is more encouraging because reduction in the period reduced the cost of calcium gluconate production.

Impact of pH

Table 22.2 indicates that optimum initial pH for maximum production was found to be 6.5 and by changing the pH below and above optimum, calcium gluconate production decreases.

Table 22.2: Impact of pH on Production of Calcium Gluconate

pH	*Glucose (g/l)*		*Calcium Gluconate (g/l)*	*Calcium Gluconate (per cent)*	*Mycelial Dry Weight (g/l)*
	Residual	*Used*			
3.5	130	20	22.42	16.69	3.2
4	100	50	29.50	23.26	27.2
4.5	90	60	48.74	28.38	29.6
5	75	75	58.29	48.41	34.4
5.5	64	86	88.76	54.26	37.6
6	57	93	92.26	65.80	48.2
6.5	35	115	94.56	75.00	68.4
7	62	88	57.05	50.58	63.4

At pH 6.5, the consumption of glucose as well as calcium gluconate production was improved significantly. It may be due to the fact that at pH 6.5 the mycelium has maximum glucose oxidase activity. The mycelial dry weight was also poor at pH below and above optimum. Above results are in accord with the work of Tripathi *et al.* (1999). However, Kundu and Das (1994) got maximum yield of calcium gluconate at initial pH: 5.5. This may be due to type of strain or its physiological conditions.

Impact of Volume of Inoculum

Figure 22.2 shows that extracellular calcium gluconate synthesis was reached at maximum level with 12 per cent v/v of inoculum after 3 days.

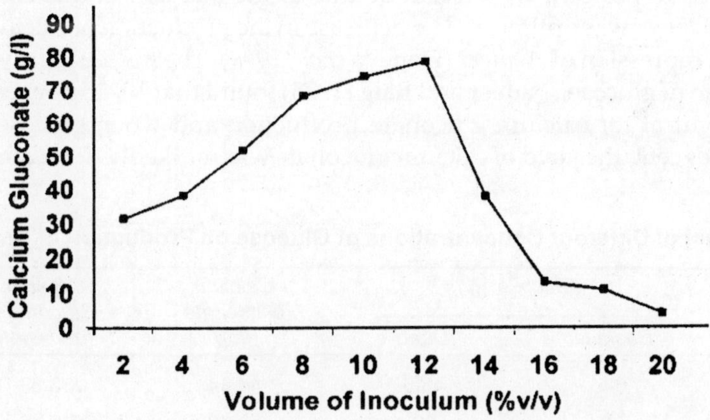

Figure 22.2: Impact of Volume of Inoculum on Production of Calcium Gluconate

Impact of Carbon Sources

The results in Figure 22.3 have been shown among the various carbon sources used for production of calcium gluconate, glucose was found to be the best substrate for maximum production of calcium gluconate. It may be due to simple chemical nature of glucose as compared to others.

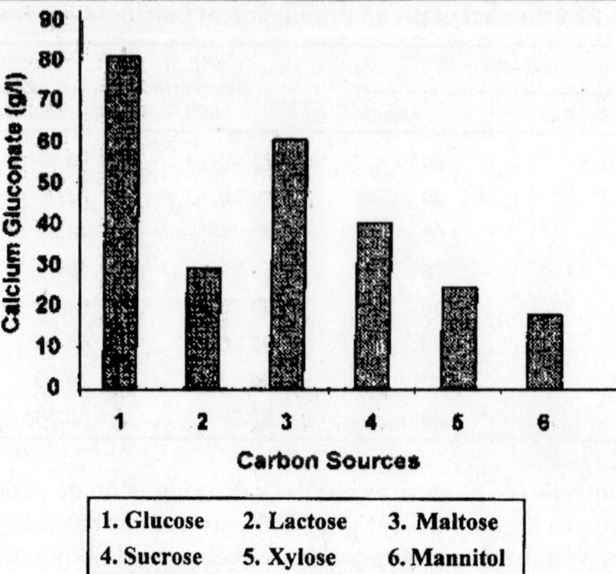

Figure 22.3: Impact of Carbon Sources on Production of Calcium Gluconate

Impact of Different Concentrations of Dextrose

Result shown in Table 22.3 indicates that maximum amount of calcium gluconate was obtained in the medium containing 10 per cent w/v dextrose. However, further increase in glucose concentration resulted in lowering both glucose utilization and calcium gluconate formation. Maximum yield of calcium gluconate at 10 per cent w/v might be due to the fact that at this concentration, glucose oxidase activity in *A. niger* was optimum. The decrease in the production of calcium gluconate may be due to the catabolic repression of *A. niger* (Doneva *et al.*, 1999). The mycelial growth was increased at higher concentration of glucose, Qadeer and Baig (1975) found that 10–15 per cent of glucose is more suitable and economical for calcium gluconate production and when glucose concentration was increased unto 30 per cent, the yield of calcium gluconate was markedly decreased (Qadeer and Baig, 1975).

Table 22.3: Impact of Different Concentrations of Glucose on Production of Calcium Gluconate

Concentration of Dextrose (% w/v)	Glucose (g/l)		Calcium Gluconate (g/l)	Calcium Gluconate (per cent)	Mycelial Dry Weight (g/l)
	Residual	Used			
5	6.5	43.5	44.84	33.38	22.4
10	20	80	76.22	56.75	28.0
15	77	73	56.05	41.72	36.8
20	142	58	53.81	40.05	40.4

Impact of Different Concentrations of Calcium Carbonate ($CaCO_3$)

Figure 22.4 shows that extracellular calcium gluconate synthesis reached at maximum value in the fermentation medium with $CaCO_3$ concentration 3.2 per cent w/v. At higher and lower

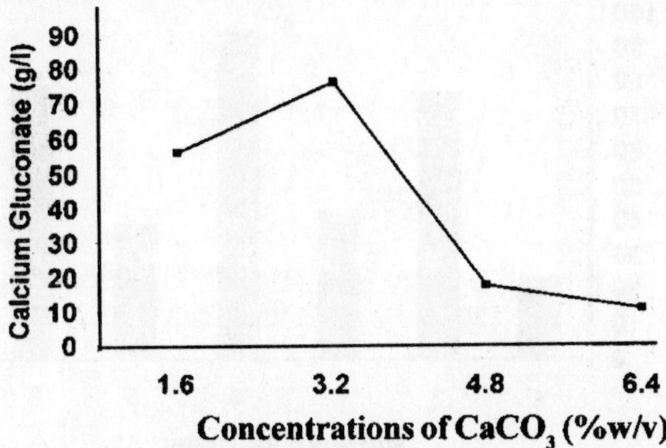

Figure 22.4: Impact of Different Concentration of Calcium Carbonate ($CaCO_3$) on Production of Calcium Gluconate

concentration, the yield decreases. The optimum yield of calcium gluconate at 3.2 per cent w/v $CaCO_3$ concentration in production medium is due to increased production of glucose oxidase and catalase by *A. niger* and metabolic shift from glycolytic pathway to gluconic acid (Liu *et al.*, 2001).

Impact of Volume of Fermentation Medium

It was found that 25 ml of production medium gives maximum yield of calcium gluconate. As the volume of production medium used for fermentation increases, the production of calcium gluconate decreases (Figure 22.5).

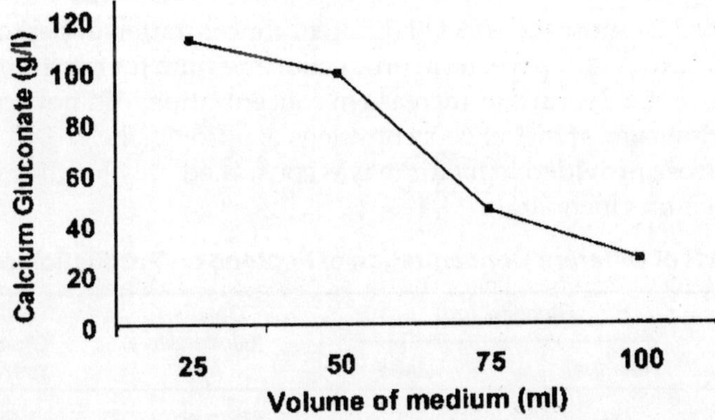

Figure 22.5: Impact of Volume of Fermentation Medium on Production of Calcium Gluconate

It may be due to the fact proposed by Dawson and Jolok (1983) that as the volume of fermentation medium increases there may be death of organisms due to suffocation.

Impact of Different Nitrogen Sources

Nitrogen sources are an important amendment that affects calcium gluconate production. The isolates revealed higher production of calcium gluconate in production medium with peptone. However, ammonium nitrate resulted in minimum yield of calcium gluconate production.

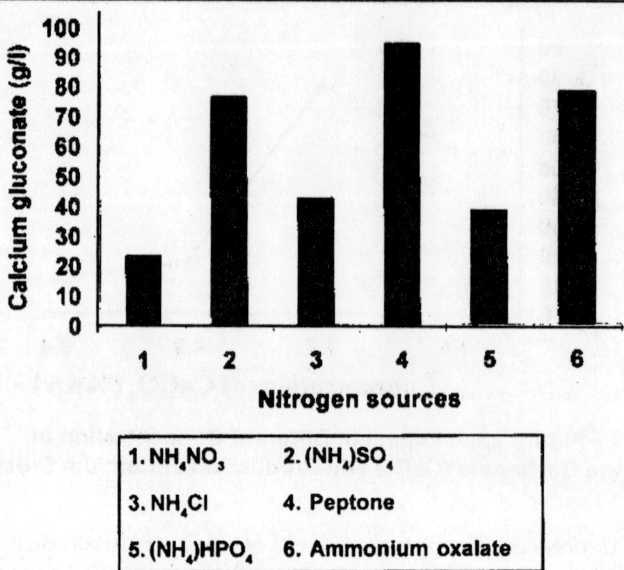

Figure 22.6: Impact of Different Nitrogen Sources on Production of Calcium Gluconate

Elnaghy and Megalla (1995) also reported peptone as best source for calcium gluconate production. Mariam and Nagra (2002) reported $(NH_4)_2SO_4$ as the best sources of nitrogen at concentration 2.0 per cent g/l in production medium for maximum production of calcium gluconate.

Impact of Different Concentration of Peptone (Nitrogen Limitation)

The data in Table 22.4 show the effect of different concentration of peptone on calcium gluconate formation. The optimum level of peptone in production medium for maximum production of calcium gluconate was 2 per cent w/v. further increase in concentration, did not enhance the production of calcium gluconate. However, at higher concentrations of nitrogen the growth of fungus was increased as the amount of glucose provided in medium was consumed only for the growth of *A. niger* and not for production of calcium gluconate.

Table 22.4: Impact of Different Concentration of Peptone on Production of Calcium Gluconate

Peptone Concentration (% w/v)	*Glucose (g/l)*		*Calcium Gluconate (g/l)*	*Calcium Gluconate (per cent)*	*Mycelial Dry Weight (g/l)*
	Residual	*Used*			
0.15	33	67	53.80	40.05	34.0
0.2	12	88	94.16	70.10	40.8
0.25	21	79	71.74	53.41	42.4
0.3	25	75	69.50	51.74	46.8
0.35	28	72	51.56	38.38	47.6

Effect of Temperature

The calcium gluconate production and glucose utilization were maximal at 30°C. Below and above optimum temperature, the production of calcium gluconate decreases but the growth of mold constantly increases as the temperature increases unto 35°C. When the temperature of medium was low, the enzyıne activity was also low, giving no significant impact on enhancement of calcium gluconate production and when temperature of medium was increased above 30°C, the biosynthesis of calcium gluconate was decreased. It may be due to that high temperature causes deactivation of enzyme (glucose oxidase). Similar findings were made by Mandal and Chatterjee (1986).

Table 22.5: Impact of Temperature on Production of Calcium Gluconate

Temperature (°C)	*Glucose (g/l)*		*Calcium Gluconate (g/l)*	*Calcium Gluconate (per cent)*	*Mycelial Dry Weight (g/l)*
	Residual	*Used*			
20	48	52	47.08	35.05	10.0
25	32	68	71.74	53.41	19.2
30	15	85	100.89	75.11	22.4
35	20	80	89.68	66.76	22.8

Conclusion

In conclusion, the isolated mesophillic strain of *A. niger* is capable of producing large amount of calcium gluconate (100.89 g/l) from lesser amount of glucose at the end of 3 days of fermentation period. The composition of fermentation medium for maximum production of calcium gluconate (g/l) formulated is dextrose 100, peptone: 2.0, $MgSO_4.7H_2O$: 0.5, KH_2PO_4: 1.0, $CaCO_3$: 32 and optimum initial pH required for maximum yield is 6.5 before sterilization. There is scope to reduce the cost of production by improvement of production strain and the use of cheap carbon and nitrogen sources.

References

Ali, S. and Qadeer, M.A., 2001. Biosynthesis of citric acid by locally isolated *A. niger* using sucrose salt media. *Online Journal of Biological Science*, 1(4): 178–181.

Available Online at http://www.health.ivillage.com [What is Calcium gluconate].

Available Online at http://www.healthtouch.com [Calcium supplement (systemic)].

Available Online at http://www.nrdcindia.com [Uses and process].

Dawson, Z. and Joloka, A., 1983. Effect of volume of medium on citric acid fermentation. *Acta Biotechnology*.

Doneva, T. *et al.*, 1999. Catalytic and biocatalytic oxidation of glucose to gluconic acid in a modified three-phase reactor. *Biotechnol. Lett.*, 21: 1107–1111.

Dubey, R.C. and Maheshwari, D.K., 2000. *Practical Microbiology*. S. Chand and Company, New Delhi.

Elnaghy, M.A. and Megalla, S.E., 1975. Gluconic acid production by *P. puberculum. Folia Microbiol.*, 20: 504–508.

Johnson, L.P., E.A. Curl, J.H. Bond and H.A., 1959. Fribourg methods for studying soil micro flora plant disease relationship. Burgess Public. Minneapolis Minn., U.S.A

Kundu, P.N. and Das, A., 1984. Utilization of cheap carbohydrate sources for production of calcium gluconate by *P. funiculosum* mutant MN-2338. *Ind. J. Exp. Biol.*, 49: 9–15.

Liu, J.Z. Huang, Y.Y., Liu, J. Weng, L.P. and Ji, L.N., 2001. Effect of metal ions on simultaneous production of glucose oxidase and catalase by *A. niger. Letters in Applied Microbiology*, 32(1): 16–19.

Mandal, S.K. and Chatterjee, S.P., 1986. Effect of some factors influencing calcium gluconate by *P. janthinellum. Folia Microbiol.*, 31, 1, 15–18.

Mariam, I. and Nagra, S.A., 2002. Production of calcium gluconate by *A. niger* in shake flask. *Pakistan Journal of Biological Science*, 5(3): 321–324.

Miller, G.L., 1963. Use of DNS reagent for the measurement of reducing sugar. *Analytical Chemistry*, 31: 426–428.

Panda, T. and Subbarao, D., 1999. Comparative analysis of calcium gluconate and sodium gluconate technique for the production of gluconic acid by *A. niger. Bioprocess Engineering*, 9: 203–207.

Pharmacopoiea, U.S., 1990. Assay of Calcium Gluconate.

Qadeer, M.A. and Baig, M., 1975. Production of calcium gluconate by *A. niger* in 50L fermentor. *Pakistan Journal of Science and Industrial Research*, 18: 227–228.

Shah, D.N. and Kothari, R.M., 1993. Glucose oxidase rich *A. niger* strain an economical substrate for the preparation of tablet grade calcium gluconate. *Biotechnol. Lett.*, 15: 35–40.

Tripathi, C.K.M. and Rastogi, S., Bihari, V. and Basu, S.K., 1990. Production of calcium gluconate by fermentation. *Indian J. Exp. Biol.*, 37, 7, 731–733.

Chapter 23

Neem Derivatives as a Larvicide for the Management of Country Bean Pest *Pericallia ricini* F. (Lepidoptera : Arctidae)

N. Revathi*[1] *and S. Kingsley*[2]

[1]*Department of Advanced Zoology and Biotechnology, Loyola College, Chennai – 600 034, T.N.*

[2]*Entomology Research Institute, Loyola College, Chennai – 600 034, T.N.*

ABSTRACT

Larvicidal property of neem oil, neem cake extract and monocrotophos at different concentrations was evaluated on fourth, fifth and sixth instar larvae of *Pericallia ricini*. It was found that neem oil was highly effective. Larval mortality was found to be lesser on the larvae of older age. An increase in the concentration of the test solutions resulted with an increase in the rate of larvicidal effect.

Keywords: *Azadirachta indica, Bioefficacy, Country bean, Larvicidal effect, Pericallia ricini.*

Introduction

Due to the increased awareness on indiscriminate use of synthetic pesticides and their persistent harmful effects, much interest has been evinced in the use of botanical products to control the

* Corresponding Author.

agricultural pests. Of the numerous plants investigated, the neem tree *Azadirachta indica* A. Juss, has shown promising results for the control of agricultural pests. More than 400 species of insects, including key pests of agriculture are susceptible to various behavioural and physiological effects of neem (Schmutterer and Singh, 1995). The practical utility of neem as a pest control agent was first demonstrated by Pradhan *et al.* (1962). Patnaik *et al.* (1987) observed larvicidal effect of *Crocidolomia binotolis* with neem oil. Larvicidal effect of neem oil was also observed earlier by Azam (1991) on *Liriyomyza trifolii,* Jhansi and Singh (1993) on *Helicoverpa armigera,* Brar *et al.* (1994) on *Earias vittella,* Jeyakumar and Uthamasamy (1997) on *Liriyomyza trifolii,* Chari *et al.* (1999) on *Spodoptera litura,* Ramamurthy and Rajaram (2000) on *Amrasca biguttula* and Vastrad *et al.* (2002) on *Plutella xylostella.* Pest control methods are aimed to combat the different stages of insects of which the larval stage is vulnerable. Controlling the pest population in the larval stage itself decreases the next generation development. This factor is one of the key aspects of the present investigation for choosing the larval stage of *P. ricini* for the evaluation of bioefficacy of neem derivatives.

The larva of lepidopteron moth *Pericallia ricini* is a serious polyphagous pest of agricultural crops. The black hairy caterpillar is a voracious leaf feeder of country bean *Dolichos lablab,* which is one of the main vegetable crops of Tamil Nadu. During the initial stages, the larvae feed on the lower side of the leaves, and when they reach third instar stage they begin to feed voraciously and defoliate the entire plant. When the larval population was uncontrolled, the crop yield loss may extend up to 70 per cent to 80 per cent. Botanical pesticides offer a viable alternative to synthetic insecticides as they are comparatively less toxic, biodegradable and are safe to non-target organisms (Verma and Pandey, 1981). This consideration was behind this study in evaluating the larvicidal effect of neem derivatives on the country bean pest *P. ricini.*

Materials and Methods

The larvicidal effect of neem oil 1.0 per cent, 1.5 per cent and 2.0 per cent, neem cake extract 1.0 per cent, 3.0 per cent and 5.0 per cent along with monocrotophos 0.05 per cent concentrations were evaluated against fourth, fifth and sixth instar larvae of *P. ricini.* The eggs of *P. ricini* were collected along with the leaves from the field during morning hours from 6.00 to 8.30 am and evening from 5.00 to 6.30 pm. The eggs were placed in a well ventilated plastic container under room temperature 28±2°C. After hatching, the larvae were transferred to plastic basin covered with nylon net. Larvae were provided with fresh leaves of country bean. The larvae were released in a container and neem derivatives were sprayed with a sprayer. The topically treated larvae were transferred into well ventilated plastic boxes, with filter paper covered bottom. Fresh leaves were supplied for treated larvae. Three replications were maintained for each treatment including control. Observations on the lalval mortality were recorded after 24 hours of treatment and percentages were calculated.

Results and Discussion

Larvicidal activity of neem oil, neem cake extract and monocrotophos at different concentrations tested against the fourth, fifth arid sixth instar larvae of *P. ricini* were presented in the Table 23.1. In the present investigation neem oil had higher larvicidal effect and it was maximum with 79.99 per cent, 71.10 per cent and 66.66 per cent when fourth, fifth and sixth instar larvae were treated with neem oil 2.0 per cent concentration, respectively. An increase in the concentration of the test solutions resulted in an increase in the rate of larval mortality of *P. ricini.* Similar observation was also reported earlier on *Helicoverpa armigera* (Behera and Satapathy, 1996); *Liriomyza trifoli* (Jayakumar and Uthamasamy, 1997); *Epilachna vigintioctcpunctata* (Shanmugapriyan and Kingsley, 2001); *Helicoverpa armigera*

(Sahayaraj and Paulraj, 2001); *Plutella xylostella* (Vastrad *et al.*, 2002); *Spodoptera litura* (Leatemia and Isman, 2004); *Helicoverpa armigera* (Pushpa *et al.*, 2004) and *Tribolium castaneum* (Gupta, 2006).

Table 23.1: Efficacy of Neem Oil, Neem Cake Extract and Monocrotophos on IV, V and VI Instar Larvae of *Pericallia ricini*

Treatment	Concentration %	Larval Mortality %		
		IV Instar	V Instar	VI Instar
Neem Oil	1.0	52.22 (46.29) bc	48.88 (44.35) bc	44.44 (41.71) bc
Neem Oil	1.5	61.10 (51.44) b	59.99 (50.87) abc	52.22 (46.29) ab
Neem Oil	2.0	79.99 (63.64) a	71.10 (58.02) a	66.66 (54.80) a
Neem Cake extract	1.0	39.99 (39.12) cd	32.21 (34.54) d	31.11 (33.77) cd
Neem Cake extract	3.0	55.55 (48.19) b	45.55 (42.42) cd	41.10 (39.84) bc
Neem Cake extract	5.0	75.55	64.44	63.33
Neem Cake extract		(60.58) a	(53.43) ab	(52.93) a
Monocrotophos	0.05	34.44 (35.86) d	29.99 (33.11) d	21.10 (27.25) d
Control	0.00	0.00 (0.19) e	0.00 (0.19) e	(0.19) e

Values mean of three replications.

Mean followed by a common letter are not significantly different at the 5 per cent level by DMRT.

Larvicidal effect of neem derivatives on the sixth instar larvae was lower than fifth and fourth instar larvae. It was observed that with an increase in the age of the larvae, there was a corresponding decrease in the larvicidal effect of the test solutions. Shanmugapriyan and Kingsley (2001) also recorded a similar decrease of the larvicidal effect, when they treated second, third and fourth larval instars of *Epilachna vigintioctopunctata* with neem oil 1.5 per cent concentration.

In the present evaluation it was observed that the neem oil had higher mortality rate followed by neem cake extract and monocrotophos. These results suggest that neem oil could be used as a larvicide for the management of *P. ricini*. However, bioefficacy ofneem derivatives will be confirmed only after field study.

References

Azam, K.M., 1991. Toxicity of neem oil against leafminer, *Liriyomyza trifolii*, Burgess on cucumber. *Pl. Port. Qrtly.*, 6: 196–197.

Behera, U.K. and Satapathy, C.R., 1996. Screening indigenous plants for their insecticidal properties against *Spodoptera litura* Fb. *Insect Envi.*, 2(2): 43–44.

Brar, K.S., Arora, S.K. and Ghai, T.R., 1994. Efficacy of neem oil, dipel and insecticides for the control of cotton jassid and fruit borer on Okra. *Pestology*, 18(2): 14–16.

Chari, M.S., Ramaprasad, G., Sitaramaiah, S. and Murthy, P.S.N., 1999. Laboratory and field evaluation of neem (*Azadirachta indica* A. Juss), Pongamia (*Pongamia pinnata* L.) Chinaberry (*Melia azedarach* L.) extracts and commercial neem formulations against tobacco caterpillar *Spodoptera litura* F. In: *Azadirachta indica A. Juss,* (Eds.) Singh, R.P. and Saxena, R.C., pp. 111–129.

Gupta, S., 2006. Performance of *Mentha piperita* against *T. castaneum* Herbst (Coleoptera : Tenebrionidae). *Curr. Sci.*, 9(1): 393–398.

Jeyakumar, P. and Uthamasamy, S., 1997. Bioefficacy of some synthetic insecticides and botanicals against *Liriyomyza trifolii. Indian J. Ent.*, 59(4): 347–350.

Jhansi, B.R. and Singh, R.P., 1993. Identification of effective and inexpensive neem (*Azadirachta indica* A. Juss) seed kernel extract for management of *Helicoverpa armigera* Hubner. Abstract. *World Neem conference*, February, Bangalore, India, pp. 7.

Leatemia, J.A. and Isman, M.Bv2004. Insecticidal activity of crude seed extracts of *Annona* spp., *Lanium domesticum* and *Sandoricum koetjape* against Lepidopteran larvae. *Phytoparasitica*, pp. 30–37.

Patnaik, N.C., Panda, N., Patro, E.R and Misra, B.K., 1987. Effect of neem oil (*Azadirachta indica* A. Juss) on mustard web-worm, *Crocidolomia binotalis* zell. (Lepidoptera : Pyralidae). *Neem News Letter*, 4(2): 15–17.

Pradhan, S., Jotwani, M.G. and Rai, B.K., 1962. The neem seed deterrent to locusts. *Indian Fmg.*, 12: 7–11.

Pushpa, P., Mary, J., Vijay, T. and Swati, J., 2004. Bioactivity of non-edible oil seed extracts and purified extracts against *Helicoverpa armigera* (Hubner). *Indian J. Expt. Biol.*, 42: 91–95.

Ramamurthy, R. and Rajaram, V., 2000. Effect of botanicals and insecticides against rice leaf folder complex. *Bio. Sci. Res. Bull.*, 16(1): 43–44.

Sahayaraj, K. and Paulraj, M.G., 1998. Relative toxicity of some plant extracts to groundnut leaf miner, *Aproaerema modicella* Dev. *Int. Arach. Newslett.*, 18: 27–29.

Sahayaraj, K. and Gabriel Paulraj, M., 2001. Efficacy of chosen plants against gram pod borer *Helicoverpa armigera* Hubner (Lepidoptera : Noctuidae). *J. Adv. Zool.*, 22(1): 8–14.

Schmutterer, H. and Singh, R.P., 1995. List of insect pests susceptible to neem products. In: *The Neem Tree Azadirachta indica A. Juss and Other Meliaceous Plants,* (Ed.) H. Schmutterer. VCH Publications, Weinheim, Germany, pp. 326–365.

Shanmugapriyan, R. and Kingsley, S., 2001. Bioefficacy of neem oil on larvae of bitter guard beetle *Epilachna vigintioctopunctata* (Coccinellidae : Coleoptera). *J. Ecotoxicol. Environ. Monit.*, 11(3): 215–219.

Vastrad, A.S., Lingappa, S. and Basavangoud, K., 2002, Multi tool approach for the management of Diamond backmoth, *Plutella xylostella* Linn. Biopesticides and neem products. *Pestology*, 24(8): 47–49.

Verma, G.S. and Pandey, U.K., 1981. Studies on the effect of *Acorus calamus*, *Cimicifuga foetida* and *Glynandropsis gynandra* extracts against insect pests of cruciferous vegetables painted bug *Bagrada cruciferum* Kirk (Hemiptera : Pentatomidae). *Zeitschriftfur Angew Zool.*, 68: 109–113.

Chapter 24

Impact of Aqueous Litter Leachate of *Eucalyptus* on Germination and Seedling Growth of Red Gram Seed

P.K. Gantayat[1], *B. Padhy*[2], *M. Behera*[3] *and S.K. Padhy*[3]

[1]*Department of Botany, Rayagada College, Rayagada – 1, Orissa*

[2]*Department of Botany, Berhampur University, Berhampur – 7, Orissa*

[3]*Department of Botany, T.S.R. and T.B.K College, Gajuwaka, Visakhapatnam, A.P.*

ABSTRACT

Experiments were conducted in plastic trays by sand culture method in laboratory conditions to find out the allelopathic effects of different concentrations *viz.*, 4, 8, 12, 16 and 20 per cent aqueous leachate of *Eucalyptus globulus* leaf litters on germination and seedling growth of the red gram (*Cajanus cajan* cv UP AS-120). Leachate of lower concentrations (4, 8 and 12 per cent) showed better seed germination and seedling growth. Whereas higher concentrations (16 and 20 per cent), although showed late germination, there was no seedling growth. The total chlorophyll content in the first leaves of the seedlings raised in the 4 per cent leachate on 14 DAS (days after sowing) was found to be higher, while first leaves of seedlings grown in 8 and 12 per cent leachate exhibited decrease in chlorophyll content. Maximum carotenoid content was estimated from leaves of the seedlings grown in 4 per cent leachate during the same period of incubation. The shoot and root fresh weights as well as dry weights followed more or less same trends as was marked in the case of chlorophyll content.

Keywords: *Allelopathic effect, Eucalyptus, Red gram (Cajanus cajan), Seed germination, Seedling growth.*

Introduction

Plant foods mainly legumes and cereals have consistently been listed as major food sources in the third world countries, where larger segment of the people suffer from protein malnutrition. In developing countries where protein malnutrition is widespread, legume seeds play vital role in bridging the protein gap and meeting the demand for protein. Hence, the legume seeds are considered as protein tablets. Increasing the yield of such crops can fill the gap between the population growth and protein supply. As a whole the principal aim in such endeavor is to boost the production of legume seeds with increasing interest.

In India forests are being destroyed in large scale. Hence, government has started tree plantation in order to increase the growing demands of green coverage of the globe Thus, the cultivation of *Eucalyptus* is gradually spreading in more and more areas of the country to meet the over increasing demand for plantations. But these trees are rich sources of secondary metabolites, which play a vital role in regulating the pattern of vegetation distribution in plant communities. Thus, *Eucalyptus* has been selected to investigate their Allelopathic effect on agricultural crops.

Any direct or indirect, inhibitory or stimulatory effect of one plant on other through the production of various chemical compounds has been termed as allelopathy (Rice, 1984). The physiochemical involved in allelopathic interactions are known as 'allelochemicals', a term originally coined by Whittaker and Feeny (1971).

Furaya *et al.* (1987) pointed out that the presence of allelochemicals are responsible for poor under story vegetation.

Eucalyptus has occupied a prominent role in various forest plantation schemes such as social forestry. Indian foresters considered it, for a long time to be the almost perfect plant architect to suit many afforestation and reforestation programmes.

Agrochemical Features of Test Crops

Red gram [*Cajanus cajan* (L) Mill Sp. *Syn. C.indicus* Spreng, cv UPAS-120]

Sl.No.	Variety	Pedigree	Year of Identification	Year of Release	Yield (q/ha)	Area of Adoption and Season	Salient Feature
1.	Red gram UPAS-120	Selection from P-47 58	1974	1976	22.4.	South zone, North zone of India (Rabi)	Moderately susceptible to sterility mosaic virus and pod hover

Materials and Methods

The senesced and freshly fallen down leaves (litter) were collected from the under story of *Eucalyptus globulus* trees planted in university campus of Bhanja Bihar, washed thoroughly with running tap water, followed by distilled water to remove the dust and other particles sticked on the surfaces of the leaves.

The surface water was removed by soaking with blotting papers. There after the leaves were dried in an oven at 60±2°C for three days 20 gm. Of such washed and dried leaves were soaked in 100ml of distilled water for three days at 30±2°C to get aqueous leaf leachate following the methods adapted by Padhy *et al.* (2000). After three days of soaking the leachate was filtered through Whatman filter paper

No. 1, which served as 20 per cent concentration. The filtrate, thus collected, was diluted with distilled water to get 4, 8, 12 and 16 per cent concentration of leachate for further use as per the experimental schedule.

The visually selected healthy seeds of uniform colour, size and shape were surface sterilized with 0.3 per cent formalin solution for 10 minutes and washed thoroughly with distilled water till complete removal of formalin. These seeds were then placed in 0.5 cm below from the top sand level and allowed to germinate @ 20 seeds/plastic tray ($3 \times 9 \times 12$ cm size) containing equal volume of sterilized sand wetted with equal volume of 4, 8, 12, 16 and 20 per cent concentration of leachate. The seeds sown in sand wetted with distilled water equal volume to leachate, served as control. Each treatment and control set had 6 replicates for accuracy of the experiment. The trays having seeds were kept in BOD incubator (Reico make) maintained at 30±1°C for germination. Care was taken to add equal quantity of distilled water or leachate as per the experimental desire to maintain wetness of the sand. Emergence of plumule from the sand bed was taken as criterion of seed germination. Germination was recorded at 24, 48 and 72 hours after sowing.

After germination the trays containing the germinated seeds of both controlled and treated sets were transferred from BOD incubator to seedling growth chambers provided with illumination of 20.8 Klux light intensity from two fluorescent electric tube lights maintained at 30±2°C and 12 hrs. photoperiods.

The shoot and root lengths of the seedlings were measured by collecting 15 seedlings at random from each tree on 8, 10, 12 and 14 days after sowing (DAS).

The seedlings are separated into shoot and root portions and fresh weights of both shoot and root were taken separately excluding the cotyledons. After recording the fresh weight of the shoot and root, the same materials were kept in an oven maintained at 60±1°C for 24 hrs. and then dry weight of both shoot and root were recorded.

The first leaves were collected, surface dried, cut into pieces and mixed thoroughly. Then 50 mg of such cut leaf tissues were homogenized with 1–2 ml of 80 per cent (v/v) acetone with a pre-chilled mortar and pestle. The homogenates were centrifuged at 5000 rpm for 15 minutes at 28±1°C. The supernatants were collected and the pellets were again macerated with 1–2 ml of 80 per cent acetone and centrifuged again. The procedures were repeated till colorless pellets were obtained. All the supernatants were pooled together and a definite volume was prepared with 80 per cent acetone, which served as chlorophyll and carotenoid extract.

The absorbancy (A) or optical density (OD) of the pooled supernatants was measured at 475, 645 and 663 nm with help of Systronics spectrophotometer (Systronics model 105). The amount of chlorophyll (Chl-a, Chl-b and total chlorophyll) and carotenoid contents were calculated following the formula by Arnon (1949) and Liaaen-Jensen and Jensen (1971) respectively. The values were expressed in mg/gm.fr.wt.

Results and Discussion

All the concentration of leachate showed inhibitory effect on seedling growth resulting decrease in shoot and root length as well as their fresh and dry weights. Although 25 and 5 per cent germination were marked in seeds treated with 16 and 20 per cent leachate respectively on 72 hours after sowing (HAS), no seedling growth was established later on by the influence of the leachate.

The shoot lengths exhibited +ve correlations with increase in the growth period and –ve correlations with increase of leachate concentration.

Table 24.1: Effect of Different Concentrations of Aqueous Leaf Litter Leachate of *Eucalyptus globulus* on Germination (per cent) of Red Gram Seeds (Each datum is the mean of 5 replicates±SE)

Name of the Crop Plant and Variety	*Concentrations of the Leachate (%)*	*Hours After Sowing (HAS)*		
		24	*48*	*72*
Red gram (*Cajanus cajan*) cv UPAS-120	Control	70±1.2	90±1.4	100
	4	50±1.4	75±0.8	90±2.3
	8	45±1.2	65±1.2	85±1.2
	12	38±1.6	61±1.0	75±1.0
	16	0	7±0.6	25±0.2
	20	0	0	5±0.4

Table 24.2: Effect of Different Concentrations of Aqueous Leaf Litter Leachate of *Eucalyptus globulus* on Seedling Growth of Red Gram (*Canjanus cajan* cv UPAS-120) (Each datum is the mean of 15 replicates±SE)

Conc. of Leachate (%)	*Days After Sowing (DAS)(cm)*	*Shoot Length (cm)*	*Root Length (mg)*	*Shoot Fresh Weight (mg)*	*Root Fresh Weight (mg)*	*Shoot Dry Weight (mg)*	*Root Dry Weight*
Control	8	8.3±0.05	3.4±0.02	30.1±0.04	4.6±0.06	4.5±0.08	1.1±0.02
	0	9.2±0.03	4.1±0.04	32.2±0.06	5.1±0.01	5.2±0.04	1.8±0.08
	12	10.1±0.02	4.7±0.03	33.0±0.08	5.5±0.08	5.7±0.06	2.2±0.08
	14	11.4±0.05	5.6±0.09	33.9±0.04	5.9±0.04	6.3±0.01	2.8±0.02
4	8	6.7±0.04	2.6±0.04	28.7±0.04	3.1±0.08	3.8±0.06	0.8±0.08
	10	7.5±0.02	3.1±0.02	29.2±0.08	3.9±0.04	4.3±0.02	1.2±0.08
	12	8.4±0.04	3.6±0.04	30.2±0.02	4.6±0.04	4.6±0.08	1.4±0.02
	14	9.3±0.01	4.2±0.08	31.1±0.0 1	5.2±0.09	5.1±0.01	1.9±0.007
8	8	5.6±0.02	2.4±0.02	26.0±0.08	2.3±0.04	2.8±0.06	0.4±0.02
	10	6.4±0.02	2.8±0.08	27.0±0.05	3.1±0.08	3.0±0.08	0.8±0.04
	12	7.3±0.04	3.2±0.02	28.9±0.08	4.0±0.02	3.9±0.02	1.0±0.04
	14	8.2±0.08	4.0±0.09	29.5±0.01	4.7±0.08	4.5±0.07	1.5±0.01
12	8	4.9±0.06	2.0±0.02	24.4±0.04	2.1±0.02	2.4±0.04	0.3±0.02
	10	5.8±0.02	2.5±0.02	26.1±0.02	3.0±0.06	2.9±0.02	0.5±0.08
	12	6.9±0.02	2.9±0.08	28.8±0.06	3.5±0.08	4.0±0.08	0.8±0.06
	14	7.4±0.02	3.3±0.09	29.2±0.08	4.1±0.09	4.2±0.07	1.2±0.01
16	8	*	*	*	*	*	*
	10	*	*	*	*	*	*
	12	*	*	*	*	*	*
	14	*	*	*	*	*	*
20	8	*	*	*	*	*	*
	10	*	*	*	*	*	*
	12	*	*	*	*	*	*
	14	*	*	*	*	*	*

*: Seeds germinated but seedlings growth was not established.

Table 22.3: Effect of Different Concentrations of Aqueous Leaf Litter Leachate of *Eucalyptus globulus* on Chl-a, Chl-b and Total Chlorophyll Content in the First Leaf of Red Gram (*Canjanus cajan* cv UPAS-120) Seedlings (Data are the mean of 5 replicates ± SE expressed in mg/g. fresh wt.)

Conc. of Leachate (%)	*Days After Sowing (DAS)*	*Chl-a*	*Chl-b*	*Total Chlorophyll*
Control	8	0.56±0.08	0.14±0.07	0.70±0.08
	10	0.62±0.09	0.17±0.08	0.79±0.09
	12	0.74±0.05	0.19±0.01	0.93±0.02
	14	0.58±0.02	0.15±0.02	0.73±0.07
4	8	0.48±0.08	0.11±0.06	0.59±0.07
	10	0.52±0.06	0.12±0.02	0.64±0.08
	12	0.62±0.02	0.15±0.06	0.77±0.07
	14	0.50±0.09	0.10±0.08	0.60±0.08
8	8	0.40±0.06	0.07±0.06	0.47±0.53
	10	0.44±0.08	0.09±0.07	0.53±0.08
	12	0.48±0.02	0.11±0.06	0.59±0.06
	14	0.40±0.01	0.08±0.02	0.48±0.05
12	8	0.28±0.08	0.06±0.06	0.34±0.07
	10	0.34±0.04	0.09±0.04	0.43±0.01
	12	0.40±0.01	0.10±0.08	0.50±0.07
	14	0.30±0.01	0.08±0.04	0.38±0.08
16	8	*	*	*
	10	*	*	*
	12	*	*	*
	14	*	*	*
20	8	*	*	*
	10	*	*	*
	12	*	*	*
	14	*	*	*

*: Seeds germinated but seedling growth was not established

The root lengths showed similar trends as was noticed in case of shoot lengths.

The shoot fresh weights also exhibited +ve correlations with increase in growth period and –ve correlations with increase in leachate concentrations.

The root fresh weights exhibit +ve correlations with increase in growth period and –ve correlations with increase in leachate concentrations.

The shoot dry weights showed –ve correlations with increase in concentration of leachate and +ve correlations with the increase of growth period.

The root dry weights exhibited similar trends as was noticed in case of shoot dry weight, with increase of leachate concentrations and seedling growth period.

Table 22.4: Effect of Different Concentrations of Aqueous Leaf Litter Leachate of *Eucalyptus globulus* on Carotenoid Content in the First Leaf of Red Gram Seedlings at Different Developmental Stages (Each datum is the mean of 5 replicates±S.E)

Name of the Crop Plant and Variety	Concentrations of the Leachate (%)	Carotenoid Content (mg/g.fr.wt.) on			
		8DAS	10 DAS	12 DAS	14DAS
Red Gram (*Cajanus*	Control	0.11±0.02	0.10±0.02	0.08±0.01	0.15±0.03
cajan)	4	0.13±0.01	0.11±0.04	0.10±0.02	0.18±0.03
cv UPAS-120	8	0.15±0.01	0.14±0.03	0.11±0.01	0.20±0.05
	12	0.18±0.04	0.16±0.02	0.14±0.01	0.22±0.04
	16	*	*	*	*
	20	*	*	*	*

*: Seeds germinated but seedling growth was not established.

The different chlorophyll synthesis in the first leaves of red gram seedlings was considerably checked by all the concentrations of test leachate. The chlorophyll-a content exhibited +ve correlations with increase of plant age up to 12 DAS and –ve correlations with-increase of leachate concentrations. The chl-b content also exhibited similar trends as was noticed in case of Chl-a content. The total chlorophyll content in the first leaves of both control and treated seedlings gradually increased with the increase in the age of the seedlings upto 12 DAS and then declined

The carotenoid content in the first leaves of test gram seedlings gradually decreased from 8 DAS till 12 DAS and then increased on 14 DAS.

The delay or inhibition in the process of germination caused by different phenolic compounds present in the leachate might have directly or indirectly checked the amylase and protease activities which are produced or secreted by the aleurone layer and or indirectly by inhibiting the synthesis of gibberlin which usually induce and activate the synthesis of hydrolyzing enzymes responsible for better germination.

Del Moral and Muller (1970) and Muller and Chou (1971) reported that during the process of *Eucalyptus* litter decomposition the allelochemicals form complexes with inorganic soil constituents which alter the physicochemical nature of the soil by increasing their effective toxicity. The germinations and seedling growth of the test seeds were significantly affected by water soluble phytotoxic substances leached out from *Eucalyptus* leaves.

The growth and development of both shoot and root systems are controlled by various metabolic activities carried out in leaves. Hence changes in shoot and root length of the seedlings and their fresh and dry weights, changes in certain macro-molecular content etc. are considered as important parameters of seedling establishment. The impact of different concentrations of aqueous leal litter leachate of *Eucalyptus globulus* is given below.

The retardation of seedling growth by the influence of different concentrations of aqueous leaf litter leachate of *Eucalyptus* might be due to the inhibitory effect of allelochemicals present in the leachate on cellular metabolic processes, enzyme and/or cytological activities.

The fresh and dry weights of the seedlings were significantly reduced by the influence of all the concentrations of leachate. Since all the concentrations of leachate checked shoot and root length, the reduction in seedling fresh weight might be the reflection of seedling growth. As the leaf litter of Eucalyptus contain several phenolic compounds, terpenoids etc. such compounds might have arrested or checked the various metabolic pathways of the seedlings during their growth resulting decrease in the seedling fresh weight as well as dry weight.

The inhibition of chlorophyll synthesis might be due to interaction of phytotoxins and other phenolic compounds present in the leachate with photo and oxidative phosphorylation pathways and/or inhibition of light activated Mg+ and ATPase activity (Mooreland and Nuvitzsky, 1998).

It was further noticed in all the test seedlings that the chlorophyll content considerably decreased on and after 14 DAS. This indicates that the leaves might have entered into the age of senescence where the metabolites degrade and tissues became inactive for carrying out different vital activities.

The carotenoid content gradually decreased on increase of growth period up to 12 DAS and then increased. Positive correlation was established between carotenoid content and increase of concentration of leachate. So, it seems that the phenolics and other phytotoxic compounds present in the leachate might have caused degradation or denatures of chlorophyll content which lead to increase in carotenoid content. The phytol tail get detached from the chlorophyll and converted into carotenoid.

Conclusion

From the above results and discussions it is concluded that the leaf litters of *Eucalyptus* are harmful for pulses like red gram. Hence such pulse crops should not be raised under the canopy of *Eucalyptus* or in other words the *Eucalyptus* seedling should not be planted in and around the crop fields for timber or any other purposes.

In a nutshell it can be concluded that further in detailed work is needed to draw any definite conclusions.

Acknowledgements

Thanks are due to Head, P.G. Department of Botany, Berhampur University, Berhampur for facilities; Deputy Secretary, UGC, Regional Office, Kolkata for awarding Teacher Fellowship to Mr. Gantayet and Officer-in-Charge, Regional Agricultural Research Station of Orissa University of Agriculture and Technology (OUAT), Berhampur for providing pure-line seeds for this experiment.

References

Arnon, O.J., 1949. Copper enzymes in isolated chloroplasts. Polyphenol oxidase in *Beta vulgaris*.

Del Moral, R. and Muller, C.H., 1970. The allelopathic effect of *E. camaldulensis*. *Amer. Midland Naturalist*, 83: 254–282.

Furaya, T.V., Orihara, P. and Hayashi, C., 1987. Triterpenoids from *Eucalyptus paramiana* cultured cells. *Phytochemistry*, 26: 715–719.

Liaaen-Jensen, S. and Jensen, S., 1971. Quantitative determination of carotenoids in photosynthetic tissues. In: *Methods in Enzymology*, Vol. 23, (Eds.) S.P. Colowick, and N.D. Kaplan. Academic Press, New York, London, pp. 586–602.

Mooreland, D.E. and Notitzky, W.P., 1988. Interference by flavone and tlavonols with chloroplasts mediated electron transport and photophosphorylation. *Phytochemistry*, 27: 3359–3366.

Muller, C.H. and Chou, C., 1971. Phytotoxins: An ecological phase of phytotoxicity. In: *Phytochemical Ecology*, (Ed.) J.B. Harbourne. Academic Press, London, p. 201–216.

Padhy, B., Patnaik, P.K. and Tripathy, A.K., 2000. Allelopathic potential of eucalyptus leaf litter leachate on germination and seedling growth of finger millet. *Allelopathy Journal*, 7(1): 69–78.

*Rice, E.L. 1984. In: *Allelopathy*, Second Edition, (Ed.) E.L. Rice. Academic Press, Orlando, Floride, pp. 422.

Whittkar, R.H. and Feeny, P.P., 1971. Allelochemicals: Chemical interactions between plant species. *Science*, 171: 757–770.

* Originals not seen.

Chapter 25

Weed Flora, Relative Density and Summed Dominance Ratio of Weeds, as Influenced by Weed Management Practices in Direct Sown Wet Seeded Rice

B. Rajendra Kumar[1] and A. Christophrer Lourduraj[2]*

[1]*Agricultural Officer, Agricultural Research Station, Amadalavalasa, ANGRAU, A.P.*

[2] *Associate Professor, Department of Agronomy, Tamil Nadu Agricultural University, Coimbatore*

ABSTRACT

Wet seeded rice was infested with composite weed flora pre-emergence application of Pretilachlor + safener 0.45 kg a.i. ha^{-1} at 4 DAS + 1 hand weeding at 35 DAS registered better control of weeds, as compared to other herbicide treatments. The grasses, sedges and broad leaved weed density and dry weight were less in this treatment and the effective weed control provided at the early period of crop growth by Pretilachlor + safener 0.45 kg a.i. ha^{-1} at 4 DAS persisted up to maturity due to one hand weeding at 35 DAS.

Keywords: Relative density, Summed dominance ratio, Weeds, Wet seeded rice.

* *Corresponding Author.*

Introduction

Weed spectrum depends upon the efficiency of the control measures adopted. Weeds are the major problem in the early stages of rice cultivation. Weeds belonging to various species *viz.*, grasses, sedges, broad leaved and aquatics are associated with rice cultivation. Among the grasses, *Echinochloa crus-galli* is predominant, followed by *Cyperus difformis* in sedges and *Ammania baccifera, Eclipta prostrata* and *Sphenoclea zeylanica* in broad leaved weeds right from initial stages in direct seeded rice

Materials and Methods

Field experiment was conducted at Tamil Nadu Agricultural University, Coimbatore during *rabi* season 2002–2003 to study the weed management of anaerobic tolerant cultivars raised under wet seeding in low land direct sown rice.

The experiment was laid out in split plot design with three varieties ASD 19, ADT 38 and CO 43 in the main plot and weed management practices (Pretilachlor + safener 0.45 kg a.i. ha^{-1} at 4 DAS + 1 hand weeding at 35 DAS; Butachlor I kg a.i. ha^{-1} at 8 DAS + 1 hand weeding at 35 DAS; Butachlor + Propanil 0.84 + 0.84 kg a.i. ha^{-1} at 10 DAS (tank mix) + 1 hand weeding at 35 DAS; two hand weedings at 20 and 35 DAS; unweeded check) in the subplots with three replications.

Relative density was calculated using the number of individual weed species and total number of weeds. The summed dominance ratio (s.d.r) of the weed species was computed using the formula suggested by Janiya and Moody (1989).

Results and Discussion

Among the different types of weeds, grasses were predominant followed by broad leaved weeds and sedges. The dominant weed flora observed were *E. crusgalli* (L.) Beauv (Barnyard grass), *E. colonum* (L.) Link (Jungle grass) and *B. platyphyla* (Rises) nash (Broad leaved Signal grass) under grasses. The Sedges were *C. difformis* L. (Small flower umberella plant, *C. iria* L. (Rice flatsedge) and *Fimbristylis miliacea* (L.) vahl (Hoorah grass); while broad leaved weeds were *E. alba* (L.) Hassk (False daily compositae), *Ammania baccifera* (Redstem), *M. quardrifoliata* (L.) (water fern); *Ludwiga purviflora* Roxb (Neer Karambu) and *M. vaginails* (Burn) perso (Neer Thamarai). A pectrum of dominant weeds in direct seeded puddle rice with similar results was reported by Bayan (2000).

The relative density of grasses was higher at 20 and 35 DAS (76.4 and 74.2 per cent respectively) and decreased at 50 DAS (66.2 per cent). However in case of sedges and broad leaved weeds, reverse trend was observed, with lower relative density at 20 and 35 DAS and higher density at 50 DAS. Among the varieties, ASD 19 (M_1) recorded lower relative density of grasses and sedges at 20, 35 and 50 DAS and higher relative density of broad leaved weeds was observed in this variety at all the stages as compared to ADT 38 (M_2) and CO43 (M_3). In case of broad leaved weeds at 20 DAS, lower relative density was noticed in ADT 38 (16.1) while at 35 and 50 DAS lower value was observed in CO 43 (11.1 and 21.6, respectively). In all the weed control treatments, at all three stages of observation, the relative density of grasses was higher followed by broad leaved weeds and sedges. Out of the total weed population, grass weeds recorded 70–80 per cent, and broad leaved weeds 15–20 per cent sedges 5–10 per cent. Among the varieties tested, ASD 19 (M_1) recorded lowest per cent of grasses and sedges at all the three stages of crop growth. Pretilachlor + safener 0.45 kg a.i. ha^{-1} + 1 hand weeding at 35 DAS (S_1) recorded lowest summed dominance ratio for grasses (74.5 per cent) and sedges (3.7 per cent), but recorded higher value for broad leaved weeds (21.7 per cent) at 20 DAS. Highest summed dominance ratio of grasses was recorded in Butachlor + Propanil 0.84 + 0.84 kg a.i. ha^{-1} + 1 hand weeding at 35

DAS (S_3) (82.9). At 35 DAS, the summed dominance ratio of grasses and sedges was lowest in unweeded check (S5) (69.0 and 9.5, respectively) and highest in S_3 (76.2 and 10.8, respectively), while in case of broad leaved weeds, reverse trend was observed with S_3 recording lowest value (12.7) and S_5 with highest value of 21.0. At 50 DAS, the summed dominance ratio of grasses was lowest in S_5 (67.8) and highest in S_3 (75.5). In case of sedges, S, recorded lowest Summed dominance ratio of 9.9 and highest value of 12.5 in S_3. In case of broad leaved weeds, S_5 recorded highest value of 21.3 compared to 11.6 in S_3, which recorded lowest summed dominance ratio.

Relative density of weeds in unweeded check indicates the dominance of grasses in weed flora constituting 72 per cent of the total weeds at 20, 35 and 50 DAS. Pretilachlor + safener effectively checked the grass population and thereby relative density of grasses (Table 25.1). The field observations revealed that, when there was lesser weed density, the hand weeding could check the weed completely, but when there was high density as in the case of Butachlor (S_2) and Butachlor + Propanil applied plots (S_3), the hand weeding could control the weeds partially or only the apical portion of weeds which resulted regeneration in a short span. Similar results of ineffectiveness of hand weeding due to higher density of weeds in wet seeded rice was earlier reported by Moody and Cordova (1985). Effective control of sedges in the herbicide treated plots favoured the development of grasses causing their dominance. Srinivasan (1989) also reported dominance of grasses in herbicide treated rice crop.

Table 25.1: Effect of Weed Control Treatments on Relative Density of Grasses, Sedges and Broad Leaved Weeds in Direct Seeded Wet Land Rice

	Grasses			*Sedges*			*Broad Leaved Seeds*		
	20 DAS	*35 DAS*	*50 DAS*	*20 DAS*	*35 DAS*	*50 DAS*	*20 DAS*	*35 DAS*	*50 DAS*
M_1	74	74	63	6.5	6.5	11.3	19.6	18.1	25.2
M_2	79	75	65	6.2	6.7	11.8	16.1	20.8	22.7
M_3	78	76	66	6.6	7.5	11.3	16.6	19.1	21.6
C.D	N.S	N.S	N.S	N.S	N.S	N.S	N.S	N.S	N.S
C.V%	12.5	12.0	11.0	13.5	11.5	13.0	14.0	12.5	13.0
S_1	75	75	67	4.8	8.1	11.0	20.5	18.1	21.5
S_2	76	70	61	4.9	8.5	13.6	18.7	20.8	24.8
S_3	82	78	69	7.3	10.5	16.8	10.0	11.1	13.6
S_4	75	73	69	6.5	9.7	14.7	18.0	17.6	26.1
S_5	74	75	65	7.4	12.2	15.5	18.3	18.9	24.1
Mean	76.4	74.2	66.2	6.12	8.38	13.24	17.1	17.3	22.02
C.D	N.S	N.S	N.S	N.S	1.8	3.6	4.3	6.0	5.5
C.V.%	11.5	12.0	11.5	13.0	11.5	14.0	13.0	14.5	13.0

M_1: ASD 19; M_2: ADT 38; M_3: CO 43

S_1: Pretilachlor + safener 0.45 kg a.i. ha^{-1} + 1 hand weeding at 35 DAS

S_2: Butachlor 1 kg a.i. ha^{-1} + 1 hand weeding at 35 DAS

S_3: Butachlor + Propanil 0.84 + 0.84 kg a.i. ha^{-1} + 1 hand weeding at 35 DAS

S_4: Two hand weedings at 20 and 35 DAS

S_5: Unweeded check.

Table 25.2: Effect of Weed Control Treatments on Summed Dominance Ratio of Grasses, Sedges and Broad Leaved Weeds in Direct Seeded Wet Land Rice

	Grasses			*Sedges*			*Broad Leaved Seeds*		
	20 DAS	*35 DAS*	*50 DAS*	*20 DAS*	*35 DAS*	*50 DAS*	*20 DAS*	*35 DAS*	*50 DAS*
M_1	75.5	70.8	65.0	5.9	9.8	10.6	17.45	18.4	20.8
M_2	78.7	72.3	68.8	6.0	9.5	10.8	17.00	19.3	20.4
M_3	79.1	72.6	69.4	6.1	9.4	11.7	15.20	15.2	21.2
C.D.	N.S	N.S	N.S	N.S	N.S	N.S	N.S	N.S	N.S
C.V.%	11.5	11.0	11.0	13.0	11.5	14.0	12.0	12.5	13.5
S_1	74.5	72.9	73.2	3.7	9.9	9.9	21.70	17.6	16.6
S_2	76.9	71.4	71.0	4.6	9.9	10.8	18.20	18.2	17.8
S_3	82.9	76.2	75.5	6.6	10.8	12.5	10.50	12.7	11.6
S_4	75.4	73.3	75.1	6.5	10.1	11.4	17.60	16.6	18.2
S_5	75.4	69.0	67.8	7.0	9.5	10.3	17.30	21.0	21.3
C.D.	N.S	N.S	N.S	1.8	N.S	1.5	2.1	3.5	3.1
C.V.%	13.0	11.5	14.5	13.5	12.0	11.0	11.0	12.5	13.0

M_1: ASD 19; M_2: ADT 38; M_3: CO 43

S_1: Pretilachlor + safener 0.45 kg a.i. ha^{-1} + 1 hand weeding at 35 DAS

S_2: Butachlor 1 kg a.i. ha^{-1} + 1 hand weeding at 35 DAS

S_3: Butachlor + Propanil 0.84 + 0.84 kg a.i. ha^{-1} + 1 hand weeding at 35 DAS

S_4: Two hand weedings at 20 and 35 DAS

S_5: Unweeded check.

References

Bayan, H.C., 2000. Studies on integrated weed management under split application of nitrogen in wet seeded rice. *Ph.D. Thesis*, Tamil Nadu Agricultural University, Coimbatore.

Janiya, J.D. and Moody, K., 1989. Weed populations in transplanted and wed seeded rice as affected by weed control method. Tropic. *Pest Management*, 35: 8–11.

Moody, K. and Cordova, V.G., 1985. Wet seeded rice. In: *Women in Rice Farmings.* International Rice Research Institute, Los Banos, Philippines, pp. 467–480.

Srinivasan, G., 1989. Influence of integrated weed management on weed dynamics in rice based cropping system. *Ph.D. Thesis*, Tamil Nadu Agricultural University, Coimbatore.

Chapter 26

Effect of Sugar Mill Effluent on Biochemical Changes in the Liver of Freshwater Fish *Cirrinus mrigala*

K. Shanthi[1] *and V. Dhanalakshmi*[2]

[1]*Post Graduate Department of Zoology, Kongunadu Arts and Science College, Coimbatore – 641 029, T.N.*

[2]*Department of Zoology, N.G.M. College, Pollachi – 642 001, T.N.*

ABSTRACT

Usage of water in industries is increasing daily. The important raw material of industries is water and it has been discharged after utilization as an effluent as a major concern in water pollution. Discharge of industrial effluent such as distillery, tannery, pulp, paper, textile and sugar mill effluents have been of major concern in water pollution abatement. Presently an attempt has beer made to study the effect of sugar mill effluent on biochemical changes in the liver of fresh water fish *Cirrinus mrigala.* Liver enzymes and metabolites play a significant role in metabolism and body functioning of the fish.

Keywords: *Biochemical parameters, Cirrinus mrigala, Sugar mill effluent.*

Introduction

The healthy functioning of the biosphere in our planet of life depends entirely on the water flow and steady state phenomenon. Water is not only a vital environmental factor to all form of life, but it has also a great role to play in socio-economic development of human population (Park, 1997). It is estimated that three fourth of the waste water generated is from municipal sources, industrial waste contribute to over half of the total pollution and the major portion of this is coming from large and medium industries (Neena, 1993). Among the industries that contribute the aquatic pollution in India are pulp, paper and textile industries. Almost every river system of India is polluted perennial stream Khala by the sugar factory effluent near Laksar have been done by (Verma *et al.*, 1969). (Senthil Kumar *et al.*, 2001) have worked on physico-chemical characteristics and toxicity of sugar mill effluent. Sugar industries play an important role in the economic development of country, but the effluent released by them produce high pollution especially in the aquatic medium particularly fish population. (Avasnmaruthi and Ramakrishna Rao, 2001). Similar study was carried by (Konar *et al.*, 1991 and Manikavasagam, 1997). Economically important fishes having nutritive value are under threat due to discharge of sugar mill effluent into ecosystem. Anithakumari and Sree Ram Kumar (1996) have studied the effect of water pollution on the biochemical parameters of selected tissues of *Channa striatus.*

Materials and Methods

The effluent from Ponni Sugars and Chemical Ltd which is located in Pallipalayam near Erode in Tamil Nadu, was collected at the discharge point and analysed for the physico-chemical parameters as per the APHA, AWWA, and WPCFA (1985). The experimental fish *Cirrinus mrigala* was procured from Aliyar fish farm, Tamil Nadu Fisheries Corporation Ltd., Aliyar, South India. The fishes were acclimatized in laboratory condition for a month. Ten batches of healthy fishes were exposed to different concentrations (2.6 per cent, 3.6 per cent and 4.6 per cent) of the sugar mill effluent and survivability was determined in an interval of 24, 48, 72 hrs. One set of fishes were maintained as control in tapwater. LC_{50} value was calculated by the method of Finney (1971). Controls and treated fishes were sacrificed at the end of the exposure period. Samples of liver tissue were collected for the analysis of total protein, carbohydrates and lipid. The liver tissue weighing about 10 gms were homogenized in 80 per cent ethanol, centrifuged at 3500 rpm and clear supernatant was used for the analysis. Biochemical studies like Liver Glycogen, Total Protein, Cholesterol, Glutamate Oxaloacetate Transaminase (GOT), Glutamate Pyruvate Transaminase (GPT), Acid Phosphatase (ACP), Alkaline Phosphatase (ALP) were analysed. The data of the results obtained from the above studies were statistically analysed.

Results and Discussion

The results of the effect of sugar mill effluent on biochemical changes of *Cirrinus mrigala* are given in Table 26.1.

The liver glycogen, total protein, cholesterol, liver acid phosphatase and alkaline phosphatase were declined throughout the experimental period when compared to the control. The 't' test analysis showed significant variations between treatment and their controls throughout the exposure period. Liver glutamate oxaloacetate tranaminase and glutamate pyruvate tranaminase were increased in the experimental groups. The aquatic medium is very efficient solvent for many chemical compounds. Aquatic organisms are extremely vulnarable to toxic effect resulting from absorption or oral intake of these contaminants from the immediate environment. Various chemical compounds have been investigated to determine their potential toxicity to certain organisms, especially fish (Rand and Petrocelli, 1985).

Liver is the central metabolic clearing house of the body and it maintains blood glucose concentrations by the process of glyconeogenesis and glycogen breakdown (Voet and Voet, 1995). In

Table 26.1: Biochemical Changes in the Liver of *Cirrinus mrigala* Exposed in Different Sublethal Concentrations of Effluent in Different Periods

Concentration	2.60%			3.60%			4.60%		
Exposure Period (hrs)	24	48	72	24	48	72	24	48	72
Glycogen	81.51±1.161	81.33±1.175	80.44±1.095	81.51±1.161	81.33±1.175	80.44±1.095	81.51±1.161	81.33±1.175	80.44±1.0.95
	78. 24±1. 125	74. 20± 1.134	70.40±1.304	67.29±1.075	65.30±1.175	63.20±1.150	61.24±1.125	59.23±1.130	57.77±1.902
	3.48106*	8.43964*	10.164*	15.5545*	16.709*	18.7952*	21.702*	23.4728*	25.8832*
Total Protein	83.27±1.100	82. 26±1.200	82. 05± 1. 051	83. 27±1.1 00	82.26±1.200	82.08±1.051	83.27±1.100	82.26±1.200	82.08±1.051
	78.34±1.263	73.20±1.250	68.60±1.250	64.27±1.125	61.36±1.200	59.18±1.160	53.22±1.140	46.25±1.060	39.28±1.135
	3.15554*	9.02479*	13.2276*	20.9155*	21.325*	23.7115*	32.851*	38.9436*	45.0809*
Cholesterol	78.21±1.115	77.27±1.1 00	77.13±1.095	78.21±1.115	77.27±1.100	77.13±1.095	78.21±1.115	77.27±1.1 00	77.13±1.095
	75.66±1.340	72.47±1.280	69.44±1.341	67.17±1.1 00	64.26±1.212	62.8±1.924	58.15±1.010	55.20±1.215	52.74±1.297
	2.63104*	4.9255*	8.59148*	13.9446*	13.7687*	17.0674*	22.8919*	23.3219*	24.876*
GOT	41.24±1.110	41.42±1.015	41.64±1.1 05	41.24±1.110	41.42±1.015	41.64±1.1 05	41.24±1.110	41.42±1.015	41.64±1.1 05
	44.34±1.185	48.39±1.266	52.25±1.250	57.33±1.165	60.55±1.195	64.15±1.135	69.43±1.298	73.85±1.060	77.42±1.197
	3.41634*	7.44159*	5.12569*	17.3248*	21.1298*	24.7095*	7.5892*	38.2737*	38.2882*
GPT	63.25±1.120	63.50±1.090	64.55±1.073	63.25±1.120	63.50±1.090	64.55±1.073	63.25± 1.120	63.50±1.090	64.55±1.073
	68.48±1.145	72.44±1.235	76.20±1.195	80.27±1.180	84.24±1.145	88.46±1.218	94.35±1.1 05	98.58±1.205	102.90±1.080
	5.66252*	9.50667*	12.5604*	18.1149*	22.9796*	24.0346*	26.9789*	37.7855*	48.2285*
ACP	163.42±1.220	163.1 0±1.053	162.13±1.1 01	163.42± 1.220	163.10±1.053	162.13±1.1 01	163.42±1.220	163.10±1.053	162.13±1.101
	159.15±1.II0	155.24±1.1 05	151.21±1.105	147.42±1.317	143.76±1.274	139.66±1.200	135.78±1.110	131.41±1.270	127.35±1.260
	4.86098*	9.38669*	5.64565*	15.1029*	18.5242*	23.8999*	29.1434*	35.9616*	48.2314*
ALP	103.34±1.197	103.12±1.110	103.27±1.146	103.34±1.197	103.12±1.110	103.27±1.146	103.34±1.197	103.12±1.110	103.27±1.146
	102.21±1.176	98.37±1.215	100.13±1.205	91.62±1.205	87.53±1.165	89.15±1.130	80.33±1.180	75.22±1.120	77.32±1.200
	1.16972*	4.92388*	3.43452*	3.59044*	16.4215*	15.3411*	23.7112*	29.8682*	22.2666*

First row entries control values (mean±SD of five individual observations).

Second row entries experimental values (mean±SD of five individual observations).

Third row entries t-test values (values are significant at 5 ± level with degrees of freedom 4).

the present investigation sharp decline in glycogen content of liver was noticed, when the fishes were exposed to the effluent for 24, 48 and 72 hours exposure at 2.6, 3.6 and 4.6 per cent concentration. The significant decrease in glycogen content in the liver indicated the excess utilization of the reserve glycogen to withstand heavy metal toxicosis. Similar observations were made by (Amutha *et al.*, 2002), when the tissues like gill, liver, muscle of fresh water teleost fish *Oreochromis mossambicus* was exposed to diary effluent.

The decrease of protein content of liver may be due to lowered protein synthesis indicating a lower metabolism. Fall in protein content in tissue under stress of pollutants may be due to altered enzyme activities (Jadhav and Lomte, 1982). The present investigation showed significant decrease in total protein content in liver tissue which snowed maximum in control fish which was found to decrease during exposure period.

The liver plays a central role in the regulation of cholesterol balance of the body (Champe and Harvey, 1994). Decline in liver cholesterol was associated with depressed de novo synthesis along with greater mobilization of liver cholesterol for the synthesis of corticostroids or their conversion to bile acids that form the main route for the excretion of cholesterol (Kondal *et al.*, 1988). In the present study the cholesterol level was found to be decreased in liver when exposed to sugar mill effluent.

Glutamate Oxaloacetate Transaminase (GOT) and Glutamate Pyruvate Transaminase (GPT) activity in the present study showed significant increase in liver. This indicates an increased rate of proteolysis in the tissues (Shukla and Sastry, 1998).

Decreased activity of both the acid phosphatase and alkaline phosphatase as a result of effluent treatment showed that the physiological functions of the tissues were affected. Similar decline in the enzyme activities observed when catfish *Heteropneustes fossilis* was exposure to cadmium (Sastry and Subhadra, 1985). The present investigation showed the alteration of carbohydrate, protein, cholesterol and enzymes can be taken as biochemical indices of Sugar mill effluent toxicity to assess effluent pollution of the aquatic environment. This made additional contribution to the information already available in the field of toxicology.

References

Amutha, P., G. Sangeetha and Mahalingam, S., 2002. Dairy effluent induced alterations in the protein, carbohydrate and lipid metabolidsm of a fresh water teleost fish *Oreochromis mossambicus. Poll. Res.*, 21(1): 51–53.

APHA, AWWA and WPCFA, 1985. *Standard Methods for the Examination of Water,* 16th Ed., Amer. Public Health Association Inc.

Avasnmaruthi, Y. and Ramakrishna Rao, S., 2001. Effect of sugar mill effluent on organic reserves of fish. *Poll. Res.*, 19(3): 391–393.

Anithakumari, S. and Sree Ram Kumar, N., 1996. Effect of water pollution on the biochemical parameters of selected tissues of *Channa striatus. J. Ecobiol.*, 8(3): 221–227.

Champe, P. and Harve, R.A., 1994. *Illustrated Reviews: Biochemistry,* 2nd Edition, (Eds.) Pamela C. Champe and Richard A. Harve. J.B. Lippincott Company, Philadelphia.

Chawdhury, N., 1997. Worst polluted areas India. *Ind. J. Environ. Hlth.*, 21: 73.

Finney, D.J., 1971. *Probit Analysis,* 3rd edition. Cambridge University Press, London, p. 20.

Jadhav, M.L. and Lomte, V.S., 1982. Biochemical composition of the fresh water bivalve *Lamellidens marginalis,* Riv. *Di. IdrobioL.*, 21: 1–3.

Konar, S.K. Ghosh, T.K. and Dey, M., 1991. Management of complex pollutional problems Indian reservoirs to increase fish yield. *Environ. Ecol.*, 9(2): 460–472.

Kondal, J.K., Saxena, P.K and Soni, 1988. Effect of vegetable oil factory effluent on the liver of a fresh water teleost *Heteropneusatus fossilis* (Bloch): A biochemical study. *Biol. Wastes*, 25: 17–24.

Manikavasagam, 1997. *Industrial Effluent: Origin, Characteristics, Effect, Analysis and Treatment*. Sakthi Publications, pp. 99–106.

Neena, V., 1993. Survey of the environment: Government stance of hoped and reality. *The Hindu*, pp. 1–196.

Park, K., 1997. *Park's Textbook of Preventive and Social Medicine*. Banasidas Bharat Publishers, Jabalpur.

Rand, G.M. and Petrocelli, S.R., 1985. *Fundamentals of Aquatic Toxicology*. McGraw-Hill International Book Company, New York, p. 666.

Sastry, K.V. and Subhdra, 1985. *In vivo* effects of cadmium on some enzyme activities in tissues of the fresh water catfish. *Environ. Res.*, 36: 32 – 45.

Senthil Kumar, R.D., Narayanaswamy, R. and Ramachandran, K., 2001. Pollution studies on sugar mill effluent: Physiochemical characteristics and toxic metals. *Poll. Res.*, 20(1): 93–97.

Shukla, Vineeta and Sastry, K.V., 1998. Acute toxic effects of cadmium on various parameters in the fresh water fish. *Proc. Acad. Environ. Biol.*, 7(2): 155–160.

Verma, S.R. and Shukla, 1969. Pollution in a perennial stream Kholar by sugar factory effluent near Lacksav. *Ind. J. Environ. Hlth.*, 11: 145–162.

Voet, D. and Voet, J.G., 1995. *Biochemistry*. John Willy and Sons Inc., New York, pp. 1360.

Chapter 27

Studies on Physico-chemical and Heavy Metal Accumulation in Selected Sites along the Coastal Belt of Kerala

K. Murugan
Department of Botany, University College, Thiruvananthapuram, Kerala – 695 034

ABSTRACT

Heavy metals are one of the most hazards of environmental pollution. These, depending on their oxidation states, can be highly reactive and, as a consequence, toxic to most organisms. They are produced by an expanding variety of andropogenic sources suggesting increasingly important for this form of pollution. The present investigation was carried out to assess pH, turbidity, DO, BOD, salinity, chlorinity, NO_3, NO_2, NH_3 and heavy metals such as cadmium, lead, mercury, manganese, zinc and copper along the selected coastal sites of Chetua, Cochin, Kollam in Kerala. The study revealed fluctuation of values of different parameters in the three sites. The profile of heavy metals in the coastal waters from Cochin and Kallam clearly showed that Cu, Mn, Cd, Pb and Hg are the major contaminants pollute water. Heavy metal accumulation was coupled with the low level of DO and high load of BOD confirm the polluted state of the sites. The high pollution at Cochin *may* be due to the discharge of industrial effluents and loading of ships whereas Kallam recorded major oil leakage in the past. Chetua a small island nearly uninhabited record poor concentrations of heavy metals making the place comparatively nonpolluted. Heavy metal pollution is a major threat to all biota living or depending on it. Further, the variability of major ecological parameters has also leaded a major impact on the presence and the degree of effect of the pollutant on both the organism and the water body.

Keywords: *Heavy metal, Kerala, Pollution.*

Introduction

Toxic heavy metals in air, soil, and water are global problems that are a growing threat to humanity. Major anthropogenic sources of heavy metals sound are industrial and urban waste, wastewater' discharges and shipping activity (Lionetto, 2001). In response to the growing problems, state governments have instituted environmental regulations to protect the quality of surface and groundwater from heavy metal pollutants, such as Cd, Cu, Pb, Hg, Cr, and Fe. Developed and developing countries are suffering from different form of water pollution. Developed countries, which have water pollution problem due to industrial proliferation and modernised agricultural technologies, are now on the way of combating the problems through improving wastewater treatment technique. But developing countries with lack of technical know how, weak implementation of environmental policies and with limited financial resources are still facing problems. In India different lakes receive a heavy influx of sewage, industrial effluents, domestic and agricultural waste which consists of varying hazardous chemical and causing deleterious effects on fish and other aquatic organism (Allen and Moore, 2004). In addition fishing and recreational activities in lake also pollute its water. Coastal lagoons receive a variety of pollutants from land drainage. Information on the distribution of heavy metals in coastal lagoon water is essential to assess the accumulation levels in the organism and their possible transfer to food chain, which governs the fishery potential. The primary sources of heavy metals pollution in coastal lagoons are input from rivers, sediments and atmosphere. Highly reactive free radicals are produced with the exposure of plants to a range of abiotic stresses which include water, salt, heavy metal toxicity, etc. These reactive oxygen species have been implicated directly with molecular damage in plant cells. Heavy metals are integrated components of the ecosystem (Franklin *et al.*, 2002). Though some heavy metals are essential as micronutrients, uptake of higher concentrations of heavy metal is found to be toxic for plants. Certain metals are known to produce/act as catalysts for the production of free radicals in biological systems. Many heavy metals like Fe, Cu, Cd, Cr, Zn, etc. have been shown to cause oxidative damage in various higher plants. Conventional aquatic toxicity testing generally assesses the toxic effects of contaminants in their dissolved state (Kennish, 1996). Many contaminants entering the marine environment, however, have the ability to move into various solid components of an ecosystem. For example, heavy metals may be bound within organic matrices or accumulate in sediments and macroalgae. In this way, organisms may be affected by contaminants not only through direct exposure to metals in the dissolved state but also indirectly through the physical environments they inhabit and the food they consume. It is therefore essential for ecotoxicologists to examine the concentration of metals dissolved in the water along with other parameters (Nayar *et al.*, 2003). With this background the objective of this work was to evaluate the heavy metal concentration coupled with physico-chemical parameters in selected sites along the coastal belt of Kerala.

Materials and Methods

The polluted kalamukke in Cochin (9°97′N–76°23′E), Ashtamudi lake in Kallam 8°53′–9°52′N and 76°31′–76°41′E) and non polluted Munamakalle in Chetua were the sites selected for the study. Water samples were collected in 250ml acid washed polythene bottles and were kept in 4°C during transportation. At the laboratory, the samples were immediately filtered using acid washed and pre-weighed millipore filters to separate particulate metals from the dissolved metals. Samples were analyzed for copper (Cu), zinc (Zn), lead (Pb), manganese (Mn) and cadmium (Cd) by acid digestion and quantified using atomic spectroscopy (Bharghava and Raghupathy, 1993). Sample digestion for the determination of mercury (Hg) was done by cold atomic spectrophotometer in a mercury analyzer. Physico-chemical parameters of surface and sediment water sampling such as salinity, silicates, phosphates, ammonia, turbidity, pH, temperature, NO_3 and NO_2 were also analyzed. For DO analysis sample was collected in a 250 ml reagent bottle and fixed at the spot and analysis was done in the laboratory as per the methods of APHA (1995).

Results and Discussion

Environmental pollution by metals became extensive as mining and industrial activities increased in the late 19th and early 20th century (Togna *et al.*, 2001). Table 27.1 displays the physico-chemical parameters of surface water sampling. Temperature varied from 27.85 to 31.7°C in the three sites. pH ranges from 7.8 to 8.1, highest pH value was recorded at Kallam. pH is the scale of intensity of alkalinity. DO values ranged from 2.89 to 4.8 mg/L DO was comparatively higher at Chetua and lowest in Cochi. Decrease in DO is an index of increased organic pollution which is mainly due to addition of waste (Santos *et al.*, 2002). The decrease in DO can be correlated with the higher BOD suggesting the higher accumulation of sediment waste in the area. Turbidity value varies from 18.9–35. Concentration of NO_3 varied from 5.4 to 36.1 µmol/L. NO_2 content ranged from 2.4–8.0 µmol/L. Similarly NH_3 5.2 to 10.1 µmol/L. NO_2 occurs in water as an intermediate product in the biological breakdown of N_2, being produced either through the oxidation of NH_3 or the reduction of NO_3 (Ying *et al.*, 2002). The higher NO_2 content in Kollam may be due to the higher decomposition of nitrogenous compounds. The results for sediment analysis are represented in Table 27.1. The physico-chemical characters of sediments were not much different from the values of surface water. Thus, the initial results show that Cochi and Kollam coastal belts are more polluted than Chetua.

Table 27.1: Physico-chemical Analysis of Surface Water and Sediments from Cochin, Kollam and Chetua Sites

	Surface Water			*Sediment*		
	Cochin	*Kollam*	*Chetua*	*Cochin*	*Kollam*	*Chetua*
pH	7.8	8.1	7.2	7.7	8.3	7.4
Temp.	27.9	29.5	32	28.7	28.9	30.1
Salinity (ppt)	2.5	26.3	0.65	13.4	36.3	5.65
Chlorinity (µmol/l)	1.1	15.8	0.35	6.1	17.8	3.35
Silicate (µmol/l)	209.6	34.5	29.8	246.6	38.5	32.8
NO_3 (µmol/l)	8.5	5.4	36.1	7.5	4.9	25.1
NO_2 (µmol/l)	8	2.4	5.9	8	1.98	4.9
NH_3 (µmol/l)	10.1	8.3	5.2	9.1	10.3	4.82
PO_4 (µmol/l)	28.6	6.2	5	24.5	5.4	12
Turbidity (NTU)	32.4	35	18.9	35.7	38.4	14.6
DO (mg/L)	2.89	3.1	4.8	2.21	2.6	4.45
BOD (ppm)	45.3	43	16.4	43.2	44.2	14.4

The current world-wide mine production of copper, cadmium, lead and mercury is considerable (Smirnoff *et al.*, 2001). The amount of heavy metals in the water bodies from three localities selected for this study clearly showed that copper remains as one of the major pollutants in sites caused by the offshore and onshore loading and cleaning of ships (Table 27.2). High concentration of copper was observed in polluted sites (Cochin and Kollam) and was 86.3 ppm compared to the non-polluted site

(Chetua). The level of copper in the polluted site was higher than the maximum permitted by United Nations Environmental Protection Agency (USEPA, 1999) regulations, for marine ecosystems. Concentration of zinc did not show sound difference between the polluted and non-polluted site. The case for cadmium, mercury, manganese and lead showed a somewhat proximity with copper. For the past several years heavy metals have perceived greater impact because they provide a coded history of marine ecosystem (Pawar and Sathe, 2006). In the present investigation the heavy metals like Cd, Pb and Hg are comparatively higher in Cochin and Kollam whereas in Chetua the concentration was low. Heavy metal toxicity affects aquatic organisms and consequently to the other consumers (Grepe *et al.*, 2000). The interactions of heavy metals in aquatic systems are complicated because of the possible changes due to many dissolved and particulate components and non equilibrium conditions. Often heavy metals experience a change in the chemical form or speciation as a result of these processes and their distribution, subsequently affecting the bioavailability and other interactions in the environment. For example the speciation of heavy metals is controlled not only by their chemical properties but also environmental variables like pH, redox potential, DO, ionic strength, alkalinity, hardness, concentration and nature of dissolved organic chelating agents such as organic acids, humic materials, peptides and polyamino carboxylates. As environmental factors change the chemical reactions and speciation of heavy metals, they influence not only the mobilization, transport and bioavailability, but also the toxicity of heavy metal ions towards biota in both fresh water and marine ecosystems (Pinto *et al.*, 2003). Metal ions commonly bond with sulfhydryl and carboxyl acid groups in aminoacids and inturn affect proteins. The change in physiochemical conditions can lead to metylation of Hg to develop methyl mercury which is one of the most toxic chemicals harmful to the ecosystem, particularly the bottom feeding biota. The biomagnification of these toxic compounds is also possible which endangers the whole marine ecosystem (Hu *et al.*, 1996). Thus, the preliminary study reveals a serious threat of heavy metal pollution due to discharge of effluents from industries and shipping. This can cause a serious pollution problem in aquatic ecosystem or otherwise and may disturb the primary food chain. Further studies are warranted to assess the impact of these in the marine algae and animals.

Table 27.2: Concentrations of Heavy Metals (ppm)

	Cadmium	*Copper*	*Manganese*	*Zinc*	*Lead*	*Mercury*
Cachin	19.2	86.3	389.9	323	67	0.21
Kallam	19.7	62.7	412	289	81	0.18
Chetua	9.5	41.1	270.5	267	32	0.019

References

Allen, J. and Moore, M., 2004. Environmental prognostics: Is the current use of biomarkers appropriate for environmental risk evaluation? *Marine Environmental Research*, 58: 227–232.

APHA, 1975. *Standard Methods for the Examination of Water and Wastewater*, 14th edition. American Public Health Association, Washington, D.C.

Bharghava, B.S. and Raghupathy, H.B., 1993. Analysis of plant material for macro and micro nutrients, In: *Methods of Analysis of Soils, Plants, Water and Fertilizer*, (Ed.) H.L.S. Tandon. FDCO, New Delhi, p. 53.

Franklin, N.M., Stauber, J.L., Apte, S.C. and Um, R.P., 2002. Effect of initial cell density on the bioavailability and toxicity of copper in microalgal bioassays. *Environ. Toxicol. Chem.*, 21: 742.

Hu, S., Tang, C.H. and Wu, M., 1996. Cadmium accumulation by several seaweeds. *Sci. Total Environ.*, 187: 65.

Kennish, M.J., 1996. *Practical Handbook of Estuarine and Marine Pollution*, CRC Press, New York.

Lionetto, M.G., 2001. Biomonitoring of heavy metal contamination along the Salento coast (Italy) by metallothionein evaluation in *Mytilus galloprovincialis* and *Mullus barbatus*. *Aquatic Conservation: Marine and Freshwater Ecosystems*, 11: 305–310.

Nayar, S., Goh, B.P.L. and Chou, L.M., 2003. *In situ* microcosms to study the impact of heavy metals resuspended by dredging on periphyton in a tropical estuary. *Aquat Toxicol.*, 54: 293.

Pinto, E., Teresa, C.S., Sigaud-Kutner, Leitao, M.A.S., Okamoto, O.K., Morse, O. and Colepicolo, P., 2003. Heavy metal induced oxidative stress in algae. *J. Phycol.*, 39: 1008.

Santos, M.M., Garrido, I.M., Goncalves, F., Soares, A.M.V.M. and Ribeiro, R., 2002. An *in situ* bioassay for estuarine environments using the microalga *Phaeodaetylum trieornutum*. *Environ Toxicol. Chem.*, 21: 567.

Smirnoff, N., Conklin, P.L. and Loewus, F.A., 2001. Biosynthesis of ascorbic acid in plants: A renaissance. *Ann. Rev. Plant Physiol. and Plant Mole. Biol.*, 52: 37.

Togna, M.T., Kazumi, J., Sabine, A., Kirtay, V. and Young, L.Y., 2001. Effect of sediment toxicity on anaerobic microbial metabolism. *Environ. Toxicol. Chem.*, 20: 2406.

USEPA, 1999. National recommended water quality Criteria: Correction EPA 822–Z–99001. Office of water. US Environmental Protection Agency, Washington, DC, USA.

Ying, G., Murray, P., Sauve, S. and Hendershot, W., 2002. Low metal bioavailability in a contaminated urban site. *Environ. Toxicol. Chem.*, 21, 954.

Chapter 28

In vitro Plant Regeneration from Different Explants of *Centella asiatica* (L.)

Surjeet S. Bisht, Snehlata Bhandari and N.S. Bisht

Department of Botany, H.N.B. Garhwal University, Campus Pauri, Pauri (Garhwal) – 246 001, Uttarakhand

ABSTRACT

In vitro organogenesis study of *C. asiatica* (L.) member of family Apiaceae was tried considering its medicinal importance. MS medium containing Kn (1.0–3.0 mg/l^{-1}) and 2,4-D (1.5–4.0 mg/l^{-1}) is best for callusing in leaf and petiole explants. 1.5–2.0 mg/l^{-1} concentration of Kn and 2,4-D or IBA proved best for callus induction in nodal explant. BA (1.5 mg/l^{-1}) and IBA (1.0 mg/l^{-1}) combination in MS medium induced maximum number of shoots, while 1.0 mg/l^{-1} of auxin in MS full was suitable for higher root development. Controlled conditions and regular irrigation with nutrient media at initial stages are desirable for higher survival of plantlets.

Keywords: *Apiaceae, Callusing, Organogenesis, Rooting, Shooting.*

Introduction

Centella asiatica L. (Apiaceae), a weakly scented medicinal plant is occurring in part of India, Sri Lanka, Indonesia, Malaysia and Southern and Central Africa. Plant is used in several Ayurvedic preparations for the treatment of leprosy, varicose veins, ulcers, lupus and certain eczemas and of mental retardation since ancient times (Kartnig, 1988). The drug is also reported to possess anti-epileptic activity (Moharana and Moharana 1994). Clinical trials have shown that extracts of *C. asiatica* heal wounds, bums and ulcerous abnormalities of the skin, cure stomach and duodenal ulcers. Plant possesses anti-leprotic, anti-filarial, adaptogenic, antiviral and antibacterial and insecticidal properties. The requirement of *C. asiatica* is met from natural vegetation, leading to its gradual depletion. Tissue culture technique can play an important role in the rapid multiplication as well as germplasm conservation of the plant. Standardisation of an *in vitro* multiplication is a crucial pre-requisite for this. Shoot regeneration and somatic embryogenesis from different explants of *Brahmi* (*Bacopa monniera* L.) have been reported by Tiwari *et al.* (1998). Patra *et al.* (1998) had successfully regenerated *C. asiatica* from callus cultures. *In vitro* multiplication of *C. asiatica* from leaf explant has been studied by Banerjee *et al.* (1999) and Tiwari *et al.* (2000).

Material and Methods

Centella asiatica explants (leaf, petiole, node) collected from nature (Oak pine forest of Garhwal) are initially washed with tap water followed by a wash with 1 per cent (v/v) Labolene detergent for 15 minutes and then in running tap water for 30 minutes. The explants are then surface sterilized with an aqueous solution of 0.1 per cent $HgCl_2$ (w/v) for 2–3 min, followed by a final 5–6 rinses with sterilized double distilled water. The cut surfaces exhibiting mercuric chloride damage were asceptically trimmed with sharp, sterile surgical blade.

Murashige and Skoog (1962) basal medium (MS) supplemented with 3 per cent sucrose and 8 per cent agar is used in the present investigation. Auxins (Indole 3-acetic acid, Indole 3-butyric acid, Napthalene acetic acid or 2,4-Di chlorodiphenoxy acetic acid) and cytokinins (6-Benzyl Adenine or Kinetin) are incorporated into the basal medium in varying concentrations and combinations as indicated in the result. The pH of all the media combinations is adjusted to 5.8±0.1 using 0.1 N NaOH or 0.1 N HCl. Autoclaving is done at 1.06 kg cm^{-2} at 121° for 25 to 30 minutes. Cultures are incubated at 24°±2°C temperature and 60 per cent relative humidity with 16 : 8 hrs, light: dark photoperiod. All treatments having 4 to 8 replicates. Each culture flask contained 40 ml of culture medium and 2–4 explants. The experiments are repeated thrice.

Rooted plantlets removed from the culture media were thoroughly and gently washed under running tap water to remove agar from the roots, and planted in thermocol cups (5 cm. dia) having three different types of soil *viz.* (*i*) sterilized mixture of garden soil and organic manure in the ratio of 1:1 (v/v), and (*ii*) sterilized mixture of sand and soil in the ratio of 1:1 (v/v), (*iii*) sterilized forest soil. Planted thermocol cups were kept in growth chamber for two to four weeks. Potted plantlets were irrigated either with (*i*) water only or (*ii*) 1/4 strength of growth regulator free MS medium subsequently with water or (*iii*) 1/2 strength of growth regulator free MS medium and subsequently with water during the plants growth in Growth Chamber. Plantlets were covered with 250 ml transparent glass beakers or polythene bags to provide high humidity. Temperature of growth chamber was maintained at 20°C±1°C with 16hr photoperiod.

Plantlets were transferred to earthenware pots (25 cm. dia) containing different types of soil mixture. Planted earthenware pots were kept in shade for two weeks and watered regularly. All the potted plants were then placed outdoor with the gradual increase in the periodicity under full sun. The plants transferred to full sun were irrigated regularly for a week and number of plants surviving till the end were counted and survival percentage was computed.

Results and Discussion

Callusing Response

MS medium containing either Kinetin or 2,4-D (0.5 mg/l^{-1}) did not show any callusing response. Poor callusing is observed when MS medium is fortified with auxin and 0.5 mg/l^{-1} Kn. Higher concentration of Kn (1.0–3.0 mg/l^{-1}) combined with higher concentration of 2,4-D (1.5–4.0 mg/l^{-1})

exhibited best callusing in leaf and petiole explants. Kn (1.0–1.5 mg/l^{-1}) and IBA (1.5 mg/l^{-1}) combination in MS proved most appropriate in the production of best callus mass in leaf explant.

Higher concentration of another cytokinin BA (2.0–2.5 mg/l^{-1}) produced best calli when combined with 0.5 to 1.5 mg/l^{-1} of NAA in leaf explants while 1.5–2.0 mg/l^{-1} concentration of both BA : NAA and Kn : 2,4-D or IBA are found to be most suitable of the development of best calli in petiole and nodal explants respectively.

Shooting Response

Development of petiole from the stolon part is considered as shooting in this plant. Periodic variations in the number of petioles were observed from 15th day after inoculation and at the interval of 15 days onwards up to 120 days. BA was combined with IBA in MS medium for shooting purpose. BA used was in the concentration range of 1.0 to 3.5 mg/l^{-1} and IBA in the range of 0.5 to 2.0 mg/l^{-1} and both were combined separately in different concentrations. Number of petiole ranged from 2.0±0.82 to 3.55±1.28 initially at 15 day stage. The rate of appearance of new petioles in different concentrations varied during study period. At 120-day stage maximum number of petioles developed is 1.5 : 1.0 mg/l^{-1} followed by 2.0 : 1.0 mg/l^{-1} combination (31.1±3.3). Higher concentrations of BA and IBA in MS medium resulted in the slowest growth rate where the petiole number was 10.0±1.63.

Lowest concentration of these chemicals used in the present study also exhibited lower number of petioles (11.6±2.28). 2.5 : 1.5 mg/l^{-1} combination was found to be better than the combinations of either lowest or highest concentrations of cytokinin and auxin. Thus most favourable combination for best growth was 1.5 : 1.0 mg/l^{-1} of BA and IBA (Table 28.1).

Table 28.1: Periodic Shooting Variation in *C. asiatica*

MS Basal + Growth Regulators (mg/l^{-1})	*Number of Days*							
Cytokinin : Auxin	*15*	*30*	*45*	*60*	*75*	*90*	*105*	*120*
1.0 (BA) : 0.5 (IBA)	3.33±1.24	5.0±1.63	5.66±1.24	7.0±1.63	7.12±1.63	8.0±1.63	10.0±1.63	11.6±2.28
1.5 (BA) : 1.0 (1BA)	3.0±0.82	8.33±1.25	13.33±2.28	19.0±1.63	21.3±2.05	26.6±2.15	32.6±1.71	37.0±1.63
2.0 (BA) : 1.0 (IBA)	3.0±0.81	7.33±1.24	12.33±2.05	16.6±2.54	21.0±1.63	24.6±2.05	26.3±2.05	31.3±3.32
2.5 (BA) : 1.5 (IBA)	2.0±0.82	4.66±1.69	7.0±l.63	9.66±1.69	12.3±1.88	15.3±2.05	17.3±3.09	20.6±2.62
3.0 (BA) : 1.5 (IBA)	3.12±1.20	5.4±1.76	6.0±1.61	6.29±1.32	7.0±1.63	7.43±1.62	8.0±1.63	10.0±1.63
3.5 (BA) : 2.0 (IBA)	3.55±1.28	5.7±1.82	5.79±1.32	7.12±1.61	7.33±1.63	8.12±1.69	8.52±1.75	10.0±1.63

±: SD.

Rooting Response

MS medium was fortified with 0.5 to 1.5 mg/l^{-1} of auxin (NAA or IBA) in separate sets of experiment to induce rooting in the small fragments having well developed shoots that were transferred in the fresh medium for rooting purpose. Root number was periodically recorded at the interval 15 days upto 120 days.

Rooting behaviour was studied in MS medium fortified with 0.5 or 1.0 mg/l^{-1} of NAA or IBA as well as in MS half strength combined with 1.5 mg/l^{-1} of auxins. Requirement of 1.0 mg/l^{-1} auxin in MS full for the development of greater number of roots is well evident. However, good results were noticed when half strength of MS medium was used with 1.5 mg/l^{-1} of auxin. Lower number of roots were observed in the nutrient medium containing lower concentration of auxins (Table 28.2).

Table 28.2: Periodic Variation in Number of Roots of *C. asiatica*

MS Auxins (mg/l^{-1})	*Number of Days*							
	15	*30*	*45*	*60*	*75*	*90*	*105*	*120*
MS full + 0.5 NAA	3.33±1.24	7.33±2.05	11.0±3.74	16.0±3.74	18.0±2.94	20.0±0.82	34.3±3.3	38.6±2.87
MS full + 1.0 NAA	6.0±0.82	10.3± 1.0 1	14.6±2.05	22.0±1.63	27.6±2.05	32.0±2.16	43.3±3.39	50.6±3.09
MS half + 1.5 NAA	2.66± 1.69	8.33±1.24	16.0±2.94	21.0±2.94	26.3±5.74	35.0±4.08	41.0±2.94	45.6±4.94
MS full + 0.5 IBA	3.0±0.8	6.66± 1.69	11.0±2.44	15.0±4.32	17.0±2.94	22.0±4.96	33.0±3.26	37.6±2.89
MS full + 1.0 IBA	3.66± 1.24	9.33±1.2	13.3±1.07	21.3± 1.24	26.3±2.05	35.3±2.62	42.3±2.86	49.0±2.16
MS half + 1.5 IBA	4.66± 1.69	9.33±2.05	15.3±2.05	24.0±2.94	32.0±2.49	35.3±5.31	41.3±4.19	43.3±4.78

±: SD.

Field Transfer

Observation indicates that higher percentage of survival can be achieved by keeping the plantlets under controlled conditions (Growth Chamber) for a longer period with the supply of nutrie:1ts (half or quarter MS). Irrigation with water only increased the mortality rate thus reducing survival percentage.

It was important to maintain low temperature of 18°C to 20°C because at 25°C±2°C most of the plantlets died. Similar findings have been reported for *Valeriana wallichii* (Mathur *et al.*, 1988) and *Swertia chirata* (Wawrosch *et al.*, 1999), the Himalayan plants which grow at an altitude of 1500–3000 m.

Thus it indicates the requirement of sufficient nutrients during initial stages is important for enhancing the survival rate. Results indicate that hardening of the plants weaned away from culture vessels at least for a minimum period of 2 weeks was a pre-requisite for successful establishment, because all plantlets transferred directly to earthen pots could not survive while those hardened for 2–4 weeks in the Growth Chamber with high humidity showed higher percentage of survival.

Irrigation soon after planting and intermittent misting at 4 hour intervals was the most suitable to achieve high percentage establishment of the plants (Sreekumar *et al.*, 2000). These workers have reported the highly desirable condition to rear the hardened plants at least for two weeks in the net house with 50 per cent shade and regular irrigation before field planting.

80 to 85 per cent survivality of *C. asiatica* was reported by Banerjee *et al.* (1999) during transfer from growth chamber to glass house. Field transfer trials made in the present study indicate that garden soil with organic compost potted plants kept in controlled environment for a indicate that garden soil with organic compost potted plants kept in controlled environment for a longer duration, irrigation with half strength MS solution and water are the desirable conditions to achieve high survival percentage for *C. asiatica*.

No callusing has been seen without cytokinins and auxins in the culture media, irrespective of the explant used. Kukreja (1996) had reported that there was no morphogenetic response on MS medium and similarly the leaf explant of *Azadirachta indica* had no callusing in MS medium alone (Eeswara *et al.*, 1998).

Saikia *et al.* (2001) have reported the best concentration and combination for callus induction was 1.0–1.0 mg/l^{-1} of 2,4-D and BAP with increase in the concentration of both the phytohormones resulted in poor growth of friable callus in *Hydrocotyle rotundifolia*. Leaf explant of *C. asiatica* in the present study also reflected best growth of callus in 1.0 : 1.0 mg/l^{-1} of Kn and IBA but similar response was noticed with 1.5 : 1.5 mg/l^{-1} combination as well. Higher concentration of Kn (1.5–2.5 mg/l^{-1}) in combination with higher concentration of 2,4-D (1.5–3.5 mg/l^{-1}) also responded towards best callusing in the present case. The difference in the response in our study may be due to the age of explant as they were collected from the plants growing in the field. Reports on callus induction in *Psoraiea corylifolia* using mature leaves and stem in MS medium supplemented with combination of cytokinins (BA and Kn) and auxins (NAA and 2,4-D) all at 0.0–4.0 mg/l^{-1} by Saxena *et al.* (1997) are substantiating our findings.

Lower concentration of BA (0.5 mg/l^{-1}) in combination with 0.5–2.5 mg/l^{-1} of NAA showed poor callusing while higher concentration of BA (2.0–2.5 mg/l^{-1}) with lower concentration of NAA (0.5–1.5 mg/l^{-1}) reflected fast callus proliferation resulting into best callus mass during growth period. Similar finding were earlier reported by Matsubara *et al.* (1996) while studying *in vitro* callus formation of *Ocimum basilicum* cultivars and *Perilla* spp.

Generally a cytokinin or a combination of cytokinin and auxin is required for *in vitro* shoot proliferation (Thrope and Patel, 1984). Fracaro and Echeverrigaray (2001) also observed that the presence of cytokinin on the culture media positively influence the micropropagation of *Cunila galioides*, and that BA gave significantly more shoot per explants than the other cytokinins tested (Tiwari *et al.*, 1998; Banerjee *et al.*, 1999; Saikia *et al.*, 2001). The combination of cytokinins and auxins stimulated the *in vitro* multiplication and the growth of shoots of several plant species (George 1993).

Islam *et al.* (1996) had observed that number of shoot/seedling increased with increase in BA concentration and 2.0 mg/l^{-1} BA showed the maximum response. Maximum multiple shoots were noticed in 1.5 mg/l^{-1} BA concentration with IBA (1.0 mg/l^{-1}) in the present study.

Inhibitory effect of increased auxin concentration on shoot bud formation was reported by Nandawani and Ramawat (1991). Increase in the concentration of auxin suppresses the bud differentiation. This may be explained by assuming that the amount of auxin synthesized endogenously was more than sufficient to interact with added cytokinin to achieve the balance necessary for bud induction. The present findings have the supports of observation made by Mandal and Gadgil (1979).

Banerjee *et al.* (1999) have reported that the initial sprouting required the presence of BAP at 2.0 mg/l^{-1} and IBA at 0.1 mg/l^{-1} concentration. Tiwari *et al.* (2000) also observed that the maximum frequency and optimum number of shoots was attained on MS medium containing high BA and low NAA concentrations. High shoot numbers obtained in the present investigation in high BA (1.5–2.5 mg/l^{-1}) concentrations find support from above findings.

Excised shoots were transferred to rooting medium. Earlier workers have reported the use of full or half or quarter strength of MS medium supplemented with auxin (IAA, IBA, NAA etc.). The use of half or full strength of MS media for root induction can be attributed to the favourable effect of low concentrations of macro and micro nutrients on rooting is probably due to the decreased requirement of nitrogen for rhizogenesis in some plants, while it may not be the case with other plant species. Low level of sucrose with auxin was reported optimum in *C. asiatica* by Patra *et al.* (1998) and in *Punica granatum* by Nataraja and Neelambika (1996).

Nandwani and Ramawat (1991) have observed the increased rooting with increased concentration of IBA up to 3.0 mg/l^{-1} and NAA up to 5.0 mg/l^{-1} incorporated singly in half strength of MS. In the present study also maximum number of roots were induced on MS half with higher concentration of auxin.

Banerjee *et al.* (1999) while studying *in vitro* micropropagation using leaf explant of *Centella asiatica* have reported root initiation on full or half strength MS medium with or without an auxin. The average number of roots was good on half strength MS + 1.5 mg/l^{-1} auxin. In the present study, however, maximum number of roots were observed on MS full + 1.0 mg/l^{-1} NAA or IBA (Table 28.2).

References

Banerjee, S., Mehra, Z. and Kumar, S., 1999. *In vitro* multiplication of *Centella asiatica*: a medicinal herb from leaf explant. *Current Science*, 76(2): 147–148.

Eeswara, J.P., Stuchbury, T., Allan, E.J. and Mordue, A.J. (Luntz), 1998. A standard procedure for the micropropagation of the neem tree (*Azadirachta indica* A. Juss). *Plant Cell Reports*, 17: 215–219.

Fracaro, F. and Echeverrigaray, S., 2001. Micropropagation of *Cunila galioides*: A popular medicinal plant of South Brazil. *Plant Cell Tissue and Organ Culture*, 64: 1–4.

George, E.F., 1993. *Plant Propagation by Tissue Culture, Part 1: The Technology.* Exegetics Ltd., Edington.

Islam, R., Hossain, M., Reza, M.A., Mamun, A.N.K. and Jorder, O.I., 1996. Adventitious shoot regeneration from root tips of intact seedlings of *Aegle marmelos. J. Hort. Sci.*, 11(6): 995–1000.

Kartnig, T., 1988. Clinical application of *Centella asiatica* (L.) Urb. In: *Herb Spices and Medicinal Plants*, (Eds.) Craker, L.E. and Simson, J.E. Oxyx Press, Phoenix, p. 145–173.

Kukreja, A.K., 1996. Micropropagation and shoot regeneration from leaf and nodal explant of peppermint (*Mentha pipperita* L.). *Journal of Spices and Aromatic Crops*, 5(20): 111–119.

Matsubara, S., Ino, M., Murakami, K., Kamada, M. and Ishihara, I., 1996. Callus formation and plant regeneration of herbs in *Perilla* family. *Scientific Report of the Faculty of Agriculture*, Okayama University, 85: 23–30.

Mandal, S. and Gadgil, V.N., 1979. Differentiation of shoot buds and roots in fruit callus culture of *Solanum nigrum* L. *Indian Journal of Experimental Biology*, 17: 876–878.

Mathur, J., Ahuja, P.S., Mathur, A., Kukreja, A.K. and Shah, N.C., 1988. *In vitro* propagation of *Valeriana wallishii. Planta Medica*, 39: 187, 277, 1–3.

Moharana, D. and Moharana, S., 1994. A clinical trial of mental in patients with various types of epilepsy. *Probe* 33: 160–162.

Murashige, T. and Skoog, F., 1962. A revised medium for rapid growth and bioassys with tobacco tissue cultures. *Physiol. Plant*, 15: 473–497.

Nandwani, D. and Ramawat, K.G., 1991. Callus culture and plantlets formation from nodal explants of *Prosopis juliflora* (Swartz) DC. *Indian Journal of Experimental Biology*, 29(5): 523–527.

Nataraja, K. and Neelambika, G.K., 1996. Somatic embrogenesis and plantlet from petal cultures of pomegranate, *Punica granatum* L. *Indian Journal of Experimental Biology*, 34(7): 719–721.

Patra, A., Rai, E., Rout, G.R. and Das, P., 1998. Successful plant regeneration from callus culture of C. *asiatica* (Linn.) Urban. *Plant Growth Regul.*, 24: 13–16.

Saikia, F.R., Baruah, C.C., Geka, A.C. and Kalita, M.C., 2001. Micropropagation of *Hydrocotyle rotundifolia*: An indigenous medicinal plant of north east. *Indian Journal of Medicinal and Aromatic Plant Sciences*, 22(4A), 23(1A): 291–293.

Saxena, C., Palai, S.K., Samantaray, S., Rout, G.R., and Das, P., 1997. Plant regeneration from callus cultures of *Psoralea corylifolia* Linn. *Plant Growth Regulation*, 22(1): 13–17.

Sreekumar, S., Seeni, S. and Pushpangadan, P., 2000. Micropropagation of *Hemidesmus indicus* for cultivation and production of 2-hydroxy,4-methoxy benzaldehyde. *Plant Cell Tissue and Organ Culture*, 62: 211–218.

Thorpe, T.A. and Patel, K.R., 1984. Clonal propagation: Adventitious buds. In: *Cell Culture and Somatic Cell Genetics of Plant Cells, Vol. 1: Laboratory Procedures and Their Application*, (Ed.) IX. Vasil. Academic Press, Orlando, Florida, p. 49–60.

Tiwari, K.N., Sharma, N.C., Tiwari, V. and Singh, B.D., 2000. Micopropagation of C. *asiatica* (L.): A valuable medicinal herb. *Plant Cell Tissue and Organ Culture*, 63: 179 –185.

Tiwari, V., Singh, B.D. and Tiwari, K.N., 1998. Shoot regeneration and somatic embryogenesis from different explants of Brahmi [*Bacopa monniera* (L.) Wettsty]. *Plant Cell Rep.*, 17: 538–543.

Wawrosch, C. Maskay, N. and Kopp, B., 1999. Micropropagation of the threatened Nepalese medicinal plant *Swertia chirata* Buch.-Ham. Ex Wall. *Plant Cell Report*, 18: 997–1001.

Chapter 29

Influence of Weather Factors on Dynamics of Bud Fly Incidence on Three Common Varieties of Linseed

Rabindra Prasad, Uday Kumar Prasad and Devendra Prasad

Department of Entomology, Birsa Agricultural University, Kanke, Ranchi – 834 006 (Jharkhand)

ABSTRACT

The results revealed that the budfly (*Dasyneura lini* Barnes) emerged as severe pest on linseed. Budfly incidence started during 4th week of February, 2005 on linseed, C.V. Neelum, Shubhra and T-397 to the tune of 2.60, 1.20 and 1.80 per cent bud infestation. The pest incidence gradually increased with rise in the ambient temperature and advancement of the crop growth. In between the period of 2nd week of March to end of March and 1st week of April, 2005, the insect pest incidence picked up very fast. As such, ultimately maximum bud-infestation of 50.55, 21.60 and 17.50 per cent was noticed on Neelum, Shubhra and T-397 at dough stage of the crop respectively. Optimum and congenial condition for bud fly's growth and multiplication was found to be the higher temperature and milder R.H. As such, range of 29.70°C to 33.0°C (max. temp.) and 12.70°C to 15.80°C (min. temp.) as well as a range of 78–89 per cent (max. R.H.) and 39 to 42 per cent (min. R.H.) were found to be favourable for the population and incidence build up of the linseed budfly. There was significant positive correlation between the pest incidence and atmospheric temperature (maximum and minimum). Relative humidity (max. and min.) and rainfall had the negative impact on the pest incidence. From the degree of resistance point of view, the three linseed varieties were found to be in order of T-397 > Shubhra > Neelum.

Keywords: *Dasyneura lini, Dynamics of incidence, Effect, Linseed, Weather parameters.*

Introduction

Linseed (*Linum usitatissimum* Linn.) is one of the important *rabi* oilseed crops grown in both irrigated and rainfed conditions. Despite, the significant area under linseed in Jharkhand, its productivity and production is very low. One of the significant factors for lower yield of the crop, attack of bud fly (*Dasyneura lini* Barnes) is major one. Budfly's maggots cause damage to the flower buds to the tune of 15 to 50 per cent (Pruthi and Bhatiya, 1937) and some times upto 80 per cent (Gupta, 1995). As such, reduction in seed yield of linseed is ranged from 50–70 per cent (Prasad, 1998). In order to suppress the pest intensity, one should have knowledge of the impact of meteorological parameters on the pest species, which will be utilized in the adjustment in shifting in dates of sowing to enable the crop to escape from the peak period of the pest incidence. This may be able to minimize the crop loss. With this objectives the present investigation was under taken.

Materials and Methods

A field experiment was conducted in *rabi* season with three linseed varieties T-397, Neelum and Shubhra in randomized block design with plot size of 5×3 m^2. Seeds were sown on 30th November, 2004. The crop was raised under the recommended agronomical packages of practices. No protection measures were provided to enable the crop to be naturally infested with the pest. Observation on the incidence of budfly (*D. lini*) was recorded at weekly intervals starting from the bud initiation stage. Number of total floral buds (healthy and damaged) and budfly infested buds were recorded on 10 randomly selected plants in all the three linseed varieties replication-wise for calculating percentage of bud damage caused by *D. lini.* As such, mean percentage of bud damage was calculated variety-wise for each of the weekly observation. Finally, correlation coefficient (r) was worked out between atmospheric temperature and relative humidity and rainfall to draw the conclusion.

Results and Discussion

Results (data) pertaining to budfly (*D. lini*) infestation in terms of percentage of bud damage are presented in Table 29.1 and correlation matrix between bud damage and weather factors are shown in Table 29.2.

Dynamics of the Pest Incidence

The experimental results revealed that budfly infestation started during 4th week of February (*i.e.* 8th standard meteorological week, SMW) to the tune of 2.60, 1.20 and 1.80 per cent in Neelum, Shubhra and T-397 respectively. Since then, the pest incidence exhibited a progressive enhancement in terms of intensity of bud-infestation with the increase in atmospheric temperature and advancement in the age of the crop plants and decrease in the atmospheric relative humidity. Accordingly, bud-infestation gradually increased in all the three linseed varieties almost in order of T-397 > Shubhra > Neelum. It was interesting to note that, in between the period of 2nd week of March to end of March and 1st week of April, 2005, the degree of pest incidence picked up very pest which showed the faster multiplication rate of budfly (*D. lini*). Accordingly, 14.50, 17.60 and 34.80 as well as 15.80, 18.70 and 44.50 per cent bud damage were recorded during 12th and 13th SMW in case of T-397, Shubhra and Neelum respectively. As such, upto the dough stage of the crop, the bud damage reached upto its maximum level, *i.e.* 17.50, 21.60 and 50.55 per cent in the respective linseed varieties during the 13 SMW that fell in the first week of April 2005. A close perusal of the results (Table 29.1) revealed that maximum temperature range of 29.70°C–33°C and minimum temperature range of 12.70°C to 15.80°C along with relative humidity ranging from 78–89 per cent (maximum RH) and 39-42 per cent (minimum RH) were found to be congenial for the faster multiplication of bud fly and higher damage of floral buds of linseeds, caused by the pest, in the present study. The present findings are more or less in agreement with that of observations registered by Malik *et al.* (2000).

Table 29.1: Dynamics of Incidence and Infestation of Budfly (*Desyneura fini*) on Three Linseed Varieties as Affected by Weather Factors

Standard Meteoro-logical Week (SMW)	Date of Obser-vations	Bud Fly Infestation in Terms of Percentage of Bud Damage (mean of three replications)			Whether Factors				
					Temperature °C		Relative Humidity (RH%)		Rainfall (mm)
		Neelum	Shubhra	T-397	Max.	Min.	Max.	Min.	
5	03-02-05	0.00	0.00	0.00	18.70	9.60	90	63	23.20
6	10-02-05	0.07	0.04	0.02	25.20	11.30	86	56	39.00
7	17-02-05	1.90	0.06	0.04	28.70	13.60	88	54	0.00
8	24-02-05	2.60	1.20	1.80	28.90	12.80	90	52	10.80
9	03-03-05	4.50	1.70	1.95	27.90	12.70	89	40	0.00
10	10-03-05	12.72	2.50	2.74	32.20	15.80	89	39	0.00
11	17-03-05	24.40	12.55	2.76	29.30	26.30	78	42	16.60
12	24-03-05	34.80	17.60	14.50	33.10	16.20	89	39	5.60
13	31-03-05	44.50	18.70	15.80	33.00	15.60	88	40	2.00
14	07-04-05	50.55	21.60	17.50	33.5	16.80	86	38	0.00

(Dough stage of the crop).

Table 29.2: Correlation Matrix of Weather Factors and Budfly Infestation (SFI) in Three Linseed Varieties

Sl.No.	Weather Factors Correlation 'r' Between	Value of Correlation Coefficient 'r' in Three Linseed Varieties		
		Neelum	Shubhra	T-397
1.	Max. temperature Vs bud infestation (%)	0.802**	0.816	0.835**
2.	Min. temperature Vs bud infestation (%)	0.785**	0.779**	0.766**
3.	Min. RH (per cent) Vs bud infestation (%)	–0.816**	–0829**	–0.836**
4.	Min. RH (per cent) Vs bud infestation (%)	–0.769**	–0.776**	–0.728**
5.	Rainfall (mm) Vs bud infestation (%)	–0686*	–0.705**	–0.719**

*: Significant; **: Highly significant.

Relationship of Budfly Incidence with Weather Factors

Correlation studies of the present investigations (Table 29.2) revealed that there was highly significant positive correlation between maximum atmospheric temperature and bud fly incidence *i.e.* r = 0.835**, 0.816** and 0.802** in case of T-397, Shubhra and Neelum. Similarly, correlation coefficient. between minimum temperature and bud fly infestation (*i.e.* 0.766**, 0.779** and 0.785**) also remained positive and significant. Relative humidity (both maximum and minimum) had significant and negative correlation with bud fly incidence having value of correlation coefficient, r = –0.836**, –0.829**, –0.816** (for maximum RH) and r= –0.728**–0.776** and –0.769** (for minimum RH). Rainfall also had negative impact on the incidence of bud fly having values of r = –0.719**, –0.705** and –0.686* in the respective varieties in the present study. Almost, similar results were obtained by earlier workers (Kumar *et al.*, 1989, Malik *et al.*, 1998).

Degree of Bud Infestation as Affected by Varietal Reaction

Results of the present studies (Table 29.1) also suggested that inspite of delayed sowing (*i.e.* on 30-11-2004), linseed varieties like T-397 and shubhra suffered from substantially lesser bud damage *i.e.* 17.50 and 21.60 per cent as compared to that of a susceptible variety, Neelum which showed the highest bud damage of 50.55 percent. As such, T-397 and Shubhra exhibited as moderately resistant against budfly. This findings is almost in agreement with that of results obtained by Prasad (1998).

Conclusion

It may be concluded that atmospheric temperatures had significant and positive effect and relative humidity and rainfall had significantly negative influence on the dynamics of incidence of (*D. lini*) on the three linseed varieties. However, degree of bud damage was found to be influenced by the extent of resistance inherited by different linseed varieties

References

Gupta, R.L., 1995. Linseed gall fly, *Dasyneura lini* Barnes and its control. In: *Proc. 42nd Indian Science Congress*, 3: 294.

Kumar, D., Singh, S.V., Kanvar, R.P. and Singh, Y.P., 1989. Incidence of various insect pests of linseed in relation to environmental factors. *Farm Sc. J.*, 4(1&2): 87–91.

Malik, Y.P., Singh, B., Pandey, N.D. and Singh, S.V., 1998. Infestation of linseed bud fly, *Dasyneura lini* Barnes in relation to weather factors under irrigated condition in central Uttar Pradesh. *Ann. Pl. Prot. Sci.*, 6(1): 206–208.

Malik, Y.P., Hussain, K., Singh, S.V. and Srivastava, R.L.,2000. Development of management module for bud fly, *Dasyneura lini* in linseed. *Indian J. Ent.*, 62(3): 260–269.

Prasad, R., 1998. Low cost integrated management of the bud fly for sustainable production of linseed. In: *Abstracted Res. Papers: P. 260 Inter. Nat. Conf. of Pest and Pesticide Management for Sustainable Agriculture*, 11–13 Dec, held at C.S.A. Univ. of Ag. & Tech. Kanpur (India).

Pruthi, H.S. and Bhatiya, H.L., 1937. A new cecidomyid pest of linseed in India. *Indian J. Agric. Sci.*, 7: 797–806.

Chapter 30

Integrated Ecofriendly Management of Solid Wastes of Pulp Industry: Review

P. Jothimani, A. Bhaskaran and S. Vidhya

Department of Crop Management, Anbil Dharmalingam Agricultural College and Research Institute, Tiruchirappalli – 620 009, T.N

Introduction

Increased industrialization has raised the pressure on raw materials and the waste output to a point where natural reclamation pathways cannot cope up. Hence, it is clear that new methods of waste management and reuse must be developed. Among the major industries in India, pulp industry is one, which contributes to water and soil pollution to a significant level. Pulp industries are discharging huge quantities of solid and liquid wastes, the proper disposal of which is a problem for such industries. The wastewater is discharged on land or into the stream, which makes the watercourses polluted and often renders the water useless for miles down stream. Large quantities of sludge are also generated during the mechanical and activated sludge treatment of these wastewaters. Activated sludge and bio-sludge accounts for 10–50 per cent of total sludge produced from the industry. As the manurial value of pulp industrial sludge is generally rated higher than farmyard manure, its primary usage as an organic manure and soil conditioner is an attractive proposition. Bio-sludge and other solid wastes from pulp industries can be very well organically recycled to enrich the soil fertility and to substitute part of chemical fertilizers and ameliorants without causing environmental hazards. Recycling of organic wastes for the maintenance of soil fertility and crop productivity is of economic importance. Hence, the alternate eco-friendly use of industrial solid wastes for crop production and land reclamation is gaining importance in recent years.

Characteristics and Management of Pulp Industrial Solid Wastes

The treatment plant of any pulp industrial effluent produces solid wastes. During pulping and rayon manufacturing 30 to 50 per cent materials are converted as solid wastes (Chatterjee and Bhargava, 1982). The quantum and ingredients of solid wastes generated from each plant varies. Disposal of solid wastes has been a problem since the availability of landfill sites has decreased and the treatment of solid wastes before their disposal at landfill sites consumes high energy. Utilization of the municipal waste treatment facilities, combining the high BOD, low nutrient pulp industrial wastes with high nutrient loaded city liquid wastes is another option for its disposal. A third and probably the sound approach from the economic and ecologic stand point over the long run, is the disposal of such organic wastes on land. Land application of solid wastes provides an effective and environmentally acceptable option of waste disposal, which also recycles valuable nutrients into the soil plant system (Chattetjee and Bhargava, 1982). Applying pulp industrial sludges to agricultural land as a soil amendment or as a source of lime has been suggested as more desirable alternative to land fill (Gabriels, 1998). Due to high content of organic matter in pulp industrial sludges, composting and land application have become an attractive method of their disposal. Further, the results of detailed sludge characterization work, greenhouse studies, decomposition work and column leaching studies showed that land application was feasible (Macyk, 1999)

Nutrient Content of Pulp Industrial Solid Wastes

The quantities and characteristics of solid waste generated from pulp industries are highly dependent on raw materials, processes and process conditions used in the industry. The principal organic constituents in pulp industrial primary sludge include cellulose, hemi cellulose, lignin and other components of wood fibre, process chemicals and compounds produced during the pulping (Xiao *et al.*, 1999). Industrial sludges contain considerable amount of macro- and micro-nutrients and it can be substituted for commercial mineral fertilizers. Primary sludge consisted of organic matter mainly in the form of cellulosic or wood fibre and had low N contents of 0.1 to 0.25 per cent on dry weight basis (Bellamy *et al.*, 1995). Simard (2001) reported that the combined primary and secondary pulp industrial sludge was rich in nitrogen.

The paper and pulp industrial lime sludge had higher concentrations of Ca and Mg. Pulp industrial sludge was reported to contain various micronutrients such as Fe, Cu, Mn and Zn. The copper content in sludge was reported to be within acceptable limits (Bellamy *et al.*, 1995). The C/N ratio of the sludges ranged from 18 : 1 (Elvira *et al.*, 1996) to 150 : 1 (Bellamy *et al.*, 1995).

Impact of Pulp Industrial Solid Wastes on Soil Properties

Physical Properties

Waste recycling through land application has begun in the 1960's as an alternative to land filling and as a way to amend degraded and nutrient depleted soils. The high fibrous content of pulp industrial solid wastes makes these materials ideally suited to improve the soil physical properties. The application of pulp industrial sludge to mine spoils stabilized the soil structure, and improved the infiltration and water-holding characteristics. Paper industrial sludges increased the soil moisture holding capacity (Bellamy *et al.*, 1995) and improved the general soil physical properties (Arfaouri *et al.*, 2001). Physical characters of soil can best be maintained using small yearly application or a single large application of sludge. Solid wastes had a favorable effect on soil physical conditions and tilth due to their high organic matter content. At high levels of application, soil properties were reported to be impaired due to the presence of high concentration of metals and toxic constituents. Improved soil structure and total porosity due to pulp industrial sludge application was also observed by Sun and Lie (1992). The clay ash fraction of the pulp and paper industrial sludge could be beneficial to soils less in clay content.

Chemical Properties

Application of pulp industrial sludge to soil resulted in a significant increase in soil pH, electrical conductivity and organic matter content (Hameed Sulaiman, 1997). Several studies have shown that applying pulp industrial sludge with a high C/N ratio to agricultural land will initially cause a net immobilization of nitrogen. Continuous use of pulp industrial sludge over a period of 15 years to a sandy soil increased the soil EC, exchangeable Na, Ca, Mg and K, available P, K, Fe, Mn, Zn and Cu contents. Paper industrial sludge enhanced the nutrients such as Ca, K, Mg, N and P when applied to mine soil, which was usually low in all these nutrients. Pulp and paper industrial sludge when applied to soil increased the cation exchange capacity and nutrient retention capacity (Bellamy *et al.*, 1995).

Biological Properties

Sludge amendments increased the number and activity of microorganisms and available plant nutrients and developed a soil environment conducive for plant growth. Application of decomposed pulp industrial waste at the rate of 10 to 20 t ha^{-1} improved the soil biological properties. Continuous use of pulp industrial sludge over a period of 15 years to a sandy soil increased the activity of urease and phosphatase enzymes. Invertase enzyme activity was higher in soils amended with pressmud and pulp industrial sludge (Palaniswami and Sree Ramulu, 1994). Hameed Sulaiman (1997) observed that among the amendments, ETP sludge recorded the highest microbial population, whereas actinomycetes and fungi were totally absent under lime sludge incorporation.

Impact of Pulp Industrial Solid Wastes on Crop Productivity

Application of solid wastes at lower doses could overcome the heavy metal toxicity problem and cause a positive effect on yield and quality of crops. Bellamy *et al.* (1995) conveyed that application of pulp and paper industrial sludge benefited general agriculture, tender fruit and container nursery production. Better shoot length, root length, collar diameter and dry matter production, were observed under ETP sludge whereas seedlings completely withered in case of *Acacia nilotica* and *Casuarina equisetifolia* due to lime sludge application (Hameed Sulaiman, 1997). Simard (2001) presented that combined primary and secondary pulp industrial sludge could be used for horticultural crop production.

The treatments receiving pulp industrial sludge plus twice the amount of recommended rate of fertilizer produced the highest yield of Bermuda grass (*Cyanodon dactylon*) on mine spoil land. The pulp industrial sludge could be used to add organic carbon to mine spoils and obtain efficient growth of *Cyanodon dactylon*. Pressmud and pulp industrial ETP sludge were identified as effective ameliorants for paddy crop. Bagasse pith and activated sludge in 1 : 1 ratio with 100 per cent NPK was found to increase the growth parameters of groundnut (Udayasoorian *et al.*, 1999). Addition of organic matter has been a method for reducing the heavy metal uptake by plants and this was due to the sorption capacity of organic matter. The plants recovered 10 per cent of the sludge borne heavy metals and nearly 90 per cent of the heavy metals were retained in the top 0–15 cm depth of soil horizon. The Zn and Cu contents were found to be greatly elevated in the tissues of sludge treated vegetable crops. Zinc concentrations in grass were reported to be quite high (98.53 ppm) due to high mobility of this metal in sludge-amended soils.

Composting of Industrial Solid Wastes

The recycling of organic wastes as fertilizers for maintenance of soil health by hygienic method is vital for increasing crop production and the welfare of mankind. Composting is a microbiological, nonpolluting and safe method for disposal and recycling of organic wastes by bioconversion to manures (Gaur, 1987). Compo sting of organic wastes and its utilization in agriculture reduces the requirement of mineral fertilizers and builds up soil fertility on a permanent basis at less cost.

Composting is considered as one of the oldest bio-technological approach of solid waste disposal known to man. The oldest existing reference on composting dates back to the Akkadian Empire of Mesopotamian valley a thousand years before Moses (Gaur, 1987). Historical records show that ancient Greeks and Romans were familiar with the decay process that occurred spontaneously in stocks of plant or animal wastes (Zibiliske, 1998). In India, the use of dung as manure appears to have been practised since the *Rig* Vedic age (2500–1500 BC) (Gaur, 1987). The scientific study of composting was initiated by Hutchinson and Richards at Rothamsted.

Composting is a method of solid waste management whereby the organic component of the solid waste is biologically decomposed and stabilized under controlled conditions to a state where it can be handled, stored and applied to the land to supply essential nutrients without adversely affecting the environment. Haug (1993) defines composting as the biological decomposition and stabilization of organic substrates under conditions that allow development of thermophilic temperatures as a result of biologically produced heat, to produce a final product that is stable, free of pathogens and can be beneficially applied to land. According to Kalaiselvi and Ramasamy, (1996) composting is the biological conversion of heterogeneous organic substrates, under controlled conditions into a hygienic, humus rich and relatively bio-stable product that conditions soil and nourishes plant. Composting reduces the offensive odors, reduces the difficulties in handling due to stickiness, inactivates pathogens and sterilizes the organic constituents, thus producing uniform organic material suitable for soil application. The goals of composting range from sanitation, eliminating pathogenic organisms, reducing the bio-available energy content of the wastes and reducing the volume of wastes (Zibiliske, 1998).

Decomposition of Pulp Industrial Sludges

Pulp industrial sludges are difficult to get decomposed because of their high lignin, cellulose and hemicellulose contents. Therefore, efficient cellulolytic and ligninolytic microbial cultures can playa major role in the bioconversion of pulp industrial sludge. *Pleurolus sajor-caju*, an edible mushroom has been recognized as one of the most efficient lignin degrading organisms known. Cellulose degradation by *Pleurolus sajor-caju* has been rapid at the initial stage of growth on pulp industrial sludge, but the degradation of lignin was very lower at the initial stage of growth. *Pleurolus sajor-caju* degraded 78 per cent lignin and 73 per cent cellulose in pulp industrial sludge after 30 days of inoculation. Cellulose degradation is the growth associated primary metabolic event in *Pleurolus sajor-caju* and lignin degradation is a secondary metabolicevent which is not growth associated. Lignolytic degradation is carried out by bacteria and various group of fungi. White rot fungi is the most promising lignin degrader among all the microorganisms. The activities of exoglucanase and β-glucosidase were minimum at 12 and 26 days of growth, respectively during the decomposition of sludge by *Pleurolus sajor-caju*. Sesay *et al.* (1997) composted the pulp industrial sludge using the relatively low cost, aerated static pile (ASP) method. They suggested a compo sting strategy of a two weeks aeration phase, and a further four to five weeks maturation.

The relationship between nutrient addition and temperature control on composting of pulp industrial sludges was studied by Jackson and Line (1997). The time to produce matured compost could be shortened by 30–50 days if composting was undertaken at higher temperature. The method of nutrient addition had no effect on the respiratory activities of compost microbial population but had a major influence on pH, which determined the intensity and period of ammonia volatilization. If pH was controlled, then incremental nutrient addition could be advantageous from the perspective of nutrient conservation. Composting of pulp industrial solid waste was attempted by Mini (1998). Mixing bagasse pith with activated sludge and effluent treatment plant (ETP) sludge in the ratio of 2: 1 : 1 produced better quality compost. The potential for composting of combined primary and secondary dewatered pulp industrial sludge was assessed and the suitability of the finished product as a potting soil substitute was evaluated by Evanylova and Daniels (1999). They concluded that the pulp industrial sludge compost may best be used as an organic fertilizer, soil amendment, or supplemental nutrient source for potting media, rather than as a potting medium alone.

Vermicomposting of Pulp Industrial Sludges

Pulp industrial sludge is proved as a suitable medium for optimum growth and reproduction of earthworms. They further opined that presence of earthworms in decomposing pulp industrial sludge accelerated the mineralization of organic matter, favoured the break down of structural polysaccharides and increased humification rate. Consequently the C/N ratio and the degree of extractability of heavy metals were lower in the worm-composted product. In Spain, partially composted sludges from a pulp industry was used as a feed stock for *Eisenia andrei* which transformed the substrate by a process of aerobic and mesophilic hydrolysis to a good compost suitable for land application. Ceccanti and Masciandaro (2000) evaluated the potential of vermicomposting as an economical or environmental alternative for pulp industrial sludge management. The experiments were performed on mixtures of aerobic sewage sludge and anaerobically treated paper industrial sludge. Following physical and biological conditioning through an accelerated aerobic composting at the end of the vermicomposting process, a high quality humic product was obtained for use as a soil organic amendment.

Impact of Land Application of Compost on Soil Fertility and Productivity

The compost, comparable to natural humus, is an attractive product for farmers because of the basic economic and technical advantages. Compost preserves the soil nutrients at minimum cost and protects the soil from being washed away by rainwater and improves its structure. The water retaining capacity of the soil is improved by the compost application thus providing a considerable advantage in the event of drought. Compost improved soil porosity, soil structure, water holding capacity, and nutrient availability. The bio-sludge compost proved to be a valid alternative to organic soil conditioners as well as in terms of the recovery of degraded areas. Recycling waste resource has value associated with returning organics to soil, and it helps communities to solve the problem of waste management. The most logical way to preserve or restore the soil fertility is addition of organic manure or compost. The pulp and paper industrial sludge compost may be used as an organic fertilizer and soil amendment (Evanylova and Daniels, 1999). Humic substances and bio-chemicals produced by the decomposition of organic matter not only have the ability to transform insoluble form of iron, zinc and manganese to soluble metal complexes available to plants, but also the insoluble humic compounds in the decomposing organic matter fraction as a sink for poly valent cations such as cadmium and lead. Compost application increased the soil organic matter level, total and residual nitrogen, buffer the pH, increase the micro nutrient status. Scientific tests show that compost can be an effective biomanure, though, compared to mineral fertilizers, the nutritive value of compost is less important.

Application of city waste compost increased the organic carbon, total and available nitrogen, phosphorus, potassium, Zn, Fe and Mn of the soil. Total porosity was greater in the plots treated with urban waste compost than the control plots and enzymatic activities (dehydrogenase, alkaline phosphomono esterase and asparaginase) were high with no inhibiting influence of the heavy metals present. Compost application also helped to bio-remediate polycyclic aromatic hydrocarbons and pesticides contaminated soils. Application of composts from pulp board improved the structure of the plough horizon, decreased its density, and enriched the soil with organic matter. It promoted the development of strong root systems and stimulated biological activity in the soil. Emmerling *et al.* (2000) reported that soils treated with varying amounts of composted sewage sludge and compost resulted in an increase in microbial respiration and enzyme activities like invertase and alkaline phosphatase and also recorded an improvement in soil physical properties such as water and nutrient retention-capacity.

Among the enzymes, catalase, cellulase, dehydrogense, phosphatase and amylase participate in the mineralisation of nutrients. Amylase and cellulase hydrolyse macromolecules and the degradation products become available as nutrient for the organisms. The measurement of dehydrogenase activity in the soil gives information on the activities of microbial population in the soil. Any management practice that influences the microbial population of the soil would be expected to produce changes in soil enzyme activity and the level of enzyme activity can be used as an indicator of soil fertility (Burns, 1982).

The activities of various soil enzymes are influenced by the nature and age of the crop, addition of fertilizers and manures (Burns, 1982). Perucci (1990) observed that phosphatase activity was stimulated by the addition of compost. The enzyme phosphatase hydrolyses organic phosphorus into inorganic form that was taken up by the plants. It was concluded that quantity and type of organic matter added were highly important in influencing the activity of soil enzymes. The phosphatase activity was inhibited up to 50 cm depth due to the addition of vermicomposted sludge but stimulated the activity of dehydrogenases and protease. The increase in soil enzyme activity depended on the rate of sludge applied, the time, which had elapsed since application and on the depth in the soil profile. Masciandaro *et al.* (2000) observed the increase in soil dehydrogenase activity due to *in situ* vermicomposting of biological sludges.

Effect of Compost Application on Crop Growth and Productivity

A balanced application of composts and mineral fertilizers is essential to obtain optimum yield and to maintain soil fertility. All crops respond to organic manuring and the extent of response depends on several factors such as degree of humification, maturity of the compost, its carbon and nitrogen ratio, the time and method of application and moisture regime of soil during the growth of the crop (Gaur, 1987). The results of more than thirty experiments with potato showed an increase in yield with application of organic manure varying from four to thirty per cent over the control (Gaur, 1987). The final result of a six-year pot experiment on the organic fertilization of sunflower crop using compost prepared. from animal waste. The improved crop production was related to the persistence of improved conditions of residual fertility. The use of compost can lead to improved soil fertility, even after several years, which in turn results in a quantitative improvement in crop yield and product quality. Gaur and Singh (1995) found that the incorporation of municipal waste compost at 10t ha^{-1} with 50 per cent recommended NPK recorded normal yield of finger millet crop. Application of N through compost accelerated the metabolic activities, which enhanced the synthesis of amino acids, protein and carbohydrates resulting in higher assimilation of these contents in grains of black gram and wheat.

Lulakis and Pestas (1996) reported an improvement in the quality of products with compost, over those obtained with more conventional fertilizer practices. Application of composted pulp and paper industrial sludge improved the growth and foliar nitrogen of *Pine radiata*. Application of mushroom spent rice straw compost has resulted in an enhanced grain yield of rice to the tune of 20 per cent over 100 per cent recommended NPK. Plant uptakes of nutrients were also more in the compost-amended plots. The number of nodules, nitrogenase activity of nodules and grain and straw yield of green gram were higher in compost-amended soil. Compost application also suppressed numerous plant diseases, plant nematodes and weed population. Composted pulp industrial sludge contains higher proportion of more resistant lignin residues and humates, thus accounting for the lower shrinkage when used as a medium for plant growth while fresh sludges reduced the plant growth due to shrinking in the rooting medium (Bellamy *et al.*, 1995).

Effect of Lime Sludges as Ameliorant for Acid Soil on Soil Properties and Crop Productivity

Alternative sources of liming material and nutrients could be useful to correct soil acidity. In several wastewater treatment plants, hydrated lime is increasingly used to stabilize sludge. These stabilized lime sludges have high $CaCO_3$ equivalencies and can alleviate Al toxicity in acidic soils. In addition, the use of good nutrient sources such as compost and fertilizers could improve the nutrient balance of the acid soil. The pulp industrial sludge amendment increased the soil pH, organic matter content, exchangeable sodium per cent (ESP) and the Ca/Mg ratio. The combined application of sludge with amendments would be of greatest benefit on acid soils and no pollution problems would arise if such applications were suitably managed.

Soil electrical conductivity increased with higher lime-sludge application rates due to soluble salts in the soil. It is concluded that lime-treated municipal sludge can be applied to agricultural land with no detrimental effect on the environment, while maintaining good crop yield. Vityakon and Seripong, (1995) evaluated the paper industrial lime sludge as a liming material for the growth of maize in an Ultisol. Application of lime sludge increased soil pH, exchangeable Ca and Bray II extractable P and decreased exchangeable Al, H and Al saturation of the effective CEC, but had no effect on exchangeable K and Mg. Soil pH increased with lime-containing amendments and reduced the soluble Mn. The largest decreases in exchangeable Al were associated with lime-containing amendments.

Land application of lime-stabilized bio-solids has been reported to cause Mn deficiency in wheat (*Triticum aestivum*) and soybean (*Glycine max*). Soil pH increased significantly with lime-cake addition, reducing the availability of trace metals. Bueno *et al.* (1997) conducted a greenhouse experiment to evaluate the effects of limestone and sewage,sludge on the dry matter accumulation and nutrient absorption by grain sorghum, and the contents of some heavy metals in plants. The application of limestone reduced the absorption of Zn and Mn, the levels of heavy metals extracted from the soil were low, although the plants had not absorbed these metals. Cereal and potato crops recorded a significant increase in yield due to application of primary sludges at the rate of 22.5 t ha^{-1} for land reclamation purposes under which the soil quality and crop response was satisfactory. Simard (2001) used the limed sludge for ameliorating the acid soil along with cement kiln dust as nutrient source. They reported that these ameliorants increased the soil micronutrient content and had no significant effect on soil heavy metal. The lime addition did not influence the growth or heavy metal accumulation in clover.

Okamoto (2000) studied the form and behavior of some heavy metals in a soil with longterm applications of limed sewage sludge. The sludge was conditioned with calcium hydroxide as a coagulant. The results indicated that sludge application increased the Zn, Cd, Ni and Cu contents in the topsoil. The increase in the rate of the Fe-Mn oxides fraction of Zn and Ni, and Cd carbonate fraction started to level off after the 4th application, when the soil pH also leveled off. This suggested that an excessive amount of calcium over the soil cation exchange capacity is a factor in suppressing the sorption of heavy metals to soil.

References

Arfaouri, M.A., R.R. Simard, G. Blanger, M.R. Laverdiere and Chabot, R., 2001. Mixed paper mill residues affect yield, nutritive value and nutrient use of a grass–*Alfalfa sward. Can. J. Soil Sci.*, 81: 103–111.

Bellamy, K.C., C. Chong and Cline, R.A., 1995. Paper sludge utilization in agriculture and container nursery culture. *J. Environ. Qual.*, 24: 1074–1082.

Bueno, M.F., P.A. Bellingieri and Freitas, J.C., 1997. Effects of liming and of organic and mineral fertilization in the culture of grain sorghum [*Sorghum bicolor* (L.) Moench]. *Cientifica (Jaboticabal)*, 25: 163–176.

Burns, R.G., 1982. Enzyme activity in soil location and possible role in microbial ecology. *Soil Biol. Biochem.*, 14: 423–427.

Ceccanti, B. and Masciandaro, G., 1999. Researchers study vermicomposting of municipal and paper mill sludges. *Biocycle*, 40: 71–72.

Chatterjee, B. and Bhargava, K.S., 1982. Reuse of backwater and recovered fibre from pulp and paper machine effluent at Orient Paper Mill, Brajrajnagar. In: *Proc. Int. Sem. Mgt. Envtl. Prob. Pulp Pap. Ind.*, (Ed.) N.L. Ramanathan. Delhi Productivity Council, New Delhi, pp. 177–181.

Elvira, C., M. Goicoechea, L. Sanpedro, S. Mato and Nogales, R., 1996. Bio-conversion of solid paper pulp mill sludge by earthworms. *Bioresource Technol.*, 57: 173–177.

Emmerling, C., C. Liebner, M.H. Rosar, J. Katzur and Schroder, D., 2000. Impact of application of organic waste materials on microbial and enzyme activities of mine soils in the Lusatian coal mining region. *Pl. Soil*, 220: 129–138.

Evanylova, G.K. and Daniels, W.L., 1999. Paper mill sludge composting and compost utilization. *Compost Sci. and Utilization*, 7: 30–39.

Gabriels, D., 1998. Use of organic waste materials for soil structurisation and crop production: Initial field experiment. *Soil Tech.*, 12: 89–92.

Gaur, A.C., 1987. Recycling of organic wastes by improved techniques of composting and other methods. *Resources and Conservation*, 13: 157–174.

Gaur, A.C. and Singh, G., 1995. Recycling of rural and urban wastes through conventional and vermicomposting. In: *Recycling of Crop, Animal, Human and Industrial Waste in Agriculture*, (Ed.) H.L.S. Tandon. Fertilizer Development and Consultation Organization, New Delhi, India, 150 p.

Hameed Sulaiman, S.M., 1997. Influence of paper mill (TNPL) solid wastes as amendments for forest nursery and their impact on soil properties. *M.Sc.(Ag.) Dissertation*, Department of Environmental Sciences, Tamil Nadu Agricultural University, Coimbatore.

Haug, R.T., 1993. *The Practical Handbook of Compost Engineering*. Lewis Publishers, Boca Raton, Fla., 388 p.

Jackson, M.J. and Line, M.A., 1997. Composting pulp and paper mill sludge: Effect of temperature and nutrient addition method. *Compost Sci. and Utilization*, 5: 74–81.

Kalaiselvi, T. and Ramasamy, K., 1996. Compost maturity: Can it be evaluated? *Madras. Agric. J.*, 83: 609–618.

Lulakis, M.D. and Pestas, S.I., 1996. Effect of humic substances from vine canes mature compost on tomato seedling growth. *Biores. Technol.*, 54: 179–182.

Macyk, T.M., 1999. Land application of mechanical pulp mill sludges in Alberta: Research and operational activities. *Pulp and Paper Canada*, 100: 34–37.

Masciandaro, G., B. Ceccanti and Garcia, C., 2000. *In situ* vermicomposting of biological sludges and impacts on soil quality. *Soil Biol. Biochem.*, 32: 1015–1024.

Mini, K., 1998. Impact of composted bagasse pith and paper mill effluent irrigation on soil environment and groundnut crop. *M.Sc. (Environmental Sciences) Dissertation*, Tamil Nadu Agricultural University, Coimbatore.

Okamoto, T., 2000. Form and behavior of some heavy metals in a soil with long-term applications of limed sewage sludge. *Japanese J. Soil Sci. Pl. Nutri.*, 71: 231–242.

Palaniswami, C. and Sree Ramulu, U.S., 1994. Effects of continuous irrigation with paper factory effluent on soil properties. *J. Indian Soc. Soil Sci.*, 42: 139–140.

Perucci, P., 1990. Effect of the addition of municipal solid waste compost on microbial biomass and enzyme activities in the soil. *Biol. Fertil. Soils*, 10: 221–226.

Sesay, A.A., K. Lasaridi, E. Stentiford and Budd, T., 1997. Controlled composting of paper pulp sludge using the aerated static pile method. *Compo Sci. and Util.*, 5: 96.

Simard, R.R., 2001. Combined primary or secondary paper mill sludge as a nitrogen source in a cabbage-sweet corn cropping sequence. *Can. J. Soil Sci.*, 81: 1–10.

Sun, H.G. and Lie, W.Q., 1992. The effects of straw paper pulping wastes using ammonium sulfate on the microstructural characteristics of calcareous soil and paddy soil. *Acta Agric.*, 18: 181–188.

Udayasoorian, C., K. Mini and Ramaswami, P.P., 1999. Bagasse pith compost: An effective ameliorant in paper mill polluted soil environments. In: *Proc. of the Workshop on Bioremediation of Polluted Habitats*, Tamil Nadu Agricultural University, Coimbatore, March 22–23, pp. 69–71.

Vityakon, P. and Seripong, S., 1995. Evaluation of paper mill lime sludge as an acid soil amendment in northeast Thailand. In: *Proc. Plant-Soil Interactions at Low pH: Principles and Management*, (Eds.) R.A. Date, N.J. Grundon, G.E. Rayment and M.E. Probert. The Third International Symposium, Brisbane, Queensland, Australia, pp. 595–599.

Xiao, C., L.Q. Ma and Sarigumba, T., 1999. Effects of soil on trace metal leachability from paper mill ashes and sludge. *J. Environ. Qual.*, 28: 321–333.

Zibiliske, L.M., 1998. *Composting of Organic Wastes*. Lewis Publishers, Boca Raton, Florida, 402p.

Chapter 31

Mycoflora of the Soil Sediment of Two Contrasting Lentic Systems, Tiruchirappalli, Tamil Nadu

R. Sivakami[1], R. Ramya[1], V. Guru[1], M. Suganya[1], and G. Premkishore[2]

[1]*PG and Research Department of Microbiology, Kurinji College Arts and Science College, Tiruchirappalli – 620 002, Tamil Nadu*

[2]*Alan Feldman Public School and Junior College, Kazhakuttam, Tiruvandram, Kerala*

ABSTRACT

A study was carried out on the fungal population of two contrasting lentic systems in Tiruchirappalli, Tamil Nadu, which are utilized for polyculture. Among the 24 species of fungi identified, 15 belonged to mesophilic group and 9 to the thermophilic group. Species diversity was higher in the river fed perennial pond when compared to the temporary rain-fed pond. The river fed pond as well as the rain fed pond through recorded unique species both were dominated by the genus *Aspergillus*.

Keywords: *Mesophilic fungi, Soil sediment, Thermophilic fungi.*

Introduction

There has been an ever-increasing demand for fish because they form the best source of easily digestible proteins and fatty acids that are good for cardiovascular system (Katoch *et al.*, 2003). However, in the past two decades, there has been a tremendous stress on the native fish in India in the wake of overall development and over exploitation of the aquatic ecosystems. This is because aquaculture is more complex than terrestrial animal husbandry because in the aquatic environment, the physical, chemical and biological changes are unpredictable and may change rapidly. Ponds in India receive a large input of organic matter and so they act as a store for a variety of substances. It also harbours a rich microbial community whose activity of leads to anaerobic conditions and development of microbial flora which in turn utilizes organic compounds (Boyd, 1995). Soil affects the nutrient level in water, which in turn influences the productivity. The out break of diseases is mainly due to deterioration in the quality of water and soil condition. Therefore, it is necessary to know about the role of microbes in aquaculture productivity. Even though there are several investigations on the role of bacteria in soil fertilization and growth of aquatic organisms (Maeda and Liao, 1992; Maeda and Liao, 1994). There appears to be only very few reports on the mycoflora of the freshwater aquaculture systems (Sivakami *et al.*, 2003; Paulraj *et al.*, 2003). Hence an attempt was made to analyse the mesophilic and thermophilic fungi in the soil sediments of two contrasting lentic systems in culturing carps.

Materials and Methods

Soil sediments samples were collected from two different ponds–a rain fed astatic pond and a perennial river-fed pond situated at Ayilapettai Region (10 km away from the city) near the city of Tiruchirappalli, Tamil Nadu. Both of them are within one kilometer radius and so have almost the same climatic regime and are used for carp culture regularly. The samples were taken every month from October 2005 to March 2006 and were analyzed using the dilution plate method for the isolation of mycoflora.

One gram of soil from each sample was serially diluted and 1 ml of 10^{-2} dilution was transferred into triplicate plates. Potato dextrose agar was used for the isolation of mesophilic fungi and the plates were maintained at room temperature of 30±2°C. Emerson's yeast phosphate soluble starch agar (YpSs) was used for the solution of thermophilic fungi and the plates were incubated at 47±2°C (Emerson, 1941). After 3–5 day of incubation, the colonies were isolated, identified and enumerated using standard procedures (Cooney and Emerson, 1964; Gilman, 1967; Quions *et al.*, 1981 and Subramanian, 1971).

Results

The mesophilic and thermophilic fungi that belonged to both astatic and the perennial lentic systems used for carp culture are represented in Table 31.1. A total of 24 species could be identified in both the ponds during the study period. Of this, 15 species belonged to mesophilic and 9 to thermophilic group. Among the two ponds, the perennial river fed pond recorded 12 mesophilic and 8 thermophilic species. The astatic rain-fed pond recorded 10 mesophilic and 6 thermophilic species. Of the 12 mesophilic species recorded in the perennial river-fed pond, 5 species were present throughout the period of study while in the astatic pond 6 aspects were recorded during the period of study. Among the thermophilic groups, 4 species were recorded throughout study period in perennial pond and only 3 in the rain-fed pond.

The most common geneus that was found to occur in both the ponds belonged *Aspergillus* (6 species followed by *Pencillium* (2 species) as the whole, species diversity was found to be higher in the perennial pond (20 species) when compared to the astatic pond (16 species). Among the species, five were unique to perennial river fed pond and four to the astatic pond.

Table 31.1: Mesophilic and Thermophilic Fungi Isolated from the Two Contrasting Lentic Systems at Tiruchirappalli

Sl.No.	*Fungi*	*River Fed Pond (Perennial)*			*Rain Fed Pond (Astatic)*		
		*8**	*2**	*Cfu*	*8**	*2**	*Cfu*
	Mesophilic Fungi						
1.	*Aspergillus chevalieri*	+	–	21.4	+	–	3.5
2.	*Aspergillus flavipes*	+	+	132.4	–	–	
3.	*Aspergillus flavus*	+	–	150.2	+	–	126.0
4.	*Aspergillus nudulans*	+	+	136.2	–	–	–
5.	*Aspergillus niqer*	+	+	65.1	+	+	150.2
6.	*Aspergillus terreus*	–	+	11.0	–	–	
7.	*Curuvularia lunata*	+	–	37.2	+	–	1.4
8.	*Cunninghamella echinulatus*	–	–	–	–	+	5.5
9.	*Drechslera australieses*	–	–	–	–	+	48.4
10.	*Fusarium oxysporum*	–	+	41.5	+	–	25.2
11.	*Monila siophila*	–	+	0.6	–	+	20.3
12.	*Penicillium funciculosum*	–	+	16.4	–	–	–
13.	*Penicillium oxalicum*	+	–	15.2	+	–	35.8
14.	*Paecilomyces variotii*	–	–	–	–	+	0.08
15.	*Mucor racermosus*	+	+	55.2	+	–	29.6
	Non sporulating colonies	+	+	52.6	+	+	115.2
	Thermophilic Fungi						
1.	*Aspergillus fumigatus*	+	–	102.2	+	–	137.4
2.	*Cheatoimium* var. *corprophile*	–	+	15.6	–	+	36.25
3.	*Cheatomium* var. *dissitum*	+	–	1.0	–	–	–
4.	*Cheatomium* var. *thermophilie*	–	+	7.6	–	–	–
5.	*Humicola grisea*	–	–	–	–	+	10.6
6.	*Mucor pusillus*	+	–	92.0	+	–	41.9
7.	*Rhizopus stolonifer*	+	+	0.62	–	–	–
8.	*Thermomyces lanuginosus*	+	–	2.8	+	–	110.2
9.	*Torula thermophila*	–	–	–	–	+	8.5
	Non Sporulating colonies	+	+	17.9	+	–	42.2

Where 8*, 2* occurrence of fungal species during study period.

Discussion

Both the ponds are regularly used for polyculture of both native and exotic carps. Feed given to carp usually consist of rice barn, ground nut oil-cake and cow-dung in various ratios. Hence, both the ponds received a large amount of nutrients, the availability of oxygen and the nutrients would have provided an optimal medium suitable for the growth of fungi. Singh and Wadhwani (1989) reported that the fungi were abundant in stagnant waters when compared to flowing water. The high amount of fungal flora present in both the ponds could be attributed to the stagnancy of water. Which in turn provided better chances for germination, growth and survival of fungi on dead and decaying organic matter lying at the bottom (Manoharachary and Ramarao, 1983). Similar reports on the presence of a wide variety of fungi in fresh water mud ponds (Subramanian, 1971), and estuaries (Johnson and Sparrow, 1961) peat soil (Girivasan *et al.*, 1998) and cave soil (John Koilraj *et al.*, 1999) have also been reported.

In the present study *Aspergillus* was the most dominant genera found in both the ponds. The dominance of *Aspergillus* has also been reported in Tilapia and cat-fish culture ponds from Nigeria (Manoharachary and Ramaroa, 1983) as well as prawn culture ponds in India (Paularaj *et al.*, 2003). Among the genus *Aspergillus*, *A. flavus* was found to be dominant in the river fed pond and *A. niger* in rain fed pond.

Whenever a fungal or bacterial species grow along with other orgainism, becomes dominant and the growth of other organisms get retarded due to competition. This phenomena was also reported jn the growth of *Aspergillus niger* in sewage samples (Hiremath *et al.*, 1990) and prawn culture ponds (Paulraj *et al.*, 2003).

Among the two groups, the mesophilic fungi was found in larger diversity than thermophilic species. The presence of large number of mesophilic fungi was observed in our study has also been reported earlier (Hiremath *et al.*, 1990; Paulraj *et al.*, 2003). Some of the fungi present in the study have also been isolated from air and cowdung (Cooney and Emerson, 1964). Thus, the presence of fungi in the soils of fresh water ponds might be due to aerial dissemination or manuring the pond with cowdung or both.

Conclusion

In the present study fungi were more in the river-fed perennial pond when compared to the rain-fed astatic pond. This might probably be due to the entry of water from the canal which would be carrying surface run-off from nearby agricultural fields containing fungi. The variability in counts (Cfu's) between similar species between both the ponds could be due to the differences in the nutrient contents between the two ponds as well as the differences in the characteristics of the soil.

Thus the presence of fungi in aquaculture ponds requires further investigations to analyse their impact on the water, soil as well as in the productivity of aquatic ecosystem of this region.

Acknowledgements

The authors (R. Sivakami) is grateful to the Chairman Dr. P. Selvaraj, J.J.E.T., Vice-Chairman A.K.K. Ravichandran the Principal Dr. K. Santhanam, Kurinji College of Arts and Science, Tiruchirappalli-2 for providing financial assistance, enormous encouragement and motivation.

References

Boyd, C.E., 1995. *Bottom Soil, Sediment and Pond Aquaculture.* Chapman Hill Publ., pp. 1–9.

Cooney, D.G. and Emerson, R., 1964. *Thermophilic Fungi: An Account of Biology, Activities and Classification,* (Ed.) W.H. Freeman. San Francisco, Californiz, 188 pp.

Emerson, R., 1941. An experimental study of the life cycle and taxonomy of Allomyees. *Lioydia*, 4: 77–144.

Gilman, J.L., 1967. *A Manual of Soil Fungi*. IBH Publ. Co., Calcutta, pp. 97.

Girivasan, K.P., Rajagopal, K., Muruganandan, V.R and Suryanarayanan, T.S., 1988. Isolation of fungi from tropical peat soil of southern India. *Curr. Sci.*, 74: 359–360.

Hiremath, A.B., Prabhakarn, M., Nimbaraj and Jayaraj, M., 1990. Activities of fungi and bacteria in domestic waste purification. In: *Recent Trends in Limnology*, (Ed.) R.S. Mishra. Daya Publishing House, New Delhi, pp. 77–85.

John Koilraj, A., Marimuthu, G., Natarajan, K., Saravanan, S., Maran, P. and Minna, J.H.S., 1999. Fungal diversity inside caves of south India. *Curr. Sci.*, 77: 1081–1089.

Johnson, T.W. and Sparrow, F.K., 1961. *Fungi in Oceans and Estuarias*. Hafner Publ. Co., New York, 688 pp.

Katoch, R.C., Mandeep Sharma, D., Pathania, Subhash V.R.C. and Aravind Mahajan, 2003. Recovery of Bacterial and mycotic fish pathogens from carps and other fish in Himachal Pradesh. *Ind. J. Microbiol.*, 43(1): 65–66.

Maeda, M. and Liao, I.C., 1992. Effect of bacterial population on growth of prawn larvae, *Penaeus monodon. Bull. Natl. Res. Inst. Aqua.*, 21: 25–29.

Manoharachary, C. and Ramarao, P., 1983. Microbial ecology of two fresh water mud ponds of Hyderabad district, India. *Bibliotheca Mycologia*, 91: 411–420.

Paulraj, A., Muthmary, K., G., Kathirvaran and Udaya Prakash, N.K., 2003. Fungal flora of the soil sediment of giant prawn culture ponds. *Ind. J. Microbiology*, 43(1): 61–63.

Qnions, A.H.S., Allopp, D. and Eggins, H.O.W., 1981. *Introduction to Industrial Mycology*. Edward Arnold Publ, Londan, pp. 168–209.

Singh, A. and Wadhwani, K., 1989. On the correlation of fungal population in following and stagnant aquatic habitats. *J. Environ Biol.*, 10: 149–157.

Sivakami, R., Joseph Sajith, and Prem Kishore, G., 2003. Fungal flora of soil sediment of Ponds at Vypin area, Kerala. In: *Paper Presented in UGC Sponsored International Symposium on Fresh Water Aquaculture: Potentials and Prospects,* held on 12th–13th of September at P.G. Research Department of Zoology and Biotechnology, A.V.V.M. Sri Pushpam College, Poondi, Thanjavur, Tamilnadu, pp. 55.

Subramanian, C.V., 1971. *Hypomycetes: An Account of Indian Spices Except Cercosporae*. Indian Council of Agricultural Research, New Delhi, 86 pp.

Chapter 32

Estimation of Economic Feasibility of Integrated Management of Sucking Insect Pests of Brinjal Through Inter-cropping and Insecticides

Uday Kumar Prasad and Rabindra Prasad

Department of Entomology, Birsa Agricultural University, Kanke, Ranchi – 834 006, Jharkhand

ABSTRACT

The results of a field experiment, conducted in Birsa Agricultural University, Ranchi with the objective to formulate inter-disciplinary and eco-toxicological pest management strategy to manage sucking insect pests menace, infesting brinjal (var. Mukta Keshi), revealed that efficacy of soil application of the tested insecticides were found to be in order of phorate > carbofuran > no insecticide and that of inter-crops remained to be in order of : marigold> French bean>sole crop of brinjal in terms of suppression of incidence of jassid (*Amrasca bigutulla bigutulla* Ishida), aphid (*Aphis gossypii* Glover) and white fly (*Bemisia Tabaci* Genn.). The findings indicated that the interactive effect between the insecticides proved to be significantly higher (more) as compared to that of their individual impact in terms of quantum of suppression of incidence of the three insect pest species. Significantly higher reduction in the incidence of *A. bigutulla bigutulla*, *A. gossypii* and *B. tabaci* on account of joint (*i.e.* interactive) effect of the insecticide(s) and the inter-crop(s) resulted to higher yield of marketable fruits of brinjal ranging from 170.07 to 231.73 q/ha. Yield of fruits obtained from untreated inter-planted crop of brinjal with marigold and French bean were 144.20 and 135.07 q/ha respectively. Encouraging yield of marketable fruits was obtained from sole crop of brinjal under the protection of cover of phorate, carbofuran and quinalphos remained to be 222.60, 190.87 and 156.73 q/ha respectively, as against the lowest yield of 107.53 q/ha obtained from untreated sole crop of brinjal. As such, compatible amalgamation of olericultural and floricultural crops as inter-crops with the need based desirable modifications in time and space might be emerged as an ecofriendly pest control module with need based application of granular insecticides around the root zone of the plants.

Keywords: *Brinjal, Interactive effects, Inter-cropping, Insecticides, Net profit, Sucking pests, Yield.*

Introduction

Brinjal (*Solanum melongena* L.) is the second most important vegetable crop next to tomato in India. Sucking pests fauna such as jassid (*Amrasca bigutulla bigutulla* Ishida), aphid (*Aphis gossypii* Glover), white fly (*Bemisia tabci* Genn.) and red spider mite (*Tetrancychus urticae* Koch) are some of the devastating pest complex which is responsible for tremendous loss in yield varying from 15 to 52 per cent with change in the time and space. Frequent reports of inability of chemical insecticide(s) used alone in combating the pest problem are nowadays emerging in vegetable crops in general and brinjal agro ecosystem in particular. In general, soil application of granular insecticides are considered not only more effective but also appeared to be relatively less hazardous as compared to that of foliar spraying of insecticides particularly for green vegetables. Keeping in view the complexity of the pest problem, the present study was undertaken, comprising of growing brinjal in the inter cropping system under the protection cover of judicious soil application of granular insecticides.

Materials and Methods

An experiment was conducted in the field of Birsa Agricultural University, during late *kharif* (*i.e.* early *rabi*) season with altogether twelve treatments replicated thrice in the factorial randomized block design (FRBD). There were, three granular insecticides and three inter-crops (Tables 32.1–32.3). The experiment was conducted with brinjal (Variety: Mukta Keshi) with only one soil application of granular insecticides (phorate 10G, carbofuran 3G or quinalphos 5G) applied@ 1.0 kg a.i./ha, amply supplemented with promising inter-cropping *e.g.* marigold (*Tagetus erecta* L.) French bean (*Phaseolus vulgaris* L.) with brinjai and brinjal grown alone. Seedling of brinjal (Var. Mukta Keshi) was transplanted in inter-cropping system in last week of August in 1 : 1 row ratio indicating one row of brinjal and one row of the inter-crops *viz.*, marigold and French bean (Var. Sel-9) and brinjal was grown as sole crop also. Granules of the insecticides *viz.*, phorate 10, carbofuran 5G and quinalphos 5G was applied around the root zone of brinjal plants @ 1.0 kg a.i./ha at 30 days after transplanting. A need based uniform foliar spraying was given with chlorpyriphos 20EC @ 1000 ml/ha at 85 days after transplanting to take care of peak incidence of tissue feeding and tissue boring insect pests. Number of jassids. (*A. bigutulla bigutulla*), aphids (*A. gossypii*) and white flies (*B. tabaci*) were recorded on three leaves basis from lower, middle and top canopy region of five randomly selected plants at weekly intervals to calculate mean number of the respective insect pest(s) per 15 leaves for pooling them in terms of mean of all the observations for the statistical analysis. Similarly, weights of marketable fruits of brinjal and French bean and flower of marigold were recorded at each picking and ultimately pooled together to determine the mean value for the statistical analysis. Thus, net return was calculated on the basis of money value of out-come of yield advantage borne by the addition of yields of sole crop and the subsidiary crop in the respective crop combinations to find out the economic feasibility of the pest management module comprising crop association, insecticide and their interaction in terms of net economic return realized per unit area.

Table 32.1: Effect of Inter-cropping on the Incidence of Sucking Insect Pests and their Ultimate Impact on Yield of Brinjal Fruits

Treatments (Inter-crop-combination)	*No. of Jassids per 15 Leaves*	*No. of Aphids per 15 Leaves*	*No. of Whiteflies per 15 Leaves*	*Yield of Healthy Fruits of Brinjal (q/ha)*
Brinjal + Marigold	7.92 (63.23)	7.17 (52.22)	2.14 (3.68)	199.55
Brinjal + French bean	8.41	7.64 (71.29)	2.34 (59.36)	182.14 (4.65)
Brinjal alone	9.11	8.35 (85.65)	2.84 (72.90)	169.43 (8.23)
SE m (±)	0.15	0.22	0.08	7.06
C.D. (P = 0.05)	0.43	0.63	0.22	21.65
C.Y. (%)	6.05	9.67	10.56	8.12

Table 32.2: Effect of Root Zonal Application of Insecticide on the Incidence of Sucking Insect Pests and their Resultant Impact on Yield of Brinjal Fruits

Treatments (Inter-crop-combination)	*No. of Jassids per 15 Leaves*	*No. of Aphids per 15 Leaves*	*No. of Whiteflies per 15 Leaves*	*Yield of Healthy Fruits of Brinjal (q/ha)*
Brinjal grown in soil treated with phorate	6.84 (46.16)	5.86 (33.91)	1.89 (2.64)	227.58
Brinjal grown in the soil treated with carbofuran	7.73 (58.99)	6.99 (48.29)	2.13 (3.56)	199.47
Brinjal grown in the soil treated with quinalphos	8.76 (76.02)	8.25 (67.34)	2.34 (4.54)	169.51
Brinj al grown in the soil left untreated (Control)	10.59 (112.39)	9.77 (96.43)	3.37 (11.35)	128.93
SE m (±)	0.17	0.25	0.09	8.34
C.D. (P = 0.05)	0.50	0.73	0.25	25.60
C.Y. (%)	6.05	9.67	10.56	8.18

Figures under the parentheses correspond to original values and that of out side figures are $\sqrt{n+1}$ transformed values.

Results and Discussion

Results pertaining to the individual impact of inter-cropping and insecticides(s) on the incidence of the sucking insect pests infesting brinjal and ultimate effect on the fruits yield are shown in Tables 32.1 and 32.2 above respectively. Interactive effects of inter-cropping and insectides in this regard are presented in Table 32.3.

Table 32.3: Interactive Effects of Inter-cropping and Insecticides on the Incidence of Sucking Insect Pests and their Ultimate Impact on Yield of Healthy Brinjal Fruits, and Net Monetary Return

Treatment Combinations		*No. of Jassids per 15 leaves*	*No. of aphids per 15 leaves*	*No. of white flies per 15 leaves*	*Yield of healthy fruits of brinjal (q/ha)*	*Net monetary return (Rs./ha)*
Inter-cropping	*Insecticides*					
Brinjal + Marigold	Phorate	5.48 (29.57)	6.02 (35.73)	1.79 (2.24)	231.73	56,746
	Carbofuran	6.64 (43.41)	7.33 (53.15)	2.07 (3.31)	212.53	49,166
	Quinalphos	8.00 (63.23)	8.49 (71.36)	2.25 (4.09)	181.73	38,084
	No. insecticides	8.56 (72.67)	9.04 (81. 05)	2.46 (5.08)	144.20	25,334
Brinjal + French bean	Phorate	5.82 (33.49)	6.48 (41.47)	1.89 (2.64)	228.40	55,546
	Carbofuran	7.09 (49.53)	7.66 (58.11)	2.13 (3.60)	195.00	42,204
	Quinalphos	8.33 (68.63)	8.78 (76.33)	2.38 (4.72)	170.07	33,666
	No. insecticides	9.31 (85.79)	9.73 (93.86)	2.94 (7.65)	135.07	21,698
Brinjal grown alone	Phorate	6.27 (38.68)	6.87 (46.61)	1.99 (3.03)	222.60	37,494
	Carbofuran	7.26 (51.92)	7.73 (59.10)	2.18 (3.78)	190.87	25,702
	Quinalphos (70.15)	8.41 (78.68)	8.92 (4.76)	2.40	156.73	14,346
	No. insecticides (130.85)	11.45 (139.36)	11. 84 (21.33)	4.71	107.53	–
SE m (±)		0.08	0.32	0.15	6.03	–
C.D. (P = 0.05)		0.26	0.98	0.46	19.03	
C.Y (per cent)		6.05	9.67	10.56	8.18	

Figures in the parentheses are original values, outside figures are $\sqrt{n+1}$ transformed values.

Impact of Inter-cropping

Results (Table 32.1) reveal that brinjal inter-cropped with marigold (*T. erecta*) recorded the minimum population of jassid, *A. bigutulla bigutulla* (63.23/15 leaves), aphid, *A. gossypii* (52.22/15 leaves), and white fly, *B. tabaci* (3.68/15 leaves) almost being at par with brinjal inter-cropped with French bean but both of these inter-cropping combinations proved significantly superior over brinjal crop grown alone in reducing the incidence of these pest species (Table 32.1). Sole crop of brinjal harboured significantly higher number of jassid (85.65/15 leaves), aphid (72.90/15 leaves) and white fly (8.23/15 leaves) in the present study. Prasad *et al.* (1998) also found similar results in inter-cropping systems. Present findings indicates that the inter-crops caused pronounced reduction in the pest incidence which gave rise to higher yield of brinjal. As such, fruits yield of brinjal were found to be 192.55, 182.14 and 169.43 q/ha when it was grown in association with marigold, French bean and it was grown alone respectively. Findings of Prasad *et al.* (1998) and Prasad and Prasad (2005) are also in agreement with the result of the present study. Thus, performance of the crop association in terms of quantum of the pest suppression and their positive impact on fruit-yield was found in order of brinjal grown with marigold>brinjal grown with French bean > brinjal grown alone, in the present investigations (Table 32.1). Presence of row of non-host plants in between rows of host plants may influence pest incidence in the inter cropping system probably on account of creation of physical barrier, and repellent action contained in the inter-crop in host finding capacity of the pest complex of the crop (Prasad, 2001 and Prasad and Prasad, 2005).

Impact of Insecticides

A perusal of Table 32.2 reveals that all the three granular insecticides *viz.*, phorate, carbofuran and quinalphos applied, around the root zone of brinjal, after 30 days after transplanting, @ 1.0 kg a.i./ha proved significantly superior in reducing incidence of jassid, aphid and white fly which were found to be responsible for corresponding enhancement in the yield of marketable yield of brinjal, ranging from 169.51 to 227.58 q/ha as compared to the untreated sole crop of brinjal (Table 32.2). As such, bio-efficacy of the granular insecticides were recroded to be in order of: phorate > carbofuran > quinalphos > no application of insecticide: in reducing the attack of the sucking insect pests of the crop and also in increasing the fruit yield (Table 32.2).

Interactive Impact of Inter-cropping and Insecticides

It was found that all the treatments were found to be superior in suppressing the incidence of the insect pests *viz.*, *A. bigutulla bigutulla*, *A. gossypii* and *B. tabaci* and in enhancing the yield of marketable fruits of brinjal over the untreated sole crop of brinjal (Table 32.3). Lower pests incidence was recorded in different treatment combinations which significantly varied among the insecticidal treatments combined with either of the three crop combination as compared with untreated sole crop which harboured the highest incidence of population of: jassid (130.85/15 leaves); aphid (139.26/15 leaves) and white fly (21.33/15 leaves). Both the inter-crop *viz.*, marigold and French bean when grown with brinjal, significantly reduced the population of the three pest species, which was again reduced by supplementation with the insecticidal treatments in order of: phorate > carbofuram > quinalphos > no insecticide application (Table 32.3). As such, the interactive effect of inter-cropping and insecticide were found to be significantly effective in suppressing the pest attack and in encouraging the yield enhancement in the present study. Highly pronounced pest reduction in case of brinjal inter-planted with either marigold or French bean were observed probably due to certain volatile substance emitted by both the inter-crop(s) which repelled and obstructed the entry and incidence of the insect pest species in the field/plot and ultimately upon the plant(s) up to the significant extent. Pronounced pest suppression was caused as the inter-crops were responsible for preventing the pests from locating their food plants on account of their repellant properties as well as their role played as physical barrier (Jimenez and Carrilo, 1979; Prasad, 2001). The interactive effects between the inter-cropping and insecticide(s) were found to be significantly effective in enhancing the marketable yield of brinjal furits ranging from 170.07 to 231.73 q/ha probably on account of better protection received by the crop against the insect pests. Brinjal grown in association with marigold receiving protection also by root zone soil application with phorate gave rise to the highest yield of fruit (23 1.73 q/ha) which remained at par with that of brinjal grown in association with either French bean or even when it was grown alone provided the soil was treated with phorate it self.

Economic Feasibility of the Crop-Combination and the Insecticides

The highest net monetary return of Rs. 56,746/ha was realized by brinjal grown in association with marigold in the soil treated with phorate (Table 32.3). This was closely followed by brinjal inter-planted with French bean and soil treated with phorate itself. Though, carbofuran and quinalphos when applied as soil application in the inter-cropping system of brinjal grown either with marigold or French bean was also found to be quite encouraging in terms of fetching higher net monetary return ranging from Rs. 33,666/ha to Rs. 49,166/ha. Sole crop of brinjal grown in the soil treated with either phorate, carbofuran or quinalphos as soil application around the root zone gave rise to relatively lower level of net profit of Rs. 37,494, Rs. 25,702 and Rs. 14,346 per hectare respectively which remained little higher to that of the respective insecticidal treatments alongwith the supplementation with the inter-cropping systems (Table 32.3). Prasad *et al.* (1998) obtained almost similar results.

Conclusion

Conclusively, interactive effect of inter-cropping of brinjal either with marigold or French bean and root zone soil application of the insecticides *viz.*, phorate or carbofuran appeared to be ecologically feasible and economically viable which may find a good place in the integrated management of sucking pests for realizing encouraging yield of brinjal fruits.

Acknowledgements

The first author is thankful to the Vice Chancellor, Dean (Agric.) and Dean PGS of Birsa Agricultural University for granting the study leave.

References

Jimenez Aragon, J.G. and Carrilo Senchez, J.L., 1979. Beneficial insect fauna on cotton inter sown with maize as compared with cotton alone. *Agriculture Teachnica en Mexico*, 4(2): 143–156.

Prasad, R., Chand, P. and Prasad, D., 1998. Studies on the interactive effects of insecticides and crop-associations on the pest complex of brinjal (*Solanum melongena* L.) for its sustainable cultivation in Ranchi (India). In: *International Conference on Pest and Pesticide Management for Sustainable Agriculture* (ICPPMSA), pp. 346–347. Held at C.S.A. Univ. of Agric. and Technology, Kanpur, on 11–13th December.

Prasad, R., 2001. Effect of crop association and insecticide on the incidence of shoot and fruit borer, *Leucinodes orbonalis* Guen. of brinjal. *Shashpa*, 8(2): 195–197.

Prasad, R. and Prasad, D., 2005. Management of shoot and fruit borer (*Leucinodes orbonalis* Guen.) of brinjal through crop-association. In: *National Symposium on Recent Advances in Integrated Management of Brinjal Shoot and Fruit Borer*, pp. 185–186. Held at IIVR, Varanasi on 3–4 October, 2005.

Chapter 33

Succession and Status of Insect Pests Infesting Linseed

Rabindra Prasad, Uday Kumar Prasad and Devendra Prasad
Department of Entomology, Birsa Agricultural University, Kanke, Ranchi – 834 006, Jharkhand

ABSTRACT

Overall findings of pest monitoring and survey conducted in different rural areas of Ranchi district of Jharkhand for four consecutive years revealed that linseed crop was found to be infested with as many as 13 insect pest species during the different growth stages of the crop. The pest species was found to occur almost in overlapping manner. Termite (*Microtermes* spp.) was found to infest in seedling stage of the crop. The incidence of termite remained upto low to mild level over the observational periods. Grass hoppers (*Chrotogonus trachypterous* Blanch and *Oxya velox* F.) occurred in seedling to vegetative stage, the status of the these two pests remained as minor pest with low incidence. Bihar hairy caterpillar (*Spilosoma obliqua* Walker) appeared upto low to mild pest incidence in stray mode of incidence. Semi looper (*Plusia orichalcea* Fab.), aphid (*Aphis craccivora* Koch; *Macrosiphum* sp.), green stink bug (*Nezara virudula* Linn.) and pentatomid bug (*Piezedorus hybneri* Gmel) appeared in stray form of occurrence as minor pest on linseed. Capsule borer (*Heficoverpa armigera* Hubner) and bud fly (*Dasyneura lini* Barnes) were noticed as regular pests in the reproductive phase of the crop as major pest, whereas leaf miner. (*Phytomyza h6rticola* Gour) emerged occasionally in capsule formation stage upto very low level of incidence. Among the major pests, bud fly (*D. lini*) emerged as severe and key pest of linseed. So far as the natural enemies of pest fauna of linseed is concerned, yellow wasp (*Polistes* spp.) spiders (unidentified), black ant (*Acantholepsis simplex* Forel), Rove beetle (*Paederus* spp.), coccinelid beetle (*Coccinela septempanctata* L.) and preying mantid (unidentified) were found to occur in the linseed agro-ecosystem.

***Keywords**: Insect pests, Linseed, Natural enemies, Status.*

Introduction

Linseed (*Linum usitatissimum* Linn.), an oilseed of various industrial utilities, is grown during *rabi* both in rain fed and irrigated situations in India. It is going to be emerged as the crop of immense medicinal values and usages. The extent of loss in seed yield caused by insect pest species is found to be ranging from 50 to 80 per cent (Gupta, 1995; Prasad, 2003). However, pest fauna of a crop varies from region to region. Familiarization with pest spectrum of a crop in the particular zone or locality is the first and foremost pre-requisite for formulating a suitable pest management strategy. Keeping this view in mind, the present investigation was undertaken.

Materials and Methods

Pest survey and surveillance was conducted during 1995–96 to 1998–99 in different rural areas of Ranchi district of Jharkhand for visualizing with the real, pests-situation occurring on linseed by random sampling in roving survey procedure made at fortnightly intervals between 7.00 am to 10.30 am. Absolute population per five plants per plot per location were recorded in case of leaf hopper, aphid, Bihar hairy caterpillar, semi looper, green stink bug, pentatomid bug, leaf miner and capsule borer whereas in case of grass hopper their numbers were counted per square meter (Deshmukh *et al.*, 1992). The occurrence of bud fly (*Dasyneura lini* Barnes) at green bud stage as well as dough stage of the crop was recorded by counting the total number of floral buds as well as damaged buds, due to bud fly, to work out percentage of bud fly infestation. Almost similar process was followed to record percentage of capsule borer infestation. Ultimately, based on the numerical abundance, the pest fauna were placed in accordance with the crop phonological phase (Table 33.1). Finally, based on the mean value of the abundance and severity of the infestation of the respective pest species, they were classified as pest of low, mild, minor, major, severe and key pest of linseed.

Incidence of natural enemies (NEs) complex was recorded by a net sweep method by employing five sweeping of insect collecting net per location for calculating mean numerical value per, sweep. On the basis of mode of occurrence, over the years, they were categoriesed as stray, regular and occasional mode of incidence.

Results and Discussion

The overall results, recorded during 1995–96 to 1998–99 are shown in Table 33.1.

Pest Incidence and their Succession

A perusal of results (Table 33.1) indicated that as many as thirteen insects pests appeared to infest the crop. The first insect to appear as pest was termite species, *Microtermes* spp. and *Odototermes* spp. which usually caused damage to the germinating seeds of linseed. Their status was observed to be as low to mild pest. Deshmukh *et al.* (1992) observed termite and cut worm (*Agrotis* spp.) as pest of linseed in its seedling stage. The crop in its stages of seedling to early vegetative form was found to be infested by four insect pest species *viz.*, grass hoppers, *Oxya velox* F. and *Chrotogonous tranchypterous* Blanch, Bihar hairy caterpillar, *Spilosoma obliqua* Walker and some times by *Spodoptera litura* with their low and stray incidence, though S. *oblique* appeared as major pest (*i.e.* low to mild) for early sown crop and minor pest for late sown crop. The another pests infesting the crop as minor and stray pest in the vegetative and reproductive phase of the plants were aphids, *Aphis craccivora* Koch, *Microsiphum* sp. green stink bug, *Nezara virudula* Linn; pentatomid bug (*piezodorus hybreri* Gmel.) leaf hopper; *Amrasca Kerri* Pruthi, semi looper, *Plusia orichalcea* F. and capsule borer, *Helicoverpa armigera* Hub. More or less similar observations were recorded by some earlier workers like Deshmukh *et al.* (1992), Malik (1998) and Patel and Thakur (2005) in the agro climatic conditions of Delhi, Kanpur and Orissa respectively.

Table 33.1: Succession and Status of Insect Pests of Linseed Under the Agro-climatic Conditions of Ranchi, Jharkhand (1995–96 to 1998–99)

Sl. No.	Name of Insect	Crop Stage	Mode of Occurrence	Status
	A. Insect Pest Species			
1.	Grasshopper (*Chrotogonus trachypterus* Blanch)	Seedling and vegetative	Stray	Low
2.	Grasshopper (*Oxya velox* F.)	Seedling and vegetative	Stray	Low
3.	Leaf hopper (*Amrasca kerri* Pruthi)	Vegetative	Stray	Minor
4.	Semi looper (*Plusia orichalcea* Fab.)	Vegetative, flowering and capsule formation	Stray	Minor
5.	Aphid (*Aphis craccivora* Koch.)	Flowering and capsule formation	Stray	Minor
6.	Aphid (*Macrosiphum* sp.)	Flowering and capsule formation	Stray	Minor
7.	Green stink bug (*Nezara viridula* Linn.)	Vegetative and reproductive phase	Stray	Minor
8.	Pentatomid bug (*Piezedorus hybneri* Gmel)	Vegetative and reproductive	Stray	Low
9.	Bihar hairy caterpillar (*Spilosoma obliqua* Wlk.)	Seedling and early vegetative	Regular	Low to mild
10.	Capsule borer (*Helicoverpa armigera* Hub.)	Bud, flowering and capsule formation	Regular	Mild–major
11.	Bud fly (*Dasyneura lini* Barnes)	Bud and flowering, dough stage	Regular	Severe-key
12.	Termites (*Microtermes* spp.) (*Odontotermes* spp.)	Seedling	Regular	Low to mild
13.	Leaf miner (*Phytomyza hoticola* Gaur)	Capsule formation	Stray, Occasional	Low
	B. Natural Enemies (Predatory fauna)			
1.	Yellow wasp (*Polistes* spp.)	Vegetative and reproductive	Stray, Occasional	Low
2.	Spiders (unidentified)	Vegetative and reproductive	Regular	Mild
3.	Black ant (*Acantholepis simplex* Forel)	Vegetative and reproductive	Regular	Mild
4.	Rove beetle (*Paedorus* spp.)	Late vegetative phase	Regular	Low
5.	Cocciflelid beetle (*Coccinela septempuctata L.*)	Late vegetative and reproductive phase	Regular	Low
6.	Preying rnantid (unidentified)	Late vegetative and reproductive phase	Regular, Occasional	Low

However, in case of late sown crop of linseed, *H. armergera* appeared as major (mild to major) pest in the capsule formation stage which was generally not reported earlier for this crop. As the most injurious pest of the crop, bud fly, *Dasyneura lini* Barnes appeared in the bud formation stage in the severe forms in almost all the years of the investigation resulting to unopening of the infested buds that in turn gave the insect its status as *key pest*. Early sown (*i.e.* middle October) crop was found to be suffered with relatively lower infestation and late sown crop (middle November and onward) suffered the most. Almost similar information was generated by Malik (1998) and Prasad *et al.* (2005).

It was found that incidence of *D. lini* and *H. armigera* was observed to be increased with delaying in sowing time of the crop. Literature survey indicated that in the agro-climatic condition of Delhi, linseed suffered with as many as twenty one insect pests out of which only three pests *viz.*, leaf miner, *P. horlicola*, cutworm, *Agrotis ipsilion* Hufn. and capsule borer, *H. armigera* exhibited as major pests and budfly *D. lini* appeared as the extremely severe pest (*i.e.* key pest) of linseed (Deshmukh *et al.*, 1992). In the present study, thirteen insect pest species appeared on linseed; among them, termites and Bihar hairy caterpillar occurred in low to mild form whereas capsule borer appeared as major pest. Bud fly, *D. lini* caused the severe and extremely servere damage to the floral buds and hence this was categorized as *key pest* of linseed. Experimental findings of Prasad *et al.* (2004) is almost in the conformity with that of the present results.

Incidence of Natural Enemies of Pest Fauna

So far as the prevalence of the bio-agents (natural enemies, NEs) regulating the incidence of insect pests and occurring in the linseed ecosystem is concerned, as many as six species of NEs namely, (*i.e.* preying mantids (unidentified), spiders (unidentified), rove beetle (*Piezedorus* spp.) yellow wasp (*Polistes* spp.) black ant (*Ancantholepis simplex* Forel) coccinnelid beetle (*Coeeinela septempunetata* L.) were noticed during the period of the investigations in the agro-climatic conditions of Ranchi. Information generated by Malik (1998) and Patel and Thakur (2005) are more or less similar to the findings of the present study.

Conclusion

It may be concluded that linseed crop plants was found to be infested with as many as 13 insect pests species in overlapping manner. Out of those pests, the bud fly (*D. lini*) appeared as severe and most damaging pest (key pest) and capsule borer (*H. armigera*) behaved as newly emerging major pest on linseed. Among the natural enemies of the pest fauna, six arthropod fauna were found to occur in linseed agro-ecosystem as predators on the pest species.

References

Deshmukh, S.D., Singh, K.M. and Singh, R.N., 1992. Pest complex and their succession in linseed, *Linum usitatissimum* L. *Indian J. Ent.*, 54(2): 168–173.

Malik, Y.P., 1998. Insect pest complex and their phenology in linseed, *Linum usitatissiumum* L. *Indian J. Ent.*, 60(4): 364–370.

Patel, R.K. and Thakur, B.S., 2005. Screening of different cultivars of linseed against the budfly, *Dasyneura lini* Barnes. *J. Pl. Protec. Environ.*, 2(2): 99–101.

Prasad, D., Kumar, S. and Prasad, R., 2004. Succession and incidence of major insect pests infesting different genotypes of linseed. *Ann. Pl. Protec. Sci.*, 12(2): 312–315.

Prasad, R., Kandulna, A. and Prasad, D., 2005. Efficacy and economics of insect pest management of linseed grown under farmers and improved agronomical practices. *J. Pl. Protec. Environ.*, 2(2): 49–54.

Chapter 34

Water Quality Monitoring in Respect to Physico-chemical Characteristics of a Tropical Lake of Udaipur City of Rajasthan

Rajesh Kumar[1] *and Kamal Kapoor*[2]*

[1]*Lecturer in Environmental Science, Seth G.B. Podar College, Nawalgarh – 333 042, Rajasthan*
[2]*Professor, Department of Botany, M.L.S. University, Udaipur – 313 002, Rajasthan*

ABSTRACT

The water of a tropical fresh water body *i.e.* Lake Bari were analyzed about fourteen physico-chemical characteristics of Udaipur city. The water samples were collected at monthly intervals for two consecutive years from November, 2000 to October, 2002. The water of this lake was found suitable for drinking as well as other purposes after comparison with standard limits. In thoese years, Udaipur zone or whole state of Rajasthan received lesser rainfall. The lake Bari received water from so many seasonal *nullah* and genesis of river Bedach. This lake also received run-off from agricultural lands.

Keywords: *BOD, COD, Lake city, Monitoring of water quality, TDS, Tropical Lake.*

* *Corresponding author.* E-mail: rajeshyadav_2006@rediffmail.com

Introduction

Water is an essential ingredient of life. It has been rightly describes as the "elixir of life". 98 per cent of the planet Earths water is in the Oceans, remaining 2 per cent is fresh water, but 98 per cent of which is stored in ice caps at the poles. In other words only 0.04 per cent of water is available for human being use. The civilized world depends largely upon surface water for its domestic, irrigational, industrial requirements and many other purposes besides potable water. Since the quality of water affects our lives in many ways, it must be of good quality along with its aesthetic value in the scenic environment, which must be appreciated.

The quality of water depends on its temperature, clarity and colour, which in any aquatic habitat at a given time is the sum total of the effect of various physical, chemical and biological factors. The undesirable impurities are termed as pollutants, which are expressed with the help of various physico-chemical characteristics such as BOD (Biochemical Oxygen Demand), COD (Chemical Oxygen Demand), TDS (Total Dissolved Solids), pH, turbidity, Chloride, Alkalinity, etc.

Materials and Methods

Udaipur city, situated in southern Rajasthan known as "city of lakes" on world tourism map. There are so many lakes in and around this urban area. In these, lake Bari was selected for present investigation. It was comparatively clean, pollution free and virtually has no pollutant input. It is located at 24°34′00″–24°35′13″ north latitude and 73°36′00″ to 73°36′16″ east longitude and situated 14 km north of the Udaipur city near village Bari. Sometimes, called as Jiyan Sagar after the queen Jiyan Devi. It was built in the year AD 1664. It's water is used primarily for irrigation and secondly for municipal supplies in Udaipur city. It recharges from the seasonal *nullah* Bedach (Mwarchiro, 1993). The detailed informations about morphometric features of lake Bari is shown in Table 34.1.

The climate of Udaipur city and nearby areas is more or less tropical with occasional high temperature and rainfall gradient. A glimpse of meteorological data reveal that there had been scanty rainfall in the past four-five consecutive years in Udaipur, and all over Rajasthan state. The average annual rainfall was 439.9 mm (January, 2000–December, 2000), 531.2 mm (January, 2001–December, 2001) and 365.3 mm (January, 2002–December, 2002). The variations in the average annual rainfall and the number of rainy days during study period are very significant as relative humidity was maximum in rainy season only (Table 34.2). The study area's and sampling site map is shown in Figure 34.1

Observations on various physico-chemical characteristics of study site *i.e.* lake Bari were carried out from November, 2000–October, 2002 *i.e.* for a two-year period. Initially two sampling sites of the lake were selected, but as there was no significant difference in parameters recorded, only one site was reserved for the observations. The water samples were collected during morning hours between 9:30 AM–10:00 AM every month at a depth of 30 cm with sufficient precautions, so as to prevent any vertical disturbances, the water samples were collected in acid washed five litre capacity plastic cans and physico-chemical characteristics were determined, soon after on the same day for the help of related literature (Trivedi and Goel, 1984; Manivaskam, 1996 and APHA, 1998).

Results and Discussion

The physico-chemical characteristics of the lake Bari are summarised in Tables 34.3 and 34.4. According to observations, during first year of study maximum temperature was recorded in June, 2001 (38.9°C) and minimum in January, 2001 (15.2°C) around lake Bari, while maximum water

Table 34.1: Meteorological Data of Udaipur City and its Environs (Year, 2000–2002)

	Jan.	*Feb.*	*March*	*April*	*May*	*June*	*July*	*Aug.*	*Sept.*	*Oct.*	*Nov.*	*Dec,*	*Mean (X)*
Year–2000													
Temperature (°C)													
Maximum	25.8	25.4	31.5	38.6	36.8	36.8	30.9	30.2	28.1	35.4	30.8	28.0	31.5
Minimum	4.6	7.1	12.4	20.4	25.6	26.1	23.6	23.4	20.8	14.4	10.7	5.9	16.2
Mean	15.2	16.2	21.9	29.5	31.2	31.4	27.2	26.8	24.4	24.9	20.7	16.9	23.8
Rainfall (mm)	0.0	0.0	0.0	0.8	52.4	66.4	173.4	93.6	48.3	5.0	0.0	0.0	36.6
Relative Humidity (per cent)													
Maximum	73.8	61.9	49.8	46.7	59.4	61.8	75.8	82.1	80.7	69.7	64.1	77.5	66.1
Minimum	32.4	29.7	17.6	20.3	34.7	40.5	52.6	59.4	56.4	31.9	29.8	29.9	36.2
Mean	53.1	45.8	33.7	33.5	47.0	51.1	64.2	70.7	68.5	50.8	46.9	53.7	51.5
Year–2001													
Temperature (°C)													
Maximum	24.7	28.5	31.7	36.0	36.8	33.6	28.2	29.7	33.1	34.0	30.1	27.4	31.1
Minimum	3.8	8.1	14.9	18.4	24.7	25.0	23.5	22.6	19.7	16.5	9.4	5.7	16.0
Mean	14.2	18.3	23.3	27.2	30.7	29.3	25.8	26.1	26.4	25.2	19.7	16.5	23.5
Rainfall (mm)	0.0	0.0	0.0	8.6	0.0	120.3	250.8	132.7	18.8	0.0	00	00	44.2
Relative Humidity (per cent)													
Maximum	72.2	52.4	43.5	43.0	52.8	75.0	88.8	88.6	76.2	79.0	71.0	75.9	68.2
Minimum	29.2	19.2	15.2	17.8	27.5	51.7	79.5	73.8	49.1	30.9	22.6	22.4	36.5
Mean	50.7	35.8	29.4	30.4	40.2	63.3	84.1	81.2	62.6	55.0	'+6.8	'+9.1	52.3
Year–2002													
Temperature (°C)													
Maximum	24.2	26.2	32.6	37.5	39.4	36.7	31.7	29.3	32.1	31.6	30.4	27.9	31.6
Minimum	4.4	5.9	14.0	19.6	26.2	26.3	25.0	23.2	20.8	15.0	10.5	7.4	16.5
Mean	14.2	16.0	23.3	28.5	32.8	31.5	28.3	26.2	26.4	23.4	20.4	17.6	24.0
Rainfall (mm)	0.0	0.0	0.0	2.0	0.0	112.7	2.4	192.6	53.6	0.0	0.0	2.0	30.4
Relative Humidity (per cent)													
Maximum	79.9	67.8	50.9	38.1	50.0	66.1	70.8	86.3	82.a	66.3	63.7	78.3	66.6
Minimum	31.8	24.2	19.8	13.8	22.8	40.8	50.8	70.4	49.7	19.3	23.3	36.6	33.6
Mean	55.8	46.0	35.3	25.9	36.4	53.4	60.8	78.4	65.8	42.8	43.5	57.5	50.1

Source: Department of Soil and Water Conservation, College of Technology and Agricultural Engineering, Maharana Pratap University of Agriculture and Technology (MPUAT), Udaipur, Rajasthan.

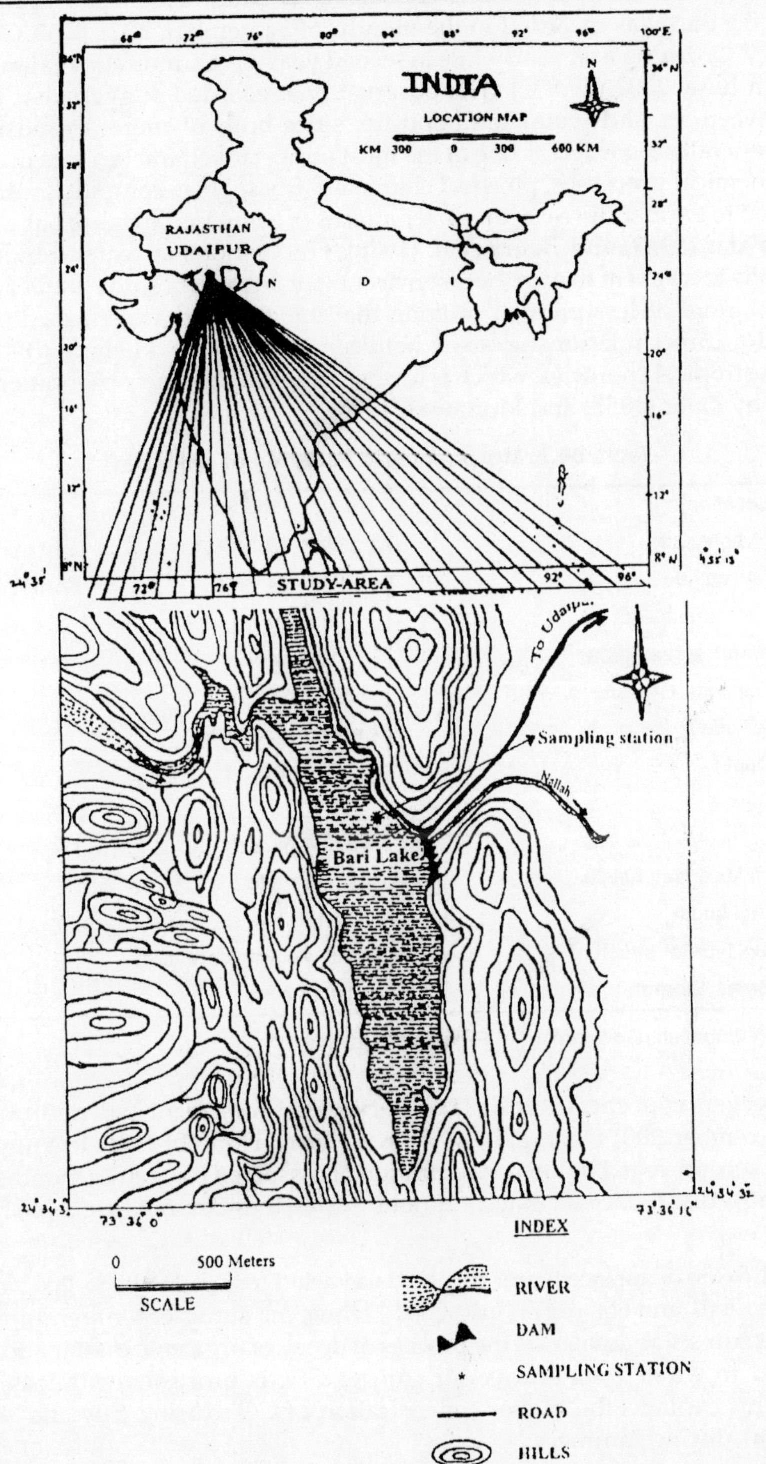

Figure 34.1: Location and Sampling Site of Study Area *i.e.* Lake Bari of Udaipur City

temperature of lake Bari was recorded in the months of September, 2001 (27.3°C) and minimum in January, 2001 (15.9°C) during first year, while in second year it was minimum in January, 2002 (18.2°C) and maximum in June, 2002 (29.8°C). The observations recorded suggest that there exits a close relationship between air and water temperature, since both of there are positively correlated. Temperature is generally regarded as one of the most important characteristics in aquatic ecology. In addition many chemical processes, physical characteristics such as solubility, visibility and density are directly related to temperature *e.g.* thermal patterns of a few Indian lakes have been described by Ganpati (1960), Zafar (1968) and Bohra *et al.* (1978). Fluctuations in water temperature of aquatic habitat are generally less violent than in the air environment. In present study water and air temperature were found to go more or less parallel proving that the atmospheric temperature governs water temperature and a direct relationship exist between the bright sun-shine. It's duration and air temperature in the tropical countries. which is further supported by our observations (Tables 34.3 and 34.4) as reported by Zafar (1955) and Munawar (1970).

Table 34.2: Morphometric Features of Lake Bari

1.	Location		
	(*i*) Longitude	73°36'00"–73°36'16"	East Longitude
	(*ii*) Latitude	24°34'00"–24°35'13"	North Latitude
2.	Catchment area	9.6 km²	
3.	Water spread area	1.46 km²	
4.	Capacity Gross/Live	112.82/72 mm³	
5.	Altitude	587 meter above MSL	
6.	Depth	23 meter	
7.	Dam		
	(*i*) Height of the dam	27.34 meter	
	(*ii*) Maximum height	29.38 meter	
	(*iii*) Length	0.128 km	
	(*iv*) Type of structure	Both sides face wall with earth filling	
8.	Gross Storage	31.15 mm³	

Source: Department of Irrigation, Government of Rajasthan, Udaipur.

Dissolved oxygen concentration in lake Bari water was recorded minimum in July, 2001 (6.8 mg/l) and November, 2001 (4.6 mg/l), while maximum in March, 2001 (10A mg/I) and February, 2002 (9.8 mg/l) in second year. Higher concentration of dissolved oxygen in months of February and March were probably due to the fact that conditions during these months are more favorable for high rate of photosynthesis.

Free carbon dioxide of surface water of lake Bari varied from nil in June, 2001 and October, 2002 to 124 mg/l in July, 2001 and 600 mg/l (July, 2002). However, summer temperature which is closely related with bright sun-shine accelerate the process of decay of orgasmic matter and large quantity of free CO_2 is added to water thus an overall gain of CO_2 is more than its loss of utilization in photosynthesis. This explains the higher concentration of CO_2 during July and August inspire of phytoplankton peak during summers.

Table 34.3: Statistical Data of Different Physico-chemical Characteristics of Lake Bari Water (November, 2000–October 2001) of Udaipur City

Sl.No.	Characteristics Studied	Minimmum Value	Maximum Value	Range	Mean±S.D.	C.V.	S.E.
1.	Temperature						
	Air	15.2	38.9	23.7	29.0±7.650	26.30	2.208
	Water	16.1	27.3	11.2	23.1±3.897	16.84	1.124
2.	Dissolved Oxygen	6.8	10.4	3.6	8.63±1.008	11.68	0.290
3.	Free CO_2	0.0	124.0	124.0	36.5±36.391	99.70	10.505
4.	pH	7.92	8.79	0.87	8.41±0.282	3.353	0.081
5.	Total dissolved Solids	309.0	521.0	212.0	398±56.151	14.11	16.209
6.	Conductivity	479.0	692.0	213.0	546.88±64.88	11.86	18.730
7.	Chloride	42.55	96.74	54.19	74.80±15.319	24.79	4.422
8.	Total hardness (as $CaCO_3$)	146.0	199.0	53.0	169.95±17.892	10.52	5.165
9.	Total alkalinity	96.0	192.0	96.0	147±29.753	20.24	8.589
	Phenolphthalein	0.0	9.0	9.0	2.08±3.776	181.53	1.909
	Methyl orange	88.0	192.0	104.0	144.9±30.897	21.32	8.919
10.	Phosphate	0.073	0.388	0.315	0.612±0.092	56.79	0.026
11.	Nitrate	6.527	11.772	5.245	9.98±1.350	13.52	0.389
12.	Sulphate	6.0	47.0	41.0	27.33±14.392	52.66	4.154
13.	Biochemical oxygen demand	1.6	5.6	4.0	3.07±1.207	39.31	0.348
14.	Chemical oxygen demand	14.20	35.06	20.86	20.9l±6.431	30.75	1.856

Units: mg/l except temperature (°C), conductivity (μmho/cm) and pH.

Abbreviations: S.D.: Standard Deviation; C.V.: Coefficient of Variance; S.E.: Standard Error.

The water of lake Bari was sufficiently alkaline during November, 2000–October, 2002. Values of pH were recorded minimum in November, 2000 and September, 2002 (7.92 and 7.73), while maximum in September, 2001 and February, 2002 (8.79 and 8.76).

Total dissolved solids in water of lake Bari were minimum in July, 2001 and September, 2002 (309 mg/l and 270 mg/l) and maximum in June, 2001 and May, 2002 (521 mg/l and 503 mg/l). According to Beeton (1965) noted that the concentration of TDS in oligotrophic lakes less than 100 mg/l and in eutrophic lakes the concentration of TDS more than 100 mg/l. Conductivity in lake Bari water minimum in months of July, 2001 and September, 2002 (479 μmho/cm and 427 μmho/cm), while maximum in months of March, 2001 and August, 2002 (677 μmho/cm and 792 μmho/cm). It has also seen that there was a positive correlation between concentration of total dissolved solids and conductivity.

Chloride concentration in water sample was minimum in August, 2001 and 2002 (42.55 mg/l and 46.79 mg/l), while maximum in December, 2000 and 2001 (96.74 mg/l and 107.8 mg/l). Total hardness as observed was minimum in August 2001 and April, 2002 (146.0 mg/l and 148.0 mg/l), while maximum in months of January, 2001 and August, 2002 (199.0 mg/l and 238.0 mg/l). Total alkalinity was mainly due to bicarbonates, because carbonate alkalinity was negligible.

Table 34.4:Statistical Data of Different Physico-chemical Characteristics of Lake Bari Water (November, 2001–October 2002) of Udaipur City

Sl.No.	*Characteristics Studied*	*Minimmum Value*	*Maximum Value*	*Range*	*Mean±S.D.*	*C.V.*	*S.E.*
1.	Temperature						
	Air	19.0	43.7	24.7	31.8±8.552	26.89	2.468
	Water	18.2	29.8	11.6	25.1±4.395	17.50	1.268
2.	Dissolved Oxygen	4.4	9.8	5.4	7.33±1.906	26.00	0.550
3.	Free CO_2	0.0	600.0	600.0	113.58±179.4	158.34	51.917
4.	pH	7.45	8.46	1.01	8.11±0.289	3.56	0.083
5.	Total dissolved solids	278.0	503.0	225.0	415.83±73.16	17.59	21.120
6.	Conductivity	427.0	792.0	365.0	623.41±115.7	18.57	33.423
7.	Chloride	36.86	107.8	70.94	n.72±22.955	31.13	6.626
8.	Total hardness (as $CaCO_3$)	148.0	238.0	90.0	184.3±24.733	13.41	7.140
9.	Total alkalinity	88.0	174.0	86.0	138.66±30.1 1	21.721	8.694
	Phenolphthalein	0.0	30.0	30.0	2.5±8.660	346.40	2.500
	Methyl orange	64.0	174.0	110.0	136.16±35.01	25.70	10.107
10.	Phosphate	0.0336	6.281	6.247	0.639±1.778	278.24	0.513
11.	Nitrate	1.019	17.971	16.95	10.222±6.644	64.99	1.923
12.	Sulphate	6.0	30.7	24.7	24.89±8.165	32.80	2.357
13.	Biochemical oxygen demand	1.27	6.0	4.73	2.86±1.751	61.22	0.505
14.	Chemical oxygen demand	5.76	49.16	43.40	20.80±14.61	70.24	4.217

Units: mg/l except temperature (°C), conductivity (μmho/cm) and pH.

Abbreviations: S.D.: Standard Deviation; C.V.: Coefficient of Variance; S.E.: Standard Error.

Phosphate-phosphorous was found to be minimum in February, 2001 and October, 2002 (0.099 mg/l and 0.033 mg/l), while maximum in September, 2001 and April, 2002 (0.388 mg/l and 6.201 mg/l). Utilization of these nutrients for the growth and reproduction of phytoplankton has been emphasized (Michael, 1969) and as primary limiting nutrient in lakes and ponds (Schindler, 1971).

Nitrate–nitrogen concentration in lake Bari was minimum in months of October, 2001 and July, 2002 (6.527 mg/l and 1.01 mg/l) and maximum in August, 2001 and October, 2002 (11.772 mg/l and 20.37 mg/l). Nitrates-nitrogen enters in an aquatic body from various sources like erosion of natural body or soil as well as artificially fertilized soil, through rainfall and sewage (Kapoor and Bamniya, 2001). Sulphate ions concentration was minimum in June, 2001 and October, 2002 (6.0 mg/l), while maximum in November, 2000 and 2001 (47.0 mg/l and 36.0 mg/l). Munawar (1970) observed that partially sulphates are reduced to sulphides and dissociate in water gradually.

Bio-chemical oxygen demand in water samples was observed minimum in September, 2001 and March, 2002 (1.6 mg/l and 1.2 mg/l), while maximum in January, 2001 and March, 2002 (4.7 mg/l and 5.6 mg/l). The concentration of chemical oxygen demand (COD) was observed minimum in April, 2001 and 2002 (14.2 mg/l and 5.76 mg/l), while maximum in February, 2001 and 2002 (35.06 and 39.04 mg/l).

The basic information and data on the fresh water aquatic habitats are required to be generated in order to evolve effective and appropriate strategies for management of our natural and man-made reservoirs.

Acknowledgements

The authors are thankful to the Head, Department of Botany, Mohanlal Sukhadia University, Uudaipur (Rajasthan) for providing necessary laboratory facilities for this work. Also thankful to Dr. B.R. Bamniya, Assistant Professor, Department of Environmental Sciences, Mohanlal Sukhadia University, Uudaipur, Rajasthan.

References

APHA, AWWA and WPCF, 1998. *Standard Methods for the Examination of Water and Wastewater*, 20th Edition. American Public Health Association, Washington, D.C., 20005–2605.

Beeton, A.M., 1965. The eutrophication problem. *Journal of Fisheries, Research Board of Canada*, 29: 673–682.

Bohra, O.P., Misra, S.D. and Bhargav, S.C., 1978. Occurrence of a thermocline in two lakes of the semi-arid region of Western Rajasthan, India. *Hydrobiologia*, 59(3): 203–211.

Ganpati, S.V., 1960. Ecology of tropical waters. In: *Proceedings of Symposium on Algology*, ICAR, New Delhi, p. 204–218.

Kapoor, K. and Bamniya, B.R., 2001. Water, eutrophication and algal blooms. In: *Ecology and Conservation of Lakes, Reservoirs and Rivers*, (Ed.) A. Kumar. ABD Publishers, Jaipur, p. 42–57.

Manivaskam, N., 1996. *Physico-chemical Examination of Water, Sewage and Industrial Effluents*. Pragati Prakashan, Meerut (India), pp. 227.

Michael, R.G., 1969. Limnological investigation on pond plankton, macrofauna and chemical constituents of water and bearing of fish production. *Ph.D. Thesis*, Calcutta University, Calcutta, India.

Munawar, M., 1970. Limnological studies on fresh water certain polluted and unpolluted environments. *Hydrobiologia*, 39: 105–138.

Mwarchiro, E.C., 1993. Limnological evaluation of the Lake Bari, Udaipur. *Ph.D.* Thesis, Rajasthan College of Agriculture, Udaipur, Rajasthan.

Schindler, D.W., 1971. Carbon, nitrogen, phosphorous and eutrophication of fresh water lakes. *Journal of Phyology*, 17: 321–329.

Trivedi, R.K. and Goel, P.K., 1984. *Chemical and Biological Methods for Water Pollution Studies*. Environmental Publishers, Karad, Maharashtra, India.

Zafar, A.R., 1955. Studies on the periodicity and distribution of algae in certain fish ponds in the vicinity of Hyderabad. *Ph.D. Thesis*, Osmania University, Hyderabad, India.

Zafar, A.R., 1968. Studies on the ecology of algae in certain fish ponds of Hyderabad, India–I. Physico-chemical complexes. *Hydrobiologia*, 23: 195–197.

Chapter 35

Variations in Different Biochemical Parameters of Commercial Pearl Millet [*Pennisetum glaucum* (L.) R.Br.] Hybrids Under Irrigated and Terminal Water Stress Conditions

Harshal E. Patil*

State Level Biotechnology Centre, Mahatma Phule Agricultural University, Rahuri – 413 722, Distt. Ahmednagar, Maharashtra

ABSTRACT

Seven pearl millet hybrids were grown under irrigated and terminal water stress conditions to study the variations in different biochemical parameters with response to drought stress and natural condition. The significant decrease in chlorophyll content and increase in carotenoid content in both the stages was observed under terminal water stress condition as compared to irrigated condition in all the hybrids tested. All the hybrids accumulated higher proline under stress condition as compared to irrigated condition. Free proline accumulation and higher grain yield under drought condition revealed that the hybrids accumulating higher proline under stress besides having high yield stability index. Hence, possibly proline accumulation under water stress helps the hybrids to resists drought.

* E-mail: harry_biotech@yahoo.com

The hybrid GHB 538 will prove to be much superior due to significantly higher accumulation of total chlorophyll, carotenoid and proline content with high yield stability under stress condition as compaired with the performance of the others hybrids *viz.*, GHB 577, GHB 558, GHB 526, MH 169, GHB 720 and GHB 664 under terminal drought condition.

***Keywords**: Carotenoid, Chlorophyll, Proline, Terminal water stress and Yield stability index.*

Introduction

Pearl millet [*Pennisetum glaucum* (L.) R.Br.] is an indispensable cereal crop of arid and semi-arid tropics. Due to erratic rainfall the crop is generally exposed to water stress at different stages, which decreases growth and yield of plants in many ways. It is an important food crop for less fortunate population inhabiting drier regions of the country where drought stress limits grain production. In rainfall crops, drought stress can occur at any time during crop growth and it is a well-known fact that hybrids perform better than the varieties but the physiology of hybrids is poorly understood (Goyal *et al.*, 2001). Vigour in productivity of a hybrid mainly depends upon photosynthetic efficiency of plant and hybrids are characterized to possess higher concentration of chlorophyll pigments (Fleming and Palmer, 1975). Hence, to examine the extent of genotypic differences, information is needed for various biochemical and yield parameters in drought stress conditions. The present investigations were carried out to ascertain effect of terminal water stress for different biochemical parameters as compaired with irrigated condition on released pearl millet hybrids.

Materials and Methods

The investigations were carried out on seven pearl millet hybrids (released and promising) *viz.*, GHB 577, GHB 558, GHB 526, MH 169, GHB 538, GHB 720 and GHB 664 for analysis of in response to terminal drought stress for different biochemical parameters as compared with irrigated condition at plant breeding research farm of Anand Agricultural University, Anand. The whole set will be individually subjected to two diverse environments *viz.*, irrigated and terminal water stress, with two replications each, so that each environment will be a separate experiment of its own. The experiment was conducted in randomized block design with control (irrigated) and terminal drought stress condition on promising hybrids as two treatments with four replications. Each sub-plot consisted of four rows of 4.5 m length with 50 cm apart. The crop was irrigated as necessary except for terminal drought treatment in which irrigation was withheld after boot leaf stage to physiological maturity (Bidinger *et al.*, 1987). Sampling was done at 60, 70, 80 days after sowing (*i.e.* 0, 10 and 20 days after withholding irrigation).

Photosynthetic pigments were estimated according to the method of Hiscox and Israelstam (1979) using DMSO. Calculations for chlorophyll a, chlorophyll b, and total chlorophyll in μ mol/g dry weight were made according to the equation developed by Anderson and Boardman (1964). Carotenoid was calculated in terms of μ mol/g dry weight by using equation of Goodwin (1955). The estimation of proline was done by Bates *et al.* (1973). Samples were analysed in triplicate and data were analysed using simple CRD (Panse and Suchatma, 1967).

Results and Discussion

Table 35.1 represents the data for total chlorophyll content in μ mol/g DW. The results show that up to ten days of water stress there was increase in total chlorophyll, but after that it decreased under control (irrigation). Under stress conditions, hybrid GHB 538 maintained higher amount of chlorophyll as compared to other hybrids. Significant decrease in chlorophyll content was observed under stress. However, the per cent reduction was less in case of hybrid GHB 538, and this was to the tune of 14 per sent. A similar trend was observed for chlorophyll a and b in hybrid GHB 538.

Table 35.1: Effect of Irrigation and Terminal Water Stress on Total Chlorophyll (μ mol/g dry wt.) and Carotenoid (μ mol/g dry wt.) Contents of Various Hybrids of Pearl Millet

Sl.No.	Hybrids	0 (Day After Drought)		10 (Days After Drought)		20 (Days After Drought)		Mean
		Irrigated	Drought	Irrigated	Drought	Irrigated	Drought	
				Total Chlorophyll (μ mol/g dry wt.)				
1.	GHB 577	3.67	3.67	4.48	3.72	4.06	2.98	4.07
2.	GHB 558	2.94	2.94	3.39	3.25	3.30	2.54	3.21
3.	GHB 526	2.84	2.84	3.85	3.21	3.45	2.63	3.38
4.	MH 169	3.25	3.25	4.40	3.73	3.31	2.40	3.65
5.	GHB 538	4.52	4.52	4.87	4.13	4.21	3.33	4.53
6.	GHB 720	3.87	3.87	4.42	3.78	3.12	2.34	3.80
7.	GHB 664	4.12	4.12	4.28	3.59	3.62	2.58	4.00
	Mean	3.60	3.60	4.24	3.63	3.58	2.68	3.80
	C.D.± (5%)	V x T x S = 0.069		V x S = 0.057		V x T = 0.032		
				Carotenoid (μ mol/g dry wt.)				
1.	GHB 577	0.90	0.90	1.23	1.45	1.38	1.63	1.17
2.	GHB 558	1.13	1.13	1.22	1.46	1.48	1.59	1.27
3.	GHB 526	1.26	1.26	1.53	1.63	1.63	1.71	1.47
4.	MH 169	1.21	1.21	1.39	1.56	1.62	1.82	1.40
5.	GHB 538	1.33	1.33	1.49	1.78	1.61	1.93	1.47
6.	GHB 720	1.16	1.16	1.49	1.62	1.52	1.72	1.39
7.	GHB 664	1.32	1.32	1.55	1.65	1.63	1.56	1.40
	Mean	1.18	1.18	1.41	1.59	1.55	1.70	1.38
	C.D.± (5%)	V x T x S = 0.073		V x S = 0.06 1		V x T = 0.033		

The 20 days after stress (Table 35.2), rapid decrease in concentration of chlorophyll a was observed, whereas chlorophyll b showed less decrease. However, in both the cases the fall was significant. Results reveal that interaction between genotypes × treatment was significant.

Chlorophyll content in all the genotypes decreases significantly under stress in plants grown in field conditions. However, the reverse trend was observed for carotenoid content. The decreased chlorophyll content under water stress may be due to decreased synthesis and increased degradation of chlorophyll content in leaves. These findings are in accordance with the earlier reports of Hasio (1976). Further, leaves of hybrids contained greater chlorophyll content than its parental lines and male parent was found to be better than female parent. Similar results were also reported in pearl millet by Joshi *et al.* (1986), Goyal *et al.* (2001) and Chanda *et al.* (1988).

Table 35.2: Effect of Irrigation and Terminal Water Stress Condition on Chlorophyll-'a' (μmol/g dry wt.) and Chlorophyll-'b' (μmol/g dry wt.) Contents of Various Hybrids of Pearl Millet

Sl.No.	*Hybrids*	*0 (Day After Drought)*		*10 (Days After Drought)*		*20 (Days After Drought)*		*Mean*
		Irrigated	*Drought*	*Irrigated*	*Drought*	*Irrigated*	*Drought*	
				Chlorophyll-'a' (μmol/g dry wt.)				
1.	GHB 577	2.72	2.72	3.62	3.18	3.10	2.12	3.14
2.	GHB 558	2.23	2.23	2.62	2.94	2.67	1.92	2.50
3.	GHB 526	2.38	2.38	2.96	2.37	2.67	1.97	2.67
4.	MH 169	2.12	2.12	3.02	2.56	2.53	1.82	2.55
5.	GHB 538	3.52	3.52	3.98	3.45	3.19	2.34	3.56
6.	GHB 720	2.76	2.76	3.19	2.68	2.47	2.06	2.80
7.	GHB 664	3.02	3.02	3.44	2.91	2.98	2.17	3.14
	Mean	2.67	2.67	3.26	2.87	2.80	2.05	2.90
	C.D.± (5%)	V × T × S = 0.078		V × S = 0.067		V × T = 0.060		
				Chlorophyll-'b' (μmol/g dry wt.)				
1.	GHB 577	0.75	0.95	0.86	0.54	0.96	0.86	0.92
2.	GHB 558	0.71	0.71	0.77	0.31	0.63	0.62	0.70
3.	GHB 526	0.46	0.46	0.89	0.84	0.78	0.66	0.71
4.	MH 169	1.13	1.13	1.38	1.17	0.78	0.58	1.09
5.	GHB 538	1.00	1.00	0.89	0.68	1.03	0.99	0.97
6.	GHB 720	1.11	1.11	1.23	1.10	0.65	0.28	0.99
7.	GHB 664	1.10	1.10	0.84	0.68	0.64	0.41	0.86
	Mean	0.92	0.92	0.98	0.76	0.78	0.62	0.89
	C.D.± (5%)	V × T × S = 0.058		V × S = 0.045		V × T = 0.031		

There was a significant rise in carotenoid content under drought stress conditions. Elevation rates were higher in hybrid GHB 577, GHB 538 and MH 169 as compared to other hybrids. At both, 10 and 20 days after stress all the hybrids had optimum carotenoid content. The increase in carotenoid content is in accordance with Joshi *et al.* (1986) and Goyal *et al.* (2001). The increase in carotenoid content was more in GHB 538, thus enabling to escape drought as the increased carotenoid content hindered the photo-inhibition and hence prevented the oxidation of chlorophyll.

Proline content in all the hybrids increased significantly under terminal drought conditions as compared to irrigated condition (Table 35.3). The proline content was found to be increased at both stages 10 and 20 days of water stress in all the hybrids. In 10 and 20 days after water stress increased in proline content from 1.5 to 2 times as compaired with the irrigated conditions. Maximum accumulation observed in hybrid GHB 538 and GHB 664 that was 1.89 and 1.36 fold more in 20 days after water stress as compared to initial stage of water stress. However, hybrid GHB 538 showed highest increase in proline (426 and 484 μg/g fresh wt.) at both the stages than all other hybrids. The increases in proline level suggest greater osmoprotection under drought stress. It was probably owing to proline oxidation in irrigated condition as the free proline is converted to glutamic acid and in other compounds under non-stress condition. Proline works as a source of energy, carbon and nitrogen and also protects several enzymes, against the inactivating effects of heat during water stress. Hence possibly proline accumulation under water stress helps the hybrids to resists drought. Similar results were also reported in pearl millet by Goyal *et al.* (2001).

Table 35.3: Effect of Irrigation and Terminal Water Stress Condition on Proline (µg/g fresh wt.) Contents and Grain Yield per Plant (kg/plant) of Various Hybrids of Pearl Millet

Hybrids	0 (Day After Drought) Proline (µg/g fresh wt.)		10 (Days After Drought) Proline (µg/g fresh wt.)		20 (Days After Drought) Proline (µg/g fresh wt.)		Mean Proline (µg/g fresh wt.)		Grain Yield/Plant (kg/ Plant)	
	Irrigated	Drought	Irrigated	Drought	Irrigated	Drought	Irrigated	Drought	Irrigated	Drought
GHB 577	234	234	269	344	316	422	273	333	132	82
GHB 558	320	320	342	412	346	438	336	390	128	78
GHB 526	242	242	302	386	328	462	290	363	122	86
MH 169	210	210	248	356	312	424	256	330	162	98
GHB 538	256	256	312	426	328	484	298	338	148	80
GHB 720	272	272	292	372	307	440	290	361	141	74
GHB 664	221	221	262	380	310	472	264	357	170	94
Mean	251	251	290	382	321	448	286	360	143	84
C.D. ± (5%)	V × T × S = 0.089		V × S = 0.067		V × T = 0.052					

Free proline accumulation with higher grain yieid stability index under terminal water stress condition for the hybrid GHB 538, indicated better adaptability under drought conditions. The resistant hybrids accumulated higher levels of free proline than susceptible ones under stress conditions. Similarly, Narayan and Mishra (1989) found out proline accumulation in bread wheat and durum wheat under stress condition with high yield stability index.

Conclusion

The study indicated that the hybrid GHB 538 will prove to be much superior due to significantly higher accumulation of total chlorophyll, carotenoid and proline content with high yield stability index under terminal drought condition as com paired with the performance of the others hybrids *viz.*, GHB 577, GHB 558, GHB 526, MH 169, GHB 720 and GHB 664.

Acknowledgments

The authors are thankful to the Millet scientist, Millet Research Station, Jamnagar for supplying the seed material and authorities of Anand Agricultural University, Anand (Gujarat) for providing facilities for conducting biochemical research on terminal water stress condition.

References

Anderson, J.M. and Boardman, N.K., 1964. Studies on the greening of dark grown bean plants. II. Development of photochemical activity. *Aust. J. Biol. Sci.*, 17: 93–101.

Bates, L.S., Waldren, R.P. and Teare, I.O., 1973. Rapid determination of free proline for water stress studies. *Plant and Soil*, 39: 205–207.

Bidinger, F.R., Mahalakshmi, V. and Rao, G.D.P., 1987. Assessment of drought resistance in pearl millet–I. Factors affecting yield in stress. *Aust. J. Agric. Res.*, 38: 37–48.

Chanda, S.V., Joshi, A.K., Vaishnav, P.P. and Singh, Y.D., 1988. Comparative analysis of photosynthetic parameters and its evaluation in pearl millet hybrids and its parents. *Agron. Crop Sci.*, 160: 217–233.

Fleming, A.A. and Palmer, J.H., 1975. Variation in chlorophyll content of maize lines and hybrids. *Crop Sci.*, 15: 617–620.

Goodwin, J.W., 1955. Carotenoids. In: *Modern Methods of Plant Analysis*, (Eds.) Peach, K. and M.V. Tracey. Springer Verlag, 3: 272–311.

Goyal, V., Bansal Parveen, Goel Anuradha and Bishnoi, N.R., 2001. Effect of terminal drought stress on different pigments of various genotypes of pearl millet. *Crop Res.*, 21(3): 271–273.

Hasio, T.L., 1976. Stomatal ion transport. In: *Encyclopedia of Plant Physiology*, New Series Vol. 2, (Eds.) Luttigeand, U. and M.G. Pitman. Springer Verlag, Berlin, pp. 194–221.

Hiscox, J.D. and Israelstam, G.F., 1979. A method for the extraction of chlorophyll from leaf tissue without maceration. *Can. J. Bot.*, 57: 1332–1334.

Joshi, A.K., Chanda, S.V., Krishnan, P.N., Vaishnav, P.P. and Singh, Y.D., 1986. Chlorophyll content, hill reaction, PEP case activity in relation to hybrid vigour in pearl millet during early seedling growth. *Crop Sci.* 157: 1–12.

Narayan, D. and Mishra, R.D., 1989. Free proline accumulation and water stress resistance in bread wheat (*Triticum aestivum*) and drum wheat (*T. durum*). *Indian J. Agric. Sci.*, 59: 176–178.

Panse, V.G. and Sukhatme, P.V., 1967. *Statistical Methods for Agricultural Workers*, 2nd edition. ICAR, New Delhi.

Chapter 36

Life-Form Analysis and Biological Spectrum of the Macrophytes of the Laisoipat Lake, Manipur

Laishram Kamala Devi and B. Manihar Sharma

Ecology Research Laboratory, Department of Life Sciences, Manipur University, Canchipur – 795003, Imphal, Manipur

ABSTRACT

The present study focussing on the life-form and biological spectrum of the macrophyte community in the Laisoipat lake, Manipur have been carried out with details of the floristic study. Laisoipat lake is a freshwater eutrophic lake situated at Keinou village in Bishnupur district. The lake covers an area of about 0.3640 sq.km and lies at an intersection of 93°48′0.54″ to 93°48′17.78″E longitude and 24°38′27.24″ to 24°38′54.24″N latitude and at an altitude of 770 m above the mean sea level. The study has been carried out for a period of two years at monthly intervals and includes details of floristic study *i.e.* habitats, heights and natures of perennating buds together with seasonal changes in the diversity of the plant species. Altogether 43 macrophytic plant species were recorded showing six (6) different life forms. Out of the 43 species, 15 species were Therophyte (34.88 per cent), 11 species were EVH (25.58 per cent), 9 species were Geophytes (20.93 per cent) 6 species were Hemicryptophytes (13.95 per cent), 9 species each were chamaephytes (2.33 per cent) and Lianas (2.33 per cent). The biological spectrum of the Laisoipat lake revealed the occurrence of Thero-Errant Vascular Hydrophytic phytoclimate.

***Keywords**: Biological spectrum, Laisoipat lake, Life forms, Macrophytes, Phytoclimate.*

Introduction

The life-form reveals the characteristics of the adaptation of the plant to the climate. The life form of the vegetation reflects the major feature of climate providing a sound basis for a natural ecological classification (Clements and Shelford, 1939). Conversely the major vegetation formation is also delimited by climatic data (Holdridge, 1947). Macrophytic vegetation not only represents the sum total of plants but also reflects other characteristics such as life forms, density, coverage etc. (Weaver and Clements, 1938).

Macrophytic communities play a significant role in regulating the structure of a lake ecosystem. The structure and dynamics of the vegetation provide an indication of the nature of the whole environment, biological and physical. The structural attributes through which a plant community is studied are conveniently viewed as analytic and synthetic characters. The analytic characters may be qualitative and quantitative. The qualitative analysis comprises the floristic composition, life-form analysis and growth form classification. The plant species can be grouped into life-forms and growth form classes on the basis of structure and function.

The percentage distribution of species among the various life-forms of a flora is called the biological spectrum of that place. Thus, through calculation of percentage value of each life-form, biological spectrum of the area is obtained. The dominant life forms on the basis of percentage contribution of the species is termed as phytoclimate.

The bio-resources of north-eastern region in the Eastern Himalayas of India including Manipur, being one of the 25 "Biodiversity hotspots" of the world (Myers *et al.*, 2000) are much degraded which in turn affect the phytoclimate of an area.

Materials and Methods

The study site is located in the Keinou village, Bishnupur District, Manipur. It extends between 93°48′0.54″ to 93°48′17.78″E longitude and 24°38′27.24″ to 24°38′54.24″ N latitude and at an altitude of 770 m above Mean Sea Level (MSL). The area covered by lake is 0.3640 sq.km.

Vegetational analysis were carried out by regular periodical survey and sampling of vegetation at monthly interval during January 2004 to December 2005 in 4 different sites by using quadrat method described by Curtis (1959) and Misra (1968).

Life forms of the macrophytes were determined after detailed floristic studies. The habit, form, height and nature of perennating buds of each species were studied in the field. The life form classification of macrophytes on physiognomic basis has been done as per Raunkiaer's system modified by Ellenberg and Mueller-Dombois (1967) and Mueller-Dombois and Ellenberg (1974). Raunkiaer's classification (1934) of five life-forms have been expanded into 23 major life-forms. The biological spectrum for the study area has been compared with the Raunkiaer's normal spectrum as well as with the spectra available for other localities.

Results and Discussion

Out of the 43 macrophytic plant species recorded from the freshwater lake of Laisoipat lake, the dominant macrophytic species like *Alternanthera philoxeroides, Atylosia scarabaeoides, Azolla pinnata, Brachiaria mutica, Chara zeylanica, Eichhornia crassipes, Kyllinga triceps, Mikania micrcintha* etc. are found to be present in all the study sites.

The macrophytic plant species were found belonging to 6 different life forms *viz.*, Therophytes, Errant Vascular Hydrophytes, Geophytes, Hemicryptophytes, Chamaephytes and Lianas. The

Therophytes included 15 species constituting the maximum species percentage of 34.88 per cent, Errant Vascular hydrophytes had 11 species (25.58 per cent) which was successively followed by Geophytes with 9 species (20.93 per cent), Hemicryptophytes with 6 species (13.95 per cent), Chamaephytes and Lianas with I species (2.33 per cent) each. The distribution of different life forms in the different sites of the study area has been shown in Table 36.1 and the life-form classification of the Laisoipat lake has been shown in Table 36.2.

Table 36.1: Distribution of Different Life Forms in the Study Area (+) and (–) signs Indicates the Presence and Absence of a Particular Species in the Study Sites

Sl.No.	Name of Species	Life-form	Site I	Site II	Site III	Site IV
1.	*Ageratum conyzoides* Linn.	Th	–	–	+	–
2.	*Alternanthera philoxeroides* (Mart.) Griseb.	Th	+	+	+	+
3.	*Atylosia scarabaeoides* (Linn.) Benth.	Th	+	+	+	+
4.	*Azolla pinnata* R.Br.	EVH	+	+	+	+
5.	*Brachiaria mutica* (Forsk.) Stapf	Th	+	+	+	+
6.	*Carex cruciata* Wahlenb	Th	–	+	+	+
7.	*Ceratophyllum demersum* Linn.	EVH	+	+	+	+
8.	*Charazeylanica* Willd.	EVH	+	+	+	+
9.	*Colocasia esculenta* (Linn.) G.Don	G	–	–	+	+
10.	*Commelina bengalensis* Linn.	H	+	+	+	+
11.	*Crassocephallum crepidioides* (Benth.) Moore	Th	+	+	+	+
12.	*Diplazium porrectum* Wall.	Ch	–	+	–	–
13.	*Echinochloa stagnina* (Retz) P. Beauv.	Th	+	+	+	+
14.	*Eichhornia crassipes* (Mart) Solms	EVH	+	+	+	+
15.	*Enhydrafluctuans* Lour.	H	+	+	+	+
16.	*Euryale ferox* Salisb.	G	+	+	+	–
17.	*Fagopyrum dibotrys* (D. Don) Hara	Th	+	+	+	+
18.	*Hydrilla verticillata* (Linn. f.) Royle	H	+	+	+	+
19.	*Hygroryza aristata* (Retz.) Nees ex. Wight and Am	H	+	+	+	+
20.	*Ipomoea aquatica* Forsk.	H	+	+	+	+
21.	*Kyllinga triceps* Rottb.	Th	+	+	+	+
22.	*Ludwigia adscendens* (Linn.) Hara	Th	+	+	+	+
23.	*Mikania micrantha* Kunth	L	+	+	+	+
24.	*Nymphaea pubescence* Wilid.	G	–	–	+	–
25.	*Nymphoides cristatum* (Roxb.) O. Kuntze	G	+	–	+	+
26.	*Oenanthe javanica* (Bl.) DC.	Th	+	+	+	+
27.	*Pennisetum glaucum* (Linn.) R:Br.	Th	+	+	+	+
28.	*Pistia stratiotes* Linn.	EVH	–	–	–	+
29.	*Polygonum orientale* Linn.	Th	+	+	+	+
30	*Potamogeton crispus* Linn.	G	+	+	+	+

Contd...

Table 36.1–Contd...

Sl.No.	Name of Species	Life-form	Site I	Site II	Site III	Site IV
31.	*Potamogeton natans* Linn.	G	+	+	+	+
32.	*Pseudoraphis spinescens* (R.Br.) Vickery	H	+	+	+	+
33.	*Riccia natans* Gorda	EVH	–	+	–	+
34.	*Saccharum procerum* Roxb.	G	+	+	+	+
35	*Salvinia cucullata* Roxb.	EVH	+	+	+	+
36.	*Salvinia natans* Hoffm	EVH	+	+	+	+
37.	*Schoenoplectus supinus* Linn.	G	–	+	+	+
38.	*Senecio saxatilis* Wall.	Th	+	+	+	+
39.	*Trapa natans* Linn.	Th	+	+	+	+
40.	*Utricularia aurea* Lour.	EVH	+	+	+	+
41.	*Utricularia exoleta* R.Br.	EVH	+	+	+	+
42.	*Vallisneria spiralis* Linn.	EVH	+	+	–	+
43.	*Zizania latifolia* (Griseb.) Stapf.	G	+	+	+	+
			35	38	39	39

Table 36.2: Life Form Classification of the Macrophytes in Laisoipat Lake (After Ellenberg and Mueller-Dombois, 1967 and Mueller-Dombois and Ellenberg, 1974)

Life-form	Name of Species	No. of Species	Percentage Composition
Therophytes (Th)	1. *Ageratum conyzoides* Linn.	15	34.88 per cent
	2. *Alternanthera philoxeroides* (Mart.) Griseb.		
	3. *Atylosia scarabaeoides* (Linn.) Benth.		
	4. *Brachiaria mutica* (Forsk.) Stapf		
	5. *Carex cruciata* Wahlenb		
	6. *Crassocephallum crepidioides* (Benth.) Moore		
	7. *Echinochloa stagnina* (Retz) P. Beauv.		
	8. *Fagopyrlum dibotrys* (D.Don) Hara		
	9. *Kyllinga triceps* Rottb.		
	10. *Ludwigia adscendens* (Linn.) Hara		
	11. *Oenanthe javanica* (Bl.) DC.		
	12. *Pennisetum glaucum* (Linn.) R.Br.		
	13. *Polygonum orientale* Linn.		
	14. *Senecio saxatilis* Wall.		
	15. *Trapa natans* Linn.		

Contd...

Table 36.2–Contd...

Life-form	Name of Species	No. of Species	Percentage Composition
Geophytes (G)	1. *Colocasia esculenta* (Linn.) G.Don	9	20.93 per cent
	2. *Euryale ferox* Salisb.		
	3. *Nymphaea pubescence* Willd.		
	4. *Nymphoides cristatum* (Roxb.) O. Kuntze		
	5. *Potamogeton crispIls* Linn.		
	6. *Potamogeton natans* Linn.		
	7. *Saccharum procerum* Roxb.		
	8. *Schoenoplectus supinus* Linn.		
	9. *Zizania latifolia* (Griseb.) Stapf.		
Hemicryptophytes (H)	1. *Commelina bengalensis* Linn.	6	13.95 per cent
	2. *Enhydra fluctuans* Lour.		
	3. *Hydrilla verticillata* (Linn. f.) Royle		
	4. *Hygroryza aristata* (Retz.) Nees ex. Wight and Am		
	5. *Ipomoea aquatica* Forsk.		
	6. *Pseudoraphis spinescens* (R.Br.) Vickery		
Chamaephytes (Ch)	1. *Diplazium porrectum* Wall.	1	2.33 per cent
Errant Vascular Hydrophytes (EVH)	1. *Azolla pinnata* R.Br.	11	25.58 per cent
	2. *Ceratophyllum demersum* Linn.		
	3. *Chara zeylanica* Willd.		
	4. *Eichhornia crassipes* (Mart.) Solms		
	5. *Pistia stratites* Linn.		
	6. *Riccia natans* Gorda		
	7. *Salvinia cucullala* Roxb.		
	8. *Salvinia natans* Hoffm		
	9. *Utricularia aurea* Lour.		
	10. *Utricularia exoleta* R.Br.		
	11. *Vallisneria spiralis* Linn.		
Lianas (L)	1. *Mikania micrantha* Kunth	1	2.33 per cent

The present findings are in conformity with the findings of Devi and Sharma, (1998) in Utrapat lake who reported 30.80 per cent species in Therophytes, 23.00 per cent species in Geophytes, 15.40 per cent in hemicryptophytes and 30.80 per cent in Errant Vascular Hydrophytes; Bebika and Sharma (2002) from Sanapat lake who reported 29.41 per cent of Therophytes, 23.53 per cent of Geophytes, 17.65 per cent of Hemicryptophytes and 29.41 per cent of Errant Vascular Hydrophytes and Devi, Kh. U. 2004 from Poiroupat who reported 36.7 per cent of Therophytes, 20.00 per cent of Geophytes, 13.3 per cent of Hemicryptophytes and 30 per cent of Errant Vascular Hydrophytes.

A comparative account of percentage composition of study area and Normal Biological Spectrum has been shown in Table 36.3. When the life-form spectra of a particular plant community is compared to the normal life-form spectrum (Raunkiaer, 1934), the adaptational capabilities of the plants to the adverse climate conditions are revealed.

Table 36.3: Biological Spectrum of the Flora of Laisoipat Lake

Parameters	*Life-form*							
	Th	*G*	*H*	*Ch*	*Ph*	*EYH*	*L*	*Total Species*
Total number of species	15	9	6	1	–	11	1	43
Life-form percentage	34.88	20.93	13.95	2.33	–	25.58	2.33	–
Raunkiaer's normal spectrum and composition (per cent)	13.00	6.00	26.00	9.00	46.00	–	–	–

Abbreviation: Th: Therophytes; G: Geophytes; H: Hemicryptophytes; Ch: Chamaephytes; Ph: Phanerophytes; EYH: Errant Vascular hydrophytes; L: Lianas.

The highest percentage to the biological spectrum of the study area was contributed by Therophytes (34.88 per cent) which was almost two or three times the percentage value of Therophytes (13 per cent) in the 'Normal Spectrum' of Raunkiaer. It was followed by Errant Vascular Hydrophytes (25.58 per cent).

The predominance of Therophytes in the present study is indicative of warm climate in conformity with the Bioclimate diagram of Dansereau. (1957). This is due to the fact that the Therophytic climate is favourable for the growth of Therophytes which are best adapted to tide over the unfavourable period in the form of seeds.

The co-predominance of Errant Vascular Hydrophytes is the indication of setting of eutrophication in the present study area since the Errant Vascular plants are mostly found in eutrophic water bodies.

The phytoclimate of a region reflects the dominant life-form which comprises the largest number of plant species occurring in the community. Therefore, based on the various life-form classes of the Laisoipat lake, this region may be assigned to the "Thero-Errant Vascular Hydrophytic" type of phytoclimate as the Therophytes and Errant Vasuclar Hydrophytes represent the dominant life forms comprising the higher number of plant species.

Similar types of phytoclimate have also been reported earlier by Rao (1968) from Karamnasa watershed, Devi and Sharma (1998) from Utrapat lake, Manipur Bebika and Sharma (2002) from Sanapat lake, Manipur and Devi, Kh. U. 2002 from Poiroupat lake, Manipur.

A comparison of life-form spectra and phytoclimate of some wetlands of Manipur are incorporated in Table 36.4.

In the Waithou lake, Manipur, Okram *et al.* (1996) observed Geo-Therophytic phytoclimate while Devi, Ch.U. (2000) reported Thero-Geophytic type of phytoclimate in the freshwater bodies (4 ponds and a stream) in Canchipur, Manipur. Errant Vascular Hydrophytic type of phytoclimate was reported by Devi and Sharma (2002) from non-Phumdi region of Loktak lake.

The rich growth of macrophytic species belonging to the Therophytic and Errant Vascular Hydrophytic plant species indicate the eutrophic nature of the lake. The eutrophic nature of the present study area is due to human activities and agricultural works in and around the study area. It

Table 36.4: Comparison of Life-form Spectra and Phytoclimate of Some Wetlands of Manipur

Wetlands		*Life-form (%)*							*Phyto-Climate*	*References*
		Th	*G*	*H*	*Ch*	*Ph*	*EVH*	*L*		
Loktak	Non-phumdi	15.38	20.07	07.69	–	–	53.86	–	Errant Vascular Hydrophytes	Devi and Sharma 2002
	Phumdi	46.57	10.96	20.55	09.59	1.37	–	10.96	Thero-Hemicryptophytes	Devi and Sharma 2002
Utrapat		30.80	23.00	15.40	–	–	30.80	–	Thero-Errant Vascular Hydrophytes	Devi and Sharma 1998
Waithoupat		28.60	57.10	–	–	–	14.30	–	Geo- Therophytes	Okram *et al.*, 1996
Wetlands of Canchipur		38.00	22.00	18.00	2.00	–	20.00	–	Thero-Geophytes	Devi, Ch.U. 2000
Sanapat		29.41	23.53	17.65	–	–	29.41	–	Thero- Errant Vascular Hydrophytes	Bebika and Sharma 2002
Ikop		24.10	31.00	13.80	3.5	–	27.6	–	Geo-Errant Vascular Hydrophytes	Devi, Ch. N. and Sharma 2003
Poiroupat		36.7	20.00	13.3	–	–	30.00	–	Thero- Errant Vascular Hydrophytes	Devi, Kh. U. 2002

is necessary to take up remedial measures in time to check further eutrophication in the present study area.

References

Bebika, Ch. and Sharma, B.M., 2002. Floristic distribution and Life-form analysis of Sanapat lake, Manipur. *Indian J. Environ. and Ecoplan.*, 6: 155–162.

Clements, Frederic E. and Shelford, V.E., 1939. *Bioecology*. John Wiley and Sons, Inc., New York.

Curtis, J.T., 1959. *The Vegetation of Wisconsin*. Univ. Wisconsin Press, Madison.

Dansereau, P., 1957. *Biogeography: An Ecological Perspective*. Ronald Press, New York, USA.

Devi, Ch. N. and Sharma, B.M., 2003. Ecological study of the macrophytes of Ikop lake, Manipur: Morphometry and qualitative analysis. *Indian J. Environ. and Ecoplan.*, 7(2): 243–250.

Devi, Ch. Umabati, 2000. Phytosociology and primary production of the macrophytes in the freshwater ecosystems of Canchipur, Manipur. *Ph.D. Thesis*, Manipur University, Manipur.

Devi, K.I., 1998. Ecological studies of freshwater macrophytes in Utrapat lake, Manipur. *Ph.D. Thesis*, Manipur University, Manipur, India.

Devi, K.I. and Sharma, B.M., 1998. Vegetation ecology of macrophytes of Utrapat lake, Manipur, India. *J. Freshwater Biol.*, 10: 1–9.

Devi, Kh. U., 2002. Macrophyte ecology of Poiroupat lake, Manipur. *Ph.D. Thesis*, Manipur University, India.

Devi, N.B. and Sharma, B.M., 2002. Life-form analysis of the macrophytes of the Loktak lake, Manipur, India. *Indian J. Environ. and Ecoplan.*, 6(3): 451–458.

Ellenberg, H. and Mueller-Dombois, P., 1967. A key to Raunkiaer's plant life form with revised subdivisions. *Ber. Geobot. Inst, ETH Stiftg. Rubel. Zurich*, 37: 56–73.

Holdridge, L.R., 1947. Determination of world plant formations from simple climatic data. *Science*, 105: 267–368.

Misra, R., 1968. *Ecology Work Book*. Oxford and IBH Publ. Co., New Delhi.

Mueller-Dombois, D. and Ellenberg, H., 1974. *Aims and Methods of Vegetation Ecology*. John Wiley and Sons, New York.

Myers, Norman, Russell, A.M., Cristina, G.M., Gustavo, A.B. da fonseca and Jenifer, K., 2000. Biodiversity hotspots for conservation priorities. *Nature*, 3: 853–858.

Okram, I.D., Sharma, B.M. and Singh, E.J., 1996. Ecological study of Waithou lake, Manipur: Morphometry and qualitative analysis of macrophytic vegetation. *J. Freshwater Biol.*, 8: 177–189.

Rao, C.C., 1968. Biological spectrum of Karamanasa watershed flora (Varanasi, India). *Proc. Symp. Recent. Adv. Trop. Ecol.*, 11: 458–465.

Raunkiaer, C., 1934. *The Life-forms of Plants and Statistical Plant Geography*. Clarendon Press, Oxford.

Sharma, B.M. and Devi, N.S., 2002. Ecological assessment of macrophytes in Loktak lake, Manipur: Phytosociology: In: *Ecology of Polluted Waters*, (Ed.) Arvind Kumar. A.P.H. Publishing Corporation. New Delhi, pp. 869–876.

Weaver, J.E. and Clements, E.E., 1938. *Plant Ecology*. McGraw Hill Inc., pp. 286.

Chapter 37

Integrated Nutrient Management System for Productivity Potential for Sorghum-Wheat Sequence Under Irrigated Conditions

H.M. Patil[1] *and A.G. Wani*[2]

[1]*Junior Research Assistant*, [2]*Professor of Agronomy*

Department of Agronomy, Mahatma Phule Krishi Vidyapeeth, Rahuri – 413 722, Distt. Ahmednagar, M.S.

ABSTRACT

An experiment on, "Permanent plot experiment on integrated nutrient supply system in a cereal based crop sequence" was carried out at the main center for Cropping Systems Research Project, Mahatma Phule Krishi Vidyapeeth, Rahuri for two consecutive years during 2004–05 and 2005–06. Ten treatments comprised of use of chemical fertilizers alone and in conjuction with different organic manures *viz.*, farm yard manure (FYM), wheat cut straw (WCS) and glyricidia leaves were laid out in randomized block design. An attempt was made to develop suitable integrated nutrient supply system for a cereal based crop sequence involving more efficient use of fertilizers in conjunction with judicious combination of organic manures by effective recycling technique without detriment to long term soil fertility and by improving productivity. The experiment was conducted on the same site without changing the randomization of treatments to asses the residual effect of fertilizers on crop yield after 21 crop cycles of crop rotation. The experiment was initiated during 1984–85.

The soil of experiment plot was clayey in texture with low in available nitrogen (153.0 kg ha^{-1}), medium in available phosphorus (14.0 kg ha^{-1}) and rich in available potash (496.0 kg ha^{-1}) with slightly alkaline in reaction (pH 8.2).

The results revealed that in *kharif* sorghum crop, higher grain yield was (28.67 q ha^{-1}) and in *rabi* season wheat grain yield was (39.29 q ha^{-1}) recorded in T_6 where 50 per cent RDF + 50 per cent N substitution through FYM was applied. It was followed by treatments T_{10} and T_{11} in which 50 per cent RDF, 50 per cent N through green manuring and 75 per cent RDF + 25 per cent N through green manuring was applied in *kharif* season. Addition of FYM in combination with chemical fertilizers had pronounced positive effect on mean grain productivity of sorghum and wheat in sorghum-wheat crop sequence over years. Sustainability yield index (SYI) and mean grain productivity on pooled mean basis in sorghum and wheat crop emanated the necessity of use of FYM, glyricidia leaves and wheat cut straw along with chemical fertilizers to sustain crop and soil productivity in long run.

Incorporation of FYM, wheat, cut straw and glyricidia leaves for 25 to 50 per cent N substitution in conjunction with balanced dose of NPK fertilizers increased the infiltration rate, water stable aggregates and organic matter content of soil and decreased the bulk density due to improvement in structural status of soil.

Thus, it could be safely recommended that in the crop sequence of sorghum-wheat where both the crops are exhaustive feeders, it is necessary to incorporate the organic manures such as FYM, glyricidia leaves or wheat cut straw as per availability in substitution of N to the extent of 25 to 50 per cent in order to have not only the reasonably high sustainability, profitability of the productivity of sequence but also for maintaining the soil fertility and microbial activity in the soil.

Introduction

India is the largest country in respect or population. To meet out demands of foodgrains forever increasing population of our country, it is necessary to improve the productivity and production of the food crops in the country. In order to increase production further there is no other option except to pay greater emphasis on increasing productivity and enhancing the cropping intensity by using available resources most efficiently. System approach is applied to agriculture for efficient utilization of all resources, maintaining stability in production and obtaining higher returns. Cropping system is the most important component in the integrated farming system management.

The area under *kharif* sorghum occupied 15.11 m ha in India with production of 15.96 million tonnes. Out of 20.97 m ha arable area of the Maharashtra state. The area under *kharif* sorghum in Maharashtra is 30.8 per cent which rank first in *kharif* crops. Sorghum-wheat cropping system has become every important production system and contributes to about one third of cereal production. Because of the high productivity of both these crops in this system, the average monetary returns are also high. New technologies which could help to achieve a quantum jump in productivity and production are yet to sight.

Judicious use of fertilizer is no doubt, the king-pin in agricultural production under different agro-climatic as well as farming situations. The use of organic manures, crop residues and green manuring with chemical fertilizers is receiving greater attention in the intensive farming under integrated nutrient management system concept. Combined use of fertilizers with manures, crop residues and green manuring is known to stimulate mineralization and immobilization of organic nitrogen and can intluence the forms and availability or phosphorus and potassium in soil in many ways.

Several studies carried out in India have shown the beneficial effects of addition of organic manures on the physical properties. Biswas *et al.* (1971) reported that the treatments consisting of the inorganic fertilizers alone and in various combinations with organic manures at different levels, the structural status, water retention and bulk density of the soil were significantly improved by the treatments with organic manures. The decrease in bulk density and increased water stable aggregates were also observed under long term use of chemical fertilizers in combination with organic manures in a permanent manorial experiments by Singh and Sadhu (1980).

In intensive cropping system, fertilizer has more importance for increasing yield of individual crop. But use of only organics were precious than inorganic. Therefore, there is a need to substitute part of it through organic wastes and organic residues. For improvement of soil fertility status for sustainable crop production only organic source can be used. Because of their long-range effects, soil physical, chemical and biological conditions can be improved. Therefore, this experiment was taken for study on integrated nutrient system on crop sequence. The objectives behind this experiment are to investigate the long term effects of organic and inorganic sources of fertilizers on the productivity of sorghum and wheat and to develop suitable integrated nutrient supply system for higher productivity of sorghum-wheat crop sequence with better soil health.

Materials and Methods

The present investigation was carried out by laying out field experiment on, "Integrated nutrient management system for productivity potential of sorghum-wheat sequences under irrigated conditions," during 2004–05 and 2005–06 at the Main Centre for Cropping Systems Research Project, Mahatma Phule Krishi Vidyapeeth, Rahuri – 413 722 (M.S.). Ten treatments comprised of use of fertilizers alone and in conjunction with different organic manures *viz.*, farm yard manure (FYM), wheat cut straw (WCS) and glyricidia leaves were laid out in randomized block design.

The soil of the experiment plot was clayey in texture with low in available nitrogen (150.0 kg ha^{-1}), medium in available phosphorus (14.0 kg ha^{-1}) and rich in available potash (496.0 kg ha^{-1}) with slightly alkaline in reaction (pH 8.2).

The details of the treatments adopted during *kharif* and *rabi* season are given in Table 37.1.

		N		P_2O_5		K_2O kg ha^{-1}
Recommended dose of fertilizer (optimum level)	(a) Sorghum:	120	:	60	:	60
	(b) Wheat	120	:	60	:	60
Spacing	(a) Sorghum: 45 × 12 cm					
	(b) Wheat: 22.5 cm between line					

Sustainability yield index calculated by formulae:

$$SYI = \frac{Y - \sigma}{Y_{max}}$$

where,

Y: Estimated arrange yield of a treatment over years.

σ: Estimated standard deviation (SD)

Y_{max}: Observed maximum yield of experiment over years.

Treatment Details

Treatment	*Kharif/Sorghum*	*Rabi Wheat*
T_1	No fertilizer, no organic matter (control)	No fertilizer, no organic matter (control)
T_2	50 per cent recommended NPK dose through fertilizers	50 per cent recommended NPK dose through fertilizers
T_3	50 per cent recommended NPK dose through fertilizers	100 per cent recommended NPK dose through fertilizers
T_4	75 per cent recommended NPK dose through fertilizers	75 per cent recommended NPK dose through fertilizers
T_5	100 per cent recommended NPK dose through fertilizers	100 per cent recommended NPK dose through fertilizers
T_6	50 per cent recommended NPK dose through fertilizers + 50 per cent N through FYM	100 per cent recommended NPK dose through fertilizers
T_7	75 per cent recommended NPK dose through fertilizers + 25 per cent N through FYM	75 per cent recommended NPK dose through fertilizers
T_8	50 per cent recommended NPK dose through fertilizers + 50 per cent N through wheat straw	100 per cent recommended NPK dose through fertilizers
T_9	75 per cent recommended NPK dose through fertilizers + 25 per cent N through wheat straw	75 per cent recommended NPK dose through fertilizers
T_{10}	50 per cent recommended NPK dose through fertilizers + 50 per cent N through Green Manuring	100 per cent recommended NPK dose through fertilizers
T_{11}	75 per cent recommended NPK dose through fertilizers + 25 per cent N through Green Manuring	75 per cent recommended NPK dose through fertilizers
T_{12}	Farmer's practice (Conventional)	Farmer's practice (Conventional)
T_{13}	100 per cent RDN through FYM	100 per cent RDN through FYM

Other Details

(*a*) Type of farming: Irrigated

(*b*) Fertilizers: As per treatments

(*c*) Varieties: Sorghum CSH-9
Wheat HD 2189

Note: Nitrogen was applied in two splits to both the crops half at sowing and remaining half at 30 and 21 days after sowing of sorghum and wheat, respectively.

Results and Discussion

Yield Studies

This is the 22nd year of experimentation. The experiment was initiated during 1984-85. Total 22 crop cycles have been completed on permanent fixed site. The data on grain and fodder/straw yields along with sorghum grain equivalent are presented in Table 37.1.

Table 37.1: Grain, Fodder/Straw and Grain Equivalence of Sorghum (2005–06)

Treatments	Sorghum (q/ha)		Wheat (q/ha)		Sorghum Grain Equivalence (q/ha)
	Grain	Fodder	Grain	Straw	
T_1	1.29	12.09	6.30	10.91	14.30
T_2	23.26	66.20	26.61	34.81	78.21
T_3	24.70	71.59	31.27	43.55	89.28
T_4	26.12	75.32	30.40	42.40	88.90
T_5	27.17	80.25	35.70	48.50	100.89
T_6	**28.67**	**85.13**	**39.29**	**50.64**	**109.81**
T_7	26.18	71.32	33.31	46.53	94.97
T_8	25.13	71.99	34.04	46.96	95.43
T_9	26.50	77.07	31.70	42.13	91.97
T_{10}	28.32	82.19	35.90	48.87	101.01
T_{11}	27.54	77.01	32.30	44.15	94.24
T_{12}	20.80	53.46	24.60	32.73	71.60
T_{13}	20.96	54.21	23.10	30.80	68.67
S.Em±	0.57	0.78	0.33	0.47	–
CD at 5%	1.65	2.25	0.94	1.35	–

Grain and Fodder/Straw Yield

It is revealed from the data that in *kharif* sorghum crop, higher grain yield was (28.67 q/ha) recorded in T_6 where 50 per cent RDF + 50 per cent N substitution through FYM was applied. It was followed by treatments T_{10} (28.32 q/ha) and T_{11} (27.54 q/ha) in which 50 per cent RDF + 50 per cent N through green manuring and 75 per cent RDF + 25 per cent N through green manuring was applied to *kharif* sorghum, respectively. Which shows that he substitution of 50 per cent N through FYM and *in situ* green manuring of dhaincha gave slightly highest grain yield than the 100 per cent RDF to *kharif* sorghum. Substitution of 50 per cent N through wheat straw significantly reduced the sorghum grain. yield as compared to other organic sources *viz.*, FYM and green manuring. The sorghum grain yield in control plot (T_1) was only 1.29 q ha^{-1}. More or less similar trend was observed for sorghum fodder yields. The treatment involving 50 per cent N substitution through FYM (T_6) recorded highest fodder yield (85.13 q/ha).

These findings are in conformity with these reported by Jadhav (1994), Mohammad *et al.* (1995), Kumpawat and Rathore (1995), Singh *et al.* (1996), Patil *et al.* (1999) and Patidar (2000).

During *rabi* season 50 per cent N substituted through FYM + 50 per cent RDF to *kharif* sorghum and 100 per cent RDF to wheat crop (T_6) recorded the highest wheat grain yield (39.29 q/ha^{-1}) which was significantly superior over remaining treatments. The wheat grain yield in control treatment (T_1) was only 6.30 q ha^{-1}, whereas treatment T_{10} where 50 per cent N substitution through organics to *khunl* sorghum followed by 100 per cent RDF to wheat crop gave significantly higher grain yield (35.90 q ha^{-1}) among the all treatments except T_6. It may be due to residual effect of organics source *i.e.* green manuring, which was applied in *kharif* season.

Similar trends were observed when the data on sorghum and wheat were pooled over the years from 1984–85 to 2005–06 (Table 37.2).

Table 37.2: Average Grain, Fodder/Straw Yield (q/ha) of Sorghum-Wheat (q/ha) Crop Sequence (1984–85 to 2005–06)

Treatments	*Sorghum (q/ha)*		*Wheat (q/ha)*		*SYI*	
	Grain	*Fodder*	*Grain*	*Fodder*	*Sorghum*	*Wheat*
T_1	2.60	18.76	5.04	10.02	–0.05	0.50
T_2	30.85	62.22	25.23	37.22	0.62	0.71
T_3	35.65	65.47	31.83	44.86	0.59	0.71
T_4	38.86	69.01	30.07	41.69	0.54	0.68
T_5	44.27	76.55	33.39	47.43	0.50	0.72
T_6	**46.55**	**78.61**	**35.69**	**49.03**	**0.65**	**0.72**
T_7	41.17	72.08	31.28	45.10	0.64	0.71
T_8	35.17	65.64	34.00	47.99	0.53	0.69
T_9	37.99	69.75	30.53	43.78	0.45	0.71
T_{10}	42.28	73.06	33.98	47.34	0.53	0.72
T_{11}	40.32	69.83	30.53	43.38	0.54	0.69
T_{12}	24.00	52.13	21.16	30.02	0.54	0.47
S.Em±	0.84	1.37	0.68	0.89	–	–
CD at 5%	2.34	3.85	1.91	2.45	–	–

Similar findings have also been reported by Sonar and Zen de (1984), Bhat *et al.* (1996), Patil *et al.* (1999) and Singh *et al.* (1999) under integrated nutrient management in cereal sequence.

Economics

Economics of sorghum-wheat sequence under irrigated nutrient supply system was computed on the prevailing market prices and are presented in Table 37.3. Highest gross monetary returns (Rs. 54,932/ha), net returns (Rs. 28,080/ha) with B : C ratio of 2.05 were observed in the treatment involving use of 50 per cent RDF + 50 per cent N substitution through the FYM to *kharif* sorghum and 100 per cent RDF to wheat (T_6). Also, the second highest gross monetary returns (Rs. 50,827/ha), net returns (Rs. 25,253/ha) with B : C ratio of 1.99 were observed in the treatment involving use of 50 per cent RDF + 50 per cent N substitution through the green manuring and it was at par with the T_6. However, the T_6 is found at par with T_{10} and T_5.

The findings are in conformity with the results obtained by Patidar (2000). Kurlekar *et al.* (1994) and Dubey *et al.* (1997). The growth, yield attributing characters, grain and fodder yield of sorghum and sorghum grain equivalence were at higher magnitude due to integrated use of 50 per cent recommended dose of fertilizer along with 50 per cent N substituted through FYM which was at par with 100 per cent recommended dose of fertilizers *i.e.* 120 : 60 : 60 kg NPK ha^{-1}. Among the organic sources FYM perforated better followed by glyricidia and wheat straw in respect of yield potential of sorghum.

Table 37.3: Economics of Sorghum-Wheat Sequence (2005–06)

Treatments	*Monetary Returns (Rs/ha)*				*Gross Returns (Rs/ha)*	*Cost of Cultivation (Rs/ha)*	*Net Returns*	*B:C Ratio*
	Kharif Sorghum		*Rabi Wheat*					
	Grain	*Fodder*	*Grain*	*Wheat*				
T_1	593	7276	20261	218	7276	20261	–12985	0.36
T_2	10697	39317	22905	696	39317	22905	16412	1.72
T_3	11362	44803	24227	871	44803	24227	20576	1.85
T_4	12016	44759	24227	847	44759	24227	20532	1.85
T_5	12499	50597	25548	970	50597	25548	25049	1.98
T_6	**13190**	**54932**	**26852**	**1012**	**54932**	**26852**	**28080**	**2.05**
T_7	12041	47472	25539	930	47472	25539	21933	1.86
T_8	11561	47714	27867	939	47714	27867	19847	1.71
T_9	12188	46230	26047	843	46230	26047	20183	1.77
T_{10}	13028	50827	25574	977	50827	25574	25253	1.99
T_{11}	12669	47313	25001	83	47313	25001	22312	1.89
T_{12}	9568	35726	22860	654	35726	22860	12866	1.56
T_{13}	9642	34367	30761	616	34367	30761	3606	1.12
S.Em±	199.40	417.16	–	9.49	417.16	–	416.93	0.08
CD at 5%	571.70	1196.02	–	27.20	1196.02	–	1195.38	0.22

Selling Rates (Rs./ha)

	Sorghum	Wheat
Grain:	Rs. 460/-	Rs. 950/-
Straw/fodder	Rs. 40/-	Rs. 20/-

Table 37.4: Year-wise Grain Fodder/Straw Yield (q/ha) of Sorghum-wheat Crop Sequence

Year	T_1	T_2	T_3	T_4	T_5	T_6	T_7	T_8	T_9	T_{10}	T_{11}	T_{12}
					Kharif Sorghum Grain Yield (q/ha)							
84–85	17.56	35.66	39.88	50.63	66.85	56.58	53.12	49.66	61.59	61.21	56.95	30.68
85–86	3.34	32.18	47.57	55.70	59.65	52.31	45.61	41.10	44.83	47.57	47.47	28.69
86–87	3.37	35.75	3637	45.03	46.16	56.89	48.12	41.44	37.41	50.88	50.84	26.43
87–88	2.27	32.53	37.28	39.53	41 16	46.94	43.38	39.26	40.58	47.40	43.50	25.58
83–89	2.03	30.71	38.66	39.07	40.48	43.50	42.38	38.28	35.74	47.40	41.22	25.41
89–90	2.01	29.43	40.48	40.22	46.12	48.23	40.36	31.45	39.03	41.72	42.97	20.116
90–91	1.49	23.34	23.23	27.09	31.28	35.89	30.06	21. 13	21.37	3 1.28	17.36	16.09
91–92	2.97	40.77	43.43	48.31	54.34	62.15	51.82	43.53	44.48	53.04	48.95	32.75
92–93	2.42	30.34	37.41	40.93	50.98	54.34	46.37	42.05	43.56	45.62	40.26	20.05
93–94	2.34	30.52	32.55	34.23	39.20	48.78	38.19	34.78	37.87	36.11	34.89	17.50

Contd...

Table 37.4–Contd...

Year	T_1	T_2	T_3	T_4	T_5	T_6	T_7	T_8	T_9	T_{10}	T_{11}	T_{12}
94–95	2.33	32.44	40.86	43.42	46.87	48.78	44.91	40.71	42.56	45.57	44.94	26.08
95–96	0.19	36.25	41.70	38.93	45.96	47.47	41.99	34.38	33.57	38.04	37.93	30.32
96–97	0.43	36.62	39.12	46.72	5215	53.12	51.14	41.61	4509	47.94	47.99	33.24
97–98	1.62	3734	43.35	50.19	55.69	57.30	54.53	45.88	53.69	5328	51.61	30.96
98–99	1.07	29.50	32.75	35.40	40.95	44.69	38.72	36.50	37.38	39.68	36.50	23.41
99–00	1.54	32.78	34.79	43.69	50.94	54.23	46.94	39.08	45.41	48.74	48.21	23.38
00–01	2.84	3621	43.74	39.26	48.96	49.03	46.11	36.62	41.45	46.32	46.37	28.10
01–02	0.29	23.52	33.17	34.39	42.81	42.58	37.24	21.81	31.90	39.34	32.91	15.14
02–03	1.01	19.1	22.28	24.16	31.25	34.72	24.44	18.08	19.81	2604	25.06	14.03
03–04	1.38	25.37	25.92	26.37	27.02	28.53	26.66	25.24	26.47	26.65	26.57	19.58
04–05	1.25	25.05	25.09	25.53	28.85	29.37	27.43	26.00	25.58	28.06	27.08	19.1
05–06	1.29	23.26	24.7	26.12	27.17	28.67	26.18	25.13	26.51	28.32	27.54	20.8
Mean	2.60	30.85	35.65	38.86	44.27	46.23	41.17	35.17	37.99	42.28	40.32	24.00
Kharif (Sorghum Fodder Yield (q/ha)												
81–85	27.84	51.85	50.64	57.99	61.75	57.70	55.32	5.17	62.56	68.23	61.17	39.29
85–86	27.14	72.74	69.44	72.11	80.21	75.23	76.97	75.69	70.95	73.78	73.67	53.70
86–87	23.38	47.11	47.45	46.01	51.50	55.15	46.30	45.60	43.29	49.48	49.02	40.80
87–88	29.51	54.40	56.42	56.13	71.18	66.84	63.08	57.00	54.98	71.76	63.95	51.22
88–89	28.07	55.41	57.44	62.79	71.76	70.46	63.08	62.27	58.30	75.09	63.41	48.47
89–90	21.70	67.71	74.94	72.34	88.33	83.62	81.60	63.37	72.05	74.07	79.28	48.03
90–91	21.37	72.96	71.74	78.09	88.71	85.80	84.70	60.69	79.53	74.70	70.67	51.63
91–92	24.88	70.60	70.89	75.81	77.26	87.67	77.55	70.60	76.68	76.97	73.21	62.79
92–93	33.42	64.81	78.13	82.18	100.69	95.78	88.83	83.33	84.49	89.12	81.60	59.32
93–94	22.71	64.96	81.16	86.23	93.46	92.74	87.96	84.64	90.13	80.30	79.28	58.16
94–95	17.58	65.83	66.98	68.14	74.07	72.48	71.47	66.78	68.78	72.92	69.59	58.16
95–96	19.82	67.85	68.14	69.01	75.95	79.43	72.19	71.04	68.46	74.22	71.76	60.19
96–97	10.42	68.58	76.10	80.15	84.78	89.99	88.54	80.44	86.61	80.15	84.78	67.13
97–98	12.59	71.18	70.46	74.36	78.85	81.45	76.97	78.70	86.37	83.48	79.57	62.93
98–99	7.29	63.66	67.42	72.19	78.85	76.97	75.52	75.38	80.73	79.14	75.2.1	57.29
99–00	11.30	69.53	72.02	84.11	96.63	106.66	81.92	72.19	79.83	87.21	83.39	55.73
00–01	21.64	62.09	66.32	63.60	75.14	74.74	78.47	67.56	70.34	64.76	69.27	48.4 7
01–02	20.85	53.60	60.04	68.69	63.09	78.08	57.43	42.43	57.04	66.26	60.76	40.65
02–03	4.45	39.45	40.47	44.56	56.5	63.53	47.61	40.26	41.29	49.13	44.13	38.54
03–04	2.87	50.73	53.30	55.74	53.28	60.29	56.4 7	47.59	49.67	51.48	50.16	37.23
04–05	12.86	67.57	69.19	72.58	81.60	89.62	82.51	72.41	75.37	82.88	75.24	53.7
05–06	12.09	66.2	71.59	75.32	80.25	85.13	71.32	71.99	71.07	82.19	71.01	53.46
Mean	18.76	62.22	65.47	69.01	76.55	78.61	72.08	65.64	69.75	73.06	69.83	52.13

Contd...

Table 37.4–Contd...

Year	T_1	T_2	T_3	T_4	T_5	T_6	T_7	T_8	T_9	T_{10}	T_{11}	T_{12}
Wheat Grain Yield (q/ha)												
84–85	4.47	25.10	31.87	30.45	35.79	34.56	32.34	39.89	32.36	3837	29.56	19.50
85–86	5.09	27.89	35.53	31.16	36.18	36.39	34.81	38.89	35.45	37.30	32.44	20.23
86–87	2.67	23.65	33.98	30.77	32.17	35.86	33.33	37.22	31.13	3714.	32.89	20.69
87–88	3.17	21.00	29.87	28.98	29.92	32.36	26.85	33.09	27.36	31.51	26.57	17.12
88–89	4.59	28.54	31.42	34.83	34.45	29.42	34.27	41.65	33.32	34.64	32.65	20.09
89–90	5.57	24.13	30.64	33.28	33.35	31.63	31.80	32.19	34.03	31.37	30.87	18.11
90–91	5.03	25.38	34.40	30.96	35.90	37.41	31.65	33.65	28.71	32.87	29.20	17.70
91–92	5.85	27.52	38.75	33.85	40.22	40.39	35.50	40.16	34.35	39.33	33.30	22.88
92–93	4.50	25.03	31.91	29.58	35.85	37.32	32.76	37.79	32.99	36.38	32.00	17.80
93–94	4.92	24.48	27.83	28.15	30.13	36.45	28.45	28.74	27.66	33.22	27.37	16.90
94–95	4.83	28.06	37.59	13.54	39.75	42.87	34.49	37.25	33.97	40.02	33.22	23.74
95–96	4.32	22.06	30.91	28.16	32.87	32.63	31.48	29.18	28.29	30.60	30.17	20.84
96–97	6.37	28.13	34.54	32.56	36.58	40.76	33.68	34.41	33.48	37.12	33.21	25.70
97–98	6.29	24.36	26.55	25.06	33.76	32.73	28.01	28.06	25.90	28.50	27.82	20.60
98–99	5.24	22.01	30.63	29.17	34.26	36.36	30.73	32.11	30.90	33.18	29.15	20.84
99–00	6.91	27.70	36.13	35.72	38.54	44.47	37.78	41.16	35.81	39.31	35.84	36.06
00–01	7.63	31.41	38.73	37.47	3278	42.64	36.50	37.85	36.44	39.93	38.62	24.65
01–02	4.82	20.69	24.90	23.27	26.38	28.87	26.24	27.37	26.26	26.98	25.71	17.54
02–03	3.84	23.14	28.02	23.57	28.29	29.95	24.58	28.18	23.74	27.32	24.94	18.04
03–04	4.31	21.52	23.40	19.58	20.19	28.10	19.83	23.28	18.22	24.64	24.10	18.19
04–05	4.26	26.65	31.37	29.01	31.59	34.78	29.82	31.74	2955	32.19	29.7	23.62
05–06	6.3	26.61	31.27	30.4	35.7	39.29	33.31	34.04	31.7	35.2	32.3	24.6
Mean	5.04	25.23	31.83	30.07	33.39	35.60	31.28	34.00	30.53	33.98	30.53	21.16
Wheat Straw (q/ha)												
84–85	10.87	44.59	51.85	44.69	54.28	52.81	50.65	51.71	50.30	54.40	47.83	21.01
85–86	7.35	37.50	43.52	39.38	48.08	44.40	44.62	52.89	45.75	47.60	40.44	25.49
86–87	8.91	36.54	35.82	32.30	36.41	39.08	37.10	45.83	41.63	43.32	41.18	23.87
87–88	9.96	30.30	35.89	39.38	38.31	43.42	41.86	46.83	41.45	43.41	37.04	29.65
88–89	7.51	33.15	34.90	37.83	26.88	40.25	3615	43.71	34.42	37.09	34.94	23.49
89–90	7.87	26.22	35.04	34.61	37.25	35.50	35.34	34.94	35.13	34.92	34.52	22.69
90–91	9.43	38.86	44.01	39.70	44.18	45.34	42.42	42.45	38.70	43.23	40.25	31.13
91–92	9.78	41.06	5703	48.70	54.02	54.22	50.64	56.28	46.96	56.16	47.28	31.25
92–93	8.95	37.20	51.69	45.99	54.63	57.52	49.24	58.19	50.51	55.48	47.61	27.40
93–94	9.55	37.59	36.26	36.37	44.24	43.11	40.13	41.58	34.40	43.32	37.44	27.66
94–95	9.58	37.42	52.78	50.57	63.30	62.97	56.64	66.26	56.59	58.97	60.44	37.56

Contd...

Table 37.4–Contd...

Year	T_1	T_2	T_3	T_4	T_5	T_6	T_7	T_8	T_9	T_{10}	T_{11}	T_{12}
95–96	6.25	30.61	37.79	33.45	40.91	44.17	39.54	40.08	36.31	41.58	35.46	26.37
96–97	13.31	45.94	53.72	50.19	51.09	49.81	52.55	57.60	55.35	54.32	55.91	40.56
97–98	11.86	36.46	36.55	42.10	47.25	44.70	47.16	37.85	41.46	38.54	40.10	30.24
98–99	13.08	31.41	39.31	35.82	48.08	40.83	36.40	39.25	37.83	42.26	36.70	25.45
99–00	17.51	38.19	46.01	44.42	53.53	57.78	46.01	50.64	46.06	50.41	45.14	32.61
00–01	10.66	43.89	63.05	53.23	61.27	68.17	63.98	62.86	56.58	60.60	55.98	39.35
01–02	8.46	29.23	38.47	32.86	41.91	41.74	36.83	38.89	36.81	34.91	35.59	23.28
02–03	8.14	32.37	34.23	31.40	34.33	36.48	32.55	34.26	32.77	33.92	33.10	21.84
03–04	12.91	54.98	66.53	58.16	64.53	70.31	61.05	58.74	56.42	68.57	57.58	51.53
04–05	7.60	40.53	48.92	43.71	50.39	55.50	44.56	48.00	45.63	50.59	45.44	35.38
05–06	10.91	34.81	43.55	42.4	48.5	50.64	46.53	46.96	42.13	48.87	44.15	32.73
Mean	10.02	37.22	44.86	41.69	47.43	49.03	45.10	47.99	43.78	47.34	43.38	30.02

Incorporation of FYM, wheat cut straw and glyricidia leaves for 25 to 50 per cent N substitution in conjunction with balanced dose of NPK fertilizers increased the infiltration rate, water stable aggregate and organic matter content of soil and decreased the bulk density because of this INM system and strengthened the chemical, biological and physical properties of soil.

References

Bhat, A.K., Beri, V. and Sidhu, B.S., 1996. Effect of long term recycling of crop residues on soil productivity. *J. Indian Soc. Soil Sci.*, 39: 380–382.

Biswas, T.D., Jain, B.L. and Mandal, S.C., 1971. Cumulative effect of different levels of manures on the physical properties of soil. *J. Indian Soc. Soil Sci.*, 19(1): 31–37.

Dubey, S.K., Sharma, R.S. and Vishwakarma, S.K., 1997. Integrated nutrient management for sustainable productivity of important cropping systems in Madhya Pradesh. *Indian J. Agron.*, 42(1): 13–17.

Hegde, D.M., 1996. Effect of integrated nutrient supply on crop productivity and soil fertility in rice (*Oryza sativa*): Rice system. *Indian J. Agron.*, 41(1): 1–8.

Jadhav, A.S., 1994. Growth, yield and quality of sorghum under double cropping systems with reference to fertilizer management. *Madras Agric. J.*, 81(3): 127–131.

Kumpawat, B.S. and Rathore, S.S., 1995. Response of maize (*Zea mays*), wheat (*Triticum aestivum*) cropping sequence to fertilizer application. *Indian J. Agron.*, 40(1): 26–29.

Kurlekar, V.C., Kale, V.C., Nagarguje, S.R., Navhat, D.A. and Topadia, B.B., 1994. Production, potential and economics of different crop sequences. *J. Maharasthra Agric. Univ.*, 18(3): 397–399.

Mohammad, S., Hason, M.V. and Bajelam, B., 1995. Genotypic responses of fertility variation in a scarce rainfall-shallow soil ecosystem. *Ann. Agric. Chem.*, 19(1): 369–373.

Patidar, M., 2000. Integrated nutrient management in sorghum and its residual effect on wheat. *Ph.D. Thesis*, Agricultural University, Udipur (Rajasthan).

Patil, C.B., Patil, B.P., Patil, B.N., Patil, S.R. and Tuwar, S.S., 1999. Twenty five years of research in cropping systems (1970–1995). *MPKV/Res/Pub.*, 27: 35–38.

Singh, Gurukirpal and Sandhu, B.S., 1980. Studies on multiple cropping. II. Effect of crop rotations on physical and chemical properties of soil. *Indian J. Agron.*, 25(1): 57–60.

Singh, A.P., Pal, M.S. and Malik, H.P.S., 1996. Performance of grain sorghum (*Sorghum bicolor*) genotypes under rainfed conditions of varying fertility-level. *Indian J. Agron.*, 41(2): 256–260.

Singh, R.P., Mundra, M.C., Gupta, S.C. and Agrawal, S.K., 1999. Effect of integrated nutrient management on productivity of pearlmillet (*Pennisetum glaueum*), wheat (*Triticum aestivum*) cropping system. *Indian J. Agron.*, 44(2): 250–255.

Sonar, K.R. and Zen de, G.K., 1984. Influence of crop sequence and fertilizer levels on crop yields. *J. Maharashtra Agric. Univ.*, 9(2): 121–128.

Chapter 38

Comparison of Energy Interaction Parameters for the Complexation of Pr(III) with Guanosine in Absence and Presence of Zn(II) in Aqueous and Aquated Organic Solvents Using 4f-4f Transition Spectra as PROBE

*T. Kriyananda Singh, Th. David Singh, N. Yaiphaba, Ch. Sumitra and N. Rajmuhon Singh**

Department of Chemistry, Manipur University, Canchipur – 795 003, Manipur

ABSTRACT

Pr(III) is hard acceptor in HSAB (hard and soft acid base) sense and it shows strong preference for oxygen and nitrogen donor sites of guanosine. In our work, we have focus on the studies of the difference in energy parameters and comparative absorption spectrophotometry involving 4f-4f transitions on Pr(III) and guanosine in absence and presence of Zn(II) in aqueous and different aquated organic solvents of CH_3OH, CH_3CN, DMF and dioxane and their equimolar

* E-mail: rajmuhon@yahoo.co.in

mixtures at pH 6 and at 298 K. Variaions in the spectral energy parameters-slater-condon (F_k) factor, Lande spin-orbit coupling constant (ξ_{4f}), nephelauxetic ratio (β), bonding parameter ($b^{1/2}$) and per cent covalency (δ) are calculated and correlated with binding of Pr(III) with guanosine in absence and presence of Zn(II).

Keywords: *Absorption spectra, Guanosine, Hypersensitive, Nephelauxetic effect, Pseudo-hypersensitive.*

Introduction

In view of the extensive application of lanthanide elements in agriculture, industries and biomolecular reactions, it has become more important to understand the behaviours of lanthanide ions in biological system (Brittain *et al.*, 1978 and Evan, 1990). So lanthanide co-ordination chemistry in solution state is of great importance with the increasing use of lanthanids as probes in the exploration of the structural functions of biomolecular reactions (Misra, 1985; Bukeitynska and Ngoethey, 1987; Holz and Thompson, 1988 and Devlin *et al.*, 1988). This is particularly due to their ability to replace Ca(II) ions in specific manner (Einsphar and Bugg, 1984 and Martin, 1983). Shah and Shah (2001) studied comparative 4f-4f transition spectra of Pr(III) with lysozyme by using energy interaction parameters to explain the behaviour of binding between them. Mehta *et al.* (2001) also studied the mode of binding between Pr(III) and Nd(III) with lysozyme in presence of Zn(II), a soft metal ion, by employing intensity parameters. The ligand we choose, guanosine (Figure 38.1) is a naturally occurring nucleoside. It has both oxygen and nitrogen donor sites *viz.*, carbonyl, amino groups and peptide linkage of purine nucleus and hydroxylic groups of ribose sugar of guanosine. Nucleosides and nucleotides, the building blocks of nucleic acidd, form a very important class of organic molecules in molecular biology.

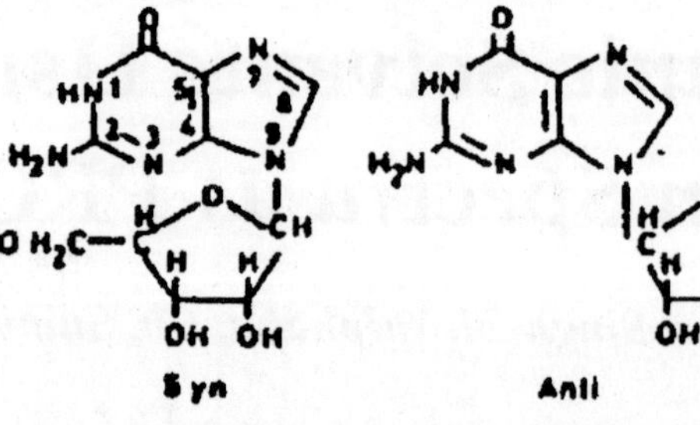

Figure 38.1: *Syn* and *Anti* Conformations of Guanosine

Guanosine can exist in two forms *viz.*, syn and anti-forms (Pinkerton and Earl, 1979). The syn conformation brings sugar oxygen and base nitrogen a bit closer; hence, metal ion can form inner sphere complexes while the 'anti' form places sugar oxygen and base nitrogen far off, so that both cannot be simultaneously involved in complexation with the metal ion.

Electronic spectral studies on the conformation of the guanosine molecule and its interaction with lanthanide cations in aqueous medium have led to an understanding of the nature and reactivity of the biochemical reactions involving such species (Birnbaum *et al.*, 1976). Complexes with ligands like polyaminepolycarboxylates and 8-hydroxyquinoline (nitrogen and oxygen donors) were studied (Moeller and Hessen, 1963; Cefola *et al.*, 1962 and Freasire *et al.*, 1956) Since Pr(III) resembles to Ca(II), its complexation can provide information about the co-ordination characteristics of diamagnetic molecule, *i.e.* Ca(II) with biomolecules during biochemical reactions. Hence, paramagnetic lanthanides are good spectral probes for exploring the biological roles of Ca(II) by isomorphous substitution. The present work discusses the quantitative spectral energy interaction parameters of Pr(III) complex with guanosine in absence and presence of Zn(II) in aqueous and aquated organie solvents at pH6 and 298 K. The present work reports the sensitivity of ligand mediated pseudohypersensitivity of $3H_4 \rightarrow 3p0$, $3H_4 \rightarrow 3P_1$, $3H4 \rightarrow 3P_2$ and $3H_4 \rightarrow {}^1D_2$ transitions of Pr(III) and utilized the magnitude and variation of slater-condon (F_k), Lande spin orbit coupling (ξ_{4f}), nephelauxetic ratio (β), bonding ($b^{½}$). Racah (E^k) and covalency (δ) parameters for the complexation of Pr (III) with guanosine in absence and presence of Zn(II).

Materials and Methods

Praseodymium (III) nitrate of 99.9 per cent purity was purchased from M/s Indian Rare Earths Ltd. and guanosine from M/s Sisco Pvt. Ltd., Mumbai. The solvents used are CH_3CN, CH_3OH, DMF and dioxane of A/R grade from E. Mark. The saturated solutions of ligands and Pr(III) nitrate of 10^{-2} M were prepared in different solvents. Absorption spectra of each solution at pH6 and at temperature 298 were recorded on Perkin Elmer Lamda-35UV-vis spectrophotometer upgraded with high resolution and expansion of scale in the region 350–900 nm and using water circulating thermostat model of HAAKE DC 10.

Nephelauxetic ratio has long been regarded as a measure of covalency (Peopock, 1964; Jorgensen and Judd, 1964 and Henrie *et al.*, 1976). The Nephelauxetic effect has been interpreted in terms of Slater-Condon and Racah parameters (inter electronic repulsion parameters) as well as by the ratio of the complex ion and free ion (Sinha and Schmidtke, 1965; Sinha, 1966; Sinha and Mehata, 1972 and Condon, 1963)

$$\beta = \frac{F_k^c}{F_k^f} \ or \ \beta = \frac{E_c^k}{E_f^k} \tag{1}$$

where,

F_k (K = 2,4,6) is the Slater-Condon parameter and E_K the Racah parameters C and f stand for complex and free ions respectively. The bonding parameter ($b^{½}$) is inter-related to Nephelauxetic effect as,

$$b^{½} = \left[\frac{1-\beta}{2}\right]^{½} \tag{2}$$

The electrostatic term E_0 is expressed in terms of the product of Slater radial integral known as Slater-Condon parameter, F_k and is given by

$$E_0 = \sum_{k=0}^{k=6} K^k F_k \tag{3}$$

The Slater-Condon parameters are also known as direct-integral and is decreasing function of K as given by the relation,

$$F_1^k = \int_0^\infty\int_0^\infty \frac{r_<^k}{r_<^{k+1}} R_i^2(r_i) R_j^2(r_j) r_i^2 r_j^2 dr_i dr_j \tag{4}$$

where,

R is the 4f-radial wave function; $r_<$ and $r_>$ are the radii of near and more distant electrons; and *i* and *j*, the ith and j^{th} electrons under consideration. Condon and Shortley (1963) redefined F^K integrals in terms of reduced integral F_K related to each other and the relation is

$$F_k = \frac{F^k}{D_k} \quad (5)$$

Combining relation (4) and (5), the reduced Slater-Condon integral can be written as:

$$F_k = \frac{1}{D_k}\int_0^\infty\int_0^\infty r_<^k r_>^{k+1} R_i^2(r_i)R_j^2(r_j)T_i^2 r_j^2 dr_i dr_j \quad (6)$$

where,

D_K is the denominator and F_K are coefficient of linear combination and represent the angular part of the interaction F_K is the expectation value of scalar product $(C_1^{(K)}C_2^{(K)})$.

The Racah energy interaction pararmeter E^K's are linear combinations of F_K given by

$$E^1 = \frac{70F_2 + 231F_4 + 20.04F_6}{9}$$

$$E^2 = \frac{F_2 - 3F_4 + 7F_6}{9} \quad (7)$$

$$E^3 = \frac{5F_2 + 6F_4 - 9F_6}{3}$$

The energy, E_{s0} arising from the most important magnetic interactions, which are spin orbit interactions may be written as:

$$E_{s0} = A_{s0}\xi_{4f} \quad (8)$$

where,

A_{s0} is the angular part of spin orbit interaction and ξ_{4f} is the radial integral and is known Lande's parameter. By first order approximation the energy E_j of the jth level is given by Wong (1961) as

$$E_j(F_k, \xi_{4f}) = E_{0j}(F_k^0, \xi_{4f}) + \frac{\partial E_j}{\partial E_k}\Delta F_k + \frac{\partial E_j}{\partial \xi_{4f}}\Delta\xi_{4f} \quad (9)$$

where,

E_{0j} is the zero order energy of the jth level. The value of F_K and ξ_{4f} are given by

$$F_k = F_k^0 + \Delta F_k$$

$$\xi_{4f} = \xi_{4f}^0 + \Delta\xi_{4f} \quad (10)$$

The difference between the observed E_j value and the zero order values, ΔE_j is evaluated by

$$\Delta E_j = \sum_{k=2,4,6} \frac{\partial E_j}{\partial E_k} \Delta F_k + \frac{\partial E_j}{\partial \xi_{4f}} \Delta \xi_{4f} \tag{11}$$

By using the zero order energy and partial derivatives of Pr(III) ion given by Wong (1963), the above equation can be solved by least square technique and the value of ΔF_2 and $\Delta \xi_{4f}$ can be found out. From these the value of F_2 and ξ_{4f} are found out by using relation *i.e. Eq.* (10). The estimated values of F_4 and F_6 are calculated by the relations:

$$\frac{F_4}{F_2} = 0.1380 \text{ and } \frac{F_4}{F_2} = 0.0150 \tag{12}$$

The per cent covalency parameter (δ) representing the Nephelauxetic effect was given by the relation

$$\delta = \frac{1-\beta}{\beta} \times 100 \tag{13}$$

Results and Discussion

From the Figures 38.2(a–d) we see that there is a red shift as gaunosine is added to Pr(III) and further longer wave length is observed on addition of Zn(II). Table 38.1 shows the variation of the magnitude of energy interaction parameters like Slater-Condon(F_K), Lande factor (ξ_{4f}), Racah energy (E^K), Nephelauxetic ration (β), bonding ($b^{1/2}$) and percentage co-valency (δ) for Pr(III), Pr(III): gaunosine and Pr(III): gaunosine: Zn (II) in aqueous and different aquated organic solvents. There is a slight decrease in the value of F_K, ξ_{4f} and E_K as the complexation goes on which lead to increase in the values of Nephelauxetic ratio.

For spectral studies on the structures of co-ordination compounds of lanthanides in solution, any evidence of the relationship between the Nephelauxetic band shift and the structure is of special interest. Jorgensen and Ryan (1966) noticed the dependence of Nephelauxetic effect on the co-ordination number. It was suggested that shortening in the metal ligand distance occurs with decrease in the co-ordination number. Langhlin and Conway (1963) studied the absorption spectra of Pr(III) ion in the iso-structural crystals $CeCl_3$, $NdCl_3$, $SmCl_3$ and $GdCl_3$ and they concluded that with identical structures of complexes (CN = 9 and D_{3h} symmetry), there exists a linear relationship between wave no. of $^3H_4 \rightarrow {}^3P_{0,1,2}$ transitions in the spectra of Pr^{3+} and distance between praseodymium and ligand, *R(Ln–O)*[*R(Ln–O)* is the mean Ln-oxygen distance in the complexes with ligands coordinated via oxygen atom]. It was observed that increase in the value of R increases the band shift towards the short wave region. Yatsimirskii and Davidenko (1970) and Yatsimirskii *et al.* (1966) pointed out a close linear correlation between R(ln–O) and wave number of some f-f transitions in the absorption spectra of some praseodymium and neodymium complexes having different co-ordination number and symmetries. To interpret the correlation and analysis of the relationship between Nephelauxetic effect and geometry, energy parameters have been derived and evaluated for complex compound using the angular overlap model, the value of '*n*' is proportional to the Nephelauxetic effect as

$$n = \left[\frac{1-\beta^{1/2}}{\beta^{1/2}}\right] \tag{14}$$

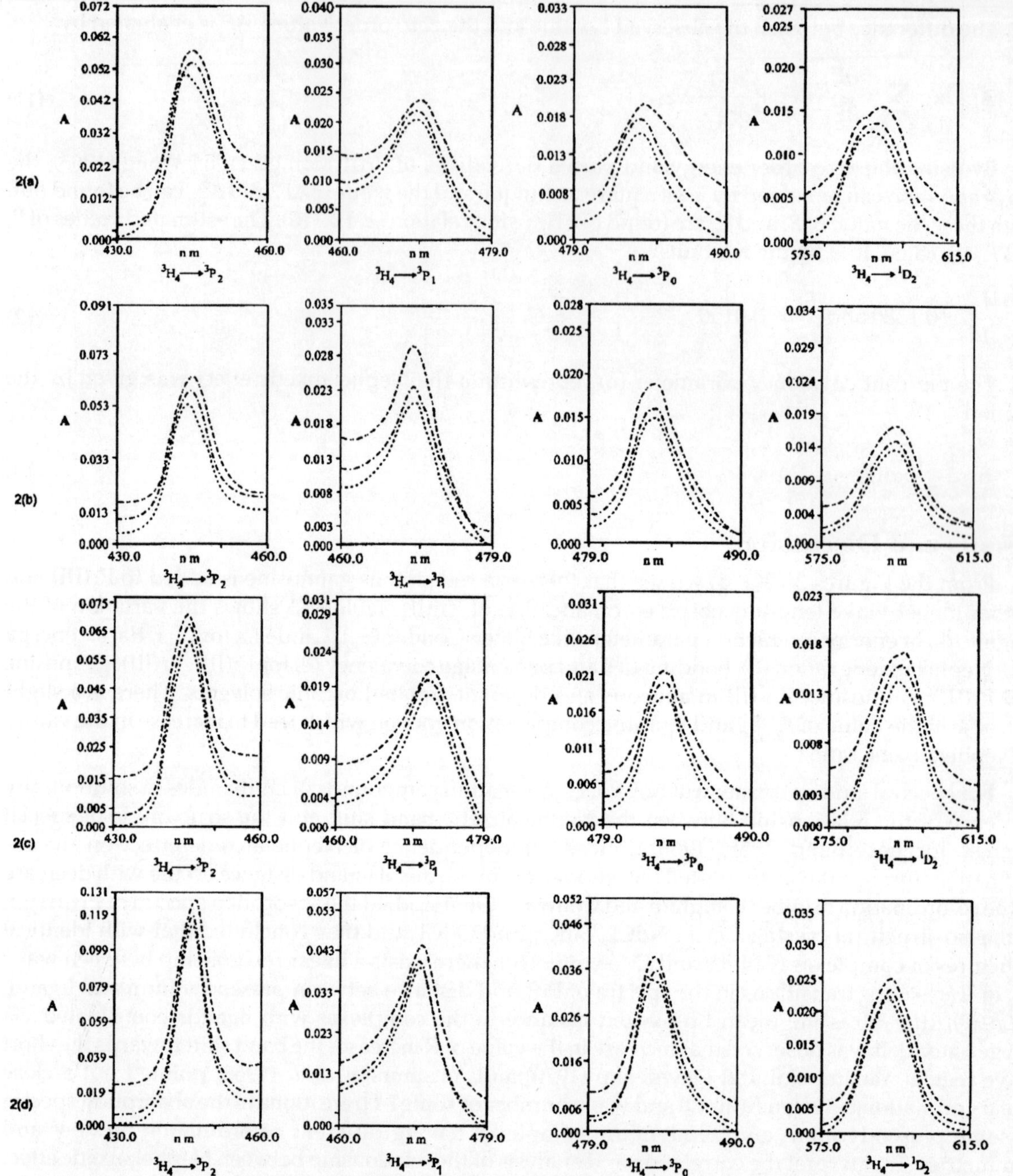

Figure 38.2: Comparative Absorption Spectra of (a) Pr(III) (- - -), P. (-.-.-) and Pr(III) : guanosine : Zn(II) (1:1:1) (-..-..) in CH_3CN. (b) Pr(III) (- - -), Pr(III) : guanosine (1:1) (-.-.-) in Pr(III) : guanosine : Z (II) (1:1:1) (. .-. .-) in dioxane. (c) Pr(III) (-—), Pr(III) : guanosine (1 : 1) (-. -. -) and Pr(III) : guanosine : Zn(II)(1 : 1 : 1) (-. . -' . -) in CH,oH. (d) Comparative absorption spectra of Pr(III) (-—), Pr(III) : guanosine(1 : 1) and Pr(III) : guanosine : Zn(II) (1:1:1) (-..-..-) in DMF

Table 38.1: Computed Value of Energy Interaction Slater-Condon F_k (cm^{-1}), Spin Orbit Interaction ξ_{4f} (cm^{-1}), Racah energy E^k (cm^{-1}), Nephelauxetic Ratio (β), Bonding ($b^{1/2}$) and Covalency (δ) Parameters of Pr(III), Pr(III): guanosine (1 : 1) and Pr(III): guanosine: Zn(II) (1:1:1) Systems in Aqueous and Different Aquated Organic Solvents (50 : 50) at pH = 6 and 298K are Given Below

Sl.No.	System	F_2	F_4	F_6	ξ_{4f}	E^1	E^2	E^3	β	$b^{1/2}$	δ
1.	Solvent–water										
	Pr(III)	309.07	42.66	4.66	721.8 8	3509.5478	23.7495	586.4776	0.9461	0.1637	5.6778
	Pr(III): guanosine	309.06	42.66	4.65	721.7 0	3509.3318	23.7481	586.4415	0.9459	0.1639	5.6948
	Pr(III): guanosine: Zn(II)	309.02	42.65	4.64	721.3 6	3508.9318	23.7454	586.3747	0.9456	0.1644	5.7263
2.	Solvent–CH_3OH										
	Pr(III)	308.96	42.64	4.66	721. 08	3508.1918	23.7403	586.2510	0.9453	0.1648	5.7587
	Pr(III): guanosine	308.94	42.64	4.66	721. 05	3508.0528	23.7394	586.2277	0.9453	0.1648	5.7601
	Pr(III): guanosine: Zn(II)	308.91	42.64	4.66	720.92	3507.7248	23.7372	586.1730	0.9453	0.1649	5.7626
3.	Solvent–CH_3CN										
	Pr(III)	309.01	42.65	4.66	721.3 9	3508.8208	23.7446	586.3560	0.9456	0.1644	5.7260
	Pr(III): guanosine	309.00	42.65	4.66	721.3 7	3508.7308	23.7440	586.3411	0.9456	0.1645	5.7291
	Pr(III): guanosine: Zn(II)	308.97	42.64	4.66	721.26	3508.3738	23.7416	586.2814	0.9455	0.1646	5.7422
4.	Solvent–DMF										
	Pr(III)	308.95	42.64	4.66	720.47	3508.1908	23.7403	586.2509	0.9449	0.1654	5.8054
	Pr(III): guanosine	308.93	42.64	4.65	720.31	3507.8548	23.7381	586.1947	0.9448	0.1657	5.8227
	Pr(III): guanosine: Zn(II)	308.89	42.63	4.65	720.28	3507.0178	23.7324	586.0547	0.9448	0.1657	5.8217
5.	Solvent–Dioxane										
	Pr(III)	308.98	42.65	4.66	721.21	3508.4278	23.7419	586.2905	0.9455	0.1646	5.7455
	Pr(III): guanosine	308.93	42.64	4.66	720.66	3507.9118	23.7384	586.2042	0.9450	0.1653	5.7947
	Pr(III): guanosine: Zn(II)	308.89	42.63	4.66	720.61	3507.4968	23.7356	586.1349	0.9449.	0.1654	5.8050
6.	Solvent–CH_3GH:CH_3CN										
	Pr(III)	308.97	42.64	4.66	720.64	3508.3068	23.7411	586.2702	0.9451	0.1652	5.7910
	Pr(III): guanosine	308.96	42.64	4.66	720.55	3508.2388	23.7407	586.2588	0.9450	0.1653	5.7945
	Pr(III): guanosine: Zn(II)	308.88	42.63	4.66	720.44	3507.3498	23.7346	586.1103	0.9450	0.1653	5.7981

Contd...

Table 38.1–Contd...

Sl.No.	System	F_2	F_4	F_6	ξ_{4f}	E^1	E^2	E^3	β	$b^{1/2}$	δ
7.	Solvent–CH_3OH:DMF										
	Pr(III)	308.94	42.64	4.66	720.48	3508.0208	23.7392	586.2224	0.9449	0.1655	5.8074
	Pr(III): guanosine	308.93	42.64	4.66	720.33	3507.7385	23.7385	586.2050	0.9448	0.1656	5.8200
	Pr(III): guanosine: Zn(II)	308.90	42.64	4.65	720.24	3507.6768	23.7365	586.1550	0.9447	0.1658	5.8315
8.	Solvent-CH_3OH:Dioxane										
	Pr(III)	308.98	42.65	4.66	721.50	3508.4348	23.7440	586.2916	0.9457	0.1643	5.7227
	Pr(III): guanosine	308.97	42.64	4.66	721.42	3508.3068	23.7411	586.2702	0.9456	0.1644	5.7309
	Pr(III): guanosine: Zn(II)	308.94	42.64	4.66	721.19	3507.9998	23.7390	586.2184	0.1647	0.1647	5.7531
9.	Solvent–CH_3CN:DMF										
	Pr(III)	309.03	42.65	4.66	721.2 7	3509.0118	23.7459	586.3880	0.9456	0.1644	5.7318
	Pr(III): guanosine	309.01	42.65	4.66	721.08	3508.8228	23.7446	586.3564	0.9454	0.1647	5.7488
	Pr(III): guanosine: Zn(II)	308.97	42.64	4.66	721.13	3508.3158	23.7412	586.2717	0.9454	0.1647	5.7523
10.	Solvent–CH_3CN:Dioxane										
	Pr(III)	309.01	42.65	4.66	721.1 7	3508.7698	23.7443	586.3476	0.9455	0.1646	5.7429
	Pr(III): guanosine	308.93	42.64	4.66	720.74	3507.8558	23.7381	586.1948	0.9451	0.1652	5.7885
	Pr(III): guanosine: Zn(II)	308.85	42.63	4.65	720.63	3507.0078	23.7323	586.0531	0.9450	0.1652	5.7871
11.	Solvent–DMF:Dioxane										
	Pr(III)	308.92	42.64	42.65	720.62	3507.8008	23.7377	586.1857	0.9450	0.1654	5.7997
	Pr(III): guanosine	308.83	42.63	4.65	720.07	3506.7728	23.7138	586.0138	0.9445	0.1661	5.8565
	Pr(III): guanosine: Zn(II)	308.41	42.57	4.65	718.10	3506.0368	23.6987	585.2225	0.9425	0.1690	5.8762

Table 38.2: Computed and Observed Values of Energies (cm^{-1}) and RMS Values for Pr(III), Pr(III): guanosine (1 : 1) and Pr(III): guanosine: Zn(II) (1:1:1) in Aqueous and Different Aquated Organic Solvents (50 : 50) at pH = 6 and 298 K are Given Below

Sl.No.	System	$^3H_4$ → E_{obs}	$^3P_2$ E_{cal}	$^3H_4$ → E_{obs}	$^3P_1$ E_{cal}	$^3H_4$ → E_{obs}	$^3P_0$ E_{cal}	$^3H_4$ → E_{obs}	$^1D_2$ E_{cal}	RMS
1.	Solvent–water									
	Pr(III)	22498.19	22452.66	21321.04	21241.99	20742.14	20682.21	16959.20	17134.75	103.35
	Pr(III): guanosine	22496.67	22450.45	21320.59	21239.59	20741.71	20680.53	16954.31	17133.34	105.39
	Pr(III): guanosine: Zn(II)	22491.11	22446.39	21318.77	21236.13	20740.85	20677.42	16947.71	17130.75	107.64
2.	Solvent–CH_3OH									
	Pr(III)	22494.14	22440.56	21309.68	21230.44	20738.27	20672.30	16937.37	17126.93	111.16
	Pr(III): guanosine	22493.63	22439.77	21308.32	21229.61	20737.41	20671.46	16937.09	17126.38	110.99
	Pr(III): guanosine: Zn(II)	22492.12	22437.96	21306.05	21227.70	20735.26	20669.49	16936.22	17125.14	110.78
3.	Solvent–CH_3CN									
	Pr(III)	22488.58	22445.85	21316.95	21235.55	20739.99	20676.78	16949.72	17130.37	106.15
	Pr(III): guanosine	22487.57	22443.20	21316.04	21234.90	20739.56	20676.18	16949.43	17129.94	106.03
	Pr(III): guanosine: Zn(II)	22484.53	224422.55	21314.22	21232.28	20737.84	20673.77	16946.27	17128.19	106.85
4.	Solvent–DMF									
	Pr(III)	22489.59	22437.48	21320.13	21228.01	20671.14	20671.32	16925.15	17125.15	120.00
	Pr(III): guanosine	22483.52	22434.67	21315.59	21225.29	20739.56	20668.75	16921.04	17123.32	118.81
	Pr(III): guanosine: Zn(II)	22478.97	22430.87	21311.95	21221.72	20738.70	20665.79	16915.88	17120.88	120.20
5.	Solvent–Dioxane									
	Pr(III)	22490.60	22442.61	21314.68	21232.41	20738.70	20674.01	16941.68	17128.26	116.61
	Pr(III): guanosine	22487.06	22436.79	21313.77	21227.05	20738.27	20669.78	16927.62	17124.58	115.67
	Pr(III): guanosine: Zn(II)	22485.04	22434.04	21308.78	21224.27	20736.12	20667.11	16926.19	17122.74	115.25
6.	Solvent–CH_3OH: dioxane									
	Pr(III)	22491.61	22439.01	21318.32	21229.39	20740.85	20672.17	16925.91	17126.10	117.75
	Pr(III): guanosine	22489.59	22438.18	21316.95	21228.63	20740.85	20671.59	16925.62	17125.58	117.49
	Pr(III): guanosine: Zn(II)	22486.05	22433.82	21312.86	21223.88	20734.40	20666.45	16923.61	17122.52	117.07

Contd...

Table 38.2–Contd...

Sl.No.	System	$^3H_4$ E_{obs} →	$^3P_2$ E_{cal}	$^3H_4$ E_{obs} →	$^3P_1$ E_{cal}	$^3H_4$ E_{obs} →	$^3P_0$ E_{cal}	$^3H_4$ E_{obs} →	$^1D_2$ E_{cal}	RMS
7.	Solvent–CH_3OH:DMF									
	Pr(III)	22489.59	22436.50	21314.68	21226.99	20739.99	20670.09	16923.04	17124.48	118.28
	Pr(III): guanosine	22485.54	22435.16	21313.32	21225.77	20739.99	20669.18	16922.47	17123.64	117.98
	Pr(III): guanosine: Zn(II)	22483.52	22432.90	21309.68	21223.54	20738.70	20667.15	16920.75	17122.15	117.96
8.	Solvent–CH_3OH:Dioxane									
	Pr(III)	22493.63	22444.14	21312.41	21233.62	20737.41	20674.61	16947.13	17129.14	106.91
	Pr(III): guanosine	22490.60	22442.96	21311.05	21232.51	20736.98	20673.67	16946.84	17128.39	106.53
	Pr(III): guanosine: Zn(II)	22484.53	22439.95	21308.78	21229.67	20736.12	20671.31	16943.97	17126.45	106.93
9.	Solvent–CH_3CN:DMF									
	Pr(III)	22498.19	22446.43	21307.41	21236.28	20742.57	20677.75	16950.0 I	17130.82	105.62
	Pr(III): guanosine	22492.12	22444.37	21306.96	21234.37	20742.14	20676.22	16948.28	17129.51	105.75
	Pr(III): guanosine: Zn(II)	22489.59	22441.60	21305.60	21231.44	20738.70	20673.19	16945.70	17127.60	106.27
10.	Solvent–CH_3CN:Dioxane									
	Pr(III)	22492.62	22444.49	21319.68	21234.39	20740.85	20676.06	16940.53	17129.55	116.25
	Pr(III): guanosine	22491.61	22436.92	21319.22	21227.07	20737.41	20669.60	16922.18	17124.62	119.43
	Pr(III): guanosine: Zn(II)	22489.59	22432.77	21313.32	21222.54	20731.39	20664.70	16920.75	17121.70	118.63
11.	Solvent–DMF:Dioxane									
	Pr(III)	22488.07	22435.91	21314.22	21226.19	20737.84	20669.01	16924.47	17124.00	117.27
	Pr(III): guanosine	22474.43	22427.04	21299.24	21217.67	20734.40	20661.61	16920.18	17118.25	115.56
	Pr(III): guanosine: Zn(II)	22464.16	22408.94	21278.40	21205.61	20725.08	20648.59	16908.70	17112.34	118.36

It may also be expressed as

$$n = \frac{H_L^2}{(H_M - H_L)^2}(S \times R)^2 N \tag{15}$$

where,

N is the co-ordination number, H_M and H_L are coulomb integrals of atomic orbital; S the overlap integral; R the radius of the orbit. For compounds with ligand coordinated through identical donor atoms, the term of Eq. (15) is a constant and Eq. (15) becomes

$$n = \text{constant } (S \times R)^2 N \tag{16}$$

The above Eq. (16) represents the nephelauxtic effect as a function of two variables, S × R and N which vary with changes in lanthanides-ligand distance in opposite directions. However, any variation in the value of R leads to a larger change in $(S \times R)^2$ compared to that in N. As a result, the Nephelauxetic effect increases when the coordination number decreases. The Ln-Q distance shortens in spite of the addition nature of β and decrease in the number of the co-ordinating ligand.

The variation in the value of E^K ($K = 2,4,6$); corresponds to that in the value of F^K, since they are inter-related. Misra; Joshi, 1979 and Misra and Singh, 1981, observed a general decrease in the values of F_K, E^K and ξ_{4f} parameters as compared to the corresponding parameters of the free ion. The $^3H_4 \rightarrow {}^3P_0$, $^3H_4 \rightarrow {}^3P_1$, $^3H_4 \rightarrow {}^3P_2$, $^3H_4 \rightarrow {}^1D_2$ transitions of Pr(III) do not obey selection rules but they have found to show substantial sensitivity reflected through wide variation of oscillator strengths and energies with even minor changes in the immediate coordination environment around it even in the presence of structurally related ligand (Misra, Joseph, 1992 and Misra, 1990) and they are termed as ligand mediated pseudohypersensitive or pseudohypersensitive transitions. Due to extremely fast water exchange rate and very small crystal field stabilization energy the conversion from one geometry to another is very convenient and facile. Karraker (1967) showed that the shape, energy and oscillator strength of hypersensitive or pseudohypersensitive transitions can be correlated with coordination number and diagnostic of immediate coordination environment around lanthanides ions. The comparative absorption spectra of Pr(III) ion, Pr(III): guanosine (in 1 : 1 molar ratio) and Pr(III): guanosine: Zn(II) (in 1 : 1 : 1 molar ratio) in different aqueous and aquated organic solvents (50 per cent) are given in Figures 38.2(a–d). The addition of guanosine in Pr(III) resulted in a marginal red shift of 4f-4f transition bands and the further addition of Zn(II) to Pr(II): guanosine gave further shift towards the longer wavelength. This shows the stimulated effect of Zn(II) towards the complexation of Pr(III) with guanosine. Table 38.1 gives the computed values of Slat or Condon (F_K), Racah (E^K) and Lande (ξ_{4f}), Nephelauxetic parameter (β), bonding parameter ($b^{½}$) and covalency (β) of Pr(III), Pr(III): guanosine and Pr(III): guanosine: Zn(II) in aqueous and aquated organic solvents. Table 38.2 gives the computed and observed values of energies for the various transition bands and root mean square deviation (RMS) showing the correctness of the various energy parameters. One can clearly see that the variation of solvent nature and composition has significant effect on the energy parameters of the different 4f-4f transition bands. Further work on the evaluation of intensity parameters is going on.

Acknowledgements

Authors are thankful to the Department of Chemistry, Manipur University, Canchipur, India for giving facilities for doing research work.

References

Birnbaum, E.R., Gomez, J.E., Abou, F. and Darnall, D.W., 1976. *Biochemistry*, 15: 5017.

Brittain, H.B., Richarsar, F.S. and Martin, R.B., 1978. *Am. J. Chem. Soc.*, 98: 8255.

Bukeitynska, K. and Ngoethey, P., 1987. *Inorg. Chern. Acta*, 131: 215.

Cefola, M., Tompa, A.S., Celiano, A.Y. and Gentile, P.S., 1962. *Inorg. Chem.*, 1: 290.

Condon, S.U. and Shortly, G.H., 1963. *The Theory of Atomic Spectra.* Cambridge University Press.

David Singh, Th., Mohondas Singh, N., Yaiphaba, N., Debecca Devi, H., Sumitra, Ch., Kriyananda Singh, T., Indira Devi, M. and Rajrnuhon Singh, N., 2003. *Chem. Environ. Res.*, 12: 3–4.

Debecca Devi, H., David Singh, Th., Yaiphaba, N., Surnitra, Ch., Indira Devi, M. and Rajmuhon Singh, N., 2004. *Asian J. Chem.*, 16: 412–428.

Devlin, M.T., Stephens, E.M. and Richardson, F.S., 1988. *Inorg. Chem.*, 27: 1517.

Einsphar, H. and Bugg, C.B., 1984. Crystal structure studies of calcium complexes and implication for biological system. In: *Metal Ions in Biological System*, (Ed.) H. Singel. Marcel Dekker, New York.

Evan, C.H., 1990. *Biochemistry of Lanthanides*. Plenum Press, New York, Chapters 2: 4.

Freasire, B.F., Oberg, A.O. and Wendlandt, W.W., 1956. *J. Phys. Chem.*, 62: 700.

Henrie, D.E., Fellow, R.L. and Choppin, G.R., 1976. *Co-ord. Chem. Rev.*, 18: 429.

Holz, R.C. and Thompson L.C., 1988. *Inorg. Chem.*, 27: 464.

Jorgensen, C.K. and Judd, B.R., 1964. *Mol. Phys.*, 8: 281.

Jorgensen, C.K. and Ryan, L., 1966. *J. Phys. Chem.*, 70: 2845.

Karrakar, D.G., 1967. *Inorg. Chem.*, 6: 1863.

Langhlin, R.D. and Conway, J.G., 1963. *J. Phys. Chem.*, 38: 1037.

Martin, R.B., 1983. Structural chemistry of calcium: lanthanides as probes, in calcium. In: *Bioloby*, (Ed.) T.O. Spiro. Marcel Dekker, New York.

Mehta, P.J., Prashant Bhatt, N. and Misra, S.N., 2001. *Bull. Pure Appl. Sci.*, 20: 2–93.

Misra, S.N. and I. Devi, M., 1997. *Spectro Chem. Acta. Part A Mol. Biomolecular Spectrosc.*, 53: 1941–1946.

Misra, S.N., 1985. *J. Sci. Ind. Res.*, 44: 366.

Misra S.N., 1990. *J. Biochem. Biophys.*, 27: 285.

Misra, S.N., I. Devi, M. and Surveer Kumar, C.M., 1977. *Indian, J. Pure Appl. Phys.*, 35: 17–22.

Misra, S.N., Joseph, Anjaiah, O.K. and Abdi, S.H.R., 1992. *Indian J. Biochem. Biophys.*, 29: 7.

Misra, S.N., Joshi, O.K. and Vaaishnav, P.P., 1979. *Indian J. Pure. Appl. Phys.*, 16: 553.

Misra, S.N., Ramchandriah, Gagnani, G., Shukla, M.A.R.S. and I. Devi, M., 2003. *Appl. Spectrosc. Rev.*, 38: 433–493.

Misra, S.N., Singh, M. and Joshi, O.K., 1981. *J. Inorg. Nucl. Chem.*, 43: 207.

Moeller, T. and Hessen, T.M., 1963. *J. Inorg, Nucl. Chem.*, 24: 1635.

Peopock, R.D., 1964 *Chem. Phys.*, 8: 281.

Pinkerton, A.A. and Earl, W.L., 1979. *J. Chem. Soc., Dalton*, p. 1347.

Shah, K.J. and Shah, M.K., 2001. *Bull. Pure Appl. Sci.*, C20: 81.

Sinha, S.P., 1966. *Spectrochim. Acta*, 22: 57.

Sinha, S.P. and Schmidtke, H.H., 1965. *Mol. Phys.*, 38: 2190.

Sinha, S.P., Mehata, P.C. and Surana, S.S.L., 1972. *Mol. Phys.*, 23: 807.

Wong, E.Y., 1961. *J. Chem. Phys.*, 35: 544.

Wong, E.Y., 1963. *J. Chem. Phys.*, 38: 976.

Yatsimirskii, K.B. and Davidenko, N.K., 1970. *Theoret. Exp. Chem.*, 6: 620.

Yatsimirskii, K.B., Davidenko, N.K., Doki, Akad and Nauk, 1966. *SSSR*, 191: 120.

Chapter 39

Mangroves: An Ecosystem in Peril

A. Saleem Khan and M. Sheik Ali
Department of Biotechnology, The New College, Chennai – 600 014, Tamil Nadu, India

ABSTRACT

Mangroves are salt tolerant plants that grow in the tropical and subtropical latitude. They thrive in salty environment and act as physical barriers in the coast, reduce soil erosion, enhance fishery production and add natural beauty to the sea coast. Oceanic Mangroves are highly productive ecosystems and enhance the food chain of coastal regions. This valuable flora contributes to economical, ecological, scientific and cultural values. But, the mangrove forests are destroyed without knowing their importance and converted to a most threatened habitat in the world. The main aim of this article is to create awareness among people about mangroves in the aftermath of Tsunami and to highlight its possible role in disaster management.

Keywords: Biomes, Mangrove, Tsunami.

Introduction

The natures's gift which stood bravely against tsunami other than the humanitarian services is the mangrove of the coastal regions. Mangroves are the forests of coastal wetlands existing in the intertidal zones of sheltered shores, estuaries, tidal creeks, backwaters, lagoons, marshes and mud-flats of the tropical and subtropical regions of the world. They form an important ecological asset and economic resources of the coastal environment. The mangroves are the most valuable biomes (Odum, 1971), which can effectively fertilize the sea and potentially protect the coastal zones and often support the development of microbial mats (Stal, 2001) and vitally serve as the breeding and feeding grounds of fishes. The word "MANGROVES" is used to refer to the plants and also to the forest community. It is believed that the term mangrove may have also been derived from a combination of the Malay word, "Manggi-Manggi". Mangrove forests are sometimes called as "Tidal forests", "Oceanic rain forest" and "Coastal woodlands" (Huang and Shedakar, 1984). Mangrove exists under very hostile and inhospitable conditions. The plants which grow there have to encounter higher salinity, tidal extremes, wind velocity, high temperature and muddy anaerobic soils. The plants have peculiar adaptations such as support roots, viviparous germination, salt-excreting leaves, breathing roots, knee roots, etc., by which the plants are well-adapted to water logged, anaerobic saline soils of coastal environments. Also the mangrove plants have great potential to adopt to the change in climate and the rise in sea level.

Distribution

Mangroves grow luxuriantly in alluvial soil substrate, which are fine textured, loose mud or silt, rich in humus and sulphides, and they develop in low lying and broad coastal plains where the topographic gradients are very small and the tidal amplitude is large. Their distribution is limited by temperature and they prefer moist atmosphere and freshwater inflow, which brings in abundant nutrients and silt from terrestial sources. Some of the important species are *Avicenia marina, A. germinans, A. bguiera, A. officinallis, Sonneratia apetala, S. caseolaris.* Mangroves are distributed circumtropically, occurring in 112 countries and territories. Total mangrove coverage is 18 million hectares and it is just about 0.45 per cent of world forests and woodlands of the total mangrove coverage, 42.4 per cent exist in south and southeast Asia and about 23 per cent of the forest occurs in Indonesia. It has been estimated that 4,69,000 acres of mangrove forests contribute to the overall health of the southern Florida in United States of America. The Sunderbans (West Bengal) in India and Bangladesh put together form the largest single block of mangroves in the world. Indian mangroves are distributed in about 6,740 sq.km which consist 7 per cent of the total Indian coastline. There are different types of mangroves in India *viz.*, deltaic, backwater estuarine type of mangroves existing in the west coast (Arabian), characterized by typical funnel-shaped estuaries of major rivers or backwaters, creeks and neritic inlets and lagoons which support a rich mangrove flora.

Ecological Importance of Mangroves

The relationship between mangroves and their associated marine life cannot be overemphasized. Mangroves provide protected nursery areas for fishes, crustacean and shelfish. They provide food for a multitude of marine species like snook, snail, crabs, snapper, tarpan, jack, sheepshead, reddrum, oyster and shrimp. Many animals find shelter either in the roots or branches of mangroves. Mangrove branches are rookeries or resting areas for beautiful coastal birds such as brown pelicans and roseate spoonbills. The forest is a natural habitat for the proboscis monkeys which feed on the leaves of mangrove trees. The floor of the mangrove forest is a habitat for extensive numbers of invertebrates. Each square meter of sediment can hold up to a hundreds of them. The most common are crabs and molluscs. Three commercially important crustaceans, king prawns, blue swimmer crabs and mudcrabs, live in mangroves. The abundant plant material produced by mangroves is a source of food for estuarine animals. However, most animals do not graze directly on the mangroves, instead they feed on the detritus (decomposting plant material). A single square kilometre of mangrove forest contributes about 600 tonnes of plant material each year to the estuarine food chain. The important nutrient content of mangroves are Nitrogen, Phosphorus, Potassium, Calcium, Sodium etc., and it greatly supports Biogeochemical cycle of mangrove ecosystem (Akasornkoae and Khemnark, 1994). Mangroves grow between the estuary and the surrounding land and act as visual screen along developed shorelines.

But they also buffer the waterway from the direct effect of run off waters and helps in maintaining estuarine water quality and reduce siltation in the waterways.

Values of Mangroves

Mangroves provide essential goods and ecological services for the livelihood of coastal people. The ecological services of mangroves include mainly protection and stabilization of coasts, enrichment of coastal waters and supporting of coastal fisheries. The mangrove forests are among the most productive ecosystem on this planet. Their daily rate or organic matter production is about 70 times more than that of tropical oceanic waters. Because of this high rate of primary production and diversified infloresence of vegetation and it acts as a great natural stabilizer of the coastal belt by dissipating the effect of cyclones, tidal intrusions, winds, storms etc. The submarine earthquake with the epicenter off Sumatra in Indonesia was the cause of the recent Tsunami which with a speed of 800 km per hour wrought havoc on 26th December, 2004 in many Asian countries. Among these, Indonesia, Sri Lanka and India were the worst affected. However, here, the extent of loss was minimal where mangroves are presently acting as a physical barrier mangroves prevented loss of life and property. So it is necessary to develop mangrove plants wherever possible along the coast. By virtue of boat building as well. Tannin is extracted from the bark of mangroves belonging to the family Rhizophoraceae which have a complex root system, the mangroves also prevent coastal erosion. A study conducted on sea defences along the coast of Vietnam showed that mangroves as "live seawalls", are more cost effective. Mangrove wood has a high content of tannin and therefore it is used as timber because of its durability. Mangrove wood is used for tanning leather, fishing nets and fish traps. The current utilization of mangrove soil includes shrimp farming, salt production, coconut palm cultivation, Nipa palm gardening, paddy growing and mangrove plantation (Changprai, 1984).

Degradation and Destruction of Mangroves: The Greatest Threats

Degradation of mangroves is caused by nature induced changes. Tropical storms and tsunami are common in the Bay of Bengal and the Caribbean sea respectively. Diseases also cause devastating damages to mangroves. To cite an example, top dying disease has damaged about 45 million *Heritierafomes* (sundari) trees (Rajendran, 2004). Biological pests and parasites also have serious impacts on mangroves. Significant damage is caused by grazing of animals in dry coastal areas of Asia and the Middle East. Young plants are damaged by barnacles and leaf eating crabs of the fruits of Rhizophora and these inhibit seed germination. Naturally, mangrove forests have withstood severe storms and changing tides for many millennia, but they are now being devastated by modern encroachments. Today, mangrove forests are among the most threatened habitats in the world, disappearing at an accelerating rate, yet with little public notice. With continuing degradation and destruction of mangroves, there is a critical need to understand them better (Kathiresan, 2002). Over time, environmental stress can kill large numbers of mangrove trees. In addition, the charcoal and timber industries have also severely impacted mangrove forests, as well as tourism and other coastal development. In sufficient mangrove recovery or replanting following clear cutting for wood products, watershed influences such as chemical pollution by industry, agriculture and freshwater diversion either in to or away from mangroves (Fast and Menasvete, 2003). The rapidly expanding aquaculture industry poses the greatest threat to the world's remaining mangroves. This highly volatile enterprise has grown exponentially over the last 15 years, leaving devasting ruin in its wake. Until recently, mangrove forests have been classified by many governments and industries alike as "wastelands" or useless swamps. This erroneous designation has made it easier to exploit mangrove forests as cheap and unprotected sources of land and water for shrimp farming. The amount of mangrove forest destruction is alarming.

Conservation and Management

Awareness to the local people has to be created especially on the importance of mangroves and their management. Very importantly opinion from local communities is essential before implementing the projects related to the mangrove ecosystem. Sufficient amount of upstream water should be allowed so as to reduce the high salt content in the soil. The large element pools stored in dead roots may serve as a conservation mechanism (Alongi *et al.*, 2004). Operation of mechanized boats in the forest area should not be encouraged should not encouraged and illegal cutting of forest trees for firewood and house constructions should be prevented. Strict enforcement in banning the operation of small mesh size net in the forest area especially during the developmental periods of juveniles. In association with forest department mangrove seedlings should be properly used (Achari *et al.*, 2002). Construction of aqua farms inside the mangrove forest as the soil is not favorable for farming should be avoided. Morever, discharge of toxic materials should not be allowed in the forest areas. The importance and the values of mangroves should be understood keenly and save the human being from natural disaster like Tsunami. Realizing the importance of mangroves, government should take initiative to conserve and plant mangroves. Very importantly people must be involved during the mangrove tree plantation to save the Mankind.

References

Achari, *et al.*, 2002. Biodiversity, ecology and conservation of mangroves. In: *Proceedings of National Seminar on Creeks, Estuaries, Mangroves and Conservation*, pp. 72–77.

Aksornkoae, S., Khemmark, C., 1984. Nutrient cycling in mangrove forest of Thailand. In: *Proceedings of the Asian Symposium on Mangrove Environment Research and Management*, pp. 545–557.

Alongi, D.M. *et al.*, 2004. Nutrient capital in different aged forests of mangrove. *Botanical-Marina*, 47(2): 116–124.

Changprai, C., 1984. Mangrove soil and their utilization in Thailand. In: *Proceedings of the Asian Symposium on Mangrove Environment Research and Management*, pp. 521–533.

Fast, A.W., Menasveta, P., 2003. *World Aquaculture*, 34(3): 6–9.

Huang, J.H., Shedakar, S., 1984. In: *Marine Geology and Oceanography of Arabian Sea and Coastal Pakistan.* Van Norsland Reinhold Co., New York, pp. 255.

Kathiresan, K., 2002. Why mangroves are degrading. *Current Science*, 83(10): 1246–124.

Odum, E.P., 1971. *Fundamentals of Ecology.* W.B. Sanders Company and Toppan Company, pp. 574.

Rajendran, T., 2004. *Mangroves.* Environmental Information System (ENVIS), pp. 6–16.

Stal, L.J., 2001. Coastal microbial mats. *South African Journal of Botany*, 67(3): 399–410.

Chapter 40

Plant Regeneration and *In vitro* Flowering of *Solanum trilobatum* L.: An Important Medicinal Plant

S.M. Nagarajan[1], *T.M. Sathees Kannan*[1], *S. Rajasekaran*[1] *and S. Kulothungan*[2]

[1]*Plant Tissue Culture Laboratory, Department of Botany, A.V.C. College (Autonomous), Mannampandal – 609 305, Mayiladuthurai, Tamil Nadu*

[2]*Department of Botany, AVVM Sri Pushpam College (Autonomous), Poondi – 613 503, Thanjavur, Tamil Nadu*

ABSTRACT

Direct organogenesis and *in vitro* flowering was obtained in the medicinal plant *Solanum trilobatum* L. High frequency and maximum number of multiple shoots were obtained from shoot tip explants on MS-B5, medium supplemented with BAP (2.0 mg/l). Regenerates when transferred to rooting medium (with IAA or IBA), they initiated flowering and rooting. Rooted plantlets were hardened and transferred to plastic pots with sterilized soil and vermiculate (1 : 1 ratio) and subsequently transferred to green house with 90 per cent survivability.

Keywords: In vitro flowering, Multiple shoots, Phytohormones, Rooting, Solanum trilobatum.

Introduction

Solanum trilobatum L. (family Solanaceae) is an important medicinal plant widely used in cough, chronic bronchitis. The leaves of this plant are bitter and astringent. The leaves are roasted along with onion and gingely oil and used to cure breathing problem (Shanmugam, 1989). The leaves contain rich amount of calcium, iron, phosphorous, protein, fat, crude fibres, minerals and other carbohydrates (Subramanian and Madhavan, 1983). β-Solamarine and solasodine have been found in *Solanum trilobatum* leaves.

Flowering, which ensures sexual method of reproduction involves many steps, usually starting with differentiation of three-dimensional structure the flower primordia. *In vitro* flowering may help in obviating the intricate interactions present in the whole plants and thus throwing light on nature of factors influencing flowering. Initiation of *in vitro* flowering may become a valuable research tool in *Solanum trilobatum* L. due to its ploidy nature (Anonymous, 1972). Hence, it becomes imperative to establish a suitable protocol for *in vitro* propagation and also peculiar occurrence of precocious flowering in vitro could be of great interest in study of developmental process and improvement of ploidy level in *S. trilobatum.* The present paper reports a highly reproducible and efficient method of *in vitro* propagation and *in vitro* flowering in S. *trilobatum.*

Materials and Methods

Shoot tips were excised from *in vitro* cultured 17 days old seedlings, used as explants cultured on MS–B5 (MS medium + B5 vitamins) medium (Murashige and Skoog, 1962; Gamborg *et al.*, 1968) fortified with different concentrations of auxins (IAA and NAA) and cytokinins (BAP and Kn) (Figure 40.1a). The pH of the medium was adjusted to 5.8 before autoclaving at 1.4 kg cm^2 for 20 minutes. For root initiation and flowering, the *in vitro* raised shoots were excised and cultured on MS Medium with various concentrations of NAA, IAA and IBA separately. Cultures were incubated in light-dark (16–8h) conditions of cool fluorescent light of *Ca.* 3000 lux at 25±2°C. For each experiment a minimum of 20 replicates were used and each experiment was repeated thrice. The number of shoots, roots and flowers were counted and analysed statistically. Rooted plantlets were hardened and transferred to plastic cups containing sterilized soil and vermiculate in the ratio of 1 : 1 and kept in green house for their survivability.

Results and Discussion

Multiple shoots were induced from shoot tip explants on medium supplemented with different concentrations of BAP (1.0–3.0 mg/l). Frequency and number of multiple shoots was more on medium with 2.0 mg/1 of BAP (Figures 40.1b,c) when compared to other concentrations of BAP and Kon tested (Table 40.1). Induction of shoots using BAP has been well documented (Bhat *et al.*, 1995; Bhattacharya and Bhattacharya, 2001 and Catapan *et al.*, 2000). But the NAA and BAP combinations induced callusing at the cut basal ends of explants. Of the three auxins (NAA, IAA and IBA) tested, IBA at 1.0 mg/l induced maximum number of roots (Figure 40.1d; Table 40.2) than the other two. Similar results of IBA inducing roots were previously reported (Manickam *et al.*, 2000 and Pattnaik and Chand, 1996).

In vitro flowering was observed within 40 days of cultures, when the shoots were transferred to rooting medium. Flowering was considered to be a complex process regulated by both internal and external factors and its induction under *in vitro* culture is extensively rare (Stephen and Jayabalan, 1998). However *in vitro* flowering has been reported in several species (Ignacimuthu *et al.*, 1997; Patil *et al.*, 1993 and Vandana *et al.*, 1995). For *in vitro* flowering the response of IBA is better than IAA (Table 40.3). High frequency and maximum number of flowers was more on MS medium supplemented with 0.5 mg/l of IBA (Figure 40.1e) compared to other concentrations of IBA and IAA tested. Thus auxins supports *in vitro* flowering. This is in accordance with the recent report on *Solanum nigrum* L. (Jabeen *et al.*, 2005). Cytokinins did not support *in vitro* flowering either singly or in combination with auxin.

Figure 40.1

(a) Inoculated shoot tip explant; (b) Multiple shoot induction on medium containing 2.0 mg/l BAP; (c) Multiple shoots and elongation (after 4 weeks of culture); (d) Rooting of *in vitro* raised shoots; (e) *In vitro* flowering.

The flowers obtained during the present work shows peduncles short, strong recurved prickles, corolla blue, lobes triangular and lanceolate.

Table 40.1: Morphogenetic Responses of Shoot Explants of *Solanum trilobatum* L.

Growth Regulators (mg/l)		*Shoot Sprouting Response (%)*	*Shoot No./Explant*	*Shoot Length (cm)*
BAP	0.5	85	3.2±0.4	3.0±0.5
	1.0	86	7.5±0.8	2.6±0.5
	1.5	92	18.4±2.6	2.3±1.2
	2.0	95	28.6±3.5	2.0±0.0
	2.5	85	8.5±1.8	1.7±0.3
	3.0	–	–	–
Kn	0.5	72	2.1±1.2	2.3±0.8
	1.0	76	4.6±0.5	2.5±0.3
	1.5	85	8.3±2.8	2.1±0.5
	2.0	87	14.0±1.8	1.6±0.3
	2.5	70	6.5±1.5	1.1±0.5
	3.0	–	–	–

Table 40.2: Effect of Various Concentrations of IBA on Root Induction of *In vitro* Raised Shoot

Growth Regulator (mg/l)	*No. of Roots/Shoots*	*Root Length (cm)*
0.5	3 2±0.2	2.7±1.2
1.0	5 0±0.8	5.5±0.5
1.5	2±0.3	1.7±0.5
2.0	2±0.3	1.3±0.7

Table 40.3: Effect of Auxins on *In vitro* Flowering of *Solanum trilobatum* L.

Growth Regulator (mg/l)		*% of Flowering Shoots*	*No. of Flowers/ Plantlets*	*Days Taken for Opening of Flowers*
IAA	0.5	0	–	–
	1.0	0	–	–
	2.0	80	1.0±0.7	52±0.2
	3.0	80	1.0±0.5	33±1.3
	4.0	–	–	–
IBA	0.5	90	3.0±0.16	40±0.2
	1.0	90	2.0±0.12	45±0.3
	2.0	80	1.0±0.08	52±0.1
	3.0	–	–	–
	4.0	–	–	–

Although in many species flowering is strictly controlled by some environmental conditions or stimulated by growth substances, this developmental step is also associated phase change to maturity. The interaction of carbohydrates and other nutritional factors with endogenous growth regulators can influence some biological parameters which are altered when plant changes from juvenile to mature phase (Zimmerman *et al.*, 1985). This protocol can be extended to plant breeding studies and for the purpose of quick flowering and fruit formation under *in vitro* conditions. In addition, though *in vitro* seed production is very low, its genetic potential is of high value.

References

Anonymous, 1972. *The Wealth of India, Raw Materials*. CSIR, New Delhi, 9: 395–396.

Bhat, S.R., Chandel, K.P.S. and Malik, 1995. Plant regeneration from various explants of cultivated piper species. *Plant Cell Rep.*, 14: 395–402.

Bhattacharya, R. and Bhattacharya, S., 2001. High frequency *in vitro* propagation of *Phyllanthus amarus* Schum and Thann by shoot tip culture. *Indian J. Exp. Biol.*, 39: 1184–1187.

Catapan, E., Otuki, M.F. and Viana, A.M., 2000. *In vitro* culture of *Phyllanthus caroliniensis. Plant Cell Tiss. Org. Cult.*, 62: 195–202.

Gamborg, O.L., Miller, R.A and Ojima, K., 1968. Nutrient experiments of suspension culture of soybean root cells. *Exp. Cell Res.*, 80: 150–158.

Ignacimuthu, S., Franklin, G. and Melchias, G., 1997. Multiple shoot formation and *in vitro* fruiting from cotyledonary nodes of *Vigna mungo* (L.) Hepper. *Curr. Sci.*, 73(9): 733–735.

Jabeen, F.T.Z., Venugopal, R.B., Kiran, G., Kaviraj, C.P. and Rao, S., 2005. Plant regeneration and *in vitro* flowering from leaf and nodal explants of *Solanum nigrum* L.: An important medicinal plant. *Plant Cell Biotech. and Mol. Biol.*, 6(1&2): 17–22.

Manickam, V.S., Elango, R. and Antonisamy, R., 2000. Regeneration of Indian ginseng plantlets from stem callus. *Plant Cell Tiss. Org. Cult.*, 62: 181–185.

Murashige, T. and Skoog, F., 1962. A revised medium for rapid growth and bioassays with tobacco tissue culture. *Physiol. Plant.*, 15: 473–497.

Patil, M.S., Ramaswamy, N.M. and Sree Rangasamy, S.R., 1993. *In vitro* flowering in sunflower (*H. annus*). *Curr. Sci.*, 65: 565–566.

Pattnaik, S.K. and Chand, P.K., 1996. *In vitro* propagation of the medicinal herbs *Ocimum americanum* (L.) Syn. (*Ocimum canum* Sim. (Loary Basil) and *Ocimum sanctum* L. (Holy basil). *Plant Cell Rep.*, 15: 846–850.

Stephen, R. and Jayabalan, N., 1998. *In vitro* flowering and seed setting formation of coriander (*Coriandrum sativum* L.). *Curr. Sci.*, 74(3): 195–197.

Shanmugam, N.K., 1989. *Dictionary of Medicinal Plants*. Nakkeran Publ. House, Chennai, pp. 452–453.

Subramanian, S.V. and Madhavan, V.R., 1983. *Heritage of the Tamil Siddha Medicine*. International Institute of Tamil Studies, Chennai.

Vandana A. Kumar, Atul Kumar and Jitendra Kumar, 1995. *In vitro* flowering and pod formation in cauliflower (*Brassica oleracea* var. Botrytis). *Curr. Sci.*, 69: 543–545.

Zimmerman, R.H., Hackett, W.P. and Paris, R.P., 1985. Harmonal aspects of phase change and precocious flowering. In: *Encyclopaedia of Plant Physiology*, New Series, (Eds.) Paris, R.P. and D.M. Reid. Springer-Berlin, pp. 79–115.

Chapter 41

Postharvest Soil Fertility Status and Nitrogen Use Efficiency of Hybrid Rice as Influenced by Nitrogen Management

S. Ravi[1], S. Ramesh[1], S.D. Sundar Singh[2] and B. Chandrasekaran[3]

[1]Department of Agronomy, [2]Registrar, [3]Director of Research, Tamil Nadu Agricultural University, Coimbatore – 641 003, T.N., India

ABSTRACT

Field experiments were conducted to evaluate the effect of different hybrids and nitrogen management approaches on the postharvest soil fertility status and the various nitrogen use efficiency parameters during *Kharif* season of 2000 and 2001 respectively in the wetland research farm of Tamil Nadu Agricultural University, Coimbatore. Treatments consisted of two hybrids in the main plot and seven levels of N management practices during *Kharif* 2000 and eight sub-plot treatments during *Kharif* 2001. The results revealed that postharvest soil organic carbon content and nitrogen were significantly higher with green manure @ 6.25 t ha^{-1} combined with LCC critical value 5 based N application whereas the available P were higher in 20 kg N ha^{-1} as basal along with LCC critical value 3 based N application. However, it was on par with application of green manure @ 6.25 t ha^{-1} combined with LCC critical value 3 based N applications in both the years. Regarding available K status of the soil, no significant difference was observed between the treatments in both the seasons. Higher agronomic efficiency and physiological efficiency were achieved with green manure @ 6.25 t ha^{-1} combined with LCC critical value 5 based N application (N_4) in both the hybrids in both the years. The partial factor productivity showed that treatment combination of green manure@ 6.25 t ha^{-1} plus LCC critical value 4 based N applications (N_3) gave higher partial factor productivity and it was closely followed by N_4 treatment in both the years of investigation.

***Keywords**: Hybrid rice, LCC, Nitrogen use efficiency, Soil fertility.*

Introduction

At present, there are about 2700 million people in the world, half of the total population of the globe, whose major staple food is rice. The population of the world will increase by about 2000 million during the next two decades and over half of these will live in Asia and be dependent on rice (Greenland, 1997). By 2015 it is necessary to increase rice production by about 50 per cent to maintain present nutritional standards. To increase the production and productivity, fertilizer nutrients especially nitrogen has received prime attention by rice researchers. Fertilizer N use efficiency is relatively low in lowland rice systems (De Datta and Buresh, 1989) and N use efficiency is less than 40 per cent in farmers field (Craswelt and Vlek, 1979; Cassman *et al.*, 1993). The increase in grain yield through N fertilization is a function of both N absorption and the efficiency with which it is utilized for grain production. Hence, maximizing the N absorption during critical phases of plant growth was essential for, increased grain yield by the rice crop (Yoshida, 1981). Appropriate N management has to be adopted for maximizing the rice production with high production efficiency for applied N (Dhyani and Mishra, 1994). Green manuring is an inexpensive, eco-friendly alternative to mounting prices of fertilizer nitrogen and has become an effective technology in economizing the agricultural production system, ensuring productive capacity of soil without causing any environmental problem (Bana and Pant, 2000). Organic carbon, total N, P, K and water holding capacity increased with green manure application (Chaphale *et al.*, 2000). Improvement in N management involves alternative water management, modified urea sources, alternative N timing and rates and integration of organic and inorganic N fertilizers. In Japan measuring the leaf green colour was widely used technique as the index of nitrogen status of the crop (Takebe and Yoneyama, 1989). To determine the relative accuracy of LCC in assessing the leaf N status, it should be compared and correlated with chlorophyll-meter reading (IRRI, 1996) and calibrated properly with cultivar groups. NUE is inferred through the parameters like partial factor productivity, agronomic efficiency and physiological efficiency. In this context, the present study was undertaken to investigate the effect of different N management strategies involving Leaf colour chart (LCC) based N application and integration of green manure on the N use efficiency and postharvest soil fertility status of rice hybrids.

Materials and Methods

Field experiments were conducted during *Kharif* (June-September) 2000 and 2001 in wetland research farm of Tamil Nadu Agricultural University, Coimbatore to study the effect of green manuring and fertilizer application through LCC grades on the postharvest soil fertility status and nitrogen use efficiency of rice hybrids. The soil of the experimental field was moderately drained, deep clay loam, classified taxonomically as typic haplustalf. The soils were low in available soil N, medium in available soil P and high in available soil K respectively. The detailed physico-chemical characteristics of the soil are given in Table 41.1. The experiment was laid out in split plot design with three replications. The main plot treatments consisted of two hybrids PA6201 and CoRH2 and the sub-plots with seven levels of N management practices during *Kharif* 2000 whereas during *Kharif* 2001 one more sub-plot treatment was included which is the existing recommended practice under integrated nitrogen management (INM) approach (T_8). The rice hybrid PA6201 was released in 2000 by Central Release Committee (CVRC). It is a medium duration (125–130 days) semi dwarf, erect type having long panicle with long slender grain. The rice hybrid CoRH2 released in 1998 by Paddy Breeding Station, Tamil Nadu Agricultural University, Coimbatore is also a medium duration semi-dwarf type having compact panicle with medium grain type.

Table 41.1: Physico-chemical Characteristics of the Experimental Field

Sl.No.	*Particulars*	*Kharif 2000*	*Kharif 2001*
1.	Mechanical analysis		
	Clay (per cent)	39.1	44.2
	Silt (per cent)	18.7	19.0
	Fine sand (per cent)	24.4	18.1
	Coarse sand (per cent)	16.9	17.2
	Textural class	Clay loam	Clay loam
2.	Chemical analysis		
	Organic carbon (per cent)	0.73	0.64
	Available Nitrogen (kg ha^{-1})	271	244
	Available Phosphorus (kg ha^{-1})	16.3	17.2
	Available Potassium (kg ha^{-1})	532	505

The details of the treatments and notations used are as under:

A. Main plot: Hybrid Rice

H_1: PA6201
H_2: CORH$_2$

B. Sub-Plot: Nitrogen management

N_1: Basal + 3 Splits of top dressing of recommended dose
N_2: Green manure at 6.25 t ha^{-1} + N as per LCC-3
N_3: Green manure at 6.25 t ha^{-1} + N as per LCC-4
N_4: Green manure at 6.25 t ha^{-1} + N as per LCC-5
N_5: 20 kg N ha^{-1} as basal + N as per LCC-3
N_6: 20 kg N ha^{-1} as basal + N as per LCC-4
N_7: 20 kg N ha^{-1} as basal + N as per LCC-5
N_8: Green manure 6.25 t ha^{-1} + 3 Splits top dressing of recommended dose

An absolute control was maintained separately in the experimental field to find out the efficiency of soil applied nitrogen. The above ground biomass of *Sesbania aculeata* at the age of 40 days was harvested, weighed and applied to the respective plots as per the treatment schedule. The green manure was spread uniformly and incorporated one week before transplanting. The fertilizer nitrogen was applied to the rice crop as per the treatments. In LCC based N management treatments, the LCC values were recorded as per the standard procedure (IRRI, 1996) at weekly intervals starting from 14 DAT to flowering. Leaf colour chart consists of six strips from light yellowish green (No. 1) to dark green (No. 6). The rice leaf colour was compared with standard LCC under the same environmental condition. The colour of the single leaf was measured by holding the leaf colour chart vertically and placing the middle part of leaf, 1 cm in front of colour strip for comparison (IRRI, 1996). Ten readings were taken for each plot and average was computed by rounding off to the nearest 0.5 to determine the need for N top dressing. Whenever the LCC values fall below the fixed critical level the amount of N to be applied at different growth stages was as per the schedule given below (IRRI, 1996).

Crop Growth Stages	*N to be Applied*
Early growth stage (14–21 DAT)	20 kg N ha^{-1}
Rapid growth stage (28–42 DAT)	30 kg N ha^{-1}
Late growth stage (45 DAT–Flowering)	20 kg N ha^{-1}

The total quantity of N applied as per the treatments are furnished in Tables 41.2 and 41.3. All the treatments including control received a uniform dose of 50 kg P_2O_5 and 50 kg K_2O ha^{-1}. The entire dose of P as Single Super phosphate (16 per cent P_2O_5) was applied as basal before transplanting. Potassium in the form of Muriate of Potash (60 per cent K_2O) was applied in three equal splits *viz.*, 50 per cent as basal and 25 per cent each at active tillering and panicle initiation stages. Zinc sulphate @ 25 kg ha^{-1} was applied to the crop during both the seasons. Postharvest soil samples were collected plot wise to a depth of 20 cm for estimating available N, P, K and organic carbon content. The methods adopted in the analysis of the soil samples are as follows:

Particulars	*Methods Adopted*	*Author*
Organic carbon	Wet chromic acid digestion	Walkley and Black (1934)
Available N	Alkaline permanganate	Subbiah and Asija (1956)
Available P	Olsen's	Olsen *et al.* (1954)
Available K	Neutral normal ammonium acetate	Standford and English (1949)

Nitrogen Use Efficiency (NUE)

The efficiency with which the applied N is used by the crop towards grain production was evaluated through following parameters.

1. Partial factor productivity (Pfp) = Y/N_t

where,

Y is the mean grain yield (kg ha^{-1})

Nt is the total amount of N applied (kg ha^{-1} in treatment)

2. Agronomic efficiency (Ea) = Δ y/Nt

where,

Δ is the incremental mean grain yield (kg ha^{-1}) for a N applied treatment compared to control (Zero N)

Nt is the total amount of N applied (kg ha^{-1}) in that treatment

3. Physiological Efficiency (Ep) = dy/dNu

where,

dy is the incremental in grain yield (kg ha^{-1}) for a N applied treatment compared to zero N

dNu is the increase in crop N uptake (kg ha^{-1}) that results from N application

Table 41.2: Leaf Color Chart Based N–application–At Weekly Intervals–*Kharif* 2000

Time of Fertilizer Application (Days After Transplanting)		H_1N_1	H_1N_2	H_1N_3	H_1N_4	H_1N_5	H_1N_6	H_1N_7	H_2N_1	H_2N_2	H_2N_3	H_2N_4	H_2N_5	H_2N_6	H_2N_7
Basal (0)		50	–	–	–	20	20	20	50	–	–	–	20	20	20
Early growth stages (EGS)	14	–	–	20	20	–	20	20	–	–	20	20	–	20	20
	21	50	20	–	20	20	–	20	50	20	–	20	20	–	20
Rapid growth stages (RGS)	28	–	–	30	–	–	30	–	–	–	30	–	–	30	–
	35	–	30	–	30	30	–	30	–	30	–	30	30	–	30
	42	50	–	30	–	–	30	–	50	–	30	–	–	30	–
Late growth stages (LGS)	49	–	20	–	20	20	–	20	–	20	–	20	20	–	20
	58	–	–	20	20	–	20	20	–	–	20	20	–	20	20
	63	50	20	–	20	20	–	20	50	20	–	20	20	–	20
	70	–	–	20	20	–	20	20	–	–	20	20	–	20	20
Total/kg/ha		200	90	120	150	110	140	170	200	90	120	150	110	140	170

Table 41.3: Leaf Color Chart Based N–application–At Weekly Intervals–*Kharif* 2001

Time of Fertilizer Application (Days After Transplanting)		H_1N_1	H_1N_2	H_1N_3	H_1N_4	H_1N_5	H_1N_6	H_1N_7	H_1N_8	H_2N_1	H_2N_2	H_2N_3	H_2N_4	H_2N_5	H_2N_6	H_2N_7	H_2N_8
Basal (0)		50	–	–	–	20	20	20	–	50	–	–	–	20	20	20	–
Early growth stages (EGS)	14	–	–	20	20	–	20	20	–	–	–	20	20	–	20	20	–
	21	50	20	–	20	20		20	50	50	20	–	20	20	–	20	50
Rapid growth stages (RGS)	28	–	–	30	–	–	30	–	–	–	–	30	–	–	30	–	–
	35	–	30	–	30	30		30	–	–	30	–	30	30	–	30	–
	42	50	–	30’	–	–	30	–	50	50	–	30	–	–	30	–	50
Late growth stages (LGS)	49	–	20	–	20	20	–	20	–	–	20	–	20	20	–	20	–
	58	–	–	20	20	–	20	20	–	–	–	20	20	–	20	20	–
	63	50	20	–	20	20	–	20	50	50	20	–	20	–	–	20	50
	70	–	–	20	20	–	20	20	–	–	–	20	20	20	20	20	–
Total/kg/ha		200	90	120	150	110	140	170	150	200	90	120	150	110	140	170	150

Results and Discussion

Postharvest Soil Fertility Status

Soil Organic Carbon Content

There is no significant difference in the available organic carbon content of the soil between the two hybrids in both the years of experiment. As regard to N management practices application of green manure @ 6.25 t ha^{-1} combined with LCC critical value 5 based N(N_4) recorded significantly higher organic carbon content of 0.85 and 0.84 per cent in H_1 and H_2 during *Kharif* 2000 and 0.75,0.74 per cent respectively (Table 41.4). The least organic carbon content was recorded in LCC critical value 3 based N application combined with 20 kg N ha^{-1} as basal (N_5) treatment in both the hybrids during the investigation. At the end of crop harvest, there was slight improvement in organic carbon at the highest level of N addition either through fertilizer or green manure compared to initial content. This might be due to the contribution of organic carbon by *Sesbania aculeata* and residue addition by rice roots. Building up of organic carbon in soil has also been reported by Somani (1987), Swarup (1986), Sharma and Mittra (1988) and Gurung and Sherchan (1996) in rice soil ecosystem.

Table 41.4: Postharvest Soil Fertility Status (kg ha^{-1}) as Influenced by Hybrids and Nitrogen Management Strategies

Treatments	Organic Carbon %		Available N		Available P		Available K	
	Kharif 2000	Kharif 2001	Khar 2000	Kharif 2001	Kharif 2000	Kharif 2001	Kharif 2000	Kharif 2001
Hybrids								
H_1	0.81	0.69	245.32	378.21	17.37	18.28	511.86	479.75
H_2	0.80	0.69	242.27	374.68	19.36	19.81	508.43	476.37
SEd	0.002	0.025	0.64	0.92	0.08	0.07	1.78	1.49
CD (0.005)	NS	NS	2.77	3.43	0.35	0.31	7.68	6.44
'N' Management								
N_1	0.79	0.70	252.10	382.55	17.21	18.11	500.0	472.0
N_2	0.82	0.72	229.93	362.57	19.53	20.20	521.0	489.0
N_3	0.83	0.74	244.90	376.54	18.27	19.42	509.0	477.0
N_4	0.84	0.68	267.37	396.81	16.68	17.41	494.50	466.0
N_5	0.77	0.66	227.19	358.97	20.04	20.88	523.0	492.0
N_6	0.78	0.67	236.74	366.78	18.95	19.97	516.0	483.0
N_7	0.79	0.69	248.33	379.35	17.84	18.72	506.50	475.0
N_8	–	0.72	–	386.99	–	17.69	–	469.50
SEd	0.016	0.065	4.86	7.58	0.36	0.38	10.16	9.60
CD (0.05)	0.033	0.135	10.04	15.52	0.75	0.78	20.97	19.68

Available Soil N

The results revealed that hybrid rice PA6201 recorded significantly higher soil available N of 245.50 and 378.24 kg ha^{-1} in both the seasons. As regard to N management the application of green manure @ 6.25 t ha^{-1} combined with LCC critical value 5 based N applications (N_4) registered significantly higher available N status than the rest of treatments during *Kharif* 2000 (Table 41.4). But during *Kharif* 2001 the treatment N_4 was on par with application of green manure @ 6.25 t ha^{-1} combined with 150 kg N ha^{-1} (T_5). The application of 20 kg N ha^{-1} as basal with LCC critical value 3` based N applications (N_5) exhibited the lowest available N in both the seasons. Application of inorganic and organic N resulted in positive influence on soil available N. Higher N availability in green manured plot was due to the slow and steady release of N from the green manure. Build up of soil available N with green manuring is in consonance with the findings of Salu Reddy *et al.*, 1995 and Tiwari *et.al* (1995). The addition of organic N from green biomass and N fixation by rhizobial nodules and subsequent sustained release of N throughout the crop growth period and even after growing period would have contributed to mineralization of N which naturally increased the available N content in soil. Higher N levels increased the N status of the soil with higher N availability (Budhart, 1994; Thakur, 1993).

Available P and K

The available P and K of the soil as influenced by the two hybrids revealed that the PA6201 was significantly superior to CoRH2 in the available soil P status in both the seasons whereas the available level of soil K was found to be comparable with each other in first and second season of experimentation (Table 41.4). The application of 20 kg N ha^{-1} as basal with LCC critical value 3 based N application (N_5) treatment registered significantly higher available P status of the soil at postharvest stage in both the seasons. As far as available K status of the soil was examined, no significant difference was observed between the treatments in both the seasons. Decline in the soil available P and K was recorded with the increased inorganic and organic N application due to the higher amount of P and K uptake. Better growth of the crop with higher N supply might have resulted in increased uptake of P and K, thereby reducing soil available P and K content.

Nitrogen Use Efficiency (NUE)

Partial Factor Productivity (Pfp)

The efficiency of absorption of applied N is expressed as Pfp. The hybrid PA6201 recorded higher mean partial factor productivity of 48.05 compared to CoRH2 with Pfp of 43.40 kg kg^{-1} N (Table 41.5). Maximum partial factor productivity was recorded with application of 20 kg N ha^{-1} as basal combined with LCC 3 based N application (N_5). The partial factor productivity was least with application of 200 kg N ha^{-1} alone (N_1). Similar trend was observed in both the hybrids during both the years. Krishnakumar and Subramanian (1991) reported similar results.

Agronomic Efficiency (Ea)

Generally, the agronomic use efficiency decreased with graded levels of N application. With increase in N level, the rate of response or rate of change in production increased with decreasing rate. The hybrid PA6201 registered higher mean agronomic efficiency of 25.63 as compared to CoRH2 with agronomic efficiency of 24.24 kg^{-1} N during both the years (Table 41.5). Regarding N management practices, in both the hybrids it was observed that the agronomic efficiency was higher during *Kharif* 2000 with application of green manure @ 6.25 t ha^{-1} combined with LCC critical value 4 based N application (N_3) and low with 200 kg N ha^{-1} alone (N_1). This is in conformity with the works of many research workers on N management (Morries *et al.*, 1986 and Krishnakumar and Subramaniyam, 1991). During *Kharif* 2001, PA6201 recorded maximum agronomic efficiency (31.32 kg grain/kgN) with application of green manure @ 6.25 t ha^{-1} combined with LCC critical value 5 based.

Table 41.5: Nitrogen Use Efficiency (kg N ha^{-1}) as Influenced by Hybrids and Nitrogen Management Strategies

Treatments	Partial Factor Productivity		Agronomic Use Efficiency		Physiological Efficiency	
	Kharif 2000	Kharif 2001	Kharif 2000	Kharif 2001	Kharif 2000	Kharif 2001
			Hybrids			
H_1	48.05	47.96	25.63	26.43	64.50	66.02
H_2	43.40	43.38	24.24	25.14	59.65	61.80
			'N' Management			
N_1	37.64	37.84	19.60	19.68	62.24	62.39
N_2	48.80	49.37	29.63	30.41	60.15	62.59
N_3	45.43	45.72	29.96	30.31	63.87	65.63
N_4	42.61	43.69	29.53	30.71	65.68	68.03
N_5	54.10	54.77	21.32	21.76	56.07	57.11
N_6	48.34	48.83	22.57	22.89	62.63	64.09
N_7	43.16	43.49	21.95	22.13	64.03	64.78
N_8	–	41.77	–	28.42	–	65.96

References

Bana, O.P.S. and Pant, A.K., 2000. Green manuring for ecologically sound crop production. *Indian Farmer's Digest*, pp.19–20.

Budhart, M.N., 1994. Effect of conjunctive and individual application of fertilizer and green manure nitrogen on low land rice (*Oryza sativa* L.). *Ph.D. Thesis*, Tamil Nadu Agricultural University, Coimhatore.

Cassaman, K.G., M.J. Kropfft, J. Gaunt and Peng, S., 1993. Nitrogen use efficiency of rice reconsidered. What are the key constraints. In: *Plant Nutrition from Genetic Engineering to Field Practice*, (Ed.) N.J. Barrow. Kluwert Dord rec, pp. 471–474.

Cassmannt, K.G., M.J. Kropff and De Yanzhem, 1994. A conceptua1 framework for nitrogen management of irrigated rice in high-yield environments. In: *Hybrid Rice Technology: New Developments and Future Prospectus*, (Ed.) Virman, p. 81–96.

Chaphalet, S.D., B.S. Chaphalet, A.Z. Yerne and Lanjewar, A.D., 2000. Effect of green manuring on soil properties and yield performance of rice. *J. Soils and Crops.*

Crasswelt, E.T. and Vlek, P.L.G., 1979. Fate of fertilizer nitrogen applied to the wetland rice. In: *Nitrogen and Rice*. International Rice Research Institute, Los Banos, Laguna, Philippines, pp. 175–192.

De Dattat, S.K. and Buresh, R.J., 1989. Integrated nitrogen management in irrigated rice. In: *Adv. Soil Sci.*, (Ed.) B.A. Stewart, Springer-Verlag, New York, pp. 143–164.

Dhyani, B.P. and Mishra, B., 1994. Scheduling nitrogen application for rice in mollisols. *Oryza*, 31: 202–205.

Greeniand, D.J., 1997. *The Sustainability of Rice Farming*. CAB International, IRRI, pp. 1–18.

Gurung, G.B. and Sherchan, D.P., 1996. Study on biomass production of dhaincha when relayed under maize and its green manuring effect on rice. PAC–Technical–paper pakhribas. Agricultural center, p. 170.

IRRI, 1996. Use of leaf colour chart (LCC) for N management in Rice. Int. Rice Res. Inst., P.O. Box. 933, Manila 1099, Philippines.

Kalpana, R., 1992. Studies on intercropping of grain legumes and green manures for multiple use and their effect on succeeding rice. *M.Sc. (Ag.) Thesis*, Tamil Nadu Agric. Univ., Coimbatore.

Krishnakumar, V. and Subramanian, S., 1991. Influence of meteorological parameters on growth and yield of lowland rice. *Oryza*, 28: 49–54.

Morris, R.A., R.E. Furoc and Dixon, M.A., 1986. Rice responses to a short duration green manure. I: Grain yield. *Agron. J.*, 78(7): 409–412.

Olsen, S.R, C.V. Cole, F.S. Watanabe and Dean, A.L., 1954. Estimation of available phosphorus in soils by extraction with sodium carbonate. Circular No. 939, USDA.

Salu Reddy, N., G. Prabhakara Reddy and Reddy, Yellamanda, 1995.Nutrient uptake and yield of semi-dry rice (*Oryza sativa*) as influenced by preceding crops. *J. Res. APAU*, 23: 97–99.

Sharma, A.K. and Mittra, B.N., 1988. Effect of green manuring and mineral fertilizer on growth and yield of crops in rice based cropping on arid laterite soil. *J. Agric. Sci. Camb.*, 110: 605–608.

Somani, L.L., 1987. Decomposition of some organic materials in hilly soils of Rajasthan. *Acta Agronomica Hungarica*, 36: 103–119.

Stanford, G. and English, L., 1949. Use of Flame Photometer in rapid soil tests for K and Ca. *Agron. J.*, 41: 446.

Subbiah, B.V. and Asija, G.L., 1956. A rapid procedure for the estimation of available nitrogen in soils. *Curr. Sci.*, 25: 259–260.

Swarup, A., 1986. Effect of pre-transplanting submergence and green manure on yield and sodic soil improvement. *IRRN*, 11: 39.

Takebe, M., T. Yonekama, K., 1989. Measurement of leaf colour scores and its implication to nitrogen nutrition of rice plants. *J.A.R.Q.*, 23: 86–93.

Thakur, R.B., 1993. Effect of split application of nitrogen on growth yield and N uptake in wetland rice. *Indian J. Agron.*, 38(2): 191–194.

Tiwari,C., Y. Sing and Singh, Dheer, 1995. Effect of Sesbania green manure and nitrogen fertilizer in rice-wheat cropping system. *Indian J.Agric.Sci.*, p. 65.

Walkley, A. and Black, C.A., 1934. An examination of the degtare zmethod for determining soil organic matter and a proposed modification of the chromic and titration method. *Soil Sci.*, 40: 233–243.

Yoshida, S., 1981. *Fundamentals of Rice Crop Science*. Int. Rice Res. Inst., Los Banos, Philippines.

Chapter 42

Structure and Optical Properties of DC Reactive Magnetron Sputtered Zinc Oxide Films

M. Lavanya[1], *Sa K. Narayandass*[2] *and D. Mangalaraj*[3]

[1]*Department of Physics, P.S.G.R Krishnammal College for Women, Coimbatore, T.N.*

[2]*Professor and Head (Retd.), Department of Physics, Bharathiar University, Coimbatore, T.N.*

[3]*Reader, Department of Physics, Bharathiar University, Coimbatore, T.N.*

ABSTRACT

Zinc oxide films were deposited on glass substrates in argon and oxygen atmosphere by dc reactive magnetron sputtering using a metallic zinc target. The influence of annealing temperature on the structural and optical properties of the films were systematically investigated and optimized the deposition parameters to prepare single phase zinc oxide films with preferred (002) orientation. The annealed films exhibited a good transmittance of around 99 per cent with a band gap of 3.28 eV.

Keywords: *DC reactive magnetron sputtering, Zinc oxide.*

Introduction

Zinc oxide is an n-type wide band gap semiconducting material that has wide applications in optics and optoelectronics. In FPD applications ZnO serves as a transparent electrode. Application of these films in each branch of technology demands certain specific values of conductivity, transmittance, and band gap etc. Due to high optical band gap, ZnO films have been used as window layers in copper indium selenide based heterojunctiuon solar cells to enhance the short circuit current (Kojima *et al.*, 1998). Another important advantage of ZnO is its chemical stability in the presence of hydrogen plasma, which enable for use in amorphous silicon solar cell fabrication by plasma enhanced chemical vapour deposition (Granqvist, 1990). It has also been recognized as a promising alternative material to transparent conducting indium tin oxide because of its low cost and non-toxicity (Stolt *et al.*, 1993). Apart from the above ZnO films have been used in many devices such as gas sensors, high temperature solid lubricant in gas turbine engines and electro chromic devices (Subramanyam *et al.*, 2000). ZnO films have large piezoelectricity, wide band gap and good optical transparency in the visible region and displays large conductivity changes on contact with different gas atmospheres. Among the various thin film deposition techniques, dc reactive magnetron sputtering received much attention, in recent years, because of its high deposition rates onto a large area offers good control over the composition of the film. In the present investigation an attempt was made in the preparation of ZnO films by dc reactive magnetron sputtering and studied the influence of annealing effect on the structural and optical properties of the films.

Materials and Methods

The zinc oxide films were deposited onto ultrasonically cleaned glass substrates by dc reactive magnetron sputtering method. A stainless steel vacuum chamber, 250 mm was pumped using diffusion pump–rotary pump combination which could give an ultimate vacuum of the order of 3×10^{-5} mbar. The pressure was measured using a Pirani-Penning gauge combination. High purity (IOLAR 2 grade) argon and oxygen were used as the sputtering and reactive gases respectively. The target to substrate distance maintained was 8 mm. The voltage applied during the deposition was 500 volts and the ion current was 250 mA. Before deposition of an oxide film the target was presputtered in pure argon atmosphere for about 15 minutes in order to remove the surface oxide layer of the target. The structure of the films was studied by X-ray diffraction using (model SUNTAG–USA, A = 1.5406). The optical transmittance of the films was measured using UV-VIS-NIR double beam spectrophotometer. The room temperature deposited films were subjected to open air annealing at 200°C and 400°C respectively. Film depositions were carried out at different thicknesses. The thicknesses of the films were measured by Multiple Beam Interference (MBI) method.

Results and Discussion

Figure 42.1 shows the XRD pattern of the films formed at various thicknesses and for the as deposited and 673 K annealed films. The XRD spectra were recorded in the 20 ranges from 10° to 100° in 0.02° steps. However, the diffraction peaks were observed between 30° and 60°. The presence of diffraction peaks indicates that the films were polycrystalline in nature with hexagonal wurtzite structure (Jin Ma *et al.*, 1996 and Schulz *et al.*, 1979). Figure 42.1(a) represents the XRD pattern of the as deposited film of thickness 148 nm. It revealed that the as deposited films were nearly amorphous in nature exhibited a small single peak at 54.9° corresponding to (102) orientation of un oxidized zinc. Figure 42.1(b and c) represents the XRD pattern of 673 K annealed films of thickness of 125nm and 193 nm respectively. From Figure 42.1(c) it was observed that the films of thickness 193 nm exhibited a single phase of zinc oxide with hexagonal (002) orientation. The intensity of (002) peak is a direct indication of the c-axis orientation perpendicular to the substrate surface (Menner *et al.*, 1998). The intensity of the (002) peak was found to increase with increase in thickness and no variation in their peak position was observed.

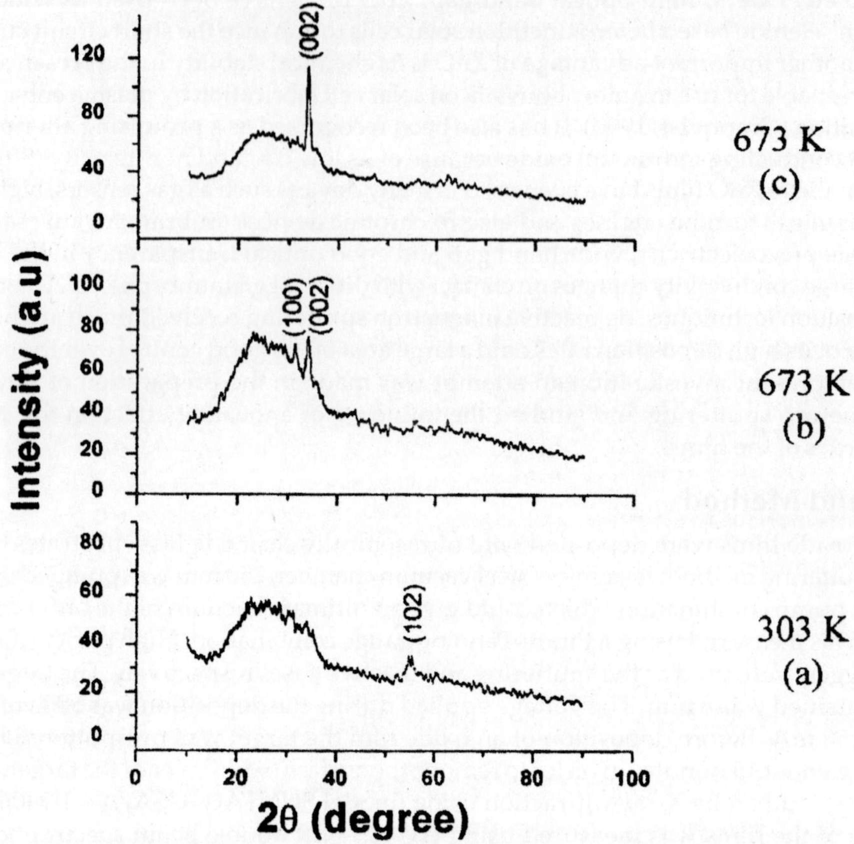

Figure 42.1(a–c): X-ray Diffraction Pattern of ZnO Film at Various Thicknesses

The structural parameters such as lattice constants (a), grain size (D), strain (ε_s), dislocation density (δ) and the interplanar distance (d) of the films were calculated. The evaluated lattice parameters a = 3.26 A° and c = 5.24 A° are compared with the earlier reports and it is in good agreement with them (Jenkins *et al.*, 1997 and Sundaram and Khan, 1997). The calculated structural parameters are shown in the Table 42.1.

Table 42.1: Micro Structural Parameters Associated with the Sputtered ZnO Thin Films

Film Thickness Nm	Annealing Temperature K	Grain Size (D) Nm	Dislocation Density (δ) 10^{15} lines/m²	d Spacing A°	Starin 10^{-3}
125	673	8.9	12.37	2.620	3.47
193	673	14.3	4.83	2.620	2.207

From Table 42.1 it can be seen that there is no variation in interplanar distance with thickness, accordingly the crystallite size was found to increase with the thickness. Variation of grain size compromises the change in interplanar distance and also there was no variation found in the peak positions of (002) planes. Very small crystallite size of 8.9 nm and 14.3 nm were observed in the films and no change in the *d* value shows the formation of nanoparticles like nature. Higher dislocation density and strain was observed in 125 nm thick films. This may be due to the structural distortion induced by the (100) plane-oriented crystallites.

Figure 42.2(a–c) shows the optical transmittance of the sputter deposited zinc oxide films of various thicknesses of that annealed at various temperatures. The plots reveal very low transmission for as deposited films of different thickness. It was observed that the transmittance increase from 23 per cent of 125 nm thick to 37 per cent for 148 nm and then decrease to 31 per cent for 193 nm thick films. The increase in transmission for 148 nm films is the consequence of higher base pressure (6×10^{-5} torr) during deposition. In this pressure there is a change of higher degree of oxidation leading to increase in transmission. For this sample the XRD results shows zinc rich films. Thus we believe that the lower transmitting 125 nm and 193 nm thick films obey the same rule. Overall low transmission in the as deposited films may be due to higher zinc concentration, which is the nature of relatively thick metal rich films at visible wavelengths.

In the case of 473 K annealed samples no significant variation is noted in lower thickness film, a decrease in transmittance is observed in 193 nm and a notable increase in transmittance due to oxide formation is observed in 148 nm films. The decrease of transmittance in higher thickness film may be due to temperature induced oxygen desorption in the surface creating oxygen deficient centers. Good transmission around 99 per cent and a sharp absorption edge was observed for 673K annealed samples of thickness 148 and 193 nm. The higher transmission shows the formation of stoichiometric ZnO. XRD results revealed the same. For lower thickness film in contrast to the other two low transmissions about 45 per cent and gradual absorption was observed. This may be due to the presence of a large number of defect centers in the film. The higher dislocation density and strain in this sample as calculated from the XRD confirms the interpretation. The gradual absorption edge and lower

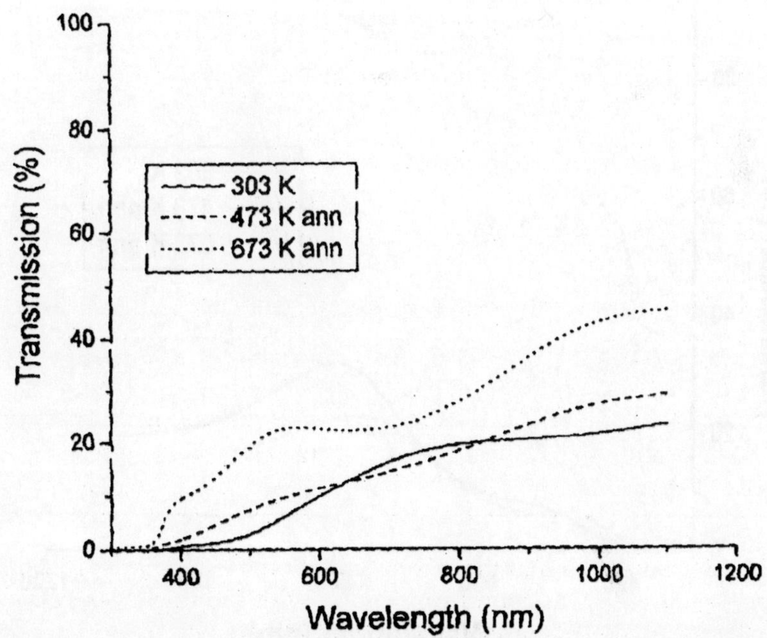

Figure 42.2(a): Transmittance Spectra of the ZnO Films of 125 nm Thickness

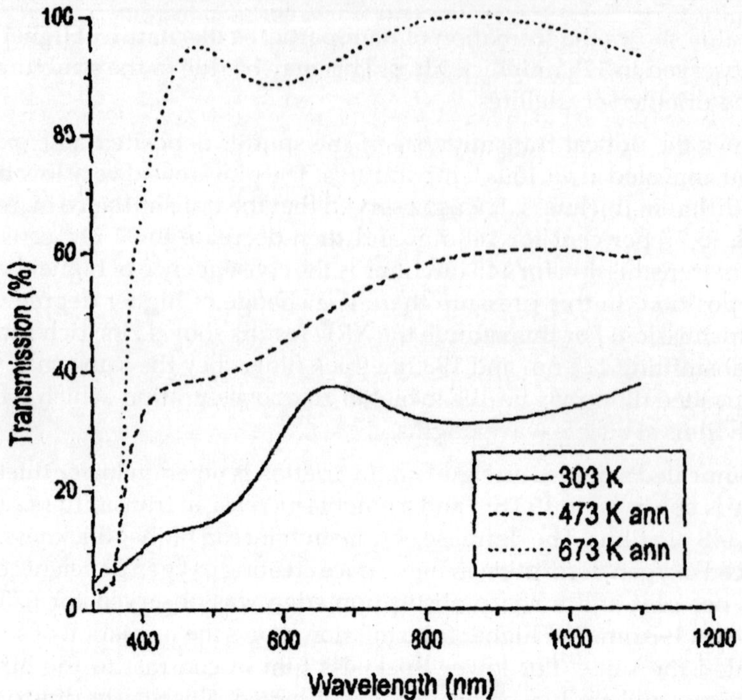

Figure 42.2(b): Transmittance Spectra of the ZnO Films of 148 nm Thickness

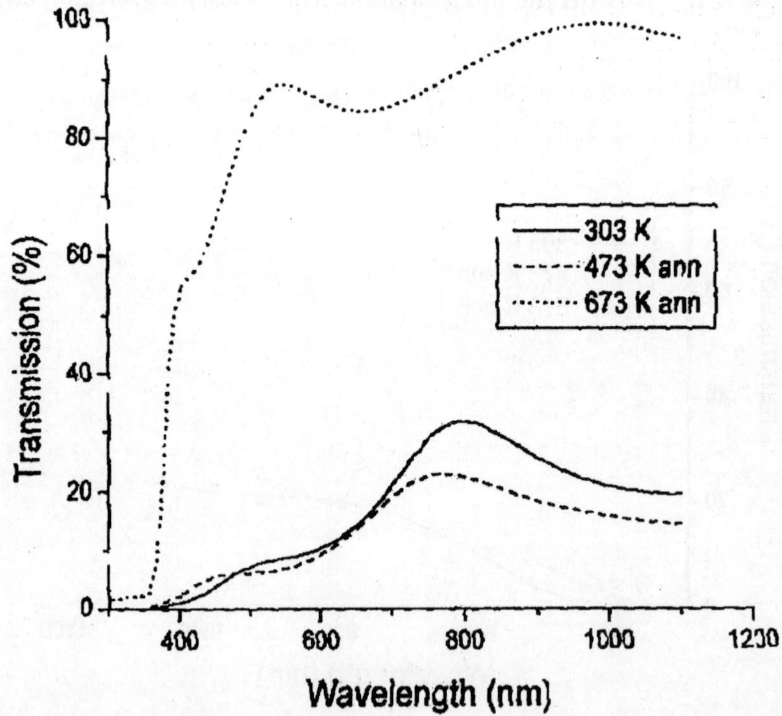

Figure 42.2(c): Transmittance Spectra of the ZnO Films of 193 nm Thickness

transmission observed in the plots for as deposited and 473 K annealed samples may be due to the presence of interband trap states. The interband trap states may have been induced due to the amorphous nature, oxygen deficiency and structural distortion in the sample.

The optical band gap value is found to increase with the increase in annealing temperature. This may be due to the increase in crystallinity and improvement in stoichiometry. For the films annealed at 673 K the band gap value varies from 3.19 eV to 3.29 eV, which is good agreement with the earlier reports (Zang *et al.*, 1994). The low value of optical band gap 2.69 eV for the as deposited films of thickness 148 nm may be due to the metal rich nature and oxygen deficiency.

Conclusion

Thin films of zinc oxide have been deposited onto the glass substrate by dc reactive magnetron sputtering technique at different annealing temperatures with different thickness. The annealing effect played a major role on the structure and optical properties of the films. 673 K annealed films exhibited single phase of ZnO with preferred (002) orientation, with an optical transmittance of 99 per cent with band gap of 3.28 eV.

References

Granqvist, C.G., 1990. *Thin Solid Films*, 193/194730.

Jenkins, D.F.L., Cunningham, M.J., Velu, G. and Remiens, D., 1997. *Sensors and Actuators*, A63: 135.

Jin Ma, Feng Ji, Hong-lei Ma and Shu-Yung, Li, 1996. *Thin Solid Films*, 279: 213.

Kojima, T., Koyanagi, T., Nakamura, K., Yanagisawa, T., Takahisa, K., Nishitami, M. and Wada, T., 1998. *J. Solar Energy Matter. Solar Cells*, 50: 5117.

Menner, R., Dimmler, B., Mauch, R.H. and Schook, H.W., 1998. *J. Cryst. Growth*, 86: 906.

Schulz, H. and Thiemann, K.H., 1979. *Solid State Commun.*, 32: 783.

Stolt, L., Hedstrom, J., Kessler, J., Velthaus, K.U. and Schock, H.W., 1993. *Appl. Phys. Lett.*, 62: 597.

Subramanyam, T.K., Srinivasulu Naidu, B. and Uthanna, S., 2000. *Cryst. Res. Technol.*, 35: 1193.

Sundaram, K.B. and Khan, A., 1997. *Thin Solid Films*, 295: 87.

Zang, D.H. and Brodie, E.E., 1994. *Thin Solid Films*, 38: 95.

Chapter 43

Antibacterial and Antifungal Activity of Marine Bacteria from Gulf of Mannar Resources

N.K. Asha Devi[1], R. Gopal[2], K. Balakrishnan[3] and S. Padmavathy[1]

[1]*Department of Zoology and Microbiology, Thiagarajar College, Madurai – 9, Tamil Nadu, India*
[2]*Reader and Head, Department of Zoology, Yadava College, Madurai – 14, Tamil Nadu, India*
[3]*Reader, Department of Biotechnology, Bharathidasan University, Tiruchirapalli – 24, Tamil Nadu, India*

ABSTRACT

Marine microorganisms of bacterial species were isolated from coastal water of South-east Zone of Rameshwaram, Dhanushkodi and Thondi, Ramanathapuram District, Tamil Nadu, India. Out of 23 bacterial isolates, 6 were identified and selected for antimicrobial activity against selected human bacterial pathogens of *Staphylococcus aureus, Pseudomonas aeruginosa, Salmonella typhi, Vibrio cholerae, Klebsiella pneumoniae* and *Aspergillus niger.* Of the 6 marine bacterial species, *Planococcus* sp. and *Bacillus* sp. showed strong antimicrobial activity. But all the marine bacterial samples showed antifungal activity against *Aspergillus niger.* The results of the present investigation reveal that the marine bacteria from coastal environment are a potent source of novel antibiotics.

Keywords*: Antibacterial and Antifungal activity, Human pathogens, Marine bacteria.*

Introduction

Marine ecosystems represent 95 per cent of the biosphere and coastal regions are particularly promising, because of the highly adapted species found in these harsh environments. Marine microorganisms have become important in the search of novel microbial products. Many microorganisms such bacteria and fungi had been reported to produce antibacterial, antifungal, antiviral and antitumor substances (Bernan *et al.*, 1997; Fenical, 1997; Molinski, 1993; Okami and Okazaki, 1972; Rosenfeld and Zobell, 1947). Marine bacteria have recently been identified as a source of new bioactive metabolites (Bernan *et al.*, 1997 and Jaruchoktaweechai *et al.*, 2000). There are few reports on antimicrobial activity from marine microorganisms from East Coast of India. Hence, a detailed study on the antimicrobial potential microorganisms of bacteria from coastal water of South East zone of Rameshwaram, Dhanushkodi and Thondi, Ramanathapuram District, Tamil Nadu, India.

Materials and Methods

The water samples were collected in clean, sanitized and autoclaved bottles and transported to the laboratory for further analysis. Marine agar (Hi Media) was used to culturing the bacterial colonies, which were incubated at 37°C for 24–48 hours. The isolated bacteria were identified based on the colony morphology, gram staining and biochemical tests (Cappuccino, and Sherman, 1999; Holt *et al.*, 1994). Marine bacterial samples isolated were cultured in 300 ml in Marine broth for the production of secondary metabolites in 500 ml Erlenmeyer flasks. Flasks were incubated on a rotary shaker at 220 rev/min at 25°C. After 7 days of cultivation, the culture broth was centrifuged at 10000 rpm for 20 minutes to remove the cell and filtered through Millipore filter (0.45 µm) to get cell free extract. Antimicrobial activities were assayed in duplicate, using 50 µl of marine bacterial samples in well diffusion assay (Schillinger and Lucke, 1989) and the test organisms were *Staphylococcus aureus, Pseudomonas aeruginosa, Salmonella typhi, Vibrio cholerae, Klebsiella pneumoniae* and *Aspergillus niger.*

Results

Among 23 marine bacteria isolates, 6 were selected for further studies on the basis of antimicrobial activity and growth characteristics. Based on morphological, gram staining and other biochemical characteristics, they were identified as *Marinonascens* sp., *Mesophilobacter* sp., *Marinoncoccus* sp., *Planococcus* sp., *Bacillus* sp. and *Alteromonas* sp. (Table 43.1). Selected 6 bacterial filtrated samples were tested against selected human pathogens. *Planococcus sp.* showed significant antimicrobial activity and it did not show zone of inhibition in *Staphylococcus aureus* and *Klebsiella pneumoniae.* Except the gram negative bacteria of *Klebsiella pneumoniae,* other three gram negative bacteria were sensitive to *Planococcus* sp. *Bacillus* sp. also showed strong antibacterial activity against gram negative bacteria but not against *Salmonella typhi.* Of the 6 marine bacterial samples, only 2 species of *Mesophilobacter* sp. and *Marinococcus* sp. showed antibacterial activity against gram positive bacteria of *Staphylococcus aureus.* All 6 marine bacterial samples, including *Alteromonas* sp. and *Marinonascens* sp. showed significant antifungal activity against *Aspergillus niger* (Table 43.2).

Discussion

The first antibiotic from a marine bacterium was identified and characterized in 1966 (Burkholder *et al.*, 1996). The results of the growth inhibition assays against five species of pathogenic fungi provide very promising results. In this present work, *Planococcus* sp. and *Bacillus* sp. showed stronger antimicrobial activity compared to other marine bacterial samples. In recent studies wide arrays of marine bacteria were tested for bacterial antagonistic effects and it was demonstrated that this trait appears to be a widespread feature in marine habitats and present in bacterial groups (Burgess *et al.*, 1999; Grossart *et al.*, 2004; Long and Azam, 2001).

Table 43.1: Morphological and Biochemical Characteristics of Selected Marine Bacterial Species

Isolated Organisms	*Morphology*	*Gram Stain*	*Shape*	*Motility*	*Oxidase*	*Catalase*	*Indole Production*	*Methyl Red*	*Voges Proskaur*	*Citrate Utilisation*	*Nitrate Reduction*
Marinonascens sp.	Smooth, circular, raised yellow in colour	+	Cocci or short rods	Nonmotile	+	+	–	–	–	–	+
Mesophilobacter sp.	Irregular, smooth, pale yellow in colour	–	Cocci	Nonmotile	+	+	–	+	–	–	+
Marinococcus sp.	Circular, smooth, convex colonies, yellow orange in colour	+	Rod	Motile	+	+	–	–	–	+	+
Planococcus sp.	Wrinkled, yellow orange colonies	+	Cocci	Motile	–	+	–	+	–	+	–
Bacillus sp.	Irregular Wrinkled, orange colonies	+	Rod	Motile	+	+	–	+	–	+	–
Alteromonas sp.	Irregular, colourless wrinkled colonies	–	Cocci	Motile	+	–	–	+	–	–	–

+: Indicates positive; –: Indicates negative.

Table 43.2: Antimicrobial Activity of Marine Bacteria Against Human Pathogens

Tested Organisms	*Marine Bacteria (Zone of Inhibition in mm)*					
	Marinonascens sp.	*Mesophilobacter sp.*	*Marinococcus sp.*	*Planococcus sp.*	*Bacillus sp.*	*Alteromonas sp.*
Staphylococcus aureus	–	11	11	–	–	–
Pseudomonas aeruginosa	–	–	–	27	31	–
Salmonella typhi	–	–	–	17	–	–
Vibrio cholerae	–	–		27	17	–
Klebsiella pneumoniae	–	–	–	–	15	–
Aspergillus niger	17	11	13	13	17	15

–: Indicates absence of antimicrobial activity.

In the present study, 33 per cent of the isolates were found to inhibit *Salmonella typhi.* This is a very encouraging and fruitful finding in the context of resistance of this pathogen to most of the widely used antibiotics. Normally, *Pseudomonas* sp. are even capable of growth in some antibiotics, such as quaternary ammonium compounds. Their resistance to more antibiotics has also been a source of medical concern (Tortora *et al.*, 2000). It is interesting to note that *Bacillus* sp. and *Planococcus* sp. response represents antibiotic potential competing microorganisms against *Pseudomonas aeruginosa* in the marine environment.

In the current study, 100 per cent of the tested marine bacterial species exhibited antifungal activity. Many marine microorganisms showed antifungal activity against *Aspergillus niger,* but not against *Candida albicans* (Kokare *et al.*, 2004). Thus the research on these marine bacteria for newer products is a timely one as the ocean appears to be a good alternative source of unique materials. For complete characterization of an antibiotic it should be isolated in pure form as a single component, but this is impractical in a screening programme like this work.

This present investigation shows that the marine bacteria from the coastal environment could be a great source for screening and isolating new antibiotics to combat many human pathogens particularly the most resistant *Pseudomonas aeruginosa* and *Salmonella typhi,* in a successful manner in the future.

Acknowledgements

Thanks are due to the Department of Zoology and Microbiology, Thiagarajar College, Madurai, Tamil Nadu, India for providing all the facilities.

References

Bernan, V.S., Greenstein, M. and Maiese, W.M., 1997. Marine microorganisms as a source of new natural products. *Adv. Appl. Micro. Biol.*, 43: 57–90.

Burgess, J.G., Jordan, E.M., Bregu, M., Means-Spragg, A. and Boyd, K.G., 1999. Microbial antagonism: A neglected avenue of natural products research. *J. Biotechnol.*, 70: 27–32.

Burkholder, P.R., Plflister, R.M. and Leitz, F.P., 1996. Production of a pyrrole antibiotic by marine bacterium. *Appl. Microbiol.*, 14: 649–653.

Cappuccino, J.G. and Sherman, N., 1999. *Microbiology: A Laboratory Manual*, 4th ed, pp. 179–182.

Fenical, W., 1997. New pharmaceuticals from marine microorganisms. *Trends Biotechnol.*, 15: 339–341.

Grossart, H.P., Schligoff, A., Bernhard, M., Simom, M. and Brinkhoff, T., 2004. Antagonistic activity of bacteria isolated from organic aggregates of the German Wadden Sea. FEMS: *Microbial Ecol.*, 70: 2560–2565.

Holt, J.G., Krieg, N.R., Sneath, P.H.A., Staley, J.T. and Williams, S.T., 1994. *Bergey's Manual of Determinative Bacteriology*, 9th ed. Williams and Wilkins, Baltimore, pp. 518–537.

Jaruchoktaweechai, C., Suwanborirux, K., Tanasupawatt, S., Kittakoop, P. and Menasveta, P., 2000. New macrolactins from a marine *Bacillus* sp. Sc. 026. *J. Nat. Products*, 63(7): 984–986.

Kokare, C.R., Mahadik, K.R. and Kadam, S.S., 2004. Isolation of bioactive marine actinomyces from sediments isolated from Goa and Maharashtra coast lines (West Coast of India). *Indian Journal of Marine Sciences*, 33(3): 248–256.

Long, R.A. and Azam, F., 2001. Antagonistic interactions among marine pelagic bacteria. *Appl. Environ. Microbiol.*, 67: 4975–4983.

Molinski, T.F., 1993. Developments in marine natural products receptor-specific bioactive compounds. *J. Nat. Prod.*, 58: 1–8.

Okami, Y. and Okazaki, T., 1972. Studies on marine microorganisms (1) Isolation from the sea. *J. Antibiot.*, 25: 456–460.

Rosenfeld, W.D. and Zobell, C.E., 1947. Antibiotic production by marine microorganisms. *J. Bacteria*, 154: 393–398.

Schillinger, U. and Lucke, F., 1989. Antibacterial activity of *Lactobacillus* strain isolated from meat. *Applied and Environmental Microbiology*, 55: 1901–1906.

Tortora, G.J., Funke, B.R. and Case, C.L., 2000. *Microbiology*, 6th ed. The Benjamin/Cummings Publishing Company Inc., pp. 300, 659–668.

Chapter 44

Estimation of Understorey Grass Production in *Acacia tortilis–Cenchrus ciliaris* Silvopastoral System through Principal Component Regression Analysis

T.A. Khan, R.K. Bhatt and G. Suresh
Indian Grassland and Fodder Research Institute, Jhansi

ABSTRACT

This study deals with the establishment of relationships of understorey grass production in a 4 × 4m spaced trees of *Acacia tortilis–Cenchrus ciliaris* based silvopastoral system. Data recorded on understorey grass production, aboveground (Photosynthetically active radiation) and belowground (soil temperature, moisture and tree roots at three depths, 0–15, 15–30 and 30–45 cm depths) factors were used in the study. Principal component regression analysis was employed for establishing relationships of understorey grass production with respect to the formulated principal components as predictors. While formulating the prediction models in estimation of understorey grass production, the predictor variables particularly the below ground variables (soil moisture, soil temperature and tree roots), were found to be highly correlated. The principal components removed the multicollinearity and served as predictors to meet the statistical criterion of being independents and provide best fitted relationships.

***Keywords**: Principal components, Understorey grass production.*

Introduction

Silvopastoral systems have much to offer in checking land degradation trend on the one hand and in meeting the multiple demands of food, fibre, fodder, fuel wood, timber, medicine etc., on the other. It is a low input technology for rehabilitation of degraded lands (Deb Roy and Pathak, 1974). Estimation of understorey grass production is a part of silvopastoral system modelling to compare systems at different stages of growth using system analysis methodologies (Khan *et al.*, 1994). Silvopastoral models require the development of the relationship between understorey pasture production and overstorey tree production (Knowles, *et al.*, 1999). The understorey grass production in a silvopastoral system is affected by many factors like light availability for the ground vegetation, effect of soil moisture, temperature, tree-roots etc. Principal component interactions in the silvopastoral system also revealed the importance of belowground factors (soil moisture, temperature and tree-roots in the estimation of understorey grass production (Khan *et al.*, 2000). The log-transformations most often used to change the multiplicative effects to additive effects and meet the assumptions in the model. However in silvipastoral systems the transformed values also have high degree of correlations amongst themselves. This could be due to the presence of multi-collinearity among the predictor variables. Multiple regressions could be erroneous if independent variables are highly correlated. To avoid the problem of mutli-collinearity, new variables called Principal Components could be used instead of predictor variables (above ground and below ground). These components represent the same amount of variation explained in the original variables and would be independent to each other *i.e.* correlations among them would be zero. The idea of principal components was first discussed by Pearson (1901). He gave theoretical solution to the problem of principal axis in terms of least squares. The description of practical computing method came later by Hotelling (1931). Regression analysis can be carried out using principal components as regressor in place of original variables. This paper deals with the development of relationship of understorey grass production by principal component analysis in a silvipastoral system.

Material and Methods

The experiment was carried out for three consecutive years (1998–2000) at Central Research Farm of Indian Grassland and Fodder Research Institute, Jhansi which is situated at 25°27′N, 78°37′E and at an altitude of 275 msl. The climate is dry tropical with an annual rainfall of 900 mm of which 90 per cent is received during June–September. Mean maximum and minimum temperatures are 32.5° and 17.7°C with temperature as high as 48°C and as low as 1.5°C during May and June respectively. The soil of the experimental area has pH (6.89–7.69), C/N ratio (39.6–55.4).

A silvopastoral system (*Acacia tortilis–Cenchrus ciliaris*) was studied for estimation of understorey grass production through principal component regression analysis. In this experiment trees were planted during 1982 and grasses were re-established in 1996. The data from four-tree plots, (divided into nine sub-grids) of 4m tree to tree spacing, were recorded for both above ground *viz.*, Photosynthetic Active Radiation (PAR) at hourly interval starting from 8:00 to 15:00 hrs and below ground factors *viz.*, soil temperature, soil moisture and tree roots at three depths (0–15, 15–30, 30–45cm) from the ground surface along with the understorey grass production, in each grid at the grass harvest (September and December). The PAR was recorded with Quantum Radiometer (Licor-1200) using quantum sensor. Soil moisture and temperature were recorded by gravimetric method and temperature sensors (Century Multi Channel Temperature Indicator, Model CT-804), respectively. Correlations amongst the variables both above ground and belowground were studied. The principal component were formulated with respect to belowground and aboveground factors, using SPSS package ignoring the eigen values less than unity to fulfils the requirement of the model that the predictors should be independent. To find out the relationship of understorey grass production with these variables as predictor variables, the model of the type $Y = X\beta + \varepsilon$, has been evolved. *Where*, Y ($n \times 1$) is understorey grass production and X ($n \times p$) is matrix of p-predictor variables in the form of principal components, β is the matrix represents regression coefficients to be determined through regression analysis.

Step-wise procedure of Regression analysis has been used for evolving the best-fitted relationships of grass in the three years pooled databased formulated principal components (Draper and Smith 1981). To see the serial correlations amongst the errors Durbin-Watson Test has been applied to test the hypothesis that there exist no serial correlations amongst the residuals. Analysis of residuals has been done plotting the standardized errors, with the Estimated values of grasses to see the outliers (-2,2) (Draper and Smith 1981).

Results and Discussions

The correlation among the different variables showed high degree of correlations amongst all the predictor variables particularly amongst the belowground factors (soil moisture, soil temperature, tree roots) at the three depths (Table 44.1). The principal components, formulated from both above and belowground factors. The first five principal components could explain 79.054 per cent variation of the original variables (Table 44.2). Therefore, out of seventeen predictor variables, only five can serve as predictor variables, ignoring only less than 21 per cent variation of the original variables. The relationships for estimation of understorey grass production with respect to these five principal components as predictor variables could retain fourth components (W4) for the 1st cut of the grass, 1st and the 2nd components (W1 and W2) for 2nd cut and 1st components (W1) for total grass estimation. The standardized coefficients of variables have been presented in Table 44.3. Where 'Ws' are transferred back components to original variables coefficients.

Based on pooled data for three years the following relationships of understorey grass production (DM basis) of *Cenchrus ciliaris* under *Acacia tortilis*, with principal components were established:

Dry grass (Cut–I–g/m^2) = EXP ($0.325^* - 0.282^* W_4$), Adj. $R^2 = 0.062$, D–W Statistics = 0.853

Dry grass (Cut–IInd–g/m^2) = EXP ($2.649 - 0.315^* W_1 + 0.286^* W_2$), Adj. $R^2 = 0.148$, D–W Statistics = 0.354, and

Dry grass (Total–g/m^2) = EXP ($0.780^{**} - 0.269^{**} W_1$), Adj. $R^2 = 0.054$, D–W Statistics = 0.4 7 6

The values of regression though estimated to be less but the insignificant values of D- W statistic in all the three estimations (Grass production at 1st cut, 2nd and Total) shows the strength of the relationships. The insignificance of the D–W statistic approved the null hypothesis and basic assumptions of the model. So, out of five principal components, only three components were retained for estimation of grass at 1st cut, 2nd cut and total grass estimation and rest were excluded, as they did not satisfy the criterion of step-wise procedure of regression analysis.

The relationships evolved with principal components as predictors were best fitted models as they satisfy the step-wise procedure of regression analysis and insignificance of Durban-Watson (DW) statistic, which shows the absence of serial correlations amongst the errors.

The difference in observed and predicted understorey grass production shows the gap between the observed and estimated production of grass is less and the variability in the estimated grass with respect to grids is less compared to observed production (Figure 44.1). Principal components can help in analysis where no. of variable is more than the number of observations.

Table 44.1: Correlations Amongst Different Components in *Acacia tortilis–Cenchrus ciliaris* Silvipastoral System (based on pooled data)

	PAR 08	*PAR 09*	*PAR 10*	*PAR 11*	*PAR 12*	*PAR 13*	*PAR 14*	*PAR 15*	*ST 0–15*	*ST 15–30*	*ST 30–45*	*SM 0–15*	*SM 15–30*	*SM 30–45*	*RWT 0–15*	*RWT 15–30*
PAR 09	0.602**															
PAR 10	0.647**	0.553**														
PAR 11	0.185	0.177	0.343*													
PAR 12	0.272*	0.338*	0.354**	0.548**												
PAR 13	0.422**	0.317*	0.598**	0.266	0.350**											
PAR 14	0.218	0.287*	0.170	0.399**	0.657**	0.487*										
PAR 15	0.406**	0.289*	0.378**	0.141	0.321*	0.484*	0.301*									
ST 0–15	–0.171	–0.117	–0.333*	–0.359**	–0.376**	–0.189	–0.388**	0.073								
ST 15–30	–0.085	–0.040	–0.278*	–0.295*	–0.334*	–0.141	–0.329*	0.090	0.982**							
ST 30–45	–0.012	0.002	–0.213	–0.244	–0.277*	–0.098	–0.288*	0.120	0.942**	0.983**						
SM 0–15	–0.362**	–0.332*	–0.547**	–0.557**	–0.608**	–0.480**	–0.513**	–0.413**	0.712**	0.679**	0.627**					
SM 15–30	–0.192	–0.247	–0.545**	–0.549**	–0.641 **	–0.465**	–0.538**	–0.383**	0.627**	0.611 **	0.575**	0.882**				
SM 30–45	–0.254	–0.259	–0.553**	–0.508**	–0.657**	–0.452**	–0.553**	–0.379**	0.526**	0.481**	0.433**	0.808**	0.877**			
RW 0–15	–0.014	0.207	0.093	–0.203	–0.077	–0.098	–0.294*	–0.162	0.036	0.003	0.005	0.179	0.118	0.266		
RWT 15–30	–0.201	–0.202	–0.161	0.132	–0.093	0.187	0.101	–0.038	0.002	–0.018	–0.027	–0.052	–0.069	0.018	–0.110	
RWT 30–45	0.065	0.108	0.069	0.361**	0.340*	0.185	0.092	0.063	0.032	0.048	0.054	–0.108	–0.186	–0.029	0.410**	0.079

*: Significant ($P < 0.05$); **: Significant ($P < 0.01$).

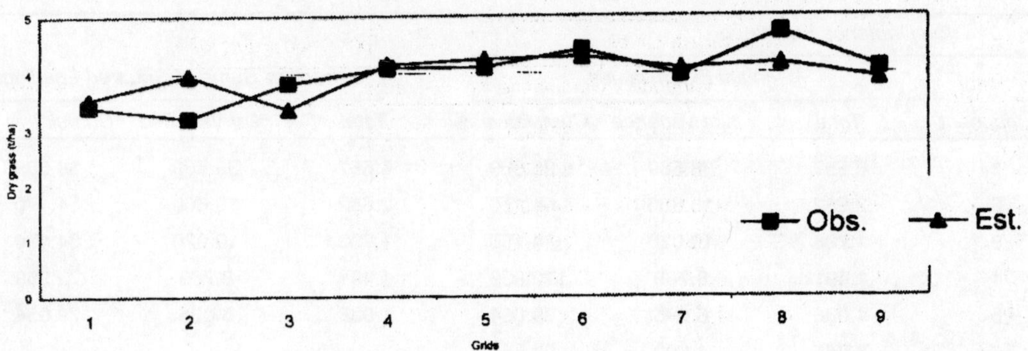

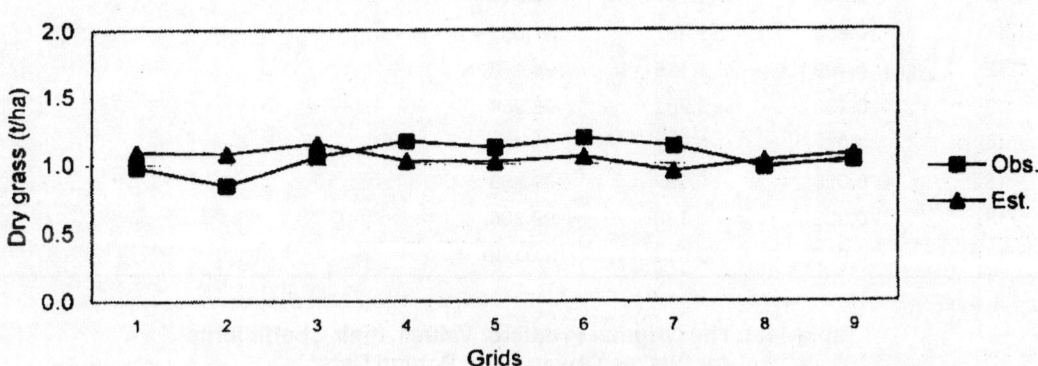

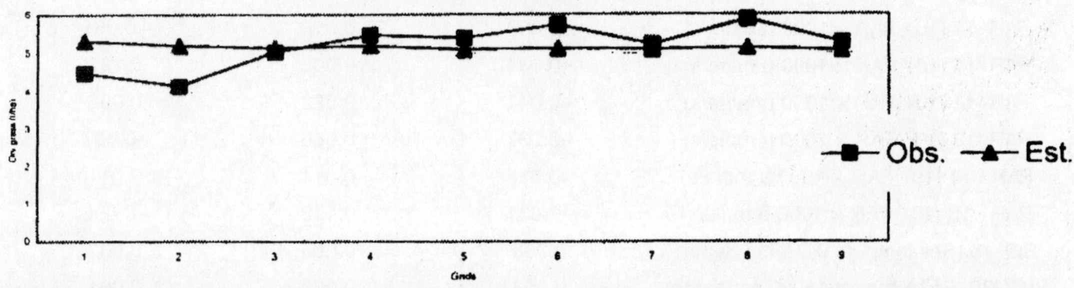

Figure 44.1: Difference Between Observed and Estimated Understorey Dry Grass Production (1st cut, 2nd cut and total cuts–98-00) in *Acacia tortilis–Cenchrus ciliaris* Silvipastoral System

Table 44.2: Variation (%) Explained by Different Principal Components in *Acacia tortilis* Based Silvopastoral System (based on pooled data)

Total Variance Explained

	Initial Eiqen values			*Extraction Sums of Sauared Loadings*		
Component	*Total*	*%of Variance*	*Cumulative %*	*Total*	*%of Variance*	*Cumulative %*
1	6.557	38.569	38.569	6.557	38.569	38.569
2	2.652	15.600	54.170	2.652	15.600	54.170
3	1.703	10.020	64.190	1.703	10.020	64.190
4	1.491	8.769	72.959	1.491	8.769	72.959
5	1.036	6.095	79.054	1.036	6.095	79.054
6	0.765	4.500	83.554			
7	0.731	4.301	87.855			
8	0.526	3.094	90.949			
9	0.485	2.856	93.805			
10	0.300	1.767	95.572			
11	0.253	1.487	97.059			
12	0.189	1.115	98.173			
13	0.133	0.781	98.954			
14	0.087	0.511	99.466			
15	0.056	0.327	99.793			
16	0.033	0.193	99.986			
17	0.002	0.014	100.000			

Table 44.3: The Original Predictor Values, their Coefficients for 'Ws' as Observed in Pooled Data

Original Predictor Variable	*Standardized Coefficients* W_1	*Standardized Coefficients* W_2	*Standardized Coefficients* W_4
PAR at 08:00hrs (10+PAR × 1000 µ mole/s)	–0.048	0.148	–0.812
PAR–09 (10+ PAR × 1000 µ mole/s)	–0.032	0.104	–0.031
PAR–10 (10+ PAR × 1000 µ mole/s)	–0.038	0.058	–0.061
PAR–11 (10+ PAR × 1000 µ mole/s)	–0.034	0.005	0.127
PAR–12 (10+ PAR × 1000 µ mole/s)	–0.041	0.022	0.096
PAR–13 (10+ PAR × 1000 µ mole/s)	–0.037	0.066	–0.007
PAR –14 (10+ PAR × 1000 µ mole/s)	–0.038	0.011	0.011
PAR –15 (10+ PAR × 1000 µ molels)	–0.031	0.112	–0.096
St0–15 (Soil temp. at 0–15 cm depth)	0.003	0.005	0.002
St15–30 (Soil temp. at 15–30 cm depth)	0.003	0.006	0.001
St30–45 (Soil temp. at 30–45 cm depth)	0.003	0.007	0.001

Contd...

Table 44.3–Contd...

Original Predictor Variable	*Standardized Coefficients W_1*	*Standardized Coefficients W_2*	*Standardized Coefficients W_4*
Sm0–15 (Soil moisture per cent at 0–15 cm depth)	0.005	0.001	0.002
Sm15–30 (Soil moisture per cent at 15–30 cm depth)	0.006	0.001	–0.004
Sm30–45 (Soil moisture per cent at 30–45 cm depth)	0.005	–0.0003	0.004
Rw0–15 (√ 0.5+ Tree root weight (g/m³) at 0–15 cm depth)	0.003	0.004	0.066
Rw15–30 (√0.5+ Tree root weight (g/m³) at 15–30 cm depth)	0.0001	–0.003	0.011
Rw30–45 (√ 0.5+ Tree root weight (g/m³) at 30–45 cm depth)	–0.002	0.006	0.060

It is concluded that principal components regression analysis could be used for estimation of understorey grass production in silvopastoral system. The principal component regression analysis in a situation where predictor variables are correlated should be employed to avoid the problem of multi-collinearity.

Acknowledgements

Authors are thankful to Director IGFRI for providing facilities. We are thankful to ICAR, New Delhi for providing financial assistance under AP Cess Fund Project.

References

Deb Roy, R. and Pathak, P.S., 1974. Silvipastoral management. *Indian Farming*, 24: 41–42.

Hotelling, H., 1933. Analysis of a complex of statistical variables into principal components. *Journal of Educational Psychology*, 24: 417–441, 498–520.

Draper, N.R. and Smith, H., 1981. *Applied Regression Analysis.* John Wiley and Sons, New York, pp. 709.

Khan, T.A., P.S. Pathak and Gupta, S.K., 1994. Modelling *Acacia tortilis* based silvipastoral System using system analysis methodologies. *Range Mgmt. and Agroforestry*, 15(2): 203–210.

Khan, T.A., G. Suresh and Bhatt, R.K., 2000. Principal component interactions in silvopastoral systems. *Indian Journal of Agroforestry*, 2: 79–85.

Knowles, R.L., G.C. Horvath, M.A. Carter and Hawke, M.F., 1999. Developing a canopy closure model to predict overstorey/under storey relationships in *Pinus radiata* silvopastoral systems. *Agroforestry Systems*, 43: 109–119.

Pearson, K., 1901. On lines and planes of closest fit to a system of points in space. *Philosophical Magazine*, 2: 557–572.

Chapter 45

Enumeration and Analysis of Weed Flora of Agricultural Fields in Warangal District, Andhra Pradesh

M. Dharmender Rao
Department of Botany, Lal Bahadur College, Warangal – 506 007, A.P., India

ABSTRACT

The present study aimed at the enumeration of weed flora of different agricultural fields of Warangal District, Andhra Pradesh and to analyse their distribution in different families, genera and species. A total number of 122 weed species of 37 angiospermic families were collected from wet and dry crops in three revenue divisions of Warangal District. They were grouped in to three categories on the basis of habitat: (1) wet field weeds, (2) dry field weeds and (3) weeds common to dry and wet fields. It was observed that 32 families belongs to dicotylodons five belongs to monocolyledons. The dominant families are Amaranthaceae, Asteraceae, Euphorbiaccae and Fabaceae among dicots and Cyperaceae, Poaceae among monocots accounts for 57 per cent of weed species. Acanthaceae, Capparidaceae, Malvaceae, Rubiaceae, Tiliaceae, Scrophulariaceae and Solanceae together accounts for 19 per cent of weed species. All the above families constitute 76 per cent of the weed species. The remaining families were found to be poorly represented. The varying distribution ratio in dry and wet fields was also analysed.

Keywords: *Dry field weeds, Enumeration, Weed flora, Wet field weeds.*

Introduction

Any plant other than the crop sown is a weed (Brenchley, 1920). A plant which is wild or intrusive in cultivated ground and not valued for its usefulness or its beauty is termed as weed. It is a plant growing where it is not desired (Shaw, 1956). The weeds have developed a tremendous power of aggression, persistence, reproductive capacity during the course of their evolution and equipped with several mechanisms of their propagule dispersal. Thus, they became troublesome and undesirable elements of earths vegetation. The diversity and composition of weed flora in different areas depends on various factors such as type of irrigation, cropping pattern and environmental factors. The survey and enumeration of weed flora is important in the weed management programme. De and Mukhopadhyay (1984) carried out the study of weed flora in sub-lateritic belt of West Bengal. Similarly, Tripathi *et al*. (2002) worked out the composition of weed flora in *Kharif* grain crops of eastern Uttar Pradesh. Warangal District of Andhra Pradesh is situated on the globe with the geographical coordination of 17°–19′ and 18°–36′ N, and 78°–49′ and 80°–43′E, covering the geographical area of 12.875 sq.km. consisting 1,098 villages (Figure 45.1). Eventhough, this largest district of Telangana region proved to be a potential agriculture centre, the information pertaining to weed flora of Warangal District (A.P.) is lacking in the weed science database. Therefore, an exclusive enumerative study on weed flora of dry and wet field crops was carried out in this district to record the composition and structure of weeds.

Materials and Methods

A total number of 122 weed species belonging to 37 angiosperm families were collected in two consecutive seasons for the present study. Weed plants were collected from different localities of Warangal city and from all the three revenue divisions by visiting various monsoon (*Kharif*) and winter (*Rabi*) crops at an interval of 15–30 days depending upon the flowering stages of weeds. The plant specimens were numbered, treated with Mercuric Chloride, pressed, mounted on herbarium sheets and preserved. All the collected specimens were identified with the help of standard floras, (Gamble and Fisher 1957; Cooke 1958; Saldana and Nicolson 1976; Nair and Henry 1983; Bor 1960). The naming of the weed species was updated as per the International Code of Botanical Nomenclature (ICBN). The weeds are grouped into three categories on the basis their habitat: (1) wet field crop weeds (2) dry field crop weeds, (3) Weeds common to both dry and wet field crops. The botanical names of weeds and their respective families enlisted Table 45.1. Further, the distribution of weeds in different families, genera and species was studied on the basis of their habitat or field.

Results and Discussion

The weed species investigated for the present study and their distribution patterns are shown in the Tables 45.1–45.5. The places of collection of these weeds is shown in the Figure 45.1. A perusal of Tables 45.2 and 45.5 reveals that the 122 weed species recorded in Warangal district, belong to 94 genera and 37 angiospermic families. Of which, 32 families represent Dicotyledons and five Monocotyledons. The ratio of the Monocots and Dicots is 1 : 7.4 of families, 1 : 3.27 of genera and 1 : 2.59 of species. The dominant weed families are Amaranthaceae, Asteraceae, Euphorbiaceae and Fabaceae among Dicots; and Cyperaceae, Poaceae among Monocots represented by 11, 10, 11, 7, 14, 16 species respectively. These together account for 57 per cent of weed species. The dominant families of the second group *viz.*, Acanthaceae, Capparidaceae, Malvaceae, Rubiaceae and Tiliaceae are represented by three species each and Scrophulariaceae, Solanaceae are represented by four species each. These together account for 19 per cent of the weed species. The dominant families of first and

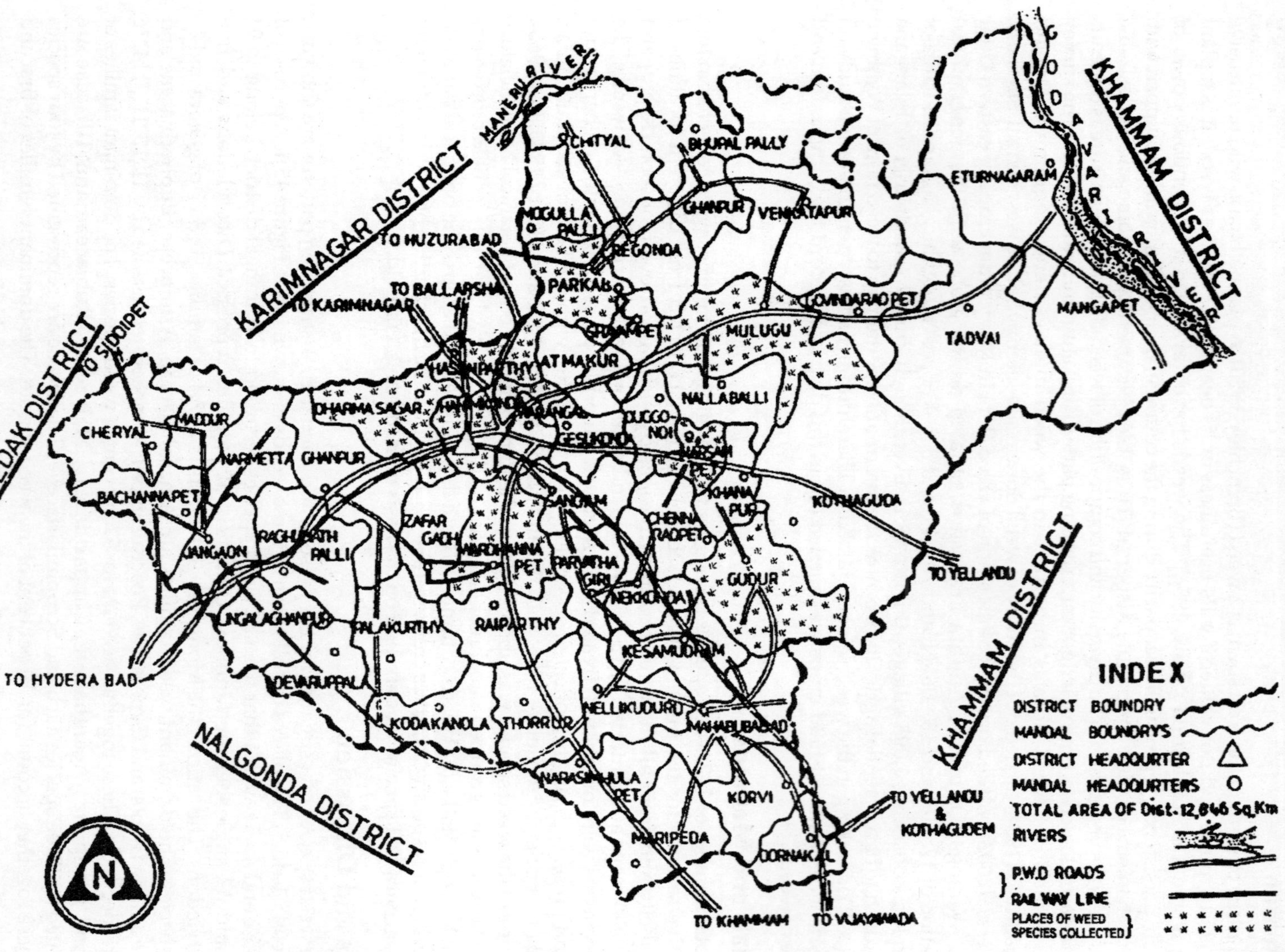

Figure 45.1: Warangal District with Divisions and Mandals Showing the Places of Weed Species Collected

second group together constitute about 76 per cent *i.e.* more than 3/4 of the weed species in different fields of Warangal district studied. The remaining families are poorly represented while many others are totally absent. However, the distribution ratios vary with the dry and wet fields.

Table 45.1: List of Weed Species Investigated in Warangal District, A.P.

Sl.No.	*Name of the Species*	*Family*
	Wet Field Weeds	
1.	*Hygrophila auriculata* (*Asterocanthan longifolia*)	Acanthaceae
2.	*Limnophytonobstusifolium*	Alismataceae
3.	*Alternathera sessilis*	Amaranthaceae
4.	*Caesulia axillaries*	Asteraceae
5.	*Eclipta alba*	Asteraceae
6.	*Sphaeranthus indicus*	Asteraceae
7.	*Heliotrorium ovalifolium*	Boraginaceae
8.	*Bacopa monnieri*	Scrophulariaceae
9.	*Sphenoclea zeylanica*	Sphenocleaceae
10.	*Bergia capensis*	Elatinaceae
11.	*Ammania baccifera*	Lythraceae
12.	*Ludwigia perennis* (*Ludwiga parviflora*)	Onagraceae
13.	*Bulbostyls barbata*	Cyperaceae
14.	*Cyperus rotundus*	Cyperaceae
15.	*Cyperus difformis*	Cyperaceae
16.	*Cyperus compressus*	Cyperaceae
17.	*Cyperus exalatus*	Cyperaceae
18.	*Fimbristylis bisumbellate*	Cyperaceae
19.	*Fimbristylis ovata*	Cyperaceae
20.	*Fimbristylis quinquangularis*	Cyperaceae
21.	*Fuirena ciliaris*	Cyperaceae
22.	*Kyllinga bulbosa*	Cyperaceae
23.	*Schoenoplectus articulatus*	Cyperaceae
24.	*Schoenoplectus lateriflorus*	Cyperaceae
25.	*Alloteropsis cimicina*	Poaceae
26.	*Bothriochloa pertusa*	Poaceae
27.	*Paspalum distichum*	Poaceae
28.	*Sataria pumila*	Poaceae
29.	*Urochloa panicoides*	Poaceae
30.	*Monochoria vaginalis*	Pontederiaceae

Contd...

Table 45.1–Contd..

Sl.No.	Name of the Species	Family
	Dry field weeds	
31.	*Justicia procumbens*	Acanthaceae
32.	*Indoneesiella echioides* (*Andrographis echioides*)	Acanthaceae
33.	*Trianthma portulacastrum*	Aizoaceae
34.	*Trianthma triquetra*	Aizoaceae
35.	*Achyranthes aspera*	Amaranthaceae
36.	*Aerva lanata*	Amaranthaceae
37.	*Allmania nodiflora*	Amaranthaceae
38.	*Alternanthera pungens*	Amaranthaceae
39.	*Amaranthus roxburghianus* (*Amaranthus polygamus*)	Amaranthaceae
40.	*Amaranthus spinosus*	Amaranthaceae
41.	*Amaranthus viridis*	Amaranthaceae
42.	*Celosia argentea*	Amaranthaceae
43.	*Digera muricata*	Amaranthaceae
44.	*Gomphrena celosioides*	Amaranthaceae
45.	*Catharanthus pusillius*	Apocynaceae
46.	*Aristolochia bracteolate*	Aristolochiaceae
47.	*Acanthospermum hispidum*	Asteraceae
48.	*Ageratum comyzoides*	Asteraceae
49.	*Lagasca mollis*	Asteraceae
50.	*Parthenium hysterophorus*	Asteraceae
51.	*Tridax procumbens*	Asteraceae
52.	*Vernonia cinerea*	Asteraceae
53.	*Xanthium strumarium*	Asteraceae
54.	*Trichodesma sedgwickianum* (*Trichodesma amplexicaule*)	Boraginaceae
55.	*Cleome aspera*	Cappariadaceae
56.	*Cleomo gynandra*	Cappariadaceae
57.	*Cleome viscose*	Cappariadaceae
58.	*Chenopodium alhum*	Chenopodiaceae
59.	*Evolvulus alsinoides*	Convolvulaceae
60.	*Acalypha lanceolata*	Euphorbiaceae
61.	*Chamaesyce hirta*	Euphorbiaceae
62.	*Chamaescye indica*	Euphorbiaceae
63.	*Chamaescye serpens*	Euphorbiaceae
64.	*Chrozophora rottleri*	Euphorbiaceae
65.	*Croton bonplandianus*	Euphorbiaceae
66.	*Euphorbia heterorphyla*	Euphorbiaceae

Contd...

Table 45.1–Contd..

Sl.No.	*Name of the Species*	*Family*
67.	*Phyllanthus amarus* (*Phyllanthus niruri*)	Euphorbiaceae
68.	*Phyllanthus madaeraspatensis*	Euphorbiaceae
69.	*Phyllanthus virgatus* (*Phyllanthus simrlex*)	Euphorbiaceae
70.	*Traga plukenetii*	Euphorbiaceae
71.	*Alysicarpus monilifer*	Fabaceae
72.	*Goniogyna hirta*(*Heylandia latibrosa*)	Fabaceae
73.	*Indigoera tinctoria*	Fabaceae
74.	*Psoralea coryifolia*	Fabaceae
75.	*Tephrosia purpurea*	Fabaceae
76.	*Tephrosia villosa* (*Tephrosia hirta*)	Fabaceae
77.	*Vigna trilobata* (*Phasedus trilobus*)	Fabaceae
78.	*Cassia absus*	Caesalpinaceae
79.	*Cassia occidentalis*	Caesalpinaceae
80.	*Enicostema axillare*	Gentianaceae
81.	*Leucas aspera*	Lamiaceae
82.	*Abutilon indicum*	Malvaceae
83.	*Malachra capitata*	Malvaceae
84.	*Sida cordifolia*	Malvaceae
85.	*Arygemone nexicana*	Papavaraceae
86.	*Polygala arvensis*	Polygalaceae
87.	*Portulaca oleracea*	Portulacaceae
88.	*Portulaca quadrifida*	Portulacaceae
89.	*Hedyotis puberula* (*Oldenlandia umbellate*)	Rubiaceae
90.	*Spermacoce articularis* (*Borreria hispida*)	Rubiaceae
91.	*Spermacoce pusilia*	Rubiaceae
92.	*Cardiospermum halicacabum*	Sapindaceae
93.	*Stemodia viscose*	Scrophulaiaceae
94.	*Striga angustifolia*	Scrophulaiaceae
95.	*Pyhysalis minima*	Solanaceae
96.	*Solanum nigrum*	Solanaceae
97.	*Solanum surattense*	Solanaceae
98.	*Withania somnifera*	Solanaceae
99.	*Waltheria indica*	Sterculiaceae
100.	*Corchorus tridens*	Tiliaceae
101.	*Triumfetta rhomboidea*	Tiliaceae
102.	*Triumfetta rotundifolia*	Tiliaceae

Contd...

Table 45.1–Contd..

Sl.No.	Name of the Species	Family
103.	*Tribulus terrestris*	Zygophyllaceae
104.	*Brachiria eruciformis*	Poaceae
105.	*Bracharia ramose*	Poaceae
106.	*Bracharia reptans*	Poaceae
107.	*Dactyloctenium aegyptium*	Poaceae
108.	*Digitaria ciliaris*	Poaceae
109.	*Dinebra retroflexa*	Poaceae
110.	*Tragus roxburghii*	Poaceae
	Weeds Common to Dryland Wet Field Crops	
111	*Biophytum sensitivum*	Oxalidaceae
112.	*Oxalis corniculata*	Oxalidaceae
113.	*Sopuhia delphinifolia*	Scrophulariaceae
114.	*Phyla nodiflora*	Verbanaceae
115.	*Amiscophacelus axillaries* (*Cyanotis axillaries*)	Commelin-aceae
116.	*Commelina benghalensis*	Conunelin-aceae
117.	*Cyperus iria*	Cyperaceae
118.	*Fimbristylis miliacea*	Cyperaceae
119.	*Cynodon dactylon*	Poaceae
120.	*Echinochloa colona*	Poaceae
121.	*Echinochloa crus-galli*	Poaceae
122.	*Echinochloa procera*	Poaceae

Table 45.2: The Number of Families, Genera and Species of Angiosperm Weed Studied in the Agricultural Fields of Warangal District

Sl.No.	Taxa	Dicotyledons		Monocotyledons		Total
		Number	Percentage	Number	Percentage	
1.	Families	32	86.49	5	13.51	37
2.	Genera	72	76.60	22	23.40	94
3.	Species	88	72.13	34	27.87	122

Table 45.3: Distribution of Wet Field Weeds as per their Families, Genera and Species Under Dictos and Monocots

Sl.No.	Taxa	Dicotyledons		Monocotyledons		Total
		Number	Percentage	Number	Percentage	
1.	Families	11	68.75	5	31.25	16
2.	Genera	15	46.88	17	53.13	32
3.	Species	15	35.72	27	64.29	42

Table 45.4: Distribution of Dry Field Weeds as per their Families, Genera and Species under Dictos and Monocots

Sl.No.	Taxa	Dicotyledons		Monocotyledons		Total
		Number	Percentage	Number	Percentage	
1.	Families	28	90.33	3	9.67	31
2.	Genera	64	85.34	11	14.66	75
3.	Species	77	83.70	15	16.30	92

Table 45.5: Synopsis of Weed Distribution in Different Families, Genera and Species

Sl.No.	Families	Wet Field			Dry Field			Total		
		No. of Genera	No. of Species	%of Species	No. of Genera	No. of Species	%of Species	No. of Genera	No. of Species	%of Species
					Dicotylons					
1.	Acanthaceae	1	1	2.38	2	2	2.17	3	3	2.46
2.	Aizoceae	–	–	–	1	2	2.17	1	2	1.64
3.	Amaranthaceae	1	1	2.38	10	10	10.87	8	11	9.01
4.	Apocyanaceae	–	–	–	1	1	1.09	1	1	0.81
5.	Aristolochiaceae	–	–	–	1	1	1.09	1	1	0.81
6.	Asteraceae	3	3	7.15	7	7	7.61	10	10	8.20
7.	Boraginaceae	1	1	2.38	1	1	1.09	2	2	1.64
8	Caesalpiniaceae	–	–	–	1	2	2.17	1	2	1.64
9	Capparidaceae	–	–	–	1	3	3.26	1	3	2.46
10.	Chenopodiaceae	–	–	–	1	1	1.09	1	1	0.81
11.	Convolvulaceae	–	–	–	1	1	1.09	1	1	0.81
12.	Elatinaceae	1	1	2.38	–	–	–	1	1	0.81
13.	Euphorbiaceae	–	–	–	7	11	11.96	7	11	9.01
14.	Fabaceae	–	–	–	6	7	7.61	6	7	5.74
15.	Gentianaceae	–	–	–	1	1	1.09	1	1	0.81
16.	Lamiaceae	–	–	–	1	1	1.09	1	1	0.81
17.	Lythraceae	1	1	2.38	–	–	–	1	1	0.81
18.	Malvaceae	–	–	–	3	3	3.25	3	3	2.46
19.	Onagraceae	1	1	2.38	–	–	–	1	1	0.81
20.	Oxalidaceae	2	2*	4.76	2	2*	2.17	2	2	1.64
21.	Papavaraceae	–	–	–	1	1	1.09	1	1	0.81
22.	Polygalaceae	–	–	–	1	1	1.09	1	1	0.81
23.	Portualaccaceae	–	–	–	1	2	2.17	1	2	1.64
24.	Rubiaceae	–	–	–	2	3	3.26	2	3	2.46
25.	Sapindaceae	–	–	–	1	1	1.09	1	1	0.81

Contd...

Table 45.5–Contd...

Sl.No.	Families	Wet Field			Dry Field			Total		
		No. of Genera	No. of Species	%of Species	No. of Genera	No. of Species	%of Species	No. of Genera	No. of Species	%of Species
26.	Scrophulariaceae	2	2(1)*	4.76	3	3(1)*	3.26	4	4	3.28
27.	Solanaceae	–	–	–	3	4	4.35	3	4	3.28
28.	Sphenocleaceae	1	1	2.38	–	–	–	1	1	0.81
29.	Sterculiaceae	–	–	–	1	1	1.09	1	1	0.81
30.	Tiliaceae	–	–	–	2	3	3.26	2	3	2.46
31.	Verbanaceae	1	1*	2.38	1	1*	1.09	1	1	0.81
32.	Zygophyllaceae	–	–	–	1	1	1.0	1	1	0.81
	Monocotyledons									
33.	Alismataceae	1	1	2.38	–	–	–	1	1	0.81
34.	Commelinaceae	2	2*	4.76	2	2*	2.17	2	2	1.64
35.	Cyperaceae	6	14(2)*	33.34	2	2*	2.17	6	14	11.48
36.	Poaceae	7	9(4)*	21.43	7	11(4)*	11.96	12	16	13.12
37.	Pontederiaceae	1	1	2.38	–	–	–	1	1	0.81

Note: *, and (–)* indicate the number of species common to both dry and wet field crops.

Weeds Distribution in Wet Field

The weeds inhabiting the wet field crops amount to 42 species belonging to 32 genera and 16 families. Of these 11 families are of Dicots and five are of Monocots (Table 45.3). The ratio of Monocots and Dicots is 1 : 2.2 of families, 1 : 0.88 of genera and 1 : 0.55 of species. This indicates the dominance of monocot weeds.

The dominant families Asteraceae, Cyperaceae and Poaceae are represented by 3,14 and 9 species respectively together account for 62 per cent of weed species. The dominant families of the second group Scrophulariaceae, Oxalidaceae and Commelinaceae are represented by two species each account for 14 per cent of wet field weed species. The dominant families of the first and second groups constitute 76 per cent of weed population in wet field. However, 12 species, belonging to 10 genera comprising six families *viz.*, Oxalidaceae, Scrophulariaceae, Verbanaceae, Commelinaceae, Cyperaceae and Poaceae are common to both dry and wet field crops (Table 45.5).

Weeds Distribution in Dry Field

The total weed taxa present in dry field crops are 92 species comprising 75 genera and 31 families (Table 45.4). Of these, 28 families belong to Dicots and three to Monocots. The ratio of Monocots to Dicots is 1 : 9.34 of families, 1 : 5.82 of genera and 1 : 5.14 of species. The clear dominance of Dicots over Monocots is quite obvious contrary to wet field weeds. The dominant families Amaranthaceae, Asteraceae, Euphorbiaceae, Fabaceae and Poaceae are represented by 10, 7, 11, 7 and 11 species respectively. These all account for 50 per cent of dry field weed species. The dominant families of the second group weed families *viz.*, Acanthaceae, Aizoaceae, Caesalipinaceae, Oxalidaceae, Portulacaceae,

Solanaceae are represented by four species. All these together account for 36 per cent of dry field weed taxa. The dominant weed families of the first and second groups together make 86 per cent which is more than 3/4 of the dry and wet field weed species.

In comparison to that of Sugha and Shukla (1977) the present investigation reveals that Cyperaceae, Poaceae and Scropulariaceae and Asteraceae are common and represented by more than 50 per cent of the total weeds. Further, in the dry fields Amaranthaceae, Asteraceae, Fabaceae, Euphorbiaceae are commonly represented. The significance of this peculiar distribution of weeds is their reproductive strategy which is the part of their adaptive mechanism. Weeds show tremendous reproductive capacity equipped with several methods of their propagule dispersal. A thorough understanding of the biology of the weeds as a precondition for the control measures has been emphasized by King (1966), Harper (1958), Tripathi (1977). The present investigations provide a valuable information to understand the aggressive nature of the weed and to design an integrated weed management programme.

References

Bor, N.L., 1960. *The Grasses of Burma, Ceylon, India and Pakistan (excluding Bambaseae)*. Pergmen Press, Oxford.

Brenchley, W.E., 1920. *Weeds of Farm Land*. Longmans, Green, London, pp. 239.

Cooke, T., 1958. *The Flora of the Presidency of Bombay*, Vols. I–III. BSI, Calcutta, (Repr. ed.).

De, G.C. and Mukhopadhyay, S.K., 1984. Weed flora in sub-humid lateritic belt of West Bengal, India. *J. Weed. Sci.*, 16: 101–115.

Gamble, J.S. and Fischer, C.E.C., 1957. *Flora of the Presidency of Madras*, Vols. I–III. BSI, Calcutta, (Repr. ed.).

Harpar J.L., 1958. *Herb. Abst.* 28: 151.

King, L.J., 1966. *Weeds of the World: Biology and Control*. Wiley Eastern Private Limited, New Delhi, pp. 525.

Nair, N.C. and Henry, A.N., 1983. Flora of Tamil Nadu, India. Series I Analysis, BSI, Coimbatore.

Saldanha, C.J. and Nicolson, H., 1976. *Flora of Hassan District, Karnataka, India*. Amerind Publishing Co. Pvt. Ltd., India and New York.

Shaw, W., 1956. Terminology Committee Report, Weed Society of America, Weeds. 4: 278.

Sugha, S.K. and Shukla, S.P., 1977. Angiospermic weed flora of Rice (*Oriza sativa* L.) fields in Kangra. *Indian J. Weed. Sci.*, 9: 1–8.

Tripathi, R.S., 1977. *Trap. Ecol.*, 18: 138.

Tripathi J.S.S., Vaishya, R.D., Khan, A.H., Singh, S.S. and Tripathi, R., 2002. Composition of weed in Kharif grain crops of Eastern Uttar Pradesh. *Indian J. Envir. and Ecoplan.*, 6: 309–312.

Chapter 46

Evaluation of Biopesticides Against *Spodoptera litura*, Fab. Infesting *Kharif* Groundnut

M.M Sanap, S.T. Aghav, P.K. Dharne, D.A. Shambharkar, A.J. Patil and R.B. Jadhav

Oilseeds Research Station, M.P.K.V., Jalgaon – 425 001, Maharashtra, India

ABSTRACT

The field experiments conducted during three consecutive seasons, *Kharij* 2003, 2004 and 2005 at Oilseeds Research Station, M.P.K V., Jalgaon revealed that, SNPV @ 1.5 × 10^{12} POB/ha (34.21 per cent) and NSE 5 per cent (30.2 per cent) were equally effective with chemical insecticide, Chiorpyriphos @ 200g.a i./ha (22.9 per cent) in reducing the foliage damage caused by *Spodoptera litura.* Maximum dry pod yield was obtained from the treatment with Chiorpyriphos @ 200g.a i./ha (14.37 q/ha) followed by NSE 5 per cent (14.24 q/ha) and SNPV @ 1.5 × 10^{12} POB/ha (14.17 q/ha).

Keywords: *Bio-pesticides, Days after spraying, Kharif, Spodoptera litura.*

Introduction

Groundnut or peanut (*Arachis hypogae* L.) the soil enriching legume grown over an area of 6.7 million ha with the production and productivity of 6.5 million tonnes and 967 kg/ha in India. In the state of Maharashtra it is cultivated over an area of 4.16 lakh ha with the production of 4.59 lakh m tonnes and productivity of 1103 kg/ha (2004–05). About 83.41 per cent area is under *Kharif* Groundnut and the productivity is highly variable due to vagaries of monsoon, incidence of pest and diseases.

Among the various pests attacking groudnut crop the defoliator, *Spodoptera litura*, Fab. is one of the notorious pest occurring from early vegetative stage till harvest of the crop causing considerable yield losses in *Kharif* Groundnut in Maharashtra state. The chemical pesticides are found to be effective against the pest, however, due to their ill effects the bio-pesticides like *Nomuraea rileyi*, SNPV and NSE were evaluated to combat the pest which were reported to be effective against the pest (Pokharkar *et al.*, 2001; Loganathan *et al.*, 2002; Kulkarni and Huger, 1999). *Nomuraea rileyi* in combination with botanicals (Patil *et al.*, 2002) and chemical insecticides (Murali Krishna *et al.*, 2004) were observed to be more effective against the pest.

Material and Methods

Field experiments for three consecutive *Kharif* seasons 2003, 2004 and 2005 were carried out at Oilseeds Research Station, Jalgaon, Maharashtra in Randomised Block Design replicated four times in the plot size of 5.0 × 3.60 m with the spacing of 30 × 10 cm on Groundnut cv.JL-286. The crop was sown during first fortnight of July each season and was raised by following the recommended agronomic practices. The biopesticides *viz.*, *Nomuraea rileyi* @ 5×10^{13} conidia/ha, and SNPV @ 1.5×10^{12} POB/ha were evaluated in comparison with NSE 5 per cent, chemical insecticide Chiorpyriphos @ 200g.a i./ha and untreated control. Only one spray at the peak activity of pest with hand operated knapsack sprayer @ 500L spray fluid/ha was given. Observations on per cent foliage damage caused by *Spodoptera litura* were recorded 7 DAS and were transformed to arcsin values for statistical analysis. Similarly dry pod yields obtained from each treatment was recorded seperately and analysed statistically.

Results and Discussion

Foliage Damage

The data on foliage damage caused by *Spodoptera* (Table 46.1) indicated that the differences in foliage damage among the treatments were significant during *Kharif* 2003 and 2004 and were non significant during *Kharif* 2005. Significantly the lowest damage of 22.80 per cent and 16.50 per cent during 2003 and 2004, respectively was recorded in the treatment with Chlorpyriphos @ 200g a.i./ha which was significantly superior over rest of the treatments except NSE 5 per cent which was at par with it during *Khartf* 2003. However, during *Khartf* 2005, the treatment with SNPV @ 1.5×10^{12} POB/ha recorded numerically the lowest damage (29.30 per cent), followed by Chiorpyriphos @ 200g a.i./ha (29.60 per cent).

The pooled mean foliage damage differences among the treatments were also significant and all the insecticide treatments recorded significantly less damage than untreated control. However, the lowest damage of 22.90 per cent was recorded in the treatment with Chlorpyriphos @ 200 g a.i./ha which was significantly lower than *Nomuraea rileyi* whereas it was at par with SNPV and NSE 5 per cent, thereby indicating that the bio-control agents, *viz.*, SNPV @ 1.5×10^{12} POB/ha and NSE 5 per cent spray were equally effective in reducing the *Spodoptera* foliage damage to that of chlorpyriphos, aside from being safer to environment. The effectiveness of chlorpyriphos in reducing the *Spodoptera litura* damage is in agreement with that of Kulkami and Lingappa (2002).

Dry Pod Yield

The dry pod yield differences among the treatments were significant during *Kharif* 2003 however, the same were non significant during 2004, 2005 season and for pooled mean. During *Kharif* 2003 the treatment with *Nomuraea rileyi* recorded significantly higher yield of 17.50 q/ha which out yielded rest of the treatments. However, the treatments with chlorpyriphos, SNPV and NSE were equally effective in obtaining the yield. Numerically the highest yield of 12.65, 15.42 and 14.37 q/ha was harvested from the treatment with chlorpyriphos @ 200g a.i./ha during *Kharif* 2004, 2005 and pooled mean yield for three seasons, respectively. All the bio-control treatments recorded numerically higher yield over untreated control and were more or less same to that of chlorpyriphos treatment.

Table 46.1: Evaluation of Biopesticides Against *Spodoptera litura*, Fab. Infesting *Kharif* Groundnut

Sl.No.	*Treatments*	*Per cent Foliage Damage (7 DAS)*				*Dry Pond Yield (q/ha)*			
		2003	*2004*	*2005*	*Mean*	*2003*	*2004*	*2005*	*Mean*
1.	*Nomuraea rileyi* @ 5× 10^{13} conidia/b.a	40.6 (39.57)	34.6 (35.96)	35.2 (36.47)	36.8 (37.33)	17.50	11.82	13.26	14.16
2.	SNPV @ 1.5 × 10^{12} POB/ha	38.8 (38.45)	34.4 (35.87)	29.3 (32.31)	34.2 (35.54)	15.93	11.04	15.56	14.17
3.	NSE 5 per cent	24.7 (29.73)	32.7 (34.89)	33.3 (35.09)	30.2 (33.23)	15.51	12.24	14.97	14.24
4.	Chiorpyriphos @ 200g a.i./ha	22.8 (27.56)	16.5 (23.79)	29.6 (32.59)	22.9 (27.98)	15.06	12.65	15.42	14.37
5.	Untreated control	68.4 (55.81)	45.0 (42.06)	48.2 (43.97)	53.7 (47.28)	12.39	10.74	13.20	12.11
	SE±	3.02	1.95	2.72	2.58	0.47	0.59	1.19	0.64
	CD at 5%	9.29	6.01	N.S.	8.43	1.45	N.S.	N.S.	N.S.
	CV%	15.78	11.30	15.06	12.34	6.14	10.01	16.42	8.02

DAS: Days after sowing.

From the results it is evident that the biopesticides, SNPV @ 1.5 × 10^{12} POB/ha and NSE 5 per cent spray are quite effective in suppressing the *Spodoptera* damage and in obtaining the highest, dry pod yield in groundnut.

References

Kulkarni, G.G. and Huger, P.S., 1999. Persistence of nuclear polyhedrosis virus on different host plants of *Spodoptera litura* Fab. *Haryana Agricultural Univ. J. Res.*, 29: 3–4.

Kulkarni, N.S. and Lingappa, S., 2002. Bioefficacy of entomopathogenic fungus, *Nomuraea rileyi* (Farlow) Samson on *Spodoptera litura* and *Cydia ptychora* in soybean and on *Spodoptera litura* in potato. *Karnataka J. Agril. Sci.*, 15(1): 47–55.

Loganathan, M., Babu, P.C.S., Balasubramanian, G. and Kailasam, C., 2002. Crop-pest damage model for groundnut infested with *Spodoptera litura* under field conditions. *Indian J. Ento.*, 64(4): 484–492.

Murali Krishna, T., Ramaiah, M., Rao, A.R.K. and Reddy, B. Chenchu, 2004. Efficacy of certain biopesticides along with chemical pesticides against *Spodoptera litura* in groundnut. In: *Current Trends of Res. on Groundnut in India (Supplement)*. NRCG, ICAR, Junagadh, p. 33–35.

Patil, R.K., Lingappa, S., Hiremath, I.G., Hiremath, R.V., Tandon, P.L., Ballal, C.R., Jalali, S.K. and Rabindra, R.J., 2002. Efficacy of botanicals with *Nomuraea rileyi* (Farlow) Samson in groundnut ecosystem. Biological control of lepidopteran pests. In: *Proceedings of the Symposium of Biological Control of Lepidopteran Pests*. July, 17–18, 2002, Bangalore, India, p. 313–317.

Pokharkar, D.S., Charati, S.N., Bade, B.A. and Ghorpade, S.A., 2001. Evaluation of biocontrol agents for managing *Spodoptera litura* damaging potato (*Solanum tuberosum*). *Indian J. Agril. Sci.*, 71(12): 797–799.

Chapter 47

Euglenophyceae of Hatnur Dam, Jalgaon District, Maharashtra

P.N. Narkhede

P.G. Research Centre, Department of Botany, Dhanaji Nana Mahavidyalaya, Faizpur – 425 503, M.S., India

ABSTRACT

Present article deals with the description of eleven taxa of euglenoids reported from Hatnur dam, Jalgaon district of Maharashtra State, situated on Tapi river at 21°05′10″ N latitute and 75°57′30″ E longitude. The water samples were collected monthly for a period two years. Total 63 taxa are reported from these samples. The present paper deals with the taxonomic enemuration of 11 taxa with 2 genera, 7 species, 1 subspecies, 1 variety and 2 forma. All these 11 taxa are reported first time from Hatnur dam and 7 taxa namely *Euglena intermedia* (Klebs) Schmitz, *Euglena ovum* (Ehr.) Lemm. f. *caudata* Defl., *Phacus pleuronectes* (O.F.M), *Phacus dangeradii* Lemm. *var. oblonga* (Skv) H. P., *Phacus undulatus* (Skv.) Pochm., *Phacus longicauda* (Ehr.) Duj. *var coradata* Pochm (*Phacus longicauda*) subsp. *cordata* Pochm, *Phacus bravicaudatus* (Klebs) Lemm.f. *minor* f.nov. are reported first time form North Maharashtra.

Keywords: *Euglenoids, Hatnur dam, Maharashtra.*

Introduction

In Jalgaon district of Maharashtra the work on Euglenophyceae are known through Barhate and Tarar (1985), Bhoge and Raghothaman (1986), Jawale and Kumawat (2000). The hydrobiology of Hatnur dam was studied first time. The present communication deals with 11 taxa, 2 genera, 7 species, 1 subspecies, 1 variety and 2 forma are reported first time from Hatnur dam.

Materials and Methods

The water samples were collected fortnightly in the morning between 8.00 to 10.00 am. monthly throughout the two years since June 2001 to May 2003. Algal samples were examined fresh as far as possible and preserved 4 per cent formalin. For further study. Sketching of euglenoids with the help of Camera Lucida. Identification is based on the monographs of Gojdics (1953), Huber- Pestalozzi (1955), Kamat (1962, 1976), Asual (1975), Bold and Wyne (1978), Patel and Waghodekar (1980), Patel (1990) and research papers of Pochman (1942).

Systematic Enumeration

***Euglena intermedia* (Klebs) Schmitz.**

Sladecek and Perman, 1978, 57 (1): 58; Waghodekar, 1980, P. 86, Pl. 4, Figure 1.

Cells elongate, cylindrical, anterior end blunt, posterior end short pointed, pellicle paintly spirally striated, cytoplasm densely granulated, chromatophore many, discs without pyrenoids, 210 paramylum bodies, nucleus central, slightly posteriors cells 77.7–113.2 µ long, 11–12.1 µ broad.

Observed length 78.54; breadth 13.9µ, f.1, (Colle No. 54)

***Euglena caudata* Huebn.**

Sladecek and Perman, 1978, 57(1): 58; Waghodekar, 1980, P. 93, Pl. 5, Figure 6.

Cells broadly spindle-shaped, tapering pasteriorly, rounded anteriorly, spirally striated pellicle, many chromatophores, discoid, sheated pyrenoid, paramylum many, small, oval, granules, central nucleus, cells 71.3–99.6µ; long, 34.5–48.1µ broad.

Observed length 71.06µ; breadth 37.4µ, f. 2. (Colle No.6)

***Lepocinclis ovum* (Ehr.) Lemm. *F.ecaudata* Defl.**

Popova and Safonova, 1976, 20, Pl. 2, f. 10, 11; Waghodekar, 1980, Pl. 15, Figure 30.

Present form differs from the type in having broadly ovate cell; narrow rounded at both the ends, posterioir end without any caudus, cells 22.1–35µ long, breadth 19.25µ.

Observed length 28.5µ; breadth 18.7µ, f. 3 (Colle No. 24)

***Phacus Pleuronectes* (O.F.M) Duj.**

Wynne. 1978,264, f. 5, 10; Waghodekar, 1980, P. 179, Pl. 7, Figures. 20, 21, Pl. 12, Figures 9–10; Pl. 20, Figure 19.

Cells ovoid to suborbicular, anterior end rounded, posterior end sharp pointed, curved tail, pellicle striated, paramylum 1–2 ringlike discs, cells 26.4–37.7µ, long, breadth 21.29µ.

Observed length 28.5µ; breadth 22.41µ, f. 4 (Colle No. 36)

***Phacus dangeardii* Lemm. Var. *oblonga* (Skv.) H.P.**

Popova and Safonova, 1976, 52, Pl. 10: Waghodekar, 1980, P. 158, Pl. 7, Figure 2.

Cells elongate, oval, flattened, about twice as long as broad, rounded at both ends, slightly narrow posteriorly, pellicle sprially striated, paramylum 1–2 rings, central, cells 20–22.6µ, long 8.7–11.7µ broad.

Observed length 15µ; breadth 9.35µ, f.5, (Colle No. 44)

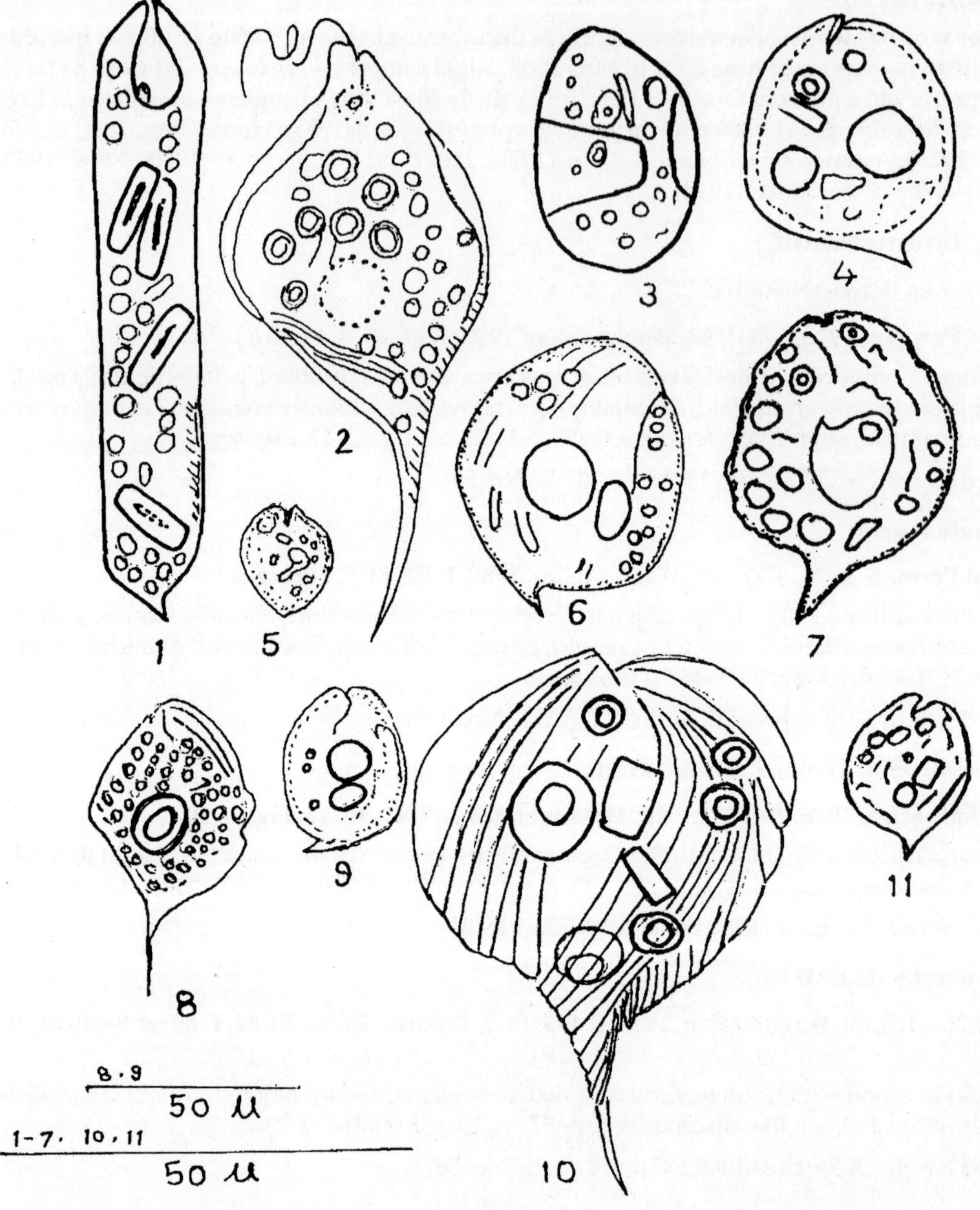

Plate 47.1

(1) ***Euglena intermedia*** **(Klebs) Schmitz; (2)** ***Euglena caudata*** **Hubn.; (3)** ***Lepocinclis ovum*** **(Ehr.) Lemmr. f.** ***ecaudata*** **Deft.; (4)** ***Phacus pleuronectes*** **(O.F.M.) Duj.; (5.)** ***Phacus dangeardii*** **Lemm var.** ***oblonga*** **(Sku.) H.P.); (6)** ***Phacus triqueter*** **(Ehr.) Duj.; (7)** ***Phacus undulatus*** **(Skv.) Pochm.; (8)** ***Phacus longicauda*** **(Ehr.)Duj Var.** ***cordata*** **Pochm. (*Phacus longicauda* subspecies** ***cordata*** **Pochm.); (9)** ***Phacus orbicularis*** **Huebner.; (10)** ***Phacus tortus*** **(Lemm.) Sky.; (11)** ***Phacus bravicaudatus*** **(Klebs) Lemm. f.** ***minor*** **f.nov.**

Phacus triqueter **(Ehr.) Duj.**

Yamagishi and Hirano, 1973, 24 (2): 74, Pl. 5, f.8; Popova and Safonova, 1976, P. 69–70, Pl. 17, f. 1–2: Waghodekar, 1980, P. 205, Pl. 17, Figure 24; Pl. 25, Figure 20.

Cells ovoid, anteriorly rounded, bilobed, narrow posteriorly sharp pointed, slightly deflected tail, triangular cell, deeply furrow on ventral side, pellicle striated extending into tail, paramylum two to several rings, cells 40–49.3μ long, 26–28.7μ broad.

Observed length 33.66μ; breadth 26.18μ, f.6. (Colle No. 20)

Phacus undulatus **(Skv.) Pochm.**

Asaual, 1975,245, Pl. 144, f. 3–9; Waghodekar, 1980, f. 192, Pl. 25, Figure 14.

Cells broadly avoid, rounded anterior end, slightly bilobed, posterior end short, strongly sharp–pointed, slightly deflected tail, pellicle striated, paramylum 1–2 large central discs or rings, cells 34–62.3μ long, 26–39.6μ broad.

Observed length 35.28μ; breadth 28.5μ, f. 7. (Colle No. 76)

Phacus longicauda **(Ehr.) Duj. *Var. cordata* Pochm. (*Ph. longicauda*) subsp. *cordata* Pochm)**

Suxena *et al.*, 1973, Pochmann, 1942, 95(2): 199, f. 108–110; Waghodekar, 1980, P. 221, Pl. 8, Figures 8–9, 13.

Cells broadly ovoid, cordate shaped, anteriorly broad; posteriorly tapering into a long, pointed straight or slightly curved tail–piece, 30–40fl in length, pellicle striated with crossstriae, paramylum 1–2 rings or discs, cells 72–134μ long, 30–55 broad.

Observed length 73.26μ; breadth 29.97μ, f. 8. (Colle No. 56)

Phacus orbicularis **Huebner**

Duthie and Socha, 1976, 103: 105; Hajdu, 1976, 11:21, g. 48, 61; Waghodekar, 1980, P. 200, Pl. 8, Figure 1,3, Pl. 13, Figures 1–2, 5, 6.

Cells ovoid, orbicular, dorsal side convex, ventral side flat or little concave, anterior end broad, posterior end narrow or broad ending in a short, sharp, oblique caudus, pellicle longitudinally striated, with cross seriae, paramylum 1–2 discs. unequal, cells 26.4–72.1μ, long, 20.8–48.1μ broad.

Observed length 39.96μ; breadth 29.97μ, f. 9. (Cell No. 16.)

Phacus tortus **(Lemm.) Skv.**

Duthie and Socha, 1976, 103: 105; Waghodekar, 1980, P. 229, Pl. 8, Figure 10, PI. 9, Figure 2.

Body twisted about the longitudinal axis through 11/2 to 2 complete turns, oval fusiform or rectangular shape, bilobed, sharply pointed caudus posteriorly, pellicle longitudinally striated, striations with twists extending slightly into caudus, paramylum 1–2 large disc or rings, nucleus in lower háif of cell, cells 56–102.3μ long, 26.47μ broad.

Observed length 71.06μ; breadth 43.01μ, f. 10. (Colle No. 32)

Phacus bravicaudatus **(Klebs) Lemm. f. *minor* f. nov.**

Waghodekar, 1980, P. 160, Pl. 12, Figures 7–8

Cells broadly oval, anterior end rounded, posterior end with dull point, margins entire, dorsal fold upto posterior end, pellicle longitudinally striated, paramylum single ring or 1–2 discs, nucleus central, cells 20.5–21μ long, 16–17μ broad.

Observed length 18.7μ; breadth 14.96μ, f. 11. (Colle No. 88)

References

Asaul, Z.L., 1975. Euglenophyta of the Ukrainian RSR, Pub. Naukova 2 Dumka Kiev, pp. 1–408.

Bharhate, V.P. and Tarar, J.L., 1985. Addition to the algae flora of Maharashtra Euglenophyceae form Khandesh–I. *Phycos*, 24: 184–185.

Bhoge, O.N and Ragothaman, G., 1986 Studies on the Euglenophyceae from Jalgaon, Maharshtra. *Phycos*, 25: 132–135.

Bold, H.C. and Wyne, M.J., 1978. *Introduced to the Algae Structure and Reproduction*. Prentice Hall of India Private Ltd, New Delhi.

Gojdics, M., 1953. *The Genus Euglena*. Pub. Medison, The Univ. Wisconsin Press. pp. 1–288.

Gonzalves, E.A. and Joshi, D.B., 1949. Fresh water algae near Bombay–15: The seasonal succession of the algae in a tank of Bandra. *J. Bombay Nat. Hist. Soc.*, 46(1): 155–158.

Huber-Pestalozzi, G., 1955. Die Binnenegewasser von Prof Dr August hienemann Das phytoplankton des Susswassers. Systematic and Biologie Euglenophyceae, pp. 1–606.

Jawale, A.X. and Kumavat, D.A., 2000. Study of physico-chemical character and algae sugar factory waste. *J. Adv. Sci. Tech.*, 3(1 and 2): 83–90.

Kamat, N.D., 1962b. The Euglenophyceae of Ahmedabad, India. *J. Univ. Bombay*, 31(3 and 5): 20–27.

Kamat, N.D. and Freitas, J.F., 1976 Checklist of Euglenophyceae and Chlorophyceae of Nagpur, Maharashtra. *Phycos* 15: 121–125.

Manuwar, M., 1970. Limnological studies on fresh water ponds of Hyderabad, India. I: Biotype. *Hydrobiologia*, 35: 127–162.

Patel, R.J. and Waghodekar, V.H., 1981. The euglenophyceae of gujarat, India–I. Genus Phacus Dujardin. *Phycos*, 20(1 and 2): 24–33.

Patel, R.J., 1990. Algae Novo, Pub. Avichal Science Foundation Science Series.

Pochmann, A., 1942. Synopsis der Gattung Pbacus. *Arch Protistenk*, 95(2): 81–252.

Rishi, V. and Kachroop, 1984. Euglenophy Doodhgangastream, Kashmir. *Phycos*, 23(1 and 2): 65–70.

Chapter 48

Larvicidal Activity of Plant Oils Against the Mosquito, *Culex quinquefasciatus* (Say)

K. Manimegalai, S. Agalya, S. Sajani and R. Uthayakumari

Department of Zoology, Avinashilingam University, Coimbatore – 641 043, T.N., India

ABSTRACT

The larvicidal activities of *Ricinus communis*, *Pongamia glabra* and *Madhuca longifolia* oil were tested against fourth instar larvae of the mosquito, *Culex quinquefasciatus.* To select the effective dose of the three oils, *Pongamia glabra* oil was found to be most effective at higher concentrations followed by *Ricinus communis* oil.

Keywords: Culex quinquefasciatus, Larvicidal activity, Madhuca longifolia.

Introduction

Mosquitoes are an ancient group of insects which has persisted for millions of years. Mosquitoes are very successful species found in different types of environment and they breed within a short period of time. Oil was the first material used to control mosquito larvae. Effects of various plant oil such as *neem* oil, eucalyptus oil, lemon grass oil, castor oil, pongamia oil, marikolundhu oil, Cinchona oil have been tested on the larvicidal and pupicidal activities on dengue vector, *Aedes albopictus* (Yang and Ma, 2005).

Materials and Methods

Ricinus communis oil, *Pongamia glabra* oil and *Madhuca longifolia* oil were collected from the local shop. Hay infusion method was adopted for culturing mosquito larvae. The experimental set up consisted of six treatments each with three applications. 250ml beakers each containing 200 ml of water were used for the experiments. A control with three replications was also maintained simultaneously. Twenty newly emerged fourth instar larvae of *Culex quinquefasciatus* were introduced into the beakers for the bioassay studies. The results obtained in the bioassay studies were subjected to Probit Analysis (Finney, 1971) and the data were analysed using One way Analysis of Variance (ANOVA) as described by Panse and Sukhatme (1985).

Results and Discussion

The results of larvicidal activity of *Madhuca longifolia, Ricinum communis, Pongamia glabra* are presented in Table 48.1. The 24 h LC_{50}, regression equation, 95 per cent confidence limits DCL, LC_{50} and LCL LC_{50} of *Pongamia glabra* oil were 269.35 ppm, y = 2.92+0.85x, 1961.23 ppm and 36.98 ppm. Based on 24 hr LC_{50} value, the order of toxicity was *Pongamia glabra* (269.35 ppm) > *Ricinum communis* (293.88 ppm) > *Madhuca longifolia* (500.58 ppm).

Table 48.1: Lethal Concentration of *Ricinus communis, Pongamia glabra* and *Madhuca longifolia* Oils Against *Culex quinquefasciatus*

Oils Used	Hours	LC_{50} (ppm)	LC_{70} (ppm)	LC_{90} (ppm)	Regression Equation	95% Confidence Limit			
						UCL ppm		LCL ppm	
						LC_{50}	LC_{70}	LC_{50}	LC_{70}
Ricinus communis	24	293.88	715.65	2587.00	Y=1.65 + 1.35 x	508.11	22488.16	169.91	297.60
	48	130.79	715.84	8327.14	Y=3.49 + 0.71 x	373.91	1.05	45.75	6.58
	72	126.00	599.97	1.22	Y=5.12 +0.14 x	1.70	1.70	5.87	5.87
	96	23.48	59.35	226.3	Y=3.21 + 1.3 x	123.04	364.99	4.48	140.31
Pongania glabra	24	269.35	1110.07	8576.04	Y=2.92 + 0.85 x	1961.23	2.08	36.98	3.52
	48	95.89	1112.07	39087.05	Y=4.02 + 0.49 x	3.48	1.04	0.48	1.46
	72	75.17	128.81	280.28	Y=0.79 + 2.24 x	173.78	680.27	32.52	115.48
	96	36.13	74.69	213.08	Y=2.4 + 1.66 x	96.18	298.85	13.57	151.89
Madhuca longifolia	24	500.58	2355.26	22030.7	Y=2.89 + 0.77 x	133440.7	1.27	1.87	3.95
	48	120.71	365.59	1810.20	Y=2.73+ 1.09 x	158.48	17712.29	91.94	185
	72	74.25	188.68	725.07	Y=2.57 + 1.29 x	489.75	55862.2	11.25	9.41
	96	47.77	94.94	255.85	Y=2.06 + 1.78 x	368.78	821.78	6.19	79.66

LCL: Lower Confidence Limit; UCL: Upper Confidence Limit.

Sujatha, *et al.* (2001) observed the various degrees of larvicidal activity against fourth instar larvae of *Culex quinquefasciatus* with neem products and Palmarosa oil at higher concentrations more than 50 per cent mortality occurred within 24 hr. Trobousil *et al.* (2002) analysed the mixture of different essential oil against mosquito larvae with varying degrees of mortality with *Mentha communis,* and *Mentha microphylla.* Thomas *et al.* (2004) studied the essential oil obtained from the *Ipomea carica* plant could induced 50 per cent mortality in the larvae of *Anopheles stephensi* at a very low concentrations of 14.9 ppm, which is 2 to 7.6 times lower than these of plants. Mansour *et al.* (2000) studied the *Citrus reticulata* oil resin for the fourth instar larvae of *Culex quinquefasciatus* and isolated main ingredient, carvacrol, from the Egyptian plant, *Thymus capitatus* where lethal concentration was 100 ppm. *Neem* oil (*Azadirachta indica*) is well known for its activities against various insect vectors and has been reported as an effective larvicide against *C. fatigans wiedemann* (Attri and Prasad, 1980). *Cedrus deodara roxburge* oils has been reported to exhibit adulticidal and larvicidal action against *A. stephensi* (Singh *et al.*, 1984, Kumar and Dutta, 1987). Present study revealed that more than fifty per cent mortality was observed in 140ppm at 24 hr in *M. longifolia* oil.

This study indicates that the plant oil has remarkable larvicidal properties and its use as larvicide against mosquitoes.

References

Attri, B.S. and Prasad, G.R., 1980. Neem oil extractive and effective larvicide. *Ind. J. Entomol.*, 42: 371–374.

Finney, D.J., 1971. *Probit Analysis*. Cambridge University Press, p. 266–310.

Kumar, A. and Dutta, C.P., 1987. Indigenous plant oil as larvicidal agent against *Anopheles stephensi* mosquitoes. *Curr. Sci.*, 56: 959–960.

Mansour, S.A., Messeha, S.S. and El-Gengaihi, S.S.E., 2000. Botanical biocides. Mosquitocidal activity of certain Thymus capitatus constituents. *J. Nat. Toxins.*, 9:49–63.

Panse, V.G. and Sukhatma, P.V., 1985. *Statistical Methods for Agricultural Workers*, 4th edition. ICAR, New Delhi, p. 37.

Singh, D., Rao, S.M. and Tripathi, A.K., 1984. Cedar wood oil as a potential insecticidal agents against mosquitoes. *Naturwissen Schaften*, 71(5): 265–266.

Sujatha, C.H., Vasuki, V., Mariappan, T., Kalyana Suntharam, M. and Das, P.K., 1991. Evaluation of plant extracts for biological activity against mosquitoes. *Int. Pest. Control*, 30(5): 120–124.

Thomas, T.G., Sharma, S.K., Prakash, A. and Sharma, B.R., 2004. Insecticidal properties of essential oil of *Cannabis sativa* against mosquito larvae. *Entomon.*, 25(1): 21–24.

Trabousil, A.F., Taoubi, K., El-Haj, S., Besseire, J.M. and Rammal, S., 2002. Insecticidal properties of essential plant oils against the mosquito *Culex pipiens molistus* (Diptera: Culicidae). *Pest Manag. Sci.*, 58: 491–495.

Yang, P. and Ma, Y., 2005. Repellent effect of plant essential oils against *Aedes albopicus*. *J. Ver. Ecol.*, 30(2): 231–234.

Chapter 49

Shigometric Studies in Regard to Heartwood Formation

S.V. Deore[1] and P.V. Ramaiah[2]

[1]Jijamata College, Department of Botany, Nandurbar – 425 412, Maharashtra, India
[2]Centre for Post-Graduate studies and Research in Botany, G.T.P. College Campus, Nandurbar – 425 412, Maharashtra, India

ABSTRACT

In the present study moisture content of some tropical trees has been measured using a shigometer, in the living condition of the tree. The advantages of using shigometer are its instant read out, easy to operation and minimum injury to the tissue. The electrical resistance, being lowest in the barks of all the three species studied, gradually increased towards inside and showed sudden increase in the transition zone. Heartwood showed highest resistance due to low moisture content.

Keywords: *Angiosperms, Heartwood formation, Moisture content, Shigometer.*

Introduction

Loss of moisture content in the secondary xylem is an important factor involved in heartwood formation (Rudman, 1966, Zimmermann and Brown, 1971). Water deficit can cause hydrolysis of starch to sugar and this process reduces the oxygen content of the cells while increasing the concentration of carbon dioxide. Very likely, this increased gas pressure causes the death of the tissue (Rudman, 1966). As the moisture content increases strength, electrical resistivity and decay resistance of the wood decrease (Panshin and de Zeeuw, 1980).

Most of the available information on moisture content of wood is based on oven dry weight. The major disadvantage of the oven dry method is that it is a destructive test; it can take up to several days to complete; and a few species contain volatile components, other than water that can be driven off in the oven dry process, resulting in an incorrectly high moisture indication (Haygreen and Bowyer, 1982). The advantages of using a shigometer are its instant readout, easy of operation and minimum injury to the tissue. Therefore, to measure the relative humidity in various zones of the wood in *Acacia auriculiformis, Albizia lebbeck* and *Lagerstroemia parviflora,* a shigometer was used.

Materials and Methods

In the present investigation moisture content, indicated by the electrical resistance was measured using a shigometer, Model 7950, in different zones of wood in three tree species *viz., Acacia auriculiformis, Albizia lebbeck* and *Lagerstroemia parviflora.* A very small hole (3.5 mm in diameter and 15–30 cm long) was made into the trunk of healthy trees, with the help of battery powered driller. In to this hole, the probing wire electrode was slowly inserted. While inserting the probing wire slowly (pushed about 1.0 cm into the hole every time) the readings on the shigometer were noted. Electrical resistivity of the bark was measured using needle probes, which were inserted into the bark and readings on the shigometer were noted.

Results and Discussion

The high electrical resistance indicates low moisture content. In all the three species, bark showed low electrical resistance. The readings shown by the shigometer were 002 k ohms in, *Lagerstroemia parviflora,* 003 k ohms in *Albizia lebbeck* and 004 k ohm in *Acacia auriculiformis* respectively. Electrical resistance in all the three species gradually increased from back through sapwood to inwards. The readings (in ohms) obtained at different distances from the periphery to inwards are given in Table 49.1.

Usually the electrical resistance offered by the outer sapwood in all the three species was below 100 k ohms, and there was a sudden increase in the electrical resistance in the zone near the heartwood. The readings are: 230 k ohms in *Acacia auriculiformis,* 258 k ohms in *Albizia lebbeck* and 348 k ohms in *Lagerstreomia parviflora.* Heartwood of all the species showed high electrical resistance. The highest values obtained are: 462 k ohms in *Acacia auriculiformis,* 415 k ohms in *Albizia lebbeck,* 396 k ohms in *Lagerstroemia parviflora.*

Electrical resistance directly depends on the moisture content. As the moisture content increases, resistance to a pulsed current decreases. Bark showing low values for electrical resistance indicate high moisture content present in it. From outer sapwood to inwards, moisture content gradually decreases. A sudden decrease in the moisture content occurs at the transition zone. Heartwood showed low moisture content indicated by high electrical resistance.

Water relations are considered to be the main factor determining plant organ behavior and longevity (Levitt, 1972). Water stress, even when temporary, has been shown to accelerate senescence and reduce longevity of cut flowers (Borochov *et al.,* 1976; Mayak and Halevy, 1971). A sudden decrease in the moisture content in the transition zone in all the three species may be attributed to the senescing state of the parenchyma cells in this zone. Further more, loss of water, which is an important factor involved in heartwood formation (Zimmermann and Brown, 1971), can cause hydrolysis of starch to sugar. This process reduces the oxygen content of the cells causing concentration of carbon dioxide which brings about the death of the cells (Rudman, 1966). The moisture content decreased in the inner sapwood and transition zone, in all the three species studied. Moreover, loss of water is reported to trigger the production of ACC and ethylene (Borochov *et al.*, 1982). Therefore, it may be concluded that the sudden decrease in the moisture content in the transition zone in all the species indicates the production of ethylene which triggers the synthesis of heartwood phenolics. Ethylene is known to be involved in the synthesis heartwood phenolics (Nelson, 1978).

Table 49.1: Pattern of Electrical Resistance in Three Different Zones of Wood

Species Name	*Distance Traveled by Probing Wire*	*Diameter of the Wood*	*Probable Region of Wood*	*ER in k Ohms*
Acacia auriculiformis	1.00 cm	46.00 cm	Bark	004
	1.60 cm	46.00 cm	Outer Sapwood	085
	2.00 cm	46.00 cm	Middle Sapwood	122
	2.50 cm	46.00 cm	Inner Sapwood/ Transition Zone	192
	3.00 cm	46.00 cm	Heartwood	230
	3.50 cm	46.00 cm	Heartwood	312
	4.00 cm	46.00 cm	Heartwood	379
	6.50 cm	46.00 cm	Heartwood	462
Albizia lebbeck	1.00 cm	20.00 cm	Bark	003
	1.40 cm	20.00 cm	Outer Sapwood	084
	2.00 cm	20.00 cm	Outer Sapwood	122
	2.50 cm	20.00 cm	Middle or Inner Sapwood	142
	3.00 cm	20.00 cm	Inner Sapwood/ Transition Zone	258
	5.00 cm	20.00 cm	Heartwood	308
	6.50 cm	20.00 cm	Heartwood	415
Lagerstroemia parviflora	1.00 cm	42.00 cm	Bark	002
	1.50 cm	42.00 cm	Outer Sapwood	059
	2.00 cm	42.00 cm	Middle Sapwood	162
	2.50 cm	42.00 cm	Inner Sapwood	215
	3.00 cm	42.00 cm	Transition Zone	348
	3.50 cm	42.00 cm	Heartwood	380
	6.50 cm	42.00 cm	Heartwood	396

Heartwood in all the three species showed high electrical resistance indicating low moisture content this may due to deposition of phenolics in the lumen and displacement of water molecules by the phenolic groups in the cell walls. Such a phenomenon is also reported by Pandya (1986) and Ramaiah (1986).

Acknowledgements

Dr. Ramaiah is thankful to the University Grants Commission, New Delhi, India for financial support.

References

Borochov, A., Mayak, S. and Broun, R., 1982. The involvement of water stress and ethylene in senescence of cut carnation flowers. *Exp. Bot.*, 33(137): 1202–1209.

Borochov, A., Tirosh, T. and Halevy, A.H., 1976. Abscisic acid content of senescing petals of cut rose flowers as affected by sucrose and water stress. *Pl. Physiol., Lancaster* 58: 175–178.

Haygreen, J.G. and Bowyer, J.L., 1982. *Forest Products and Wood Science.* The IOWA State University Press, Ames.

Levitt, J., 1972. *Responses of Plants to Environmental Stresses.* Chapter 13. Academic Press, New York, pp. 322–353.

Mayak, S. and Halvey, A.H., 1971. Water stress as the cause of failure of flower bud opening in iris. *J. Amer. Soc. Hort. Sci.*, 96: 482–483.

Nelson, N.D., 1978. Xylem ethylene, phenol-oxidizing enzymes and nitrogen and heartwood formation in walnut and cherry. *Can. J. Bot.*, 56: 626–634.

Pandya, S.A., 1986. Sap wood and heart wood. *Ph.D. Thesis*, Sardar Patel University, V.V. Nagar, Gujarat, India.

Panshin, A.J. and Carl de Zeeuw, 1980. *Textbook of Wood Technology*, pp. 1–722. McGraw-Hill Book Company, New York.

Ramaiah, P.V., 1986. Studies in heart wood formation. *Ph.D. Thesis*, Sardar Patel University. V.V. Nagar, Gujarat, India.

Rudman, P., 1966. Heartwood formation in trees. *Nature*, 210(5036): 608–610.

Zimmermann, M.H. and Brown, C.L., 1971. *Trees: Structure and Function.* Springer-Verlag, New York.

Chapter 50

Impact of Health Education on Knowledge Regarding Hepatitis B Disease Vaccine and Acceptance of Hepatitis B Vaccine Among Non-immunized Nurses in Civil Hospital

Bharat H Patel[1], *Deepak Saxena*[2*], *Deepika Singhal*[1], *Amul B. Patel*[2], *V.K. Sharma*[4**], *and M.M. Prakash*[3]

[1]*Department of Ophthalmology, SMIMER, Surat – 395 001, Gujarat*
[2]*Department of Community Medicine, Government Medical College, Surat, Gujarat*
[3]*Department of Biotechnology, Government Model Autonomous Holkar Science College, Indore – 452 017, M.P.*
[4]*Department of Zoology, PMB Gujarati Science College, Indore – 452 001, M.P.*

ABSTRACT

Hepatitis B is a well-recognized major occupational hazard among health care workers. Safe and effective vaccines against Hepatitis B have been available for many years, yet surveys of vaccine uptake in health care workers have shown lower rates all over the world. A study was conducted to find out vaccination coverage among hospital nurses, to assess impact of health education on knowledge regarding Hepatitis B disease vaccine and acceptance of Hepatitis B

* E-mail: ddeepak72@yahoo.com, **E-mail: vipulkeerti@yahoo.com

vaccination among non-immunized nurses. Present study revealed 66 per cent Hepatitis B vaccination coverage among hospital nurses. There was overall significant improvement in awareness regarding Hepatitis B disease and Hepatitis B vaccination after intervention. But it could not end up in marked acceptance of Hepatitis B vaccine (23.6 per cent).

Keywords: *Hepatitis B Disease, Health Education, Intervention Study, Non-immunized nurses.*

Introduction

Hepatitis B is clearly a major infectious occupational hazard among health care workers who come into contact with blood and blood products (Suresh, 1999 and Lase, 1998). It is responsible for much morbidity, mortality, economic loss and human suffering. Risk of transmission of Hepatitis B virus through percutaneous exposure to health care workers is 9-30 per cent (Kirby, 1995). The International Labor Office (ILO), Viral Hepatitis Prevention Board (VHPB) and Indian Academy of Pediatrics Committee on Immunization recommend Hepatitis B vaccination in all health care workers at risk (Mathew, 1995). Safe and effective vaccines against Hepatitis B have been available for many years, yet surveys of vaccine uptake in health care workers have shown lower rates all over the world (Suresh, 1999). This study explores impact of health education on knowledge regarding Hepatitis B disease, vaccine and acceptance of Hepatitis B vaccine among non-immunized nurses.

Materials and Methods

The study was carried out from September to December 2003 among non-immunized nurses. The hospital had a total 412 nurses. By personal inquiry 140 non-immunized nurses were found out. Exclusion criteria decided, were subjects having at least one dose of Hepatitis B vaccine. They were equally divided into four groups for convenience of intervention. Intervention method identified for study was health education, including a lecture of 30 minutes followed by distribution of pamphlets and interactive session.

Lecture comprised of 30 minutes of informative speech regarding various aspects of Hepatitis B disease and Hepatitis B vaccination. A pamphlet was prepared to narrate all information discussed in lecture, so they could use afterwards. In interactive session, all nurses were encouraged to participate and their queries were responded. The pre-test questionnaire was administered to two groups in morning and two groups in afternoon. Three months after intervention, post-test questionnaire was administered to assess improvement in awareness regarding Hepatitis B disease and vaccination. Acceptance of at least one dose of Hepatitis B vaccine was also assessed among study subjects. Data were analyzed using EPI Info version 6.04 software.

Results and Discussion

A total of 140 non-immunized nurses were in the study. It was found that, approximately 81 per cent of the subjects knew Hepatitis B virus as causative agent of disease. Only 4 per cent could not identify symptoms and signs of Hepatitis B. Very few subjects (21.4 per cent) were aware that Hepatitis B is incurable. Except some subjects (6 per cent), majority of them were clear about high risk of acquiring Hepatitis B infection in health staff including nurses. After health education, overall significant improvement was observed in awareness about Hepatitis B disease (Table 50.1).

Table 50.1: Awareness Regarding Hepatitis B Disease (n = 140)

Items	*Correct Response*		*χ^2 Value*	*P Value*
	*Before HE. * No (per cent)*	*After HE. No (per cent)*		
Causative agent	113 (80.7)	140 (100)	–	–
Symptoms and signs	135 (96.4)	140 (100)	–	–
Can Hepatitis B lead to live cancer?	25 (17.8)	102 (72.9)	85.44	<0.001
Is Hepatitis B curable?	30 (21.4)	62 (44.3)	16.58	<0.001
High risk groups	132 (94.3)	138 (98.6)	3.73	>0.05

*H.E.: Health Education.

Three fourth of subjects could answer correctly about transmission of Hepatitis B through blood and blood products. 70 per cent, 46.4 per cent and 37.1 per cent subjects reported parental transmission, sexual route and child-to-child transmission as the modes of spread of Hepatitis B respectively. More than 80 per cent subjects were found using at least three universal precaution measures. On post test assessment, knowledge was significantly raised in all the above-mentioned aspects of disease (Table 50.2).

Table 50.2: Knowledge About Modes of Transmission and Practice of Universal (n=140)

Items	*Correct Response*		*χ^2 Value*	*P Value*
	*Before HE. * No (per cent)*	*After HE. No (per cent)*		
Through blood and blood products	105 (75.0)	135 (96.4)	26.25	<0.001
Perinatal transmission	98 (70.0)	125 (89.3)	16.06	<0.001
Sexual route	65 (46.4)	117 (83.6)	42.45	<0.001
Horizontal transmission	52 (37.1)	107 (76.4)	44.03	<0.001
Three protection measures they take	115 (82.1)	135 (96.4)	14.93	<0.001

*H.E.: Health Education.

More than 75 per cent of subjects were aware about 3-dose schedule (0,1 and 6 months) of hepatitis B vaccination. Knowledge regarding intramuscular route of administration and deltoid muscle as site of injection was 62.1 per cent and 57.1 per cent respectively. More than half subjects could able to tell two side effects of vaccine. Almost half of the subjects knew that, this vaccine can also be used for post exposure prophylaxis. After intervention, knowledge about hepatitis B vaccination was significantly increased except dose and schedule of vaccine (Table 50.3).

Out of 140 nurses, only 33 (23.6 per cent) accepted hepatitis B vaccination. There was no significant association found between acceptance of vaccination and influencing factors like age, place of training, duration of job, family history of liver disease and health related reading.

Results revealed that pre-test knowledge was good about clinical features of Hepatitis B (96.4 per cent). McGrane and Staines (2003) have also observed good understanding about Hepatitis B in nursing staff. Ayyát *et al.* (2000) found poor level of knowledge and addressed need for health education. They could identify themselves as high-risk groups (94.3 per cent). Chekbunyasit *et al.* (1995) reported that nurses were at a higher risk than other categories of worker within same age and working year groups. It was annoying to find their ignorance about severity of disease (17.8 per cent). Even after health education, more than half subjects did not agree that Hepatitis B is incurable (55.7 per cent). O'Boyle *et al.* (1994) concluded in their study that Hepatitis B virus has not generated the heightened concern in health care workers that possible infection with HIV has. Because of the high mortality of HIV infection and daily media coverage about HIV transmission has often overshadowed the risk of Hepatitis B.

Table 50.3: Awareness Regarding Hepatitis B Vaccination (n = 140)

Items	*Correct Response*		*χ^2 Value*	*P Value*
	*Before HE. ** *No (per cent)*	*After HE.* *No (per cent)*		
Dose and schedule	110 (78.6)	122 (87.1)	3.62	> 0.05
Route of administration	87 (62.1)	115 (82.1)	13.93	< 0.001
Site of injection	80 (57.1)	128 (91.4)	43.00	< 0.001
Two side effects	72 (51.4)	112 (80.0)	23.36	< 0.001
Use as post exposure prophylaxis	68 (48.6)	118 (84.3)	40.04	< 0.001

*H.E.: Health Education.

The pre-test knowledge about mode of transmission of HBV through blood and blood products (75 per cent) and parental transmission (70 per cent) was sound enough. But awareness about transmission through sexual route (46.4 per cent) and horizontal transmission (37.1 per cent) was poor. Reason for this may be that their knowledge would be based only on workplace experience. Stein *et al.* (2003) reported that 44 per cent nurse had knowledge of risk of HBV transmission from infected patient after needle stick injury. Compliance to protective measure was good (82.1 per cent). Several studies pointed out that there were less than optimal levels of compliance with universal precautions among health care workers. Reasons for that were lack of time, low HIV, HBV risk patient and interference with technical skill (O'Boyle, 1994 and Henry *et al.*, 1994).

It was discouraging to observe that awareness regarding route (62.1 per cent), site of administration (57.1 per cent) and side effects (51.4 per cent) was not so good. Santos *et al.*, 1989) also discovered the same that there was a low information leve] about vaccine (81.5 per cent) and a low intention to be vaccinated in future (34.6 per cent), mostly due to poor information. Knowledge about use of Hepatitis B vaccine as post-exposure prophylaxis was fair (48.6 per cent). Lim *et al.* (1990) found that hospital personnel working less than 5 years pay more attention to this kind of event. In our study, no such difference was found.

Present study disclosed that Hepatitis B vaccination coverage among hospital nurses was 66 per cent. Hersey and Martin (1994) reported 55 per cent of health care staff receiving at least one shot of recommended in HBV vaccination series. Duseja *et al.* (2002) have observed that 15 per cent of health care workers received only one dose of HBV vaccine. McGrane and Staines (2003) and Talaat *et al.* (2003) reported vaccination coverage 83 per cent and 15.8 per cent respectively. After Health education, only 33 (23.6 per cent) accepted Hepatitis B vaccine (at least one dose). Ling *et al.* (2000) concluded that health education and hospital wide Hepatitis B immunization programme implementation resulted in number of health care workers with no previous history of Hepatitis B immunization decreased significantly from 17 in 1996 to 9 in 1997 ($p < 0.001$).

In conclusion, though there was a significant improvement in awareness, it could not end up in marked acceptance of Hepatitis B vaccine. It needs longer and persistent efforts to convert knowledge into attitude and behaviour. To get maximum acceptance, workplace vaccination programmes should be introduced as part of an overall preventive and educational programme on the transmission of blood borne pathogens.

References

Ayyat, A.A., Syed, H.A. and Abou, Had A.M., 2000. A KAP study among staff and student nurses about infection control in Theodor Bilharz Hospital. *J Egypt Soc Parasitol.*, 30(2): 511–522.

Checkbunyasit, N., Potacharoen, O. and Sirisanthana, T., 1995. Prevalence of HBV infection in nurse and manual workers in Maharaj Nakarn Chiang Mal Hospital. *J. Med. Assoc. Thailand*, 78 (Suppll.): 519–520.

Duseja, A., Arora, L., Masih, B., Singh, H, Gupta, P. and Behera, D., 2002. Hepatitis B and C virus–prevalence and prevention in health care workers. *Trop. Gastroenterol.*, 23(3): 125–126.

Henry, K., Campbell, S., Collier, P. and O'Boyle, Williams C., 1994. Comdiance with universal precautions and needle handling and disposal practices among emergency department staff of two community hospitals. *American Journal of Infection Control*, 22(3): 129–137.

Hersrey, J. and Martin. L., 1994. Use of infection control guidelines by workers in health care facilities to prevent occupational transmission of HBV and HIV: Results from a National Survey. *Infection Control and Hospital Epidemiology*, 15(4): 243–252.

International Labor Conference. 1985. Recommendation 171, Occupational Health Services Convention.

Kirby, M., 1995. AIDS and the law: opportunities and limitations. *AIDS Asia*, 7: 9–10.

Lase, J.M., 1998. *Public Health and Human Ecology*, 2nd edn. CT: Appleton and Lange, pp. 139–140.

Lim, K.S., Chou, M.Y., Hsu, G.J., Lu, M.Y. and Chan, I.W., 1990. Investigation of relation between HBV knowledge and needle-stick injury personnel. *Zhonghua Xuc Za Zhi*, 46(2): 104–108.

Ling, M.L., Wee, M. and Chan, Y.H., 2000. Sharps and needle-stick injury: The impact of Hepatitis B vaccination as an intervention measure. *Ann. Acar. Med., Singapore*, 29(1): 86–9.

Mathew, A., 1995. Immunization dialogue. *Indian Pediatrics*, 32: 835–836.

McGrane, J. and Staines, A., 2003. Nursing staff knowledge of the Hepatitis B virus including attitudes and acceptance of Hepatitis B vaccination: Development of effective program. *AAOHN J.*, 51(8): 347–352.

O'Boyle, Williams C., Campbell, S., Henry, K. and Collier, P., 1994. Variables influencing worker compliance with universal precautions in emergency department. *AJIC*, 22(3): 138–148.

Santos, J.L., Seisdedos, T., Sancher, J.M. and Moreno, J., 1989. Analysis of attitude of hospital staff regarding hepatitis B vaccine. *Enterm Intece Microbiol Clin.*, 7(3): 257–260.

Stein, A.D., Makarawo, T.P. and Ahmad, M.F., 2003. A survey of doctors and nurses knowledge, attitude and compliance with infection control guidelines in Birmingham teaching hospital. *J. Hosp. Infect.*, 54(1): 68–73.

Suresh, T.P., 1999. Hepatitis B: An occupational hazard. *Nurs. J. India*, 90(6): 137.

Talaat, M., Kendeel, A., Elshoubary, W., Badenschatz, C., Khairy, I. and Oun, S., 2003. Occupational exposure to needle–stick injuries and Hepatitis B.

Chapter 51

Analytical Study of Molecular Interactions in Binary and Ternary Liquid Mixtures

D.K. Gupta and G. Meenakshi

Department of Physics, Kanchi Mamunivar Centre for Post Graduate Studies, Lawspet, Pondicherry – 605 008

ABSTRACT

Ultrasonic velocity measurements are extensively used to study the physico-chemical behaviour of liquid mixtures. In recent years, a good deal of work has been done to correlate the experimental ultra sonic velocities with those of computed theoretically using empirical and semi-empirical relations. The ultrasonic velocities have been evaluated in the binary systems of Bromobenzene with Methanol, Ethanol, Propan-1-ol and Butan-1-ol at 303.15 K. Sound velocity has also been evaluated in ternary mixtures of Aniline + Benzene + Acetic acid, Aniline + Benzene + Propionic acid and Aniline + Benzene + Butryic acid at a temperature of 303.15 K. Alpha, excess molar volume, excess velocity are also evaluated and are used to explain the interaction in the mixtures.

***Keywords**: Binary and ternary liquid mixtures, Molecular interactions, Sound velocity.*

---

* E-mail: ashwinikdixit@yahoo.co.in

Introduction

Liquids and liquid mixtures have found wide applications in the chemical textiles, leather and nuclear industries (Rowlinson and Swinson, 1982; Acree, 1984 and Prausnitz *et al.*, 1986). The measurement of ultrasonic velocity in the liquid mixtures has been found to be an important tool to study the physico-chemical properties of the mixtures. The ultrasonic velocity is also useful to study the molecular interactions in variety of liquid mixtures. The main objective of the present investigation is to explore molecular interactions in binary and ternary liquids, because there is no much data available in such systems in literature. For this purpose four different alcoholic liquids with Bromobenzene for binary system and three different acids with Aniline and Benzene for ternary system are selected. The theoretical evaluation of sound velocity and excess parameters are important as the deviation of experimental values from the ideal values may yield the information on the nature of molecular interactions. The calculations are based on the experimental results of Nikam *et al.*, 2000 for binary mixtures and Kannappan *et al.*, 1992 in the case of ternary mixtures.

Materials and Methods

Theoretical evaluation of sound velocity for binary and ternary liquid mixtures are done using the following empirical formulae (Pandey *et al.*, 1991; Gupta and Pandey, 1988; Srivastava, 1992 and Ali *et al.*, 2000):

$$U_{mix} = \left[\frac{X_1R_1 + X_2R_2}{X_1V_1 + X_2V_2}\right]^3 \quad (1)$$

where,

R_1, R_1 are the molar sound velocities, X_1, X_2 are mole fractions and V_1, V_2 are the molar volumes.

$$\left[\frac{1}{X_1M_1 + X_2M_2}\right]\left[\frac{1}{U^2{}_{im}}\right] = \left[\frac{X_1}{M_1V_1{}^2} + \frac{X_2}{M_2V_2{}^2}\right] \quad (2)$$

$$U_{mix} = \left[\frac{K}{L_{f(mix)}\rho^{1/2}{}_{mix}}\right] \quad (3)$$

where,

K is the temperature constant and ρ_{mix} is the density of the mixture.

$$U_{mix} = \left[\frac{U_\infty(X_1S_1 + X_2S_2)(X_1B_1 + X_2B_2)}{V_m}\right] \quad (4)$$

where,

S_1, S_2 are collision factors, $V_\infty = 1600\ ms^{-1}$, B_1, B_2 are actual volumes of the molecule per mole and V m is molar volume.

Results and Discussion

Bromobenzene in Methanol, Ethanol and Propan-1-ol and Butan-1-ol at the temperature 303.15 K are discussed. The experimental value of ultrasonic velocity and the percentage deviation of the theoretical velocity from the experimental values in the above mixtures of different mole fractions of Bromobenzene are reported in Table 51.1. The calculated values of alpha (α), excess velocity (U^E) and excess volume (V^E) in different alcohols with mole fraction of Bromobenzene are also given in the same table.

Table 51.1: Percentage Deviation of the Calculated Value of Ultrasonic Velocity from the Experimental Value and Excess Parameters of Binary Liquid Mixtures at 303.15 K

Name of Com-pounds	*Mole Fraction of Bromo-benzene*	*Experimental Value of 'U'ms⁻¹*	*Alpha (α)*	*Percentage Deviation* $\left(\frac{\Delta U}{U}\right)\%$				*Excess Velocity (UE) ms⁻¹*	*Excess Volume (V^E) $10^{-3}m^3mol^{-1}$*
				Nomoto	*Van–Dael*	*FLT*	*CFT*		
Bromo	0.0222	1070.7	–0.0425	–2.33	1.16	–3.16	–3.67	–23.51	–0.4212
benzene	0.0805	1068.3	–0.0514	–3.11	7.66	–6.75	–7.30	–28.55	–0.1913
in Methanol	0.1697	1066.4	–0.0617	–3.97	14.36	–10.50	–11.08	–34.49	0.0459
	0.3226	1076.0	–0.0566	–3.91	21.31	–12.07	–13.08	–31.81	–0.0245
	0.6473	1096.1	–0.1468	–3.14	22.88	–9.77	–10.15	–27.42	–0.1812
Bromo	0.0316	1134.2	–0.0008	–0.05	2.48	–0.57	–0.81	–0.43	–0.1691
benzene in	0.1118	1126.4	–0.0156	–0.78	6.83	–2.47	–3.15	–8.55	–0.3135
Ethanol	0.2269	1120.3	–0.0264	–1.39	11.11	–4.60	–5.26	–15.11	–0.3962
	0.4064	1115.0	–0.0368	–1.95	14.28	–5.33	–6.72	–21.13	–0.4128
	0.7255	1121.6	–0.0276	–1.44	12.45	–3.97	–4.73	–15.80	–0.2068
Bromo	0.0407	1197.8	0.0095	–0.55	2.25	–1.50	–0.06	–5.68	–0.3627
benzene in	0.1410	1172.2	–0.0240	–0.99	4.04	–2.76	–2.48	–14.32	–0.6305
Propan-1-ol	0.2768	1155.5	–0.0393	–1.66	6.24	–3.56	–4.03	–23.43	–0.6283
	0.5173	1142.2	–0.0396	–1.61	7.78	–4.05	–4.36	–23.28	0.3287
	0.7751	1131.0	–0.0346	–1.50	5.21	–3.28	–3.28	–20.07	0.3774
Bromo	0.1683	1191.2	–0.0366	–1.70	1.56	–1.86	–2.50	–22.40	–0.1048
benzene in	0.3208	1172.2	–0.0455	–2.08	2.78	–2.59	–3.31	–27.63	–0.2699
Butan-1-ol	0.4145	1162.2	–0.0484	–2.20	3.10	–3.10	–3.57	–29.17	–0.3276
	0.5241	1149.2	–0.0539	–2.49	2.87	–3.87	–3.89	–32.27	–0.2939
	0.6538	1137.1	–0.0551	–2.59	2.23	–2.95	–3.85	–32.66	–0.2100

From the Table 51.1, it is observed that the percentage deviations of computed ultrasonic velocity using various theories, like Nomoto's method, Van Dael ideal mixing relation, collision factor theory and FLT for the mixture of Bromobenzene with different alcoholic group show that the Nomoto's method is most suitable for the calculation of ultrasonic velocity. The interaction effect can be explained from the calculated excess thermodynamic parameters. Both Bromobenzene and alcohols are polar liquids which can interact by forming hydrogen bonding through dipole-dipole interaction. Alcohols are liquids, associated through hydrogen bonding and in the pure state exhibit equilibrium between monomers and multimer species. In the pure state the halogenated compounds present weak dipole-dipole interaction. When Bromobenzene is mixed with alcohols the halogenated solvents can interact with or groups (Arul and Palaniappan, 2001; Kannappan and Rajendran, 1992). The aromatic derivatives set up an interaction between the electron cloud and the hydroxyl groups. Of course this interaction is of minor intensity compared with hydrogen bonding but they can lead to the formation of intermolecular complexes.

In the present study α, excess velocity and excess volume are found to be negative. The negative values of excess parameters indicate that the strong interaction in the liquid mixtures as well as interstitial accommodation of Bromobenzene molecules is in aggregates of alcohol. The negative excess volume indicates the formation of molecular clusters and complexes. This may involve in charge transfer complexes. The negative excess sound velocity vary with mole fraction may be attributed to the breaking of both the hydrogen bond and dipolar interaction in mixing. The average percentage deviation of ultrasonic velocity in ideal mixing relation decreases in the order Methanol > Ethanol > Propan-1-ol > Butan-1-ol (13.78 > 9.84 > 5.33 > 1.77). This trend suggests that the hetro association and homo association of molecules decreases with increase of chain length of carbon atoms in alcohols. This is probably due to less proton donating tendency of higher alcohols. The decrease of average percentage deviation in the ideal mixing relation with the increasing chain length of the carbon atoms in alcohols suggests that the ideal mixing relation can give good results with higher alcohols.

The ternary systems chosen for this study are Aniline and Benzene with carboxylic acids namely Acetic, Propionic and Butryic. The percentage deviation of ultrasonic velocity for the systems Aniline and Benzene in Acetic acid, Propionic acid and Butryic acid and the calculated values for α, excess volume and excess velocity are listed in Table 51.2. It is observed that the collision factor theory agrees better with the experimental values.

Table 51.2: Percentage Deviation of the Calculated Values of Ultrasonic Velocity from the Experimental Values and Excess Parameters of Temary Liquid Mixtures at 303.15 K

X_1 and X_3 Refers to	Name of Compounds	Mole Fraction X_1	Mole Fraction X_3	Experimental Value of (U^E) ms^{-1}	Alpha (α)	Percentage Deviation $\left(\frac{\Delta U}{U}\right)\%$ Nomoto	Van-Dael	FLT	CFT	Excess Velocity (U^E) ms^{-1}	Excess Volume (V^E) 10^{-3} m^3 mol^{-1}
Aniline	Aniline +	0.0980	0.5041	1234.1	–0.4167	–3.30	1.49	–3.69	–3.09	–297.58	–0.4795
and	Benzene +	0.2002	0.4014	1304.0	–0.4531	–1.35	4.80	–1.26	–0.98	–399.82	–0.743
Acetic	Acetic Acid	0.2988	0.3000	1352.2	–0.4890	–0.86	5.82	–0.76	–0.45	–284.53	–0.7433
Acid		0.3990	0.2003	1399.2	–0.5221	–0.34	6.07	0.24	0.00	–430.01	–0.6579
		0.4967	0.1023	1440.4	–0.5519	–0.02	5.26	0.15	0.17	–474.76	–0.5466
Aniline	Aniline +	0.0694	0.5266	1224.0	–0.4398	–3.24	–1.61	–3.58	–3.11	–317.03	–0.8575
and	Benzene +	0.1954	0.4100	1293.0	–0.4766	–1.75	1.35	–1.89	–1.54	–362.87	–0.8235
Propionic	Propionic	0.2911	0.3053	1344.3	–0.4983	–0.83	2.89	–0.80	–0.62	–394.59	–1.3683
acid	acid	0.3950	0.2058	1395.8	–0.5277	–0.08	3.91	0.47	0.08	–436.59	–1.0608
		0.4972	0.1044	1439.9	–0.5544	0.19	3.97	0.27	0.26	–478.25	–0.7915
Aniline	Aniline +	0.1007	0.4975	1249.0	–0.4624	–1.18	–0.14	–1.46	–1.28	–338.27	–0.7555
and	Benzene +	0.2015	0.3951	1295.7	–0.4921	–0.72	1.12	–0.98	–0.86	–376.81	–0.5751
Butryic	Butryic acid	0.3008	0.2989	1344.8	–0.5109	–0.10	2.34	–0.37	–0.26	–406.98	–1.0725
acid		0.3999	0.1979	1390.0	–0.5384	0.09	2.94	–0.07	–0.05	–447.61	–0.6365
		0.5058	0.1012	1436.9	–0.5567	0.19	3.26	0.26	0.08	–481.47	–0.9028

Benzene is a non-polar solvent and the carboxylic acids are polar in nature. Therefore they are associated with unsymmetrical molecules. In view of the high dipole moments, Aniline can be classified as dissociating solvent as it has the ability to dissociate the molecules of associative liquids like water, alcohols and carboxylic acids (Kannappan and Rajendran 1992) etc. The α, excess velocity and excess molar volume are found to be negative for all the concentrations studied. The negative value of a indicates the presence of complex formation. As explained in the case of binary mixtures, in ternary mixture also show negative values of excess sound velocity which may be attributed to the breaking of both hydrogen bond arid bipolar interaction on mixing. The negative value of excess volume indicates the formation of molecular clusters and complexes and may involve in charge transfer complexes. The various results show the existence of strong hydrogen bonding interactions as well as interaction due to the formation of charge transfer complex between acids and benzene.

Conclusion

Generally in liquid mixtures the possible interactions are between like molecules as well as unlike molecules. These interactions are of two types; namely, long range and short range. The long range interaction includes electro static induction and dispersion forces. These forces arise without the overlap of the electron clouds due to the closer approach of interacting molecules. Normally the long range interactions are highly directional. On the other hand, short range interactions such as dipole-dipole, dipole-induced dipole, charge transfer, complex formation and hydrogen bonding interactions arise when the molecules come closer together, resulting in a significant overlapping of electron clouds. Therefore, when two or more liquids are mixed (in the case of binary and ternary mixtures), then the mixture is not ideal. Thus the deviation from the ideality is explained on the basis of molecular interactions between the components of the liquid mixtures.

Acknowledgements

The authors are thankful to Dr. S. Mayilavelan, Head of the Department of Physics, Dr. Ashwini Kumar Dixit, Department of Plant Science and the Director, Kanchi Mamunivar Centre for Post Graduate Studies, Lawspet, Pondicherry, for their constant encouragement and valuable suggestions while preparing this manuscript. The authors are also wish to thank the Staff members of Physics Department, for their constant encouragement during the preparation of this manuscript.

References

Acree, W.E., 1984. *Thermodynamic Properties of Non-electrolytic Solutions.* Academic Press, New York.

Ali, A., Tiwari, K., Nain, A.K. and Chakravortty, V., 2000. Ultrasonic study of molecular interaction in ternary mixtures of dimethylsulphonamide + corbomtetrachloride + aromatic hydrocarbons. *Ind J. Physics*, 74B(5): 351–355

Arul, G. and Palaniappan, C., 2001. Molecular studies in the ternary liquid mixture of cyclohexene + tolune + 2 proponol. *Ind. J. Pure and Appl. Phys.*, 39: 561.

Gupta, D. and Pandey, J.D., 1988. Sound velocity in reciprocalfused salt pairs. *Acoustic Letters*, 12(3): 55–58

Kannappan, A.N. and Rajendran, V., 1992. Acoustic parameters of some ternary liquid mixtures. *Ind. J. Pure and Appl. Phys.*, 30: 240.

Kannappan, A.N. and Rajendran, V., 1994. Molecular interaction studies in ternary liquid mixtures from ultrasonic data. *Ind. J. Pure and Appl. Phys.*, 29: 465.

Nikam, P.S., Kapade, V.M. and Hasen, M., 2000. Molecular interaction in binary mixtures of bromo benzene with normal alkanols: An ultrasonic study. *Ind. J. Pure and Appl. Phys.*, 38: 170–173.

Panday, J.D., Dubey, G.P., Shukla, B.P. and Dubey, S.N., 1991. Prediction of ultrasonic velocity and intennolecular tree length and their correlation with interaction in binary liquid mixtures. *Pranama J. Phy.* 37(6): 497–503.

Prausnitz, J.M., Linchenthalr and Azeved, E.G., 1986. *Molecular Thermodynamics of Fluid-phase Equilibriam*, 2nd edn. Eagle Wood Cliffs, Prentice Hall Inc.

Rowlinson, J.S. and Swinson, F.L., 1982. *Liquid and Liquid Mixtures*, 3rd edn. Buttor Worth Scientific, London.

Srivastava, A.P. 1992. Evaluation of sound velocity in binary systems of anisole with some aromatic compounds by different methods. *Ind. J. Chem.*, 31A: 577–580.

Chapter 52

Silver Toxicity to *Anabaena doliolum* Bharadwaja

Praveen K. Chandel*, Dilip K. Rathore and Krishna Chandra Sonie

Department of Botany, University College of Science,
Mohanlal Sukhadia University, Udaipur – 313 001, Rajasthan, India

ABSTRACT

Cyanobacteria have tremendous capacity of nitrogen fixation, as well as, capacity of adaptability of high order. Detergents and heavy metals are discharged from various industrial processes, modern agricultural practices, acid mine drainage and human wastes in aquatic habitat. Sustenance of life in such contaminated water bodies requires proper management and restoration by way of nutrient and metal removal. *Anabaena doliolum* can play an efficient role in this kind of activity. It has exhibited the survival potential up to 0.75 ppm of silver concentration. So, its growth therein will help in removal and gleaning of this particular metal ion. Thus, cyanobacteria are valuable tool for bioassay of metal toxicity and the tolerant forms can be used for water self-purification processes in natural water bodies.

Keywords: *Heavy metals, Gleaning, Toxicity, Water self-purification.*

Introduction

Cyanobacteria are fascinating group of photosynthetic prokaryotes, endowed with a tremendous capacity of nitrogen fixation and its distribution in ecologically diverse habitats proves their capacity

* E-mail: cruzshiphar@yahoo.co.in

of adaptability of high order. They constantly face stressful environment due to presence of detergents and heavy metals in aquatic habitat which are discharged from various industrial processes, modern agricultural practices, acid mine drainage and human wastes.

Metal ions may mark their presence in trace or ultra trace amounts and may be acting as micro- or macro-nutrients. Some metals are considered as heavy metals input into aquatic environment by continued anthropogenic activities. They possess a potential threat to the structure and functioning of the ecosystem. Metal can be accumulated by aquatic, organisms, they enter the food chain leading to man and thus become a human health problem (Rachlin *et al.*, 1982). The toxicity of heavy metal to eukaryotic algae has been widely acknowledged (Rai *et al.*, 1981; Stokes, 1983; Whitton, 1984). Cyanobacteria have, however, been little explored in this regard (Stratton and Corke, 1979a,b; Singh and Yadav, 1983, 1984; Rai and Raizada, 1985). Heavy metal toxicity is influenced by presence of various group due to their metal binding properties (Rai *et al.*, 1981; Rai and Raizada, 1985; Raizada and Rai, 1985).

Many works related to cyanobacterial interaction with heavy metal provide us toxicity sequence of heavy metal Ag > Ni > Cr > Pb (Rai *et al.*, 1990). Silver is widely distributed in nature but in micro amounts, silver is also universally present in soils and plants. It is well-known as less toxic element for higher organisms but probably the most toxic element for microorganism.

Exposure of *Anabaena doliolum*, a pale blue-green filamentous, heterocystous cyanobacterium with barrel shaped cells at different concentration of silver has resulted in reduced growth and heterocyst production. This also has led to decrease in the content of chlorophyll a, carbohydrates, proteins and amino acids.

Materials and Methods

A. doliolum is selected to find out the silver toxicity and physiological responses in terms of chlorophyll-a, total carbohydrate, protein and amino acids. *A. doliolum* was cultured in nitrogen-free Allen and Amon's (1955) medium.

Growth was measured using Systronics UV-VIS spectrophotometer–117 by taking optical density of pigments at 630, 645 and 665 nm and calculated as follows (Kaushik, 1987).

$$\text{Total Growth (Chl.a)} = C_a \,.\, V_a / V_c \text{ mg/l}$$

where,

Ca: Amount of Chl.a

V_a: Volume of 80 per cent acetone

V_c: Volume of culture concentrate

$$(C_a = 11.6\ OD_{665} - 1.31\ OD_{645} - 0.14\ OD_{630})$$

and the growth rate percentage was calculated using formula given below.

$$\text{Growth rate percentage} = dx/dt \,.\, 1/x \,.\, 100$$

where,

dx: Change in growth (final – initial growth)

dt: Time (days)

x: Initial growth

Specific growth rate was calculated at 663 nm using Systronics UV-VIS Spectrophotometer–117. Protein was estimated by Lowry's method (Lowry *et al.*, 1951).

Carbohydrate was estimated by acid hydrolysis Anthrone reaction method (Plummer, 1979) after the growth period of 18 days.

Heterocyst frequency is expressed as number of heterocysts per hundred vegetative cells. Observations were taken from at least 10–15 fields and an average of 10 filaments were taken as per requirement of the experiment.

Results and Discussion

Various physiological and biochemical parameters were analyzed for the interaction of *A. doliolum* with silver.

In growth measurements, specific growth rate at 0.05 ppm has shown stimulation over control and later a gradual decrease was observed with increasing concentration (Table 52.1). Similar trend was also observed for growth rate percentage (Figure 52.1).

Table 52.1: Effect of Different Concentrations of Silver on Specific Growth Rate

Concentration of Silver (ppm)	*Specific Growth Rate (ΔA_{663})*	
	Mean±SD	*Inhibition (%)*
0.00	0.541±0.002	–
0.05	0.584±0.002	–7.95
0.10	0.443±0.005	18.11
0.15	0.428±0.003	20.89
0.20	0.427±0.003	21.07
0.25	0.411±0.003	24.03
0.30	0.331±0.001	38.82
0.35	0.328±0.002	39.37
0.40	0.298±0.004	44.92
0.45	0.295±0.005	45.47
0.50	0.250±0.005	53.79
0.55	0.243±0.005	55.08
0.60	0.221±0.002	59.15
0.65	0.171±0.003	68.39
0.70	0.166±0.004	69.32
0.75	0.144±0.006	73.38

Data with negative sign show stimulation over control.

At various concentrations of silver, initially total carbohydrate content increased up to 0.10 ppm (3.4089 g/l) and then a regular decrease was seen. The maximum value for total protein content was observed in control (4.4618 g/l) and thereafter, it decreased gradually. Amino acid content of the test organism showed the peak value at 0.05 ppm (1.0194 g/l) and then a declined towards the toxic level (Figure 52.2). pH of supernatant obtained after centrifugation of cultured teat organism showed slight variation in relation to different concentrations of silver (Figure 52.3).

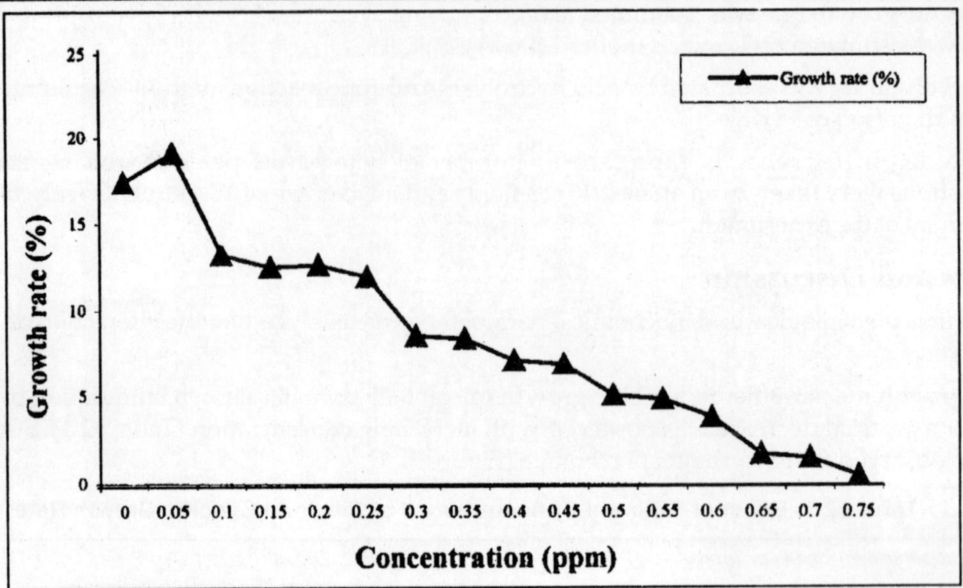

Figure 52.1: Growth Rate Percentage of *Anabaena doliolum* at Various Silver Concentrations

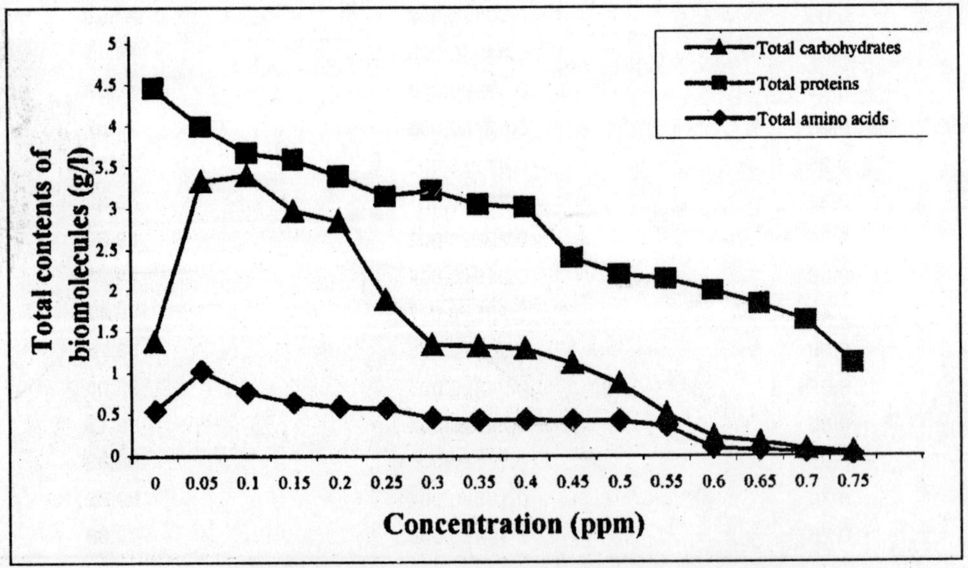

Figure 52.2: Total Content of Carbohydrates, Proteins and Amino Acids of *Anabaena doliolum* at Different Concentrations of Silver

Heterocyst frequency was maximum at control (0.00 ppm) and decreased regularly up to 0.65 ppm. At 0.70 ppm and 0.75 ppm, no significant frequency (*i.e.*, nearly zero) was observed which might be due to silver toxicity. In contrast, heterocyst dedifferentiation showed an increase up to 0.75 ppm (Figure 52.4).

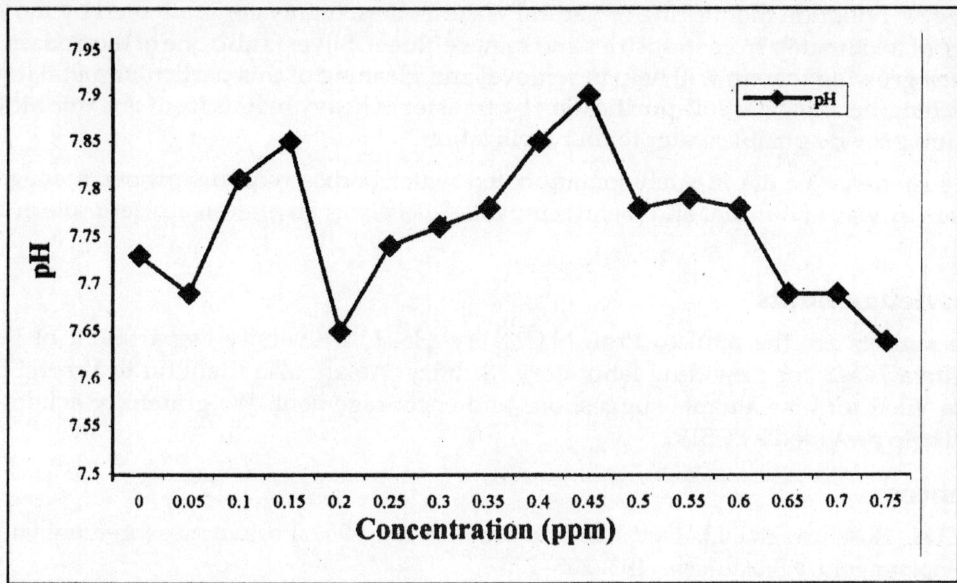

Figure 52.3: pH of the Medium after Incubation Period at Different Concentrations of Silver Ion

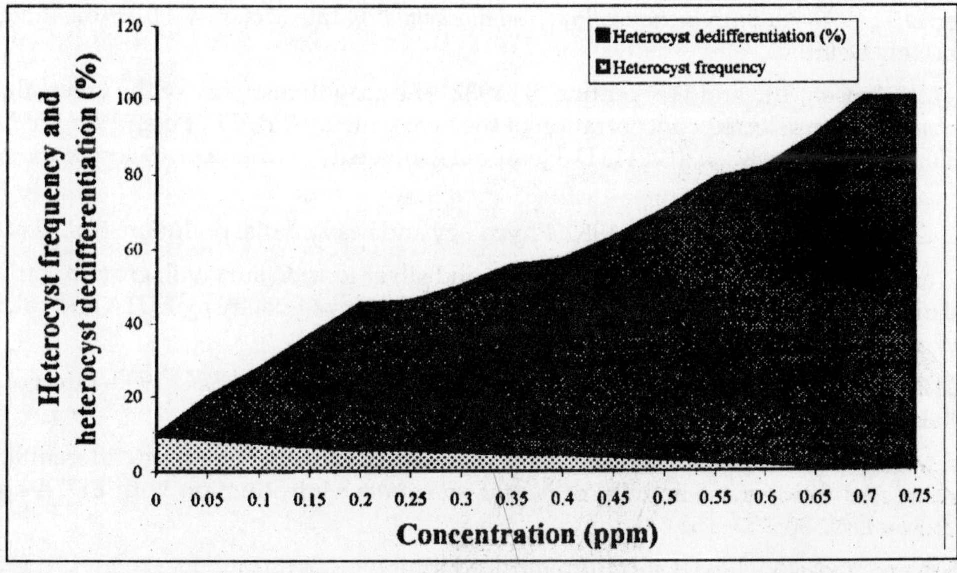

Figure 52.4: Heterocyst Frequency and Heterocyst Dedifferentiation of *Anabaena doliolum* at Different Concentrations of Silver

Conclusion

A. doliolum exhibited the survival potential up to 0.75 ppm of silver concentration. So, the present study and finding will help towards the major environmental health problems confronting both developed and developing countries related to municipal water supply and proper facilities for disposal of sewage *i.e.* pollution monitoring of natural water bodies, highly contaminated by indiscriminate discharge of toxic metals from industries and mines effluent. Silver is also one of the toxic heavy metal. *A. doliolum* growing therein will help in removal and gleaning of this particular metal ion. This will further strengthened water self-purification by transfer of heavy metals from one trophic to another, thus in turn provide potable water for our civilization.

So, sustenance of life in such contaminated water bodies requires proper management and restoration by way of nutrient and metal removal. *A. doliolum* can play an efficient role in this kind of activity.

Acknowledgements

The authors are thankful to Prof. N.C. Aery, Head, University Department of Botany, and Dr. Sandhya Tyagi, for providing laboratory facilities. We are also thankful to Preeti Sharma and Premlata Vikal for his valuable suggestions and encouragement. We gratefully acknowledge the financial help provided by CSIR.

References

Lowry, O.H., Rosebrough, N.J., Farr, A.L. and Randall, R.J., 1951. Protein measurement with the folin-phenol reagent. *J Biol. Chem.*, 193: 265–275.

Kaushik, B.D., 1987. *Laboratory Methods for Blue-green Algae.* Associated Publishing Company, New Delhi, India, p. 31–62.

Plummer, D.T., 1979. *An Introduction to Practical Biochemistry.* Tata McGraw-Hill Publishing Company Ltd., New Delhi.

Rachlin, J.W., Jensen, T.E. and Warkentine, B., 1982. The growth response of the green alga (*Chlorella saccharophila*) to selected concentration of the heavy metals Cd, Cu, Pb and Zn. In: *Environment Health, XVI: A Symposium,* (Ed.) D.D. Hemphill. University of Missouri, Columbia, U.S.A, pp. 45–54.

Rai, L.C., Gaur, J.P. and Kumar, H.D., 1981. Phycology and heavy metal pollution. *Biol. Rev.*, 56: 99–151.

Rai, L.C. and Raizada, M., 1985. Effect of nickel and silver ions on survival, growth, carbon fixation and nitrogenase activity of *Nostoc muscorum:* Regulation of toxicity by EDTA and calcium. *J. Gen. Appl. Microbiol.*, 31: 329–337.

Rai, L.C., Raizada, M., Mallick, N., Husaini, Y., Singh, A.K. and Dubey, S.K., 1990. Effect of four heavy metals on the biology of *Nostoc muscorum. Bio. Metals,* 2: 229–234.

Raizada, M. and Rai, L.C., 1985. Metal induced inhibition of growth, heterocyst differentiation, carbon fixation and nitrogenase activity of *Nostoc muscorum* : Introduction with EDTA and calcium. *Microbios Lett.*, 30: 153–163.

Singh, S.P. and Yadav, V., 1983. Cadmium induced inhibition of nitrate uptake in *Anacystis nidulans*: Interaction with copper and other divalent cation. *J. Gen. Appl. Microbiol.*, 29: 297–304.

Singh, S.P. and Yadav, V., 1984. Cadmium induced inhibition of ammonium and phosphate uptake in *Anacystis nidulans*: Interaction with other divalent cation. *J. Gen. Appl. Microbiol.*, 30: 79–86.

Stokes, P.M., 1983. Response of freshwater algae to metals. In: *Progress in Phycological Research,* (Eds.) F.E. Round and V.J. Chapman. Elsevier, Amsterdam, 2: 87–111.

Stratton, G.W. and Corke, C.T., 1979a. Effect of cadmium ion on the growth, photosynthesis and nitrogenase activity of *Anabaena inaequalis*. *Chemosphere*, 5: 277–282.

Stratton, G.W. and Corke, C.T., 1979b. Effect of mercuric ion on the growth, photosynthesis and nitrogenase activity of *Anabaena inaequalis*. *Environ. Microbiol.*, 38: 537–543.

Whitton, B.A., 1984. Algae as monitors of heavy metals in freshwaters. In: *Algae as Ecological Indicators*, (Ed.) L.E. Shubert. Academic Press, London, pp. 257–280.

Chapter 53

Water Quality Studies on Freshwater Lakes of Kolar Taluk, Kolar, Karnataka

H. Krishna Ram, M. Ramachandra Mohan, Chithanagiri and N. Latha

Water Quality Research Laboratory, Department of Zoology, Bangalore University, Bangalore – 560 056, Karnataka, India

ABSTRACT

Water is one of the five elements required for the very existence of life. Freshwaters most basic and important benefit is health of peoples. Ecosystem and societies, but tragedy of the present century in that human negligence is depleting this vital source indispensable for the very existence of human race on earth. The immediate and future well being of all countries depends on their abilities to manage water resources more effectively. In lake one the water will not be fit for domestic purpose, only for irrigation purposes and other three lakes were used for domestic and irrigational purpose. The study reveals that all the parameters are varied in nature.

Keywords: *Kolar, Nutrients, Water quality.*

Introduction

Water is considered as one of the most basic and important element for human civilization our dependence on freshwater resources has accelerated in the last century due to rapid world population growth and economic development (Postel, 1992). As a result, freshwater resources have deteriorated both in quantity and quality in many areas of the world, lakes providing the majority of easily accessible freshwater resources, have been most influenced. Many lakes in the world suffer environmental problems such man made eutrophication, chemical contamination etc. consequently causing ecosystem and human health damage.

Freshwater is man's most precious needs. Since the beginning of civilization man has tried to settle at the banks of great rivers which supplied freshwater not only for the basic physiological needs of man but also for various socio-economic requirement. The freshwater present on the surface of the earth is put to multifarious uses, it is used for drinking, domestic and municipal uses, agricultural irrigation, industries navigation, recreation etc. after being used in the above mentioned human activities the water flowing out becomes contaminated. Some of the extensive contributions on limnological studies of freshwater bodies in India are of Zutshi and Vas (1978). Harshey *et al.* (1982), Kaul Sidharth (1983), Zafar (1964), Munawar (1970), Seenayya (1971) and Rao (1991). This study, therefore, attempts to evaluate the quality of lentic waters of Kolar Taluk in South of Karnataka State with respect to domestic and irrigational purposes.

Materials and Methods

Study Site

The study was conducted in Kolar District. It is one of the districts falling a madian area of Karnataka having a geographical area of about 8240.00 sq km with the population of about 25 lakhs. It is a predominantly on agricultural district. Where in the major profession of the rural mass is agriculture the average annual rainfall of,the district, is between 700 to 750 mm. In this study we are selected Kolar lakes of Kolar District.

Kolar Amanikere–1

It is situated on the Kolar–Bangalore Road (NH4) to the East of Kolar city. It is a natural lake with a total area of one and half square kilometres and an average depth 2.20 m. Now the lake turned out to be unfit for human consumption and for agricultural and industrial purposes. It receives enormous amount of domestic sewage. At the very look the water appears greenish with periodic algal blooms of throughout the year.

Tambihally Lake–2

It is located 15 km far from district head quarters of Kolar. The lake is man-made rainfed and oftenly used by people for cattle drinking and cloth washing the catchment area is 0.64 sq.km and water spread area is 5.76 hectares. It receives enormous amount of agricultural wastes.

Harati Lake–3

It is located 20 km far from district headquarters of Kolar. The lake is man-made fed by rain. The lake is run off materials from surrounding agricultural plots. The lake used by people for cattle drinking and washing. The catchment area is 0.77 sq.km and water spread area is 1.28 hectares.

Muduvadi Lake–4

It is located 15 km far from district headquarters of Kolar. The lake is man-made, rainfed, and oftenly used by people for agricultural purposes. The catchment area is 14.74 sq.km and water spread area is 266.25 hectares.

Physio-chemical Analysis

Water samples were collected during the period January 2005 to December 2005, in the polythene bottles before 10 am for analysis of following standard methods.

Results and Discussion

The physio-chemical parameters of an aquatic ecosystem is an essential component which drastically effects the flora and fauna of it.

pH

pH represents the effective concentration (activity) of hydrogen ions (H^+) in water. In the present study of lakes the pH of the lake-1, Maximum, pH was recorded 8.3 in site-7 and minimum pH was recorded 7.0 in site-1. In lake-2, the maximum pH was recorded 8.4 in site-7 and minimum pH was recorded 7.1 in site-2 (Table 53.1). In lake-3, the maximum pH was recorded 7.8 in site-2 and minimum pH was recorded 7.1 in site-7. Inlake-4 the maximum pH was recorded 8.1 in site-5 and minimum pH was recorded 7.3 in sites-1, 6, 7 (Table 53.2). pH gives an idea to the type and intensity of pollution (Verma *et al.*, 1984 and Mishra and Saksena, 1991) and pH is one of the most important single factor which influences aquatic production (Swingle, 1967). The life process and growth of microorganisms is influenced by H-ion concentration of the water a greater extent (Shastree, 1991) observed that pH was not very significant with reference to other physico-chemical as well as biological features.

Table 53.1: Physio-chemical Characteristics of Kolar Amanikere and Tambihally Lake 2005

Parameters	*Kolar Amanikere*							*Tambihally Lake*						
	KS1	*KS2*	*KS3*	*KS4*	*KS5*	*KS6*	*KS7*	*TS1*	*TS2*	*TS3*	*TS4*	*TS5*	*TS6*	*TS7*
pH	7	7.5	7.8	7.9	7.7	7.6	8.3	7.3	7.1	7.8	7.5	7.3	7.5	8.4
Conductivity	2200	1800	2500	1010	1500	1640	1840	1980	2200	2500	4700	3000	1580	1320
Hardness	564	512	920	324	520	516	496	560	320	780	1840	932	416	292
TDS	1470	1050	1510	570	870	940	1070	1140	1280	1460	2620	1750	910	758
Carbonate (CO_3)	24	16	9	10	8	6	2	0	14	8	7	12	5	14
Bicarbonate (HCO_3)	552	534	356	343	339	362	368	421	578	306	279	181	397	196
Calcium (Ca)	208	164	228	75	126	126	114	200	93	263	547	270	142	77
Magnesium (Mg)	41	24	88	34	53	50	53	15	22	29	118	64	15	25
Sodium (Na)	176	170	142	79	98	134	188	197	357	208	222	250	173	170
Potassium (K)	23	4	9	5	5	8	8	3	0	7	14	16	0	2
Iron (Fe)	0.20	0	0	0.10	0.05	0.40	0.03	0.04	0.30	0.00	0.30	0.03	0.2	0.15
Sulphates (SO_4)	340	80	100	56	123	130	152	72	67	126	117	145	60	50
Chloride (Cl)	364	263	144	170	232	269	3i6	398	330	608	1414	120	283	294
Nitrate (NO_3)	10	35	27	33	25	11	22	13	72	20	7	73	7	40
Floride (F)	0.40	1.00	0.60	0.50	0.75	1.00	0.95	1.30	0.80	1.25	0.90	1.00	1.6	1.6

All the parameters expressed in mg/ltr except pH, conductivity, micro mhos.

Conductivity

Conductivity is a measure of how well water can conduct an electrical current. In the present investigation the lake-1, Maximum conductivity was recorded 2500 micro mhos in site-3 and minimum conductivity was recorded 1010 micro mhos in site-4. In lake-2, the maximum conductivity was recorded 4700 micro mhos in site-4 and minimum conductivity was recorded 1320 micro mhos in site-7 (Table 53.1). In lake-3, the maximum conductivity was recorded 2700 micro mhos in site-6 and minimum conductivity was recorded 580 micro mhos in site-1. In lake-4 the maximum conductivity was recorded 1240 micro mhos in site-1 and minimum conductivity was recorded 650 micro, mhos in site-4 (Table 53.2). Conductivity increases with increasing amount and mobility of ions these ions, which come from the breakdown of compounds, conduct electricity because they are negatively and positively charged when dissolved in water (CDPHE-WQCD) the increase concentration of salts because of discharge of domestic effluents and organic matter from the near by residential areas. Which bring along with in the lake. High levels of conductivity reflects on the pollution status as well as trophic levels of the aquatic body (Anitha *et al.*, 2005).

Table 53.2: Physio-chemical Characteristics of Harati Lake and Muduvadi Lake–2005

Parameters	*Hariti Lake*							*Muduvadi Lake*						
	HS1	*HS2*	*HS3*	*HS4*	*HS5*	*HS6*	*HS7*	*MS1*	*MS2*	*MS3*	*MS4*	*MS5*	*MS6*	*MS7*
pH	7.2	7.8	7.4	7.2	7.3	7.7	7.1	7.3	8.0	7.4	7.8	8.1	7.3	7.3
Conductivity	580	1000	880	1090	2000	2700	2000	1240	740	890	650	860	790	730
Hardness	200	280	244	284	588	816	640	300	240	236	212	348	304	296
TDS	350	590	520	680	1190	1610	1180	720	415	530	380	495	450	410
Carbonate (CO_3)	10	0	9	0	0	0	0	22	0	9	19	14	0	19
Bicarbonate (HCO_3)	196	377	225	270	441	252	372	304	204	225	240	240	270	220
Calcium (Ca)	51	84	60	80	198	208	197	86	56	62	54	88	54	80
Magnesium (Mg)	18	16	23	21	53	60	37	21	24	20	19	32	42	24
Sodium (Na)	40	85	89	141	128	257	163	145	68	95	52	31	40	33
Potassium (K)	2	7	3	15	8	7	3	3	5	2	2	9	2	4
Iron (Fe)	0.5	0.00	0.00	0.20	0.10	0.20	0.0	0.2	0.0	0.0	0.0	0.1	0.0	0.1
Sulphates (SO_4)	26	67	58	50	144	116	97	53	29	66	31	58	32	18
Chloride (Cl)	41	76	123	171	282	614	370	196	59	123	44	104	87	89
Nitrate (NO_3)	35	38	37	40	127	195	95	9	9	6	9	15	22	12
Floride (F)	2.00	1.10	1.70	0.50	1.00	1.00	1.30	2.00	2.40	1.80	1.80	0.40	1.60	0.6

All the parameters expressed in mg/ltr except pH, conductivity, micro mhos.

Hardness

Hardness is measure of polyvalent cations (ions with a charge greater than +1) in water. Hardness generally represents the concentration of calcium (Ca^{2+}) and magnesium (Mg^{2+}) ions because these are the most common polyvalent cations. In the present study lake-1, the maximum hardness was recorded 920 mg/L. in site 3 and minimum hardness was recorded 324 mg/L in site 4. In lake-2 the maximum hardness was recorded 1840 mg/L in site-4 and minimum hardness was recorded 292 mg/L in site-7 (Table 53.1). In lake-3 the maximum hardness was recorded 816 mg/L in site-6 and minimum.hardness was recorded 200 mg/L in site-1. In lake-4 the maximum hardness was recorded 348 mg/L in site-5 and minimum was recorded 212 mg/L in site-4 (Table 53.2). Hardness often employed as indicator of water quality depends on the concentration of carbonate and bicarbonate salts of calcium and magnesium (temporary hardness) or sulphate chloride or other anions of mineral acids (permanent hardness) (Anitha, *et al.*, 2005). The higher values depended upon geographical formation of the study area and also available minerals. The hardness exhibited positive association with calcium, magnesium chloride and sulphates.

TDS

Total dissolved solids as a measure of the amount of material dissolved in water. The present study lake-1, the maximum TDS was recorded 1510 mg/L in site-3 and minimum TDS was recorded 570 mg/L in site-4. In lake-2 the maximum. TDS was recorded 2620 mg/L in site-4 and minimum TDS was recorded 758 mg/L in site-7 (Table 53.1). In lake-3 the maximum TDS was recorded 1610 mg/L in site-6 and minimum TDS was recorded 350 mg/L in site-1. In lake-4 the maximum TDS was recorded 720 mg/L in site-1 and minimum TDS was recorded 380 mg/L in site-4 (Table 53.2). The changes in TDS concentrations can be harmful because the density of the water determines the flow of water into and out of an organism's cells (Mitchell and Stapp, 1992). However, TDS concentrations are too high or too low. The growth of many aquatic life can be limited and death may occur water with high TDS often has a bad taste and/or higher water hardness, and could result in a laxative effect.

Carbonate (CO_3)

In the present study the lake-1, the maximum CO_3 was recorded 24 mg/L in site-1 and minimum CO_3 was recorded 2 mg/L in site-7. In lake-2 the maximum CO_3 was recorded 14.0 mg/L in site-2 and 7 and minimum CO_3 was recorded in 0.0 mg/L in site-1 (Table 53.1). In lake-3 the maximum CO_3 was recorded 10 mg/L in site-1 and minimum CO_3 was recorded 0.0 in sites-2, 4, 5, 6, 7, Lake-4 the maximum CO_3 was recorded 22 mg/L in site-1 and minimum CO_3 was recorded 0.0 mg/L in sites-2 and 6 (Table 53.2). Carbonate alkalinity (CO_3) is a measure of the buffering capacity of water or the capacity of bases to neutralize acids, measuring carbonate alkalinity is important in determining a lakes ability and neutralize acidic pollution from rainfall or wastewater. Carbonates all added to a water system if the water passes through soil and rock that contain carbonate minerals. Carbonate ions in water can remove toxic metals (such as lead, arsenic and cadmium).

Bicarbonate (HCO_3)

In the present study lake-1 the maximum HCO_3 was recorded 552 mg/L in site-1 and minimum HCO_3 was recorded 339 mg/L in site-5. In lake-2 the maximum HCo3 was recorded 578 mg/L in site-2 and minimum HCO_3 was recorded 181 mg/L in site-5 (Table 53.1). In lake 3 the maximum HCO_3 was recorded 441 mg/L in site-5 and minimum HCO_3 was recorded 225 mg/L in site-3. In lake-4 the maximum HCO_3 was recorded 270 mg/L in site-6 and minimum HCO_3 was recorded 204 mg/L in site-2 (Tables 53.1 and 53.2). Bicarbonate alkalinity (HCO_3) in a measure of the buffering capacity of water. The capacity of bases to neutralize acids. Bicarbonates are added to a. water system if the water passes through soil and rock that contain Bicarbonate minerals. Bicarbonate ions in water can remove toxic metals (such as lead, arsonic, and cadmium) by precipitating the metals out of solution. Generally above pH 8.3, alkalinity is mostly in the form of carbonate (CO_3) and below 8.3 alkalinity is present mostly as bicarbonate (HCO_3).

Calcium (Ca)

In the present study the lake-1, maximum Ca was recorded 228 mg/L in site-3 and minimum Ca was recorded 75 mg/L in site-4. In lake-2 the maximum Ca was recorded 547 mg/L in site-4 and minimum Ca was recorded 77 mg/L in site-7 (Table 53.1). In Lake-3 the maximum Ca was recorded 208 mg/L in site-6 and minimum Ca was recorded 51 mg/L in site-1. In lake-4 the maximum Ca was recorded 88 mg/L in site-5 and minimum Ca was recorded 54 mg/L in sites-4 and 6 (Table 53.2). This element is an important component of the carbonic buffer system and also cycles through biotic and a biotic component of the eco system. Calcium was 3rd in the order of equivalent ionic dominance in the lakes (Anitha, *et al.*, 2005). Calcium as a main cations factors causing water hardness in nature water and originate from natural process *i.e.* on a dissolvent of mineral containing Ca and other sources can be industrial wastes and sometime agricultural waste. It is non-toxic.

Magnesium (Mg)

In the present study the lake-1, maximum Mg was recorded 88 mg/L in site-3 and minimum Mg was recorded 24 mg/L in site-2, In lake-2 the maximum Mg was recorded 118 mg/L in site-4 and minimum Mg was recorded 15 mg/L in site-6 (Table 53.1). In Lake-3 the maximum Mg was recorded 60 mg/L in site-6 and minimum Mg was recorded 16 mg/L in site-2. In lake-4 the maximum Mg was recorded 42 mg/L in site-6 and minimum Mg was recorded 19 mg/L in site-4 (Table 53.2). Magnesium as main cation are factors causing water hardness. In natural water Mg originate from natural process that is as dissolvent of mineral containing Mg and other sources can be industrial waste and some times agriculture waste Mg is non-toxic.

Sodium (Na)

In the present study the lake-1, maximum Na was recorded 188.mg/L in site-7 and minimum Na was recorded 79 mg/L in site-4. In lake-2 the maximum Na was recorded 357 mg/L in site-2 and minimum Na was recorded 170 mg/L in site-7 (Table 53.1). In Lake-3 the maximum Na was recorded 257 mg/L in site-6 and minimum Na was recorded 40 mg/L in site-1. In lake-4 the maximum Na was recorded 145 mg/L in site-1 and minimum Na was recorded 31 mg/L in site-5 (Table 53.2). Sodium is main cation in natural water in freshwater such as lake water. Sodium in collected from natural minerals dissolved in water. In low concentration sodium in drinking water is a mineral necessary for electrolyzing blood balance in the human bodies.

Potassium (K)

In the present study the lake-1, maximum K was recorded 23 mg/L in site-1 and minimum K was recorded 4 mg/L in site-2. In lake-2 the maximum K was recorded 16 mg/L in site-5 and minimum K was recorded. 0.0 mg/L in site-6 (Table 53.1). In Lake-3 the maximum K was recorded 15 mg/L in site-4 and minimum K was recorded 2 mg/L in site-1. In lake-4 the maximum K was recorded 9 mg/L in site-5 and minimum k was recorded 2 mg/L in sites-3, 4 and 6 (Table 53.2). Potassium is main cation in natural water in freshwater such as lake water. Potassium occurs in natural water salty through found in small amount. It plays a vital role in metabolism offish environment and is a important macro nutrients.

Iron (Fe)

In the present study the lake-1, maximum Fe was recorded 0.40 mg/L in site-6 and minimum Fe was recorded 0.0 mg/L in sites-2 and 3. In lake-2 the maximum Fe was recorded 0.30 mg/L in site-2 and 4 and minimum Fe was recorded 0.03 mg/L in site-5 (Table 53.1). In lake-3 the maximum Fe was recorded 0.5 mg/L in site-1 and minimum Fe was recorded 0.0 mg/L in sites-2, 3 and 7. In lake-4 the maximum Fe was recorded 0.20 mg/L in site-1 and minimum Fe was recorded 0.0 mg/L in sites-2, 3, 4 and 6 (Table 53.2). In aquatic environments metals have been termed as conservative pollutants because once added to the environment. They prevail forever. Further the seriousness and persistence of heavy metals in water are compounded by the fact that generally they are water soluble non degradable and strongly bounded to many biochemical especially poly peptides and proteins besides, other materials which loses electron rich functional groups such as sulfohydryl, amino and imidasole. Being stable compounds like heavy metals are non-reality removed by oxidation, precipitation other processes and affects the activities of the animals (Anitha, *et al.*, 2005).

Sulphates (SO_4)

In the present study the lake-1, maximum SO_4 was recorded 340 mg/L in site-1 and minimum SO_4 was recorded 56 mg/L in site-4. In lake-2 the maximum SO_4 was recorded 145 mg/L in site-5 and minimum SO_4 was recorded 50 mg/L in site-7 (Table 53.1). In Lake-3 the maximum SO_4 was recorded 144 mg/L in site-5 and minimum SO_4 was recorded 26 mg/L in site-1. In lake-4 the maximum SO_4 was recorded 66 mg/L in site-3 and minimum SO_4 was recorded 18 mg/L in site-7 (Table 53.2). Sulphur exists is a number of oxidation status from the most oxidized sulphate to the most reduced sulphide. The biological reduction of sulphur can take place in both aerobic and anaerobic conditions. A high concentration of sulphates stimulates the action of sulphate reducing bacteria. Which produce hydrogen sulphide a gas highly toxic to fish life. The high concentration of sulphate associated with the depletion of oxygen led to anoxic or anaerobic conditions in the lake waters.

Chloride (Cl)

In the present study the lake-1, maximum Cl was recorded. 364 mg/L in site-1 and minimum Cl was recorded 144 mg/L in site-3. In lake-2 the maximum Cl was recorded 1414 mg/L in site-4 and minimum Cl was recorded 120 mg/L in site-5 (Table 53.1). In Lake-3 the maximum Cl was recorded 614 mg/L in site-6 and minimum Cl was recorded 41 mg/L in site-1. In lake-4 the maximum Cl was recorded 196 mg/L in site-1 and minimum Cl was recorded 44 mg/L in site-4 (Table 53.2). This element is an important component of the carbonic buffer system and also cycles through biotic and a biotic component of ecosystem the chloride values is maximum in bottom layer and this may be due to washing down of organic matter from the surrounding catchment area (Datta *et al.*, 1984).

Nitrates (NO_3)

In the present study the lake-1, maximum NO_3 was recorded 35 mg/L in site-2 and minimum No_3 was recorded 10 mg/L in site-1. In lake-2 the maximum NO_3 was recorded 73 mg/L in site-5 and minimum No_3 was recorded 7 mg/L in site-4 and 6 (Table 53.1). In Lake-3 the maximum NO_3 was recorded 195 mg/L in site-6 and minimum NO_3 was recorded 35 mg/L in site-1. In lake-4 the maximum NO_3 was recorded 22 mg/L in site-6 and minimum No_3 was recorded 9 mg/L in sites-1, 2 and 4 (Table 53.2). It is one of the most, important elements in aquatic systems, as it helps in synthesis and maintenance of proteins and productivity of the water. Nitrate concentration of the water bodies all influenced with the geochemical conditions organic load the rate of their mineralization, the rainfall is to be responside for increasing the amount of nitrate in water. Nitrate is generated by heterotrophic microbes as a primary and product of decomposition of organic matter either directly from protein or from the organic compound. The presence of excessive nitrate in water is due to man made domestic activities and fertilizers from fields. Nitrate represent high load all cause of eutrophication of the aquatic body. Nitrates stimulate the growth of plankton and water needs that provide food for fish. It is may increase the fish population. However if argue grow too wildly, oxygen levels will be reduced and fish will die (USEPA, 1986; Knepp and Arkin, 1973).

Fluoride (F)

In the present study the lake-1, maximum F was recorded mg/L in site-2 and 6 and minimum F was recorded 0.40 mg/L in site-1. In lake-2 the maximum F was recorded 1.60 mg/L in site-6 and 7 and minimum F was recorded 0.80 mg/L in site-2 (Table 53.1). In Lake-3 the maximum F was recorded

2 mg/L in site-1 and minimum F was recorded 0.50 mg/L in site-4. In lake-4 the maximum F was recorded 2.40 mg/L in site-2 and minimum F was recorded 0.40 mg/L in site-5 (Table 53.2). Fluoride is the highest member of the halogen family is the most electronegative among all chemicals elements (Hodge and Smith, 1965) In ions inhabit a variety of enzymes often by forming complexes with magnesium ions and other metals ions (NAS, 1971). In low concentration fluoride prevents dental caries and higher concentration of fluoride also causes respiratory failure fall of blood pressure and general paralysis.

References

Anitha, G., Chandrashekar, S.V.A. and M.S. Kodarkar, 2005. Limnological, studies on Mir Alam lake, Hyderabad. *Poll. Res.*, 24(3): 681–687.

Colorado, 2002. Department of Public Health and Environment and Water Quality Control Division (DPHE– WQCD). The basic standards and. methodologies for surface water. 5CCR, 1002–31.

Datta, N.C., Choudhary, A. and S. Choudhary, 1984. Annual rhythm of some hydrobiological parameters and their interaction in a brackish water impoundment of West Bengal, India. *Camp. Physiol. Ecol.*, 9(2): 149–154.

Harshey, D.K., Pakl, S.G. and D.F. Singh, 1982. Limnological studies on a trophical freshwater tank of Jabalpur, India. I. Abiotic factors. *Geobios*, 1(2): 98–102.

Hodge, H.C. and Smith, F.A., 1965. *Fluorine Chemistry*, Vol. 4, Academic Press, New Delhi.

Kaul, Sidharth, 1983. Limnological studies of an aquatic body, Lalpari Talab at Rajkot. In: *Proc. Indian Nat. Sci. Acad.*, B45: 564–570.

Knepp and Arkin, 1973. Ammonia Toxicity Levels and Nitrate Tolerance of, Channel Catfish. *The Progressive Fish Culturist*, 35: 221.

Mishra, S.R. and Saksena, D.N., 1991 Pollutional ecology with reference to physico-chemical characteristics of morar (Kalpi) river, Gwalior (M.P.). *Current Trends in Limnology*, 1: 159–184.

Mitchell and Stapp, 1992. *Field Manual for Water Quality Monitoring.*

Munawar, M., 1970. Limnological studies on freshwater ponds of Hyderabad–II. The biocenose. *Hydrobiologia*, 36(1): 105–128.

NAS, 1971. *Biologic Effects of Atmospheric Pollutants: Fluorides.* Washington, D.C. National Academy of Science. National Research Council. Committee on Biological Effects of Atmospheric Pollutants, pp. 239.

Postel, S., 1992. *Last Oasis: Facing Water Scarcity.* W.W. Norton and Company, Inc.

Rao, V.S., 1991. An Ecological Study of three freshwater ponds of Hyderabad, India, the environment. *Hydrobiol.*, 35(1): 213–223.

Seenayya, G. and A.R. Zafar, 1979. An Ecological Study of the Mir Alam Lake. Hyderabad, India. *Ind. J. Biol.*, 2(2): 214–220.

Shastree, N.K., Islam, M.S., Pathak, S, and M. Afsan, 1991. Studies on the physico-chemical dimensions of the lentic hydrospere of Ravindra Sarovar (Gaya). *Current Trends in Limnology*, 1: 133–153.

Swingle, H.S., 1967. *Standardisation of Chemical Analysis for Water and Muds.* FAO. Fish Rep., 4(4): 397–421.

USEPA, 1986. *Quality Criteria for Water*. U.S. Environmental Protection Agency. EPA #440/5–86–001.

Verma, S.R., Sharma, P., Tyagi. A., Rani, Gupta, A.K. and R.C. Dalela, 1984. Pollution and sporshic status of Eastern Kalinadi. *Limnologica (Barlin)*, 15(1): 69–133.

Zafer, A.R., 1964. On the Ecology of algae in certain fish ponds of Hyderabad, India. I. Physio-chemical complexes. *Hydrobiologia*, 23: 179–195.

Zutshi, D.P. and Vass, K.K., 1978. Limnological Studies on Dal Lake: Chemical features. *Indian J. Ecol.*, 5(1): 90–97.

Chapter 54

Effect of Mulches on Soil Properties and Yield of Maize in Mizoram

U.K. Sahoo and Nigel J.S. Kharmalki

Department of Forestry, School of Forestry and Earth Sciences, Mizoram University, Post Box 190, Aizawl – 796 001, Mizoram, India

ABSTRACT

The effect of four mulches *viz.*, rice straw, weeds (*Eupatorium*), *Leucaena* leaves and polythene with and without alder (*Alnus nepalensis* D. Don) were studied on soil properties and yield of maize in Mizoram during 2004. The N, P and K levels in the soil were increased due to addition of *Leucaena* and *Eupatorium* leaf mulches whereas the soil pH decreased by these mulches. The maize productivity was significantly ($P < 0.05$) higher under *Leucaena* leaf mulched plots than the other treatments. The study depicted an important role played by alder in improving soil properties and enhancing maize yield.

Keywords: *Alder, Maize yield, Mizoram, Mulches, Soil properties.*

Introduction

Maize (*Zea mays* L.) is one of the most important cereal crops and is grown either as a sole crop or with other species over 8260 ha of land in Mizoram. The rapid use of chemical fertilizers was reported to decrease the yield of maize but growing of maize with different nitrogen fixing tree species in recent years have resulted in enhanced yield (Mathew *et al.*, 1992, Khola *et al.*, 1997, Sahoo *et al.*, 2005). Shifting cultivation (locally known as *jhum*) which is traditionally practised widely in the hilly terrain causes numerous problems such as heavy soil and water erosion and consequent low maize productivity. The productivity of maize can be increased by adopting improved package of practices, particularly *in situ* moisture conservation by mulching. The soil properties may also be affected by use of different mulches in the system (Sharma and Pawar, 1998; Mc Intyre *et al.*, 2000). The present investigation, therefore, seeks to determine the effect of three vegetative mulches (rice straw, weed leaves, *Leuceana leucocephala*) and one artificial mulch (polythene) on soil moisture retention, pH, nitrogen, phosphorus, potassium and organic carbon and productivity of maize during 2004 in Mizoram.

Materials and Methods

Study Site

Study was carried out at Tanhril which is located on the south-western part of Aizawl city, the capital of Mizoram and is about 22 km from Aizawl city and lies between 92°38′ to 92°42′ E longitude and 23°42′ to 23°46′ N latitude. The site is at an altitude of 350 m with an average rainfall of 250 cm. The site is moderately slope and covers an area of 167.5 sq. m. The experiment plot consists of one year old established Alder (*Alnus nepalensis* D. Don) on recently cleared forest stand. Maize (*Zea mays* L.) was selected as an agriculture crop introduced between the inter-spaces of the tree species.

Design and Layout of Experiment

Field experiment was carried out following the Randomized Block Design (RBD). The experimental plot covers a total area of 167.5 sq.m. This plot was divided into two equal parts (plot 1 and plot 2). In plot 1 maize was raised along with alder trees, while in plot 2 maize was planted without the Alder trees. In both plot 1 and plot 2, four types of surface mulch were applied to maize. A control (without mulch) was also maintained in each of the plots, thus there were a total of ten (10) treatments and each treatment was replicated thrice. The treatments included: T_1: Maize + *Alnus* + Rice Straw, T_2: Maize + *Alnus* + Weed (*Eupatorium*), T_3: Maize + *Alnus* + *Leuceana leucocephala* leaves, T_4: Maize + *Alnus* +Artificial Mulch (polythene), T_5: Maize + *Alnus*, T_6: Maize + Rice Straw, T_7: Maize + Weed (Eupatorium), T_8: Maize + *Leuceana leucocephala* leaves, T_9: Maize + Artificial mulch (polythene) and T_{10}: Only maize crop. The alder trees were maintained at a uniform spacing of 1.5 × 1.5 m. Maize was planted at a uniform spacing of 60 × 40 cm. Thus there were a total of 60 alder trees and 670 maize plants at the experimental site. Each category of mulch was incorporated at the time of maize planting at the rate of 10 tons per hectare. During the period of the experiment, three weeding were carried out. The first weeding was done in mid-April, the second in mid-June and the last in the first week of August, prior to harvesting of maize. The crop was raised purely under rainfed condition.

Soil moisture was determined on a monthly interval from the samples collected at a depth of 30 cm. Soil moisture loss on drying to a constant weight was determined for 100 g fresh soil. The other soil properties such as soil pH, nitrogen, phosphorus, potassium, and organic carbon were determined at monthly intervals as per Allens *et al.* (1974). Crop productivity was assessed at the time of crop harvest. The data were subjected to analysis of variance (ANOVA) to see the effect of mulches on soil moisture retention, soil properties and maize yield.

Results and Discussion

Soil Moisture Retention

Soil moisture retention was lowest in month of May and highest in July in all the treatments. However, straw mulch was found to conserve more moisture as compared to other mulches although there was ample variation in the soil moisture content from the various treatments (Table 54.1). The higher moisture retention by straw mulch could be attributed to its slower rate of decomposition than the other mulch materials. Least moisture conservation was recorded for the control plots without mulch. This was found to be in consonance with the findings of Mohan *et al.* (2004). The vegetable mulch materials may act like some cover crops on hill slopes (Khera and Singh, 1995, 1998) and help not only in reducing the run off and nutrient losses (Hadda and Sur, 1989) but also in improving soil water regime (Unger, 1978).

Table 54.1: Effect of Mulches on Soil Moisture Retention (%) in the Site During the Study Period

Treatments	*Months*				
	April	*May*	*June*	*July*	*August*
T_1	9.70	8.87	9.05	9.27	9.13
T_2	8.97	7.97	8.72	9.07	9.05
T_3	8.92	8.00	8.79	9.12	9.06
T_4	9.25	8.23	8.96	9.12	9.09
T_5	8.10	7.81	8.30	9.00	8.89
T_6	9.73	8.80	9.04	9.23	9.11
T_7	8.92	7.98	8.68	9.08	9.06
T_8	8.94	7.98	8.97	9.06	9.06
T_9	8.87	8.09	8.91	9.10	9.10
T_{10}	8.11	7.84	8.31	9.29	8.92
S.E. m ± n = 24	0.10	0.07	0.05	0.032	0.02
C.D. (P = 0.05)	0.35	0.25	0.20	NS	0.11

Soil Properties

The mean monthly records of soil pH of the various treatments (Table 54.2) showed highest pH in the month of April and lowest in August. The treatments with plastic mulch and control raised without Alder trees showed the highest pH. However, the lowest pH values were recorded for treatments with *Leuceana* and weed (*Eupatorium*) mulch. The soil pH gradually declined with decomposition time of mulch materials and this was true in all the cases, possibly due to release of more bases to the soil. This is clearly reflected from the fact that the pH did not change in control (Table 54.2). The highest soil nitrogen (per cent) levels were recorded in May and lowest in August although it did not vary significantly between the months. Soil nitrogen was higher in *Leuceana* and *Eupatorium* (weed) mulch applications compared to other mulch materials (Table 54.3). Incorporation of plant residues as mulch increased the concentrations of nitrogen in the soil. The foliage of *Leucaena* and *Eupatorium* containing considerable nitrogen when mixed with the soil improved the soil nitrogen status. Although there was no significant variation in soil phosphorus between treatments, a decreasing trend was observed from May to August (Table 54.4). Highest P (33.78 kg/ha) was found in the plot treatment with *Leuceana* mulch and lowest (30.47 kg/ha) in control. During May to August, the microbial population must have multiplied frenetically consuming sizeable amount of phosphorus resulting in its lower level in all treatments. The levels of potassium gradually decreased form the initiation to the end of the study period. Highest K content (146.99 kg/ha) was recorded for treatment with *Leuceana* mulch in May and the lowest (142.64 kg/ha) was for the control in August (Table 54.5). Soil organic carbon (per cent) increased significantly ($P < 0.05$) compared to the control towards the end of study period which could have been due to faster decomposition of most of the mulch materials (Table 54.6). This explains the role played by decomposing mulches in bringing some changes in organic carbon (per cent) in all treated plots especially at the end of study period. The organic carbon levels were highest in the month of April and lowest in the months of July and August. The highest organic carbon (7.08 per cent) in the month of April was for treatment with weed (*Eupatorium*) mulch and the lowest (5.35 per cent) was in control during the month of August.

Table 54.2: Effect of Mulches on Soil pH in the Site During the Study Period

Treatments	*Months*				
	April	*May*	*June*	*July*	*August*
T_1	5.77	5.50	5.20	5.06	4.98
T_2	5.42	5.11	5.03	4.92	4.76
T_3	5.41	5.25	5.03	4.87	4.81
T_4	6.07	5.71	5.53	5.01	4.98
T_5	6.07	5.72	5.56	5.08	5.00
T_6	6.13	5.59	5.30	5.09	4.94
T_7	6.04	5.23	4.96	4.93	4.87
T_8	5.65	5.27	4.93	5.03	4.85
T_9	6.25	5.77	5.15	5.06	5.00
T_{10}	6.25	6.05	6.14	6.04	6.08
S.E. m ± n = 24	0.06	0.05	0.05	0.02	0.02
C.D. (P = 0.05)	0.10	0.20	0.28	0.32	0.36

Table 54.3: Effect of Mulches on Soil Nitrogen (%) in the Site During the Study Period

Treatments	*Months*				
	April	*May*	*June*	*July*	*August*
T_1	0.20	0.19	0.19	0.18	0.18
T_2	0.21	0.20	0.19	0.18	0.19
T_3	0.20	0.19	0.19	0.19	0.19
T_4	0.20	0.19	0.18	0.16	0.15
T_5	0.20	0.19	0.17	0.16	0.15
T_6	0.20	0.19	0.18	0.18	0.19
T_7	0.21	0.19	0.18	0.18	0.18
T_8	0.20	0.20	0.19	0.19	0.19
T_9	0.20	0.19	0.17	0.16	0.16
T_{10}	0.20	0.20	0.17	0.16	0.16
S.E. m. ± n = 24	0.001	0.0013	0.0022	0.0024	0.0028
C.D. (P = 0.05)	NS	NS	NS	0.014	0.012

Table 54.4: Effect of Mulches on Soil Phosphorus (kg/ha) in the Site During the Study Period

Treatments	*Months*				
	April	*May*	*June*	*July*	*August*
T_1	33.47	32.41	32.25	31.83	31.53
T_2	33.38	33.17	32.84	32.26	32.04
T_3	33.78	33.13	32.80	32.04	31.59
T_4	33.33	32.19	32.15	31.53	30.89
T_5	33.32	32.14	32.09	31.40	30.60
T_6	33.43	32.49	32.56	31.57	31.15
T_7	33.37	33.16	32.37	31.49	32.09
T_8	33.47	32.72	32.36	32.04	31.98
T_9	33.21	32.36	31.8	30.96	30.50
T_{10}	33.19	31.95	31.95	31.00	30.47
S.E. m ± n = 24	0.05	0.12	0.12	0.10	0.12
C.D. (P = 0.05)	NS	NS	NS	NS	0.54

Table 54.5: Effect of Mulches on Soil Potassium (kg/ha) in the Site During the Study Period

Treatments	*Months*				
	April	*May*	*June*	*July*	*August*
T_1	146.78	146.19	145.47	144.65	144.60
T_2	146.82	146.31	145.92	145.27	145.03
T_3	146.96	146.42	145.59	145.16	144.97
T_4	146.34	146.07	145.06	144.04	142.89
T_5	146.26	146.09	145.17	143.44	143.13
T_6	146.72	145.90	145.45	144.57	144.52
T_7	146.76	146.00	145.83	145.00	144.76
T_8	146.99	146.87	145.57	144.92	145.02
T_9	146.29	146.18	145.56	143.57	143.09,
T_{10}	146.14	146.04	145.54	143.48	142.64
S.E. m ± n = 24	0.08	0.10	0.11	0.15	0.19
C.D. (P = 0.05)	NS	NS	NS	0.98	0.87

Crop Productivity

Maximum number of cobs was obtained under treatment with *Leuceana* mulch that was grown along with Alder trees (*Alnus nepalensis*) and minimum number of cobs was from maize grown under control without Alder trees (Table 54.7). It was observed that the mean maximum weight of grain per cob was 78.07 g under *Leuceana* mulch (with alder trees) and the lowest mean weight of grain per cob was from maize in the control plots (70.61 g). The mean maximum number of grains was 405.13 under

Leuceana mulch treatment along with Alder trees. The mean least number of grains per cob (367.73) was recorded in the control sub-plot without alder trees.

Table 54.6: Effect of Mulches on Soil Organic Carbon (%) in the Site During the Study Period

Treatments	*Months*				
	April	*May*	*June*	*July*	*August*
T_1	7.00	6.88	6.45	6.35	6.41
T_2	7.08	6.96	6.51	6.42	6.50
T_3	6.98	6.89	6.59	6.51	6.46
T_4	6.99	6.85	6.26	5.68	5.38
T_5	6.96	6.0	5.89	5.67	5.35
T_6	7.00	6.84	6.40	6.40	6.45
T_7	7.03	6.88	6.41	6.35	6.40
T_8	6.96	6.96	6.50	6.45	6.45
T_9	6.94	6.82	6.06	5.59	5.50
T_{10}	6.96	6.69	5.78	5.30	5.00
S.E. m ± n = 24	0.01	0.01	0.02	0.03	0.03
C.D. (P = 0.05)	NS	NS	0.15	0.16	0.15

Table 54.7: Effect of Mulches on Crop Productivity of Maize in the Site

Treatments	*Parameters for Crop Productivity*		
	Average No. of Cobs per Plant	*Average Weight of Grains per Cob (gm)*	*Average No. of Grains per Cob*
T_1	5.67	74.14	394
T_2	6.73	77.03	402
T_3	6.81	78.07	405
T_4	4.87	71.28	377
T_5	4.7	70.61	375
T_6	5.53	73.6	389
T_7	6.6	75.93	400
T_8	6.7	76.92	401
T_9	4.6	70.82	370
T_{10}	4.43	70.63	368
S.E. m ± n = 24	0.18	0.53	2.81
C.D. (P = 0.05)	0.59	1.38	13.99

The role of organic waste as compost and mulch in increasing various crop yield such as rice (Bawm, 1990), corn yield (Gill *et al.*, 1992, Ortega *et al.*, 2000) pigeon pea (Gajera *et al.*, 1998), potato (Acharya and Kapur, 2001) and ginger (Agarwal *et al.*, 2003) have been reported. The N availability in the soil due to stubble and vegetable waste materials may help in fertility management (Bawm, 1990,

Singh, 1995). Therefore it is suggested to practise tillage and mulching combining for improved crop productivity (Hadda and Sur, 1989, Moitre *et al.*, 1994).

The overall yield of maize was found to be significantly ($P < 0.05$) higher in the treatments with *Leuceana, Eupatorium* (weed) and straw mulch raised along with *Alnus nepalensis*. This is in accordance with the findings of Mannering and Mayor (1963) and Singh *et al.* (2002) in which growth of maize was highest under vegetative mulch treatments. The present study depicts better growth and higher grain yield of maize by application of mulch as compared to the un-mulched crops. Further, mulching has conserving effect on soil hydrothermal regime; hence it is a very important management practice for wide spaced maize in sloppy lands under rainfed conditions.

Acknowledgements

We wish to acknowledge the grant received from ICAR, New Delhi (Sanction No.5(29)197-SW&DF) for carrying out this work.

References

Acharya C.L. and Kapur, O.C., 2001. Using organic wastes as compost and mulch for potato (*Solanum tuberosum*) in low water retaining hill soils of North-West India. *Ind. J. Agric. Sci.*, 71: 306–309.

Agarwal, S., Korla, B.N. and Raina, J.N., 2003. Effects of mulch on soil hydrothermal regimes, weed incidence, yield and quality of ginger. *J. Ind. Soc. Soil Sci.*, 51: 65–67.

Allen, S.E., Grime S., Parkinsons, I.A. and Quarmby, G., 1974. *Chemical Analysis of Ecological Materials*. Blackwell Scientific Publication, Oxford.

Bawn, P.E., 1990. Effects of stubble and N fertilization management on N availability and up take under successive rice (*Oyza sativa* L.) crops.*Plant and Soil*, 121: 11–19.

Gajera, M.S., Ahlawat, R.P.S. and Ardashna, P.B., 1998. Effect of irrigation schedule, tillage depth and mulch on growth and yield of winter pigeon pea (*Cajanus cajan*). *Ind. J. Agron.*, 43: 689–693.

Gill, K.S., Arshad, M.A., Chivunda, B.K. and Gumbo, M., 1992. Influence cf residue mulch, tillage and cultural practices on weed and corn yield from three field experiments. *Soil and Tillage Res.*, 24: 211–223.

Hadda, M.S. and Sur, H.S., 1989. Effect of mulch on rate of runoff, sediment and nutrient losses from a sandy soil under untilled and tilled conditions. *J. Res. P.A.U., Ludhiana*, 26: 37.

Khera, K.L. and Singh, G., 1995. Effect of paddy straw mulch and rainfall intensity on runoff and soil loss under simulated rainfall. *Ind. J. Soil Conserv.*, 23: 20–23.

Khera, K.L. and Singh, G., 1998. Effect of crop cover and field slope on soil on soil erosion in northern plain hot sub-humid Punjab. *Ind. J. Soil Conserv.*, 26: 19–21.

Khola, O.P.S., Dube, R.K. and Sharma, N.K. 1997. Biological and economic feasibility and intercropping of legumes with maize (*Zea mays* L.) on slopping valley lands. *Ind. J. Soil Cons.* 25: 141.

Mannering, J.V. and Mayor, L.D., 1963. The effect of various rates of surface mulch on infiltration and erosion: empirical studies on soil conservation technique and design procedure. *Soil Conserv. RPC, Morgan*, 19.

Mathew, R.B., Lungu, S., Volk, J., Holden, S.T. and Solberg, K., 1992. The potential of alley cropping in improvement of cultivation systems in high rainfall areas of Zambia. II. Maize production. *Agroforestry System*, 17: 241.

Mc Intyre, B.D., Speijer, P.R., Riha, S.J. and Kizito, F., 2000. Effects of mulching on biomass, nutrients and soil water in banana inoculated with nematodes. *Agron. J.*, 91: 1081–1085.

Mohan, R., Lalsanglura, C. and Sahoo, U.K., 2004. Effect of different mulches on soil, moisture and crop productivity of maize in an agroforestry system of Mizoram. *Sci. Vis.*, 4: 14–17.

Moitra, R., Ghosh D.C., Majumdar S.K. and Sarkar, S., 1994. Influence of tillage and mulching on growth, yield and economics of rainfed yellow (*Brassica rapa* var. *glauca*) in laterite acid belt of West Bengal. *Ind. J. Agric. Sci.*, 64: 156–159.

Ortega, A.L., Sayre, K.D. and Francis, C.A., 2000. Wheat and maize yields in response to straw and nitrogen under a bed planting system. *Agron. J.*, 92: 295–302.

Sahoo, U.K., Mohan, R., Vanlalhluna, P.C. and Sangma, M.M., 2005. Effect of *Leucaena leucocephala* (Lam.) de Wit on the growth and yield of *Zea mays* L. in the humid subtropics of Mizoram, India. *J. Nat. Conserv.*, 17: 177–183.

Sharma, P.P. and Parmar, D.K., 1998. Effect of phosphorus and mulching on the uptake of secondary nutrients and productivity of wheat at different growth stages on an alfisol from Western Himalayas. *J. Ind. Soc. Soil Sci.*, 46: 67–70.

Singh, H., 1995. Nitrogen mineralization, microbial biomass and crop yield as affected by wheat residue placement and fertilizer in a semi-arid tropical soil with minimum tillage. *J. Appl. Ecol.*, 35: 588–595.

Unger, P.W., 1978. Straw mulch rate effect on soil water storage and sorghum yield. *Soil Sc. Soc. Am. J.*, 42: 486–491.

Chapter 55

Biological Treatment of Azodyes

P. Dhevagi[1] and K. Sujatha[2]

[1]*Department of Environmental Science, Tamil Nadu Agricultural University, Coimbatore – 3, T.N.*

[2]*Kongu Nadu Arts and Science College, Coimbatore – 29, T.N.*

Introduction

Textile mills use raw water about 61–646 1/kg of cloth with an average of 172 1/kg of cloth processed (50–81 per cent of water consumed, Bal 1999). Coloured effluents are continuously released into sewage and adjoining water resources which render a catastrophic effect on the environment. The wastewater generated contains dyes at concentration of 10 to 200 mg/l and about 10 to 20 per cent goes along with other organic materials such as fats, waxes, pectin solid fragment, starch derived from the cloth during its processing and constitute major source of pollution (Zollinger, 1961; Anliker, 1979; Chudgar, 1985; Lewis, 1991 and Ashoka *et al.*, 2002). Soluble reactive dyes which are being used in increasing quantities are known to hydrolyze during application without a complete fixation results in larger proportion of these dyes being released into the environment (Carliell *et al.*, 1996). The recalcitrance of the azodyes to biological degradative processes results in severe contamination of the rivers and groundwater in those areas of the world with a high concentration of dyeing industries (Namasivayan and Yamuna, 1992).

Azo dyes are largely used in textile industries, can also be found in the food, pharmaceutical, paper and printing, leather and cosmetics industries (King-Thom Chung *et al.*, 1978 and Sudhakar *et al.*, 2002). Many of these dyes find their way into the environment via wastewater facilities. These compounds retain their colour and structural integrity under exposure to sunlight, soil, bacteria and sweat; they also exhibit a high resistance to microbial degradation in wastewater treatment systems. The development of synthetic fabrics such as nylon, Lycra, rayon and polyester has required the production of new dyes that can effectively bond to these materials. Azo dyes must be continually updated to produce colors that reflect the trends dictated by changing social ideas and styles. Brighter, longer lasting colors are often necessary to satisfy this demand. There are more than 8000 chemical products associated with the dyeing process listed in the Colour Index (1998). While over 1,00,000 commercially available dyes (7×10^5 metric tons) are produced annually (Zollinger, 1987).

A multitude of dyes were used in textile industry are usually aromatic and heterocyclic (Vyas and Motiforis, 1995) and some are toxic and carcinogenic. The different classes of dye include azo, acid, reactive, metal complex, disperse, vat, mortant, direct, basic, suphur etc. (Vijaya *et al.*, 2003).

Effective and economic treatment of a diversity of effluents containing azo group has become a problem. No single treatment system is adequate for degrading the various dye structures. Physico-chemical and biological process has been investigated extensively for treating dye-bearing effluents. Even though chemical coagulation appears to be effective in treating the dye bearing effluents (Venkata Mohan *et al.*, 1999), the toxic sludge handling problem associated and the chemical cost retarded its usage. Adsorption employing activated carbon and various other adsorbents has been quite widely studied; however, the cost of carbon and the regeneration limited its application. Chemical oxidation is sometimes coupled with UV light exposure to increase the color removal. Other techniques involve electrochemical or wet oxidation, activated carbon adsorption, reverse osmosis, or coagulation/ flocculation (Edwards, 2000). Many of these technologies are cost prohibitive, therefore are not viable options for treating large waste streams. Therefore, scientists have been forced to become very creative in the development of effective treatment technologies for the Azo dye removal.

Now the science and technology comes handy with nature's waste disposer–the microbes, in cleaning up the environmental pollutants. Bioremediation of pollution is mandatory for an ecofriendly environment. Recent stress is on microbial based biological methods which are considered to be sustainable, ecofriendly, cost effective and also easy manageable effluent treatment system.

Understanding the basic composition of azo dyes and specifically fiber-reactive azo dyes is necessary to envision how these molecules can be destroyed. By understanding the general chemical structure of these compounds, consideration can be given to the toxic potential that they pose to the environment.

Azo Dyes and its Intermediates

Azo dyes contain at least one nitrogen-nitrogen (N=N) double bond, however, many different structures are possible (Cripps *et al.*, 1990 and Zollinger, 1991). Monoazo dyes have only one N=N double bond, while diazo and,triazo dyes contain two and three N=N double bonds, respectively. The azo groups are generally connected to benzene and naphthalene rings, but can also be attached to aromatic heterocyclic or enolizable aliphatic groups (Zollinger, 1991). These side groups are necessary for imparting the color of the dye, with many different shades and intensities being possible. A common example of an azo dye can be seen in Figuer 55.1 (Emitiazi, 2000). When describing a dye molecule, nucleophiles are referred to as *auxochromes,* while the aromatic groups are called *chromophores.* Together, the dye molecule is often described as a *chromogen.* The absorption and reflection of visible and UV irradiation is ultimately responsible for the observed color of the dye. Synthesis of most azo dyes involves diazotization of a primary aromatic amine, followed by coupling with one or more nucleophiles. Amino- and hydroxyl-groups are commonly used coupling components (Zollinger, 1991). Because of the diversity of dye components available for synthesis, a large number of structurally different azo dyes exist and are used in industry (McCurdy, 1991).Worldwide production of organic dyes is currently estimated at nearly 4,50,000 tons, with 50,000 tons being lost in effluents during application and manufacture (Lewis, 1999).

Figure 55.1: Example of an Azo Dye Structure (Remazol Black 5) (Emitiazi, 2000)

Eighty to ninety-five per cent of all reactive dyes are based on the azo chromogen (Zollinger, 1991 and Edwards, 2000). Reactive dyes are colored compounds that contain one or two functional groups capable of forming covalent bonds with the active sites in fibers. A carbon or phosphorous atom of the dye molecule will bond to hydroxyl groups in cellulose, amino, thiol and hydroxyl groups in wool, or amino groups in polyamides. Most fibre-reactive azo dyes are used for dyeing cellulosic materials, such as cotton and are a major source of dye waste in textile effluents. Fibre-reactive azo dyes exhibit a high wet-fastness, due to their ability to covalently bond to substrates. However, dyes that hydrolyze in solution prior to bonding to a substrate are often lost in the washing processes (Loyd, 1992). Schematic diagram of fiber reactive azo dyes is given in Figuer 55.2.

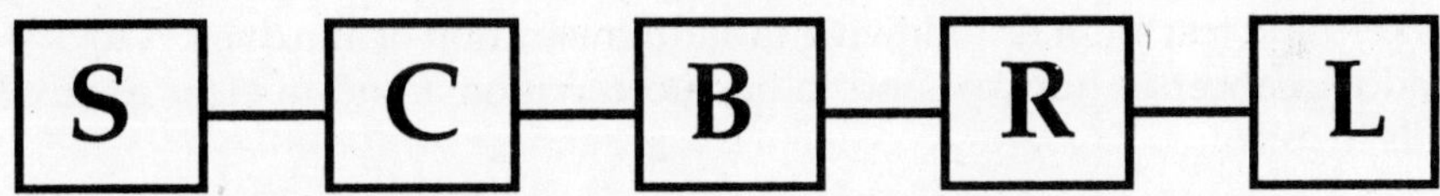

Figuer 55.2: Schematic Diagram of Fiber Reactive Azo Dye

where,

S: Water solubiling group; C: Chromogen; B: Bringing group (can be part of chromogen); R: Reactive group; L: Fiber reactive group

The bridging group serves to combine the chromogen with the reactive group of the dye molecule. The bridging group must be stable, soluble in water and exhibit a certain degree of flexibility. Amino and alkylamino groups are generally used for this purpose. The reactive group serves to bond the dye molecule to a substrate via nucleophilic substitution or addition. Mono-, di- and trichlorotriazinyl are all examples of reactive functional groups. Approximately 200 different reactive groups are listed in the patent literature. Twenty-two new reactive azo dye structures were submitted for patent in early 1998, which was nearly five times as many from other classes of azo dyes (Freeman and Sokolowska, 1999). With such a disproportionate production rate, it is not surprising that a large percentage of dye pollution problems are related to fiber-reactive azo dyes.

Toxicity Considerations

The potential for toxic effects to the environment and humans, resulting from the exposure to dyes and dye metabolites, is not a new concern. As early as 1895 increased rates in bladder cancer were observed in workers involved in dye manufacturing (Rehn, 1895 and Rehn *et al.*, 1994). Since that time, many studies have been conducted showing the toxic potential of azo dyes. A complete review of these studies is beyond the scope of this study; however, a broader understanding of the problem can be found in the works of Levine (1991) as well as Brown and DeVito (1993). Both papers indicate that the problem associated with azo dyes is created by the dye metabolites.

Azo dyes have been identified as potentially carcinogenic agents (IARC, 1996). While most azo dyes themselves are non-toxic a significantly larger portion of their metabolites are toxic (Ganesh, 1992). An investigation of several hundred commercial textile samples revealed that nearly 10 per cent were mutagenic in the Ames test (McCarthy, 1997). Another study conducted on 45 combined effluents from textile finishing plants showed that 27 per cent of the wastewater samples were mutagenic in the Ames test.

Many textile effluents contain heavy metals that are complexes in the azo dyes. Little *et al.*, 1974 reported that all cationic dyes are more toxic and vat dyes are less toxic. High concentrations of salt are often used to force fibre-reactive dyes out of solution and onto substrates (Zollinger, 1991). The azo dyes reduced by cytoplasmic reductases, results in the formation and accumulation of aromatic amines (Chung and Stevens, 1993). These reduction products may be toxic, mutagenic and/or carcinogenic to animals and humans (Levine, 1991) and reduces photosynthesis in plants (Saranaik and Kanekar, 1995). Reduced compounds can cause high electrolyte and conductivity concentrations in the dye wastewater, leading to acute and chronic toxicity problems equal to that of other metals (Venkata Mohan *et al.*, 2002).

Aromatic amines, which are known to be human carcinogens, have been found in the urine of dyestuff workers and test animals following the administration of azo dyes. Azo dye compounds are linked to bladder cancer in humans and to hepato carcinoma and nuclear anomalies in intestinal epithelial cells in mice.

Understanding the dye structures and how they are degraded is crucial to understanding how toxic by-products are created. Brown and DeVito (1993) have proposed the biological mechanisms thought to be responsible for carcinogenic activation of azo dye compounds. This list is based on an extensive review of the literature regarding azo dye toxicity. Brown and DeVito (1993) postulate that:

1. Azo dyes may be toxic only after reduction and cleavage of the azo linkage, producing aromatic amines.
2. Azo dyes with structures containing free aromatic amine groups that can be metabolically oxidized without azo reduction may cause toxicity.
3. Azo dye toxic activation may occur following direct oxidation of the azo linkage producing highly reactive electrophilic diazonium salts.

An example of a reductive pathway that is mediated by intestinal anaerobic bacteria and leads to the formation of benzidines is shown in Figuer 55.3 (Brown and DeVito, 1993).

Similar reactions and by-products would be possible in an anaerobic treatment system. Examples of toxic aromatic amines, which could be created from the degradation of azo dye compounds and its human effects (Cartwright, 1983, Levine, 1991, Brown and Devito, 1993 and Chung, 2000), were given in Table 55.1.

Treatment Options

The treatment of textile effluents is of interest due to their toxic and esthetic impacts on receiving waters. While much research has been preformed to develop effective treatment technologies for wastewater containing azo dyes, no single solution has been satisfactory for remediating the broad diversity of textile wastes. Despite being aware of the problem, many textile manufactures have failed

Congo Red

NH_2 SO_3Na

$-N=N-$ $-N=N-$

NaO_3S H_2N

Anaerobic Bacteria

Benzidines : R = H_2, CH_3, CH_3O

H_2N— —N_2H

R R

Figure 55.3: Example of Azo Dye Reduction via Anaerobic Bacterium

to adequately remove azo dye compounds from their wastewaters. Until dye and textile manufactures are able to develop efficient technologies, allowing for increased dye-fiber bonding and lower dye house losses (Lewis, 1999), the problem of treating these types of wastes will fall to the wastewater treatment facilities.

Table 55.1: Effect of Aromatic Amines and Potential Dye Metabolites

Sl.No.	*Compound*	*Human Effect*
1.	1-Napthylamine	Slight/mixed
2.	2-Napthylamine	Good
3.	2,5-Diamiotoluene	Slight
4.	3,3'-Dichlorobenzidine	Slight/mixed
5.	3,3'- Dimethoxy benzidine	Slight /mixed
6.	3,3'-Dimethylbenzidine	Slight
7.	4-Biphenylamine	Good
8.	4-Nitrobiphenyl	Slight/mixed
9.	4,4'-Methylenebis(2-chloroaniline)	Slight
10.	Auramine	Slight
11.	Benzidine	Good
12.	N,N-Bis(2-chloroetlhyl)-napthylamine	Good
13.	Magenta	Slight

To ensure the safety of effluents, proper technologies need to be used by treatment facilities when degrading fiber-reactive azo dyes. Azo dyes usually resist biodegradation in conventional treatment methods (Pagga and Brown, 1986, Shaul *et al.*, 1991 and Willmott *et al.*, 1998). Previous research efforts have focused on various biological, chemical and physical techniques for treating azo dye wastes. There is evidence that all three areas have potential for remediating dye house wastes. However, chemical treatment is often cost and application limited, while physical removal can lead to extra solid wastes and increased overhead (Yamini and Lalithakumari, 2001). Biological treatment has been effective in reducing dye house effluents and when used properly has a lower operating cost than other remediation processes. Combinations of chemical and biological or physical and biological treatment have also proven to be effective. Because this research explores the use of microbial sludges for the treatment of a textile effluent, the focus of this literature review will be on biological techniques used for degrading wastewaters containing azo dyes.

Because of their recalcitrant nature, azo dyes often pass through activated sludge facilities with little or no reduction in color (Cariell *et al.*, 1995, 1996; Pagga and Brown, 1986). Although some researchers have observed slight color reductions, their findings are largely outweighed by those who have not (Loyd, 1992, Zissi *et al.*, 1997). Reductions in the carbon content and oxygen demand of azo dye wastewaters following aerobic treatment are well cited (McCurdy, 1991, Horning 1977, Loyd, 1992, Pagga and Brown, 1986).

The anaerobic reduction of azo dyes to simpler compounds has been well researched (Chinwetkitvanich *et al.*, 2000, Razo-Flores *et al.*, 1997, Brown and Laboureur, 1983, Chung *et al.*, 1978). These and other studies, have all demonstrated the ability of anaerobic microbes and sludges to effectively reduce azo dyes to their structures, thus destroying the apparent color. Many of these intermediates are aromatic amines with constituent side groups. By reducing the dye compounds to their intermediates, the problem of aesthetic pollution is solved, but a larger and more deleterious problem may be created. Most azo dyes are non-toxic, but a higher percentage of their intermediates have been identified as carcinogens (Brown and DeVito, 1993,Wong and Yuen, 1996). Because of the toxic potential of many aromatic amines, further degradation of the dye compound is necessary if toxicity is to be eliminated or reduced (Levine, 1991 and Brown and DeVito, 1993).

Aerobic Treatment

Conventional activated sludge treatment of wastes is often an effective and highly economic system for reducing organic pollutants in wastewater. A fair amount of research has been conducted assessing the viability of using activated sludge to treat textile effluents (Pagga and Brown, 1986, Shaul *et al.*, 1991, Loyd, 1992, Brown and DeVito, 1993 and Zissi *et al.*, 1997). However, aerobic treatment of azo dye wastes has proven ineffective in most cases, but is often the typical method of treatment used today (Edwards, 2000, Yang *et al.*, 1998). Because aerobic microbes cannot reduce azo linkages, their ability to destroy dye chromogens is less than anaerobic bacterium. However, aerobic sludges have been successfully used to stabilize dye metabolites (Brown and Laboureur, 1983). Pagga and Brown (1983) conducted a study on 87 commercial dye stuffs. Some of the dye stuffs were in a technically pure form, while others were in a sales form containing organic substances such as wetting agents. The tests were performed in a reactor designed to simulate the conditions of and adapted activated sludge wastewater treatment plant. The samples were tested for color and DOC removal following 42 days of treatment. Kumar *et al.*, 2006 studied the decolorisation, biodegradation and detoxification of Direct Black-38, a benzidine based azo dye, by a mixed microbial culture isolated from an aerobic bioreactor treating textile wastewater. Pagga and Brown indicated that the primary mechanism for removal of dyes in activated sludge systems may occur by adsorption onto the cell walls (Knapp *et al.*, 1995 and Banat *et al.*, 1996). In an earlier study by Brown and Laboureur (1983a), the aerobic biodegradability of aniline, o-toludine, p-anisidine, p-phenetidine, o-dianisidine and 3,3′-dichlorobenzidine, was investigated. These compounds are all lipophilic aromatic amines and possible by-products of azo dyes. Because many aromatic structures are non biodegradable in anaerobic environments and are not hydrophilic, they can accumulate in the adipose tissues of organisms.

Brown and Laboureur (1983) indicated that azo dyes may be broken down to their intermediate structures in a reductive environment, but were not amenable to further degradation by anaerobes. Brown and Laboureur (1983a) concluded that aniline, p-anisidine, p-phenetidine and o-toludine were readily biodegradable by aerobes, while o-dianisidine and 3,3′-dichlorobenzidine were inherently biodegradable. They suggested that these compounds could be stabilized if released into the environment or directly from a dye house into a conventional wastewater treatment plant. Shaul *et al.* (1991) conducted a study on 18 dyes to determine their fate in the activated sludge process. Of these dyes, 15 were acid azo dyes and three were direct azo dyes. The dyes were spiked into pilot-scale treatment systems. Eleven of the dyes passed through the activated sludge system substantially untreated, four were significantly absorbed into the sludge and three were apparently biodegraded.

Loyd (1992) also performed activated sludge treatment tests on two wastewaters. The first was a textile dyeing and finishing process water that contained Reactive Navy 106 and the second was a municipal wastewater consisting predominately of textile effluents. Both effluents were fed to laboratory-scale activated sludge reactors. Loyd concluded that aerobic treatment of the azo dye wastewaters provided significant biodegradation with minimal decolorization and the biodegradation did not include the azo dyestuffs.

Moosvi *et al.*, 2005 developed consortium exhibited 94 per cent decolourization ability within 37 h under a wide pH range from 6.5 to 8.5 and temperature ranging from 25 to 40 °C. The bacterial consortium was able to grow and decolourize RV5 under static conditions in the presence of glucose and yeast extract and also showed an ability to decolourize in the presence of starch in place of glucose. Maximum decolourization efficiency was observed at 200 ppm (mg/l) concentration of RV5. Bacterial consortium RVM11.1 had the ability to decolourize 10 different dyes tested.

Idaka *et al.*, 1987 described the reductive fission of azo bonds by *P. cepacia* and aerobic reduction of simple azo compounds by *Aeromonas hydrophilia.* The ability of bacteria to utilize simple azo dye compounds as sole source of carbon and energy was shown by Dykes *et al.*, 1994 and Yatome *et al.*, 1993 and *Pseudomonas* K22 strain capable of utilizing carboxy orange II was well established by Kulla 1981, Zimmermann *et al.*, 1982, Kulla *et al.*, 1983 and Coughlin *et al.*, 1997 and 1999.

Finally, Zissi *et al.* (1997) investigated the biological oxidation of p-aminoazobenzene (pAAB) by *Bacillus subtilis* and results proved that *Bacillus subtilis* could cometabolize pAAB in the presence of glucose, breaking the N=N double bond and producing aniline and p-phenylenediamine. The degradation of the dye was the direct result of an oxygen insenstive azo reductase enzyme found to be present in the soluble fraction of the biomass. This enzyme was also synthesized independently of the presence of pAAB.

Bacterial cultures namely *Vibrio logei* and *Pseudomonas nitroreducens* from a wastewater treatment plant degraded a toxic azo dye (methyl red) by decolourization. Complete decolourization using a mixed-culture was achieved at pH 6, 30°C within 6 h at 5 mg/l methyl red concentration and 16 h at 20–30 mg/l (Adedayo *et al.*, 2004).

Azo Dye Degradation by Fungi

The inappropriate disposal of dyes in wastewater constitutes an environmental problem and can cause damage to the ecosystem. Alternative treatments have been reported that fungi are particularly effective in the decolorization of textile effluents.

The wood degrading whit rot basidiomycete *Phanerochaete chrysosporium* was found to degrade a number of Azo dyes (Glenn and Gold, 1983 and Paszczynski *et al.*, 1992, Spadaro *et al.*, 1992). The extracellular peroxidases produced by *P. chrysosporium,* Laccase produced by *Pyricularia orison*, Mn peroxidase produced by *Aspergillus terreus* were found to initiate azo dye degradation (Glenn *et al.*, 1986, Chivukula Cripps *et al.*, 1990 and Renganathan, 1995). The degradation pathway of copper-pthalocyanine dyes by *P. chrysosporium* was given by spectrometry and polarography (Conneely *et al.* 1999). Anthraquinone dye degradation by *Geotricum candidum* was reported by Kim *et al.*, 1995 and Kim and Shoda, 1999.

The decolorization of dyes with different molecular structures by *Cunninghamella elegans* was evaluated under several media conditions. The decolorization profile was highly dependent upon the incubation time, the molecular structure of the dye and presence or absence of co-substrates. The presence of sucrose or both sucrose and peptone significantly increased the decolorization of the solutions, however, the presence of only the nitrogen source suppressed it (Ambrosio and Campos-Takaki 2004).

Basidiomycete PV 002, a recently isolated white-rot strain from decomposed neem waste displayed high extracellular peroxidase and rapidly decolorized azo dyes. White rot strain PV 002 efficiently decolorized Ranocid Fast Blue (96 per cent) and Acid Black 210 (70 per cent) on day 5 and 9 respectively under static conditions. The degradation of azo dyes under different conditions was strongly correlated with the ligninolytic activity. The optimum growth temperature of strain PV 002 was 26°C and pH 7.0 (Pradeep Verma and Datta Madamwar 2005).

Algal Biodegradation

The commonly available algal spirogyra species was found to degrade/decolorize the aromatic basic dye-stuff containing azine group in batch scale operation (Venkata Mohan *et al.*, 2002). The decrease in aqueous phase pH below 3 during initial phase reduces the surface charge of algal cell, which sorbed the dye molecule in the aqueous environment through diffusion. This was observed in the initial stages of the study and about 10 per cent of the dye removal may be attributed to this mechanism (Biosorbtion). Once the dye molecules adsorbed onto the surface of the cell diffuses into the cell and stored in vacuole and participate in the respiration/photo conversion and subsequently releases polysaccharides and related organic compounds (bioconversion). The released metabolic intermediates possess excellent coagulation capacity and the dye remaining in the aqueous phase tend to absorb on the surface of the polymers and settles (Biocoagulation).

Actinomycete Biodegradation

A variety of actinomycetes like *Streptomyces viridosporrns* (Ramachandra *et al.*, 1988), *Thermomonospora mesophila* and *Streptomyces badius* (Godden *et al.*, 1992) and *Streptomyces chromofuscus* A11 (Pasti-Grigsby *et al.*, 1996) involving in azo dye degradation have been reported. These *Streptomycetes* sp. produce extracellular peroxidases that contribute to their ability to decolorize the dyes. Evidence does show that the aerobic biodegradation of azo dye intermediates is possible and is perhaps an effective treatment process for stabilizing these compounds after anaerobic reduction.

Anaerobic Treatment

Anaerobic reduction of azo dyes using microbial sludges can be an effective and economic treatment process for removing color from dye house effluents. Previous studies have demonstrated the ability of anaerobic bacteria to reductively cleave the azo linkages in reactive dyes (Chung *et al.*, 1978; Brown and Laboureur, 1983; Brown and Hamburger, 1987; Loyd, 1992; Ganesh, 1992; Razo-Flores *et al.*, 1997 and Chinwetkitvanich, 2000). Although this effectively alters the chromogen and destroys the observed color of the dye, many aromatic groups are not susceptible to anaerobic reduction. However, there is evidence that some azo dye metabolites may be fully stabilized in anaerobic environments (Weber and Wolfe, 1987 and Razo-Flores, 1997).

During anaerobic treatment, it is expected that the dye molecules would cleave via reduction of the azo bonds. While this is often effective for reducing the color of the dye, it does not usually lower the carbon content (Loyd, 1992; Ganesh, 1992). Brown *et al.* (1993) studied the degradability of various azo dyes in both anaerobic and aerobic systems. In the first study, Brown and Laboureur (1983) investigated the anaerobic degradability of 22 commercial dyes. Of the dyes studied, four monoazo and six diazo-dyes showed substantial biodegradation, while two polyazo dyes showed moderate to variable reductions. Later in 1987, Brown and Hamburger conducted a study on 14 azo dyes subjected to anaerobic sludge digestion followed by aerobic treatment. Brown and Hamburger's results confirmed the findings from earlier research, showing decolorization of the azo dyes. Confirming the cleavage of azo linkages, the production of metabolites was also observed, but at less than theoretical concentrations.

Brown and Hamburger (1987), Loyd (1992) and Ganesh (1992), who all reported good color losses, but low carbon removals following the anaerobic treatment of various azo dyes. It is hypothesized that the dye molecules undergo only partial degradation, limiting the carbon loss, but not the color removal. More recently, Razo-Flores *et al.* (1997) investigated the fate of Mordant Orange 1 (MO1) and Azodisalicylate (ADS) under methanogenic conditions using continuous Upflow-Anaerobic-Sludge-Blanket (UASB) reactors and demonstrated the ability of an anaerobic consortium to completely mineralize some azo dye compounds. Razo-Flores *et al.* (1997) observed the complete mineralization of ADS with and without a cosubstrate, indicating the possibility for aromatic amine destruction in methanogenic environments. The compound, 1,4-phenylenediamine was not observed to degrade in either test reactor, indicating the specificity of aromatic amine utilization by anaerobes.

In studies conducted by Loyd (1992) and Ganesh (1992), the anaerobic reduction of textile mill effluents and the azo dyes Reactive Black 5 and Navy 106 were investigated, respectively. In both cases, good decolorization with minimal nutrient removal was observed. Chinwetkitvanich *et al.* (2000) performed a study on various reactive dye bath effluents concluded that higher initial color concentrations might be deleterious to acid forming bacteria, resulting in a lower dye removal. Additionally, the authors suggest that sulfate-reducing bacteria might out-compete other anaerobic microorganisms for available organic carbon, but contribute minimally to decolorization. This could serve to limit the reduction equivalents necessary for dye degradation.

Anaerobic/Aerobic Sequential Step-Treatment

Anaerobic bacteria are often able to reduce the azo linkages, but are generally unable to further stabilize the dye metabolites; it would seem advantageous to follow anaerobic treatment processes with an aerobic treatment step. As mentioned previously, aerobic organisms can oxidize aromatic ring compounds to simpler-molecules. A substantial amount of research has been conducted on ANA/

AER sequential step-treatment systems used for degrading textile wastewaters (Brown and Hamburger, 1987; Loyd, 1992; Seshadri and Bishop, 1994; Fitzgerald and Bishop, 1995 and O'Neill *et al.*, 2000). Brown and Hamburger (1987) conducted a study on the ultimate biodegradability of various dye stuffs. Fourteen azo dyes and two other dye types were studied using lab-scale anaerobic and activated sludge reactors. Metabolite production was observed following anaerobic treatment, indicating the presence of eight identifiable aromatic amines. Results from the aerobic treatment phase corresponded with the work performed by Brown and Laboureur (1983a), showing reductions in DOC for most of the dyes. Some sulphonated aromatic amines were found to be non- biodegradable during this test.

Seshadri and Bishop (1994) investigated the fate of azo dyes Acid-Orange 7, Acid-Orange 8, Acid-Orange 10 and Acid-Red 14 in an ANA/AER sequential step-treatment system in a bench-scale Fluidized-Bed Anaerobic Reactor (FBR) followed by a bench-scale activated sludge reactor as a sequenced second stage treatment step. Their results indicated that the transformation of all the dyes to intermediates was readily achieved via anaerobic reduction; and was assumed to be the result of azo bond cleavage. Complete mineralization was not observed, however, COD and color removals were greatest at HRT's of 12 and 24 hours, with one hour being the shortest HRT tested and 24 hours the longest. However, the largest cumulative removals occurred in the first two hours of treatment. This would be expected based on the higher nutrient concentrations available for biodegradation during that period. The Acid-Orange 10 appeared to inhibit dye removal at concentrations of 15 mg L^{-1}. The authors believed the production of aromatic amines was responsible for the toxicity, citing the work of Chakrabarti *et al.* (1988). All other dyes tested did not exhibit inhibition at a concentration of 15 mg L^{-1}. COD removal was variable depending on the dye, but reductions were seen in both the anaerobic and aerobic phases. Aerobic oxidation of dye intermediates was necessary to decrease COD levels to an acceptable range.

Results from Zissi and Lyberatos (1996) confirmed that the anoxic biodegradation of p-aminobenzene, a simple azo dye, is effective in removing its apparent color. The findings from this research suggest that an ANA/AER sequential step-treatment system is the most effective method for biologically treating textile wastewaters.

Fitzgerald and Bishop (1995) used three lab-scale reactor systems to study the degradation of the azo dyes Acid-Orange 10, Acid-Red 14 and Acid-Red 18. The reactor system included an anaerobic fluidized bed system in the first stage, which was followed by an aerobic swisher reactor in the second stage. Their results indicated a high degree of degradation of the Acid Red 18 and Acid Red 14 in the anaerobic stage, with decolorization greater than 90 per cent. Acid Orange 10 was only decolorized by 70 per cent, however. Furthermore, analysis of the dye intermediates suggested a high degree of removal in the anaerobic stage. This is interesting, as most of the literature reports poor metabolite reduction in anaerobic environments. Low color loss and COD removal was measured in the aerobic stage. The authors do state that further investigation is necessary to verify if the dye intermediates biodegraded in the anaerobic reactor.

Recently, O'Neill *et al.* (2000) conducted a study on degradation of reactive azo dye Procion Red H-E7B in an ANA/AER sequential step-treatment system comprised of a lab-scale UASB reactor and an activated sludge tank. Based on this research, O'Neill concluded that Procion Red H-E7B could be qualitatively shown to degrade to aromatic amine derivatives after anaerobic treatment with subsequent oxidation of these derivatives following aerobic treatment.

Factors Affecting Azo Dye Biodegradation

Due to the highly variable nature of biological treatment systems and especially textile effluents, there are a number of factors that may affect the biodegradation rate of azo dyes. Non-dye related parameters such as temperature; pH, dissolved oxygen or nitrate concentrations, type and source of reduction equivalents, bacteria consortium and cell permeability can all affect the biodegradation of azo dyes and textile effluents. Dye related parameters such as class and type of azo dye (*i.e.*, reactive-monoazo), reduction metabolites, dye concentration, dye side groups and organic dye additives could also affect the biodegradability of azo dye wastewaters.

Wuhrmann *et al.* (1980) investigated the effects of pH, temperature, type and concentration of respiration substrates and oxygen tension on the rate of biological reduction of a variety of azo dyes. A consortium of microbes was used, including *Bacillus cereus, Sphaerotilus natans* and two others isolated from sewage-activated sludge. Also, activated sludge was used in experiments with mixed biocenoses. Temperatures, which are too high or too low, can result in the exclusion of a particular group of microorganisms. Using activated sludge, Wuhrmann *et al.* (1980) determined that temperature has an increasing liner relationship with the reduction rate of Orange II and Lanasyl violet up to 28°C. In general however, most studies have been conducted at set temperatures, offering minimal data on temperature effects. The wastewater pH can affect the proper functioning of both anaerobic and aerobic organisms (Grady *et al.*, 1999). Wuhrmann *et al.* (1980) also investigated the effect of pH on dye reduction rates, but were unable to conclusively establish a relationship. However, they did state that an exponential increase in the decolorization rate was observed by decreasing the pH, but this relationship depended on the dye being tested. Loyd (1992) observed an indirect increase in the rate of decolorization of Navy-106, with decreased pH values in anaerobic batch tests. No statistical data was performed to verify this result; however Nitrate and especially oxygen may play an important role in determining the rate of dye reduction. The presence of oxygen generally inhibits the degradation of azo dye chromogens. Interestingly, Wuhrmann *et al.*, demonstrated that obligate aerobes might actually de colorize azo compounds under temporary anoxic conditions. However, high nitrite or nitrate concentrations in the mixed liquor of activated sludge plants could significantly inhibit dye removal. Zissi and Lyberatos (1996) observed *Bacillus subtilis* to degrade p-aminoazobenzene under anoxic conditions.

Without the necessary reduction equivalents to optimize bacterial respiration and growth, dye reduction may be inhibited. Often, bacterial cultures are unable to proliferate when an azo dye is the sole carbon and nitrogen source. Therefore, additional, readily biodegradable sources may be necessary. Wuhrmann *et al.* (1980) indicated that in the absence of oxygen an azo compound will act as the sole oxidant, and its reduction rate will be governed by the rate of formation of the electron donor. This may be a problem in WWTPs that receive a high loading of dye waste with a low, readily available, organic carbon content. Gingell and Walker (1971) state that the presence of oxygen may compete with the azo dye as the preferred oxidizer of reduced electron carriers in the respiration chain and thus limit the reduction of azo linkages. The type of bacteria or consortium used for dye biodegradation will undoubtedly affect the reduction rate. In general however, aerobic microbes do not have the ability to substantially decolorize azo dyes, but can oxidize the dye metabolites. Ganesh (1992) concluded that very little of the dye added to a biological reactor will be leached from the biomass when placed in a landfill. This might suggest that the dye is effectively reduced after adsorption to the cell wall or that very little dye is actually adsorbed. Cell permeability might play an important role in dye biodegradation.

Jo-Shu Chan *et al.*, 2004 reported that the addition of supernatants containing metabolites from growth and decolorization cultures triggered an enhancement of decolorization rates. Although extracellular metabolites play a crucial role for stimulating decolorization, they still cannot enable decolorization alone without involvement of biodecolorizers (*E. coli* strain NO_3).

In the study conducted by Wuhrmann *et al.* (1980), all dyes that were not reduced by whole cells were effectively degraded by cell extracts from both facultative, anaerobes and obligate aerobes. This suggests that many cells might be capable of dye biodegradation, but are limited by the permeability of their cell walls. The azo dye structure can play a significant role in the dye biodegradation rate. Depending on the number and placement of the azo linkages, some dyes will biodegrade more rapidly, than others. In general, the more azo linkages that must be broken will cause the reduction rate to be slower. Brown and Laboureur (1983) observed that two poly-azo dyes showed only moderate to variable biodegradation as compared to four monoazo and six diazo dyes. The authors indicate that poly-azo dyes are less likely to degrade than mono- or diazo dye types. Fiber-reactive azo dyes often contain solubilizing side groups, as well as a nucleophilic reactive group. Depending on the nature of these groups, biodegradation might be inhibited.

According to Ganesh (1992), sorption of dye to sludge depends on the type, number and position of the constituents in the dye molecule. Hydrophilic sulfo groups reduce the dye removal through sorption and conversely, sorption is increased by the presence of hydroxyl, nitro and azo groups in the dye molecule. The production of toxic byproducts or the presence of toxic dye additives may also inhibit biodegradation. High, salt concentrations are not uncommon in textile effluents and may result in adverse conditions for biodegradation. Dispersing and solubilizing agents may also create inhibitory conditions for dye reduction (Carliell, *et al.*, 1994).

In various studies (Wuhrmann *et al.*, 1980; Donlagic' and Levec, 1998, and O'Neill *et al.*, 2000) the production of inhibitory dye metabolites is cited as causing a decrease in biodegradation. O'Neill *et al.* concluded from respiration-inhibition testing that anaerobic degradation of simulated textile effluent generated metabolites that were toxic to some aerobic organisms, Donlagic' and Levec concluded that the distribution of dye intermediates plays an important role in determining the aerobic biodegradability of Orange II. They further state that low dye removal may be attributed to the presence of intermediates that are less susceptible to microbial degradation or that act as inhibitory agents. Wurhmann *et al.* (1980) attributes a decline in decolorization rates to the accumulation of reduction products in the test medium. Inhibition of various microorganisms to dye metabolites is frequently cited throughout the literature and is assumed to be a key factor when treating wastewaters containing azo dyes.

A final and important factor to evaluate is the initial dye concentration of the wastewater. Seshadri and Bishop (1994) performed a study investigating the effect of different influent dye concentrations on the color removal efficiency. They concluded that elevated dye concentrations may cause a drop in per cent dye removal. Furthermore, the inhibition may be directly related to the effects of increased dye formation due to higher dye concentrations. Less pronounced reductions were seen at lower concentration levels. It should be noted that tolerable influent concentrations are likely specific to individual or related groups of dyes. Cariell *et al.* (1995) also states that toxicity assays showed that C.I. Reactive Red 141 was inhibitory to anaerobic organisms at concentrations greater than 100 mg/l, however prior biomass adaptation increased their resistance to elevated dye concentrations.

The Research Need

Biodegradation of azo dyes by a single species of microorganism requires that species to possess all the necessary enzymes involved in the pathway and also genetic information. Continual exposure of microbes to azo dyes can result in the evaluation of novel metabolic processes which enables the microorganisms to wholly or partially degrade the azo dyes. There are two possible ways by which the microbes acquire new traits. They are (1) Natural mutation and natural gene transfer and

(2) Artificial gene transfer. Natural mutations occur at the rate of 1 mutation per 100 cell divisions, mostly non beneficial to the cell. Many genes that encode key enzymes in the catabolism of azo dyes are plasmid encoded. Such plasmids can be transmitted from one cell to another by the processes of conjugation and transformation. In cells possessing plasmids which only encode for a few enzymes for azo dye degradation must be complemented with chromosomal genes so that plasmid encoded pathway is linked to a metabolic pathway. Encoding of degradative genes on plasmids gives the possibility for the construction of microbes by genetic engineering with the metabolic requirements for the treatment of azo dye containing wastes.

References

Adedayo, O.S., Javadpour, C., Taylor Anderson, W.A. and Moo-Young, M., 2004. Decolourization and detoxification of methyl red by aerobic bacteria from a wastewater treatment plant. *World J. Microbiol. Biotechnol.*, 20(6): 545–550.

Ambrosio, S.T. and Campos-Takaki, G.M., 2004. Decolorization of reactive azo dyes by *Cunninghamella elegans* UCP 542 under co-metabolic conditions. *Biores. Technol.*, 91(1): 69–75.

Anliker, R., 1979. Ecotoxicology of dyestuffs: A joint effort by industry Ecotox. *Environ. Safety*, 3: 59–74.

Ashoka, C., Geetha, M.S. and Sullia, S.B., 2002. Biobleaching of compost textile dye effluent using bacterial consortia. *Asian J. Microbiol. Biotech. Env. Sci.*, 4: 65–68.

Bal, A.S., 1999. Waste water management for textile industry: An overview. *Indian J. Env. Hlth.*, 41: 264–290.

Banat, I.M., Nigan, P., Singh, D. and Marchanto, R., 1996. Microbial decolourisation of textile dye containing effluents: A review. *Biores. Technol.*, 58: 217–227.

Brown, D. and Hamburger, B., 1987. The degradation of dyestuffs, Part III: Investigations of their ultimate degradability. *Chemosphere*, 16(7): 1539–1553.

Brown, D. and Laboureur, P., 1983. The degradation of dyestuffs, Part I: Primary biodegradation under anaerobic conditions. *Chemosphere*, 12(3): 397–404.

Brown, D. and Laboureur, P., 1983a. The aerobic biodegradability of primary aromatic amines. *Chemosphere*, 12(3): 405–414.

Brown, Mark A. and Stephenm C. DeVito, 1993. Predicting azo dye toxicity. *Crit. Rev. Environ. Sci. Tech.*, 23(3): 249–324.

Brown, M.S. and Devito, S.C., 1993. Predicting azo dye city. *Crit. Rev. Environ. Sci. Tech.*, 23: 249–342.

Cariell, C.M., Barclay, S.J. and Buckley, C.A., 1996. Treatment of exhausted reactive dye bath effluent using anaerobic digestion: Laboratory and full scaletrials. *Water SA*, 22(3): 225–233.

Cariell, C.M., Barclay, S.J., Naidoo, N., Buckley, C.A., Mulholland, D.A. and Senior, E., 1994. Anaerobic decolorisation of reactive dyes in conventional sewage treatment processes. *Water SA*, 20(4): 341–344.

Canell, C.M., Barclay, S.J., Naldoo, N., Buckley, C.A., Mulholland, D.A. and Senior, E., 1995. Microbial decolourisation of a reactive azo dye under anaerobic conditions. *Water SA*, 21(1): 61–69.

Cartwright, R.A., 1983. Historical and modern epidemiological studies on populations exposed to N-substituted aryl compounds. *Env. Health Perspect.*, (49): 13.

Chakrabarti, T., Subrahmanyam, P. and Sundarsan, B., 1988. Biodegradation of recalcitrant industrial wastes. *Biotreatment Systems Vol. II.*, (Ed.) D. Wise. CRC Press, Inc., Boca Raton, FL.

Chinwetkitvanich, S., Tuntoolvest, M. and Panswad, T., 2000. Anaerobic decolorization of reactive dye bath effluents by a two-stage UASB system with tapioca as a co-substrate. *Water Research*, 34(8): 2223–2232.

Chivukula, M. and Renganathan, V., 1995. Phenolic azo dye oxidation by laccase from *Pyricularia oryzae. Appl. Environmental Microbiol.*, 61: 4374.

Chudgar, R.J., 1985. Azo dyes. In: *Kirk–Othmer Encyclopedia of Chemical Technology*, 4th edn, (Ed.) Kroschwitz. Wiley, New York, pp. 821–875.

Chung, King-Thom and Stevens, S.E., 1993. Degradation of azo dyes by environmental microorganisms and helminthes. *Environ. Toxicol. Chem.*, 12: 2121–2132.

Chung, King-Thom, Fulk G.E. and Egan, M., 1978. Reduction of azo dyes by intestinal anaerobes. *Appl. Environmental Microbiol.*, 35(3): 558–562.

Chung, K.T., 2000. Mutagenictiy and carcinogenicity of aromatic amine metabolites produced from azo dyes. *Enviro Carcino. and Ecotox. Rev.*, 18: 51–74.

Colour Index International: Additions and Amendments. 1998. *The Society of Dyers Colourists*, 109: 2860–2895.

Conneely, A., Smyth W.F. and McMullan, G., 1999. Metabolism of the pthalocyanine textile dye turquoise blue by *Phanerochaete chrysosporium. FEMS Microbiol. Lett.*, 179: 333–337.

Coughlin, M.F., Kinkle, B.K. and P.L. Bishop, 1999. Degradation of azo-dyes containing aminonaphthol by *Spingomonas* sp. strain ICX. *Indian J. Microbiol. Biotechnol.*, 23: 341–346.

Coughlin, M.F., B.K-.Kinkle, K. Tipper and P.L. Bishop, 1997. Characterisation of aerobic azo-dye degrading bacteria and their activity in biofilms. *Water Sci. Tech.*, 36: 215–220.

Cripps, C., J.A. Bumpus and S.D. Aust, 1990. Biodegradation of azoheterocyclic dyes by *Phanerochaete chrysosporium. Appl. Environ. Microbiol.*, 56: 1114–1118.

Donlagic, Jelka and Janez Levec, 2000. Comparison of catalyzed and noncatalyzed oxidation of azo dye and effect on biodegradability. *Environ. Sci. Technol.*, 32: 1294–1302.

Dykes, G.A., Timm, R.G. and Von Holy, A., 1994. Azo reductase activity in bacteria associated with the greening of instant chocolate pudding. *Appl. Environ. Microbiol.*, 60: 3071–3074.

Edwards, Jesscia C., 2000. Investigation of color removal by chemical oxidation for three reactive textile dyes and spent textile dye wastewater. *Masters Thesis*, Virginia Polytechnic Institute and State University, 56 pp.

Emitiazi, G., 2000. Decolorisation and biodegradation of dyes by *Aspergillus terreus* grown on wheat straw with Mn peroxidase activity. *Poll. Res.*, 19: 31–35.

Fitzgerald, Sean W. and Paul L. Bishop, 1995. Two stage aerobic/aerobic treatment of sulfonated azo dyes. *J. Environ. Sci. Health*, 30(6): 1251–1276.

Freeman, Harold S. and Sokolowska, J., 1999. Developments in dyestuff chemistry. *Review of Progress in Coloration and Related Topics*, pp. 29.

Ganesh, R., 1992. Fate of azo dyes in sludges. *Masters Thesis*, Virginia Polytechnic Institute and State University. 193 pp.

Gingell, R. and Walker, R., 1971. Mechanisms of azo reduction by *Streptococcus faecalis* II: The role of soluble flavines. *Xenobiotica*, 1: 231–239.

Glenn, J.K. and Gold, M.H., 1983. Decolorisation of several polymeric dyes by the lignin degrading basidiomycete *Phanerochaete chrysosporium. Appl. Environ. Microbiol.*, 45: 1741–1747.

Glenn, J.K., Akileswara, L. and Gold, M.H., 1986. Mn (II) oxidation is the principal function of the extra cellular Mn peroxidase from *Phanerochaete chrysosporium. Arch. Biochem. Biophys.*, 251: 688–696.

Godden, B.A., S. Ball, Helvenstein, P.A., Carthy J. Mc and Penninckx, M.J., 1992. Towards elucidation of the lignin degradation pathway in actinomycetes. *J. Gen. Microbiol.*, 138: 2441–2448.

Grady, C.P.L. Jr., Daigger, G.T., Lim, H.C., 1999. *Biological Wastewater Treatment*. Marcel Dekker, Inc., New York, NY, 1076 pp.

Horning, R.H., 1977. Characterization and treatment of textile dyeing wastewaters. *Textile Chemist and Colorist*, 9(3): 73–76.

IARC, World Health Organization International Agency for Research on Cancer. 1982. Monographs on the Evaluation of the Carcinogenic Risk of Chemicals to Humans. *Some Industrial Chemicals and Dyestuffs*, Vol. 29, Lyon, France. Jager, I., *et al.*, 1996. *Melliand Textilber*, 77: 72.

Idaka, E., Ogawa, T. and Horitsu, H., 1987. Some properties of azo reductase produced by *Pseudomonas. Bull. Environ. Contam. Toxicol.*, 39: 982–989.

Jo-Shu Chan, Bor-Yann Chen and Yung Sheng Lin, 2004. Stimulation of bacterial decolorization of an azo dye by extracellular metabolites from *Escherichia coli* strain NO_3. *Biores. Technol.*, 91(3): 243–248.

Kim, S.J. and M. Shoda, 1999. Purification and characterization of a novel peroxidase from *Geotrichum candidum* December 1 involved in decolorisation of dyes. *Appl. Environ. Microbiol.*, 65: 1029–1035.

Kim, S.J., Ishikawa, K., Hirai, M. and Shoda, M., 1995. Characteristics of a newly isolated fungus, *Geotrichum candidum* December 1 which decolorizes various dyes. *J. Fermentation and Bioengineering*, 79: 601–607.

King-thom, Chung, George E. Fulk and Mary Egan, 1978. Reduction of azodyes by intestinal anaerobes. *Appl. Env. Microbiol.*, pp. 558–562.

Knapp, J.S. and Newby, P.S., 1995. Decolourization of dyes by wood rotting basidiomycete fungi. *Water Research*, 29: 1807–1809.

Kumar, K., Saravana Devi, S., Krishnamurthi, K., Gampawar, S. Nilesh Mishra, Pandya G.H. and Chakrabarti, T., 2006. Decolorisation, biodegradation and detoxification of benzidine based azo dye. *Biores. Technol.*, 97(3): 407–413.

Kulla, H.G., Klausner, U. Meyer, B., Ludeke B. and Leisinger, T., 1983. Interference of aromatic sulfo groups in the microbial degradation of the azo dyes Orange I and Orange II. *Arch. Microbiol.*, 135: 1–7.

Kulla, H.G., 1981. Aerobic bacterial degradation of azo dyes in microbial degradation of xenobiotics, recalcitrant compounds, (Eds.) T. Leisinger, A.M. Cook, J. Nuesh and R. Hutter. Academic Press, London, pp. 387–399.

Levine, Walter, G., 1991. Metabolism of azo dyes: Implications for detoxification and activation. *Drug Metabolism Reviews*, 23(3–4): 253–309.

Lewis, David, M., 1999. Coloration for the next century. *Review of Progress in Coloration and Related Topics*, 29: 23–28.

Longbottorn, J.E. and Lichtenberg, J.J., 1982. *EPA Test Methods*. EPA–600/4–82–057.

Little, L.W.J., Lamd, C., Chillingworth, M.A and Durkin, W.B., 1974. Acute toxicity of selected commercial dyes to the fathead minnow and evolution of biological treatment for reduction of toxicity. In: *Proc. 29th Purdue Ind. Waste Conf.* Ann. Arbor Science, Ann Arbor, Michigan.

Loyd, Chapman K., 1992. Anaerobic/aerobic degradation of a textile dye wastewater. *Masters Thesis*, Virginia Polytechnic Institute and State University, 184 pp.

McCarthy, B.J., 1997. Biotechnology and Coloration. *Review of Progress in Coloration and Related Topics*, 27: 26–31.

McCurdy, Michael W., 1991. Chemical reduction and oxidation combined with biodegradation for the treatment of a textile dye wastewater. *Masters Thesis*, Virginia Polytechnic Institute and State University, pp. 181.

Moosvi, S., Haresh Keharia and Datta Madamwar, 2005. Decolourization of textile dye Reactive Violet 5 by a newly isolated bacterial consortium RVM 11.1. *World J. Microbiol. Biotechnol.*, 21(5): 667–672.

Namasivayam, C. and Yamuna, R.T., 1992. Removal of congored from aqueous solution by biogas waste slurry. *J. Chem. Tech. Biotechnol.*, 53: 153–157.

O'Neil, C., Lopez, A., Estevez, S., Hawkes, F.R., Hawkes, D.L. and Wilcox, S., 2000. Azo-dye degradation in an anaerobic-aerobic treatment system operating on simulated textile effluent. *Appl. Microbiol. Biotech.*, 53: 249–254.

Pagga, U. and Brown, D., 1986. The degradation of dyestuffs, Part II: Behaviour of dyestuffs in aerobic biodegradation tests. *Chemosphere*, 15(4): 479–491.

Pasti-Grigsby, M.B., Burke, N.S., Goszezynski S. and Crawford, D.L., 1996. Transformation of azo dye isomers by *Streptomyces chromofuscus*. *Appl. Environ. Microbiol.*, pp. 1814–1817.

Paszcezynski, A., Pasti-Grigsby, M., Goszceynski, S., Crawford, R. and Crawford, D.L., 1992. Mineralization of sulfonated azo dyes and sulfanilic acid by *Phanerochaete chrysosporium* and *Streptomyces chromofuscus*. *Appl. Environ. Microbiol.*, 58: 3598–3604.

Ramachandra, M., Crawford, B.L. and Hertel, G., 1988. Characterization of extracellular lignin peroxidase of the lignocellulose actinomycetes *Streptomyces viridosporus*. *Appl. Environ. Microbiol.*, 54: 3057–3063.

Razo-F1ores, E., Luijten, M., Donlon, B., Lettinga, G. and Field, J., 1997. Biodegradation of selected azo dyes under methanogenic conditions. *Water Sci. Technol.*, 36(6–7): 65–72.

Rehn, L., 1895. Blasengeschwulste bei Fuchsin Arbeitem. *Arch. Klin. Chir.*, 50: 588.

Saranaik, S. and Kanekar, P., 1995. Bioremediation of colour of methyl violet and phenol from a dye industry waste effluent using *Pseudomonas* sp. isolated from factory soil. *J. Appl. Bacteriol.*, 79: 459–469.

Seshadri, S., Bishop, P.L. and Agha, A.M.1994. Anaerobic/aerobic treatment of selected azo dyes in wastewater. *Waste Management*, 14(2): 127–137.

Shaul, Glenn M., Holdsworth, T.J., Dempsey, C.R. and Dostal, K.A., 1991. Fate of water soluble azo dyes in activated sludge process. *Chemosphere*, 2(1–2): 107–119.

Spadaro, J.T., Gold, M.T. and Renganathan, V., 1992. Degradation of azo dye by the lignin degrading fungus *Phanerochaete chtysosporium. Appl. Environ. Microbiol.*, 58: 2391–2401.

Sudhakar, P., Palaniappan, R. and Gowrishankar, R., 2002. Degradation of azo dye (Blach E) by an indigenous bacterium *Pseudomonas* sp. BSP-4. *Indian J. Microbiol. Biotech. Env. Sci.*, 4: 203–208.

Venkata Mohan, S., Charidrasakar Rao, N. and Krishna Prasad, K., 2002. Biological decolorisation of stimulated basic dye effluents by algal Spirogyra species. *Asian J. Microbiol. Biotech. Env. Sci.*, 4: 107–111.

Verma, Pradeep and Madamwas, Datta, 2005. Decolorization of azo dyes using Basidiomycete strain PV 002. *World J. Microbiol. Biotechnol.*, 21(4): 1573.

Vijaya, P., Padmavathy, P. and Sandiya, S., 2003. Decolourisation and biodegradation of reactive azodyes by mixed culture. *Indian J. Biotech.*, 2: 259–263.

Vyas, B.R. and Molitoris, H.P., 1995. Involvement of an extracellular H_2O_2-dependent lignolytic activity of the white-rot fungus, *Pleurotus ostreatus* in the decolorization of Remazol Brilliant Blue R. *Appl. Environ. Microbiol.*, 61: 3919–3927.

Weber, Eric J. and Lee Wolfe, N., 1987. Kinetic studies of the reduction of aromatic azo compounds in anaerobic sediment/water systems. *Environmental Toxicology and Chemistry*, 6: 911–919.

Willmott, N., Guthrie, J. and Nelson, G., 1998. The biotechnology approach to colour removal from textile effluent. JSDC, 114: 38–41.

Wong, P.K. and Yuen, P.Y., 1996. Decolourization and biodegradation methyl red by *Klebsiella pneumoniae* RS-13. *Water Research*, 36: 1744–1746.

Yamini, R. and Lalithakumari, D., 2001. Aerobic treatment of textile mill effluent and decolorisation of commercial azo dyes. *Asian J. Microbiol. Biotech. Env. Sci.*, 3: 353–358.

Yang, Y., Wyatt, D.T. II and Bahorshky, M., 1998. Decolorization of dyes using UV/H_2O_2 photochemical oxidation. *Textile Chemist and Colorist*, 30: 27–35.

Yatome, C., Matsufuru, H., Taguchi, T. and Ogawa, T., 1993. Degradation of 4-dimethylamino azobenzene-2-carboxylic acid by *Pseudomonas stuzeri. Appl. Microbiol. Biotechnol.*, 39: 778–781.

Zimmermann, H., Kulla, H.G. and Leisinger, T., 1982. Properties of purified orange li azo reductase enzyme initiating azo dye degradation by *Pseudomonas* KF46. *Env. J. Biochem.*, 129: 197–203.

Zissi, U. and Lyberatos, G., 1996. Azo-dye biodegradation under anoxic conditions. *Wat. Sci. Tech.*, 34(5–6): 495–500.

Zissi, U., G. Lyberatos and S. Pavlous, 1997. Biodegradation of p-aminobenzene by *Bacillus subtilis* under aerobic conditions. *J. Industrial Microbiol. Biotechnol.*, 19: 49–55.

Zollinger, H., 1961. *Azo and Diazo Chemistry Aliphatic and Aromatic Compounds*. Interscience Publishers, New York, London.

Zollinger, H., 1987. *Colour Chemistry: Synthesis Properties and Applications for Organic Dyes and Pigments*, pp. 92–102. VCH Publishers, Inc., New York,

Zollinger, Heinrich, 1991. *Color Chemistry: Syntheses, Properties and Applications of Organic Dyes and Pigments*. 496 pp.

Chapter 56

Cyanobacterial Diversity in Polluted Lakes of Jalgaon District of North Maharashtra

S.N. Nandan[1] and S.R. Mahajan[2]

[1]*P.G. Department of Botany, S.S.V.P.S's L.K. Dr. P.R. Ghogrey Science College, Dhule – 424 005 (M.S.)*
[2]*Department of Botany, Bhusawal Arts, Science and P.O.N. Commerce College, Bhusawal – 425 201, Distt. Jalgaon (M.S.), India*

ABSTRACT

Cyanobacteria is an extremely diverse group of prokayotic organisms which make valuable contributions to soil fertility by fixing atmospheric nitrogen and play an important role in all types of water bodies whether they being heterocystous or non-heterocystous. Polluted waters harbour characteristic type of blue-green algae. Considering above their distributional pattern in polluted and mildly polluted lakes of Jalgaon has been worked out. Their role as indicators of water quality and eutrophication has been summarized. Its highest population has been recorded during summer while lowest during monsoon months. 154 species belonging to 35 genera were identified. Of these 39 species were heterocystous and 115 species were non-heterocystous members.

Key Words: Cyanobateria, Heterocystous, Lake, Non-heterocystous.

Introduction

During the recent past, studies on Cyanobacterial have emphasized their important role in ecosystems. Cyanobacteria algae grow at any place and in any environments where moisture and sunlight are available. However, specific algae grow in specific environment and therefore, their distributional pattern, ecology, periodicity, qualitative and quantitative occurrence differ widely. Now a days people are facing many problems and water pollution is one of them. Industrial wastes, domestic wastes and sewage of the area either accumulated in the form of ditches of ponds or are being let into nearby rivers. Some nuisance algae grow inside these ditches and ponds which develop bloom and ultimately water becomes unpotable even for animals and not fit for irrigation puposes as well.

To know the role of Cyanobacterial in polluted Hartala and mildly polluted Velhala lake of Jalgaon, present study has been carried out. Their role as indicators of water quality and eutrophication has been investigated. A list of Cyanobacterial collected from these lakes has also been given.

Material and Methods

Two different lakes of Jalgaon (India) have been selected for the present study and work has been carried out from October, 1998 to September, 2000. The Hartala lake is located near the village Hartala of Tal-Muktainagar lies on small tributary of river Tapti. Geographically, Hartala lake is located at latitude 21°.00′ and longitude 76°.00′ East. The drainage area of this lake is 6.61 square miles. The lake has a capacity of 140 millions of cubic feet. Three stations were selected for the collection of water and algal samples. H-I station is very near to village, at this station during the study period, the water utilized for clothes washing and bathing by villagers and also cattle wadding was seen. At this station water was much contaminated due to addition of domestic wastes etc. H-II station is near to Chakradhar Swamy temple nearly 1 k.m. away from the H-I station. At this station during the study period, the water was utilized for bathing and washing clothes by villagers. At this station fishing and cattle wadding were observed in the study period. H-II station remained away from human activities and was free from much contamination.

Another lake selected for study is Velhala lake situated near village Velhala of Bhusawal taluka. It lies at 20°.55′ North latitude and 75°.50′ East longitude. The net available storage of lake is 74 millions of cubic feet. Three stations were selected for the collection of water and algal samples. V-I station of Velhala lake received ash water from the leakage of pipe lines of waste water of Thermal Power station, Deepnagar. During the study period, at number of places through the leakage of pipelines of water, ash water leaked out into rivulets of catchment area and finally get contaminated into lake water at this station. V-II station is near to village Velhala. The water was utilized for bathing and washing clothes by villagers. Cattle wadding were also observed during the study period. V-III station is nearly 3 km away from the village. It was free from much contamination and human activities.

Monthly collection of water and algae were made for two years from October, 1998 to September, 2000. The algal samples were collected in the acid washed plastic collection bottles and immediately preserved in 4 per cent formalin for microscopic observations. Line drawings of different forms of Cyanobacteria were made by camera lucida. Cyanobacteria were identified by relevant monographs and recent available literature (Desikachary, 1959, Dixit, 1936, Gonzalves and Kamat, 1960, Kamat, 1963, Mahajan,1983, Nandan and Patel, 1985a, Rao, 1937). The characterization and identification of Cyanobacterial taxa was based on the taxonomic keys given by Desikachary (1959). Systematic enumeration of Cyanobacteria, their habitat and month of collection have been recorded.

Systematic Enumeration

In all 154 cyanobacterial forms were identified belonging to families Chroococcaceae, Entophysalidaceae, Cyanidiaceae, Chamaesiphonaceae, Pleurocapsaceae, Oscillatoriaceae, Nostocaceae, Scytonemataceae, Rivulariaceae, Mastigocladaceae and Stigonemataceae. Out of these 39 were heterocystous and 115 were non-heterocystous.

Cyanophyceae

Order - Chroococcales

Family - Chroococcaceae

***Microcystis aeruginosa* Kuetz. (Plate 56.1, F.1)**

Desikachary, 1959, P.93, Pl.17, F.L2,6 and Pl.18, F.10

Cells 3.33μ m in diam., spherical, generally with gas vacuoles.

Habiatat: Hartala lake, H-III, June, 1999.

Velhala lake, V-I, November, 1998, February, 1999.

***Microcystis flox-aquae* (Wittr.) Kirchner. (Plate 56.1, F.2)**

Desikachary, 1959, P.94. Pl.17., F.11.

Cells 3.3μ in diameter, spherical, nannocytes present.

Habitat: Velhala lake, V-I, October, 1999, July, 1999, September, 1999.

***Microcystis marginata* (Menegh.) Kuetz. (Plate 56.1, F.3)**

Desikachary, 1959, P.87, Pl.17, F.3-5

Cells 5.5μ in diam., closely arranged, with gas vacuoles.

Habitat: Velhala lake, V-I, July, 2000.

***Microcystis robusta* (Clark) Nygaard (Plate 56.1, F.4)**

Desikachary, 1959, P.85, Pl.17, F.7-10

Cells 6.1-6.6μ diameter, spherical, without gas-vacuoles.

Habitat: Velhala lake, V-I, March, 1999, V-III, July, 1999, January, 1999.

***Microcystis viridis* (A.Br.) Lemm. (Plate 56.1, F.5)**

Desikachary, 1959, P.87, Pl.18, F.1-6

Cells 3.3μ in diameter, spherical with gas vacuoles.

Habitat: Velhala lake, V-I, May, 1999, V-III, January, 1999.

***Chroococcus cohaerens* (Breb.) Nag. (Plate 56.1, F.6)**

Desikachary, 1959, P.111, Pl.26, F.3-9

Cells single or up to 2-8 in groups, without envelop 2.2μ diameter, and with sheath 2.5μ diameter, sheath thin, colourless, unlamellated.

Habitat: Hartala lake, H-I, March, 1999.

***Chroococcus hansgirgi* Schmidle (Plate 56.1, F.7)**

Desikachary, 1959, P.85, Pl.105

Cells violet-blue, single spherical or oblong 7.2μ broad, 8.3μ long sheath thin hyaline, adhering, cell contents homogeneous.

Habitat: Velhala lake, V-I, April, 2000, V-II, January, 1999.

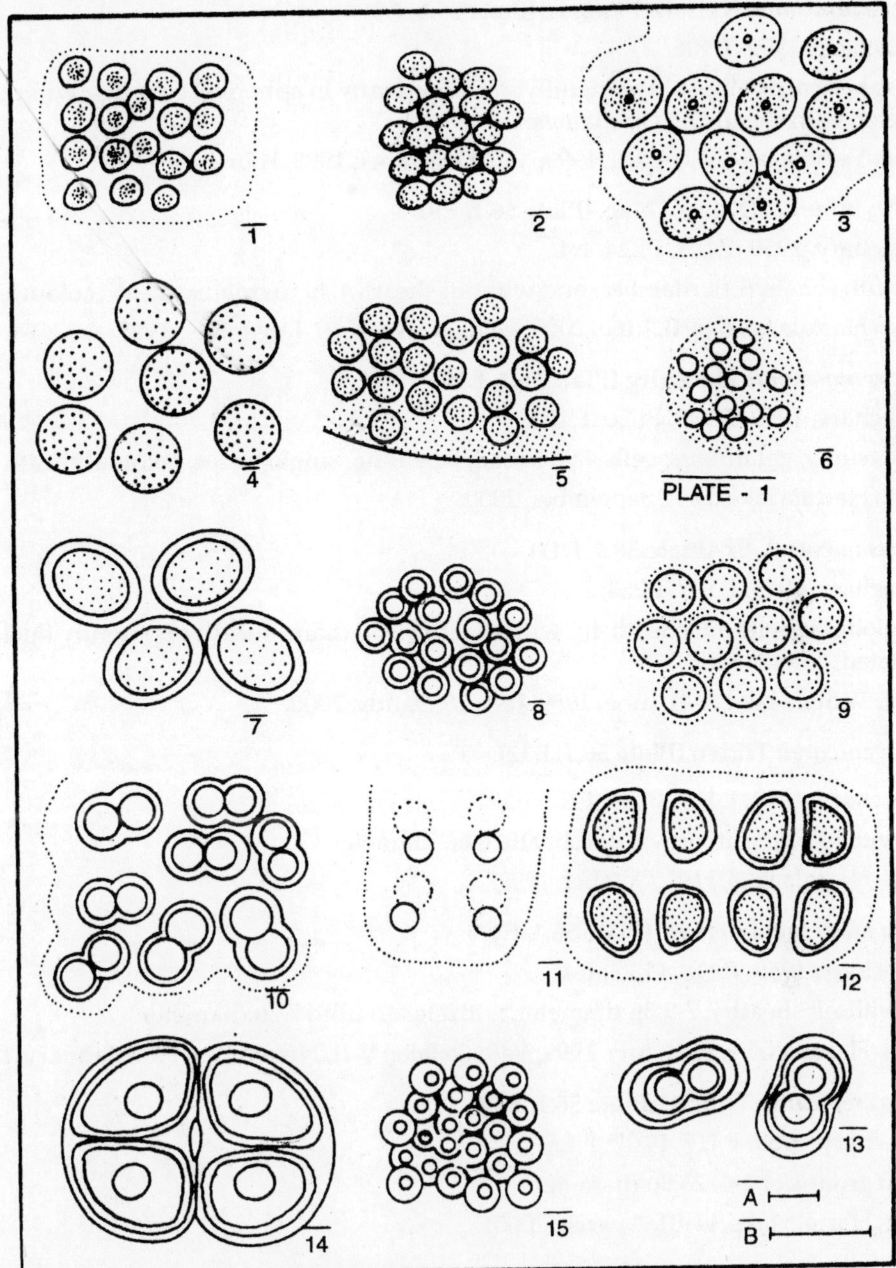

Plate 56.1: (1) *Microcystis aeruginosa* Kuetz., (2) *M. flox-aquae* (Wittr). Kirchn., (3) *M. marginata* (Mengegh.) Kuetz., (4) *M. robusta* (Clark) Nygaard, (5) *M. viridis* (A. Br.) Lamm., (6) *Chroococus coharens* (Breb.) Nag., (7) *C. hansgirgi* Schmidl, (8) *C. minimus* (Keissl.) Lamm., (9) *C. minutus* (Kuetz.) Nag., (10) *C. montanus* Hansg., (11) *C. varius* A. Br., (12) *Gloeocapsa calcarea* Tilden, (13) *G. kuetzingiana* Nag., (14) *G. crepidinum* Thuret, (15) *G. punctata* Nag.

Scale Bars A & B: 10µm Scale A: Fig. 11, Scale B. Fig. 1, 2, 3, 4, 5, 6, 7, 8, 9, 10, 12, 13, 14, 15.

***Chroococcus minimus* (Keissler) Lamm. (Plate 56.1, F.8)**

Desikachary, 1959, P.106

Cells spherical or ellipsoid, generally in two or many in spherical colonies, without sheath 2.2μ diameter, with sheath 3.8μ in diameter.

Habitat: Velhala lake, V-I, April, 1999, V-II, December, 1998, February, 1999.

***Chroococcus minutus* (Kuetz.) Nag. (Plate 56.1, F.9)**

Desikachary, 1959, P.103, Pl.24, F.4

Cells with sheath 6.1μ diameter and without sheath 4.4μ diameter, sheath colourless.

Habitat: Hartala lake, H-II, June, 2000; Velhala lake, V-III, December, 1998.

***Chroococcus montanus* Hansgirg (Plate 56.1, F.10)**

Desikachary, 1959, P.108, Pl.26, F.12

Thallus slimy, gelatinous, cells 4.9μ broad, 8.8μ long, single or 2-4, colonial sheath lamellose.

Habitat: Hartala lake, H-II, September, 2000.

***Chroococcus varius* A.Br. (Plate 56.1, F.11)**

Desikachary, 1959, P.107, Pl.24, F .5

Cells globular, without sheath 4μ, with sheath 5.95μ diam. sheath apparently thick, indistinctly lamellated, colourless.

Habitat: Velhala lake, V-II, June, 1999, July, 1999, July, 2000.

***Gloeocapsa calcarea* Tilden (Plate 56.1, F.12)**

Desikachary, 1959, P.115, Pl.24, F.6

Cells with or without individual sheath, 6.6μ diameter.

Habitat: Hartala lake, H-II, October, 1999.

***Gloeocapsa kuetzingiana* Nag. (Plate 56.1, F.13)**

Desikachary, 1959, P.118, Pl.23, F.4

Cells without sheath 2.7-3.3μ diameter, with sheath upto 7.7μ diameter.

Habitat: Hartala lake, H-II, July, 1999; Velhala lake, V-II, November, 1998, January, 1999.

***Gloeocapsa crepidium* Thuret. (Plate 56.1, F.14)**

Desikachary, 1959, P.117, Pl.27, F.4,5

Cells in groups of 2-4, 23.8μ diameter

Habitat: Hartala lake, H-III, August, 2000.

***Gloeocapsa punctata* Nag. (Plate 56.1, F.15)**

Desikachary, 1959, P.115, Pl.23, F.2

Cells without sheath 1.6μ diameter with sheath 4.9μ diameter.

Habitat: Hartala lake, H-II, May, 1999.

Velhala lake, V-II, March, 1999, April, 1999, May, 1999.

V-III, May, 1999, August, 2000.

Gloeocapsa stegophila **(Itzings.) Rabenh. v. *crassa* Rao, C.B. (Plate 56.2, F.1)**

Desikachary, 1959, P.119, Pl.25, F.3

Cells without sheath 3.8μ broad, 5.5μ long, with sheath 4.4μ broad, 6.6μ long, 2-4 cells in globose groups.

Habitat: Velhala lake, V-I, January, 1999, V-III, August, 2000.

Gloeothece rupestris **(Lyngb.) Bornet v. *maxima* West (Plate 56.2, F.2)**

Desikachary, 1959, P.127, Pl.25, F.4

Cells cylindrical with rounded ends, 9μ broad and 14.2μ long, with large granules, mucilage sheath thick, cells with sheath 11.9μ broad, 22.6μ long.

Habitat: Velhala lake, V-III, January, 1999.

Gloeothece samoensis **Wille (Plate 56.2, F.3)**

Desikachary, 1959, P.108, Pl.23, F.3

Cells ellipsoidal, without sheath 5.5μ broad, 8.3μ long, cells bluish green, in rounded colonies.

Habitat: Velhala lake, V-I, April, 2000.

Gloeothece samoensis **Wille v .*major* Wille. (Plate 56.2, F.4)**

Desikachary, 1959, P.128, Pl.23, F.6

Cells ellipsoidal, without sheath 5.5μ broad, 8.8μ long, with envelope 11.6μ broad, 16.6V long, sheath hyaline.

Habitat: Velhala lake, V-II, April, 1999.

Aphanocapsa banaresensis **Bharadwaja (Plate 56.2, F.5)**

Desikachary, 1959, P.133

Cells oval or almost spherical 4.4μ in diameter.

Habitat: Hartala lake, H-I, September, 2000.

Aphanocapsa crassa **Ghose (Plate 56.2, F.6)**

Cells spherical 8.3μ diameter.

Habitat: Hartala lake, H-I, March, 2000.

Aphanocapsa grevillei **(Mass.) Rabenh. (Plate 56.2, F.7)**

Desikachary, 1959, P.134

Cells spherical 5μ diameter closely arranged.

Habitat: Hartala lake, H-I, September, 2000.

Aphanocapsa koordersi **Strom. (Plate 56.2, F.8)**

Desikachary, 1959, P.132

Colony spherical 2-3 mm in diameter, cells loosely arranged or in groups of four, spherical 2.2-2.7μ in diameter.

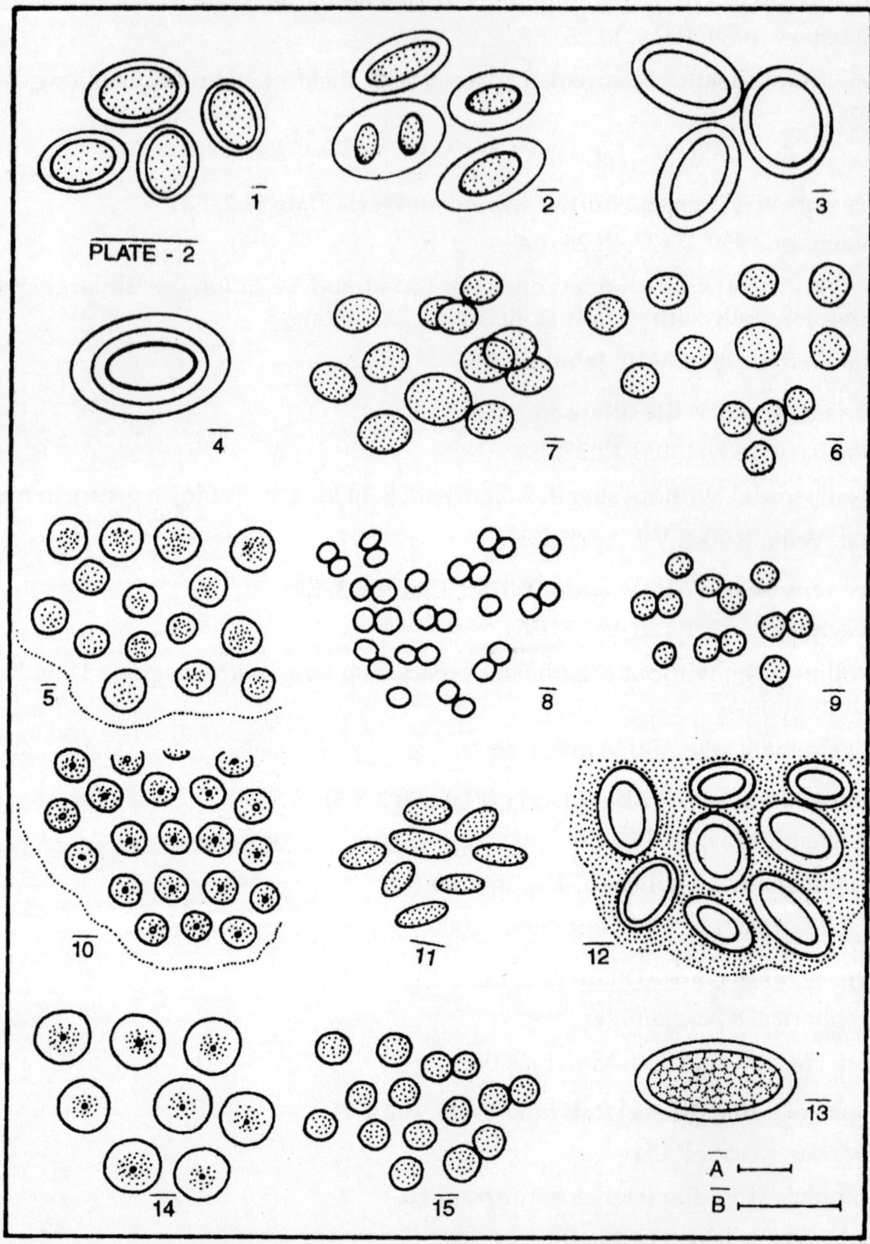

Plate 56.2: **(1)** ***Gloeocapsa stegophila*** **(Itzings.), Rabenh.** ***v. crassa*** **Rao, C.B., (2)** ***G. rupestris*** **(Lyngb.) Borb** ***v. maxima*** **West, (3)** ***G. samoensis*** **Wille, (4)** ***G. samoensis*** **Wille** ***v. major*** **Wille, (5)** ***Aphansocapsa banaresensis*** **Bharadwaj, (6)** ***A. crassa Ghose*****, (7)** ***A. grevillie*** **(Hass.) Rebenh., (8)** ***A. kooredersi*** **Storm, (9)** ***A. montana Cramer*****, (10)** ***A. roseana*** **De Bary, (11)** ***A. castagnei*** **(Breb.) Rabenh., (12)** ***A. pallida*** **(Kuetz.) Rebenh, (13)** ***Synechococcus aeruginosus*** **Nag., (14)** ***Synechosystis aqualitis*** **Sauv., (15)** ***S. pevalekii*** **Erceg.**

Scale Bars A & B: 10µm Scale A: Fig. 2, 6, 10, 13, Scale B. Fig. 1, 3, 4, 5, 7, 8, 9, 11, 12, 14, 15.

Habitat: Hartala lake, H-I, October, 1998, December, 1998, May, 1999, June, 1999, July, 1999, H-III, July, 1999.

***Aphanocapsa montana* Cramer (Plate 56.2, F.9)**

Desikachary, 1959, P.135, Pl.20, F.8

Thallus of no definite shape, gelatinous, cells 2.77-3.33μ diameter, spherical, light blue-green, single or in pair, mucilage colourless, diffluent.

Habitat: Hartala lake, H-II, May, 1999, H-III, October, 1998, November, 1998, December, 1998, January, 1999, February, 1999.

***Aphanocapsa roseana* De Bary (Plate 56.2, F.I0)**

Desikachary, 1959, P.131

Cells 5.1-8.3μ diameter.

Habitat: Hartala lake, H-III, October, 1999, March, 2000.

***Aphanothece castagnei* (Breb.) Rabenh. (Plate 56.2, F.11)**

Desikachary, 1959, P.140, Pl.21, F.8

Cells ellipsoidal to cylindrical, 2.2μ broad, 4.4μ long.

Habitat: Velhala lake, V-I, October, 1999.

***Aphanothece pallida* (Kuetz.) Rabenh. (Plate 56.2, F.12)**

Desikachary, 1959, P.140, Pl.22, F.3

Cells oblong, elliptical or cylindrical, 5.5μ broad and 8.3μ long.

Habitat: Velhala lake, V-III, December, 1998, January, 1999, December, 1999, November, 1999.

***Synechococcus aeruginosus* Nag. (Plate 56.2, F.13)**

Desikachary, 1959, P.143, Pl.25, F.6,12

Cells cylindrical, 13.3μ broad, 30.5μ long single, pale blue-green.

Habitat: Velhala lake, V-I, September, 2000.

***Synechocystis aquatilis* Sauv. (Plate 56.2, F.14)**

Desikachary, 1959, P.144, Pl.25, F.9

Cells spherical, single or in rows, 5.5-6.6μ diameter.

Habitat: Hartala lake, H-I, January, 1999, March, 1999, June, 1999. H-II, October, 1998, January, 1999.

Velhala lake: V-II, January, 1999, V-III, October, (1998).

***Synechocystis pevalekii* Ercegovic (Plate 56.2, F.15)**

Desikachary, 1959, P.145, Pl.25, F.11

Cells spherical, 2.7-3.3μ broad, single or two together.

Habitat: Hartala lake, H-I, February, 1999, May, 1999, July, 1999.

Coelosphaerium kuetzingianum **Nag. (Plate 56.3, F.1)**

Desikachary, 1959, P.148, Pl.28, F.7,8

Cells spherical or sub-spherical, 2.7μ broad, cells loosely arranged.

Habitat: Hartala lake, H-I, October, 1998.

Merismopedia convoluta **Breb. (Plate 56.3, F.2)**

Desikachary, 1959, P.152, Pl.29, F.8,12,13

Cells spherical to oblong, 4.7μ broad, 9.5μ long, forming a large 1-4mm long and broad, flat often leaf like convolute colonies, blue-green.

Habitat: Velhala lake, V-II, October, 1998, November, 1998, January, 1999, February, 1999, January, 2000.

Merismopedia minima **Beck (Plate 56.3, F.3)**

Desikachary, 1959, P.154, Pl.29, F.11

Cells pale blue-green 4 to many in small colonies, 0.5μ broad, free swimming, groups of four cells 2-3×3μ.

Habitat: Hartala lake, H-I, March, 2000, H-II, August, 2000.

Merismopedia punctata **Mayen (Plate 56.3, F.4)**

Desikachary, 1959, P.155, Pl.23, F.5

Colonies small, 4-64 cells about 60μ broad, cells not closely packed, spherical to ovoid, 3.8μ broad, pale blue-green.

Habitat: Hartala lake, H-I, March, 2000.

Velhala lake, V-II, May, 2000.

Merismopedia tenuissima **Lemm. (Plate 56.3, F.5)**

Desikachary, 1959, P.154, Pl.29, F.7

Cells 1.6μ broad.

Habitat: Hartala lake-H-I, September, 1999, August, 2000.

Dactylococcopsis raphidioids **Hansg. (Plate 56.3, F.6)**

Desikachary, 1959, P.158, Pl.29, F.1,2

Cells spindle-shaped, sigmoid or lunulately bent, 2.7-3.9μ broad, upto 29μ long, light blue-green, single to a few together in a little mucilage.

Habitat: Hartala lake, H-I, June, 1999.

Velhala lake, V-II, July, 1999, April, 2000.

Family: Entophysalidaceae

Johannesbaptistia pellucida **(Dickie) Taylor and Dr. (Plate 56.3, F.7)**

Desikachary, 1959, P.165, Pl.32, F.14-19.

Cells 5.8μ broad and 2.9μ long, mucilage homogenous.

Habitat: Velhala lake, V-III, March, 1999.

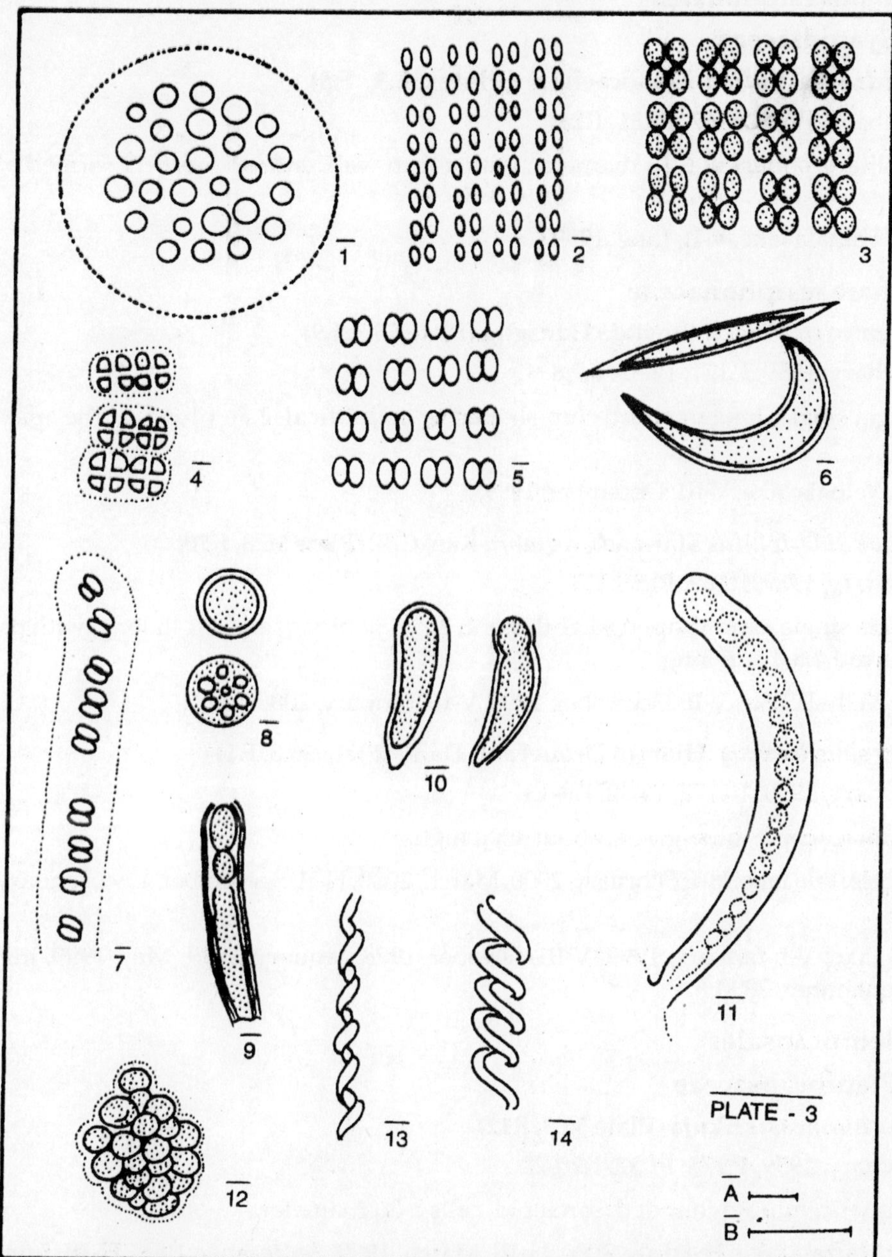

Plate 56.3: (1) *Coelosphaerium kuetzingianum* Nag., (2) *Merismopedia convoluta* Breb, (3) *M. minima* Beck, (4) *M. punctata* Mayen, (5) *M. tenuissima* Lemm., (6) *Dactylococcopis raphidiodes* Hansg., (7) *Johannesbaptistia pellucida* (Dickie) Taylor and Dr. (8) *Chroococcidiopsis indica* Desikachari, (9) *Chamaesiphon rostaffinskii* (Rostaf.) Hansg., (10) *C. sideriphilus* Starmach *v. glabra* Rao, C.B., (11) *Stichosiphon sansibaricus* (Hieron.) Dr. and D., (12) *Myxosarcina burmensis* Skuja, (13) *Spirulina major* Kuetz. ex Gom., (14) *S. meneghiniana* Zanard. ex. Gomont.

Scale Bars A & B: 10µm Scale A: Fig. 2, 3, 4, 7, 11, Scale B: Figs. 1, 5, 6, 8, 9, 10, 12, 13, 14.

Order - Chamaesiphonales

Family - Cyanidiaceae

***Chroococcidiopis indica* Desikachari (Plate 56.3, F.8)**

Desikachary, 1959, P.167, Pl.31, F.28

Cells solitary, spherical 6.1μ diameter, with a firm wall, endospores 8-12 formed by successive division.

Habitat: Velhala lake, V-II, June, 1999.

Family: Chamaesiphonaceae

***Chamaesiphon rostaffinskii* (Rostaf.) Hansgirg (Plate 56.3, F.9)**

Desikachary, 1959, P.171, Pl.32, F.7,8

Sporangia single, long or short, club-shaped or cylindrical, 2.7μ broad, at the apex-3.3μ broad, 21μ long.

Habitat: Velhala lake, V-III, December, 1999.

***Chamaesiphon sideriphilus* Starmach v. *glabra* Rao, C.B. (Plate 56.3, F.10)**

Desikachary, 1959, P.169, Pl.33, F.3

Sporangia single, club-shaped, slightly bent, 2.7-3.3μ broad, and 4-14μ long, with pseudovagina 3.3-3.8μ and 4.8-17.2μ long.

Habitat: Velhala lake, V-II, December, 1999, V-III, January, 2000.

***Stichosiphon sansibaricus* (Hieron) Drouet and Daily (Plate 56.3, F.11)**

Desikachary, 1959, P .177, Pl.32, F.9-13

Plants microscopic, blue-green, about 400μ high.

Habitat: Hartala lake, H-I, February, 2000, March, 2000, H-II, September, 1999, January, 2000, H-III, July, 2000.

Velhala lake, V-I, January, 1999, V-III, October, 1998, January, 1999, May, 1999, July, 1999, July, 2000, September, 2000.

Order - Pleurocapsales

Family - Pleurocapsaceae

***Myxosarcina burmensis* Skuja (Plate 56.3, F.12)**

Desikachary, 1959, P.178, Pl.32, F.20-22

Plant aquatic, minute rounded, sarcinoid, cells 2-3μ diameter.

Habitat: Hartala lake, H-I, July, 2000, H-III, March, 1999, September, 1999, H-III, July-2000.

Order - Nostocales

Family - Oscillatoriaceae

***Spirulina major* Kuetz. ex Gomont (Plate 56.3, F.13)**

Desikachary, 1959, P.196, Pl.36, F.13

Trichome 1.6μ broad, regularly spirally coiled, blue-green, spirals 3.3μ broad and 4.4μ distant.

Habitat: Hartala lake, H-II, September, 2000.

Spirulina meneghiniana **Zanard. ex Gomont (Plate 56.3, F.14)**

Desikachary, 1959, P.195, Pl.36, F.8

Trichomes 1.6μ broad, flexible, spirally coiled, bright blue-green, spirals 4.9μ broad and 3.8μ distant from each other.

Habitat: Hartala lake, H-I, August, 2000.

Oscillatoria brevis **(Kuetz.) Gom. (Plate 56.4, F.l)**

Desikachary, 1959, P.241.

Cells 6.1μ broad, 3.8μ long, end cell rounded.

Habitat: Hartala lake, H-I, August, 2000.

Oscillatoria calcuttensis **Biswas (Plate 56.4, F.2)**

Desikachary, 1959, P.237, Pl.42, F.20, 21

Cells 6.6μ broad, cells 6.6μ long, cross-walls with 3 granules.

Habitat: Velhala lake, V-II, April, 2000.

Oscillatoria chilkensis **Biswas (Plate 56.4, F.3)**

Desikachary, 1959, P.215, Pl.39, F.1

Cells shorter than the diameter, about 2.2μ in length.

Habitat: Velhala lake, V-II, May, 2000.

Oscillatoria curviceps **Ag. *v.angusta* Ghose (Plate 56.4, F.4)**

Desikachary, 1959, P.209, Pl.38, F.2

Trichomes bent at the end, not attenuated, not constricted at the cross walls, 5.5μ broad 2.7μ long, end cells flat rounded, not capitate.

Habitat: Hartala lake, H-I, August, 2000, H-II, April, 2000.

Oscillatoria chalybea **(Mertens) Gom (Plate 56.4, F.5)**

Desikachary, 1959, P.218, Pl.38, F.3

Cells 8.9μ broad, 3.5μ long, end wall obtuse, not capitate.

Habitat: Hartal lake, H-I, January, 2000, H-II, August, 2000.

Oscillatoria formosa **Bory f. *loktakensis* Bruhl and Biswas (Plate 56.4, F.6)**

Desikachary, 1959, P.232, Pl.40, F.15

Trichomes 4.4μ broad, bent, flexible, constricted at the cross-walls, ends slightly attenuated, capitate, cells 2-7μ long.

Habitat: Hartala lake, H-III, October, 1998.

Oscillatoria hamelli **Fremy (Plate 56.4, F.7)**

Desikachary, 1959, P.225

Cells 4.9μ broad, 7.2μ long.

Habitat: Velhala lake, V-I, May, 2000.

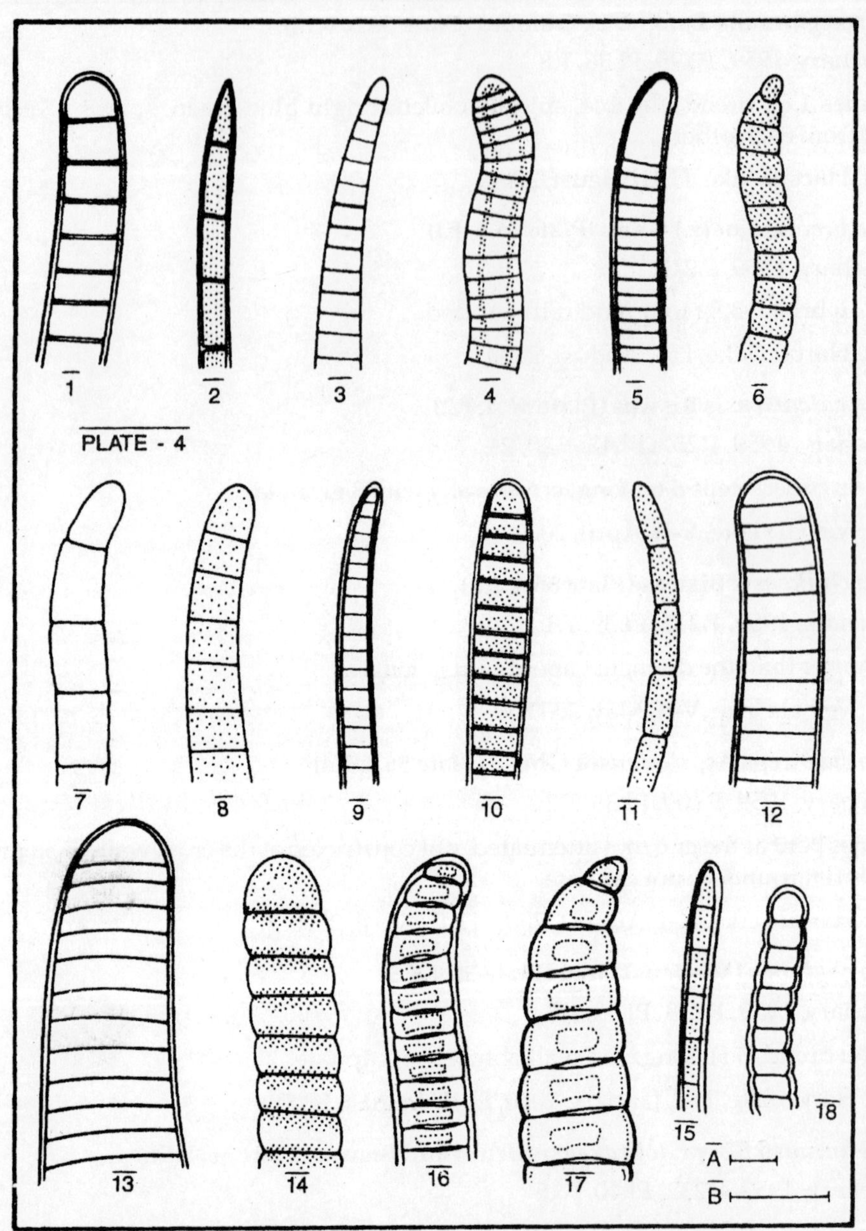

Plate 56.4: (1) *Oscillatoria brevis* (Kuetz.) Gom., (2) *O. calcuttensis* Biswas, (3) *O. chilkensis* Biswas, (4) *O. curviceps* Ag. *v. angusta* Ghose, (5) *O. chalybea* (Mertens) Gom., (6) *O. formosa* Bory f. *loktakensis* Bruhl and Biswas, (7) *O. hamelli* Fremy, (8) *O. irrigua* Kuetz., (9) *O. jasorvenisis,* (10) *O. laete-virens* Gom., (11) *O. limnetica* Lamm., (12) *O. limosa* Ag., (13) *O. margaretifera* Kuetz, (14) *O. nigro-virdis* Thwaites, (15) *O. ornata* Kuetz., (16) *O. pernornata* Skuja, (17) *O. perornata* Skuja *v. attenuata* Skuja, (18) *O. quadripanctulata* Bruhl and Biswas.

Scale Bars A & B: 10μm Scale A: Fig. 5, 12, 13, 15, 16, Scale B: Figs. 1, 2, 3, 4, 6, 7, 8, 9, 10, 11, 14, 17, 18.

Oscillatoria irrigua **Kuetz. (Plate 56.4, F.8)**

Desikachary, 1959, P.224, Pl.42, F.7,9

Cells nearly quadrate, 4.4μ long.

Habitat: Hartala lake, H-III, January, 1999.

Oscillatoria jasorvenisis **Vouk (Plate 56.4, F.9)**

Desikachary, 1959, P.221

Trichome straight, 2.7μ broad, cells as long as broad.

Habitat: Velhala lake, V-II, July, 2000.

Oscillatoria laete-virens **Gom. (Plate 56.4, F.I0)**

Desikachary, 1959, P.213, Pl.29, F.2,3

Cells 4.7μ broad, apices attenuated, 2.7μ long.

Habitat: Hartala lake, H-II, March, 2000, July, 2000.

Velhala lake, V-II, February, 1999.

Oscillatoria limnetica **Lamm (Plate 56.4, F.11)**

Desikachary,1959, P.226, Pl.37, F.3

Trichome slightly bent, 1.6μ broad, not attenuated, cells 6.6μ long.

Habitat: Hartala lake, H-II, October, 1998.

Velhala lake, V-III, January, 1999.

Oscillatoria limosa **Ag. (Plate 56.4, F.12)**

Desikachary, 1959, P.206, Pl.42, F.11

Trichome more or less straight, 15.4μ broad, cells 1/3 - 1/6 as long as broad, 4.7μ long.

Habitat: Hartala lake, H-I, August, 2000, H-II, October, 1999.

Oscillatoria margaretifera **Kuetz. (Plate 56.4, F.13)**

Desikachary, 1959, P.202, Pl.42, F.8

Cells 1/3 - 1/7 as long as broad, 4.7-5.9μ long.

Habitat: Velhala lake, V-II, September, 1998, V-III, November, 1999.

Oscillatoria nigro-virdis **Thwaites (Plate 56.4, F.14)**

Desikachary, 1959, P.202, Pl.42, F.6

Cells 1/2 - 1/4 as long as broad, 4.4μ long.

Habitat: Hartala lake, H-II, November, 1999, May, 2000.

Oscillatoria ornata **Kuetz. (Plate 56.4, F.15)**

Desikachary, 1959, P.206, Pl.37, F.12

Cells 1/2 - 1/6 as long as broad, 4.7μ long.

Habitat: Hartala lake, H-III, August, 2000.

***Oscillatoria peronata* Skuja (Plate 56.4, F.16)**

Desikachary,1959, P.205, Pl.41, F.8,9,14.

Cells commonly 1/2 - 1/5 as long as broad, 5.3μ 1ong.

Habitat: Hartala lake, H-I, August, 1999, January, 2000, H-Ill, February, 2000.

***Oscillatoria peronata* Skuja v.attenuata Skuja (Plate 56.4, F.17)**

Desikachary, 1959, P.205

Cells 1/2 - 1/3 as long as broad.

Habitat: Hartala lake, H-I, May, 2000.

***Oscillatoria quadripunctulata* Bruhl and Biswas (Plate 56.4, F.18)**

Desikachary, 1959, P.227, Pl.37, F.5

Cells 4.4μ long, 1.5μ in diameter.

Habitat: Hartala lake, H-I, January, 2000.

***Oscillatoria princeps* Voucher ex Gomont (Plate 56.5, F.1)**

Desikachary, 1959, P.210, Pl.37, F.1,10,11

Cells ½ - ¼ as long as broad, 3.6μ long.

Habitat: Hartala lake, H-II, November, 1998.

***Oscillatoria pseudogeminata* Schmid (Plate 56.5, F.2)**

Desikachary, 1959, P.228, Pl.41, F.10

Cells 2.2μ broad, cells as long as broad.

Habitat: Hartala lake, H-I, November, 1999.

***Oscillatoria pseudogeminata* Schmid *v.unigranulata* Biswas (Plate 56.5, F.3)**

Desikachary, 1959, P.228, Pl.39, F.19

Cells 3.8μ in length, cell-wall thick, distinct with one large granule situated at the centre of the partition walls on either side.

Habitat: Hartala lake, H-III, November, 1998.

***Oscillatoria raoi* De Toni, J. (Plate 56.5, F.4)**

Desikachary, 1959, P.223, Pl.42, F.16-19.

Cells 4.4μ long, 6.6μ broad.

Habitat: Hartala lake, H-I, June, 1999.

***Oscillatoria raciborskii* Wolosz (Plate 56.5, F.5)**

Desikachary, 1959, P.236, Pl.37, F.15

Cells ½ times as long as broad, 8.3μ broad.

Habitat: Hartala lake, H-I, September, 1999, H-III, April, 2000.

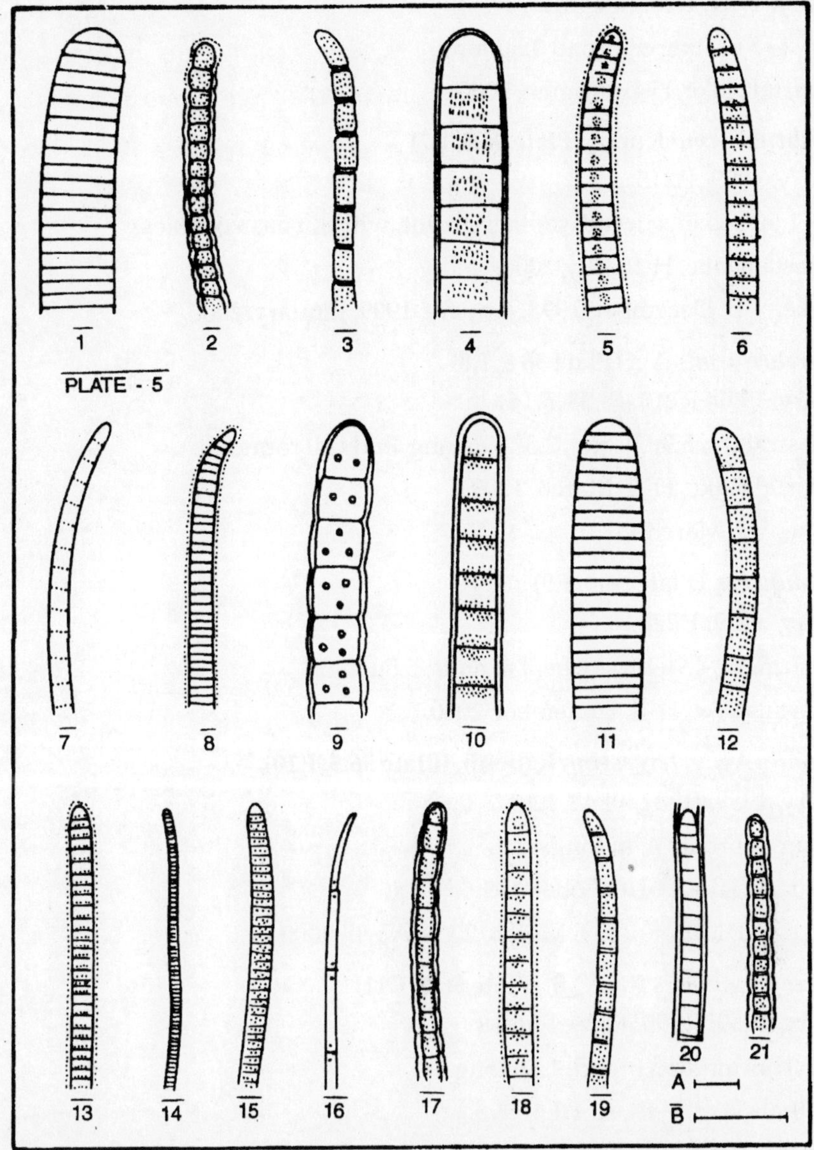

Plate 56.5: (1) *Oscillatoria princeps* Voucher ex Gom, (2) *O. pseudogeminata* Schmid., (3) *O. pseudogeminata* Schmid. *v. unigranulata* Biswas, (4) *O. raoi* De Toni, (5) *O. raciborskii* Wolosz, (6) *O. rubescens* D.C., (7) *O. subtillissima* Kuetz., (8) *O. terebriformis* Ag., (9) *O. tenuis* Ag., (10) *O. tenuis* Ag. *v. tergestina* Rabenh, (11) *O. vizagapatensis* Rao, (12) *O. willei* Gardner, (13) *Phormidium ambigum* Gom., (14) *P. angustissimum* West and West, (15) *P. autumnale* (Ag.) Gom., (16) *P. bigranulatum* Gardner, (17) *P. calcicola* Gardner, (18) *P. corium* (Ag.) Gom. *v. capitatum* Gardner, (19) *P. fragile* (Manegh.) Gom., (20) *P. incrustatum* (Nag) Gom., (21) *P. inudatum* Kuetz.

Scale Bars A & B: 10μm Scale A: Fig. 1, 5, 6, 8, 15, 17, 18, 20, 21, Scale B: Figs. 2, 3, 4, 7, 9, 10, 11, 12, 14, 16, 19.

***Oscillatoria rubescens* D.C. ex Gomont (Plate 56.5, F.6)**

Desikachary, 1959, P.235, Pl.42, F.12

Cells 1/2 - 1/3 as long as broad 3.5μ long.

Habitat: Hartala lake, H-II, August, 2000.

***Oscillatoria subtillissima* Kuetz. (Plate 56.5, F.7)**

Desikachary, 1959, P.215

Trichome, 1.5μ broad, curved, septa indistint, without gas vocuoles.

Habitat: Hartala lake, H-II, July, 2000.

Velhala lake, V-II, December, 1998, January, 1999, January, 2000.

***Oscillatoria terebriformis* Ag. (Plate 56.5, F.8)**

Desikachary, 1959, P.217, Pl.38, F.16

Trichomes straight, 5.9μ broad, 2.3-2.9μ long, end cell rounded.

Habitat: Hartala lake, H-II, March, 1999.

Velhala lake, V-I, March, 2000.

***Oscillatoria tenuis* Ag. (Plate 56.5, F.9)**

Desikachary, 1959, P.222

Trichome straight, 4.9μ broad, bent at ends, 5.5μ long.

Habitat: Hartala lake, H-II, September, 2000.

***Oscillatoria tenuis* Ag. v. *tergestina* Rabenh. (Plate 56.5, F.10)**

Desikachary, 1959, P.222, Pl.42, F.15

Cells as long as broad, 4.4μ long.

Habitat: Hartala lake, H-I, October, 1999.

Velhala lake, V-I, March, 1999, March, 2000, April, 2000.

***Oscillatoria vizagapatensis* Rao C.B. (Plate 56.5, F.11)**

Desikachary, 1959, P.205, Pl.39, F.16,18

Cells much shorter than broad, 1.6μ long.

Habitat: Velhala lake, V-III, April, 1999.

***Oscillatoria willei* Gardner (Plate 56.5, F.12)**

Desikachary, 1959, P.217, Pl.38, F.4,5

Trichome 2.7μ broad, cells 3.8μ long.

Habitat: Hartala lake, H-II, June, 2000.

***Phormidium ambigum* Gom. (Plate 56.5, F.13)**

Desikachary, 1959, P.266, Pl.44, F.16

Trichomes, 4.3-5.8μ broad, cells shorter than broad, 1.4-2.1μ long.

Habitat: Hartala lake, H-II, March, 1999.

Velhala lake, V-I, December, 1999.

Phormidium angustissimum **West and West (Plate 56.5, F.14)**

Desikachary, 1959, P.253.

Trichomes bent, 0.8μ broad, 2μ long.

Habitat: Hartala lake, H-II, December, 1999.

Phormidium autumnale **(Ag.) Gom. (Plate 56.5, F.15)**

Desikachary, 1959, P.276, Pl.44, F.24,25

Trichomes 4.3μ broad cells 2.9μ long.

Habitat: Hartala lake, H-III, December, 1999.

Phormidium bigranulatum **Gardner (Plate 56.5, F.16)**

Desikachary, 1959, P.261

Filaments 0.8μ broad, cells 9μ long, at each end with one granule.

Habitat: Velhala lake, V-II, May, 2000.

Phormidium calcicola **Gardner (Plate 56.5, F.17)**

Desikachary, 1959, P.267, Pl.43, F.4,5.

Trichomes 5.8μ broad, cells slightly longer than broad.

Habitat: Hartala lake, H-III, August, 1999.

Phormidium corium **(Ag.) Gom. v. *capitatum* Gardner (Plate 56.5, F.18)**

Desikachary, 1959, P.269, Pl.43, F.11,12

Trichomes 5.9μ broad, not constricted at the cross-walls.

Habitat: Hartala lake H-III, March, 2000.

Phormidium fragile **(Manegh.) Gom. (Plate 56.5, F.19)**

Desikachary, 1959, P.253, Pl.44, F.1-3

Trichomes 2.2μ broad.

Habitat: Hartala lake, H-II, January, 1999, H-III, February, 2000.

Phormidium incrustatum **(Nag.) Gom. (Plate 56.5, F.20)**

Desikachary, 1959, P.269, Pl.44, F.12

Trichome 4.7μ broad, cells nearly quadrate, 3.5μ long.

Habitat: Hartala lake, H-II, August-1999.

Phormidium inudatum **Kuetz. (Plate 56.5, F.21)**

Desikachary, 1959, P.271

Trichome straight, 5μ broad, 7.1μ long.

Habitat: Hartala lake, H-III, August, 1999, March, 2000.

Velhala lake, V-II, November, 1998, December, (1998).

Phormidium jadianianum **Gomont. (Plate 56.6, F.l)**

Desikachary, 1959, P.256, Pl.55, F.9

Trichome with straight long acuminate ends 4.3μ broad, 2.9μ long.

Habitat: Hartala lake, H-III, May, 1999.

Phormidium lucidium **(Kuetz.) Gomont (Plate 56.6, F.2)**

Desikachary, 1959, P.275, Pl.44, F.17,18

Trichome 7.1μ broad, cells 2.3μ long.

Habitat: Hartala lake H-III, May, 1999.

Phormidium mucosum **Gardner v .*arvense*** **Rao, C.B. (Plate 56.6, F.3)**

Desikachary, 1959, P.265, Pl.43, F.6,7

Trichome 2.2μ broad, cells 3.3μ long.

Habitat: Hartala lake, H-I, January, 2000.

Phormidium pachydermaticum **Fremy (Plate 56.6, F.4)**

Desikachary, 1959, P.267, Pl.43, F.8-10.

Trichome 4.4μ broad, cells nearly quadrate, 7.7μ long.

Habitat: Hartala lake, H-III, September, 1999.

Phormidium papyraceum **(Ag). Gom.(Plate 56.6, F.5)**

Desikachary, 1959, P.271

Trichomes 4.3μ broad, Cells 3.6μ long.

Habitat: Hartala lake, H-III, December, 1999.

Velhala lake, V-II, October, 1998,1999, V-III, October, 1999.

Phormidium purpurascens **(Kuetz.) Gom. (Plate 56.6, F.6)**

Desikachary, 1959, P.262, Pl.44, F.4

Trichome 1.6μ broad cells 2.2μ long, cross walls marked by two granuals on either side.

Habitat: Hartala lake, H-II, December, 1999.

Velhala lake, V-III, August, 1999.

Phormidium retzii **(Ag.) Gom. (Plate 56.6, F.7)**

Desikachary, 1959, P.268, Pl.44, F.13,15

Trichomes 8.3μ broad, cells 7.1μ long.

Habitat: Hartala lake, H-I, March, 2000, H-III, September, 1999.

Velhala lake, V-III, September, 1999.

Phormidium rotheanum **Itzigs. (Plate 56.6, F.8)**

Desikachary, 1959, P.258, Pl.45, F.14,16

Trichome 10.1μ broad, cells much shorter than broad, 4.3μ long.

Habitat: Hartala lake, H-III, May, 1999.

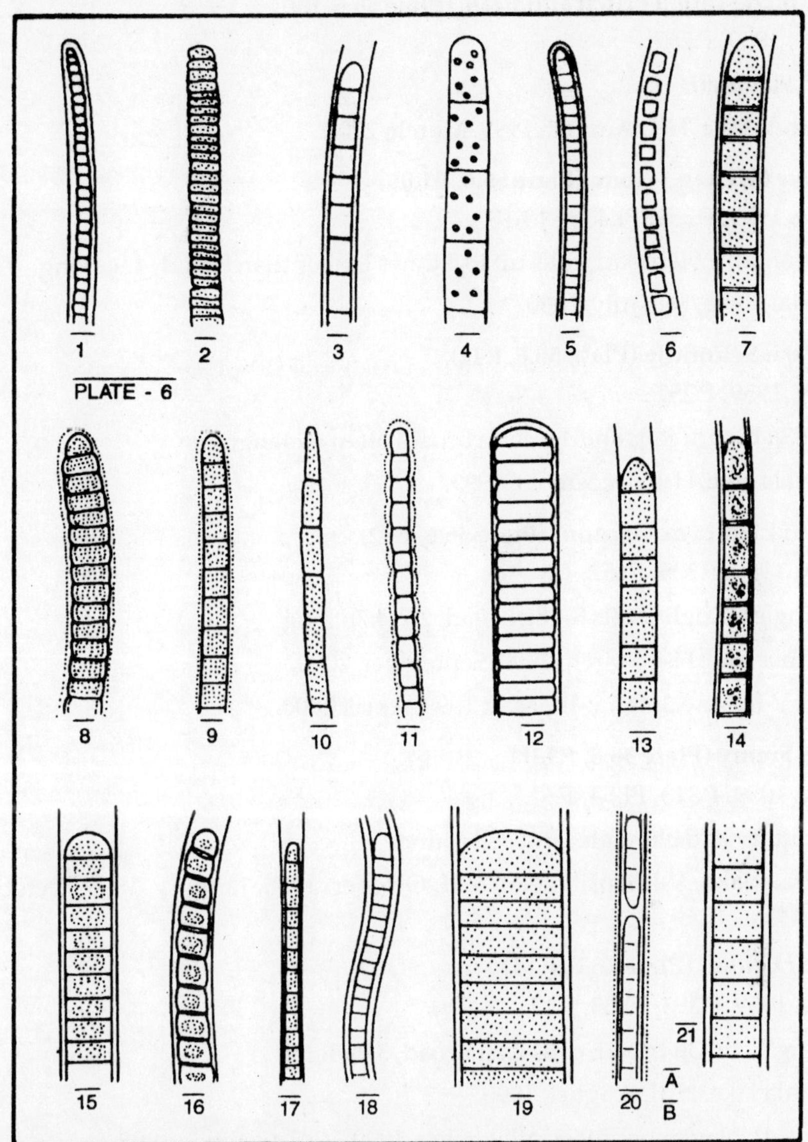

Plate 56.6: (1) *Phromidium jadianianum* Gom., (2) *P. lucidium* (Kuetz.) Gom., (3) *P. mucosum* Gardner *v. arvense* Rao, C.B., (4) *P. pachydermaticum* Fremy, (5) *P. papyraceum* (Ag.) Gom., (6) *P. purpurascens* (Kuetz.) Gom., (7) *P. retzii* (Ag.) Gom., (8) *P. rotheanum* Itzigs., (9) *P. subincrustatum* Fritch and Rich, (10) *P. tenue* (Meneg.) Gom., (11) *P. usterii* Schmidle, (12) *Lyngbya aestuarii* Liemb ex Gomont, (13) *L. allorgei* Fremy, (14) *L. borgertii* Lemm., (15) *L. ceylanica* Wille, (16) *L. cryptovaginata* Schkorbatow, (17) *L. digueti* Gom., (18) *L. epiphytica* Hieron, (19) *L. hieronymusii* Lemm., (20) *L. infixa* Framy, (21) *L. hieronymusii* Lemm. *v. crassivaginata* Ghose.

Scale Bars A & B: 10μm Scale A: Fig. 1, 2, 3, 5, 7, 8, 9, 12, 15, 16, 21, Scale B: Figs. 4, 6, 10, 11, 13, 14, 17, 18, 19, 20.

Phormidium subincrustatum **Fritch and Rich. (Plate 56.6, F.9)**

Desikachary, 1959, P.267.

Trichomes 5.9μ broad.

Habitat: Hartala lake, H-I, August, 1999, March, 2000.

Phormidium tenue **(Meneg.) Gom. (Plate 56.6, F.I0)**

Desikachary, 1959, P.259, Pl.43, F.13,15

Trichome straight, 1.9μ broad, cells up to 3 times longer than broad, 4.9μ long.

Habitat: Hartala lake, H-I, July, 2000.

Phormidium usterii **Schmidle (Plate 56.6, F.11)**

Desikachary, 1959, P.257

Trichomes 3.3μ broad, cells shorter than broad, often rectangular.

Habitat: Hartala lake, H-II, December, 1999.

Lyngbya aestuarii **Liemb. ex Gomont (Plate 56.6, F.12)**

Desikachary, 1959, P.305, Pl.52, F.8

Filaments single straight, cells 8-18μ broad, 2.7-4.7μ long.

Habitat: Hartala lake, H-I, March, 2000, September, 2000.

Velhala lake, V-II, May, 1999, V-III, May, 1999, April, 2000.

Lyngbya allorgei **Fremy (Plate 56.6, F.13)**

Desikachary, 1959, P.313, Pl.54, F.6

Filaments solitary and elongate, end cell round.

Habitat: Velhala lake, V-I, April, 2000, V-II, October, 1998, January, 1999, October, 1999, V-III, September, 2000.

Lyngbya borgertii **Lemm. (Plate 56.6, F.14)**

Desikachary, 1959, P.293, Pl.53, F.12

Filaments single, 2.6-3μ broad, cells 2.2μ broad, 3.8μ long.

Habitat: Hartala lake, H-II, August, 1999.

Velhala lake, V-II, November, 1998, November, 1999, V-III, October, 1998.

Lyngbya ceylanica **Wille (Plate 56.6, F.15)**

Desikachary, 1959, P.299, Pl.54, F.4

Filaments 10-14μ broad, cells 5.9μ long, end cell round.

Habitat: Hartala lake, H-III, July, 1999, September, 1999.

Lyngbya cryptovaginata **Schkorbatow (Plate 56.6, F.16)**

Desikachary, 1959, P.297, P1.50, F.6

Filaments straight, 7.1μ broad, cells nearly quadrate, 5.9μ long.

Habitat: Hartala lake, H-II, March, 2000.

Lyngbya digueti **Gom. (Plate 56.6, F.17)**

Desikachary, 1959, P.310, Pl.53, F.8

Filaments 2.2μ broad, cells nearly quadrate, 2.2μ long.

Habitat: Hartala lake, H-I, August, 2000.

Lyngbya epiphytica **Hieron (Plate 56.6, F.18)**

Desikachary, 1959, P.284, Pl.53, F.7

Filaments 1.5-2.2μ broad, cell 1.6μ broad, 1.6μ long.

Habitat: Hartala lake, H-II, February, 1999.

Lyngbya hieronymusii **Lemm. (Plate 56.6, F.19)**

Desikachary, 1959, P.297, Pl.48, F.4.

Filaments 12-14μ broad, cells 11μ broad, 2.7μ long.

Habitat: Hartala lake, H-III, January, 1999.

Lyngbya infixa **Fremy (Plate 56.6, F.20)**

Desikachary, 1959, P.282

Filaments 2.7μ broad, cells commonly shorter than long, 2.2μ long.

Habitat: Hartala lake, H-I, March, 2000.

Lyngbya hieronymusii **Lemm. *v.crassivaginata*** **Ghose (Plate 56.6, F.21)**

Desikachary, 1959, P.297, Pl.55, F.8

Filaments 15-20μ broad, cells 11.9μ broad, upto 10μ long.

Habitat: Hartala lake, H-II, October, 1999.

Lyngya laxespiralis **Skuja. (Plate 56.7, F.1)**

Desikachary, 1959, P .289, Pl.50, F.4.

Trichomes 9.5μ broad, cells shorter than broad, 5.9μ long.

Habitat: Hartala lake, H-III, July, 2000.

Lyngbya limmnetica **Lemm. (Plate 56.7, F.2)**

Desikachary, 1959, P.294, Pl.50, F.11

Filaments 1.2μ broad cells 1.5μ broad, 3μ long.

Habitat: Hartala lake, H-III, July, 1999.

Velhala lake, V-II, February, 1999, May, 2000.

Lyngbya martensiana **Manegh. ex Gomont (Plate 56.7, F.3)**

Desikachary, 1959, P.318, Pl.52, F.6

Trichome 6-10μ broad, cells 1.75-3.3μ in length.

Habitat: Hartala lake, H-II, July, 1999, March, 2000., H-III, May, 2000.

Velhala lake, V-I, July, 1999, June, 2000, V-II, November, 1998, April, 1999, November, 1999, V-III, January,1999.

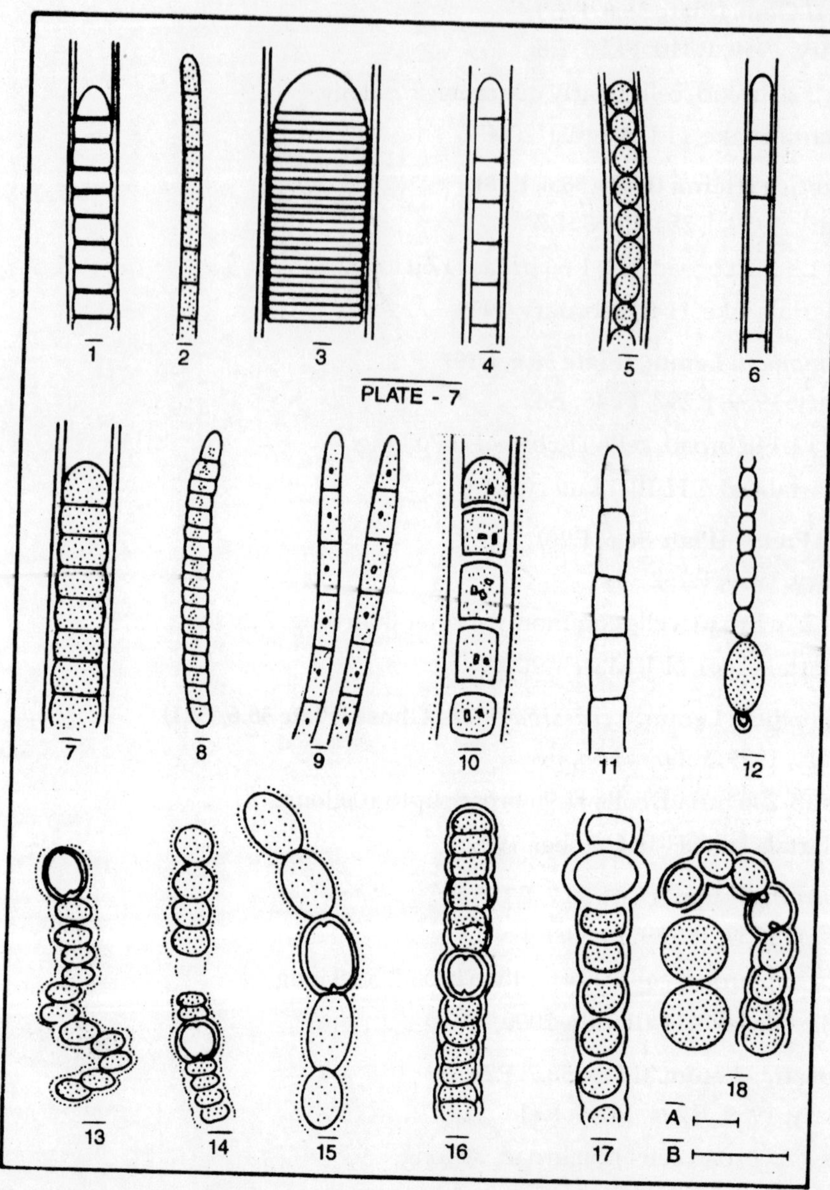

Plate 56.7: (1) *Lyngya laxespiralis* Skuja., (2) *L. limmnetica* Lemm., (3) *L. martensiana* Manegh, (4) *L. mesotricha* Skuja, (5) *L. nordgardhii* Wille, (6) *L. perelegans* Lemm., (7) *L. putealis* Mont, (8) *Schizothrix mulleri* Nag., (9) *Symploca cartilaginea* (Mont.) Gom., (10) *S. muscorum* (Ag.) Gom., (11) *Microcoleus lacustris* (Rabenh.) Farlow, (12) *Cylindrospermum indicum* Rao, C.B., (13) *Nostoc calcicola* Breb., (14) *N. coeruleum* Lyngb., (15) *N. ellipsosporum* (Desm.) Rabenh, (16) *N. linckia* (Rogh.) B. and F. *v. arvense* Rao, C.B., (17) *N. muscorum* Ag., (18) *N. piscinale* Kuetz.

Scale Bars A & B: 10μm Scale A: Fig. 1, 7, 8, 12, 14, 16, Scale B: Figs. 2, 3, 4, 5, 6, 9, 10, 11, 13, 15, 17, 18.

Lyngbya mesotricha **Skuja (Plate 56.7, F.4)**

Desikachary, 1959, P.282, Pl.50, F.1,2

Filaments 3-5μ broad, upto 1mm long, trichomes 2.5-3μ broad cell 1-2 times longer than broad, 4.6μ long.

Habitat: Hartala lake, H-II, August, 1999.

Lyngbya nordgardhii **Wille (Plate 56.7, F.5)**

Desikachary, 1959, P.287, Pl.54, F.1-3

Trichome 2.2μ broad, not attenuated, cells 1-½ as long as broad, 3.3μ long.

Habitat: Hartala lake, H-III, April, 2000, September, 1999.

Lyngbya perelegans **Lemm. (Plate 56.7, F.6)**

Desikachary, 1959, P.309, Pl.48, F.8

Trichomes 1.6μ broad, cells 7.7μ long.

Habitat: Hartala lake, H-II, November, 1999.

Lyngbya putealis **Mont. ex Gomont (Plate 56.7, F.7)**

Desikachary, 1959, P.317, Pl.52, F.12

Trichome 7.5-13μ broad, cells 8.7μ long, end cell rounded.

Habitat: Hartala lake, H-III, November, 1999.

Velhala lake, V-II, December, (1998).

Schizothrix mulleri **Nag. (Plate 56.7, F.8)**

Desikachary, 1959, P.330, Pl.57, F.9,10

Trichome slightly constricted at the cross-walls, 7.2μ broad, cells 4.4μ long.

Habitat: Velhala lake, V-I, October, 1998, V-II, September, 1999, V-III, October, 1999, September, 2000.

Symploca cartilaginea **(Mont.) Gom. (Plate 56.7, F.9)**

Desikachary, 1959, P .339, Pl.59, F.1,2

Cells 2-3μ broad, mostly longer than the broad, 3-5.8μ long.

Habitat: Hartala lake, H-II, May, 1999.

Velhala lake, V-II, October, 1998, September-October, 1998, V-III, October, December, 1998, September-October,1999, September, 2000.

Symploca muscorum **(Ag.) Gom. (Plate 56.7, F.I0)**

Desikachary, 1959, P.337, Pl.59, F.3,4

Cells 5-8μ broad, 5-11μ long.

Habitat: Hartala lake, H-II, June, 1999, November, 1999.

Microcoleus lacustris **(Rabenh.) Farlow (Plate 56.7, F.11)**

Desikachary, 1959, P.345, Pl.60, F.4,5

Cell cylindrical, 1-3 times as long as broad, 8.3μ long.

Habitat: Velhala lake, V-II, July, 1999, April, 1999, July, 2000.

Family: Nostocaceae

Cylindrospermum indicum **Rao, C.B. (Plate 56.7, F.12)**

Desikachary, 1959, P.369, Pl.64, F.4,11

Cells quadrate, 3-4.5μ long, heterocysts spherical, 2.8-5.8μ broad and 3-7.6μ long, spores 10-12μ broad and 18-22μ long.

Habitat: Velhala lake, V-I, February, 2000, January, 1999.

Nostoc calcicola **Breb. (Plate 56.7, F.13)**

Desikachary, 1959, P.384, Pl.68, F.1

Trichome 2.7μ broad, heterocysts 5.5μ broad, spores 4-5μ broad.

Habitat: Hartala lake, H-III, March, 2000.

Velhala lake, V-II, April-May, 2000.

Nostoc coeruleum **Lyngb. (Plate 56.7, F.14)**

Desikachary, 1959, P.388

Trichome 7μ broad, heterocysts 10.7μ broad, spores 8.3μ diameter.

Habitat: Hartala lake, H-III, August, 1999, December, 1999.

Velhala lake, V-I, February, 1999, February, 2000, V-II, October-December, 1998, January, 1999, V-III, October, 1998, December, 1998, December, 1999.

Nostoc ellipsosporum **(Desm.) Rabenh. (Plate 56.7, F.15)**

Desikachary, 1959, P.383, Pl.69, F.5

Trichome 4μ broad-cells 6-14μ long, heterocysts 6-7μ broad, 6- 14μ long, spores 6-8μ broad.

Habitat: Hartala lake, H-III, November, 1998, October, 1999, December, 1999.

Velhala lake, V-I, January-March-April, 2000.

Nostoc linckia **(Roth) B. and F. *v.arvense* Rao, C.B. (Plate 56.7, F.16)**

Desikachary, 1959, P.377, Pl.67, F.1

Trichomes 4-5.6μ broad, cells spherical 6.4μ long, heterocysts spherical 7.2μ broad.

Habitat: Hartala lake, H-II, September, 1999.

Velhala lake, V-I, July, 1999, June, 2000.

Nostoc muscorum **Ag. (Plate 56.7, F.17)**

Desikachary, 1959, P.385, Pl.70, F.2

Trichome 3-4 (-5)μ broad, cells short heterocysts spherical, 6.6μ broad, spores oblong, 4-8μ broad, 8-12μ long.

Habitat: Hartala lake, H-III, October, (1998).

Velhala lake, V-II, June-July, 1999, June, 2000.

Nostoc piscinale **Kuetz. (Plate 56.7, F.18)**

Desikachary, 1959, P.377, Pl.69, F.3

Trichome 3-5μ broad, heterocysts 4.5-6μ broad spores globose, 6-7μ broad, in long chains.

Habitat: Hartala lake, H-I, September, 2000, H-III, September, 1999.

Velhala lake, V-I, January-February-March-May, 1999, February-March-May, 2000, V-II, March-April, 1999, June, 2000, V-III, September, 1999, March, 2000.

Nostoc pruniformae **(L.) Ag. (Plate 56.8, F.l)**

Desikachary, 1959, P.390

Trichome 4-6μ broad, cells short barrel shaped heterocysts 6-7μ broad spores spherical upto 10μ.

Habitat: Velhala lake, V-I, March-September, 1999, March-September, 2000.

Nostoc punctiformae **(Kuetz.) Hariot (Plate 56.8, F.2)**

Desikachary, 1959, P.374, Pl.69, F.1

Trichome 3-4μ broad, heterocysts 4.4μ broad spores 5-6μ broad and 5-8μ long.

Habitat: Hartala lake, H-III, February, 1999.

Velhala lake, V-I, June, 2000.

Nostoc sphaericum **Vaucher (Plate 56.8, F.3)**

Desikachary, 1959, P.390

Trichome 4-5μ broad, cells spherical, heretocysts 5-6μ broad, spherical, spores oval 5μ broad, 7μ long.

Habitat: Hartala lake, H-I, June, 2000, H-III, November, 1999.

Velhala lake, V-I, April, 1999, April, 2000.

Nostoc sphongiaeformae **Ag. *v.varians* Rao, C.B. (Plate 56.8, F.4)**

Desikachary, 1959, P.380, Pl.68, F.2

Trichomes 3-3.5μ broad, heterocysts trichomes, 4.4μ broad and 6.6μ long, spores in long chains 4.8-6μ diameter.

Habitat: Velhala lake, V-I, April, 2000, V-II, November-December, 1998, December, 1999.

Anabaena ambigua **Rao, C.B. (Plate 56.8, F.5)**

Desikachary, 1959, P.400, Pl.76, F.2

Cells 5.5μ broad, 4.4μ long, heterocysts spherical, 7.7μ in diameter, spores 8.3μ broad and 11.6μ long.

Habitat: Velhala lake, V-I, July, 1999, June, 2000.

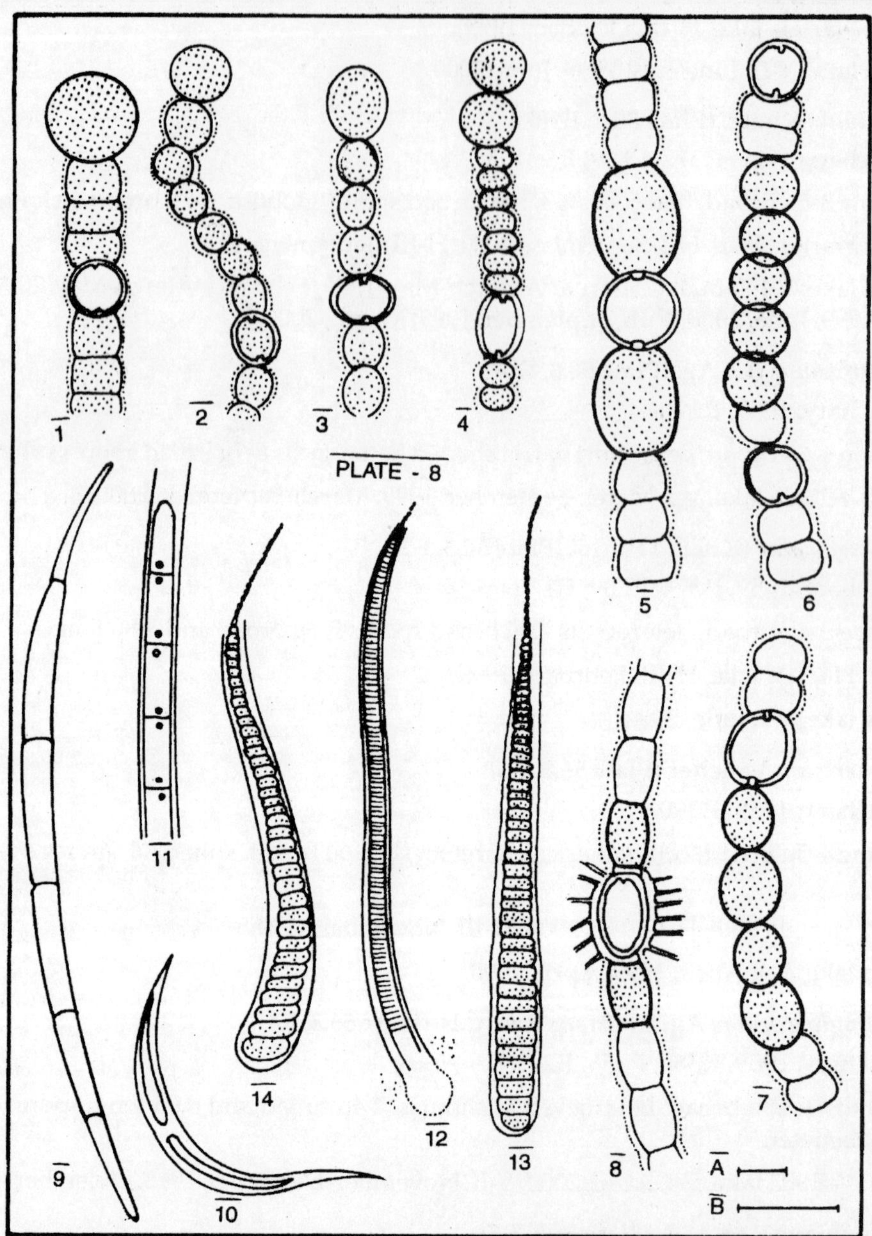

Plate 56.8: (1) *Nostoc pruniformae* (L.) Ag., (2) *N. punctiformae* (Kuetz.) Hariot, (3) *N. sphaericum* Vaucher, (4) *N. sphongiaeformae* Ag. *v. varians* Rao, C.B., (5) *Anabaena ambigua* Rao, C.B., (6) *A. fertilisimma* Rao, C.B., (7) *A. khannae* Skuja, (8) *A. spinosa* Tiwari., (9) *Raphidiopsis indica* Singh, R.N., (10) *R. curvata* Fritsch and Rich, (11) *Plectonema notatum* Schmidle, (12) *Homoethrix balearica* (Born. et Flab.) Lamm., (13) *H. hansgirigi* (Schmidle) Lemm. *v. constricta* v. nov., (14) *H. juliana* (Manegh.) Kirchn.

Scale Bars A & B: 10μm Scale A: Fig. 10, 12, 14, Scale B: Figs. 1, 2, 3, 4, 5, 6, 7, 8, 9, 11, 13.

Anabaena fertilisimma **Rao, C.B. (Plate 56.8, F.6)**

Desikachary, 1959, P.398, Pl.74, F.1

Cells barrel shaped 5.5μ broad, 4.4μ long, heterocysts 6.6μ broad, spores in long chains 6.6μ in diameter.

Habitat: Hartala lake, H-III, December, 1998, August, 2000.

Velhala lake, V-III, October, 1999.

Anabaena khannae **Skuja (Plate 56.8, F.7)**

Desikachary, 1959, P.396, Pl.75, F.2

Trichomes 2.5-4μ broad, heterocysts , 3-3.5μ broad 3.5-6μ long, spores, globose 6.6μ in diameter.

Habitat: Velhala lake, V-I, March, 2000.

Anabaena spinosa **Tiwari (Plate 56.8, F.8)**

Trichome 3.8μ broad, cells 6.6μ long, heterocysts 5μ broad, 8.8μ long, surrounded by spiny papillae, spores 3.8μ broad, 6.6μ long.

Habitat: Velhala lake, V-I, June, 2000.

Raphidiopsis indica **Singh, R.N. (Plate 56.8, F.9)**

Desikachary, 1959, P.422, Pl.79, F.1,5

Cells 2.7μ broad, cells 6-8μ times as long as broad, with gas vacuoles.

Habitat: Hartala lake, H-I, January, 2000.

Velhala lake, October, 1998, June, 1999, V-III, July, 1999.

Raphidiopsis curvata **Fritsch and Rich (Plate 56.8, F.10)**

Desikachary, 1959, P.422, Pl.65, F.5,7

Cells 5μ broad, 4-5 times as long as broad, spores 10-19.5μ long, 5μ broad, in the middle of the trichome.

Habitat: Velhala lake, V-II, July, 1999, V-III, February, 1999, May, 1999, July, 2000.

Family - Scytonemataceae

Plectonema notatum **Schmidle (Plate 56.8, F.11)**

Desikachary, 1959, P.440, Pl.83, F.5

Filaments 2.2μ broad, cells cylindrical, 2-3 times as long as broad.

Habitat: Velhala lake, V-II, May, 1999.

Scytonema coactile **Mont. (Plate 56.9, F.10)**

Desikachary, 1959, P.455, Pl.90, F.2

Filament 18-24μ broad, trichome 12-18μ broad, cells subquadrate, heterocysts sparse.

Habitat: Velhala lake, V-I, September, 2000.

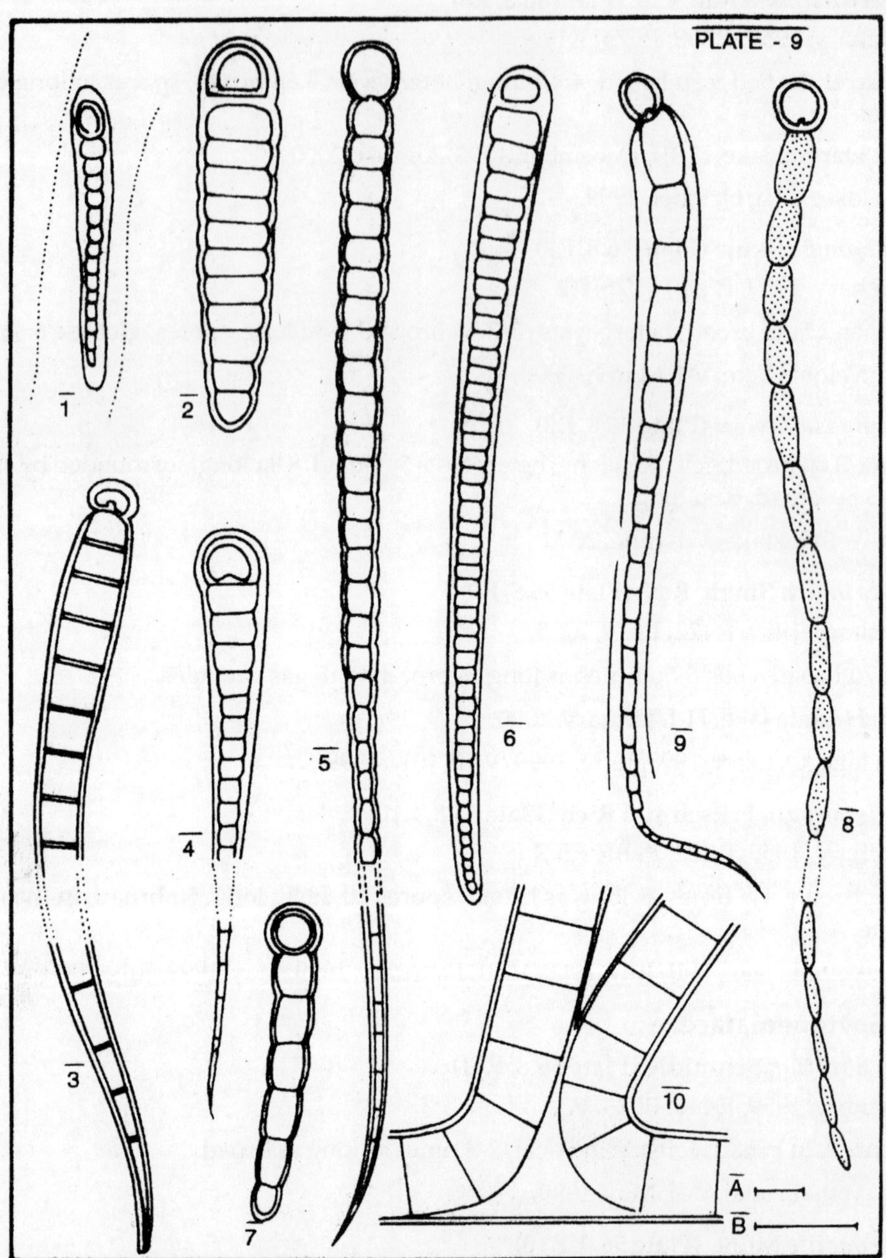

Plate 56.9: (1) *Calothrix brevissima* West G.S., (2) *C. brevissima* West, G.S. *v. chakradharensis* v. nov., (3) *C. castellii* (Massal.) B and F., (4) *C. clavata* West, G.S., (5) *C. javanic* De Wilde, (6) *C. marchica* Lamm., (7) *C. membranicea* Schmidle, (8) *Rivularia aquatica* de Wilde, (9) *Rivularia beccariana* (De. wilded) Geitler, (10) *Schytonema coactile* Mont.

Scale Bars A & B: 10µm Scale A: Fig. 1, 3, 6, 7, 8, 9, 10, Scale B: Figs. 2, 4, 5.

Family: Rivulariaceae

Homoethrix balaerica **(Born. et. Flab.) Lamm. (Plate 56.8, F.12)**

Gonzalves and Kamat, 1960, P.4, Plate 56.4, F.36 a,b.

Filaments 8.3µ broad and up to 150µ long, trichome 3.6µ broad cells 3.5µ broad and 1.5µ long.

Habitat: Hartala lake, H-III, Sept. 1999.

Homoeothrix hansgirigi **(Schmidle) Lemm.** ***v.constricta*** **v.nov. (Plate 56.8, F.13)**

Filaments 4µ broad, 66.6µ long, cross walls constricted.

Habitat: Hartala lake, H-III, November, 1999.

Homoeothrix juliana **(Manegh.) Kirchn. (Plate 56.8, F.14)**

Desikachary, 1959, P.519, Pl.107, F.7

Filaments 10-15µ broad, upto 2 mm long, trichomes 9.5-12.5µ broad, cells discoid, upto 1/3 - 1/1 as long as broad, 4.7-6µ long.

Habitat: Hartala lake, H-III, November, 1999.

Velhala lake, V-I, November, 1998, November, 1999.

Calothrix brevissima **West, G.S. (Plate 56.9, F.1)**

Desikachary, 1959, P .533, Pl.114, F.1

Filaments epiphytic 65.6µ long, 7.1µ broad, trichomes 58.3µ long, heterocysts basal, single rounded hemispherical.

Habitat: Hartala lake, H-III, July, 2000.

Calothrix brevissima **West, G.S.** ***v.chakradharensis*** **v. nov. (Plate 56.9, F.2)**

Filaments epiphytic, 39.4µ long, 6.6µ broad, trichome, 37.7µ long, Heterocysts basal, single, rounded, hemispherical.

Habitat: Hartala lake, H-III, August, 2000.

Calothrix castellii **(Massal.) B and F.** ***v.somastipurense*** **Rao, C.S. (Plate 56.9, F.3)**

Desikachary, 1959, P.529, Pl.114, F.7

Filaments 10-20µ broad, upto 232µ long cells 8.3µ long at the base, hair about 12µ long, heterocysts basal, single, hemispherical, 10.7µ broad and 5.9µ long.

Habitat: Velhala lake, V-I, April, 2000, V-II, November, 1998, November, 1999, September, 2000.

Calothrix clavata **West, G.S. (Plate 56.9, F.4)**

Desikachary, 1959, P.542, Pl.114, F.2

Filaments 60µ long, straight, prominently swollen at the base, 6.6µ broad, cells discoid at the base, heterocysts basal, single hemispherical.

Habitat: Velhala lake, V-II, September, 1999.

Calothrix javanic **De Wilde (Plate 56.9, F.5)**

Desikachary, 1959, P.525, Pl.106, F.1, 2, 5-7

Trichomes 3.8μ broad, ending in a thin hair, heterocysts basal 5.5μ broad.

Habitat: Velhala lake, V-II, June, 2000.

Calothrix marchica **Lamm. *v.intermedia* Rao, C.B. (Plate 56.9, F.6)**

Desikachary, 1959, P .543, Pl.114, F.4

Flaments epiphytic 8.3μ broad and upto 70μ long, trichomes 7.4μ broad, heterocysts single, basal, 7.1μ broad.

Habitat: Hartala lake, H-III, July, 2000.

Velhala lake, V-I, July, September, October, December, 1999, June, 2000.

Calothrix membranicea **Schmidle (Plate 56.9, F.7)**

Desikachary, 1959, P.542, Pl.106, F.10

Filaments 9.5μ broad at the base and 55μ long, trichome 8.3μ broad, cells 7.1μ broad, 9.5μ long, heterocyst single, basal, spherical, 9.5μ broad.

Habitat: Velhala lake, August, 2000.

Rivularia aquatica **De Wilde (Plate 56.9, F.8)**

Desikachary, 1959, P.552

Trichome 9.5μ broad, ending in a long thin hair, heterocysts basal, solitary, spherical, 13μ in diameter.

Habitat: Velhala lake, V-I, June-September, 2000, V-II, December, 1998, January, 1999, V-III, December, 1998, January, 1999.

Rivularia beccariana **(De. Wilde) Geitler (Plate 56.9, F.9)**

Desikachary, 1959, P.551, Pl.106, F.8,9

Trichome 8.3μ broad, cells at the base longer than broad, 13μ long, heterocyst, basal, solitary, spherical, 9.5μ in diameter.

Habitat: Velhala lake, V-I, June, 2000.

Gloeotrichia indica **Schmidle (Plate 56.10, F.1)**

Desikachary, 1959, P.558, Pl.117, F.1,7,8

Filaments radiating 200-300μ long, trichomes made of 3 cells, spores 60-70μ long.

Habitat: Velhala lake, V-II, November, 1998, January, 1999.

Gloeotrichia pilgeri **Schmidle (Plate 56.10, F.2)**

Desikachary, 1959, P.558, Pl.116, F.6,7

Filaments 150-190μ long, base with 2-3 heterocysts, trichome 8.3μ broad, spores 12-16μ broad and 29-50μ. long.

Habitat: Velhala lake, V-II, September, 2000, V-III, August, 2000.

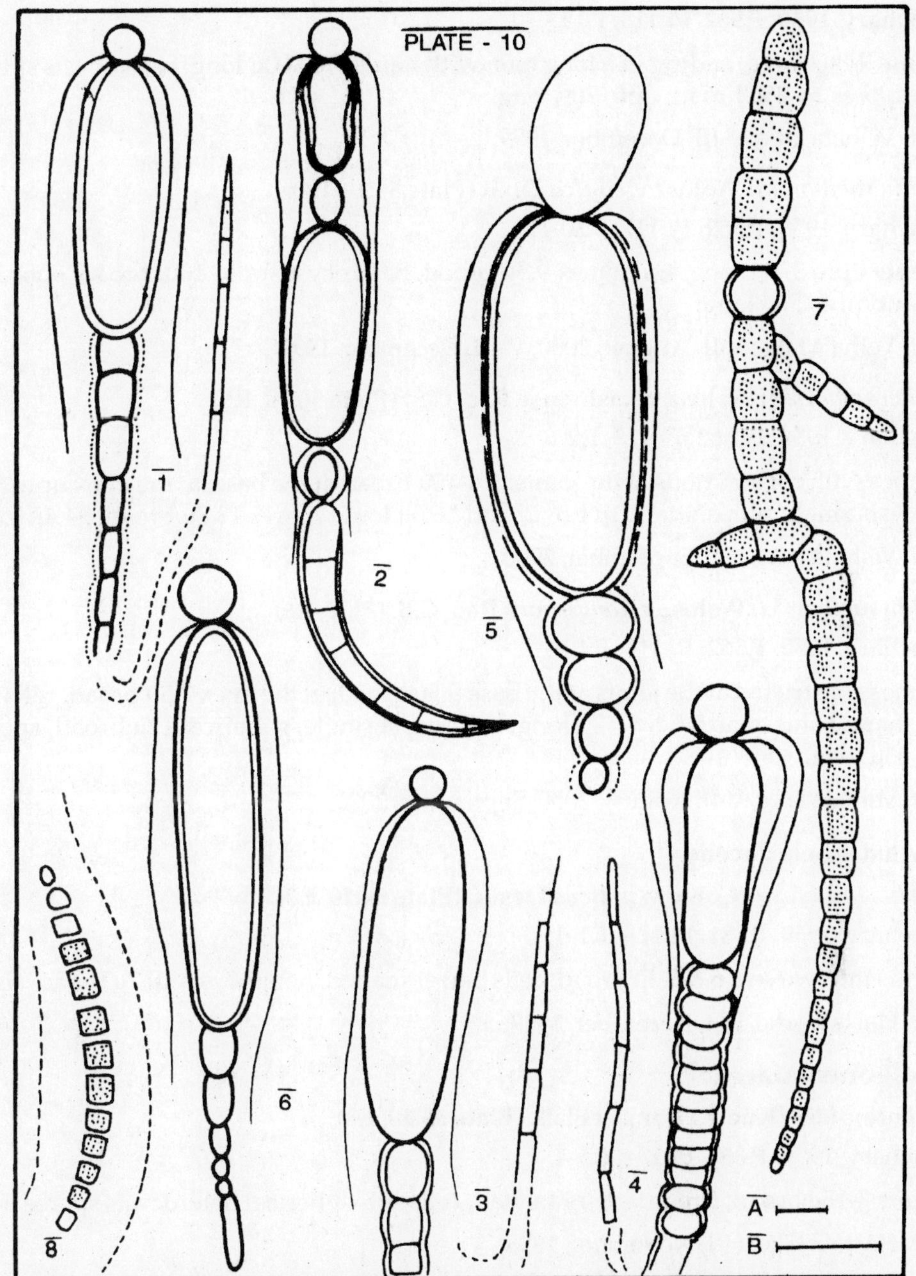

Plate 56.10: (1) *Gloeotrichia indica* Schmidle, (2) *G. pilgeri* Schmidle, (3) *G. raciborskii* Wolosz., (4) *G. raciborskii* Wolosz. *v. conica* Dixit, (5) *G. raciborskii* Wolosz. *v. kashiense* Rao, C.S., (6) *G. raciborskii* Wolosz. *v. longispora* Rao, (7) *Mastigocladus laminosus* Cohn. *v. indicus* Desikachary, (8) *Stigonem hormoides* (Kuetz.) B and F.

Scale Bars A & B: 10μm Scale A: Figs. 1, 2, 3, 4, 5, 6, 7, 8, Scale B: Fig. 5.

Gloeothrichia raciborskii **Wolosz (Plate 56.10, F.3)**

Desikachary, 1959, P.562, Pl.118, F.14

Trichome 7-8μ broad, ending in a long hair, with hair upto 800μ long, heterocysts spherical 5.9μ broad, spores 15-25μ broad, upto 70μ long.

Habitat: Velhala lake, V-III, December, 1999.

Gloeothrichia raciborskii **Wolosz v. *conica* Dixit (Plate 56.10, F.4)**

Desikachary, 1959, P.562, Pl.117, F.10-12

Filaments upto 370μ long, trichomes 9.5μ broad, heterocysts upto 10μ. broad, spores 8.5-11.5μ broad and upto 50μ long.

Habitat: Velhala lake, V-II, August, 2000, V-III, December, 1999.

Gloeotrichia raciborskii **Wolosz v. kashiense Rao, C.S. (Plate 56.10, F.5)**

Desikachary, 1959, P .562, Pl.117, F.2-6

Trichomes with constrictions at the joints, 8.4-10μ broad at the base, at the apex upto 3.5μ broad, heterocysts single, ellipsoidal, 11μ broad and 16.6μ long, spores 13.8μ broad, 34.4μ long.

Habitat: Velhala lake, V-II, September, 2000.

Gloeothrichia raciborskii **Wolosz v. *longispora* Rao, C.B. (Pl.10, F.6)**

Desikachary, 1959, P.562, Pl.118, F.4-6

Trichomes constricted at the joints at the base 9.5μ broad, at the apex 2.3μ broad, cells at the base barrel shaped and quadratic, 8-12μ long, heterocyst single, spherical 12μ broad, spores 10-12μ broad, 83μ long.

Habitat: Velhala lake, V-II, August, 2000, September, 2000.

Family: Mastigocladaceae

Mastigocladus laminosus **Cohn. v. *indicus* Desik. (Plate 56.10, F.7)**

Desikachary, 1959, P.581, Pl.126, F.1-4.

Filaments interwoven, 6.6-9.2μ broad, cells barrel shaped, heterocysts 10.7μ broad, single.

Habitat: Hartala lake, H-I, December, 1999.

Family: Stigonemataceae

Stigonema hormoides **(Kuetz.) Bom. et Flab. (Plate 56.10, F.8)**

Desikachary, 1959, P.664, Pl.134, F.1-4

Filaments 7-15μ broad, irregularly branched, cells sub-spherical heterocysts sparse intercalary.

Habitat: Hartala lake, H-I, November, 1998.

Observations and Discussion

Cyanobacterial showed maximum population during summer and a minimum during monsoon in both the lakes. Maximum blue-green algal population was observed during rainy and minimum during summer in Suraha lake. (Singh and Swarup, 1979) While Zutshi *et al.* (1984) observed their highest population during January. In the present study, comparatively a higher percentage of blue-greens was recorded at H-I station of Hartala lake. This was supported the fact that sewage contaminated water harbours more blue-green algae has also observed by Parmasivam and Sreenivasan (1981).

Blue-greens was represented by 153 species belonging to 35 genera of which 102 species occurred in Hartala lake and 87 in Velhala lake. *Oscillatoria princeps* and *Lyngbya allorgei, L. martensiana* showed a regular occurrence throughout the study period in the sewage polluted Hartala lake. This tallied with the observations made by Pandey and Tripathi (1988).

In India, *Microcystis aeruginosa* has been reported as a permanent bloom found commonly throughout the country. Singh (1953) studied the factors affecting glooms in some inland waters of India and the best single indicator of pollution, as indicated by him, is *M. aeruginosa*. During the present study *M. aeruginosa* was found commonly growing in the lake particularly during late summer in association of other algae and develop into a bloom.

Rai and Kumar (1976) observed non-heterocystous forms in the polluted waters rich in nitrogen. Present investigation also showed dominance of Cyanobacterial in these lakes with 38 heterocystous forms indicating eutrophication of the lakes.

Genus *Oscillatoria* has been found to be very tolerant to pollution which frequently inhabits the polluted waters (Rai and Kumar, 1976). Present study confirmed their observations as *Oscillatoria* was found dominating in both the lakes represented by 30 species followed by Phormidium with 20 species. In the present study the dominance of *Oscillatoria* was indicating pollutants of biological origin which agreed with the observations of Rai and Kumar (1976).

Palmer (1980) listed 14 species of Cyanobacterial as pollution tolerant algae which are common in organically enriched water bodies. During the present study 7 species from his list namely *Oscillatoria limosa, O. ChaJybea, O. brevis, O. formosa, O. princeps* and *O. tenuis, Phomidium autumale* have been observed in the lakes.

Following 12 species of Cyanobacterial which showed a wide range of adaptability can be regarded as "pollution tolerant species": *Chlorococcus minutus, Johannesbaptistia pellucida, Myxosarcina burmensis, Lyngbya allorgei, L.martensiana, Oscillatoria limosa, O. princeps, O. tenuis, Phormidium autumale, P. retzii, Stichosiphon sansibaricus, Synechocystis aquatilis.*

Acknowledgements

Authors are thankful to Mr. B.M. Patil, The Principal of S.S.V.P.S's Late Dr. P.R.Ghoghrey Science College, Dhule for providing facilities. We are also thankful to Dr. V.C. Kolhe, The Principal of Bhusawal Arts, Science and P.O. Nahata Commerce College, for encouragement during this work.

References

Desikachary, T.V., 1959. *Cyanophyta* ICAR Monographs on Algae. New Delhi, p. 686.

Dixit, S.C., 1936. The Myxophyceae of the Bombay Presidency, India. I. *Proc. Indian Acad. Sci.* 3: 93-106.

Gonzalves, E.A. and Kamat, N.D., 1960. The Myxophyceae of Karnataka III *J. Bombay Univ:* 28: 28-41.

Kamat, N.D. (1963). The Algae of Kolphapur, India. *Hydrobiologia,* 22 (3-4): 209-305.

Mahajan, A.D., 1983. Study of algal flora of Khaira district of Gujarat. *Ph.D. Thesis*, Sardar Patel Univ., Anand.

Nandan, S.N. and Patel, R.J., 1985a. Species diversities of algal flora in Vishwamitri, river Baroda *Pol. Arch. Hydrobiology*. Poland 32 (1): 1-6.

Palmer, C.M., 1980. *Algae and water pollution.* Castle House Publications Ltd., England. p.123.

Pandey, S.N. and Tripathi, A.K., 1988. Studies on algae of polluted ponds of Kanpur (India) VIII. Role of blue-green algae, *Phykos:* 27: 38-43.

Parmasivam, M. and Sreenivasan, A., 1981. Changes in algal flora due to pollution in Cauvery river, *Indian J. Environ.* Hlth. 23(3): 222-238.

Rai, L.C. and Kumar, H.D., 1976). Algal growth as a means of evaluation of nutrient status of the effluent of a fertilizer factory near Shahpuri, Varanasi, *Trop. Ecol.* 17(1): 50-56.

Rao, C.B., 1937. The Myxophycea of the Madras Presidency India. *Journal Indian Botanical Society.* 17: 81-96.

Singh, R.N., 1953. Limnological relations of Indian inland waters with special references to water blooms, *Verh. Int. Var. Rher. Anew. Limnol.* 12: 831-836.

Singh, S.R. and Swarup, K., 1979. Limnological studies of Suraha lake (Ballia). II The periodicity of phytoplankton, *J. Indian Bot. Soc.* 58: 319-329.

Zutshi, D.P., Vishin, N. and Subla, B.A., 1984. Nutrient status and plankton dynamics of a perennial pond, *Proc. Indian Natl. Sci. Acad.* 50: 577-581.

Chapter 57

Defoliation/Detasselling Studies in Maize (*Zea mays* L.): A Review

Sumit Mishra and D.S. Kler

Department of Agronomy, Punjab Agricultural University, Ludhiana – 141 004, Punjab, India

ABSTRACT

Response of defoliation to maize plants is not only affected by the extent and time of leaf area loss, but also with the position of leaf on the plant. It was also found varied with cultivar and season under which it was grown. Defoliation studies could be an excellent method for simulating various abiotic (hail, frost, lodging, etc.) and biotic (insect pest, disease, detopping, etc.) stresses and may also be a potential way for selecting cultivar suitable under these stresses with appreciably higher yield potential. Besides, effect of defoliation on dry-matter accumulation, stem reserve, nutrient absorption/distribution, phenology, yield and yield attributing characters of maize have been reviewed.

Keywords: Calorific energy, CERES-maize, Defoliation, Detasseling, EFPD, Fodder, LAGP, Maize, RDMA, Source-sink relationship,

Introduction

Many factors can cause defoliation or removal or loss of green leaves in maize. Leaf area is often lost due to attack of insect pests, wind or hail damage, or feeding by livestock. In addition, in many areas farmers remove leaves before the crop is mature to use them as fodder. The effect of defoliation on grain yield depends on the extent of leaf loss, position of leaves at plant through which leaves one affecting the plant activities and the time when leaf loss occurs. There exists a direct positive correlation between variety and leaf mass or relative growth in height but 25-50 per cent reduction in height has no effect on grain yield (Shivraj, 1972). Also, in many intercropping systems with maize the farmers deliberately defoliate the maize to allow better growth of the intercrop/primary crop.

Calorific Energy Distribution in Maize Plant

During vegetative phase, young leaf blades had a high calorific value. At silking, energy content of lower parts of stem was greater than the upper parts. This may be related to an increased lignin content of lower leaves, the upper leaves, however, had a higher energy content than leaves below ear. The ear leaf had a low energy content compared to nearby leaves. At harvest, organs below ear (lower leaves and lower stem) had a much lower energy content than at silking. Internodes and leaves near the ear had the highest energy content at maturity. Grain formation appears to draw upon the ear leaf reserves and the internode just below the ear at silking, this would account for the decrease in energy content of these plant parts. Between the sixth leaf stage and silking, the calorific value of sheath and internode increased but that of total above ground plant parts decreased. Energy accumulated in shoot at maturity was 2.7 times higher than that of accumulated at silking. One third of the combustible energy of whole plant was found in Kernels and one half in the ear (Girardin, 1985).

Contribution of Various Plant Parts in Grain Formation

The contribution of leaves on the main culm to grain production of corn has been well documented (Palmer, 1969; Tanaka and Yamaguchi, 1972). The leaves positioned close to the ear on the main culm contributed more to grain production than other leaves (Tanaka and Yamaguchi, 1972). However, limited information is available regarding contribution of laminae extending from husks (leaf sheaths) that surround ears and other plant parts. So, before discussing the effect of defoliation, it is imperative to understand the relative contribution of various leaves and other plant parts towards the grain yield of maize crop.

Culm Leaves

It is well-known in rice, barley and wheat that the stored carbohydrates in vegetative parts are translocated to grain (Yoshida, 1972). The carbohydrates preserved at presilking contributed < 10 per cent to grain production (Simmons and Jones, 1985) in the field grown maize plants. The leaves positioned close to the ear on the main culm contributed more to grain production (Tanaka and Yamaguchi, 1972) than other leaves. Partial defoliation at silking increases the photosynthetic rate of remaining leaves. This is mainly due to acceleration of photosynthate translocation from leaves by the sink. Grain production per unit leaf area increased as culm dry weight decreased with decreasing leaf area through defoliation. In addition, carbohydrates (sugar and starch) content in culm plus leaf sheath also decreased with defoliation. Contribution of the stored photosynthates in the culm to grain production may vary in relation to the photosynthetic ability during grain filling. In other words, the remobilization of the stored carbohydrate from culm to grain may be affected by photosynthetic ability during grain filling (Sawada *et al.*, 1995). Removal of upper leaves affected seed production but not their germination ability to greater extent than removal of lower leaves (Javorkova, 1990). Lower leaf removal at anthesis and 17 days after anthesis reduced yield by 15 and 9 per cent respectively. Lower ears were affected more than apical ears. Lower leaves supply carbohydrates to the root system for the maintenance of post-anthesis uptake of soil nutrients particularly nitrogen (Teyker *et al.*, 1991).

Husk Leaves

Among various leaves, those adjacent to the ear contribute more to grain production than other leaves that are farther positioned (Tanaka and Yamaguchi, 1972). Therefore, husk leaves extending from husk, in some cultivars of maize also appear to contribute to grain production to a great extent. In some cultivars of dent corn (Cantrell and Geadelmann, 1981) the photosynthesis of husk leaves contributed to grain production and the average reduction of grain yield by husk leaves removal was 2.6 per cent. However, when only husk leaves were left at the silking stage (15 per cent of the leaf area of control plant) grain yield was 45 per cent of the control plant. It is well known that the photosynthesis of leaves on the main culm of corn is the C_4 type. However, the husk leaf is reported to be C_3-C_4 intermediate type in terms of photorespiration rate and leaf anatomy (Antonielli and Venanzi, 1979). The dehusking of ear had no effect on grain yield (Tollenear and Daynard, 1978). In contrast, it has been observed (Salvador and Pearce, 1988) that decreased grain yield was responsive to husk manipulation, including removal or peeling and husk photosynthesis contributed about 2 per cent of final grain yield. The husks were the most efficient photosynthetic organ (Jain, 1971) of the plant, comprising about 7.7 per cent of total photosynthetic surface, yet contributing upto 15 per cent of the ear's dry matter. The dry matter production per unit leaf area was higher in the husk leaves than for leaves on the main culm when compared on the basis of equal area, suggesting that the husk leaves have a high potential productivity although the ability is not expressed under ordinary field conditions. The higher percentage export of C^{13} from husk leaves than from the leaves on the main culm and the partitioning of most C^{13} to the ear indicate that husk leaves may play yet an important role as a major source in the source-sink unit in maize. Although, most of C^{13} was distributed to the ear from the husk leaves examined but most of the C^{13} fed to the husk leaves was partitioned to the husk and cob plus shank than in grain, and the reverse was true for the leaves on the culm. This suggests that the husk and cob plus shank act as the major sink for photosynthates from the husk leaves than the grain. This is probably due to the greater proximity of husk leaves to the husk, compared with the grain. It has been reported that the distance between source and sink is one of the determinants of the destination of photosynthate, translocation (Tanaka and Fujita, 1979). Since, it is known that the husks act not only as barrier for CO_2 respired from the ear (Shinano *et al.*, 1993) but also as important storage site for remobilizable assimilates (Fairy and Daynard, 1978) and for maintenance of adequate moisture and temperature environment for the developing ear (Salvador and Pearce, 1988). The husk leaves may play a role not only as a source for grain production but also in maintaining the functions of the husk. It has been observed that dry matter and grain production per unit husk leaf area was higher than that of per unit culm leaf area, suggesting that husk leaves have a higher potential capacity to produce grain compared to culm leaves. It was noticed that photosynthetic rate of husk leaves increased by defoliation of culm leaves. However, such increase was not observed in culm leaves (Fujita *et al.*, 1995).

Tassel

Maize is a monoecious plant bearing male flower on the terminal inflorescence, the tassel and female flowers on the lateral axis of shoot. Manifold pollen grains are produced by a single plant while only a modicum of it is utilized for the purpose of pollination. The unutilized lot of pollen grains is sheer wastage of energy. Tassel in maize may not be serving any useful purpose after pollination, besides, they attenuate light, cause shading and continue to be metabolic drain on the plant through respiration which is about 5 per cent of the total photosynthesis (Tanaka and Yamaguchi, 1971). Hence, there is possibility of utilizing this wasteful amount of photosynthates for increasing grain yield of maize by way of tassel removal especially at increased plant population. Several workers have emphasized the advantageous effect of tassel removal (Shekhawat *et al.*, 1964; Hunter *et al.*, 1969).

Removal of the tassel may increase yield because of improved light interception by underlying photosynthetically active tissue (Duncan *et al.*, 1967; Hunter *et al.*, 1969) and more favourable translocation of photosynthate to the grain (Mostert and Marais, 1982), but results are variable depending on interplant competition, water availability and soil fertility (Hunter *et al.*, 1973). Silk length in detasselled plant was reduced to 0.5-1.0 inches (Culy *et al.*, 1992) which resulted in 0.7 to 1.1 million fewer kernels per acre and increased large and medium round kernel fractions. When tassel was not removed and silk length of > 0.25 inches had negative effect on grain yield. It is concluded that for seed production practices where female blocks are detasselled prior to pollination; silk lengths of 1.0 inch are sufficient to provide complete ear pollination. Tassel removal maintained the chlorophyll content of the ear leaf but decreased the contents of lower leaves and increased that of leaves above the ear during the period of up to 30 days after tassel emergence (Yang *et al.*, 1994). Tassel removal had no significant effect (Philip and Gautam, 1995) but contributed to as much as 19.8 per cent increase in total seed yield (Espinosa and Tadeo, 1995). However, early detasseling of the female parents as compared to normal detasseling 57 DAS (days after sowing) decreased leaf area and seed yield but did not affect seed composition or germination. The number of ears was the parameter most reduced by early detasseling, and had the greatest affect on seed yield (Menezes *et al.*, 1994). The different methods used (manual, mechanical, along with 4-5 upper leaves around tassel and male sterility) did not effect grain production, total dry matter yield or nutrient conversion into grains in the cultivars studied (Vasconullos *et al.*, 1995). However, residual nutrients increased in stalks which had leaf sheaths, husks and cobs at the time the tassel was removed. Cultural control (detasseling of 75 per cent maize plants) had a distinct effect in reducing the pest population of pyralid (*Ostrinia furnacalis*) in the field and in increasing yield (Breitchaupt, 1995). Possible reasons why detasseling (removing the tassel) can fail as a measure for controlling *Ostrinia furnacalis* on maize in the Philippines are: this technique controls only a fraction of the pest population infesting maize late in the growth stage of the plant; and the majority of larvae found on the tassel usually bore into the stalk above the ear which has only a small effect on yield. It is concluded (Velosco, 1997) that detaseling should be evaluated further particularly as natural enemies of insect pests may be deprived of pollen, which is a valuable food source.

Effect of Intensity and Timing of Defoliation

The effect on seed quality as influenced by defoliation depends upon level of defoliation, growth stage when the defoliation treatment was applied; genotype and environmental condition. Partial defoliation never affected warm germination of seeds and in only a few cases affected cold germination. Reduction in germination of seed in response to complete defoliation were associated with reduction in grain size (Vasilas and Seif, 1986). The inbred was most sensitive to defoliation and it was also the inbred which produced fewest total leaves. Twenty five per cent defoliation increased average maize grain yields from 7.7 to 8.0 t/ha and of 5 per cent defoliation at 25 DAS increased yields from 7.7 to 8.4 t/ ha, but generally defoliation decreased yields, the reduction being greatest for 60 DAS (Kelley *et al.*, 1985). Yield reduction due to defoliation was associated with decrease in grain number and weight. Grain weight was affected most by defoliation at the milking and late milking stages, whereas, grain number by defoliation at anthesis and the 14-leaf stage (Vasilas and Seif, 1985). Complete defoliation at the 4-, 5-, and 6-leaf stages did not significantly reduced grain yield, but from the 6th to 9th leaf stages, 100 per cent defoliation resulted in a linear decrease in grain yield with later leaf stage. The effect of early season complete defoliation varied with year (Shapiro *et al.*, 1986). Fifty per cent defoliation at full silk reduced yield by 2.8 per cent, losses decreased to 6 per cent in plant defoliated 4 weeks later. After 100 per cent defoliation; losses ranged from 89 per cent after defoliation at full silk to 13 per cent after defoliation 7 weeks later (Giesbrecht, 1977). Defoliation at 10-leaf stage gave higher grain yield than defoliated at 12 days after silking; which in turn gave higher yields than when plants were defoliated at 16-leaf stage. Increase in the degree of defoliation at 16-leaf stage and at 12 days after silking decreased grain yield, the effect being greatest at 16-leaf stage. Fifty per cent defoliation at 10-leaf stage increased grain yield in the *kharif* season. Grain protein content decreased with increase in plant population and increased with increase in degree of defoliation and was found higher in spring than in the *kharif* season. Leaf protein content decreased as plant approached maturity, particularly with increase in degree of defoliation. Sucrose and reducing sugar contents in the grain decrease with increase in degree of defoliation and similar trends occurred in the sugar: Protein ratio (Singh and Nair, 1975). Stem weight of undefoliated plants increased after flowering and then changed little in lower plant density but in higher density, decreased by final harvest to about the same as at flowering. Defoliation decreased final grain yields by 57 and 16 per cent in higher and by 47 and 11 per cent in lower planting densities when carried out at 122 and 143 days after sowing respectively (Allison *et al.*, 1975). Test and 100-grain weight were reduced, especially with higher defoliation rate, only by defoliation after tasseling. Defoliation before and after tasseling slightly delayed and hastened maturity (Hicks *et al.*, 1977). Defoliation at early stages significantly reduced plant height (Johnson, 1978 and M'Guettia *et al.*, 1991) due to shorter length of internodes rather than their smaller number. The defoliation treatment did not significantly affect leaf number per plant because tassel differentiation occurred 20 days after sowing, four days before the defoliation treatment started. The smaller amount of dry matter accumulated in the defoliated plants at the mid silking stage indicated that leaf area index was limiting photosynthetic activity leading to shorter plants. In general defoliation of lower parts of stem at 6 weeks stage tended to increase ear length and weight at final harvest (Remison, 1982), but otherwise early defoliation had little effect, whereas, late defoliation decreased ear and grain weight, grain moisture content and total above ground dry matter, having greatest effect with total defoliation or defoliation from upper halves of the stem. Reduced ear size was the main factor in reducing yield, while lodging, mostly due to stem breakage, increased with defoliation after mid-silking. Complete defoliation at silking produced at least 77 per cent reduction in grain yield; which lessened with delay in defoliation towards crop maturity. Less severe defoliations resulted in smaller reductions in grain yield and showed the middle and upper leaves to be more important in grain production than the lower ones. Generally, grain yield, losses exceeding 20 per cent were recorded only after two-third of the leaves were removed and than only when these included the middle one-third of the leaves. Consequently, returns from insecticide applications to control larvae after silking will decrease with crop maturity, and controls seem unnecessary unless substantial defoliation appears likely (Douglas *et al.*, 1981).

Defoliation and Dry Matter Accumulation

Considering the leaves as the source and the ears the sink. It has been stated (Tanaka and Fujita, 1970) that the photosynthesis products of the leaves above the ear were translocated to the ear and efficiency was incorporated into the grain. Improving light conditions during ripening did not increase DM production, but partial defoliation increased the NAR (Net Assimilation Rate) of the remaining leaves and promoted photosynthesis during rapid ripening, indicating that leaves do not express their full potential for photosynthesis, during ripening. Removal of ears or prevention of pollination increased sugar content of the leaves and stems which reduced photosynthesis increased respiration and in turn reduced DM production. Photosynthesis during ripening was under the control of the sink rather than the source. Leaf position had a greater effect on leaf CO_2 exchange rate (CER) after ear removal irrespective of genotype. The CER of both ear leaf and 2nd leaf was altered within 10 days following defoliation and/or ear removal. The relative source-sink ratio of the genotype was the most important factor in determining how the treatments affected CER. Leaf CER increased after defoliation only in genotypes with larger source-sink ratio in untreated plants and decreased following ear removal only in genotypes with smaller source-sink ratio in untreated plants. After treatments, weight of stalks, leaf sheaths and ears decreased as well as the stalk non-structural carbohydrates level (Barnett and Pearce, 1983). Further, it is suggested (Barnett, 1982) that translocation of assimilates out of the leaves was not affected by ear removal, possibly because of alternative sinks in the stalks. Defoliation decreased ear and grain weight and the days to physiological maturity.

Middle and upper leaves contributed almost 80 per cent of total DM to the grain and after silking photosynthates from the lower region were translocated to the roots. After total defoliation 5-6 per cent of grain DM was contributed by the stem. Grain dry matter yield and ear number were unaffected by removal of lower 33 per cent of leaves. Grain filling lasted for 76 days in normal untreated plants and was reduced by 30 and 40 days when defoliation reached 66 and 100 per cent, respectively (Diaz, 1983). The contribution of middle leaves was greatest and of lower leaves least. The effect decreased as the position moved away from the ear either upward or downwards. Transport of P32 from leaves was greater from the ear leaf than from any other leaf. Amount of chlorophyll in the various leaves differed (Zhao, 1981). Grain yields increased with delay in defoliation from 10 days after 50 per cent silking and were directly related to the amount of leaf left after defoliation. Late defoliation had little effect on yield. Removal of upper leaves had a more direct effect than removal of lower leaves. Partial defoliation increased NAR in the remaining leaves. Defoliation increased sugar and crude protein contents and decreased the starch content (Chen *et al.*, 1978). Early defoliation greatly reduced grain development and filling duration. Defoliation and ear tip removal, both resulted in an increase in N-content in the internode above the ear; soluble carbohydrate content in this internode was reduced by defoliation especially with the later defoliation, but was increased by ear tip removal. Ear tip removal increased grain N-content, but grain development and size and soluble carbohydrates concentration were unaffected (Jones and Simmons, 1983). Source silk balance was manipulated by detopping and ear removal 10 days after anthesis. Detopping increased; and ear removal decreased the net photosynthesis by about 20 per cent within one hour. The increase in net photosynthesis due to detopping was related to stomatal aperture on the lower leaf surface and the decrease due to ear removal was associated with accumulation of carbohydrates in the leaf subtending the ear. Short term changes in net photosynthesis were not related to nitrate or total N-concentration within the leaf (Pearson *et al.*, 1984).

Defoliation Relative to Absorption and Distribution of Nutrients

Removal of all the leaves led to a remarkable reduction in the accumulation of all the inorganic elements (Fujita *et al.*, 1994) indicating that leaves play an important role in absorption of inorganic elements by roots. Removal of husk leaf did not affect or only slightly affected the inorganic elements absorption, whereas removal of the main culm leaves decreased their absorption except for K. These observations signify the contribution of main culm leaves to the absorption of inorganic elements. Husk leaf removal did not affect significantly the distribution of all the inorganic elements. However, the removal of the culm leaves promoted the distribution of N, P and Mg to the ear, but did not affect the K distribution. Thus, distribution of N, P and Mg to ear is closely related to growth status of the ear. The ear growth is more active than other plant parts due to removal of main culm leaves and the distribution of N, P and Mg to ear and particularly to grain is increased. However, by further reduction of leaf area, grain growth was more severely impaired due to insufficient supply of photosynthates leading to reduction of N, P and Mg distribution to grain. Slight accumulation of N, P and Mg after complete defoliation (Fujita *et al.*, 1994) thus indicates that controlling the absorption of inorganic elements is affected more strongly by leaves on main culm. Although in an intact plant a large fraction of grain N, equivalent to 33 and 38 per cent was derived by retranslocation from culm leaves and culm plus leaf sheath, respectively yet the retranslocation of N was reduced when husk leaves alone remained. These results suggest that husk leaves play an important role in the distribution of N during the grain filling phase of crop growth.

Defoliation and Stem Reserve

Stem reserves are thought to stabilize the daily rate of grain filling against short term fluctuation in daily photoassimilation (Duncan *et al.,* 1965). They also have a role in maintaining filling rate against longer term effects of persistent post flowering stress. Inter and intra specific competition and post flowering draught have been shown (Schuller and Weslgol, 1991) to accelerate the rate of decline of stem biomass and stem non-structural carbohydrates (SNSC) in the later half of grain filling. In tropics, such stress could also develop from insect attacks, disease out breaks, nutrient deficiency, and from detopping and defoliation of plant for forage. Manipulation of source-sink ratios on fertile maize plant could strongly affect the level of post flowering stem reserves. It has been estimated (Barnett and Pearce, 1983) that normal post flowering photoassimilation required only 6 to 12 days to fill the stem of barren plant to capacity when labile carbohydrates accounted for 52 per cent of total stem weight. While others (Tollenaun and Daynard, 1982; Barnette, 1983) have reported a substantial increase in stem biomass and in stem soluble solids and sugar in barren or partially fertilized plants, despite probable feedback reductions in CO_2 exchange rate. Conversely, stem biomass and labile reserve have been substantially reduced by defoliation in the post silking period (Tollinan and Daynard, 1978) provided that complete leaf loss did not coincide with flowering, a circumstance which would normally result in barrenness. It was confirmed (Edmeades and Lafitte, 1993) that stem biomass at maturity related to undefoliated treatment was reduced at low plant density; and virtually did not at high plant density but on an average 47, 34 and 16 g plant^{-1} loss has been observed at 5, 10 and 15 plant m^{-2}, respectively. A large proportion of stem and leaf reserve might be unavailable for ear and grain growth during 2 weeks following flowering because they were being rapidly used for growth of other organs (Westgate and Boyer, 1985) nearer the source of supply. Grain set in droughted plants has been improved by infusing a sucrose-rich liquid into the stem near the ear (Boyle *et al.,* 1991) during the week following flowering. The flux of assimilates to the ear at this stage was small, perhaps < 20 per cent that was required by the stem (Edneades and Daynard, 1979). Thus where growth of the stem was reduced relative to other plant parts, there might be a slight increase in SNSC, resulting in increased ear growth. Further evidence for increased ear growth was provided by a reduced anthesis-silking interval. In a similar study of wheat cultivars (Rawsaw and Evans, 1971), it was concluded that the length of the period between the end of stem growth and the onset of rapid grain growth was a more important determinant of the effect of stress near flowering than was stem length *per se*.

Cultivar and Seasonal Response to Defoliation

Complete and 50 per cent defoliation at 14 leaf stage (Vasilas and Seif, 1985) reduced grain yield by an average of 45.5 and 2.5 per cent, respectively in FR16, FR25, PR632, FR27 rhm and MS71 and by 86.0 and 17.0 per cent respectively in FRM017 rhm. In FRM017 rhm yield was reduced because of reduced kernel size. As it produced fewest leaves, defoliation removed a greater percentage of its final leaf area. In the other lines, yield reductions with 50 per cent defoliation were due to reduced ear number, while 100 per cent defoliation reduced ear number and kernel per ear. When the part of the plant above the ear was removed (Detopping) three days after silking, PB3388, BM and CC showed 1, 29, and 49 per cent yield reductions, respectively. The pattern was similar whether plants were grown at normal (53600 plant/ha) or high (107200 plant/ha) densities. Yield of PB3388 and MB were not significantly affected by high density alone whereas that of cultivar CC was reduced by 29 per cent (Anderson *et al.*, 1979). Average yield of the full season hybrid (Crookston and Hicks, 1978) was reduced by 8 per cent while that of the short season hybrid was increased by 48 per cent after defoliation at 5 leaf stage. Yield response for 12 short season hybrids defoliated in similar fashion ranged from a 37 per cent increase to 14 per cent decrease.

Removal of the tassel only or of the tassel plus two leaves did not significantly affect yield. Removal of the tassel plus four leaves, however, reduced yield (Pinter and Kalman, 1980) to different extent according to variety. The inbred A632 and the sister line cross A632 × A635 (both are sink deficient) were unaffected by any of the treatments. Most of the material, defoliated (Conti and Landi, 1981) resulted in reduced grain yield, stalk dry weight, and moisture content in grain and stalk; these reductions were generally greater in the normal types than in the erect leafed ones. The extent to which frequency of lodging increased in response to defoliation varied among hybrids (Pinter and Kalman, 1981). Hybrids with a relatively resistance in lodging frequency were considered to have superior lodging resistance. Estimates of resistance obtained by such methods were positively correlated ($r = 0.85$) with observations of lodging resistance in the same hybrids in farm trials.

Increased yield in response to defoliation was associated with low yields and with low end-of-season available soil water. Average control yield was 48 per cent during years in which there was a positive response. In years without a positive response; average control yield was 9.31 t/ha and average response was a 10 per cent reduction (Crookston and Hicks, 1988). In spring sown maize (Elsahookia and Wuhaib, 1988) grain yield per plant increased by upto 38 per cent and 1000 grain weight also increased, when the upper leaves were removed at 6 to 8 leaf stage, whereas, in autumn maize many of the cutting treatments reduced grain yield. Extensive defoliation (Britz, 1982) during active leaf growth decreased number of grains/ear and defoliation after pollination decreased grain weight and also shortened the growth period. In general, effects increased with increasing defoliation upto 40 per cent tasseling stage and were greater in a wet than in dry season and in the east than in the west maize region.

Defoliation and Phenology

The defoliation up to 50 per cent did not affect the inbred but delayed tasselling and anthesis in the hybrid. Complete defoliation at the 7-leaf stage delayed silking of the two inbreds, and the hybrid by 3.9, 6.0 and 2.9 days, respectively. Complete defoliation at the 14-leaf stage delayed silking of FRM017 and the hybrid by 10.0 and 3.5 days, respectively, but did not affect FR27. Complete defoliation at anthesis prevented the inbred from reaching 50 per cent silking and reduced the period of pollen shed. Various inbred responded differently to the defoliation treatments. Complete defoliation increased the duration of pollen seed of FR16 but had no affect on FRM017 and decreased the duration of pollen shed of MS71, FR27, FR25 and FR632 (Vasilas and Seif, 1985).

Termination of the sucrose supply (Afuaka *et al.*, 1984) either by defoliation or by removing sucrose from the growing medium did, however, appear to induce premature black layer formation. Leaf removal at 5-leaf stage reduced grain yield, plant height number of leaves and ears but did not affect the dates of tasselling or maturity (Kayitmazbatir and Uguz, 1983). Defoliation delayed pollen shedding and silking (Johnson, 1978; Hsu, 1978) but did not change the pollen shedding to silking interval when done at 5-leaf stage. The rapid decline in soluble solid content in the stem following leaf removal (Tollena and Daynard, 1979) indicated an accelerated utilization of soluble carbohydrates from the stem for grain growth. Maturity expressed as both grain moisture content and rates of black layer formation, was significantly accelerated by defoliation. It was suggested that a decrease in the source-sink ratio during grain filling advances maturity. Defoliation also found to hasten physiological maturity of 6-11 days earlier as compared to control plant (Kajil *et al.*, 1991).

Defoliation as a Simulator of Biotic and Abiotic Stresses

Mechanical defoliation of maize plant may be a useful technique for evaluating genotypes for tolerance to inflicted leaf damage that attempts to simulate leaf area loss due to insect feeding (McMillan *et al.*, 1980). Defoliation experiment also enabled (Elmore and Doupnik, 1995) to take decision regarding the resowing or leaving the plant as such after being attacked by frost. The effect of northern leaf blight (NLB) or leaf defoliation on the yield of 9 sweet corn cultivars was studied (Solomonovitz *et al.*, 1992) in field experiments. Plants were defoliated or infected in the bottom, middle, or top third of the leaf canopy. Disease severity varied with cultivar; and almost no disease developed on leaves treated with mancozeb before inoculation. In both the defoliation and infection treatments, yield losses were greater in plants treated 55 days after planting than those treated 70 days after planting. In all cases, the highest yield loss was recorded when the middle third of the leaf canopy was treated. The relative importance of the top third of the leaf canopy increased if the treatment was done 3 weeks before harvesting. The study indicated the importance of leaf position, cultivar and plant growth stage for evaluating losses caused by NLB, and may explain the failure of various models to describe yield loss due to HLB. Ten genotypes (hybrids and lines) were tested (Palavershich *et al.*, 1990) for resistance following defoliation of the lowest ear 21 days after silking and following stem inoculation with a mixture of inoculants. Both defoliation and inoculation increased stalk rot infection, lodging and stem breakage. Defoliation of 0, 33, 66 and 100 per cent at 6-, 12-, tassel and blister stage was performed at 4 sites in Nebraska. The results generally agreed with the hail loss charts for very early season related to 100 per cent defoliation. Grain yield was reduced with increasing severity of defoliation, and with growth stages near the tassel stage. Thirty three per cent defoliation at the blister stage was the only treatment which showed less yield reduction than was predicted. Later treatments tended to show greater yield loss than the predicted one (Shapiro *et al.*, 1986). The original CERES-Maize crop growth model was modified (Piper and Weiss, 1989) for reduction of plant population and defoliation. Field data from experiments on the reduction of maize plant population during vegetative growth and defoliation (33, 66 and 100 per cent) at 6-, 12- leaf, tassel and blister growth stages were used to test the model. The model underpredicted yields for plant populations at harvest with < 5.57 plant/m^2, accurately predicted yields around 5.57 plant/m^2 and overpredicted yields for > 5.57 plant/m^2. For 1980 data, the model underpredicted yields for 33 and 66 per cent defoliation and overpredicted yields at the blister stage; while in 1981 CERES-Maize predicted higher yields for the 33 and 66 per cent defoliation than the actual yields. The CERES-Maize growth model predicted no grain yield when 100 per cent defoliation occurred during vegetative growth. Those results were, however, conflicted with field observations where 100 per cent defoliation early in the vegetative stage did not materially reduce grain yield. CERES/Maize was modified (Weiss and Piper, 1992) to realistically predicted yields when defoliation > 50 per cent (for example by hail or pest damage) occurred during vegetative growth by taking into account stem photosynthesis and DM translocation from the stem to the developing leaves. With these modifications; predictions from CERES-Maize were in good agreement with field measurements when defoliation occurred both early and later in the vegetative stages. It is suggested that this modified version of CERES-Maize could be the basis of a decision support system for defoliating pests of maize where yields can be evaluated under different management strategies and climate scenarios. Mechanical defoliation was used (Jones and Henderlong, 1981) to simulate frost damage after emergence for seedlings which were sheltered from the incident frost. Mechanical defoliation at the 5-leaf stage resulted in a small but non-significant increase in silage production for the sowing dates. The later sowing date data showed a 14 per cent reduction in grain yield compared with early sown frost damaged maize. Hill and Atkins 1982) reported that there was a small compensatory increase in weight of the upper leaves, seed number and yield at low level of defoliation (11-33 per cent). This was attributed to the observed preference of the larvae for the lower parts of the plant and upper leaves can be compensated, most efficiently for the plant damage.

Infection of the medulla of the second lowest internode by *Diplodia zeae* and *Gibberella zeae* increased with greater defoliation, while stem rind thickness decreased. The correlation between percentage infection in the medulla and lodging percentage was high. Removal of all leaves below the main ear (Pinter and Kalman, 1982) increased lodging at harvest in all varieties and the additional removal of half leaves above the ear increased it further.

Forage Yield and Quality

Total dry matter of removed tassels (315 kg ha^{-1}), tassel plus 2 (412 kg ha^{-1}) or 4 (818 kg ha^{-1}) leaves were sufficient to feed 1 livestock unit (412 kg) for 27.6, 36.1 and 71.8 days, respectively. Detasseling increased mono-culture maize grain yield by 13.7 per cent over control, while removal of tassels plus 4 leaves reduced the photosynthetic efficiency by 14 per cent compared with removal of tassel plus 2 leaves (Edje, 1984). Fresh weight of the removed forage (El-Tuhami, 1982) was maximum at the soft dough stage and declined by the hard dough stage of grain development. The crude protein (CP) content of the removed forage decreased with increasing plant maturity, while early defoliation increased the CP content of the grain. The dry weight of forage removed at the milking and soft dough stages was lower than the subsequent reduction in grain yield, whereas, defoliation at the hard dough stage resulted in dry weight of the forage removed almost equal to or higher than the reduction in grain yield. Fresh fodder yields were increased by thinning (Adhikari, 1990) at 30 days after emergence and by increasing plant density and were highest when plants were decapitated at 30 days after 50 per cent silking. Grain yield was not significantly affected by detopping. Dry matter (DM) and N-content of the removed tops decreased with delayed date of cutting. DM of forage harvested decreased from 1.69 to 1.26 t/ha with delay in detopping date (Sampet and Donthawatnadda, 1978). Detopping of maize plants above the cob at grain moisture content of 40-48 per cent, accelerated maturation and increased translocation of nutrients to grain and increased yield by 350-750 kg/ha (Zhuk and Yunaev, 1975).

Grain Yield and Yield Characters

Defoliation during flowering at 70-80 DAS caused the greatest yield reduction of 149 kg/ha compared to untreated control (5.36 t/ha). Yield components affected by defoliation during silking were grain number/plant, ear number per plant, length of upper ear and 100-grain weight. There were significant correlation between yield and ear number per plant, length of upper ear, number of kernel rows in the upper ear, grain number per plant and 100 grain weight. The most critical period of development was the flowering stage (Diaz, 1983). Removal of all the leaves from above or below the ear of a maize plant at the tasseling stage (Dhopte and Lall, 1981) decreased the number of grains per ear by 27.9 and 39.0 per cent, ear dry weight by 30 and 36.9 per cent and 1000 grain weight by 5.05 and 1.62 per cent respectively. The ear was not produced with complete defoliation. Defoliation reduced weight of ears, grains and total above ground dry matter, harvest index and grain moisture content (Remison and Omueti, 1982), however, crude protein (CP) content increased especially with complete defoliation. Removing the top five, middle three and lowest eight leaves gave grain yield of 38, 36 and 26 per cent of the control with all leaves, respectively, and these yields were equivalent to 166,488 and

86 g grain yield/m^2 of remaining leaf surface. Grain yield per unit area was greater for upper than for lower leaves. Accumulation of DM in grain increased with increasing plant age. Relative photosynthetic potential decreased as defoliation treatments were applied at weekly intervals after silking. Yield reduction due to partial defoliation was not proportionate to the leaf area removed (Sanchez and Olivere, 1973). Generally leaf or stem removal reduces grain yield by 22-44 per cent. Leaves above the ear have a greater photosynthetic efficiency than those below. Removal of lower 5 leaves during the period between tassel emergence and silking, when most plants had 15 surviving and active leaves, resulted in an average reduction of 11 per cent in grain yield. Removal of the top 5 leaves reduced grain yield by 20 per cent, when all the leaves were removed the grain yield was reduced by 24 per cent. This yield reduction was due to mainly to a reduction in the average number of years per plant (Shivraj, 1972). Removal of lower four leaves of maize after 50 to 65 days of sowing, increased the grain yield by 1.36 to 1.55 q/ha (4.4 to 5.1 per cent) over control (Kler and Singh, 1991). Further, advantage of about 5.8 per cent harvest index was found possible through detopping of maize plant after cob formation (Hari *et al.*, 1995). It is also suggested (Hari *et al.*, 1995) that besides increasing grain yield detopped part of plant can also be utilized as green fodder for animals. Though no effect of timing of defoliation on grain yield was observed (Adhikari, 1990) but lower leaf detachment 20 days after and above leaf decapitation 30 days after 50 per cent silking resulted in reduced grain yield.

Conclusion

Reduction of assimilate supply by defoliation before the start of kernel growth (12 days after mid silking) substantially reduced the number of kernels per ear. However, defoliation after the onset of linear kernel growth (24 days after mid silking) did not have such effect. It is clear that maize has capacity to alter the number of kernels per ear in order to compensate for reduction in assimilate supply that occurs during the first 3 weeks after mid silking. After this time compensation for reduced assimilates availability is shown by reduced kernel growth rate and decreased duration of grain fill and also with enhanced remobilization of soluble carbohydrates from vegetative organs such as stem. Defoliation undoubtedly reduced the supply of photosythates available for distribution to the developing grains as indicated by substantial decline in internode soluble carbohydrates concentration. However, the rate of decline was dependent upon the time of defoliation treatment. Defoliation 12 days after mid silking resulted in a slower decline of internode carbohydrates concentration, possibly due to sink demand compensation through kernel abortion and reduced growth rate, while the defoliation 24 days after mid silking resulted in a more rapid decline in internode carbohydrate concentration, since sink demand compensation by kernel abortion no longer occurred. Regardless of the time of defoliation, cessation of kernel growth appeared to be associated with the time when carbohydrate concentration of the internode reached 2 per cent. In general, mature kernel weight increased in the order of apical < mid-ear < basal, but complete defoliation reduced the magnitude of these difference, with lower photosynthate supplies, apical kernels had longer lag phase duration (LAGP) than mid-ear kernels and longer effective grain filling period (EFPD) than basal kernels, and mid ear kernels which often had higher rate of dry matter accumulation (RDMA) than basal kernels. Under stress conditions, apical, unlike basal or mid ear kernels, compensate for reduced RDMA with longer EFPD. With higher photosynthate supplies, apical kernels had shorter EFPD than basal kernel and shorter LAGP than mid ear kernels. Thus, the pattern of changes in kernel mass from base to apex of the ear in response to available photosynthates was generally consistent with the theory that the assimilate movement into the kernel is controlled by an osmotic concentration gradient. Plants defoliated during vegetative growth were able to reallocate dry matter such that within 7 days the shoot-root ratio of the defoliated plants were very close to control. As the growing season progressed, those defoliated treatments that had a shoot-root ratio similar to the control produced yields similar to the control. Increase in leaf area may be relatively little affected by a shortage of assimilates in the plant, because photosynthesis by an expanding leaf may supply much of its requirement for assimilate and because the area to dry weight ratio of leaves may increase when assimilates are in short supply. It seems clear that plants can suffer considerable destruction of leaves near the middle of the vegetative phase, may be by hail or other biotic stresses, which occurred in some seasons, without grain yield being more than moderately decreased. The results suggest the small effect due to a loss of leaf area on the growth of younger leaves and on grain number and further decreased leaf area at flowering phase resulted in a less than proportional decrease in the amount of dry matter produced after flowering. This could be explained by an increased leaf efficiency with decreased leaf area, because of decreased mutual shading of leaves. The greater leaf efficiency in defoliated than in untreated plants might also have been because of the fact that most of the removed leaf area was compensated from leaves below the ear, including leaves near the bottom of the plant which export assimilates more slowly than those higher up.

It is now well-established that the stem of the maize plant can serve as a reservoir of labile non-structural carbohydrates in the post flowering period, and that some of these stem reserves can be mobilized as sugars to contribute for grain filling. Variations in stem weight during grain filling are paralleled by fluctuations in stem soluble solid content, including sugars, which comprise a large part in the weight changes. Radio-carbon studies have confirmed that retranslocation from stem to grain actually occurs.

Acknowledgement

The first author gratefully acknowledges the financial support provided by CSIR (Human Resource Development Cell) for Extra Mural Research as Senior Research Fellowship.

References

Adhikari, K., 1990. Response of maize to different planting densities, foliage removal practices and thinning times. *J. Institute Agric. Animal Sci.* 11: 105-111.

Afuakwa, J.J., Crookstan, R.K. and Jones, R.J., 1984. Effect of temperature and sucrose availability on black layer development in maize. *Crop Sci.* 24: 285-288.

Allison, J.C.S., Wilson, J.H.H. and Williams, J.H., 1975. Effect of defoliation after flowering on changes in stem and grain mass of closely and widely spaced maize. *Rhodesian J. Agric Res.*, 13: 145-147.

Anderson, R.C., Darrah, L.L. and Zuber, M.S., 1979. Genotypic differences in tolerance to stress imposed by defoliation and plant density in three maize hybrids. *East African Agriculture and Forestry Journal*, 45: 69-73.

Antonielli, M. and Venanzi, G., 1979. Structural properties of the rachis and hypsophyll of the maize ear. *Plant Sci. latter.*, 15: 301-304.

Barnett, K.H., 1982. Physiological observations of maize genotype differing in source-sink ratio. *Dissertation. Abstracts International*, 42: 2638.

Barnett, K.H. and Pearce, R.B., 1983. Source-sink ratio alteration and its effect on physiological parameters in maize. *Crop Sci.*, 23: 294-299.

Boyle, M.G., Boyer, J.S. and Morgan, P.W., 1991. Stem infusion a liquid culture medium preventing reproductive failure of maize at low water potential. *Crop Sci.*, 31: 1246-1252.

Breithaupt, J.H., 1995. Studies on the control of the Asian corn stem borer *Ostrinia furnacalis* Guinea in the Markham Valley, Papua, New Guinea. *Mitteilungn der Deutschen Gesellschaft fur Allgemeine und Angewandte Entomologie*, 10: 291-295.

Britz, G.D., 1982. The effect of defoliation at various growth stages on maize grain yield. *Crop production*, 11: 85-91.

Cantrell, R.G. and Geadelmann, 1981. Contribution of husk leaves to maize grain yield. *Crop Sci.*, 21: 544-546.

Chen, H.Y., Chu, T.M. and Li, C.C., 1978. Studies on source-sink relationships in maize and sorghum. 1. Control of source. *J. Agric. Assoc.* of China. 104: 13-24.

Conti S. and Landi, P., 1981. Effects of defoliation in relation to leaf habit and plant density in maize hybrids. *Revista di Agronomia*, 15: 151-155.

Crookston, R.K. and Hicks, D.R., 1978. Early defoliation affects corn yields. *Crop Sci.*, 18: 485-489.

Crookston, R.K. and Hicks, D.R., 1988. Effect of early defoliation on maize growth and yield: An eleven year perspective. *Crop Sci.*, 28: 371-373.

Culy, M.D., Edwards, C.R. and Cornelius, J.R., 1992. Minimum silk length for optimum pollination in seed corn production fields. *J. Produc. Agric.*, 5: 387-392.

Dhopte, A.M. and Lall, S.B., 1981. Contribution of upper and lower leaves in cob development of maize. *Punjabrao Krishi Vidyapeeth Research Journal*, 5: 113-116.

Diaz, A.C., 1983a. Effect of defoliation on a white hybrid maize cultivar. *Revista del Instituto Colombiano Agropecuario*, 18: 1-8.

Diaz, A.C., 1983b. Response of maize ICAH-302 to defoliation. *Revista del Institute Colombiano, Agropecuaria*. 18: Extraordinario, 393-400.

Douglas, J.A., Kain W.M. and Dyson, C.B., 1981. Effect of time and extent of defoliation on grain yield of maize in relation to cosmopolitan armyworm damage. *New Zealand J. Agric. Res.*, 24: 247-250.

Duncan, W.G., Hatfield, A.L. and Ragland, J.L., 1965. The growth and yield of cron: II. Daily growth of corn kernels. *Agron. J.*, 57: 221-222.

Duncan, W.G., Williams, W.A. and Loomis, R.S., 1967. Tassels and the productivity of maize. *Crop Sci.*, 7: 37-39.

Edje, O.T., 1984. *Effect of tassel removal and defoliation of maize on yield of maize and bean grown on monoculture and in association*. Research Bulletin Bunda College of Agriculture, Univ. of Malawi, 12: 69-85.

Edmeades, G.O. and Daynard, T.B., 1979. The relationship between final yield and photosynthesis at flowering in individual maize plants. *Can J. Plant Sci.*, 59: 82-84.

Edmeades, G.O. and Lafitte, H.R., 1993. Defoliation and plant density effects on maize selected for reduced plant height. *Agron. J.*, 85: 850-857.

EI-Tuhami, M.K., 1982. Relation between removed plant organ and reduction in grain yield due to defoliation in maize plant. *Ann. Agric. Sci. Moshtohor*, 18: 360.

Elmore, R.W. and Doupnik, B. Jr., 1995. Corn recovery from early season frost. *J. Produce. Agric.*, 8: 199-203.

Elsahookia, M.M. and Wuhaib, K.M., 1988. Performance of clipped maize. *J. Agron. Crop Sci.*, 160: 10-13.

Espinosa, C.A. and Tadeo, R.M., 1995. Detasseling of maize single hybrids and its effect on seed production. *Revista Fitotecnia. Mexicans*, 18: 9-15.

Fairy, N.A. and Daynard, T.B., 1978. Quantitative distribution of assimilates in component organs of maize during reproductive growth. *Can. J. Plant Sci.*, 58: 709-717.

Fujita, K., Furuse, F., Sawada, O. and Bandara, D., 1994. Effect of defoliation and ear removal on dry matter production and inorganic elements absorption in sweet corn. *Soil Sci. Plant Nutr.* 40: 581-591.

Fujita, K., Sato H., Sawada, O. and Sendo, S., 1995. Husk leaves contributions to dry matter and grain production as well as N distribution in flint corn genotypes differ in husk leaf area. *Soil Sci. Plant Nutr.* 41: 587-596.

Giesbrecht, J.E., 1977. *Update on recommended practices for corn production. Manitoba Agronomists.* Technical and Scientific Paper presented at the Ann. Confer. of Manitoba Agronomists. Dec. 13-14, 1977, pp. 46-48.

Girardin, P., 1985. Calorific energy distribution within the corn plant. *Agron. J.*, 77: 171-174.

Hicks, D.R., Nelson, W.W. and Ford, J.H., 1977. Defoliation of corn hybrid adapted to the northern corn belt. *Agron. J.*, 69: 387-390.

Hill, M.G. and Atkins, A.W., 1982. Effect of defoliation by cosmopolitan armyworm, *Mythimna separata* Walker. on maize yield. *New Zealand. J. Agric. Res.*, 25: 251-254.

Hsu, F.H., 1978. Study on the effects of early defoliation on the agronomic traits of maize under different population densities. *J. Taiwan Livestock Res.*, 11: 63-71.

Hunter, R.B., Daynard, T.B., Hume, D.J., Tanner, J.W., Curtis, J.D., and Kannenberge, L.W., 1969. Effect of tassel removal on grain yield on corn. *Crop Sci.*, 9: 405-406.

Hunter, R.B., Martimare, G.G. and Kannenberge, L.W., 1973. Inbred maize performance following tassel and leaf removal. *Agron. J.*, 65: 471-472.

Jain, T.C., 1971. Contribution of stem, leaf leminae and ears to dry matter production of maize after ear emergence. *Indian J. Agric. Sci.*, 41: 579-583.

Johnson, R.R., 1978. Growth and yield of maize as affected by early season defoliation. *Agron. J.*, 70: 995-998.

Jones, M. and Henderlong, P.R., 1981. *The effect of frost and mechanical defoliation on corn development and yield.* Agronomy Abstracts, 73rd annual meeting, American Society of Agronomy. 106-107.

Jones, R.J. and Simmons, S.R., 1983. Effect of altered source-sink ratio on growth of maize kernels. *Crop Sci.*, 23: 129-134.

Javorkova, A., 1990. Effect of defoliation on maize seed yields. *Rostlinna Vyroba*, 36: 727-731.

Kajil, L., Sataric, I., Pesek, L., Herman, M. and Hartmann, J., 1991. *Effect of defoliation on kernel development and maturity in maize. Bio-metric in plant breeding.* Proceedings of the eight meeting of the Eucarpia Section Biometrics on Plant Breeding. July 1-6, Brno Czechoslovakia, pp. 355-366.

Kayitmazbatir, N. and Uguz, M.A., 1983. Growth and yield of irrigated maize as affected by early season defoliation in the Konya district. Konya-Bolge-Topraksu-Arastirma, *Enstitusu Yayinlari* 85: 25.

Kelley, T.G., Jackobs, J.A., Shanmuga, S. (Eds.), 1985. Yield stability of sole-crop and intercrop planting systems: *Soyabean in tropical and sub-tropical cropping systems*. pp. 49-56.

Kler, D.S. and Singh, S., 1991. Effect of source-sink relationship on maize grain yield. *Environ. Ecol.*, 9: 531-532.

McMillan, W.W., Wiseman, B.R., Widstrom, N.W., 1980. Dent and sweet corns: influence of defoliation; plant age and genotype on yield. *J. Georgia Ento. Soc.*, 15: 373-377.

Menezes, NL-de, Cicero, S.M., De-Menezer, N.L., 1994. Effect of early detasseling of maize plants on leaf area and seed yield and quality. *Pesquisa Agropecuaria Brasileira*, 29: 733-741.

Mostert, A.J. and Marais, J.M., 1982. The effect of detasseling on the yield of irrigated maize. *Crop Produc.*, 11: 163-167.

N'guettia, R.Y., Karbrah, Y. and Angbonou, K., 1991. Effect of intensity and timing of defoliation on growth, yield component and grain yield of maize. *Expt. Agric.*, 27:137-144.

Palavershich, B., Roits, M., Parlov, D. and Peinovich, M., 1990. Possibilities of utilizing induced stress in maize breeding. *Informatsionn yi Byulletin po Kukuruze*, 8: 29-43.

Palmer, A.F.E., 1969. *Translocation pattern of C^{14} labelled photosynthates in the corn plant during the ear filling stage*. Ph.D. dissertation. Cornell Univ., Ithaca, N.Y.

Pearson, C.J., Larson, E.M., Hesketh, J.D. and Peters, D.B., 1984. Development and source-sink effects on single leaf and canopy carbon-di-oxide exchange in maize. *Field Crop Research*, 9: 391-402.

Philip, J. and Gautam, R.C., 1995. Response of maize to plant population and tassel removal. *Indian J. Ecol.*, 22: 130-131.

Pinter, L. and Kalman, L., 1980. The effect of drastic detasseling on seed production in maize. *Movenytermeles*, 29: 385-391.

Pinter, L. and Kalman, L., 1981. A quick method for determining yield stability and stalk strength in maize hybrids. *Zeitschrift fur pflanzenquchtang*, 87: 139-143.

Pinter, L. and Kalman, L., 1982. Effect of leaf area reduction on stand ability and premature death of leaves in maize hybrids with different genotypes. *Acta Agronomice Scientiarum Hungaricae*, 30: 61-67.

Piper, E.L. and Weiss, A., 1989. *Extending CERES–maize to handle abiotic and biotic stresses*. 19th conference on agriculture and forest meteorology pp. 52-53.

Rawsan, H.M. and Evans, L.T., 1971. The contribution of stem reserves to grain development in a range of wheat cultivars of different height. *Aust. J. Agric. res.*, 22: 851-863.

Remison, S.U., 1982. Time of leaf blade removal on the performance of maize. *Maydica*, 27: 123-133.

Remison, S.U. and Omueti, O., 1982. Effect of N-application and leaf clipping after mid-silking on yield and protein content of maize. *Can. J. Plant Sci.*, 62: 777-779.

Salvador, R.I. and Pearce, R.B., 1988. Husk removal and its effects on maize grain yield. *Crop Sci.*, 28: 961-964.

Sampet, C. and Oonthawatnadda, T., 1978. Effect of topping time on grain yield and removed forage in maize. *Thai. J. Agric. Sci.*, 11: 165-170.

Sanchez, C.H. and Olivera, A.C., 1973. Production of dry matter and estimation of photosynthetic potential due to artificial defoliation in maize. *Anales cientificos*, 11: 189-198.

Sawada, O., Junki, I. and Fujita, K., 1995. Characteristics of photosynthesis and translocation of C13 labelled photosynthate in husk leaves of sweet corn. *Crop Sci.*, 35: 480-485.

Schusseler, J.R. and Westgate, M.E., 1991. Maize kernel set at low water potential: I sensitivity to reduced assimilates during early kernel growth. *Crop Sci.*, 31: 1189-1195.

Shapiro, C.A., Peterson, T.A. and Flowerday, A.D., 1986. Yield loss due to simulated hail damage on corn: a comparison of actual and predicted values. *Agron. J.* 78: 585-589.

Shekhawat, G.S., Singh, G.B. and Jain, R.K., 1964. Preliminary observation on effect of detasseling in maize. *Curr. Sci.*, 33: 57-58.

Shinano, T., Osaki, M. and Tadano, T., 1993. C^{14} allocation of C^{14} compounds introduced to a leaf to carbon and N-components in rice and soyabean during ripening. *Soil Sci. Plant Nutr.*, 40: 199-210.

Shivraj, A., 1972. *An Introduction to Physiology of Cereal Crops*. EmKay Publications, Delhi.

Simmons, S.R. and Jones, R.I., 1985. Contribution of presilking assimilates to grain yield in maize. *Crop Sci.* 25: 1004-1006.

Singh Hari, J., Kler, D.S., Kolar, J.S. and Kaur, H., 1995. Influence of crop geometry, fertility levels and detopping on grain yield and yield contributing character of maize. *Environ. Ecol.*, 13: 827-833.

Singh Hari, J., Kler, D.S., Kolar, J.S., Kaur, H. and Singh, S., 1995. Effect of plant geometry, fertility levels and detopping on dry matter production and harvest index of maize. *Environ. Ecol.*, 13: 680-685.

Singh, R.P. and Nair, K.P.P., 1975. Defoliation studies in hybrid maize. 1. grain yield, quality and leaf chemical composition. *J. Agric. Sci. UK.* 85: 241-245.

Solomonovitz, S., Levy, Y. and Pateky, J.K., 1992. Yield loss of sweet corn cultivars in response to defoliation and to infection by *Exserrohilum turcicum. Phytoparasitica*, 20: 113-121.

Tanaka, A. and Fujita, K., 1970. Studies on the nutrio-physiology of the maize plant. 7. Analysis of dry matter production from the source-sink concept. *Journal of the Science of Soil and Manure*, 41: 509-513.

Tanaka, A. and Fujita, K., 1979. Growth, photosynthesis and yield components in relation to grain yield of field bean. *J. Fac. Agric. Hokkaid Univ.*, 59: 145-238.

Tanaka, A. and Yamaguchi, J., 1971. *Studies on nutrio-physiology of maize plant*. 6 Respiratory rate and dry matter production during growth. *Field Crop Abstr.* 26: 6332.

Tanaka, A. and Yamaguchi, J., 1972. Dry matter production, yield components and grain yield of the maize plant. *J. Fac. Agric. Hokkaido Univ.*, 57: 71-132.

Teyker, R.H., Moil, R.H. and Jackson, W.A., 1991. Differences among prolific maize hybrid in the effects of lower leaf removal. *Maydica*, 36: 1-10.

Tollenaar, M. and Daynard, T.B., 1978. Dry weight, soluble sugar content and starch content of maize kernel during the early post-silking period. *Can. J. Plant Sci.* 58: 199-206.

Tollenaar, M. and Daynard, T.B., 1978. Effect of defoliation on kernel development in maize. *Can. J. Plant Sci.*, 58: 207-212.

Tollenaar, M. and Daynard, T.B., 1982. Effect of source-sink ratio on dry matter accumulation and leaf senescence in maize. *Can. J. Plant Sci.*, 62: 855-860.

Vasconcellos, C.A., Magalhaes, P.C., Duraes, F.O.M. and Fernandes, F.T., 1995. Detasseling practices on tropical maize and their effect on mineral nutrition and nutrition efficiency. *Pesquisa Agropecuaria Brasileira*, 30: 353-358.

Vasilas, B.L. and Seif, R.D., 1985. Effect of defoliation on the timing of anthesis and silking of maize inbred lines. *Maydica*, 30: 427-435.

Vasilas, B.L. and Seif, R.D., 1985. Pre-anthesis defoliation effects on six corn inbred. *Agron. J.*, 77: 831-835.

Vasilas, B.L. and Seif, R.D., 1985. Defoliation effects on two corn inbreds and their single cross hybrids. *Agron. J.*, 77: 816-820.

Vasilas, B.L. and Seif, R.D., 1986. Effect of defoliated corn inbred on seed quality. *J. Seed Technology*, 10: 121-128.

Velasco, L.R.I., 1992. Questions about detasseling as a control technique against the Asian corn borer *Pyralidae, Lepidoptera. Philippine Entomologist*, 8: 1103-1116.

Weiss, A. and Piper, E.L., 1992. Modifying the response of defoliation during vegetative growth in CERES–maize. *Agriculture System*, 40: 379-392.

Westgate, M.E. and Boyer, J.S., 1985. Carbohydrates reserves and reproduction development at low leaf water potential in maize. *Crop Sci.*, 25: 762-769.

Yang, J., Yu, B.G., Wang, W.C., Song, L.R. and Li, K.Y., 1994. The effect of the reproduction organs in maize on senescence of adjacent leaves under high yielding conditions. *J. Nanjing Agric. Univ.*, 17:12-16.

Yoshida, S., 1972. Physiological aspects of grain yield. *Ann. Rev. Plant Physiol.*, 23: 437-464.

Zhao, K.F., 1981. Effect of leaves in different positions in maize on the grain yield and the photosynthatic properties of leaves after the development of female flowers. *Acta Agronomica Sinica*,7: 259-266.

Zhuk, Ya.M. and Yunaev, A.A., 1975. Pre-harvesting removal of top portions of maize plants. *Kukuruza*, 2: 25-26.

Chapter 58

Wastewater Quality of Major Drains of Delhi Draining Wastewater to River Yamuna and Assessment of Water Quality of River Yamuna at Delhi Stretch

P.K. Behera[1], R.C. Trivedy[1] and P.C. Mishra[2]

[1]*Central Pollution Control Board, East Arjun Nagar, Shahadra, Delhi*

[2]*Department of Environmental Sciences, Sambalpur University, Jyoti Vihar – 768 019, Orissa, India*

ABSTRACT

A detailed monitoring of the quality of wastewater discharged to river Yamuna from five major drains at Delhi as well as quality of river water at Delhi stretch was done over a period of two years during 1993-95. Twenty-eight parameters including heavy metals and pesticides content were selected for the study. Total coliform and faecal coliforms were also estimated. Based on few important water quality parameters, water quality index of river at sampling points were calculated which reveals that the river water beyond Wazirabad is unfit for bathing, swimming and even recreation.

Introduction

Yamuna River System

The Yamuna is the largest tributary of the river Ganga. The main stream of the river Yamuna–also called Kalindi in some reaches originates from the Yamunotri glacier near Badar Punch

(30°58′ N, 78°27′ E) in the Mussoori range of the Lower Himalaya at an elevation of about 6320 metre above mean sea level (MSL) in the district of Uttar Kashi of Uttarakhand. After combining with tributaries Rishi Ganga, Kunta, Hanuman Ganga, Tons and Giri, Yamuna enters the plains at Tajewala. During its journey of nearly 1200 kms from Tajewala to its confluence with the Ganga at Allahabad the Yamuna is joined by the other tributaries *viz.* Hindon, Chambal, Sindh, Betwa and Ken.

The total catchment area of the Yamuna river system is 3,45,848 square kilometres. It covers parts of Uttarakhand, Uttar Pradesh, Himachal Pradesh, Haryana, Rajasthan and Madhya Pradesh States and the entire Delhi. On percentage basis, contributions area-wise of the Yamuna are: U.P. 21.5, Himachal Pradesh 1.6, Haryana 6.1, Rajasthan 29.8, Madhya Pradesh 40.6 and Delhi 0.4. Sub-basin-wise, the Hindon contribution is 2, the Ken 7, the Sindh 7.5, the Betwa 13.9 and the Chambal 40.5. The balance 29.1 per cent is accounted for by direct drainage into the Yamuna or the smaller tributaries.

A number of medium and large scale irrigation projects, some based on storage and some on mere diversion, draw water from the Yamuna and its tributaries. While water drawn for municipal and industrial purposes is largely, if not entirely, returned to river system itself in the form of wastewater, that drawn for irrigation in mostly consumed in evapo-transpiration. At Tajewala itself, nearly entire flow (6025 million cubic metres) of Yamuna is diverted for irrigation through the Western Yamuna canal for Haryana and the Eastern Yamuna Canal for Uttar Pradesh. The other important water abstraction points are Wazirabad, Wazirabad-Okhla stretch, Okhla-Etwah stretch and Etawah.

In terms of economic activity the Malwa region in Madhya Pradesh (M.P.) and Braj Bhoomi in Uttar Pradesh (U.P.) and Haryana have for centuries been amongst the most prosperous agricultural areas. It has large industrial complexes at Kota, Gwalior, Indore, Nugda, Khetri etc. Paper mill at Saharanpur, a large urea factory at Panipat and oil refinery at Mathura, and the Ghaziabad-Delhi-Faridabad industrial complex developments are in the Yamuna basin. The water quality of Yamuna river in the Himalayan segment and the diluted segment is comparatively good. However, due to heavy abstraction and discharge of pollutants into the river system, there are critical segments which require pollution abatement measures to improve the water quality of the river. These segments with the causes of pollution are:

Wazirabad to Okhla

Domestic and industrial wastewater of Delhi.

Okhla to Vrindavan

Domestic wastewater from Delhi and industrial effluents from Saharanpur, Ghaziabad, Noida etc.

Vrindavan to Mathura

Domestic wastewater and industrial effluent from dyeing and printing industry of Vrindavan and Mathura.

Mathura to Etwaha

Domestic wastewater from Agra and Etawah.

Environmental Status of Delhi

Area

The National Capital Delhi is located on the bank of sacred river Yamuna. It has an area of about 1485 sq km and lies between latitudes 20°14′15″ and 28°53′00″ (North) and longitudes 76°45′30″

and 77°21′30″ (East). It is situated at 250 metres above mean sea level. Delhi is surrounded on the north, south and west by the State of Haryana and on the east by the State of Uttar Pradesh.

Climate

The climate of Delhi is influenced by its inland position with the desert of Rajasthan to the west and southwest and Gangetic plains of U.P. to east. Extreme dryness with an intensely hot summer and cold winter are the characteristics of the climate. The year can be broadly divided into four seasons. The cold season starts in late November and extends to about the beginning of March. This is followed by the hot season which lasts till about the end of June when the monsoon arrives. The monsoon continues till the last week of the September. The two post-monsoon months, October and November continue, as transition period from monsoon to winter conditions. The climate of territory is semi-arid. The mean maximum temperature range from 27.6°C to 40.2°C. The mean minimum temperature range from 7.5°C to 22.2°C. The annual average rainfall is 754 mm.

Population

The population of Delhi has increased several folds after the partition (1947) making this ancient city as one of the largest cities of the world. This rapid urbanisation in turn enhanced the thrust on basic amenities to cope-up with the living standards. As per the census report of India, the population of Delhi was 9.42 million in 1991, registering an increase of 51.45 per cent in the decade 1981-1991. The population of Delhi in 1998 has been estimated to be 11.5 million, which includes 4 to 5 lakhs of floating population. Of the total population, 90 per cent is residing in urban areas and remaining 10 per cent in urban villages. It is about 13 million in year 2000.

Water Supply and Sewerage System

At present the Delhi Jal Board (DJB) is equipped to produce 2633 MLD of water and this water supply is further augmented by abstraction of groundwater for domestic consumption. Normally sewage generated is 70-80 per cent of total water supply and sewage generation works out to be 2107 in MLD. However, the flow of wastewater in various drains is observed to be 2720 MLD thus confirming the use of groundwater for drinking and other purposes. The sewage treatment capacity was 1372 MLD during 1999. The average inflow at the Sewage Treatment Plants was about 953.4 MLD and accordingly about 1766 MLD of untreated sewage finds access into various drains and ultimately into the river.

Water Supply

The habitation of Delhi is in the following categories: (a) Unauthorised/regularised colonies (b) Planned development/approved colonies (c) Re-settlement colonies (d) Urban villages (e) Unauthorised and non-regularised colonies (f) J.J. clusters. (g) Rural villages. Water supply has been provided to areas (a) to (d) and (g) above, subject to technical feasibility. Water supply is also being extended to unauthorised/non-regularised colonies by providing skeleton water supply either from the available distribution system nearby or through tube wells/deep bore hand pumps. In J.J. clusters water supply is either through public water supply or deep bore hand pumps and in the absence of either of the above through water tankers, in the far flung areas.

Sewage System

Sewerage facilities have been extended to and being extended to all planned development approved colonies whether developed by DDA, MCD, NDMC, CPWD.

1. Out of 567 unauthorised colonies sewage facilities exist in 293 colonies and work is in progress by DJB in 65 colonies.
2. Out of 44 resettlement colonies, 32 have functional sewerage system and in another 4 work is in progress.
3. Out of 126 urban villages sewerage system exists in 88, work is in progress in another 2 urban villages. The sewerage system is yet to be provided to remaining 35 urban villages.
4. There are 219 rural villages whose population is estimated around 6 lakh, where drinking water supply has been provided. However sewerage system has not been provided in these rural areas as it is not technically feasible. The wastewater from rural belts find its way into storm waste drains.
5. There are 1017 unauthorised/non-regularised colonies which have no sewerage facilities.
6. As per policy in vogue, sewerage facilities are not to be provided in J.J. clusters due to meagre rate of water supply. As such sewage generated from such areas contribute to the flow in storm water drains ultimately out-falling into river Yamuna. The daily flow of wastewater currently received and handled at the seven sewage treatment plants and some oxidation ponds serving NCT Delhi is indicated in Table 58.1. Thus more than 60 per cent of the wastewater generated in NCT Delhi is reaching the river Yamuna via the sewers or open drains without receiving any treatment.

Table 58.1: Inflow at Sewage Treatment Plants

Sl. No.	*Location*	*Capacity in MGD*	*Daily Average Flow in MGD (Year 1999)*
1.	Okhla	124	104.35
2.	Keshopur	72	59.86
3.	Rithala	40	22.06
4.	Vasant Kunj	5.2	1.95
5.	Kondli	35.00	10.09
6.	Yamuna Vihar	10.00	1.66
7.	Coronation	10.00	7.61
8.	Oxidation ponds	6.00	2.47
	Total MGD	**302.20**	**210.05**

Industries and Industrial Wastewater

The growth of large industries is restricted due to introduction of planning restrictions in Delhi, but small scale industries have expanded rapidly. There was a sharp increase in industrial units from 18500 in 1961 to 93000 in 1993. Engineering, textiles, chemicals, electronics and electrical goods are the major categories of industries located in Delhi.

Delhi has 20 large, 25 medium and approximately 93000 small scale industries located at it's 28 planned industrial estates. Large and medium scale industries constitute less than 0.05 per cent of the total industries located in Delhi, but contribute 50 per cent of the total 300 million litres of wastewater generated every day (CPCB-1994). The large and medium scale industries have some facilities for effluent treatment and emission control but the small scale industries do not have such a set-up. The planned industrial estates in Delhi were mostly developed by the DDA and a few by CPWD. Though in a large number of estates, the effluent was originally out-falling into existing sewerage system but over the years the system has degenerated with the result that industrial effluents from most of the industrial estates are now flowing through open surface drains ultimately finding access into the river Yamuna. Some of the examples are Wazirpur Industrial Area, Rajasthan Udyog, Jhilmil Industrial Area etc.

Importance of River Yamuna to National Capital

Historically the city and the river have always remained inter-twined. The Yamuna originated at Yamunotri in the Himalayas and travels through the hilly areas of Himachal Pradesh and Uttar Pradesh. After entering plains it flows along the boarder of Haryana and Uttar Pradesh to enter the National Capital Territory of Delhi at Wazirabad. Within the Delhi territory the river flows from north to south. Although the NCT Delhi lies largely on the western bank of Yamuna, the eastern bank is fast developing and has attained significance economically. The main areas of old Delhi and New Delhi are on the right (or west) bank of the Yamuna, while the old Shahdra town is located on the left (or east) bank. The total length of the river from source to its confluence with the Ganga at Allahabad is 1376 kms. The length of the reach falling within the NCT Delhi is only 48 km. At Wazirabad where the river enters the NCT Delhi, it is totally trapped for the city water supply. The river leaves the Territory of Delhi and enter into Uttar Pradesh at Okhla barrage. During its course through the city nineteen number (19 Nos.) of drains meet the river and carry into it sullage and other wastewater from various parts of city.

Methodology

Selection of Sampling Points

River Sampling Stations

The river Yamuna enters into National Capital Delhi at Wazirabad barrage and leaves at Okhla barrage. Thus the present study area covers approximately a length of 22 kilometers all over Delhi stretch, starting from upstream points at Wazirabad barrage to down-stream point at Okhla barrage. Selection of sampling sites has been done primarily on the basis of major drains joining to the river and also barrage and bridges existing on the river Yamuna at NCT Delhi stretch. Attempt was made to select the sites at cleanest point (where contamination of river water by drain water is negligible *i.e.* Wazirabad barrage and before and after joining of major drains). Thus the selected river monitoring stations and distance from Wazirabad barrage were *(i)* Wazirabad barrage 0 km., *(ii)* ISBT bridge 8 kms, *(iii)* ITO barrage 15 kms, *(iv)* Nizamuddin bridge 17 kms, and *(v)* Okhla barrage 22 kms.

Drain Sampling Points

Based on pollution load as well as flow of the drain water, only the major five (5 nos.) drains were selected for monthly collection of wastewater and sediment samples for the present investigation. The selected monitoring drains were *(i)* Najafgarh drain, *(ii)* Civil mill drain, *(iii)* Power house drain, *(iv)* Sen nursing home drain, and *(v)* Barpulla drain.

Water and Wastewater Sampling

From drains and river monitoring points samples were collected every month from December 1992 to January 1996. In case of river water, the individual grab sample were collected in triplicate from midstream in 2.5 litre plastic bottles with screw cap carefully after rinsing the sample bottles two or three times with samples. The record of every sampling collected was marked with an appropriate inscription on the container with marking pencil. Samples were collected from a depth of about 30 cm. from surface. After collection, samples were immediately transferred to the laboratory. The pH and dissolved oxygen were measured at the sampling point and for remaining parameters the water samples were immediately preserved. The suitable containers, time and chemical of preservation were selected as per APHA (1991). For pesticide analysis, separate brown glass bottles (one litre capacity) were used in sampling. The samples for heavy metals analysis were collected in polythene bottle (one litre capacity) containing concentrated HNO_3, (1 ml/bottle). For TKN, NH_3-N and nitrate samples were preserved separate bottle containing concentrated H_2SO_4, (2 ml/lt).

Analytical Methods for Estimation of Water and Wastewater

The water and wastewater samples collected from river and drains were analysed by the methods of APHA (1991) Trivedy and Goel (1984) and Saxena (1987). The following physico-chemical parameters were analysed:

1.	pH	2.	Conductivity	3.	Total Solids
4.	Suspended Solids	5.	Dissolved Solids	6.	DO (RW)
7.	Oxygen Absorbance	8.	COD	9.	BOD
10.	TKN	11.	Nitrate-Nitrogen	12.	NH_3-N
13.	Phosphate	14.	Total Alkalinity	15.	Total Sulphate
16.	Chloride	17.	Total Hardness	18.	Calcium
19.	Magnesium	20.	Sodium	21.	Potassium
22.	Total Coliforms (RW)	23.	Faecal Coliforms (RW)	24.	Iron
25.	Cadmium	26.	Chromium	27.	Copper
28.	Nickel	29.	Zinc	30.	Total BHC
31.	Total DDT	32.	Total Endosulfan		

(RW = River Water)

Before the actual analysis all the preserved samples were allowed to regain the room temperature, well shaken and filtered for individual tests. The final tabulated values represented the mean of 3 replicates.

Physico-chemical

The pH was recorded on a digital pH meter (Systronic make model No. 361). Conductivity was recorded on a conductivity meter (Systronic make No. 304). The observed values were expressed in μ mhos/cm. Total Solids was determined by evaporating a fixed volume of sample at 104°C in a hot air oven. Total Dissolved form a part of the total solids residue. In natural water, the TDS consist mainly of bicarbonates, carbonates, sulphates, chlorides, nitrates and possibly phosphates of calcium, magnesium, sodium and potassium with traces of iron, manganese and other substances. Total dissolved solids were determined by filtering the sample, drying the filtrate and determining its weight by difference. Total suspended solids (TSS) are the retained residue on a standard filter paper after filtration of a well mixed sample. The residue was dried at 103-105°C.Total suspended solids were determined by filtering the sample, drying the residue and determining its weight by difference. Dissolved Oxygen was determined immediately after collection of sample in the field by using modified Winkler's method. The samples were taken in to the ground glass stoppered 300 ml Corning make

BOD bottles preventing any bubbles. 2 ml of saturated solution of Manganous sulphate and potassium iodide and potassium hydroxide solution were added after the settlement of the precipitate. It was dissolved in sulphuric acid and titrated against 0.025 N sodium thiosulphate solution. Biochemical Oxygen Demand refers to the quantity of oxygen required by bacteria and other micro-organisms in the biochemical degradation and transformation of organic matter (OM) under aerobic conditions. It is used as an index of organic pollution in water. More the oxidisable OM present in water, more the amount of oxygen required to degrade it biologically, hence more the BOD. BOD was evaluated by measuring the DO content of the sample before and after 5 days of incubation of 20°C in the dark. Preliminary dilution and aeration of samples were usually necessary to ensure that not all the oxygen is consumed during incubation. Chemical Oxygen Demand is the measure of oxygen required in oxidising the compounds present in water by means of chemical reactions involving oxidising substances such as potassium dichromate. Digestion of sample with $K_2Cr_2O_7$ and silver sulphate-sulphuric acid solution was done by closed reflux method using a Spectralab closed reflux condenser and the excess $K_2Cr_2O_7$ was titrated with standard ferrous ammonium sulphate (0.1 N) using 4 to 5 drops of ferroin indicator. Oxygen Absorbance was measured by acidifying the sample with sulphuric acid and adding potassium permanganate to absorb oxygen. The amount of oxygen absorbed by sample was calculated from the quantities of potassium permanganate added and the amount of left after the reaction by titrating with sodium-thiosulphate solution using starch as indicator. Total Kjeldahl Nitrogen was determined by Kjeldahl Distillation Method. The water/wastewater sample was digested with concentrated sulphuric acid (H_2SO_4), potassium sulphate (K_2SO_4) and mercuric sulphate ($HgSO_4$) as a catalyst. The digest was distilled with 40 per cent sodium hydroxide and the distillate was collected in 4 per cent boric acid with mixed indicator. The contents were then titrated with 0.01 N hydrochloric acid. Ammonical Nitrogen was determined by Nesslerization Method. The sample was adjusted to pH of 7.4 with phosphate buffer and distillated. The sample containing ammonia was treated with Nessler's reagent. The yellow brown colour produced was compared with that of standards. Nitrate Nitrogen was measured by Brucine Method. The sample containing nitrate ion reacts with Brucine-sulphanillic acid solution to form a yellow colour which was measured by spectrophotometer. Total Phosphate was estimated by Stannous Chloride Method. The sample was digested with concentrated HNO_3 and perchloric acid ($HClO_4$, 70-72 per cent), and neutralized with NaOH solution. The phosphate in neutralized and filtered sample reacted with ammonium molybdate and stannous chloride and developed blue colour was measured at 690 nm. Chloride was determined by Argentometric method involving titration of the sample against silver nitrate solution of 0.014 (N) strength using potassium chromate as an indicator. Sulphate was estimated by Turbidimetric method. The sample was treated with barium chloride and conditioning reagent. The sulphate ions were precipitated as barium sulphate crystals of uniform size in acid medium. Light absorbed by precipitate was measured by using a spectrophotometer at 420 nm. Alkalinity is the quantitative capacity of aqueous media to react with hydrogen ions. The alkalinity of natural water is normally due to the presence of bicarbonate, carbonate and hydroxide compounds of calcium, magnesium, sodium and potassium. Thus, alkalinity may be expressed as total alkalinity or alkalinity due to industrial bases. Alkalinity was directly determined by titration with 0.02(N) H_2SO_4 using phenolphthalein and methyl orange indicator. Hardness due to Calcium and Magnesium were determined by EDTA titrimetric method. For total hardness erichrome black-T was used as an indicator. For calcium hardness muroxide was used as an indicator. In this test, calcium and magnesium reacted with EDTA to form soluble complex and was indicated by the colour change of suitable indicators as mentioned above. Sodium and potassium in samples were measured at wavelengths of 589 and 769 nm respectively using a flame photometer. The sodium and potassium contents were expressed in mg/l after deducing the values from the standards curve. The pesticides namely DDT, BHC and endosulfan were extracted from water by liquid extraction methods and measured by Gas Chromatography (HP-5400, Hewlett Packard, USA). Heavy metals have the tendency to form complexes readily with the organic constituents. Thus, it is essential to destroy them by digestion with strong acids. After the preliminary acid digestion with strong acids, metals were determined by atomic absorption spectrometry. Acid eliminates interference from cyanides, nitrate, sulphide, sulphite, thiosulphate and thiocyanate. The heavy metals like cadmium, copper, iron, lead, nickel, chromium and zinc present in the water sample were analysed and determined by flame atomic absorption spectrophotometry by direct aspiration into air acetylene flame. 200 ml of well shaken water sample was digested with 2 ml of concentrated HNO_3 and evaporated gently to reduce it to 50 ml. The sample after filtration was stored in polythene bottles and analysed through an atomic absorption spectrophotometer (Model 703, Perkin and Elmer), provided with baseline corrections. The metal concentrations were calculated after appropriate blank correction and expressed in mg/l.

Bacteriological Examination of Water

Total coliforms and faecal coliforms were determined by using "Membrane filter technique for member of the coliforms group". The sampling of river water was done following standard methods (APHA, 1991) using sterilized top aluminium foil packed 300 ml capacity glass bottles. The samples were brought to the laboratory within eight hours of collection and were processed immediately. Taking sample volume of 10 ml, a serial dilution series was prepared in sterile buffered dilution water. A fixed volume of diluted sample was filtered with 0.45 μm in pore size sterile filter (Gelman, Millore and Whatman). For the filtration process, the sterilized filtration assembly (Paqualab System) was used. Immediately after filtration, the membrane filter was transferred by sterile forceps on to the solid M-FC agar (kept in petridishes), incubation was carried out by immersing the petridishes, sealed with lids and inverted inside incubation chamber at 44 ± 0.2°C for 24 hrs. After incubation, the faecal coliforms count was made on a filter with a 20-80 colony range. The faecal coliforms density was determined according to the following formula:

$$\text{Faecal coliforms (CFU)/100 ml} = 100 \times \text{counted bluish colonies/Sample test volume in ml}$$

$$\text{CFU} = \text{colony forming unit}$$

For total coliforms count, filtered membrane was transferred to solid M-Endo agar in petridishes and incubated at 37°C for 24 hrs. Total coliforms developed as darker metallic spleen shaped colonies were calculated as follows:

$$\text{Coliforms (CFU)/100ml} = \frac{100 \times \text{counted dark red metallic sheen colonies}}{\text{Total Sample test volume in ml}}$$

Assessment of Water Quality

The physico-chemical environment of water functions in one or more ways and influences the biotic components. Thus, it gives a picture of the environmental suitability of water for maintaining normal life. As such, the physico-chemical factor is important in elucidating some casual influences for the faunal composition and abundance of plankton, benthos and other aquatic life. Therefore, knowledge on physico-chemical characteristics of water is essential for proper utilization of riverine wealth (Kumar, 1997).

Monthly variation of physico-chemical factors in drain water and five river sampling points (averages of three year 1993-95) under observation has been shown in tables. Two ways analysis of variance between stations and within monitoring months are given for all physico-chemical parameters. The microbial assessment of water, heavy metals and pesticides contents of the river and drain water are also presented in detail. The Water Quality Index of the river water has also been calculated.

Results and Discussion

Physico-chemical Characteristic of Drain and River Water

pH of Drain and River Water

pH is the term universally used to express the intensity of the acid or alkaline condition of a solution. It is the negative logarithm of the hydrogen-ion concentration in moles per litre. The pH of natural water, controlled by the carbon dioxide, bicarbonate equilibrium, usually ranges from 4.0 to 9.0. In majority of cases, natural waters exhibit slightly basic (pH >7) condition. It is due to the presence of bicarbonates and carbonates ions. Acidity affects the chemical and biological processes taking place in water in following manner: *(i)* Dissociation of organic and inorganic, molecules, thereby changing toxicity; *(ii)* Changing water solubility of compounds, also influencing toxicity; *(iii)* A lower pH may induce enhanced corrosion; *(iv)* Different biological species show optimum performance at different degrees of acidity, leading to shifts in species composition (D.de Zwart and Trivedy, 1994). pH of a raw water used for domestic purposes is important in that, it affects taste, corrosivity, efficiency of chlorination and treatment and processes such as coagulation etc. (Manivasakam, 1989).

Table DWM-1: Monthly Variation of pH in Drain Water (Year 1993-1995)

Month	*NAJ-1* Mean ± SD	*CIV-2* Mean ± SD	*POW-3* Mean ± SD	*SEN-4* Mean ± SD	*BAR-5* Mean ± SD	*HIN-6* Mean ± SD
Jan.	7.47 ± 0.17	7.40 ± 0.08	7.57 ± 0.24	7.47 ± 0.24	7.63 ± 0.26	7.84 ± 0.09
Feb.	7.33 ± 0.09	7.47 ± 0.05	7.28 ± 0.06	7.30 ± 0.00	7.50 ± 0.14	8.18 ± 0.13
Mar.	7.33 ± 0.12	7.57 ± 0.12	7.47 ± 0.12	7.37 ± 0.05	7.47 ± 0.31	8.12 ± 0.33
Apr.	7.37 ± 0.12	7.20 ± 0.26	7.33 ± 0.25	7.33 ± 0.09	7.53 ± 0.26	8.12 ± 0.34
May	7.60 ± 0.14	7.53 ± 0.05	7.47 ± 0.12	7.53 ± 0.12	7.75 ± 0.18	8.09 ± 0.20
Jun.	7.40 ± 0.16	7.47 ± 0.12	7.47 ± 0.09	7.33 ± 0.19	7.39 ± 0.31	8.13 ± 0.16
Jul.	7.47 ± 0.31	7.57 ± 0.19	7.40 ± 0.22	7.63 ± 0.21	7.43 ± 0.12	7.72 ± 0.28
Aug.	7.40 ± 0.28	7.27 ± 0.21	7.27 ± 0.12	7.47 ± 0.12	7.27 ± 0.19	8.01 ± 0.11
Sep.	7.37 ± 0.12	7.30 ± 0.08	7.33 ± 0.09	7.33 ± 0.05	7.40 ± 0.14	7.81 ± 0.12
Oct.	7.43 ± 0.09	7.33 ± 0.12	7.40 ± 0.14	7.40 ± 0.14	7.39 ± 0.08	7.95 ± 0.42
Nov.	7.33 ± 0.26	7.37 ± 0.19	7.57 ± 0.24	7.53 ± 0.21	7.60 ± 0.29	7.50 ± 0.30
Dec.	7.50 ± 0.28	7.33 ± 0.12	7.40 ± 0.08	7.40 ± 0.14	7.50 ± 0.22	7.71 ± 0.11

Two -way ANOVA Test

Significant difference in pH between monitoring drains, F_1=18.86 ($P<0.01$).
Significant difference in pH between monitoring months, F_2=2.85 ($P< 0.05$).

Table DWM-1 reveals a pH range of 7.33 ± 0.26 (Nov.) to 7.5 ± 0.28 (Dec.) in Najafgarh drains, 7.2 ± 0.26 (Apr.) to 7.57 ± 0.12 (Mar.) Civil mill, 7.27 ± 0.12 (Aug.) to 7.57 ± 0.24 (Nov.) in Power house, 7.3 ± 0.00 (Feb.) to 7.63 ± 0.21 (Jul.) in Sen Nursing Home, 7.27 ± 0.19 (Aug.) to 7.75 ± 0.18 (May) in

Barpulla and from 7.50 ± 0.30 (Nov.) to 8.18 ± 0.13 (Feb.) of Hindon river cut. Analysis of variance showed that significant difference exists in pH between different sampling drains (F_1 = 18.86, P< 0.01) as well as between different monitoring months (F_2 = 2.85, P<0.05). It implies that the physico-chemical nature of effluents of different drains are not similar in nature.

Seasonal and annual variation of pH of the drains shows a minimum in summer (7.43 ± 0.17) and maximum in post-monsoon (7.45 ± 0.28) in Najafgarh drains, minimum in post-monsoon (7.35 ± 0.10) and maximum in summer (7.44 ± 0.19) in Civil mill, minimum in monsoon (7.33 ± 0.16) and maximum in post-monsoon (7.50 ± 0.19) in Power house, minimum in monsoon (7.37 ± 0.17) and maximum in winter (7.5 ± 0.23) in Barpulla drain, minimum in post-monsoon (7.52 ± 0.23) and maximum in winter (7.86 ± 0.26) in Hindon river cut. Annual averages of pH values were in order of Barpulla drain (7.49 ± 0.25) > Sen nursing home (7.43 ± 0.18) > Najafgarh (7.42 ± 0.21) > Power house (7.41 ± 0.19) > Civil mill (7.04 ± 0.18). Annual average pH values of Hindon river cut was higher (7.93 ± 0.32) in comparison to other five drains.

Table RWM-1 shows monthly averages (Mean ± SD) in pH variation at the five sampling stations along the river Yamuna at Delhi stretch. The pH ranged between 7.71 ± 0.26 (July) to 8.23 ± 0.12 (Feb.) at the site WAZ-1, 7.27 ± 0.19 (Feb.) to 7.63 ± 0.12 (Apr.) at ISB-2, 7.40 ± 0 (May) to 7.63 ± 0.21 (Oct.) at ITO, 7.30 ± 0.08 (Feb.) to 7.60 ± 0.16 (Sep) at the site NIZ-4 and 7.37 ± 0.17 (Feb.) to 7.87 ± 0.25 (Sep) at sampling station OKH-5. Analysis of variance shows a significant difference in pH between sampling stations (F_1=18.86, P<0.01) as well as seasons (F_2=0.85, P<0.05). This clearly signifies the effect of pollutants on amount of hydrogen ions along the Yamuna at Delhi stretch.

Table RWM-1: Monthly Variation of pH in River Water (Year 1993-1995)

Month	*WAZ-1* *Mean ± SD*	*ISB-2* *Mean ± SD*	*ITO-3* *Mean ± SD*	*NIZ-4* *Mean ± SD*	*OKH-5* *Mean ± SD*
Jan.	8.20 ± 0.16	7.47 ± 0.25	7.43 ± 0.37	7.37 ± 0.21	7.57 ± 0.09
Feb.	8.23 ± 0.12	7.27 ± 0.19	7.33 ± 0.09	7.30 ± 0.08	7.37 ± 0.12
Mar.	8.10 ± 0.08	7.50 ± 0.08	7.63 ± 0.12	7.30 ± 0.22	7.63 ± 0.17
Apr.	8.17 ± 0.12	7.63 ± 0.12	7.43 ± 0.05	7.43 ± 0.12	7.43 ± 0.05
May	8.17 ± 0.17	7.50 ± 0.08	7.40 ± 0.00	7.47 ± 0.09	7.57 ± 0.17
Jun.	7.95 ± 0.20	7.40 ± 0.14	7.47 ± 0.12	7.50 ± 0.08	7.50 ± 0.08
Jul.	7.71 ± 0.26	7.40 ± 0.14	7.47 ± 0.17	7.43 ± 0.09	7.50 ± 0.16
Aug.	7.76 ± 0.30	7.47 ± 0.12	7.53 ± 0.12	7.53 ± 0.09	7.57 ± 0.05
Sep.	7.76 ± 0.09	7.43 ± 0.26	7.53 ± 0.09	7.60 ± 0.16	7.87 ± 0.25
Oct.	7.83 ± 0.05	7.47 ± 0.05	7.63 ± 0.21	7.57 ± 0.09	7.63 ± 0.05
Nov.	8.03 ± 0.12	7.53 ± 0.12	7.67 ± 0.12	7.57 ± 0.17	7.63 ± 0.24
Dec.	8.13 ± 0.19	7.43 ± 0.33	7.43 ± 0.26	7.43 ± 0.12	7.63 ± 0.26

Two-way ANOVA Test

Significant difference in pH between river monitoring stations, F_1=18.86 (P<0.01).
Significant difference in pH between monitoring months, F_2=2.85 (P< 0.05).

Seasonal and annual variation in pH at all sampling stations varied from 7.71 ± 0.24 (monsoon) to 8.16 ± 0.15 (winter) at WAZ-1, 7.44 ± 0.24 (winter) to 7.48 ± 0.18 (summer) at ISB-2, 7.37 ± 0.27 (winter) to 7.65 ± 0.17 (pre-monsoon) at site ITO-3, 7.37 ± 0.16 (winter) to 7.57 ± 0.14 (post-monsoon) at

NIZ-4 and 7.51 ± 0.21 (winter) to 7.64 ± 0.24 (monsoon) at monitoring site OKH-5. The annual average pH values were in order of WAZ-1 (8.0) > OKH-5 (7.58) > ITO-3 (97.50) > ISB-2 (7.46) > NIZ-4 (7.46) in all river monitoring points.

In most of time the river water at Wazirabad barrage (unpolluted stretch) shows alkaline pH (7.7 to 8.20). Murugesan *et al.* (1994) reported an alkaline pH of water in river Tampraparni with minimum pH during August and maximum during May. A slight to moderate alkaline nature of most of the Indian river waters were reported by Vass *et al.* (1977), Rajesh (1981), Mitra (1982), Bhargava (1985), Venkateswarlu (1986) in river Jhelum, Datta *et al.* (1988) in Hoogly river, Sinha (1988) in river Damodar, Sharma *et al.* (1990), Sarwar and Refat (1991) in Doodhganga river, Isaraili and Ahmed (1993), Mishra (1999) in Brahmani river.

In the present study, pH values were comparatively low at site ISB-2, ITO-3, NIZ-4 and OKH-5 in comparison to site WAZ-1, which might be due to regular additions of slightly acidic (domestic and industrial mixed) effluents from major and minor drains. Singh and Bhowmik (1985) also reported similar result in case of the river Ganga and Kumar (1999) in the river Mayurakshi.

Conductivity of Drain and River Water

Conductivity is a measure of ability of a water sample to conduct an electric current proportional to the ionic strength of the water. This mostly depends upon the nature of various dissolved ionised substances, their actual relative concentrations and temperature. The standard unit of electrical conductivity is seimence per metre (1 s/m = 10 milli-mhos/cms). In order to avoid the fractions of results in small decimal fractions, a smaller unit, the milli siemence per metre (1 MS = 10 micro-mhos/cm) is generally used.

Table DWM-2: Monthly Variation of Conductivity (μ mhos/cm) in Drain Water (Year 1993-1995)

Month	*NAJ-1*	*CIV-2*	*POW-3*	*SEN-4*	*BAR-5*	*HIN-6*
	Mean ± SD	*Mean ± SD*	*Mean ± SD*	*Mean ± SD*	*Mean ± SD*	*Mean ± SD*
Jan.	666.7 ± 98.9	917.3 ± 188.6	933.0 ± 75.0	1366.7 ± 378.1	1174.0 ± 55.7	1027.7 ± 123.0
Feb.	816.3 ± 82.2	884.6 ± 53.5	815.3 ± 168.0	1670.7 ± 905.9	1148.3 ± 74.4	1010.0 ± 99.0
Mar.	805.0 ± 107.22	1128.3 ± 106.4	1096.0 ± 53.8	2023.3 ± 968.5	1294.0 ± 99.3	802.3 ± 448.6
Apr.	862.7 ± 90.0	926.7 ± 649.0	1330.3 ± 330.5	1531.7 ± 331.0	1237.7 ± 128.4	863.7 ± 68.0
May	790.0 ± 28.3	1044.01 ± 179.2	1028.0 ± 17.2	1380.0 ± 269.2	1204.7 ± 146.4	1012.7 ± 106.6
Jun.	875.3 ± 80.0	1339.1 ± 379.1	1225.7 ± 80.6	1590.6 ± 289.7	1326.0 ± 62.4	1114.3 ± 177.7
Jul.	1332.3 ± 161.9	1037.0 ± 137.0	1113.3 ± 72.5	1527.3 ± 210.3	1142.7 ± 117.2	396.3 ± 129.01
Aug.	874.6 ± 27.1	1059.7 ± 265.3	395.0 ± 132.7	1365.0 ± 240.2	1002.0 ± 363.5	598.3 ± 237.3
Sep.	777.6 ± 151.1	852.7 ± 59.1	808.3 ± 125.1	1019.3 ± 28.9	1118.7 ± 358.3	507.3 ± 256.7
Oct.	753.3 ± 108.1	974.7 ± 74.4	1064.7 ± 61.1	1287.0 ± 89.9	1238.3 ± 51.7	636.7 ± 194.2
Nov.	785.3 ± 20.3	1198.3 ± 229.3	1058.7 ± 108.2	1356.7 ± 26.5	1270.7 ± 209.5	1185.3 ± 131.0
Dec.	771.7 ± 129.5	1022.3 ± 103.6	1140.0 ± 63.2	1321.7 ± 94.0	1255.3 ± 52.0	863.3 ± 69.4

Two-way ANOVA Test

Significant difference in conductivity between monitoring drains, F_1=20.2 (P<0.01).

Significant difference in conductivity between sampling months, F_2=2.73 (P< 0.05).

Monthly average values of conductivity (μ mhos/cm) of five drains and one river cut (Table DWM-2) ranged from 666.7 ± 98.85 to 1332.3 ± 161.94 in the NAJ-1, 852.9 ± 59.16 to 1339 ± 379.12 in CIV-2, 808.3 ± 125.19 to 1330.3 ± 330.52 in POW-3, 1019.3 ± 28.9 to 2023.3 ± 968.5 in SEN-4, 1002 ± 363 to 1326 ± 62.36 in BAR-5 and from 396.3 ± 129.6 to 1185.3 ± 131.01 in HIN-6 (Hindon river cut). Analysis of variance shows a significant difference in conductivity between the five drains (F_1=19.2, P<0.01) as well as between monitoring months, (F_2=2.74, P<0.05). This clearly signifies that the pollution load (in term of load of minerals) of different drains was not same.

Seasonal average values of conductivity varied from 714.56 ± 108.74 (winter) to 994.89 ± 274.04 (monsoon) in NAJ-1, 899.44 ± 154.12 (winter) to 1110.107 ± 198.57 (post-monsoon) in CIV-2, 972.33 ± 163.98 (winter) to 1170.00 ± 208.19 (summer) in POW-3, 1303.89 ± 281.32 (monsoon) to 1468.00 ± 585.01 (winter) in SEN-4, 1087.78 ± 308.53 (monsoon) to 1265.58 ± 123.09 (summer) in BAR-5, and from 500.67 ± 280.46 (monsoon) to 1032.33 ± 186.02 (post-monsoon) in HIN-6 (Hindon river cut). Annual average values of conductivity (μ mhos/cm) were in order of 1453.33 > 1201.03 > 1048.7 > 1032.03 > 842.58 > 837.22 on SEN-4, BAR-5, POW-3, CIV-2 NAJ-1 and HIN-6 respectively.

Monthly average values of conductivity in river water (Table RWM-2) were varied from 199.33 ± 18.21 (July) to 489 ± 313 (May) at WAZ-1, 325 ± 26.94 (Sept) to 767.67 ± 144.95 (March) at ISB-2, 271.33 ± 6.6 (Aug.) to 802.33 ± 26.66 (Feb.) at ITO-3, 297 ± 6.38 (Aug.) to 814.67 ± 38.9 (Feb.) 297 ± 6.38 (Aug.) to 814.67 ± 38.9 (Feb.) at NIZ-4 from 283.33 ± 65.62 (Aug.) to 660 ± 105.57 (Dec.) at OKH-5. Analysis of variance showed a significant difference in conductivity between the monitoring stations (F_1=20.20, P<0.01) and also in between the sampling months (F_2=2.73, P<0.05).

Table RWM-2: Monthly Variation of Conductivity (μ mhos/cm) in River Water (Year 1993-95)

Month	*WAZ-1* *Mean ± SD*	*ISB-2* *Mean ± SD*	*ITO-3* *Mean ± SD*	*NIZ-4* *Mean ± SD*	*OKH-5* *Mean ± SD*
Jan.	442.67 ± 94.24	605.00 ± 71.92	768.00 ± 103.96	742.33 ± 81.94	577.33 ± 76.36
Feb.	435.67 ± 43.86	764.33 ± 36.75	802.33 ± 26.66	814.67 ± 38.90	656.00 ± 59.26
Mar.	437.33 ± 136.80	767.67 ± 144.95	780.33 ± 140.89	803.67 ± 155.24	642.33 ± 215.09
Apr.	300.00 ± 59.30	690.00 ± 155.13	661.3 ± 175.32	700.33 ± 174.90	465.67 ± 148.45
May	489.00 ± 312.96	654.33 ± 323.04	788.33 ± 149.56	763.33 ± 117.05	476.00 ± 66.10
Jun.	208.33 ± 27.74	667.33 ± 65.57	719.33 ± 129.84	650.67 ± 107.53	434.33 ± 49.1
Jul.	199.33 ± 18.21	242.33 ± 61.67	330.67 ± 56.10	336.00 ± 65.49	343.33 ± 65.62
Aug.	249.00 ± 24.54	371.00 ± 30.63	271.33 ± 6.60	297.00 ± 6.38	283.33 ± 22.88
Sep.	254.00 ± 17.28	325.00 ± 26.94	318.33 ± 49.75	309.67 ± 69.19	298.33 ± 85.14
Oct.	376.67 ± 91.77	565.67 ± 183.84	610.00 ± 128.89	620.67 ± 149.97	597.70 ± 85.83
Nov.	461.00 ± 138.74	681.00 ± 122.63	752.67 ± 138.12	767.67 ± 135.35	837.30 ± 132.55
Dec.	484.00 ± 11	625.67 ± 119.59	771.67 ± 90.40	794.00 ± 108.71	660.00 ± 105.57

Two-way ANOVA Test

Significant difference in conductivity between river monitoring stations, F_1=20.2 (P<0.01).
Significant difference in conductivity between monitoring months, F_2=2.73 (P< 0.05).

Seasonal conductivity values ranged from 234.11 ± 31.93 to 419.89 ± 105.76 at WAZ-1, 373.44 ± 58.91 to 675.33 ± 105.29 at ISB-2, 306.78 ± 50.42 to 790.11 ± 70.37 at ITO-3, 314.22 ± 57.47 to 796.70 ± 72.58 at NIZ-4, 308.33 ± 68.38 to 629.33 ± 80.65 at OKH-5. Annual averages were in order of 633.33 at

NIZ-4 > 633.19 at ITO-3 > 595.11 at ISB-2 > 522.64 at OKH-5 355.0, at WAZ-1.

The winter season showed the highest seasonal average values at all the sampling stations followed by post-monsoon, summer and minimum in monsoon season. This could be associated with the lower water volume in the river during winter, coupled with wastewater from a number of drains entering the river Yamuna at Delhi stretch. There was no remarkable difference between the summer and post-monsoon values because almost the same volume of water was discharged from Wazirabad barrage, where as monsoon showed lower conductivity due of dilution. Imevbore (1978) and Adebisi (1980), Mishra (1999) stated the inverse relation of conductivity with water level which corroborates our findings.

Total Solids of Drain and River Water

Solids refer to the matter that remains as residue upon evaporation after drying at a definite temperature. "Total solids" is the residue that includes both dissolved solids and suspended solids. The excess of TS creates imbalance for aquatic life.

Monthly variation (Mean ± SD) of TS content (mg/l) are given in Table DWM-3. The minimum and maximum TS content were 555.0 ± 9.9 (Apr.) and 986.33 ± 136.02 (July) in NAZ-1 drain, 652.3 ± 141.2 (Jul.) and 1016 ± 372.03 (Aug.) in CIV-2, 718.67 ± 293.21 (May) and 3542.0 ± 358.27 (Nov.) in POW-3, 598.33 ± 425.25 (Oct.) and 1198.0 ± 165.89 (Mar.) in SEN-4, 732.67 ± 52.25 (Jul.) to 879.33 ± 79.93 (May) in BAR-5 and from 585.33 ± 90.49 (Jul.) to 822.3 ± 67.61 (March) in HIN-6 (Hindon river cut). Analysis of variance shows a significant difference in TS content between five drains (F_1=22.9, P<0.01) as well as between sampling months (F_2=3.25, P<0.05).

Table DWM-3: Monthly Variation in Total Solids (mg/l) of Drain Water (Year 1993-1995)

Month	*NAJ-1*	*CIV-2*	*POW-3*	*SEN-4*	*BAR-5*	*HIN-6*
	Mean ± SD	*Mean ± SD*	*Mean ± SD*	*Mean ± SD*	*Mean ± SD*	*Mean ± SD*
Jan.	586.0 ± 100.4	938.0 ± 262.3	854.0 ± 384.6	1046.6 ± 33.9	846.6 ± 32.0	702.0 ± 101.5
Feb.	654.3 ± 42.0	834.0 ± 21.4	921.3 ± 495.9	1173.0 ± 186.1	839.3 ± 92.4	724.0 ± 37.5
Mar.	588.3 ± 31.7	897.0 ± 90.2	992.6 ± 555.7	1198.0 ± 185.8	879.3 ± 74.6	822.3 ± 67.6
Apr.	555.0 ± 9.9	984.6 ± 336.6	857.3 ± 467.6	1008.6 ± 128.1	841.3 ± 76.8	665.0 ± 65.8
May	556.3 ± 102.3	829.6 ± 136.6	718.6 ± 293.2	1049.6 ± 199.6	768.3 ± 79.9	749.0 ± 66.8
Jun.	628.0 ± 84.5	741.0 ± 131.8	803.6 ± 372.8	1058.00 ± 146.9	812.0 ± 77.2	741.6 ± 112.2
Jul.	986.3 ± 136.0	652.3 ± 141.2	789.3 ± 438.4	651.3 ± 467.3	732.6 ± 118.4	585.3 ± 90.4
Aug.	745.3 ± 162.8	1016.0 ± 372.0	950.6 ± 404.4	793.3 ± 597.8	741.3 ± 131.1	785.0 ± 55.6
Sep.	746.0 ± 74.2	813.3 ± 80.5	881.0 ± 310.1	781.6 ± 564.0	756.6 ± 961.7	645.6 ± 150.9
Oct.	590.6 ± 74.4	734.6 ± 77.9	834.3 ± 374.3	598.3 ± 425.2	791.3 ± 21.3	647.6 ± 114.5
Nov.	613.6 ± 40.4	881.3 ± 173.2	3542.0 ± 358.2	731.3 ± 517.4	808.3 ± 95.6	820.3 ± 76.5
Dec.	620.6 ± 131.1	839.3 ± 93.8	1000.6 ± 443.1	821.6 ± 579.5	825.0 ± 87.9	622.6 ± 63.3

Two-way ANOVA Test

Significant difference in total solids content between monitoring drains, F_1=22.97 (P<0.01).
Significant difference in total solids content between sampling months, F_2=3.25 (P< 0.05).

Seasonal averages varied from as low as 563.83 ± 59.22 (pre-monsoon) to 825.89 ± 172.42 (monsoon) in CIV-2, 843.08 ± 138.71 (summer) to 2235.33 ± 2988.20 (post-monsoon) in POW-3, 1036.58 ± 170.61

(summer) to 1168.78 ± 214.49 (monsoon) in SEN-4, 743.56 ± 116.74 (monsoon) to 825.25 ± 87.23 (summer) in BAR-5 and 671.0 ± 94.39 (winter) to 746.50 ± 120.72 (post-monsoon) in HIN-6.

Annual averages in different drains were in the order of 1096.22 ± 1327.61 > 1072.92 ± 168.22 > 844.53 ± 213.85 > 803.53 ± 97.77 > 709.22 ± 115.9 > 655.89 ± 149.59 in POW-3, SEN -4, CIV-3, BAR-5, HIN-6 and NAJ-1 drain respectively.

Monthly averages (Mean ± SD) of total solids and the results of analysis of variance have been presented in Table RWM-3. The minimum and maximum TS contents were 137.67 ± 32.15 (Jan.) and 954 ± 481.08 (Aug.) at WAZ-l, 289.67 ± 177.90 (Apr.) 1270.33 ± 116.74 (Aug.) at ISB-2, 396.67 ± 74.7 (Oct.) and 1117.76 ± 51.8 (Aug.) at NIZ-4, 286.33 ± 58.54 (Apr.) and 1225.23 ± 225.41 (Aug.) at OKH-5. Analysis of variance shows a significant difference between sampling stations (F_1=22.94, P<0.01) as well as between sampling months (F_2=2.8, P<0.05).

Table RWM-3: Monthly Variation of Total Solids (mg/l) in River Water (Year 1993- 1995)

Month	*WAZ-1*	*ISB-2*	*ITO-3*	*NIZ-4*	*OKH-5*
	Mean ± SD	*Mean ± SD*	*Mean ± SD*	*Mean ± SD*	*Mean ± SD*
Jan.	409.0 ± 91.2	527.3 ± 7.6	546.0 ± 49.7	532.3 ± 30.2	410.0 ± 57.1
Feb.	254.3 ± 2.1	573.0 ± 65.0	527.0 ± 48.0	554.7 ± 24.6	474.3 ± 20.8
Mar.	300.0 ± 108.1	560.3 ± 103.0	508.3 ± 81.1	508.7 ± 80.7	432.0 ± 121.6
Apr.	224.0 ± 42.5	289.6 ± 177.9	393.0 ± 141.7	406.0 ± 117.1	286.3 ± 58.5
May	299.0 ± 156.5	455.3 ± 118.4	429.0 ± 84.7	507.3 ± 19.6	294.0 ± 10.0
Jun.	137.7 ± 32.1	483.6 ± 169.3	420.7 ± 32.7	406.0 ± 35.4	288.3 ± 16.8
Jul.	286.0 ± 59.5	707.6 ± 189.4	776.7 ± 411.5	522.3 ± 229.2	573.0 ± 291.7
Aug.	945.0 ± 481.0	1270.3 ± 116.7	1703.0 ± 496.3	1117.7 ± 51.8	1225.3 ± 225.4
Sep.	785.3 ± 358.1	1023.6 ± 482.6	1113.3 ± 576.3	782.3 ± 311.4	829.0 ± 456.0
Oct.	383.0 ± 142.3	500.3 ± 167.3	383.0 ± 61.6	396.7 ± 74.7	358.0 ± 38.6
Nov.	323.3 ± 70.7	456.3 ± 40.9	468.7 ± 93.6	491.3 ± 92.6	612.6 ± 179.2
Dec.	358.0 ± 65.0	497.3 ± 43.7	527.3 ± 63.9	512.3 ± 78.2	426.0 ± 54.8

Two-way ANOVA Test

Significant difference in total solids content between monitoring stations, F_1=22.94 (P<0.01).
Significant difference in total solids content between monitoring months, F_2=2.8 (P< 0.05).

Seasonal averages ranged from a minimum 240.17 ± 119.85 to the maximum 675 ± 448.91 at WAZ-l, 3743.44 ± 58.91 to 675.33 ± 105.29 at ISB-2, 306.78 ± 50.42 to 790.78 ± 72.58 at NIZ-4, 308 ± 68.38 to 629.33 ± 89.65 at OKH-5 (Okhla barrage). Exceptionally high values were recorded at all the stations of river in monsoon due to the top soil at the river bank and catchment area were carried by the stream downwards. Annual averages at different stations were in order of 649.6 > 612.08 > 561.47 > 517.42 > 392.8 at station ITO-3, ISB-2, NIZ-4, OKH-5 and WAZ-1 respectively.

Total Suspended Solids of Drain and River Water

No specific limit has been prescribed for suspended solids in water used for domestics and industrial purpose, but it indirectly controlled by the limits of turbidity (5 units). Disposal of sewage and industrial effluents contributes suspended matter to river and streams. The Indian Standards Institute has specified a maximum limit of 30 mg/l for suspended solids in wastewater dischargeable into water courses. Suspended solids are very objectionable in rivers for many reasons. Suspended sclids containing much organic matter may cause putrefaction and consequently the stream may become devoid of dissolved oxygen.

The monthly averages (Mean ± SD) of total suspended solids and results of analysis of variance in drain water have been depicted in Table DWM-4. The TSS content varied from 128.3 ± 30.2 to 276.3 ± 175.7 in NAJ-1, 130 ± 160.3 to 499.3 ± 301.3 in CIV-2 167.0 ± 39.7 to 423.0 ± 148.3 in POW-3, 242.7 ± 70.1 to 526 ± 243.5 in SEN-4, 64.7 ± 28.2 to 194.3 ± 92.4 in BAR-5 and 73.3 ± 23.6 to 472.7 ± 76.4 in HIN-6 (Hindon river cut). Analysis of variance showed a significant difference in TSS content between sampling stations (F_1=20.7, P<0.01) as well as between monitoring months (F_2=2.53, P<0.05).

Table DWM-4: Monthly Variation of Suspended Solids (mg/l) in Drain Water (Year 1993-1995)

Month	*NAJ-1*	*CIV-2*	*POW-3*	*SEN-4*	*BAR-5*	*HIN-6*
	Mean ± SD	*Mean ± SD*	*Mean ± SD*	*Mean ± SD*	*Mean ± SD*	*Mean ± SD*
Jan.	167.0 ± 65.3	257.7 ± 74.2	219.7 ± 12.4	400.0 ± 107.4	76.7 ± 17.7	97.7 ± 38.9
Feb.	168.3 ± 31.7	268.7 ± 137.2	274.7 ± 42.2	511.7 ± 142.9	113.3 ± 65.6	104.7 ± 6.2
Mar.	153.0 ± 9.4	221.0 ± 119.1	282.7 ± 58.0	470.3 ± 185.4	87.7 ± 17.9	131.7 ± 2.8
Apr.	128.3 ± 30.2	360.3 ± 85.8	326.3 ± 86.9	248.3 ± 117.8	64.7 ± 28.2	141.7 ± 18.6
May	146.3 ± 80.5	298.7 ± 84.2	167.0 ± 39.7	242.7 ± 70.1	119.7 ± 35.2	126.0 ± 11.4
Jun.	209.7 ± 43.3	130.0 ± 60.3	199.3 ± 37.3	308.3 ± 77.7	194.3 ± 92.4	73.3 ± 23.8
Jul.	276.3 ± 75.7	145.3 ± 86.4	248.0 ± 76.7	279.3 ± 36.5	113.7 ± 29.8	316.7 ± 20.5
Aug.	210.3 ± 96.9	499.3 ± 301.6	423.0 ± 48.3	526.0 ± 243.5	189.0 ± 99.3	472.7 ± 76.4
Sep.	224.3 ± 48.3	275.0 ± 88.3	386.0 ± 67.3	517.3 ± 170.7	191.0 ± 103.3	309.3 ± 75.0
Oct.	165.3 ± 33.1	228.0 ± 43.3	242.7 ± 36.3	203.3 ± 69.7	98.3 ± 24.8	204.0 ± 8.6
Nov.	142.7 ± 46.6	200.7 ± 10.1	221.3 ± 22.5	222.3 ± 67.4	113.3 ± 26.1	114.3 ± 27.5
Dec.	145.7 ± 6.6	252.3 ± 24.1	303.0 ± 32.0	355.3 ± 68.0	112.7 ± 53.4	96.0 ± 33.5

Two-way ANOVA Test

Significant difference in suspended solids content between monitoring drains, F_1=20.76 (P<0.01).
Significant difference in suspended solids content between sampling months, F_2=2.53 (P<0.05).

Seasonal TSS content ranged from 141.5 ± 361 to 237 ± 67 in NAJ-1, 228.17 ± 50.03 to 306.65 ± 82 in CIV-2, 243.83 ± 119.84 to 352.33 ± 79.6 in POW-23, 317.0 ± 125.37 to 452.56 ± 140.5 in SEN-4, 87.11 ± 46.11 to 164.56 ± 63.8 in BAR-5, and 89.67 ± 30.90 to 366.22 ± 89.8 in HIN-6. Yearly averages of TSS in different drains were in order of SEN-4 (357.08) > POW-4 (274.47) > CIV-2 (261.42) > HIN-6 (179.75) > NAJ-1 (178.11) > BAR-5 (122.86).

Seasonal TSS contents ranged from a minimum 36.22 ± 19.29 (winter) to a maximum of 505.0 ± 465.98 (Monsoon) at WAZ-1, 75.56 ± 34.31 (Winter) to 687.11 ± 437.52 (Monsoon) at ISB-2, 39.5 ± 16.28 (Summer) to 1017.56 ± 647.58 (Monsoon) at ITO-3, 98.58 ± 14.23 (Summer) to 584.44 ± 343.01 (Monsoon) at NIZ-4, 35.42 ± 16.79 (Summer) to 685.89 ± 460.92 (Monsoon) at OKH-5. Annual averages at different stations were in order of 289.19 > 243.5 >208.94 > 177.97 > 159.31 at ITO-3, ISB-2, OKH-5, NIZ-4 and WAZ-1 respectively.

Exceptionally high values were recorded at all the monitoring stations of river in monsoon months because the top soil of sandy river bank and cultivated fields of catchment area were carried by the river stream down water. In summer, the TSS values were low which may be due settling of suspended matter on the river bed, because of lean flow.

Table RWM-4: Monthly Variation of Suspended Solids (mg/l) in River Water (Year 1993-95)

Month	WAZ-1 Mean ± SD	ISB-2 Mean ± SD	ITO-3 Mean ± SD	NIZ-4 Mean ± SD	OKH-5 Mean ± SD
Jan.	38.3 ± 13.1	108.3 ± 36.8	56.7 ± 37.3	60.7 ± 18.0	44.3 ± 20.4
Feb.	35.6 ± 28.6	64.3 ± 12.0	66.3 ± 16.2	54.3 ± 23.8	70.0 ± 35.8
Mar.	35.0 ± 31.9	106.0 ± 47.1	50.0 ± 16.3	38.3 ± 6.2	40.0 ± 15.6
Apr.	31.7 ± 22.4	95.0 ± 8.2	37.3 ± 14.7	29.0 ± 4.2	29.7 ± 4.5
May	23.3 ± 12.0	54.3 ± 35.7	40.7 ± 18.8	53.3 ± 16.7	30.3 ± 7.6
Jun.	19.3 ± 13.3	128.3 ± 146.2	30.0 ± 4.2	33.7 ± 11.9	41.7 ± 26.2
Jul.	102.7 ± 3.8	464.7 ± 208.7	578.0 ± 437.1	300.0 ± 219.2	353.0 ± 314.6
Aug.	789.7 ± 540.9	790.0 ± 497.7	1536.3 ± 489.5	930.3 ± 60.8	1055.3 ± 210.0
Sep.	628.7 ± 372.0	806.7 ± 456.8	938.3 ± 599.0	523.0 ± 311.3	649.3 ± 495.6
Oct.	139.0 ± 166.3	186.0 ± 135.4	60.0 ± 12.7	56.3 ± 13.6	37.3 ± 22.4
Nov.	21.7 ± 14.4	63.0 ± 25.8	39:7 ± 19.7	32.3 ± 7.4	123.3 ± 141.9
Dec.	22.7 ± 11.6	55.3 ± 20.0	37.0 ± 20.8	24.3 ± 12.0	33.0 ± 10.2

Two-way ANOVA Test

Significant difference in suspended solids content between river monitoring stations, F_1=32.84 (P<0.01).

Significant difference in suspended solids content between monitoring months, F_2=2.17 (P<0.05).

Total Dissolved Solids of Drain and River Water

In natural water the dissolved solids exist mainly due to bicarbonates, carbonates, sulphates of calcium, magnesium, sodium and potassium with traces of iron, Manganese and other substances. The amount of dissolved solids present in water is a consideration for its suitability for domestic use. Presence of dissolved solids in industrial water is undesirable for many reasons. They form scales, cause foaming in boiler, accelerate corrosion and interfere with the colour and taste of the finished products. According to the 1962, USPHS, drinking water standards the permissible limit for total dissolved solids is 500 mg/l.

Monthly variations (Mean ± SD) of TDS content (mg/l) in river water are given in Table DWM-5. Monthly averages ranged from a minimum of 419 ± 73.27 to maximum of 535 ± 66.38 in NAZ-1, 507 ± 3.24 to 680.67 ± 119.4 in ISB-2, 495.0 ± 47.08 to 709.67 ± 91.51 in POW-3, 498.67 ± 148.95 to 832.00 ± 15.20 in SEN-4, 552.33 ± 199.74 to 791.67 ± 57.20 in BAR-5 and 266.00 ± 12.47 to 668.33 ± 276.71 in HIN-6 (Hindon river cut). Analysis of variance (Table DWM-5) shows highly significant difference in TDS content between different drains (F_1=19.1, P<0.01) as well as between sampling months (F_2=2.81, P<0.05).

Seasonal TDS content ranged from 422.67 ± 47.5 to 577.78 ± 87.78 in NAJ-1, 520.67 ± 106.56 to 617.33 ± 50.03 in CIV-2, 524.67 ± 138.3 to 643.5 ± 57.75 in POW-3, 647.44 ± 178.64 to 791.83 ± 45.41 in SEN-4, 580.11 ± 105.6 to 729.67 ± 41.30 in BAR-5, 304.89 ± 163.4 to 634.67 ± 101.56 in HIN-6. Annual averages were in order of 718.53 (SEN-4) > 673.47 (BAR-5) > 611.36 (POW-3) > 583.06 (CIV-2) > 529.53 (HIN-6) > 467.72 (NAJ-1).

Table DWM-5: Monthly Variation of Dissolved Solids (mg/l) in Drain Water (Year 1993-1995)

Month	NAJ-1 Mean ± SD	CIV-2 Mean ± SD	POW-3 Mean ± SD	SEN-4 Mean ± SD	BAR-5 Mean ± SD	HIN-6 Mean ± SD
Jan.	419.0 ± 73.3	679.7 ± 227.7	634.3 ± 84.7	695.7 ± 195.2	770.0 ± 15.3	604.3 ± 208.8
Feb.	486.0 ± 16.0	565.3 ± 47.6	646.7 ± 52.3	498.7 ± 148.9	726.0 ± 30.3	619.3 ± 253.0
Mar.	435.3 ± 39.3	676.0 ± 66.2	709.7 ± 91.5	656.1 ± 308.6	791.7 ± 57.3	684.7 ± 245.6
Apr.	426.7 ± 39.9	588.3 ± 178.4	708.0 ± 159.7	712.3 ± 35.9	693.0 ± 66.3	560.3 ± 210.2
May	430.0 ± 22.8	531.0 ± 72.8	551.7 ± 35.2	779.6 ± 158.9	348.7 ± 70.3	623.0 ± 218.4
Jun.	418.3 ± 44.8	611.0 ± 131.8	572.0 ± 38.0	716.0 ± 111.7	617.7 ± 15.7	668.3 ± 276.7
Jul.	710.0 ± 124.0	507.0 ± 73.2	550.3 ± 26.8	705.3 ± 95.8	619.0 ± 92.7	266.0 ± 12.5
Aug.	535.0 ± 66.4	516.7 ± 80.9	528.7 ± 30.7	744.0 ± 39.6	552.3 ± 199.7	312.3 ± 61.3
Sep.	488.3 ± 98.0	538.3 ± 39.9	495.0 ± 47.0	734.3 ± 49.6	569.0 ± 195.2	336.3 ± 120.5
Oct.	425.3 ± 42.0	515.7 ± 34.1	595.0 ± 78.0	733.7 ± 21.7	693.0 ± 8.2	443.7 ± 171.1
Nov.	467.7 ± 12.1	680.7 ± 119.4	647.3 ± 61.3	832.0 ± 15.3	695.0 ± 119.5	706.0 ± 308.7
Dec.	479.0 ± 136.0	587.0 ± 72.6	697.7 ± 65.7	810.7 ± 72.0	712.3 ± 37.2	530.0 ± 192.3

Two-way ANOVA Test

Significant difference in dissolved solids content between monitoring drains, F_1=19.15 ($P<0.01$).
Significant difference in dissolved solids content between sampling months. F_2=2.81 ($P<0.05$).

Table RWM-5: Monthly Variation of Dissolved Solids (mg/l) in River Water (Year 1993-95)

Month	WAZ-1 Mean ± SD	ISB-2 Mean ± SD	ITO-3 Mean ± SD	NIZ-4 Mean ± SD	OKH-5 Mean ± SD
Jan.	370.7 ± 99.9	419.0 ± 55.3	477.3 ± 65.4	471.7 ± 29.4	36.7 ± 42.2
Feb.	218.7 ± 28.8	508.7 ± 55.4	460.7 ± 32.7	500.3 ± 20.4	404.3 ± 55.7
Mar.	265.0 ± 76.2	454.3 ± 86.1	458.3 ± 76.7	476.3 ± 89.4	392.0 ± 106.5
Apr.	192.3 ± 25.6	371.7 ± 126.4	205.7 ± 138.8	377.0 ± 113.3	256.7 ± 54.4
May	275.7 ± 165.8	401.0 ± 116.4	388.3 ± 70.8	454.0 ± 34.8	263.7 ± 4.5
Jun.	118.3 ± 19.3	355.3 ± 58.9	309.7 ± 29.3	372.3 ± 25.2	246.7 ± 29.6
Jul.	183.3 ± 61.3	243.0 ± 97.3	198.7 ± 38.9	222.3 ± 68.2	220.0 ± 62.8
Aug.	164.3 ± 25.2	218.7 ± 102.3	166.7 ± 10.4	187.3 ± 16.5	170.0 ± 20.6
Sep.	156.7 ± 17.0	183.7 ± 122.1	175.0 ± 22.7	192.7 ± 37.1	179.7 ± 47.9
Oct.	244.0 ± 67.2	314.3 ± 101.3	323.0 ± 50.3	340.3 ± 82.7	322.3 ± 41.1
Nov.	301.7 ± 76.0	393.3 ± 19.7	429.0 ± 74.2	459.0 ± 93.5	489.3 ± 91.6
Dec.	335.3 ± 62.5	442.0 ± 29.8	490.3 ± 68.5	488.0 ± 78.4	393.0 ± 51.2

Two-way ANOVA Test

Significant difference in total solids content between river monitoring stations, F_1=20.38 ($P<0.01$).
Significant difference in total solids content between monitoring months, F_2=2.8 ($P<0.05$).

Table RWM-5 shows monthly averages (Mean ± SD) of total dissolved solids and the result of Analysis of variance in river water. Monthly averages varied from a minimum of 118.33 ± 19.29 (Jun.)

to a maximum of 370.67 ± 99.87 (Jan.) at WAZ-1, 183.67 ± 122.06 (Sep.) to 508.67 ± 55.43 (Feb.) at ISB-2, 166.67 ± 10.4 (Aug.) to 490.33 ± 68.519 (Dec.) at ITO-3, 187.33 ± 16.52 (Aug.) to 500.33 ± 20.42 (Feb.) at NIZ-4, 170.00 ± 20.61 (Aug.) to 393.0 ± 51.23 (Dec.) at OKH-5 (Okhla barrage). Analysis of variance shows a significant difference in TDS between sampling stations (F_1=20.38, P<0.01) as well as between sampling months (F_2=2.80, P<0.05).

Seasonal values ranged between 168.11 ± 41.06 (Monsoon) to 305.89 ± 96.63 (Winter) at WAZ-1, 215.11 ± 33.57 (Monsoon) to 463.89 ± 64.17 (Winter) at ISB-2, 180.11 ± 29.92 (Monsoon) to 484.0 ± 52.69 (Winter) at ITO-3, 200.78 ± 48.35 (Monsoon) to 499.11 ± 34.65 (Winter) at NIZ-4, 189 ± 51.87 (Monsoon) to 405.83 ± 109.60 (Pre-Monsoon) at OKH-5.

All the downstream stations of Wazirabad barrage showed peak values during winter due to lean discharge of Wazirabad barrage and no dilution of drain effluents discharge to river Yamuna at Delhi stretch. The lower values during rainy season may be due to dilution of drain effluents by rain water. Yearly averages of TDS content were in order of 378.44 (NIZ-4) > 358.75 (ISB-2) > 346.97 (ITO-3) > 308.61 (OKH-5) > 235.5 (WAZ-1).

Dissolved Oxygen of River Water

The concentration of dissolved oxygen (DO) in natural and wastewater depends on physical, chemical and biological processes in the water. Since DO is essential for the aerobic degradation of organic pollutants, the measurement of DO is important in water pollution control and in wastewater treatment. The data on DO has been provided only for river water as DO could not be detected in drain water.

Table RWM-6 shows monthly averages and the result of analysis or variance for DO. The minimum and maximum values for DO were 6.07 ± 0.34 (Jul.) and 12.23 ± 1.26 (Mar.) at WAZ-1, 0.57 ± 0.42 (Mar.) and 5.97 ± 1.01 (Sep.) at ISB-2, 0.50 ± 0.41 (Mar.) and 6.03 ± 0.42 (Aug.) at ITO-3, 0.73 ± 0.25 (Feb.) and 5.4 ± 0.24 (Aug.) at NIZ-4, 0.54 ± 0.49 (Mar.) and 6.57 ± 0.66 (Sep.) at OKH-5. Analysis of variance showed a significant difference in DO between monitoring stations (F_1=21.25, P<0.01). However, no significant difference was observed between sampling months (F_2=1.69).

Seasonal DO values ranged from 6.42 ± 1.07 (Monsoon) to 10.39 ± 1.73 (Summer) at WAZ-1, 1.07 ± 1.14 (Summer) to 4.79 ± 1.62 (Monsoon) at ISB-2, 0.80 ± 0.47 (Winter) to 4.96 ± 1.1 (Monsoon) at ITO-3, 0.81 ± 0.44 (Winter) to 4.6 ± 1.25 (Monsoon) at NIZ-4, 2.83 ± 1.32 (Winter) to 4.97 ± 1.79 (Monsoon) at OKH-5.

Summer season showed the highest seasonal average at WAZ-1 (unpolluted stretch) may be attributed to maximum penetration of incident light energy due to high transparency of water resulting in higher photosynthetic activity and phytoplanktonic production of oxygen. The winter season showed the second highest seasonal average, which may be a function of water temperature (Datta *et al.*, 1988).

The decreased water temperature during winter season has a greater capacity to hold DO than warm water and probably led to a lower rate of respiration thereby allowing maximum DO in the winter (Welch, 1951). In case of four stations (*i.e.* down stream of Wazirabad barrage) the highest DO value was observed in monsoon season, may be due to dilution of drain effluents by rain water.

In flowing polluted water, the oxygen content varied with distance and depth, and there are more than one oxygen profiles. But generally on adding sewage or any organic wastes, DO first decreases and then slowly recovers (Kumar, 1997). Same type of trend was observed in the river Yamuna at Delhi stretch, *i.e.* DO sharply decreased at ISB-2 and some what recovered at monitoring station OKH-5 (Okhla barrage).

Table RWM-6: Monthly Variation of Dissolved Oxygen (mg/l) in River Water (Year 1993-95)

Month	*WAZ-1*	*ISB-2*	*ITO-3*	*NIZ-4*	*OKH-5*
	Mean ± SD	*Mean ± SD*	*Mean ± SD*	*Mean ± SD*	*Mean ± SD*
Jan.	10.9 ± 3.0	1.8 ± 0.6	0.8 ± 0.7	0.8 ± 0.7	2.9 ± 1.6
Feb.	10.3 ± 1.8	0.7 ± 0.6	0.9 ± 0.4	0.7 ± 0.2	2.1 ± 0.6
Mar.	12.2 ± 1.3	0.6 ± 0.4	0.5 ± 0.4	0.4 ± 0.3	0.6 ± 0.5
Apr.	9.9 ± 1.4	1.1 ± 0.5	1.4 ± 1.1	2.5 ± 1.7	3.7 ± 2.3
May	10.5 ± 1.1	2.3 ± 1.5	1.6 ± 0.7	1.3 ± 0.6	2.8 ± 1.6
Jun.	8.9 ± 1.2	0.3 ± 0.2	0.8 ± 0.5	2.1 ± 2.1	3.4 ± 0.4
Jul.	6.1 ± 0.3	3.1 ± 0.6	3.9 ± 0.1	3.3 ± 1.2	2.9 ± 1.0
Aug.	6.3 ± 1.3	5.3 ± 1.4	6.0 ± 0.4	5.4 ± 0.2	5.5 ± 0.9
Sep.	6.8 ± 1.2	6.0 ± 1.1	4.9 ± 1.1	5.1 ± 0.4	6.6 ± 0.7
Oct.	9.4 ± 1.4	3.1 ± 1.9	3.6 ± 2.0	3.0 ± 1.7	4.7 ± 1.7
Nov.	8.2 ± 0.7	1.7 ± 0.7	1.6 ± 0.8	1.0 ± 0.5	4.0 ± 2.5
Dec.	10.5 ± 2.2	2.2 ± 2.2	0.9 ± 0.5	0.9 ± 0.2	2.9 ± 1.9

Two-way ANOVA Test

Significant difference in DO content between river monitoring stations, F_1=21.25 ($P<0.01$).

No significant difference in DO content between monitoring months, F_2=1.69.

Yearly averages of DO were in order of 9.18 > 3.48 > 2.36 > 2.24 > 2.21 at the station WAZ-1, OKH-5, ISB-2, ITO-3, and NIZ-4 respectively. The present study on river Yamuna at Delhi stretch showed DO content below the tolerance limits set by CPCB (1978) for class-B water (3mg/l) at all the stations (expect WAZ-1), which indicates the unsuitability of water for outdoor bathing.

Oxygen Absorbance Value of Drain and River Water

This test determines the amount of oxygen absorbed by the sample from potassium permanganate in acid solution. The test gives some indication of the nature of oxidisable substances present in the sample. Though the test is an empirical one, it is useful as a quick means to find out the pollution potential of wastewater and is useful in laboratories. When it is not feasible to conduct Biochemical Oxygen Demand (BOD) test, due to interference by toxic substances, this test is of much use. It is also used to determine the McGowan strength of polluted waters, sewage and industrial effluents. (Manivasakam, 1989).

Table DWM-6 shows monthly averages (Mean ± SD) of oxygen absorbance and the results of analysis of variance. The minimum and maximum monthly averages of oxygen absorbance values (mg/l) were 17.58 ± 4.26 (Sept.) and 31.33 ± 18.53 (Jul.) in NAJ-1, 33.2 ± 9.34 (Mar.) and 63.73 ± 7.13 (Jan.) in CIV-2, 42.6 ± 19.95 (Sep.) and 64.93 ± 4.08 (Dec.) in POW -3, 48.27 ± 11.33 (Jul.) and 69.2 ± 3.12 (Dec.) in SEN-4, 19.93 ± 2.25 (Oct.) and 34.3 ± 18.19 (Feb.) in BAR-5, 2.87 ± 0.77 (Jul.) and 8.07 ± 0.88 (Apr.) in HIN-6. Analysis of variance showed a significant difference in OA between the different drains (F_1=19.2, $P<0.01$) and also between the sampling months (F_2=2.37, $P<0.05$).

Table DWM-6: Monthly Variation of Oxygen Absorbance (mg/l) in Drain Water (Year 1993-1995)

Month	*NAJ-1* *Mean ± SD*	*CIV-2* *Mean ± SD*	*POW-3* *Mean ± SD*	*SEN-4* *Mean ± SD*	*BAR-5* *Mean ± SD*	*HIN-6* *Mean ± SD*
Jan.	25.1 ± 10.6	63.7 ± 7.1	53.6 ± 11.1	63.5 ± 5.2	32.6 ± 3.8	5.9 ± 0.7
Feb.	25.3 ± 6.0	56.1 ± 13.0	62.1 ± 3.0	67.1 ± 6.9	34.3 ± 18.2	7.1 ± 0.1
Mar.	22.4 ± 4.4	33.2 ± 9.34	57.7 ± 9.0	60.5 ± 8.1	25.5 ± 6.3	7.5 ± 4.1
Apr.	25.3 ± 4.7	59.1 ± 14.3	54.5 ± 10.5	53.7 ± 15.6	30.3 ± 12.0	8.1 ± 0.9
May	22.9 ± 7.0	51.5 ± 23.6	50.8 ± 3.3	62.9 ± 24.5	23.9 ± 3.1	4.0 ± 1.0
Jun.	20.5 ± 2.5	48.4 ± 19.5	47.7 ± 4.4	57.3 ± 8.8	29.7 ± 12.3	6.0 ± 0.5
Jul.	31.3 ± 18.53	33.2 ± 9.3	52.5 ± 6.2	48.3 ± 11.3	27.7 ± 10.8	2.9 ± 0.8
Aug.	18.3 ± 6.1	60.3 ± 21.4	52.1 ± 12.5	60.4 ± 25.7	19.3 ± 0.5	5.0 ± 2.5
Sep.	17.58 ± 4.2	51.6 ± 13.6	42.6 ± 20.0	68.4 ± 1.2	21.8 ± 1.9	3.4 ± 0.6
Oct.	18.7 ± 1.9	47.3 ± 4.1	58.5 ± 2.7	50.3 ± 8.3	19.9 ± 2.2	4.1 ± 1.2
Nov.	20.8 ± 0.7	52.7 ± 10.6	54.0 ± 4.3	60.7 ± 5.3	30.3 ± 6.7	4.4 ± 1.4
Dec.	21.1 ± 6.7	47.9 ± 6.7	64.9 ± 4.1	69.2 ± 3.1	28.0 ± 5.8	4.2 ± 1.4

Two-way ANOVA Test

Significant difference in oxygen absorbance content between monitoring drains, F_1=19.22 (P<0.01).
Significant difference in oxygen absorbance content between sampling months, F_2=2.81 (P<0.05).

Table RWM-7: Monthly Variation of Oxygen Absorbance (mg/l) in River Water (Year 1993-95)

Month	*WAZ-1* *Mean ± SD*	*ISB-2* *Mean ± SD*	*ITO-3* *Mean ± SD*	*NIZ-4* *Mean ± SD*	*OKH-5* *Mean ± SD*
Jan.	1.60 ± 0.33	10.40 ± 1.42	8.93 ± 2.85	9.60 ± 3.22	9.47 ± 3.49
Feb.	1.60 ± 0.33	10.13 ± 2.96	7.87 ± 2.45	6.93 ± 2.17	10.72 ± 4.58
Mar.	2.00 ± 0.86	13.07 ± 2.10	8.40 ± 1.50	8.80 ± 1.70	8.27 ± 0.38
Apr.	1.27 ± 0.09	9.00 ± 0.91	6.27 ± 1.32	7.73 ± 1.61	7.43 ± 2.78
May	1.60 ± 0.33	7.87 ± 2.74	6.73 ± 1.81	8.40 ± 1.50	6.73 ± 1.06
Jun.	1.87 ± 0.74	15.33 ± 9.89	8.67 ± 0.94	9.73 ± 0.68	6.67 ± 1.15
Jul.	2.33 ± 0.52	5.35 ± 0.41	5.37 ± 0.80	4.97 ± 1.27	4.07 ± 0.41
Aug.	4.53 ± 0.52	6.27 ± 0.52	5.93 ± 0.47	5.60 ± 0.33	5.73 ± 1.24
Sep.	4.13 ± 2.02	5.73 ± 2.10	6.87 ± 2.65	6.23 ± 2.12	6.40 ± 2.47
Oct.	1.87 ± 0.68	7.23 ± 3.63	5.87 ± 2.93	5.47 ± 1.80	4.80 ± 1.99
Nov.	1.27 ± 0.62	7.33 ± 1.89	5.60 ± 1.50	6.67 ± 1.15	8.80 ± 4.57
Dec.	1.43 ± 0.78	7.83 ± 2.77	6.03 ± 1.77	5.67 ± 0.96	6.80 ± 0.86

Two-way ANOVA Test

Significant difference in oxygen absorbance between river monitoring stations, F_1=20.17 (P<0.01).
Significant difference in oxygen absorbance between monitoring months, F_2=2.46 (P<0.05).

Seasonal and annual averages of oxygen absorbance varied from 18.53 ± 2.42 to 22.54 ± 5.26 in NAZ-1, 48.08 ± 20.44 to 56.71 ± 10.84 in CIV-2, 49.07 ± 14.77 to 58.09 ± 8.92 in POW-3, 58.62 ± 16.10 to

67.71 ± 6.32 in SEN-4, 22.93 ± 7.26 to 27.83 ± 12.18 in BAR-5 and 3.74 ± 1.8 to 6.42 ± 2.66 in HIN-6. Annual average oxygen absorbance values (mg/l) were 22.36, 50.43, 54.26, 60.18, 26.19 and 5.22 in NAJ-1, CIV-2, POW-3, SEN-4, BAR-5 and HIN-6 respectively.

Monthly average oxygen absorbance values (mg/l) ranged from 1.27 ± 0.09 (Apr.) to 4.53 ± 0.52 (Aug.) at WAZ-1, 7.23 ± 3.63 (Oct.) to 15.33 ± 9.89 (Jun.) at ISB-2, 5.37 ± 0.08 (Jul.) to 8.93 ± 2.83 (Jan.) at ITO-3, 4.97 ± 1.27 (Jul.) to 9.73 ± 0.68 (Jun.) at NIZ-4 and 4.07 ± 0.41 (Jul.) to 10.27 ± 4.58 (Feb.) at OKH-5 (Table RWM-7) in the river water. Analysis of variance showed a significant difference in OA between the sampling stations (F_1=20.17, P<0.01) and also between the sampling months (F_2=2.46, P<0.05).

Seasonal and annual averages of OA in Yamuna river at Delhi stretch ranged from 1.46 ± 0.46 (Winter) to 3.67 ± 1.57 (Monsoon) at WAZ-1, 5.84 ± 1.31 (Monsoon) to 11.32 ± 6.06 (Summer) at ISB.2, 5.73 ± 2.33 (Monsoon) to 7.64 ± 2.66 (Winter) at ITO-3, 5.60 ± 1.53 (Monsoon) to 7.7 ± 2.72 (Winter) at NIZ-4 and 5.4 ± 1.89 (Monsoon) to 8.64 ± 3.76 (Winter) at OKH.

It is observed that the peak values of OA were observed in summer season at all stations (except Wazirabad barrage). It may due to lean discharge of river water form Wazirabad barrage and also high concentration of oxidisable organic and inorganic pollutants present in drain effluents and discharged into river Yamuna. During the present investigation the oxygen absorbance values at Wazirabad barrage was found to be high during rainy season due to flushing of organic pollutants by rain water from river catchment areas. Annual averages of OA were in order of 8.81 >7.15 = 7.15 > 6.88 > 2.13 at station ISB-2, NIZ-4, OKH-5, ITO-3 and WAZ-1 respectively.

Chemical Oxygen Demand of Drain and River Water

The chemical oxygen demand determines the amount of oxygen required for chemical oxidation of organic matter using a strong chemical oxidant such as potassium dichromate (APHA, 1991). COD represents chemical oxidisable pollution load present in water and hence a reliable parameter for judging the quantum of organic pollution in the aquatic system.

Table DWM-7 presents the monthly average COD values of drain water. The COD ranged from 89.0 ± 20.8 to 147.3 ± 44.1 in drain NAJ-1, 284.0 ± 138.7 to 410.7 ± 196.0 in CIV-2, 278.7 ± 173.6 to 440 ± 71.2 in POW-3, 274.7 ± 69.8 to 465.3 ± 25.4 in SEN-4, 98.7 ± 23.6 to 205.0 ± 106.9 in BAR-5 and from 23.3 ± 9.6 to 35.0 ± 2.4 in Hundon river cut. Two way analysis of variance shows significant difference in COD values between the drains (F_1=19.5, P<0.01) as well as sampling months (F_2=2.33, P<0.05).

Seasonal averages of COD ranged from 122.67 ± 96.02 to 152.17 ± 66.27, 308.67 ± 68.91 to 373.0 ± 77.48, 323.11 ± 124.43 to 362.44 ± 64.22, 379.33 ± 66.59 to 449.78 ± 48.55, 136.0 ± 39.4 to 193.75 ± 47.48, 29.89 ± 22.14 to 34.11 ± 4.15 in Najafgarh, Civil mill, Power house, Sen nursing home, Barpulla drain and Hindon river cut respectively. Annual average of COD values (mg/l) were in order of 4.8028 > 349.61 > 336.31 > 166.92 > 133.89 > 31.92 in the drain SEN-4, POW-3, CIV-2, BAR-5, NAJ-1 and HIN-6 (Hindon cut) respectively. The maximum values of COD was noted in SEN-4 (Sen nursing home drain).

Monthly average COD values of river water (Table RWM-8) ranged from 5.33 ± 1.41 (Jun.) to 14.33 ± 6.94 (Aug.) at WAZ-1, 28.67 ± 15.86 (Sep.) to 65.33 ± 9.98 (Mar.) at ISB-2, 19.43 ± 5.73 (May.) to 49.33 ± 13.20 (Jan.) at, down stream station OKH-5 (Okhla barrage) observed. Analysis of variance showed a significant difference in COD values between different monitoring stations (F_1=20.45, P<0.01) and also between sampling months (F_2=2.31, P<0.05).

Table DWM-7: Monthly Variation of COD (mg/l) in Drain Water (Year 1993-1995)

Month	NAJ-1	CIV-2	POW-3	SEN-4	BAR-5	HIN-6
	Mean ± SD	Mean ± SD	Mean ± SD	Mean ± SD	Mean ± SD	
Jan.	137.3 ± 81.1	365.3 ± 26.4	364.0 ± 76.8	433.3 ± 66.3	149.3 ± 22.2	35.0 ± 2.4
Feb.	147.3 ± 44.1	415.3 ± 40.7	387.3 ± 65.8	450.0 ± 35.6	205.0 ± 106.9	34.0 ± 6.2
Mar.	133.7 ± 24.5	309.3 ± 74.5	440.0 ± 71.2	465.3 ± 25.4	184.0 ± 50.9	35.7 ± 13.2
Apr.	137.3 ± 22.9	429.3 ± 133.5	371.0 ± 44.3	374.7 ± 54.4	199.0 ± 74.3	29.0 ± 0.8
May	117.3 ± 43.5	324.0 ± 99.7	310.0 ± 51.9	440.0 ± 56.4	174.7 ± 74.6	27.7 ± 1.2
Jun.	117.3 ± 13.2	318.7 ± 135.3	318.7 ± 31.0	410.0 ± 77.9	217.3 ± 90.1	22.7 ± 0.9
Jul.	188.3 ± 141.2	284.0 ± 138.7	325.2 ± 33.5	340.0 ± 86.4	186.7 ± 55.1	39.0 ± 34.0
Aug.	90.7 ± 28.5	410.7 ± 196.0	365.3 ± 106.8	414.7 ± 198.6	145.0 ± 21.6	27.3 ± 9.4
Sep.	89.0 ± 20.8	308.3 ± 74.4	278.7 ± 173.6	531.3 ± 70.7	132.7 ± 40.7	23.3 ± 9.6
Oct.	176.3 ± 79.7	280.0 ± 65.3	345.7 ± 15.0	274.7 ± 69.8	98.7 ± 23.6	38.3 ± 12.7
Nov.	141.3 ± 24.7	317.3 ± 65.4	322.7 ± 30.3	340.0 ± 28.3	146.7 ± 37.7	38.3 ± 2.1
Dec.	130.7 ± 43.4	273.3 ± 57.3	366.7 ± 27.8	425.3 ± 62.1	164.0 ± 35.3	32.7 ± 2.5

Two-way ANOVA Test

Significant difference in COD content between monitoring drains, F_1=19.31 (P<0.01).

Significant difference in COD content between sampling months, F_2=2.31 (P<0.05).

Seasonal and annual average values of COD ranged from 5.92 ± 1.93 (Winter) to 11.78 ± 5.43 (Monsoon) at WAZ-1, 33.78 ± 12.98 (Monsoon) to 55.89 ± 11.72 (Winter) at ISB-2, 28.67 ± 8.46 (Pre-Monsoon) to 45.56 ± 8.53 (Winter) at ITO-3, 34.67 ± 6.39 (Post-Monsoon) to 57.89 ± 22.97 (Winter) at NIZ-4 and 28.36 ± 10.34 (Summer) to 39.59 ± 10.75 (Winter) at OKH-5 (Okhla barrage) respectively.

Table RWM-8: Monthly Variation of COD (mg/l) in River Water (Year 1993-1995)

Month	WAZ-1	ISB-2	ITO-3	NIZ-4	OKH-5
	Mean ± SD	Mean ± SD	Mean ± SD	Mean ± SD	Mean ± SD
Jan.	8.00 ± 0.00	54.67 ± 15.43	53.33 ± 11.78	68.00 ± 29.03	49.33 ± 13.20
Feb.	6.67 ± 1.89	64.00 ± 3.27	42.67 ± 4.99	57.33 ± 13.20	38.67 ± 3.77
Mar.	5.33 ± 1.89	65.33 ± 9.98	36.00 ± 11.78	49.33 ± 9.98	34.67 ± 3.77
Apr.	8.00 ± 0.00	48.33 ± 8.50	31.33 ± 9.67	31.67 ± 2.87	30.00 ± 11.43
May	5.33 ± 1.89	34.33 ± 16.01	19.43 ± 5.73	31.33 ± 8.99	26.10 ± 10.16
Jun.	5.00 ± 1.41	81.33 ± 64.44	21.33 ± 5.66	41.33 ± 11.47	22.67 ± 9.98
Jul.	7.67 ± 0.47	32.67 ± 10.62	26.67 ± 12.08	35.67 ± 14.06	27.00 ± 11.52
Aug.	14.33 ± 6.94	40.00 ± 8.64	29.33 ± 4.99	33.33 ± 4.99	31.33 ± 5.73
Sep.	13.33 ± 3.77	28.67 ± 15.86	28.33 ± 10.21	37.33 ± 14.73	37.67 ± 9.53
Oct.	6.83 ± 2.01	35.67 ± 8.50	24.00 ± 8.22	32.00 ± 6.53	28.00 ± 5.66
Nov.	6.67 ± 3.77	33.33 ± 6.80	28.00 ± 8.64	37.33 ± 4.99	41.33 ± 24.51
Dec.	7.33 ± 2.49	40.33 ± 6.55	37.33 ± 8.64	35.00 ± 4.08	34.67 ± 3.77

Two-way ANOVA Test

Significant difference in COD content between river monitoring stations, F_1=20.45 (P<0.01).

Significant difference in COD content between monitoring months, F_2=2.32 (P<0.05).

In station WAZ-1 the higher values of COD in monsoon months was recorded perhaps due to organic and inorganic matter content reaching the water body through run-off water. The same trend was observed in Brahmani river Orissa as reported by Pradhan and Mohanty (1991). Annual averages of COD (mg/l) in the sampling sites were in the order of ISB-2 > NIZ-4 > ITO-3 > OKH-5 > WAZ-1.

In the present investigation, annual mean COD values of five sites revealed a minimum value in the site WAZ-1 (Wazirabad) showed the minimum value. There was sharp increase in COD at ISB-2 due to the entry of allochothonus organic and inorganic material from Najafgarh and Burari drain. At site ITO-3, the value again come down due to process of sedimentation by ITO barrage. The concentration of COD increased to some extent at site NIZ-4 due to fresh addition of drain water from Sen-nursing home. But at site OKH-5, the decrease values might be due to dilution (by Hindon cut), deoxygenation, reoxygenation, coagulation and settling to the river bed while travelling approximately six km from Nizamuddin bridge to Okhla barrage.

Biochemical Oxygen Demand of Drain and River Water

Biochemical oxygen demand (BOD) is defined as the amount of oxygen required by micro-organism to decompose biologically degradable organic matter in water under aerobic conditions. BOD is an essential parameter in water pollution control management and in the evaluation of the self-purification capacity of streams (Trivedy and Goel, 1984). It represents the biologically oxidisable organic loads present in water.

Table DWM-8 shows monthly averages (Mean ± SD) and results of analysis of variance of BOD at the five sampling drains. The BOD ranged from 51.0 ± 19.70 to 108.8 ± 86.03 in NAJ-1, 139.33 ± 48.95 to 245.7 ± 135.38 in CIV-2, 77 ± 28.08 to 244 ± 76.08 in POW-3, 142.60 ± 28.4 to 265.0 ± 64.11 in SEN-4, 53.53 ± 9.67 to 99.67 ± 31.34 in BAR-5, 3.67 ± 1.53 to 12.3 ± 10.0 in HIN-6 (Hindon river cut). Analysis of variance (Table DWM-8) shows significant difference in BOD content between sampling drains (F_1=19.5, $P<0.01$) as well as between sampling months F_2=2.22, $P<0.05$).

Seasonal and annual BOD averages (Mean ± SD) varied from 58.72 ± 16.24 (Summer) to 70.6 ± 50.56 (Monsoon) in NAJ-1, 158.85 ± 26.87 (Pre-Monsoon) to 216.4 ± 50.65 (Winter) in CIV-2, 175.33 ± 63.59 (Monsoon) to 217.5 ± 35.96 (Winter) in POW-3, 170.8 ± 48.23 (Pre-Monsoon) to 272.67 ± 44.38 (Winter) in SEN-4, 63.5 ± 11.22 (Pre-Monsoon) to 78.72 ± 27.32 (Winter) in BAR-5 and 5.0 ± 0.96 (Monsoon) to 7.06 ± 2.1 in HIN-6. The average annual values were in order of 225.24 > 195.94 > 188.2 > 75.16 > 65.77 > 6.83 in the sampling drains SEN-4, POW-3, CIV-2, BAR-5, NAJ-1 and HIN-6 respectively.

Monthly variation (Mean ± SD) in BOD content and the results of analysis of variance are given in Table RWM-9. The BOD varied from 1.93 ± 0.09 (April) to 5.57 ± 3.27 (Sep.) at WAZ-1, 11.6 ± 3.61 (Sep.) to 28.17 ± 3.97 (Mar.) at ISB-2, 8.7 ± 1.67 (Nov.) to 23.87 ± 7.31 (Jan.) at ITO-3, 12.07 ± 1.31 (Jul.) to 26.5 ± 8.78 (Jan.) at NIZ-4, 10.37 ± 4.88 (Oct.) to 23.7 ± 6.38 (Jan.) at OKH-5. Analysis of variance shows the significant difference in BOD content between the monitoring stations (F_1=20.2, $P<0.01$), as well as between the sampling months (F_2=2.31, $P<0.05$).

Seasonal average values of BOD were varied from a minimum of 2.53 ± 0.71 (Winter) to 4.51 ± 2.72 (Monsoon) WAZ-1, 11.87 ± 2.51 (Monsoon) to 24.48 ± 9.19 (Summer) at ISB-2, 11.12 ± 6.86 (Pre-Monsoon) to 20.18 ± 6.35 (Winter) at ITO-3, 13.34 ± 3.81 (Monsoon) to 20.67 ± 6.93 (Winter) at NIZ-4, 12.4 ± 3.69 (Monsoon) to 18.18 ± 6.64 (Winter) at OKH-5,

Table DWM-8: Monthly Variation of BOD (mg/l) in Drain Water (Year 1993-1995)

Month	NAJ-1 Mean ± SD	CIV-2 Mean ± SD	POW-3 Mean ± SD	SEN-4 Mean ± SD	BAR-5 Mean ± SD	HIN-6 Mean ± SD
Jan.	63.4 ± 35.1	240.9 ± 84.3	199.0 ± 16.4	252.5 ± 22.3	67.3 ± 21.9	7.3 ± 1.2
Feb.	83.1 ± 12.9	225.0 ± 35.0	221.8 ± 67.5	263.0 ± 27.0	78.7 ± 37.7	8.3 ± 0.6
Mar.	61.8 ± 18.0	198.3 ± 42.5	244.0 ± 76.1	259.8 ± 54.4	78.3. ± 23.1	12.3 ± 10.0
Apr.	66.1 ± 12.2	245.7 ± 135.4	220.0 ± 70.0	178.5 ± 57.5	99.7 ± 31.3	9.7. ± 2.1
May	56.5 ± 31.3	184.2 ± 137.1	165.7 ± 25.3	265.0 ± 84.1	53.5 ± 9.7	4.7 ± 1.2
Jun.	50.5 ± 8.3	139.3 ± 49.0	170.0 ± 13.2	205.3 ± 71.3	97.7 ± 59.0	7.0 ± 1.0
Jul.	108.8 ± 86.0	149.7 ± 86.3	214.3 ± 67.7	200.5 ± 38.1	84.3 ± 65.6	3.7 ± 1.5
Aug.	52.0 ± 20.8	240.3 ± 132.1	174.7 ± 50.0	228.2 ± 144.1	72.5 ± 25.7	6.3 ± 3.5
Sep.	51.0 ± 19.7	160.7 ± 31.1	77.0 ± 78.1	255.0 ± 32.8	74.2 ± 30.3	5.0 ± 2.0
Oct.	62.1 ± 8.6	153.7 ± 6.2	161.7 ± 57.7	142.6 ± 28.4	54.0 ± 10.2	6.0 ± 2.0
Nov.	68.2 ± 4.1	162.1 ± 45.9	198.2 ± 21.3	202.3 ± 55.1	73.0 ± 2.0	6.0 ± 1.7
Dec.	65.8 ± 1.9	156.7 ± 23.6	225.0 ± 18.3	253.5 ± 120.8	87.8 ± 34.4	5.6 ± 0.5

Two-way ANOVA Test

Significant difference in BOD content between monitoring drains, F_1=19.59 (P<0.01).
Significant difference in BOD content between sampling months, F_2= 2.22 (P<0.05).

Table RWM-9: Monthly Variation of BOD (mg/l) in River Water (Year 1993-1995)

Month	WAZ-1 Mean ± SD	ISB-2 Mean ± SD	ITO-3 Mean ± SD	NIZ-4 Mean ± SD	OKH-5 Mean ± SD
Jan.	2.60 ± 0.57	26.33 ± 2.62	23.87 ± 7.31	26.50 ± 8.78	23.70 ± 6.38
Feb.	2.43 ± 0.62	24.20 ± 3.35	20.57 ± 5.92	18.37 ± 3.16	18.33 ± 4.96
Mar.	3.07 ± 1.03	28.17 ± 3.97	16.60 ± 2.34	18.70 ± 1.07	15.77 ± 4.12
Apr.	1.93 ± 0.09	20.20 ± 2.97	13.97 ± 3.15	14.20 ± 2.47	13.77 ± 3.56
May	2.77 ± 0.65	18.97 ± 6.69	17.27 ± 2.58	14.87 ± 3.67	16.33 ± 2.05
Jun.	2.33 ± 0.81	30.60 ± 12.99	19.03 ± 5.72	15.33 ± 2.87	11.87 ± 3.43
Jul.	2.67 ± 1.70	12.13 ± 1.27	11.93 ± 0.57	12.43 ± 1.19	11.47 ± 1.87
Aug.	5.30 ± 1.84	11.87 ± 2.04	10.77 ± 1.28	12.07 ± 1.31	11.37 ± 2.25
Sep.	5.57 ± 3.27	11.60 ± 3.61	15.20 ± 5.87	15.53 ± 5.76	14.37 ± 5.14
Oct.	2.63 ± 0.74	16.73 ± 9.68	13.53 ± 8.92	13.03 ± 7.79	10.37 ± 4.88
Nov.	3.27 ± 1.24	13.53 ± 3.58	8.70 ± 1.67	15.33 ± 3.97	15.07 ± 12.33
Dec.	3.17 ± 0.90	18.47 ± 8.13	15.77 ± 1.78	15.17 ± 0.79	13.63 ± 2.67

Two-way ANOVA Test

Significant difference in BOD content between river monitoring stations, F_1=20.20 (P<0.01).
Significant difference in BOD content between monitoring months, F_2=2.31 (P<0.05).

In the present investigation, the BOD values during monsoon and winter season showed higher and lower values respectively at unpolluted stretch of Yamuna river *i.e.* Wazirabad barrage. The low values of BOD in winter may be due to lesser quantity of total organic content in the water. Increase in organic matter load due to rain water may be the reason for higher BOD during rainy season at WAZ-1. In down-stream stations the trend of BOD content of river water was higher values during summer months followed by winter, seasonal and lower values during monsoon. The higher values during summer seasons may be due to lean water discharge from Wazirabad barrage and also the presence of higher concentration of biodegradable organic matter in effluents discharged into river at Delhi stretch.

In addition to this, BOD values also showed site-wise fluctuation. Annual averages were in order of 19.4 > 15.96 > 15.6 > 14.64 > 3.14 at ISB-2, NIZ-4, ITO-3, OKH-5 and WAZ-1 (Wazirabad barrage), respectively. The gradual decrease in BOD from station ISB-2 to OKH-5 supports gradual reduction due to effect of dilution and self-purification.

Total Kjeldahl Nitrogen of Drain and River Water

In water and wastewater, the forms of nitrogen of greatest interest are nitrate, nitrite, ammonia and organic nitrogen. All these compounds are inter-convertible and take part in the nitrogen cycle. Analytically, organic nitrogen and ammonia can be determined together and are referred to as "Kjeldahl Nitrogen".

Table DWM-9: Monthly Variation of TKN (mg/l) in Drain Water (Year 1993-1995)

Month	*NAJ-1*	*CIV-2*	*POW-3*	*SEN-4*	*BAR-5*	*HIN-6*
	Mean ± SD	*Mean ± SD*	*Mean ± SD*	*Mean ± SD*	*Mean ± SD*	*Mean ± SD*
Jan.	29.20 ± 7.58	49.67 ± 15.33	42.33 ± 4.19	53.67 ± 3.09	37.33 ± 3.86	4.58 ± 1.01
Feb.	31.37 ± 5.22	47.13 ± 10.34	45.00 ± 8.29	55.83 ± 7.92	37.00 ± 2.16	14.05 ± 9.04
Mar.	26.00 ± 6.57	43.37 ± 5.85	50.53 ± 12.03	54.67 ± 12.50	38.00 ± 7.48	6.90 ± 4.13
Apr.	26.63 ± 3.63	51.23 ± 20.35	44.17 ± 14.90	42.67 ± 4.64	30.33 ± 3.68	7.66 ± 3.04
May	16.27 ± 15.00	48.77 ± 26.47	33.77 ± 1.52	52.00 ± 20.61	26.00 ± 4.24	5.57 ± 1.52
Jun.	25.33 ± 8.16	34.20 ± 6.24	31.10 ± 2.76	47.67 ± 14.52	39.50 ± 11.88	7.33 ± 3.40
Jul.	47.93 ± 34.62	37.30 ± 11.13	33.83 ± 0.85	38.33 ± 8.18	36.67 ± 8.06	1.86 ± 0.47
Aug.	24.00 ± 2.48	44.17 ± 16.28	34.77 ± 3.61	43.00 ± 16.33	40.33 ± 6.34	1.46 ± 0.32
Sep.	22.60 ± 5.24	36.17 ± 7.96	30.60 ± 7.03	49.33 ± 3.40	42.00 ± 6.48	2.06 ± 1.30
Oct.	26.77 ± 7.49	45.47 ± 8.87	41.27 ± 4.85	43.83 ± 4.37	42.33 ± 4.19	4.36 ± 2.68
Nov.	29.07 ± 3.44	40.10 ± 1.64	39.53 ± 9.15	53.73 ± 3.72	40.00 ± 6.98	3.89 ± 2.38
Dec.	27.83 ± 3.30	39.60 ± 7.11	44.93 ± 7.55	62.13 ± 9.83	41.20 ± 7.44	3.50 ± 1.23

Two-way ANOVA Test

Significant difference in TKN content between monitoring drains, F_1=19.33 (P<0.01).
Significant difference in TKN content between sampling months, F_2=2.76 (P<0.05).

Table DWM-9 is present as monthly average values (Mean ± SD) and the result of analysis of variance of TKN. The minimum and maximum average values were 16.27 ± 15 and 47.93 ± 35.62 in NAJ-1, 34.20 ± 6.24 and 51.23 ± 20.35 in CIV-2, 30.6 ± 7.03 and 50.53 ± 12.03 in POW-3, 38.33 ± 8.19 and 62.13 ± 9.83 in SEN-4, 26.0 ± 4.24 and 42.33 ± 4.19 in BAR-5 and 1.46 ± 0.32 and 14.05 ± 9.04 in HIN-6, (Hindon cut). Analysis of variance shows that the variation in TKN between different sampling stations (F_1=19.3, P<0.01) as well as sampling months (F_2=2.76, P<0.05) were significant.

Seasonal and annual values in TKN of five drains (Mean mg/l ± SD) varied from a minimum of 23.56 ± 10.24 (Summer) to 29.97 ± 5.48 (Winter) in NAJ-1, 39.21 ± 12.78 (Monsoon) to 47.69 ± 12.47 (Winter) in CIV-2, 33.07 ± 4.92 (Monsoon) to 42.49 ± 8.53 (Pre-Monsoon) in POW-3, 43.56 ± 11.63 (Monsoon) to 58.17 ± 8.43 (Winter) in SEN-4, 33.46 ± 9.37 (Summer) to 40.94 ± 5.46 (Pre-Monsoon) in BAR-5, 1.79 ± 0.08 (Monsoon) to 7.10 ± 2.31 (Winter) in HIN-6 (Hindon cut). Annual average TKN values of five drains and one river cut were in the order of 49.74 (SEN-4) > 43.10 (CIV-2) > 39.32 (POW-3) > 37.56 (BAR-5) > 27.75 (NAJ-1) > 5.27 (HIN-6).

Monthly average (Mean ± SD) values and the result of analysis of variance of TKN (mg/l) in river water are tabulated in Table RWM-10. Monthly average TKN value varied between 0.79 ± 0.28 (Jul.) to 1.62 ± 0.88 (Mar.) at WAZ-1, 4.27 ± 1.67 (Jul.) to 18.73 ± 5.23 (Jan.) at ISB-2, 3.75 ± 1.83 (Aug.) to 14.83 ± 6.35(Jan.) at ITO-3, 4.58 ± 2.38 (Aug.) to 16.70 ± 5.64 (Jan.) at NIZ-4, 3.7 ± 0.91 (Aug.) to 14.73 ± 3.81 (Jan.) at OKH-5. Analysis of variance showed significant difference in TKN, between different stations (F_1=21.05, P<0.01), as well as between monitoring months (F_2=2.27, P<0.05).

Table RWM-10: Monthly Variation of TKN (mg/l) in River Water (Year 1993-1995)

Month	*WAZ-1*	*ISB-2*	*ITO-3*	*NIZ-4*	*OKH-5*
	Mean ± SD	*Mean ± SD*	*Mean ± SD*	*Mean ± SD*	*Mean ± SD*
Jan.	1.39 ± 0.09	18.73 ± 5.23	14.83 ± 6.34	16.70 ± 5.64	14.73 ± 3.81
Feb.	1.56 ± 0.56	14.12 ± 2.79	11.26 ± 2.82	11.10 ± 1.10	10.40 ± 2.53
Mar.	1.62 ± 0.88	15.89 ± 1.68	9.85 ± 0.31	11.75 ± 1.66	11.08 ± 1.29
Apr.	1.35 ± 0.16	12.37 ± 1.97	7.52 ± 2.33	9.80 ± 2.33	9.96 ± 1.20
May	1.12 ± 0.37	9.31 ± 2.62	10.57 ± 1.68	12.35 ± 3.68	12.53 ± 4.11
Jun.	0.84 ± 0.06	16.20 ± 11.08	10.28 ± 1.86	12.26 ± 3.43	6.95 ± 2.78
Jul.	0.79 ± 0.28	4.27 ± 1.67	4.20 ± 1.92	4.88 ± 2.78	4.60 ± 2.51
Aug.	1.45 ± 0.92	4.30 ± 1.77	3.75 ± 1.83	4.58 ± 2.39	3.70 ± 0.91
Sep.	1.46 ± 0.67	6.77 ± 2.35	11.63 ± 7.07	12.21 ± 8.08	10.72 ± 4.53
Oct.	1.45 ± 0.48	10.32 ± 6.23	8.31 ± 5.62	9.50 ± 7.70	8.84 ± 5.11
Nov.	1.20 ± 0.30	6.53 ± 1.65	7.11 ± 1.78	8.30 ± 1.81	10.11 ± 4.63
Dec.	1.35 ± 0.61	8.07 ± 2.40	11.94 ± 1.64	10.10 ± 1.93	8.27 ± 3.09

Two-way ANOVA Test

Significant difference in TKN content between river monitoring stations, F_1=21.05 (P<0.01).
Significant difference in TKN content between monitoring months, F_2=2.27 (P<0.05).

Seasonal and annual averages of TKN at five monitoring stations varied from 1.23 ± 0.56 (Summer) to 1.33 ± 0.42 (Winter) at WAZ-1, 5.11 ± 2.27 (Monsoon) to 14.65 ± 5.59 (Winter) at ISB-2, 6.53 ± 5.66 (Monsoon) to 12.71 ± 4.38 (Winter) at ITO-3, 7.22 ± 6.22 (Monsoon) to 13.07 ± 4.37 (Winter) at NIZ-4, 6.34 ± 4.35 (Monsoon) to 10.83 ± 4.32 (Winter) at OKH-5.

In the present study the maximum values of TKN were recorded during rainy season at Wazirabad barrage (unpolluted stretch) which might be due to influx of nitrogen rich flood water that brings large amount of autochthonous organic matter (loosely bound with soil). The other down stream stations showed peak values in summer months, which might be due to lean discharge of Wazirabad barrage, high evapo-transpiration rate and high concentration of drain effluents at polluted river stretch. The minimum values of TKN during monsoon at all down stream stations were obviously due to dilution.

Nitrate-Nitrogen of Drain and River Water

The nitrate is a stable product of aerobic oxidation of nitrogenous organic matter. Significant sources of nitrate are domestic sewage, effluent from nitrogenous fertilizer plants and agricultural runoff. Unpolluted waters contain usually small amounts of nitrate. A limit of 10 mg of nitrate-nitrogen per litre has been imposed on drinking water to prevent methaemoglobinemia in infants. In surface water, nitrate as nutrient is taken up by plants and converted in to cell protein. It is an essential nutrient for many photosynthetic autotrophs and in some cases has been identified as the growth limiting nutrients (APHA,1991).

Table DWM-10 presents monthly averages (Mean ± SD) and the results of analysis of variance of NO_3-N. The NO_3-N values varied from 2.43 ± 0.24 to 6.53 ± 0.47 in NAJ-1, 4.73 ± 1.36 to 8.87 ± 1.72 in CIV-2, 3.37 ± 0.56 to 9.33 ± 0.95 in POW-3, 3.93 ± 0.63 to 9.93 ± 2.70 in SEN-4, 2.57 ± 0.46 to 7.47 ± 1.81 in BAR-5, 1.25 ± 0.58 to 3.33 ± 1.17 in HIN-6. Results of analysis of variance shows a significant difference in NO_3-N content between monitoring drains, (F_1=19.7, P<0.01), as well as between monitoring months (F_2=2.75, P<0.05).

Table DWM-10: Monthly Variation of Nitrate-Nitrogen (mg/l) in Drain Water (Year 1993-1995)

Month	*NAJ-1*	*CIV-2*	*POW-3*	*SEN-4*	*BAR-5*	*HIN-6*
	Mean ± SD	*Mean ± SD*	*Mean ± SD*	*Mean ± SD*	*Mean ± SD*	*Mean ± SD*
Jan.	5.80 ± 0.00	8.87 ± 1.72	8.00 ± 1.07	9.47 ± 1.50	7.47 ± 1.81	2.09 ± 0.25
Feb.	6.53 ± 0.47	8.90 ± 0.99	8.57 ± 1.02	9.93 ± 2.70	6.87 ± 1.68	2.07 ± 0.37
Mar.	5.47 ± 0.47	8.23 ± 1.05	9.33 ± 0.95	8.50 ± 2.70	6.30 ± 0.49	1.76 ± 0.10
Apr.	5.57 ± 0.52	8.53 ± 2.69	7.80 ± 1.42	6.90 ± 1.28	5.20 ± 0.78	1.83 ± 0.59
May	6.40 ± 0.57	7.93 ± 2.78	7.23 ± 0.82	7.27 ± 3.15	4.63 ± 0.65	2.62 ± 0.17
Jun.	4.27 ± 0.24	6.77 ± 2.75	6.30 ± 1.02	7.33 ± 2.43	5.47 ± 1.25	3.33 ± 0.17
Jul.	2.43 ± 0.24	4.73 ± 1.36	3.37 ± 0.56	3.93 ± 0.63	2.57 ± 0.46	1.42 ± 0.49
Aug.	2.59 ± 0.44	4.90 ± 1.37	3.37 ± 0.60	4.27 ± 1.33	2.60 ± 0.16	1.25 ± 0.58
Sep.	3.97 ± 0.47	7.00 ± 2.67	4.57 ± 0.21	6.27 ± 1.07	4.80 ± 0.86	1.30 ± 0.33
Oct.	3.99 ± 0.13	7.77 ± 2.64	6.17 ± 0.45	6.30 ± 1.08	5.80 ± 0.41	2.04 ± 0.69
Nov.	4.78 ± 0.67	7.60 ± 1.91	6.63 ± 0.31	8.40 ± 1.10	6.43 ± 0.53	2.40 ± 0.54
Dec.	5.36 ± 0.37	7.80 ± 2.18	6.90 ± 0.65	8.83 ± 1.24	6.30 ± 1.08	2.11 ± 0.15

Two-way ANOVA Test

Significant difference in nitrate-nitrogen content between monitoring drains, F_1=19.77 (P<0.01).
Significant difference in nitrate-nitrogen content between sampling months, F_2=2.75 (P<0.05).

Seasonal averages shows variation with minimum of 3 ± 0.8 (Monsoon) to 6.10 ± 0.59 (Winter) in NAJ-1, 5.54 ± 2.16 (Monsoon) to 8.94 ± 1.79 (Winter) in NAJ-1, 5.54 ± 2.16 (Monsoon) to 8.94 ± 1.79 (Winter) in CIV-2, 3.77 ± 0.75 (Monsoon) to 8.03 ± 1.03 (Winter) in SEN-4, 3.32 ± 1.19 {Mon.) to 7.06 ± 1.53 {Winter) in BAR-5, 1.32 ± 0.39 (Monsoon) to 2.38 ± 0.86 (Winter) in HIN-6. Yearly averages were in the order of 7.42 > 7.28 > 6.52 > 5.37 > 4.76 > 2.02 in drains of CIV-2, SEN-4, POW-3, BAR-5, NAJ-1 and HIN-6 respectively.

Monthly average values (Mean ± SD) of NO_3-N and the results of analysis of variance in river water are tabulated in Table RWM-11.

The values ranged from 0.84 ± 0.33 (May) to 1.38 ± 0.05 (Jan.) at WAZ-1, 0.95 ± 0.03 (Jul.) to 4.64 ± 2.09 (Jun.) at ISB-2, 0.71 ± 0.16 (Jul.) to 3.61 ± 1.44 (Jan.) at ITO-3, 1.11 ± 0.23 (Jul.) to 4.06 ± 1.46 (Jan.) at NIZ-4, 1.14 ± 0.22 (Aug.) to 3.47 ± 1.01 (Jan.) at OKH-5. Analysis of variance shows significant difference in NO_3-N between the monitoring stations (F_1=20.68, $P<0.01$) as well between sampling months (F_2=2.59, $P<0.05$).

Seasonal and annual averages of NO_3-N ranged from a minimum of 0.09 ± 0.17 (Monsoon) to a maximum of 1.31 ± 0.2 (Winter) at WAZ-1, 1.21 ± 0.54 (Monsoon) to 2.94 ± 0.64 (Winter) at ISB-2, 1.29 ± 0.80 (Monsoon) to 3.13 ± 0.93 (Winter) at ITO-3, 1.64 ± 0.88 (Monsoon) to 3.59 ± 0.96 (Winter) at NIZ-4, 1.5710.88 (Monsoon) to 2.80 ± 1.00 (Winter) at OKH-5.

Table RWM-11: Monthly Variation of Nitrate Nitrogen (mg/l) in River Water (Year 1993-1995)

Month	*WAZ-1*	*ISB-2*	*ITO-3*	*NIZ-4*	*OKH-5*
	Mean ± SD	*Mean ± SD*	*Mean ± SD*	*Mean ± SD*	*Mean ± SD*
Jan.	1.38 ± 0.05	3.34 ± 0.08	3.61 ± 1.44	4.06 ± 1.46	3.47 ± 0.81
Feb.	1.32 ± 0.17	3.12 ± 0.73	3.09 ± 0.30	3.53 ± 0.26	3.00 ± 0.14
Mar.	1.15 ± 0.25	3.27 ± 0.74	2.67 ± 0.30	3.11 ± 0.53	2.52 ± 0.45
Apr.	0.95 ± 0.10	2.6.7 ± 0.33	2.74 ± 0.31	3.67 ± 1.15	3.31 ± 0.63
May	0.84 ± 0.33	2.90 ± 0.22	3.12 ± 0.72	3.92 ± 0.72	3.57 ± 0.49
Jun.	0.99 ± 0.13	4.64 ± 2.09	3.38 ± 0.50	3.56 ± 0.59	2.92 ± 0.70
Jul.	0.76 ± 0.08	0.95 ± 0.03	0.71 ± 0.16	1.11 ± 0.23	1.49 ± 0.68
Aug.	0.99 ± 0.08	0.99 ± 0.08	0.85 ± 0.14	1.18 ± 0.33	1.14 ± 0.22
Sep.	0.96 ± 0.21	1.69 ± 0.72	2.30 ± 0.58	2.63 ± 0.83	2.09 ± 0.66
Oct.	0.97 ± 0.01	2.18 ± 0.77	2.20 ± 0.24	2.38 ± 0.39	2.25 ± 0.70
Nov.	1.28 ± 0.36	2.33 ± 0.34	2.24 ± 0.18	2.67 ± 0.19	2.43 ± 0.30
Dec.	1.19 ± 0.27	2.74 ± 0.49	2.73 ± 0.12	3.26 ± 0.37	2.47 ± 1.00

Two-way ANOVA Test

Significant difference in NO_3-N content between river monitoring stations, F_1=20.68 ($P<0.01$).
Significant difference in NO_3-N content between monitoring months, F_2=2.59 ($P<0.05$).

The peak values of NO_3-N were observed during winter season because of reduced water flow at Delhi stretch due to lean discharge of Wazirabad barrage and flow of drain effluents without dilution. The lowest values during the rainy season was due to dilution effect at down-stream stations.

Ajmal *et al.* (1985) reported minimum nitrate content in winter and maximum in summer in Kali Nadi. Sunder (1988) observed minimum during July and maximum in January. Shah (1988) reported the minimum in summer and maximum in winter months in river Jhelum. Lower values of nitrate has been reported in rainy season by Trivedy *et al.* (1990).

Annual averages were in order of 2.92 > 2.57 > 2.55 > 2.47 > 1.00 at monitoring station NIZ-4, ISB-2, OKH-5, ITO-3, and WAZ-1 respectively.

Ammonical-Nitrogen of Drain and River Water

The nitrogenous organic matter present in water is degraded biologically with the production of ammonia. The most important source of ammonia in natural water is the decomposition and ammonification of organic matter and also the sewage along with the excreta of same aquatic animals

(Saxena, 1987). Presence of ammonia can be considered as the chemical evidence of organic pollution (Trivedy *et al.*, 1987; Ajmal and Razi-Ud-Din, 1988). Organically polluted waters has normally higher concentration of ammonia which is a product of ammonification of organic matter (Ellis *et al.*, 1946, Raybak and Sikarska, 1976).

Table DWM-11 shows monthly averages (Mean ± SD) and the results of analysis of variance of NH_3-N of different drains. The values (mg/l) ranged between a minimum of 13.97 ± 3.4 to a maximum of 26.53 ± 23.17 in NAJ-1, 20.10 ± 8.18 to 37.83 ± 21.76 in CIV-2, 22.57 ± 3.71 to 38.87 ± 5.33 in POW-3, 25.67 ± 4.71 to 49.33 ± 4.99 in SEN-4, 19.0 ± 4.85 to 32.67 ± 4.50 in BAR-5, 0.52 ± 0.30 to 4.49 ± 2.15 in HIN-6. Analysis of variance shows significant difference between sampling drains (F_1=19.42, P<0.01) as well as between monitoring months (F_2=2.68, P<0.05). Seasonal and annual average values of NH_3-N ranged from 16.85 ± 6.36 to 19.84 in NAJ-1, 23.47 ± 10.88 to 34.14 ± 10.0 in CIV-2, 24.81 ± 3.74 to 32.42 ± 6.7 in POW-3, 31.78 ± 8.9 to 44.61 ± 5.68 in SEN-4, 24.22 ± 9.54 to 32.0 ± 4.84 in BAR-5, 1.71 ± 0.6 to 5.63 ± 1.3 in HIN-6 respectively. Annual mean values of NH_3-N (mg/l) were 18.43, 29.04, 29.70, 38.67, 26.53, 26.53 and 4.1 in NAJ-1, CIV-2, POW-3, SEN-4, BAR-5 and HIN-6 respectively.

Table DWM-11: Monthly Variation of Ammonical-Nitrogen (mg/l) in Drain Water (Year 1993-1995)

Month	*NAJ-1*	*CIV-2*	*POW-3*	*SEN-4*	*BAR-5*	*HIN-6*
	Mean ± SD	*Mean ± SD*	*Mean ± SD*	*Mean ± SD*	*Mean ± SD*	*Mean ± SD*
Jan.	17.13 ± 2.74	33.10 ± 12.14	28.77 ± 6.27	43.17 ± 5.33	28.33 ± 1.25	2.95 ± 1.93
Feb.	22.10 ± 4.93	34.33 ± 6.76	33.00 ± 2.83	41.61 ± 3.09	25.33 ± 2.87	9.11 ± 6.30
Mar.	16.20 ± 3.96	30.17 ± 5.95	38.87 ± 5.33	43.43 ± 10.30	28.00 ± 6.68	3.15 ± 1.45
Apr.	20.17 ± 2.09	35.10 ± 16.21	34.47 ± 6.64	33.10 ± 2.02	22.33 ± 6.18	4.03 ± 0.60
May	14.40 ± 8.56	31.83 ± 21.76	29.47 ± 4.03	39.17 ± 16.88	19.20 ± 4.85	3.74 ± 1.44
Jun.	16.63 ± 7.13	22.27 ± 4.01	26.87 ± 2.29	31.67 ± 11.61	25.87 ± 10.57	4.49 ± 2.15
Jul.	26.53 ± 23.17	20.10 ± 8.18	27.20 ± 2.44	25.67 ± 4.71	19.00 ± 3.27	0.55 ± 0.08
Aug.	13.97 ± 3.40	24.27 ± 13.35	24.67 ± 3.40	30.33 ± 10.53	21.33 ± 4.99	0.63 ± 0.11
Sep.	15.80 ± 3.51	26.03 ± 9.55	22.57 ± 3.71	39.33 ± 2.87	32.33 ± 11.67	0.52 ± 0.30
Oct.	18.57 ± 5.30	30.20 ± 10.79	30.60 ± 2.41	36.00 ± 5.10	32.67 ± 4.50	1.36 ± 0.74
Nov.	20.13 ± 5.87	27.77 ± 5.12	29.13 ± 6.31	45.17 ± 1.55	31.67 ± 5.79	1.60 ± 0.95
Dec.	18.89 ± 0.94	27.33 ± 6.18	30.80 ± 7.59	49.33 ± 4.99	32.33 ± 10.66	1.89 ± 0.86

Two-way ANOVA Test

Significant difference in ammonical-nitrogen content between monitoring drains, F_1=19.42 (P<0.01).
Significant difference in ammonical-nitrogen content between sampling months, F_2=2.68 (P<0.05).

Monthly average (Mean ± SD) NH_3-N values of river stations and results of analysis of variance are tabulated in Table RWM-12. The values ranged from 0.82 ± 0.1 (Nov.) to 1.13 ± 0.55 (Mar.) at WAZ-1, 1.64 ± 0.82 (Aug.) to 9.5 ± 2.56 (Jan.) at ISB-2, 1.45 ± 0.85 (Aug.) to 8.45 ± 3.49 (Jan.) at ITO-3, 1.33 ± 0.83 (Aug.) to 7.65 ± 2.51 (Jan.) at NIZ-4, 1.4 ± 0.85 (Aug.) to 7.23 ± 1.92 (Jan.) at OKH-5. Analysis of variance showed significant difference in NH_3-N content between different stations (F_1=22.14, P<0.01) as well as between monitoring months (F_2=2.40, P<0.05).

Seasonal values (Mean ± SD) ranged from a minimum of 0.8 ± 0.49 (Monsoon) to a maximum of 0.97 ± 0.35 (Pre-Monsoon) at WAZ-1, 1.78 ± 0.81 (Monsoon) to 7.42 ± 2.77 (Winter) at ISB-2, 2.70 ± 2.62 (Monsoon) to 5.93 ± 2.01 (Winter) at NIZ-4, 2.48 ± 1.24(Monsoon) to 5.48 ± 2.32 (Summer) at OKH.-5.

Annual averages were in order of 5.05 > 4.68 > 4.63 > 0.90 at ITO- 3, ISB-2, OKH-5, and WAZ-1 river monitoring stations respectively.

Table RWM-12: Monthly Variation of Ammonical-Nitrogen (mg/l) in River Water (Year 1993-1995)

Month	*WAZ-1* *Mean ± SD*	*ISB-2* *Mean ± SD*	*ITO-3* *Mean ± SD*	*NIZ-4* *Mean ± SD*	*OKH-5* *Mean ± SD*
Jan.	0.82 ± 0.07	9.50 ± 2.56	8.48 ± 2.60	7.65 ± 2.51	7.23 ± 1.92
Feb.	1.10 ± 0.41	5.97 ± 0.97	6.49 ± 1.20	5.05 ± 1.19	4.48 ± 1.22
Mar.	1.13 ± 0.55	6.65 ± 1.79	5.15 ± 1.10	3.47 ± 0.53	4.79 ± 1.77
Apr.	1.02 ± 0.12	4.09 ± 0.81	4.32 ± 0.98	4.37 ± 0.68	5.08 ± 0.82
May	0.89 ± 0.28	5.02 ± 2.01	6.65 ± 1.36	7.12 ± 2.66	7.88 ± 1.89
Jun.	0.60 ± 0.15	7.58 ± 3.18	6.83 ± 0.70	7.20 ± 3.02	4.18 ± 2.47
Jul.	0.61 ± 0.25	1.69 ± 0.86	1.54 ± 0.40	1.50 ± 1.27	1.71 ± 1.20
Aug.	0.87 ± 0.53	1.64 ± 0.82	1.45 ± 0.20	1.33 ± 0.83	1.40 ± 0.85
Sep.	0.97 ± 0.55	2.02 ± 0.68	5.11 ± 1.30	4.42 ± 3.02	4.33 ± 1.73
Oct.	1.12 ± 0.44	3.88 ± 2.45	4.33 ± 1.00	4.55 ± 3.57	5.09 ± 2.56
Nov.	0.82 ± 0.10	3.28 ± 0.65	3.68 ± 0.90	3.64 ± 1.21	5.75 ± 2.91
Dec.	0.91 ± 0.34	4.85 ± 1.50	6.57 ± 1.30	5.27 ± 0.26	3.94 ± 1.45

Two-way ANOVA Test
Significant difference in NH_3-N content between river monitoring stations, F_1=22.14 ($P<0.01$).
Significant difference in NH_3-N content between monitoring months, F_2=2.4 ($P<0.05$).

Phosphate of Drain and River Water

Phosphorous is inevitable for all organisms as it is the chief constituent of ATP, the energy carrier acid present in nucleic and as phosphate group of nucleotides. In general, phosphorous is readily oxidised and occurs in the earth's rocks principally as orthophosphate. The phosphorous occurs in natural waters and in wastewater almost solely as phosphates rather than a free state. In water, phosphorous occurs both in inorganic and organic forms (as bound in the organic matter), only inorganic phosphorous (as orthophosphates) plays a dynamic role in an aquatic ecosystem. The most important anthropogenic sources of phosphate are the discharge of domestic sewage, detergents and agricultural run-off (Trivedy *et al.*, 1987). The amount of phosphorous in most freshwater bodies is very small and is subjected in seasonal and vertical variations (Michael, 1984). Phosphorous is essential to the growth of organisms that limits the primary productivity of a water body. The phosphate content of flowing water change slightly from time to time while that of stagnant waters it is more or less constant (Imevbore, 1978). The enrichment of this nutrient leads to the process of eutrophication.

The importance of phosphate in natural waters has been stressed by several authors (Juday *et al.*, 1928; Ohle, 1934; Phifer and Thompson, 1937; Cole, 1979; Dutta Munshi and Dutta Munshi, 1995). The total phosphate content is considered to be a nutrient of major importance in production process (Hutchinson, 1957 and Vollenwinder, 1968). The increased application of fertilizers, use of detergents and domestic sewages contribute significantly to the heavy loading of phosphorous in water (Golterman, 1975).

Table DWM-12 represents monthly average values and the result of analysis of variance of total phosphate present in the drain water in six monitoring drains. The average values ranged from a minimum of 2.12 ± 0.58 to a maximum of 4.78 ± 2.89 in NAJ-1, 3.93 ± 1.76 to 7.95 ± 1.28 in CIV-2, 3.39 ± 1.01 to 5.85 ± 0.18 in POW-3, 3.42 ± 1.17 to 5.97 ± 0.17 in SEN-4, 2.37 ± 0.31 to 4.62 ± 2.34 in BAR-5, 0.07 ± 0.05 to 1.19 ± 0.44 in HIN-6. Significant difference in PO_4-P content was found between five drains (F_1=19.43 P<0.01) as well as between monitoring months (F_2=2.69, P>0.05).

Table DWM-12: Monthly Variation of Total Phosphate (mg/l) in Drain Water (Year 1993-1995)

Month	*NAJ-1*	*CIV-2*	*POW-3*	*SEN-4*	*BAR-5*	*HIN-6*
	Mean ± SD	*Mean ± SD*	*Mean ± SD*	*Mean ± SD*	*Mean ± SD*	*Mean ± SD*
Jan.	3.55 ± 1.63	7.95 ± 1.28	4.91 ± 0.27	5.47 ± 0.74	2.94 ± 0.24	1.19 ± 0.44
Feb.	3.93 ± 0.43	7.08 ± 0.64	5.34 ± 0.65	5.90 ± 0.71	2.96 ± 0.38	0.86 ± 0.85
Mar.	3.24 ± 0.95	6.28 ± 1.37	5.85 ± 0.18	5.97 ± 0.17	3.44 ± 0.85	0.45 ± 0.27
Apr.	3.25 ± 0.36	6.49 ± 0.49	5.59 ± 0.72	4.93 ± 1.42	3.11 ± 0.44	0.71 ± 0.32
May	3.37 ± 1.18	5.97 ± 1.24	4.06 ± 1.16	5.48 ± 0.87	2.37 ± 0.31	0.63 ± 0.69
Jun.	2.12 ± 0.58	5.03 ± 1.96	4.23 ± 1.33	5.13 ± 1.13	3.92 ± 2.09	0.94 ± 1.05
Jul.	4.78 ± 2.89	3.93 ± 1.76	3.69 ± 0.72	3.97 ± 0.65	3.63 ± 2.45	0.07 ± 0.05
Aug.	2.75 ± 0.56	6.46 ± 1.71	3.63 ± 0.45	4.33 ± 0.66	2.96 ± 0.75	0.07 ± 0.08
Sep.	2.53 ± 0.91	5.12 ± 0.98	3.39 ± 1.01	5.04 ± 0.96	2.97 ± 0.90	0.16 ± 0.17
Oct.	3.18 ± 0.85	5.31 ± 1.17	4.26 ± 1.61	3.42 ± 1.17	2.89 ± 0.77	0.19 ± 0.26
Nov.	3.71 ± 0.44	5.49 ± 2.33	5.03 ± 0.90	4.40 ± 0.59	3.39 ± 0.37	0.22 ± 0.17
Dec.	3.82 ± 0.24	5.66 ± 1.23	5.47 ± 0.60	5.20 ± 0.78	4.62 ± 2.34	0.19 ± 0.26

Two-way ANOVA Test

Significant difference in total phosphate content between monitoring drains, F_1=19.43 (P<0.01).
Significant difference in total phosphate content between sampling months, F_2=2.69 (P<0.05).

Table RWM-13: Monthly Variation of Phosphate (mg/l) in River Water (Year 1993-1995)

Month	*WAZ-1*	*ISB-2*	*ITO-3*	*NIZ-4*	*OKH-5*
	Mean ± SD	*Mean ± SD*	*Mean ± SD*	*Mean ± SD*	*Mean ± SD*
Jan.	0.17 ± 0.00	1.47 ± 0.32	1.82 ± 0.34	1.81 ± 0.41	1.30 ± 0.40
Feb.	0.14 ± 0.02	1.43 ± 0.18	1.58 ± 0.52	1.30 ± 0.33	1.42 ± 0.79
Mar.	0.16 ± 0.03	1.79 ± 0.13	1.26 ± 0.01	1.27 ± 0.18	1.54 ± 1.01
Apr.	0.17 ± 0.02	1.38 ± 0.24	1.20 ± 0.09	1.18 ± 0.13	1.18 ± 0.53
May	0.15 ± 0.04	1.27 ± 0.25	1.43 ± 0.31	1.24 ± 0.15	1.19 ± 0.36
Jun.	0.11 ± 0.05	1.81 ± 0.48	1.61 ± 0.29	1.12 ± 0.11	1.00 ± 0.16
Jul.	0.08 ± 0.03	0.89 ± 0.07	0.83 ± 0.09	0.82 ± 0.13	0.64 ± 0.22
Aug.	0.08 ± 0.03	0.84 ± 0.09	0.80 ± 0.07	0.82 ± 0.21	0.66 ± 0.14
Sep.	0.16 ± 0.14	0.96 ± 0.16	1.07 ± 0.30	1.11 ± 0.14	0.91 ± 0.11
Oct.	0.10 ± 0.01	1.42 ± 0.64	1.22 ± 0.55	1.17 ± 0.50	1.21 ± 0.35
Nov.	0.14 ± 0.05	1.15 ± 0.51	0.93 ± 0.55	1.28 ± 0.48	1.66 ± 0.70
Dec.	0.13 ± 0.06	1.03 ± 0.14	1.26 ± 0.15	1.44 ± 0.30	1.30 ± 0.74

Two-way ANOVA Test

Significant difference in phosphate content between river monitoring stations, F_1=19.81 (P<0.01).
Significant difference in phosphate content between monitoring months, F_2=2.20 (P<0.05).

The seasonal average values of PO_4-P content in drain water ranged from a minimum of 2.99 ± 0.97 (Summer) to a maximum of 3.72 ± 1.02 (Winter) in NAJ-1, 5.17 ± 1.84 (Monsoon) to 7.25 ± 1.26 (Winter) in CIV-2, 3.57 ± 0.77 (Monsoon) to 5.29 ± 0.56 (Winter) in POW-3, 3.94 ± 0.97 (Pre-Monsoon) to 5.69 ± 0.62 (Winter) in SEN-4, 3.06 ± 0.63 (Pre-Monsoon) to 3.55 ± 1.57 (Winter) in BAR-5 and 0.49 ± 0.13 (Pre-Monsoon) to 1.06 ± 0.38 (Winter) in HIN-6. Annual averages were in order of 5.9 > 4.94 > 4.62 > 3.35 > 3.27 > 0.70 in CIV-2, SEN-4, POW-3, NAJ-1, BAR-5 drain and HIN-6 (Hindon river cut), respectively.

Monthly averages (Mean ± SD) and the results of analysis of variance of phosphate in river water are tabulated in Table RWM-13. The values showed variation with a minimum of 0.08 ± 0.03 (Jul.) to a maximum of 0.17 ± 0.0 (Jan.) at WAZ-1, 0.84 ± 0.09 (Aug.) to 1.79 ± 0.13 (Mar.) at ISB-2, 0.08 ± 0.07 (Aug.) to 1.82 ± 0.34 (Jan.) at ITO-3, 0.82 ± 0.21 (Aug.) to 1.81 ± 0.41 (Jan.) at NIZ-4, and 0.66 ± 0.14 (Aug.) to 1.54 ± 1.01 (Jan.) at OKH-5. Analysis of variance showed a significant difference between the sampling stations (F_1=18.81, $P<0.01$) and also between monitoring months (F_2=2.2, $P<0.05$).

Seasonal as well as annual average values of river water showed fluctuation from 0.10 ± 0.09 (Monsoon) to 0.16 ± 0.02 (Winter) at WAZ-1, 0.90 ± 0.13 (Monsoon) to 1.57 ± 0.39 (Summer) at ISB-2, 0.90 ± 0.22 (Monsoon) to 1.55 ± 0.43 (Winter) at ITO-3, 0.94 ± 0.24 (Monsoon) to 1.54 ± 0.40 (Winter) at NIZ-4, 0.74 ± 0.20 (Monsoon) to 1.19 ± 0.57 (Winter) at OKH-5, respectively.

The phosphate content in polluted stretch were maximum during winter season, possibly due to lesser flow of river, *i.e.* the percentage contribution of drain water was much higher than the river water discharge from Wazirabad barrage. The reduced level of PO_4 in monsoon season might be due to dilution of drain water by rain water.

Annual averages were in order of 1.29 > 1.25 > 1.22 > 1.17 > 0.13 at monitoring stations ISB-2, ITO-3, NIZ-4, OKH-5 and WAZ-1 respectively.

Hutchinson (1957) has reported an increase of phosphate as a result of sewage contamination. Similar findings have also been reported by Shah (1988), Rana and Palria (1982), Ghose and Sharma (1989) and Kumar (1995a) in their studies of different river system.

Sulphate of Drain and River Water

Sulphate occurs naturally in water as result of leaching from gypsum and other common minerals. The sulphate is usually second to carbonate as the principal anion in freshwaters, although chloride sometime surpasses it. Free or elemental sulphur is inactive at normal temperature. This element can combine with both metals and non-metals to form many compounds.

Atmospheric sources of sulphate in water have increased with industrial activities. Coal combustion produces maximum SO_2 and copper smelting and paper manufacturing add to it. Through precipitation and run-off sulphate level of some freshwater becomes unusually high. The sulphate content of municipal water supplies is usually increased during clarification by alum. Monthly average values (Mean ± SD) of sulphate and the results of analysis of variance tabulated in Table DWM-13.

The average values varied between a minimum of 64.5 ± 2.67 to a maximum of 77.67 ± 2.05 in NAJ-1, 62.67 ± 11.12 to 91.0 ± 8.49 in CIV-2, 70.33 ± 15.52 to 111.67 ± 17.44 in POW-3, 88.33 ± 18.12 to 124.83 ± 21.14 in SEN-4, 91.67 ± 6.6 to 112.33 ± 11.09 in BAR-5, and 17.0 ± 5.72 to 33.0 ± 2.45 at HIN-6. Results of analysis of variance (five drains) shows significant difference between sampling drains (F_1=19.03, $P<0.01$) and also between sampling months (F_2=2.77, $P<0.05$).

Seasonal values ranged from 67.87 ± 10.95 (Winter) to 71.0 ± 15.94 (Monsoon) in NAZ-1, 68.67 ± 9.68 to 88.33 ± 10.22 (Summer) in CIV-2, 76.89 ± 14.76 (Monsoon) to 90.43 ± 18.44 (Pre-Monsoon) in POW-3, 90.22 ± 8.66 (Monsoon) to 111.0 ± 14.76 (Pre-Monsoon) in SEN-4, 96.56 ± 10.24 (Monsoon) to 112.14 ± 9.39 (Pre-Monsoon) in BAR-5, 20.44 ± 2.6 (Monsoon) to 32.22 ± 3.8 (Winter) in HIZ-6. Annual average SO_4 values of five drains and one river cut were in order of 104.81 > 102.63 >, 87.61 > 80.51 > 70.99 > 29.17 in BAR-5, SEN-4, POW-3, NAJ-1 and HIN-6, respectively.

Table DWM-13: Monthly Variation of Sulphate (mg/l) in Drain Water (Year 1993-1995)

Month	*NAJ-1*	*CIV-2*	*POW-3*	*SEN-4*	*BAR-5*	*HIN-6*
	Mean ± SD	*Mean ± SD*	*Mean ± SD*	*Mean ± SD*	*Mean ± SD*	*Mean ± SD*
Jan	70.0 ± 8.8	83.0 ± 9.63	88.67 ± 8.18	105.67 ± 18.87	111.33 ± 3.40	32.00 ± 2.83
Feb	77.7 ± 5.2	76.3 ± 5.91	90.67 ± 3.68	86.33 ± 18.12	104.00 ± 6.38	33.00 ± 4.24
Mar	69.7 ± 2.1	91.00 ± 8.49	111.67 ± 17.44	93.33 ± 23.10	111.00 ± 6.48	30.67 ± 1.25
Apr	77.7 ± 2.1	87.00 ± 16.97	91.67 ± 6.18	113.00 ± 12.08	108.00 ± 14.24	34.00 ± 3.74
May	71.3 ± 8.5	89.00 ± 6.24	88.67 ± 10.34	124.83 ± 21.14	97.00 ± 4.55	31.33 ± 2.49
Jun	73.3 ± 6.8	85.67 ± 2.05	85.33 ± 10.21	106.67 ± 20.89	105.00 ± 12.96	32.00 ± 2.16
Jul	71.0 ± 20.1	67.33 ± 6.34	73.67 ± 13.07	83.33 ± 4.78	91.67 ± 6.60	23.67 ± 1.70
Aug	67.3 ± 17.8	62.67 ± 11.12	70.33 ± 15.52	88.33 ± 3.09	91.67 ± 3.09	20.67 ± 1.70
Sep	74.7 ± 4.0	76.00 ± 5.10	86.67 ± 5.79	99.0 ± 8.04	106.33 ± 10.87	17.00 ± 5.72
Oct	68.0 ± 5.7	79.83 ± 3.57	79.33 ± 17.00	98.67 ± 13.47	109.33 ± 7.36	30.00 ± 1.41
Nov	66.7 ± 2.6	86.00 ± 2.45	93.33 ± 13.10	117.67 ± 5.79	112.33 ± 11.09	33.00 ± 2.45
Dec	64.5 ± 11.1	81.67 ± 7.41	91.33 ± 18.80	112.67 ± 11.59	110.00 ± 8.16	32.67 ± 4.11

Two-way ANOVA Test
Significant difference in sulphate content between monitoring drains, F_1=19.03 (P<0.01).
Significant difference in sulphate content between sampling months, F_2=2.77 (P<0.05).

Table RWM-14 presents the monthly average (Mean ± SD) values and the results of analysis of variance of river water. The minimum and maximum sulphate content were 19.0 ± 0.82 (Jul.) and 36.33 ± 3.77 (Dec.) at WAZ-1, 21.0 ± 1.63 (Aug.) and 42.0 ± 2.94 (Feb.) at ISB-2, 25.0 ± 4.24 (Aug.) and 42.0 ± 8.49 (Mar.) at ITO-3, 26.33 ± 5.44 (Jul.) and 47.33 ± 4.19 (Nov.) at NIZ-4, 19.67 ± 1.25 (Aug.) and 54.38 ± 12.30 (Nov.) at OKH-5. Analysis of variance shows significant difference in sulphate content between monitoring stations (F_1=19.49, P<0.01) and also between sampling months (F_2=2.81, P<0.05).

Seasonal values varied between 20.0 ± 2.67 (Monsoon) to 28.78 ± 4.29 (Winter) at WAZ-1, 22.11 ± 4.31 (Monsoon) to 38.33 ± 7.96 (Summer) at ISB-2, 25.67 ± 4.47 (Monsoon) to 38.33 ± 2.81 (Pre-Monsoon) at ITO-3, 27.78 ± 4.94 (Monsoon) to 43.83 ± 5.46 (Pre-Monsoon) at NIZ-4, 23.6 ± 6.52 (Monsoon) to 46.73 ± 11.60 (Pre-Monsoon) at OKH-5. The concentration of sulphate in the river Yamuna at Delhi stretch showed some distinct fluctuation at different sites. Maximum concentration of sulphate was generally observed during summer and minimum during rainy season at Wazirabad barrage (non-polluted stretch). All down-stream stations showed peak values during winter season and lower values during monsoon season. Annual average values of SO_4 were 38.25 > 36.02 > 35.19 > 33.44 > 26.31 at NIZ-4, OKH-5, ITO-3, ISB-2 and WAZ-1 river monitoring points, respectively.

Table RWM-14: Monthly Variation of Total Sulphate (mg/l) in River Water (Year 1993-1995)

Month	*WAZ-1*	*ISB-2*	*ITO-3*	*NIZ-4*	*OKH-5*
	Mean ± SD	*Mean ± SD*	*Mean ± SD*	*Mean ± SD*	*Mean ± SD*
Jan.	28.33 ± 0.47	33.67 ± 5.44	37.00 ± 8.52	37.33 ± 6.18	36.40 ± 5.38
Feb.	24.67 ± 0.94	42.00 ± 2.94	38.00 ± 2.83	40.67 ± 6.55	45.07 ± 6.54
Mar.	27.00 ± 4.32	40.00 ± 5.66	42.00 ± 8.49	43.67 ± 8.01	43.33 ± 9.29
Apr.	24.67 ± 4.50	39.67 ± 1.70	38.67 ± 6.60	42.67 ± 4.78	36.33 ± 6.94
May	28.33 ± 2.62	40.00 ± 6.68	39.67 ± 2.62	43.00 ± 5.72	35.07 ± 5.01
Jun.	29.33 ± 11.09	33.67 ± 12.04	36.67 ± 13.57	37.00 ± 13.14	34.50 ± 5.21
Jul.	19.00 ± 0.82	22.67 ± 6.02	23.67 ± 5.19	26.33 ± 5.44	28.07 ± 6.54
Aug.	20.00 ± 1.41	21.00 ± 1.63	25.00 ± 4.24	25.33 ± 4.11	19.67 ± 1.25
Sep.	21.00 ± 4.08	22.67 ± 3.86	28.33 ± 3.30	31.67 ± 1.89	23.07 ± 6.92
Oct.	27.33 ± 3.09	32.67 ± 2.49	36.33 ± 2.05	40.33 ± 4.19	39.07 ± 0.82
Nov.	29.67 ± 2.05	36.67 ± 3.09	40.33 ± 1.89	47.33 ± 4.19	54.38 ± 12.30
Dec.	36.33 ± 3.77	36.67 ± 5.91	36.67 ± 4.99	43.67 ± 5.31	37.30 ± 3.75

Two-way ANOVA Test

Significant difference in total sulphate content between river monitoring stations, F_1=19.49 ($P<0.01$).
Significant difference in total sulphate content between monitoring months, F_2=2.81 ($P<0.05$).

Alkalinity of Drain and River Water

Alkalinity commonly results from carbon dioxide and water attaching sedimentary carbonate rocks and dissolving out some of the carbonate to form bicarbonate solution. Each carbonate ion in solution takes up hydrogen ion to become bicarbonate ion. Further hydrolysis of bicarbonate or carbonate leads to the appearance of hydroxyl ions, which will increase the value of pH. Alkalinity measures the buffering capacity of water and the content of CO_2 in its various forms are involved in this carbonate bicarbonate-carbonic and buffering system. Alkalinity value provides guidance in applying proper dose of chemicals in water and wastewater treatment processes.

Table DWM-14, represents monthly average alkalinity values (Mean ± SD) and results of analysis of variance for five drains. The average maximum and minimum values (mg/l) were, 198.67 ± 8.82 and 362.67 ± 35.48, 256.0 ± 11.78 and 366.67 ± 80.16, 301.21 ± 36.71 and 350.36 ± 29.93, 288.0 ± 98.36 and 465.33 ± 195.97, 261.33 ± 98.00 and 415.33 ± 15.17, 117.33 ± 6.24 and 259.67 ± 12.36 in NAJ-1, CIV-2, POW-3, SEN-4, BAR-5 and HIN-6 respectively. Analysis of variance shows significant difference between sampling drains (F_1=19.1, $P<0.01$) as well as between sampling months (F_2=2.77, $P<0.05$).

Seasonal averages ranged from 216.17 ± 28.94 (Summer) to 256.78 ± 43.15 (Monsoon) in NAJ-1, 271.11 ± 15.06 (Monsoon) to 340.0 ± 73.82 (Pre-Monsoon) in CIV-2, 276.33 ± 39.19 (Monsoon) to 357.56 ± 28.15 (Winter) in POW-3, 306.33 ± 70.47 (Monsoon) to 398.44 ± 31.97 (Winter) in SEN-4, 302.44 ± 95.47 (Monsoon) to 402.22 ± 30.34 (Winter) in BAR-5, 121.67 ± 20.36 (Monsoon) to 259.11 ± 28.67 (Winter) in HIN-6. Annual averages were in order of 377.58 > 361.19 > 326.17 > 320.86 > 234.25 > 199.69 in SEN-4, BAR-5, POW-3, CIV-2, NAZ-1 and Hin-6, respectively.

Table DWM-14: Monthly Variation of Total Alkalinity (mg/l) in Drain Water (Year 1993-1995)

Month	*NAJ-1*	*CIV-2*	*POW-3*	*SEN-4*	*BAR-5*	*HIN-6*
	Mean ± SD	*Mean ± SD*	*Mean ± SD*	*Mean ± SD*	*Mean ± SD*	*Mean ± SD*
Jan.	242.7 ± 40.8	348.0 ± 17.0	313.7 ± 13.1	398.7 ± 16.4	413.3 ± 23.2	259.0 ± 12.4
Feb.	274.7 ± 15.43	315.7 ± 28.6	307.5 ± 39.7	369.3 ± 24.1	415.3 ± 15.2	258.7 ± 16.8
Mar.	219.3 ± 9.3	357.3 ± 49.1	307.9 ± 36.2	397.3 ± 11.5	394.7 ± 44.1	251.0 ± 8.5
Apr.	224.7 ± 28.5	366.7 ± 80.2	301.2 ± 36.7	390.7 ± 11.5	356.7 ± 66.1	205.0 ± 27.1
May	222.7 ± 44.4	333.3 ± 45.8	302.9 ± 59.3	376.0 ± 33.9	347.0 ± 50.2	193.0 ± 35.7
Jun.	198.7 ± 8.2	301.3 ± 21.0	307.2 ± 26.7	465.3 ± 196.0	345.3 ± 38.7	181.7 ± 8.1
Jul.	302.8 ± 36.0	282.7 ± 3.8	311.5 ± 43.2	288.0 ± 98.4	362.0 ± 41.2	118.7 ± 9.0
Aug.	232.0 ± 11.3	274.7 ± 12.4	309.5 ± 32.9	320.0 ± 41.0	361.3 ± 98.0	117.3 ± 6.2
Sep.	235.7 ± 31.6	276.0 ± 22.8	311.1 ± 25.8	311.0 ± 54.8	284.0 ± 102.3	129.0 ± 3.6
Oct.	206.7 ± 35.2	310.7 ± 30.2	328.6 ± 61.7	381.3 ± 32.7	396.0 ± 11.8	186.0 ± 38.0
Nov.	204.0 ± 11.8	361.3 ± 101.5	342.1 ± 34.9	396.0 ± 60.5	369.3 ± 47.5	237.3 ± 21.7
Dec.	248.0 ± 36.4	342.7 ± 16.4	350.4 ± 29.9	437.3 ± 19.7	389.3 ± 45.8	259.7 ± 8.5

Two-way ANOVA Test

Significant difference in total alkalinity content between monitoring drains, F_1=19.11 ($P<0.01$).
Significant difference in total alkalinity content between sampling months, F_2=2.77 ($P<0.05$).

Monthly average values (Mean ± SD) and the results of analysis of variance of alkalinity are tabulated in Table RWM-15. Monthly maximum and minimum values were 81.33 ± 6.6 (Aug.) and 161.0 ± 4.24 (Dec.) at WAZ-1, 104.67 ± 16.52 (Aug.) and 208.0 ± 26.13 (Mar.) at ISB-2, 90.0 ± 15.58 (Aug.) and 234.67 ± 43.12 at ITO-3, 86.67 ± 21.0 (Aug.) and 234.67 ± 16.44 (Feb.) at NIZ-4, 92.0 ± 20.40 (Aug.) and 244.0 ± 20.4 (Feb.) at OKH-5. Analysis of variance shows significant difference in total alkalinity between different monitoring points (F_1=19.98, $P<0.01$) and also between sampling months (F_2=2.78, $P<0.05$).

Seasonal values varied from a minimum of 85.78 ± 16.42 (Monsoon) to a maximum of 147.22 ± 13.53 (Winter) at WAZ-1, 108.56 ± 17.56 (Monsoon) to 2.4.56 ± 19.6 (Monsoon) at ISB-2, 116.0 ± 52.35 (Monsoon) to 228.0 ± 26.47 (Winter) at ITO-3, 98.56 ± 26.72 (Monsoon) to 231.56 ± 23.66 (Winter) at NIZ-4, 98.89 ± 20.85 (Monsoon) to 225.33 ± 19.60 (Winter) at OKH-5 respectively.

During the present investigation carbonate alkalinity was found to be higher during winter in all stations. This may be due to lean river flow and higher contribution from drain water at Delhi stretch and lower temperature bringing down the rate of decomposition of salts to minimum there-by increasing the alkalinity. The second peak in summer may be due to low water level and presence of excess free carbon dioxide produced in the process of decomposition. Ruttener (1953), Munawar (1970), Sunder (1988) and Kumar (1994) observed the accumulation of bicarbonate in summer due to increased rate of decomposition. Hegde *et al.* (1985) and Trivedy *et al.* (1990) also noticed the same observation. The winter peak values of total alkalinity have been observed by Munawar (1970), Bhatnagar and Sharma (1975), Bilgrami and Datta Munshi (1985). The minimum values of bicarbonate alkalinity were observed during monsoon months may be due to the dilution effect and condition less favourable for the photosynthetic conversion to carbonates. Similar results were also recorded by Rao (1971), Brajnandan *et al.* (1985) and Kumar (1995). They pointed out that the dilution plays a key role in lowering the alkalinity. Adebisi (1980) also explained that the alkalinity is inversely proportional to the water level.

Table RWM-15: Monthly Variation of Total Alkalinity (mg/l) in River Water (Year 1993-1995)

Month	*WAZ-1* *Mean ± SD*	*ISB-2* *Mean ± SD*	*ITO-3* *Mean ± SD*	*NIZ-4* *Mean ± SD*	*OKH-5* *Mean ± SD*
Jan.	141.33 ± 1.89	196.00 ± 17.28	234.67 ± 43.12	225.33 ± 34.62	224.00 ± 8.64
Feb.	141.33 ± 17.99	208.00 ± 26.13	2.25.33 ± 9.98	242.67 ± 16.44	244.00 ± 20.40
Mar.	126.67 ± 9.43	206.67 ± 15.43	218.67 ± 10.50	229.33 ± 21.25	213.33 ± 11.47
Apr.	113.00 ± 9.90	178.33 ± 12.76	187.33 ± 27.15	198.67 ± 34.92	171.30 ± 19.75
May	99.67 ± 5.91	108.00 ± 65.24	176.67 ± 25.94	199.33 ± 10.87	152.03 ± 14.97
Jun.	81.33 ± 6.80	166.00 ± 15.75	179.33 ± 26.91	183.33 ± 19.62	133.33 ± 15.69
Jul.	78.67 ± 10.50	116.67 ± 20.55	103.33 ± 11.81	99.33 ± 20.68	102.67 ± 13.20
Aug.	81.33 ± 6.60	104.67 ± 16.52	90.00 ± 15.58	86.67 ± 21.00	92.00 ± 20.40
Sep.	97.33 ± 21.25	104.33 ± 11.44	154.67 ± 74.21	109.67 ± 31.75	102.00 ± 25.35
Oct.	115.03 ± 22.73	163.33 ± 35.57	162.77 ± 34.92	164.00 ± 36.81	180.00 ± 40.13
Nov.	141.33 ± 11.47	193.33 ± 24.51	202.67 ± 25.37	213.33 ± 26.40	226.67 ± 49.89
Dec.	161.00 ± 4.24	206.33 ± 8.96	225.33 ± 9.43	228.00 ± 5.66	197.33 ± 12.36

Two-way ANOVA Test

Significant difference in alkalinity between river monitoring stations, F_1=19.98 ($P<0.01$).
Significant difference in alkalinity between monitoring months, F_2=2.78 ($P<0.05$).

Annual average values of total alkalinity at Delhi stretch were in order of 181.64 > 180.60 > 169.89 > 162.64 > 114.83 at NIZ-4, ITO-3, OKH-4, ISB-2 and WAZ-1, respectively.

The present observation also showed some extra load of alkalinity at site NIZ-4, ITO-3, OKH-4 and ISB-2 which might be due to the domestic and industrial effluents. Shah (1988) also recorded higher concentration of bicarbonate alkalinity in river Jhelum due to domestic sewage. Sreenivasan and Sunderaj (1967), Munawar (1970) and Sreenivasan *et al.* (1974) have also reported similar findings.

Chloride of Drain and River Water

Contamination of water from domestic sewage can be monitored by chloride assays of the concerned water bodies. This is because human and animal excretions contains an average of 5g chloride/litre (Paramasivam and Sreenivasan, 1981). The chloride themselves do not pose any hazards as far as water quality is concerned but their concentration works out as a very good indicator of organic pollution (Trivedy *et al.*, 1990a). Chloride is influential in general due to osmotic salinity balance and ion exchange, but metabolic utilisation may not cause significant variations in the spatial and seasonal distribution with in an aquatic body. However, pollution sources of chloride *viz.* municipal wastewater, industrial sources and organic wastes can modify natural concentrations greatly (Ownbey and Kee, 1967). Chloride content also increase with the degree of eutrophication (Goel *et al.*, 1980, Sinha, 1986). Several workers like Zafar (1964) and Kumar (1995) described chloride as a pollution indicator. Thresh *et al.* (1944) also advocated that high chloride concentration are indicators of large amount of organic matter in the water which, have been further supported by Bhatnagar and Sharma (1975) and Adoni (1985).

Monthly average values (Mean ± SD) and the results of analysis of variance of chloride in drain water tabulated in Table DWM-15. The values fluctuated between 88 ± 16.33 and 182.67 ± 16.11, 88 ± 6.53 and 148.0 ± 33.15, 78.65 ± 9.98 and 136.0 ± 28.28, 128.33 ± 27.18 and 229.33 ± 78.27, 132 ± 60.78

and 192 ± 13.06, 29 ± 17.15 and 144 ± 11.05 in NAJ-1, CIV-2, POW-3, SEN-4, BAR-5, and HIN-6 respectively. Analysis of variance shows significant difference in chloride content between the drains (F_1=19.39, P<0.01) and also between monitoring months (F_2=2.7, P<0.05).

Table DWM-15: Monthly Variation of Chloride (mg/l) in Drain Water (Year 1993-1995)

Month	*NAJ-1*	*CIV-2*	*POW-3*	*SEN-4*	*BAR-5*	*HIN-6*
	Mean ± SD	*Mean ± SD*	*Mean ± SD*	*Mean ± SD*	*Mean ± SD*	*Mean ± SD*
Jan.	88.0 ± 16.3	100.0 ± 14.2	101.3 ± 24.7	145.3 ± 23.0	178.7 ± 6.8	137.0 ± 6.2
Feb.	98.7 ± 1.9	92.3 ± 23.3	97.3 ± 8.2	128.3 ± 27.2	156.0 ± 18.2	144.0 ± 11.1
Mar.	94.0 ± 12.3	133.3 ± 16.8	136.0 ± 28.3	173.3 ± 22.9	180.0 ± 21.4	101.5 ± 31.4
Apr.	108.0 ± 19.9	110.7 ± 10.0	108.0 ± 10.7	161.3 ± 3.8	130.7 ± 51.3	92.0 ± 25.7
May	101.3 ± 21.2	106.7 ± 11.5	101.3 ± 15.0	196.0 ± 6.6	182.0 ± 23.2	121.0 ± 22.1
Jun.	105.3 ± 18.0	118.0 ± 18.0	100.0 ± 16.3	229.3 ± 7.3	192.0 ± 13.1	117.0 ± 5.4
Jul.	182.7 ± 16.1	92.0 ± 13.1	84.3 ± 14.5	137.3 ± 24.5	132.0 ± 60.8	40.7 ± 9.0
Aug.	110.7 ± 11.5	89.3 ± 5.0	78.7 ± 10.0	149.3 ± 13.6	136.7 ± 51.0	29.0 ± 17.2
Sep.	112.7 ± 11.6	88.0 ± 6.5	88.7 ± 9.0	142.0 ± 10.7	137.7 ± 49.0	38.7 ± 22.1
Oct.	97.3 ± 11.5	97.3 ± 15.4	92.0 ± 17.0	149.3 ± 6.8	170.7 ± 10.0	51.0 ± 21.4
Nov.	105.3 ± 22.2	148.0 ± 33.2	124.0 ± 28.5	188.0 ± 5.7	176.0 ± 45.4	106.0 ± 15.3
Dec.	98.7 ± 34.6	116.0 ± 31.2	121.3 ± 29.3	176.0 ± 29.0	181.3 ± 26.6	100.7 ± 15.4

Two-way ANOVA Test

Significant difference in chloride content between monitoring drains, F_1=19.39 (P<0.01).
Significant difference in chloride content between sampling months, F_2=2.70 (P<0.05).

Seasonal values varied between a minimum of 86.0 ± 19.97 to a maximum of 135.33 ± 36.00 in NAJ-1, 95.89 ± 23.16 to 122.67 ± 36.2 in CIV-2, 83.89 ± 12.13 to 111.33 ± 23.75 in POW-3, 142.78 ± 26.96 to 168.67 ± 20.32 in SEN-4, 166.33 ± 19.62 to 173.33 ± 32.96 in BAR-5, 36.11 ± 5.3 to 131.89 ± 30.6 in HIN-6. Annual average values of chloride were in order of 340.0 > 296.67 > 294.53 > 268.08 > 205.0 > 185.36 in SEN-4, POW-3, BAR-5, NAJ-1 and HIN-6 respectively.

Monthly average (Mean ± SD) values and the results of analysis of variance for chloride in river water are tabulated in Table RWM-16. The chloride contents varied from 13.67 ± 1.7 (Jul.) to 45.33 ± 20.4 (Dec.) at WAZ-1, 27.33 ± 3.4 (Sep) to 105.33 ± 13.2 (Feb.) at ISB-2, 16.67 ± 3.4 (Aug.) to 109.33 ± 16.14 (Jan.) at ITO-3, 16.67 ± 4.71 (Aug.) to 109.33 ± 12.36 (Feb.) at NIZ-4, 18.00 ± 4.32 (Aug.) to 144.0 ± 69.13 (Nov.) at OKH-5 respectively. Significant difference in chloride content was found between the monitoring stations (F_1=21.93, P<0.01) and also between sampling months (F_2=2.81, P<0.05).

Seasonal averages showed a minimum of 15.78 ± 2.62 (Monsoon) to a maximum of 32.89 ± 25.0 (Winter) at WAZ-1 32.89 ± 9.57 (Monsoon) to 96.67 ± 12.1 (Winter) at ISB-2, 30.33 ± 20.9 (Monsoon) to 107.11 ± 12.56 (Winter) at ITO-3, 23.78 ± 8.13 (Monsoon) to 101.33 ± 15.10 (Pre-Monsoon) at OKH-5, respectively.

During the present investigation the chloride content showed minimum values during monsoon months. This may be certainly due to high dilution and high flow rate by rain water. Malviya *et al.* (1990) also observed a similar trend.

Table RWM-16: Monthly Variation of Chloride (mg/l) River Water (Year 1993-1995)

Month	WAZ-1 Mean ± SD	ISB-2 Mean ± SD	ITO-3 Mean ± SD	NIZ-4 Mean ± SD	OKH-5 Mean ± SD
Jan.	37.3 ± 27.4	88.0 ± 6.5	109.3 ± 16.4	102.7 ± 11.5	54.7 ± 13.2
Feb.	24.0 ± 9.8	105.3 ± 13.2	105.3 ± 10.0	109.3 ± 12.4	76.0 ± 11.8
Mar.	38.7 ± 32.2	104.0 ± 34.6	105.3 ± 26.6	101.3 ± 29.3	74.7 ± 32.7
Apr.	20.7 ± 8.4	81.7 ± 35.4	77.0 ± 32.3	74.0 ± 28.6	51.3 ± 19.6
May	27.1 ± 14.5	98.0 ± 32.8	93.3 ± 24.9	98.7 ± 20.7	40.7 ± 13.9
Jun.	20.0 ± 14.2	85.3 ± 19.7	91.3 ± 18.9	88.7 ± 15.5	47.3 ± 3.4
Jul.	13.7 ± 1.7	43.3 ± 9.6	31.7 ± 9.0	29.3 ± 8.2	32.0 ± 5.7
Aug.	18.0 ± 2.8	28.0 ± 2.8	16.7 ± 3.4	16.7 ± 4.7	18.0 ± 4.3
Sep.	15.7 ± 0.5	27.3 ± 3.4	42.7 ± 29.6	25.3 ± 5.0	28.0 ± 11.8
Oct.	30.7 ± 13.2	62.7 ± 24.1	66.7 ± 21.8	65.3 ± 20.0	58.7 ± 7.5
Nov.	42.7 ± 27.2	85.3 ± 24.7	94.7 ± 24.9	100.0 ± 22.9	144.0 ± 69.1
Dec.	45.3 ± 26.4	95.0 ± 10.8	93.0 ± 29.0	104.0 ± 17.3	81.3 ± 21.3

Two-way ANOVA Test

Significant difference in chloride content between river monitoring stations, F_1=21.93 ($P<0.0$).
Significant difference in chloride content between monitoring months, F_2=2.58 ($P<0.05$).

The present study reveals that there are two peak values in chloride content *i.e.* winter and summer seasons. The peak value during winter might be due to reduced water flow (least discharge from Wazirabad barrage). Second peak values during summer may be attributed to reduced water flow, increases rate of evaporation and concentration of solutes. Similar results were reported by Rao (1979), Billore (1981), Oberoi (1987), Shrotriya (1987), Shrotriya and Dubey (1990) and Malviya *et al.* (1990).

In the present study, site-wise fluctuations of chloride content has also been noticed. The higher values of chloride were recorded at site ITO-3, which is much higher (77.25 ± 37.28) in comparison to rest of sampling sites including WAZ-1 (27.81 ± 6.29) site which might be due to continuous influx of much contaminated domestic sewage.

Munawar (1970) suggested that higher concentrations of chloride in water is an index of pollution of animal origin and there is a direct correlation between chloride concentration and pollution level.

Total Hardness of Drain and River Water

Cations that form insoluble compounds with soap contribute to what is termed as "hardness" of water. The soap consuming power of water is lowered by heavy metals and the alkaline earth metals. The two most common bivalent cations, calcium and magnesium of water accounts for hardness.

According the Ruttener (1953), the total hardness is the total amount of alkaline earths present without reference to the particular anion to which they are bound. Hard water contains large concentrations of alkaline earths derived from the drainage of calcareous deposits (Wetzel, 1975). Most inland waters are dilute solutions of calcium and sodium carbonates but variations contributed by lithology and climate are very wide. According to Rai (1974), the total hardness is an index of water quality and is of considerable significance in connection with the discharge of domestic and municipal sewage.

Table DWM-16 presents monthly average values (Mean ± SD) and analysis of variance results of hardness of drain water. The minimum and maximum average monthly values of hardness were 188.0 ± 29.49 and 246.33 ± 4.74 in NAJ-1, 223.0 ± 22.86 and 314.33 ± 109.78 in CIV-2, 243.33 ± 27.64 and 348.67 ± 36.16 in POW-3, 265.0 ± 93.85 and 435.33 ± 113.36 in SEN-4, 218.67 ± 77.46 and 341.67 ± 50.21 in BAR-5, 110.67 ± 8.99 and 245.0 ± 12.03 in HIN-6. Analysis of variance shows significant difference in hardness between five drains (F_1=19.39, P<0.01) and also between sampling months (F_2=2.76, P<0.05).

Table DWM-16: Monthly Variation of Hardness (mg/l) in Drain Water (Year 1993-1995)

Month	*NAJ-1* Mean ± SD	*CIV-2* Mean ± SD	*POW-3* Mean ± SD	*SEN-4* Mean ± SD	*BAR-5* Mean ± SD	*HIN-6* Mean ± SD
Jan.	216.7 ± 45.7	298.7 ± 8.4	332.7 ± 12.7	360.3 ± 13.8	322.3 ± 10.7	245.0 ± 12.0
Feb.	246.3 ± 4.7	281.7 ± 17.2	322.0 ± 30.4	354.7 ± 55.8	335.7 ± 16.2	243.7 ± 15.1
Mar.	198.3 ± 8.5	282.3 ± 74.7	333.3 ± 36.9	374.3 ± 18.0	321.1 ± 23.0	227.3 ± 5.3
Apr.	191.3 ± 21.5	207.7 ± 75.2	302.7 ± 27.9	357.3 ± 19.3	274.3 ± 62.1	188.7 ± 18.0
May	194.3 ± 29.5	249.0 ± 65.7	293.3 ± 62.4	345.0 ± 24.8	260.7 ± 60.6	183.0 ± 36.7
Jun.	175.7 ± 10.3	258.7 ± 19.2	269.0 ± 36.5	435.3 ± 113.4	294.7 ± 40.9	175.3 ± 9.0
Jul.	239.7 ± 24.5	233.0 ± 16.3	267.0 ± 47.4	265.0 ± 93.9	263.0 ± 12.1	112.7 ± 10.5
Aug.	206.0 ± 11.0	232.0 ± 24.1	238.7 ± 37.1	288.3 ± 44.8	218.7 ± 77.5	110.7 ± 9.0
Sep.	203.3 ± 29.0	223.0 ± 22.9	234.3 ± 27.6	280.7 ± 56.8	259.0 ± 106.9	118.7 ± 7.4
Oct.	188.0 ± 29.4	267.0 ± 42.1	285.7 ± 64.8	326.0 ± 48.3	341.7 ± 31.2	163.0 ± 36.9
Nov.	186.0 ± 4.3	314.3 ± 109.8	332.3 ± 42.0	344.7 ± 53.6	314.0 ± 50.2	212.3 ± 25.3
Dec.	214.3 ± 36.8	299.7 ± 34.3	348.7 ± 36.2	349.0 ± 51.6	329.3 ± 42.3	244.0 ± 6.4

Two-way ANOVA Test

Significant difference in hardness content between monitoring drains, F_1=19.12 (P<0.01).
Significant difference in hardness content between sampling months, F_2=2.76 (P<0.05).

Seasonal minimum and maximum values of hardness were 187.0 ± 21.06 (Pre-Monsoon) and 217.33 ± 35.7 (Winter) in NAJ-1, 229.33 ± 21.32 (Monsoon) and 290.67 ± 86.43 (Pre-Monsoon) in CIV-2, 246.67 ± 40.89 (Monsoon) and 326.67 ± 26.33 (Winter) in POW-3, 278.0 ± 69.11 (Monsoon) and 351.22 ± 43.44 (Winter) in SEN-4, 246.89 ± 79.12 (Monsoon) and 327.83 ± 44.03 (Pre-Monsoon) in BAR-5, 114.0 ± 30.3 and 245.78 ± 31.12 in HIN-6. Annual average values of hardness were in order of 340.06 > 296.67 > 294.53 > 268.08 > 295.0 > 185.36 in SEN-4, POW-3, CIV-2, NAJ-1 and HIN-6, respectively.

Table RWM-17 shows monthly averages (Mean ± SD) of hardness (mg/l) in water of river Yamuna at five sampling stations. The range of hardness was 78.33 ± 10.5 (Jul.) to 179.0 ± 20.12 (Dec.) at WAZ-1, 111.0 ± 18.49 (Aug.) to 203.67 ± 64.01 (Nov.) at ISB-2, 97.33 ± 18.79 (Aug.) to 227.33 ± 71.26 (Nov.) at ITO-3. 93.67 ± 22.07 (Aug.) to 236.0 ± 57.15 (Dec.) at NIZ-4, 112.0 ± 30.74 (Aug.) to 259.33 ± 91.13 (Nov.) at OKH-5. Analysis of variance (Table RWM-17) showed significant difference in hardness (mg/l) between monitoring points (F_1=21.93, P<0.01) and also between sampling months (F_2=2.79, P<0.05).

Seasonal minimum and maximum values of hardness were 98.33 ± 43.20 (Monsoon) and 156.44 ± 24.06 (Winter); at WAZ-1, 117.67 ± 34.33 (Monsoon) and 195.22 ± 26.82 (Winter) at ISB-2, 120.0 ± 47.91 (Monsoon) and 184.22 ± 13.16 (Winter) at ITO-3, 107.33 ± 44.98 (Monsoon) and 208.33 ± 41.32 (Winter) at NIZ-4, 127.67 ± 48.26 (Winter) and 229.33 ± 68.09 (Winter) at OKH-5.

Table RWM-17: Monthly Variation of Total Hardness (mg/l) in River Water (Year 1993-1995)

Month	*WAZ-1* *Mean ± SD*	*ISB-2* *Mean ± SD*	*ITO-3* *Mean ± SD*	*NIZ-4* *Mean ± SD*	*OKH-5* *Mean ± SD*
Jan.	146.3 ± 14.0	169.7 ± 6.2	178.0 ± 15.6	186.3 ± 20.0	217.7 ± 72.5
Feb.	139.3 ± 13.1	198.7 ± 9.8	185.0 ± 0.8	201.3 ± 12.5	223.0 ± 75.2
Mar.	223.0 ± 10.2	182.0 ± 6.5	182.7 ± 5.2	196.0 ± 11.8	218.3 ± 51.4
Apr.	112.3 ± 6.6	157.3 ± 13.2	156.0 ± 11.3	168.0 ± 17.7	188.3 ± 48.1
May	96.3 ± 2.9	163.0 ± 13.1	161.7 ± 11.8	173.3 ± 8.7	168.3 ± 24.0
Jun.	83.3 ± 12.2	153.0 ± 5.7	158.0 ± 24.7	160.0 ± 18.7	151.0 ± 45.0
Jul.	78.3 ± 16.5	120.0 ± 9.1	104.3 ± 12.3	102.0 ± 21.4	136.0 ± 45.3
Aug.	102.7 ± 43.4	111.0 ± 18.5	97.3 ± 18.8	93.7 ± 22.0	112.0 ± 30.7
Sep.	114.0 ± 52.7	122.0 ± 55.2	158.3 ± 64.5	126.0 ± 67.4	135.0 ± 60.2
Oct.	136.0 ± 52.7	185.7 ± 52.7	186.7 ± 52.8	187.3 ± 68.1	216.7 ± 66.4
Nov.	162.3 ± 35.3	203.7 ± 64.0	227.3 ± 71.3	216.7 ± 79.7	259.3 ± 91.1
Dec.	179.0 ± 20.1	198.3 ± 38.6	184.0 ± 6.4	236.0 ± 57.2	237.3 ± 62.1

Two-way ANOVA Test

Significant difference in total hardness content between river monitoring stations, F_1=19.62 ($P<0.01$).

Significant difference in total hardness content between monitoring months, F_2=2.79 ($P<0.05$).

During the present investigation total hardness showed two peak values *i.e.* during winter and summer. The peak values during winter season might be due to lean discharge of river water from Wazirabad barrage and more contribution of drain water at Delhi stretch. The second peak values during summer might be due to higher temperature and low water level in the river as well as the addition of calcium and magnesium salts from detergents and soap used for bathing and cloth washing (through drain water). Its lower values were recorded during rainy season. This might be due to dilution on account of high rains. Ajmal and Razi-Ud-Din (1988) observed a higher hardness value in summer in river Hindon and Kali Nadi. Munawar (1970), Moss (1973), Sreenivasan *et al.* (1974), Singh (1979) and Patil (1982) found the same pattern during their studies which has been further supported by Bagde and Verma (1985). Annual average values of five monitoring stations were in the order of 188.58 > 170.56 > 164.94 > 163.69 > 122.75 at OKH-5, NIZ-4, ITO-3, ISB-2 and WAZ-1, respectively.

It was observed that the total hardness values increased sharply at ISB-2 from WAZ-1, due to input of domestic sewage through Najafgarh and Burari drain. The concentration of hardness gradually increased from station ISB-2, to OKH-5 which might be due to input of drain effluents joining to river Yamuna at Delhi stretch.

Calcium of Drain and River Water

Calcium is an essential element and human body requires approximately 0.7 to 2.0 g of calcium per day as a food element, the amount which cannot be supplied even by hard waters. Hardness are undesirable for household uses such as washing, bathing and laundering because of the consumption of more soap and other cleaning agents. Moreover calcium salts tend to cause incrustations on cooking utensil and water heaters.

Calcium is essential for normal plant growth and is desirable in water for irrigation. The presence of calcium in water is mainly due to its passage through or over deposits of lime stone, dolomite, gypsum and other gypsiferous materials (Manivasakam, 1989). The calcium content may range from zero to several hundred mg/l, depending on the source and treatment of water.

Table DWM-17, show monthly calcium average values in drain water and the results of ANOVA test for five monitoring drains. The values (mg/l) varied from a minimum of 44.2 ± 18.32 to a maximum of 68.07 ± 8.38 in NAJ-1, 72.0 ± 6.68 to 111.73 ± 39.65 in CIV-2, 68.00 ± 9.80 to 108.67 ± 16.05 in POW-3, 88.67 ± 13.70 to 130.87 ± 23.74 in SEN-4, 63.67 ± 26.16 to 84.33 ± 22.69 in BAR-5 and 34.0 ± 2.45 to 74.67 ± 8.06 in HIN-6. ANOVA (Table DWM-17) reveals significant differences in calcium content between the five sampling drains (F_1=19.11, P<0.01) and also between sampling months (F_2=2.66, P<0.05).

Seasonal mean values of calcium varied from 52.3 ± 14.1 (Summer) to 61.58 ± 8.8 (Winter) in NAJ-1, 74.44 ± 8.71 (Monsoon) to 102.7 ± 31.24 (Pre-Monsoon) in CIV-2, 74.44 ± 13.88 (Monsoon) to 104.78 ± 13.51 (Winter) in POW-3, 96.87 ± 33.81 (Monsoon) to 118.87 ± 14.5 (Summer) in SEN-4, 68.44 ± 23.08 (Monsoon) to 76.56 ± 8.96 (Winter) BAR-5, 36.0 ± 8.7 (Monsoon) to 75.0 ± 13.76 (Winter) in HIN-6. Yearly average values were in the order of 109.64 > 92.95 > 90.85 > 73.08 > 58.11 > 55.86 in SEN-4, POW-3, CIV-2, BAR-5, NAJ-1 drains and Hindon river cut respectively.

Table RWM-18 presents monthly average values (Mean ± SD) and the results of ANOVA test for calcium content (mg/l) in river water at five monitoring points. The values ranged from 20.67 ± 1.89 (Jul.) to 42.33 ± 4.92 (Nov.) at WAZ-1, 39.0 ± 7.29 (Jul.) to 52.33 ± 4.5 (Dec.) at ISB-2, 32.67 ± 2.49 (Jul.) to 56.67 ± 15.11 (Nov.) at ITO-3, 26.67 ± 4.11 (Aug.) to 64.67 ± 12.47 (Dec.) at NIZ-4, 32.33 ± 6.55 (Aug.) to 62.0 ± 17.66 (Dec.) at BAR-5. Result of ANOVA test shows significant difference in calcium content between monitoring stations (F_1=19.31, P<0.01) and also between sampling months (F_2=2.8 P<0.05).

Table DWM-17: Monthly Variation in Calcium (mg/l) of Drain Water Year (1993-1995)

Month	*NAJ-1* *Mean ± SD*	*CIV-2* *Mean ± SD*	*POW-3* *Mean ± SD*	*SEN-4* *Mean ± SD*	*BAR-5* *Mean ± SD*	*HIN-6* *Mean ± SD*
Jan.	60.0 ± 8.6	98.0 ± 5.8	106.0 ± 2.8	115.9 ± 2.9	73.6 ± 8.6	73.6 ± 8.9
Feb.	68.6 ± 8.3	95.5 ± 12.7	102.0 ± 18.5	110.0 ± 16.5	74.0 ± 7.7	74.6 ± 8.0
Mar.	44.2 ± 18.3	105.0 ± 8.2	108.6 ± 16.0	117.6 ± 4.0	71.0 ± 4.5	73.0 ± 5.8
Apr.	59.4 ± 12.3	94.0 ± 11.4	96.3 ± 14.3	113.9 ± 2.1	63.6 ± 18.1	62.6 ± 2.4
May	55.2 ± 13.2	86.0 ± 18.4	91.6 ± 22.8	113.0 ± 7.2	65.3 ± 23.8	56.3 ± 12.5
Jun.	50.4 ± 1.1	83.6 ± 4.6	90.0 ± 12.3	130.6 ± 23.7	79.6 ± 14.6	58.6 ± 6.6
Jul.	63.0 ± 5.4	72.0 ± 6.6	86.6 ± 13.1	107.9 ± 51.8	78.0 ± 4.3	34.3 ± 2.6
Aug.	46.3 ± 4.6	75.0 ± 11.5	68.0 ± 9.8	88.6 ± 13.7	63.6 ± 26.1	34.0 ± 2.4
Sep.	53.6 ± 9.7	76.3 ± 6.2	68.6 ± 9.4	94.0 ± 18.8	63.6 ± 27.5	39.6 ± 2.8
Oct.	51.6 ± 6.9	93.6 ± 14.7	83.6 ± 20.0	105.0 ± 14.1	78.3 ± 11.7	53.6 ± 11.5
Nov.	55.33 ± 6.6	111.7 ± 39.6	104.7 ± 13.5	107.0 ± 10.0	84.4 ± 22.6	64.3 ± 7.4
Dec.	62.4 ± 9.8	99.3 ± 16.4	109.0 ± 13.6	111.6 ± 12.2	81.6 ± 7.8	72.3 ± 6.9

Two-way ANOVA Test

Significant difference in calcium content between monitoring drains, F_1=19.11 (P<0.01).
Significant difference in calcium content between sampling months, F_2=2.65 (P<0.05).

Table RWM-18: Monthly Variation of Calcium (mg/l) in River Water Year (1993-1995)

Month	*WAZ-1*	*ISB-2*	*ITO-3*	*NIZ-4*	*OKH-5*
	Mean ± SD	*Mean ± SD*	*Mean ± SD*	*Mean ± SD*	*Mean ± SD*
Jan.	40.0 ± 5.7	41.3 ± 1.2	50.7 ± 3.9	56.7 ± 6.6	48.3 ± 3.7
Feb.	41.0 ± 2.2	48.6 ± 6.7	53.0 ± 1.4	62.0 ± 3.5	54.3 ± 6.9
Mar.	37.7 ± 3.0	51.0 ± 5.4	55.3 ± 3.7	60.0 ± 3.6	58.7 ± 1.3
Apr.	36.3 ± 2.9	46.3 ± 2.4	52.0 ± 3.3	52.3 ± 5.6	52.7 ± 5.7
May	33.7 ± 0.9	47.0 ± 11.1	55.0 ± 2.5	50.3 ± 4.5	48.0 ± 6.5
Jun.	25.7 ± 2.4	44.0 ± 5.9	53.3 ± 6.1	50.7 ± 6.8	39.0 ± 5.0
Jul.	20.7 ± 1.9	39.0 ± 7.3	32.7 ± 2.5	29.3 ± 3.7	33.0 ± 4.5
Aug.	25.3 ± 6.9	40.6 ± 13.1	40.3 ± 16.0	26.7 ± 4.1	32.3 ± 6.6
Sep.	28.1 ± 10.0	41.3 ± 18.9	44.7 ± 16.2	36.0 ± 15.8	39.7 + 19.5
Oct.	34.7 ± 9.8	44.6 ± 6.2	51.0 ± 13.1	50.0 ± 14.5	54.7 ± 16.9
Nov.	42.3 ± 4.9	44.3 ± 6.9	56.7 ± 15.2	58.7 ± 18.0	60.3 ± 20.9
Dec.	42.0 ± 0.8	52.3 ± 4.5	55.0 ± 5.1	64.7 ± 12.5	62.0 ± 17.7

Two-way ANOVA Test

Significant difference in calcium content between river monitoring stations, F_1=19.31 (P<0.01).
Significant difference in calcium content between monitoring months, F_2=2.8 (P<0.05).

Seasonal minimum and maximum calcium values were 24.67 ± 7.72 (Monsoon) and 41.33 ± 3.83 (Winter) at WAZ-1, 33.89 ± 6.12 (Monsoon) and 49.44 ± 5.96 (Winter) at ISB-2, 39.22 ± 14.13 (Monsoon) and 53.78 ± 3.88 (Winter) at ITO-3, 30.67 ± 10.40 (Monsoon) and 60.50 ± 8.75 (Winter) at NIZ-4, 35.0 ± 12.6 (Monsoon) and 55.89 ± 11.92 (Winter) at OKH-5. During present investigation calcium showed its peak values during winter due to lean discharge from Wazirabad barrage and high concentration of calcium in drain water. The lower values of calcium were observed during monsoon season due to dilution of drain water by rain water. Annual mean values (mg/l) of calcium at Delhi stretch were in order of 50.0 > 49.78 > 48.58 > 45.06 > 33.86 at ITO-3, NIZ-4, OKH-5, ISB-2 and WAZ-1, respectively.

Magnesium of Drain and River Water

Magnesium is an essential element for human beings and is relatively non-toxic. However at higher concentrations, magnesium salts have a laxative effect particularly when it is present as magnesium sulphate. Magnesium salts act as cathartics and diuretics among animals as well as human beings. Magnesium in industrial water is undesirable since it is the major scale forming cation. Magnesium salts occur in significant concentration in natural waters. The industrial and commercial uses of magnesium salts are numerous and the occurrence of magnesium in industrial wastewater is common. Magnesium is essential to normal plant growth. Calcium and magnesium ions in irrigation water keep the soil permeable and in good tilt (Manivasakam, 1989).

Table DWM-18 represents monthly average values (Mean ± SD) and the results of ANOVA test of magnesium content (mg/l) of five drains. The minimum and maximum mean values of magnesium were 10.48 ± 1.84 to 21.57 ± 0.83 in NAJ-1, 7.81 ± 2.73 to 18.05 ± 3.11 in CIV-2, 10.67 ± 4.35 to 16.35 ± 4.41 in POW-3, 10.93 ± 3.62 to 28.48 ± 5.49 in SEN-4, 14.73 ± 3.89 to 34.73 ± 2.68 at BAR-5 and 4.83 ± 1.84 to 15.0 ± 1.0 in HIN-6. Analysis of variance (Table DWM-18) shows a significant difference in magnesium content between five monitoring drains (F_1=2.606, P<0.05) and also between monitoring months (F_2=2.65, P<0.05).

Table DWM-18: Monthly Variation in Magnesium (mg/l) of Drain Water (Year 1993-1995)

Month	NAJ-1 Mean ± SD	CIV-2 Mean ± SD	POW-3 Mean ± SD	SEN-4 Mean ± SD	BAR-5 Mean ± SD	HIN-6 Mean ± SD
Jan.	15.8 ± 6.2	12.8 ± 3.0	16.3 ± 3.7	16.8 ± 1.8	29.9 ± 3.1	14.4 ± 3.2
Feb.	18.0 ± 3.9	10.3 ± 3.9	16.2 ± 4.7	18.1 ± 4.3	32.6 ± 3.8	15.0 ± 1.0
Mar.	13.4 ± 2.0	16.1 ± 7.3	14.2 ± 3.8	19.1 ± 5.4	34.7 ± 2.6	8.0 ± 3.5
Apr.	10.4 ± 1.8	18.0 ± 3.1	14.1 ± 5.7	18.1 ± 4.7	27.8 ± 4.1	11.3 ± 6.1
May	13.4 ± 6.7	16.1 ± 3.9	15.4 ± 9.8	15.1 ± 1.5	23.6 ± 0.4	10.0 ± 1.4
Jun.	11.9 ± 2.2	11.8 ± 1.8	10.6 ± 4.3	25.9 ± 13.0	23.3 ± 1.7	6.8 ± 1.6
Jul.	19.5 ± 2.2	13.0 ± 1.8	12.0 ± 5.2	28.4 ± 5.4	16.2 ± 0.3	6.5 ± 2.0
Aug.	21.5 ± 0.8	7.8 ± 2.7	16.5 ± 4.4	15.9 ± 4.2	14.5 ± 3.8	6.3 ± 1.7
Sep.	16.6 ± 1.8	8.3 ± 7.0	14.9 ± 5.6	10.9 ± 3.6	23.1 ± 11.2	4.8 ± 1.8
Oct.	14.2 ± 2.8	7.8 ± 1.3	18.2 ± 10.7	15.4 ± 3.8	31.6 ± 5.7	6.8 ± 3.1
Nov.	11.5 ± 3.7	8.4 ± 3.5	16.7 ± 4.5	18.4 ± 7.3	25.2 ± 3.3	9.8 ± 3.7
Dec.	14.0 ± 2.9	12.2 ± 2.2	18.2 ± 4.2	16.6 ± 5.7	30.0 ± 10.5	15.1 ± 2.7

Two-way ANOVA Test

Significant difference in magnesium content between monitoring drains, F_1=19.97 (P<0.01).
Significant difference in magnesium content between sampling months, F_2=2.65 (P<0.05).

Table RWM-19: Monthly Variation of Magnesium (mg/l) in River Water Year (1993- 1995)

Month	WAZ-1 Mean ± SD	ISB-2 Mean ± SD	ITO-3 Mean ± SD	NIZ-4 Mean ± SD	OKH-5 Mean ± SD
Jan.	10.0 ± 0.8	15.3 ± 0.9	12.0 ± 1.6	11.3 ± 1.9	12.0 ± 2.8
Feb.	8.7 ± 2.1	15.7 ± 1.2	12.7 ± 0.9	12.0 ± 1.6	13.0 ± 2.2
Mar.	6.7 ± 1.2	14.0 ± 1.4	11.0 ± 1.4	11.0 ± 0.8	8.0 ± 2.8
Apr.	7.0 ± 2.2	10.7 ± 0.9	6.3 ± 1.2	9.0 ± 1.4	6.0 ± 0.8
May	3.7 ± 0.5	8.3 ± 0.9	5.7 ± 1.2	10.7 ± 0.9	5.3 ± 0.5
Jun.	5.0 ± 1.6	9.0 ± 0.8	6.0 ± 2.2	7.8 ± 0.6	5.3 ± 0.5
Jul.	6.7 ± 3.1	8.7 ± 1.7	5.7 ± 1.2	8.3 ± 2.1	5.0 ± 1.4
Aug.	9.3 ± 6.1	7.0 ± 2.9	4.3 ± 1.2	7.0 ± 2.4	3.5 ± 1.1
Sep.	10.0 ± 7.8	10.0 ± 6.5	9.7 ± 5.3	8.7 ± 6.8	5.3 ± 4.0
Oct.	11.0 ± 6.2	15.0 ± 7.1	11.7 ± 5.9	14.7 ± 7.6	13.0 ± 5.7
Nov.	13.0 ± 6.5	17.0 ± 7.9	18.3 ± 11.1	16.5 ± 8.3	20.2 ± 10.7
Dec.	12.7 ± 4.1	11.3 ± 0.9	10.0 ± 3.6	16.0 ± 9.3	17.7 ± 11.0

Two-way ANOVA Test

Significant difference in magnesium content between river monitoring stations, F_1=21.2 (P<0.01).
Significant difference in magnesium content between monitoring months, F_2=2.90 (P<0.05).

The mean seasonal values of magnesium were varied from a minimum of 12.32 ± 4.16 (Summer) to maximum of 19.23 ± 3.00 (Monsoon) in NAJ-1, 9.73 ± 3.10 (Monsoon) to 14.50 ± 5.48 (Summer) in CIV-2, 13.92 ± 6.67 (Summer) to 17.45 ± 8.15 (Pre-Monsoon) in POW-3, 16.93 ± 6.08 (Pre-Monsoon) to 19.61 ± 8.48 (Summer) in SEN-4, 18.06 ± 7.80 (Monsoon) to 29.32 ± 5.10 (Winter) in BAR-5, and 5.89 ± 1.03 (Monsoon) to 23.04 ± 3.74 (Summer) in HIN-6 respectively. Annual mean values of magnesium were in order of 26.11 > 18.35 > 15.42 > 15.05 > 14.06 > on BAR-5, SEN-4, POW-3, NAJ-1, HIN-6 and CIV-2, respectively.

Table RWM-19 shows the monthly averages (Mean ± SD) and the results of ANOVA test for magnesium at five river monitoring points. Monthly values varied from 3.7 ± 1.5 (May) to 13 ± 6.5 (Nov.) at WAZ-1, 7.0 ± 2.9 (Aug.) to 17.0 ± 7.9 (Nov.) at ISB-2, 4.3 ± 1.2 (Aug.) to 18.3 ± 6.7 (Nov.) at ITO-3, 7.0 ± 2.4 (Aug.) to 16.5 ± 8.3 (Nov.) at NIZ-4 and 3.5 ± 1.1 (Aug.) to 20.0 ± 10.7 (Nov.) at OKH-5. Analysis of variance shows significant difference in magnesium content between monitoring stations (F_1=21.2, P<0.01) and also between monitoring months (F_2=2.9, P<0.05).

The mean minimum and the mean maximum seasonal values were 5.58 ± 2.02 (Summer) and 12.0 ± 6.51 (Pre-Monsoon), 8.56 ± 4.4 (Monsoon) and 15.0 ± 9.49 (Pre-Monsoon), 8.0 ± 4.4 (Monsoon) and 15.6 ± 8.01 (Pre-Monsoon), 4.61 ± 2.66 (Monsoon) and 16.5 ± 9.25 (Pre-Monsoon) at WAZ-1, ISB-2, ITO-3 NIZ-4 and OKH-5, river monitoring station respectively.

During present investigation magnesium showed its peak value at Wazirabad during summer season might be due to lean flow of river, high temperature, increased rate of evaporation and concentration of solutes. Lower values during Monsoon might be due to dilution with rain water. At all downstream monitoring points *i.e.* ISB-2, ITO-3, NIZ-4 and OKH-5 magnesium showed peak values during post-monsoon might be due to lean discharge of water from Wazirabad barrage and higher concentration of drain water and lower values during monsoon might be due to dilution of drain water with rain water and huge water flow from upstream.

Annual average values (1993 to 1995) of magnesium content were in order of 11.83 > 11.09 > 9.51 > 9.44 > 8.64 at ISB-2, NIZ-4, OKH-4, ITO-3 and WAZ-1, respectively.

Sodium of Drain and River Water

Sodium is present in most natural waters from negligible to appreciable concentrations. The salt of sodium are highly soluble in water and impart softness to water. It has important considerations with regard to irrigation water only. Regarding drinking and industrial purposes it is of minor importance only. However it has to be mentioned that excessive amounts of sodium in drinking water is harmful to persons suffering from cardiac, renal and circulatory diseases. Sodium when present as sodium chloride in concentration more than 500 mg/l makes the water unpalatable and causes appetite disturbances. Soil permeability has also been found to be affected by a high sodium content (more then 60 per cent) in irrigation water (Rawat, 1978).

Domestic sewage is the important source of sodium of the freshwater. Sodium ions make their way in water through weathering of rocks. Sodium is said to be utilised by blue green algae in high concentrations (Gerloff *et al.*, 1952 and Provasoli, 1969). The threshold level of 400 mg/l is required for near optimal growth of several species (Kulshrestha *et al.*, 1992).

Table DWM-19 shows monthly mean values (Mean ± SD) and the results of ANOVA test of sodium content in drains discharging their effluents to river Yamuna at Delhi stretch.

Table DWM-19: Monthly Variation in Sodium (mg/l) of Drain Water (Year 1993-1995)

Month	*NAJ-1*	*CIV-2*	*POW-3*	*SEN-4*	*BAR-5*	*HIN-6*
	Mean ± SD	*Mean ± SD*	*Mean ± SD*	*Mean ± SD*	*Mean ± SD*	*Mean ± SD*
Jan.	115.3 ± 12.1	120.3 ± 8.6	124.4 ± 3.4	163.0 ± 18.4	168.0 ± 2.9	95.3 ± 2.5
Feb.	124.7 ± 8.7	115.0 ± 13.5	121.7 ± 23.2	162.7 ± 10.4	155.0 ± 2.8	94.0 ± 5.7
Mar.	123.3 ± 9.9	146.0 ± 20.5	135.4 ± 19.1	177.0 ± 9.4	169.7 ± 17.2	94.3 ± 5.9
Apr.	145.3 ± 12.3	134.0 ± 8.8	139.0 ± 25.9	178.0 ± 8.3	157.0 ± 20.8	83.7 ± 7.1
May	139.3 ± 19.0	139.3 ± 6.6	127.0 ± 19.6	171.3 ± 9.8	150.0 ± 21.4	84.7 ± 7.3
Jun.	144.0 ± 12.4	153.0 ± 8.6	137.7 ± 20.8	179.0 ± 5.4	163.2 ± 18.4	87.0 ± 9.9
Jul.	178.2 ± 23.8	106.7 ± 18.9	115.4 ± 11.7	159.7 ± 6.5	128.2 ± 7.9	62.3 ± 4.9
Aug.	128.4 ± 13.1	106.0 ± 20.1	69.0 ± 45.7	145.7 ± 14.8	107.3 ± 34.2	61.3 ± 0.9
Sep.	130.3 ± 24.3	103.7 ± 22.6	108.3 ± 27.1	132.7 ± 7.1	119.0 ± 43.5	66.3 ± 3.7
Oct.	131.0 ± 20.5	126.3 ± 1.2	121.7 ± 19.0	153.7 ± 3.4	133.0 ± 19.4	82.3 ± 3.3
Nov.	134.0 ± 13.8	163.7 ± 26.6	138.7 ± 6.1	172.3 ± 18.8	142.7 ± 36.4	91.3 ± 1.2
Dec.	130.0 ± 21.8	144.7 ± 18.1	140.0 ± 9.6	175.0 ± 15.1	159.0 ± 9.4	92.0 ± 4.3

Two-way ANOVA Test

Significant difference in sodium content between monitoring drains, F_1=19.13 (P<0.01).
Significant difference in sodium content between sampling months, F_2=2.83 (P<0.05).

Mean values varied from 115.3 ± 12.1 to 178.3 ± 23.8 in NAJ-1 103.7 ± 22.6 to 153.0 ± 8.6 in CIV-2, 69.0 ± 45.7 to 139.0 ± 25.9 in POW-3, 132.7 ± 71 to 179.0 ± 5.4 in SEN-4, 107.3 ± 34.2 to 169.7 ± 17.2 in BAR-5, 61.3 ± 0.9 to 95.3 ± 2.5 in HIN-6. Analysis of variance shows significant difference between monitoring drains (F_1=19.13, P<0.01) as well as between monitoring months (F_2=2.83, P<0.05).

Seasonal sodium values were varied between 118.67 ± 9.46 to 145.67 ± 30.14 in NAJ-1, 105.56 ± 20.62 to 145.0 ± 26.53 in CIV-2, 97.56 ± 37.48 to 128.56 ± 16.73 in POW-3, 146.0 ± 15.03 to 176.33 ± 8.91 in SEN-4, 118.22 ± 33.38 to 163.6 ± 21.6 in BAR-5, 63.33 ± 8.1 to 94.67 ± 9.23 in HIN-6. Annual mean values were in order of 164.17, 145.16, 132.56, 129.92, 123.17 and 82.89 in SEN-4, BAR-5, NAJ-1, CIV-2, POW-3 and HIN-6 respectively.

Table RWM-20 shows monthly average values of sodium at five monitoring points of river Yamuna and the results of ANOVA test. The sodium content of river water at different monitoring stations fluctuated from 19.9 ± 7.9 to 43.0 ± 10.0 at WAZ-1, 56.33 ± 2.87 to 95.0 ± 12.03 at ISB-2, 62.67 ± 14.29 to 112.0 ± 13.95 at ITO-3, 69.67 ± 11.90 to 121.0 ± 4.24 at NIZ-4 and 56.33 ± 6.13 to 115.67 ± 19.6 at OKH-5. The ANOVA test shows significant difference in sodium content between five river monitoring stations (F_1=19.38, P<0.01) and also between sampling months (F_2=2.44, P<0.05).

The mean seasonal values varied from 21.33 ± 8.49 (Monsoon) to 37.33 ± 11.75 (Winter) at WAZ-1, 63.22 ± 8.34 (Monsoon) to 91.25 ± 14.83 (Summer) at ISB-2, 67.0 ± 12.53 (Monsoon) to 108.67 ± 9.94 (Winter) at ITO-3, 71.44 ± 17.03 (Monsoon) to 117.56 ± 6.7 (Winter) at NIZ-4, 58.89 ± 6.47 (Monsoon) to 103.5 ± 19.5 (Pre-Monsoon) at OKH-5. Seasonal lower values at all stations occurs during monsoon due to the effect of dilution. The peak in winter may be due to lean river flow (nil discharge from Wazirabad barrage) and microbial growth (Pradhan and Mohanty, 1991) mineralising the decomposed organic matter and releasing the nutrients to the sediment, increasing sodium in to the nutrient pool there by making more sodium to solubilise.

Table RWM-20: Monthly Variation of Sodium (mg/l) in River Water (Year 1993-1995)

Month	*WAZ-1*	*ISB-2*	*ITO-3*	*NIZ-4*	*OKH-5*
	Mean ± SD	*Mean ± SD*	*Mean ± SD*	*Mean ± SD*	*Mean ± SD*
Jan	38.7 ± 10.3	85.7 ± 5.4	110.7 ± 6.8	111.3 ± 7.4	82.3 ± 10.1
Feb	33.0 ± 14.5	93.0 ± 5.2	111.7 ± 14.1	121.0 ± 4.2	95.7 ± 4.8
Mar	34.0 ± 17.9	95.0 ± 12.0	19.3 ± 21.9	119.0 ± 11.1	99.0 ± 22.2
Apr	28.5 ± 7.9	91.7 ± 14.3	98.7 ± 11.6	113.0 ± 10.7	75.0 ± 10.4
May	28.7 ± 2.9	89.3 ± 21.9	112.0 ± 14.0	120.7 ± 8.6	77.0 ± 6.2
Jun	24.3 ± 5.4	89.0 ± 5.8	102.3 ± 25.3	109.0 ± 14.4	71.0 ± 5.1
Jul	19.9 ± 7.9	69.3 ± 7.4	72.0 ± 7.1	75.0 ± 19.2	62.3 ± 6.2
Aug	20.7 ± 9.2	64.0 ± 7.8	66.2 ± 13.1	69.7 ± 11.9	58.0 ± 6.1
Sep	23.1 ± 8.0	56.3 ± 2.9	62.7 ± 14.3	69.7 ± 18.6	56.3 ± 6.2
Oct	29.2 ± 9.2	76.0 ± 10.7	84.0 ± 9.3	94.7 ± 17.0	91.3 ± 9.0
Nov	38.5 ± 7.6	86.3 ± 8.4	98.7 ± 9.9	109.3 ± 20.2	115.7 ± 19.6
Dec	43.0 ± 10.0	88.7 ± 7.8	101.3 ± 0.9	111.0 ± 14.9	91.3 ± 0.9

Two-way ANOVA Test

Significant difference in sodium content between river monitoring stations, F_1=19.38 ($P<0.01$).
Significant difference in sodium content between monitoring months, F_2=2.44 ($P<0.05$).

The sodium level in river of India has been reported by many workers. In river Bharalu, sodium content ranging from 2.0 to 27.0 mg/l reported by Lal and Bhattachary (1990) 32.7 mg/l in river Cauvery (Somashekar, 1984), 9.3 mg/l in river Morar, Kalpi (Mishra and Saxena, 1991) from 2.5–3.6 mg/l in river Gelabil, Assam (Baruah *et al.*, 1993), 8.6-65.08 mg/l in river Kali (Krishnamurty and Bharati, 1995) 28.0–134.0 mg/l in river Pennar (Srikanth *et al.*, 1995).

The present study on river Yamuna at Delhi stretch showed the annual mean sodium values of 30.19 ± 12.3, 82.03 ± 15.87, 94.14 ± 22.26, 101.97 ± 23.50, 81.25 ± 20.34 at WAZ-1, ISB-2, ITO-3, NIZ-4 and OKH-5 respectively.

Potassium of Drain and River Water

Potassium is an essential nutritional element, but in excessive amounts it acts as a cathartic. Though potassium ranks seventh among the elements in the order of abundance, its concentration in most drinking water is trivial. Potassium in natural water is derived from weathering of rocks. The concentrations of potassium remain quite lower than the sodium, calcium and magnesium in aquatic ecosystems (Trivedy *et al.*, 1987). Although it is found in less quantity, it is also important in the ecology of blue-green algae (Provasoli, 1958 and Wetzel, 1966). Potassium has got a more or less similar chemistry like sodium and remains mostly in solution without undergoing any precipitation (Trivedy *et al.*, 1987).

Monthly averages (Mean ± SD) and the results of ANOVA test for potassium (mg/l) in the drain water are presented in Table DWM-20. The mean values ranged from 14.0 ± 0.82 to 28.33 ± 0.94 in NAJ-1, 11.0 ± 2.83 to 40.0 ± 21.28 in CIV-2, 12.67 ± 2.87 to 28.67 ± 3.2 in POW-3, 10.0 ± 1.41 to 22.0 ± 0.82 in SEN-4, 9.33 ± 2.36 to 20.33 ± 1.70 in BAR-5 and 7.67 ± 0.94 to 13.0 ± 0.82 in HIN-6. Result of ANOVA test shows a significant difference in potassium content between monitoring drains (F_1=20.36, $P<0.01$) and also between sampling months (F_2= 2.92, $P<0.05$).

Table DWM-20: Monthly Variation in Potassium (mg/l) of Drain Water (Year 1993-1995)

Month	*NAJ-1* Mean ± SD	*CIV-2* Mean ± SD	*POW-3* Mean ± SD	*SEN-4* Mean ± SD	*BAR-5* Mean ± SD	*HIN-6* Mean ± SD
Jan.	14.0 ± 0.8	19.0 ± 2.1	26.6 ± 1.7	19.6 ± 4.1	20.3 ± 1.7	12.0 ± 1.4
Feb.	15.3 ± 1.7	20.2 ± 1.4	26.6 ± 2.6	10.0 ± 2.4	16.6 ± 2.4	11.3 ± 3.3
Mar.	15.0 ± 0.8	22.6 ± 4.0	28.6 ± 3.3	22.0 ± 2.9	19.0 ± 4.3	12.3 ± 2.3
Apr.	19.0 ± 3.7	24.6 ± 3.4	26.6 ± 4.7	22.0 ± 0.8	18.0 ± 4.0	9.0 ± 1.4
May	17.3 ± 1.7	24.0 ± 3.2	26.0 ± 2.5	21.0 ± 1.6	16.3 ± 4.5	8.3 ± 0.4
Jun.	20.0 ± 4.2	26.6 ± 1.4	26.0 ± 4.9	18.6 ± 1.2	17.3 ± 3.6	9.3 ± 2.0
Jul.	28.3 ± 0.9	11.6 ± 2.0	13.6 ± 2.6	13.3 ± 0.4	11.6 ± 1.2	7.3 ± 1.2
Aug.	19.0 ± 4.0	11.0 ± 2.8	12.6 ± 2.8	11.0 ± 0.8	9.3 ± 2.3	7.6 ± 0.9
Sep.	15.6 ± 4.5	11.0 ± 3.7	14.0 ± 5.3	10.0 ± 1.4	13.0 ± 5.7	7.6 ± 0.4
Oct.	16.0 ± 4.9	17.6 ± 0.4	15.2 ± 7.5	14.0 ± 2.5	13.6 ± 2.0	10.0 ± 0.0
Nov.	18.0 ± 2.1	25.6 ± 4.5	23.0 ± 7.2	16.6 ± 0.9	15.0 ± 4.5	11.6 ± 0.4
Dec.	16.6 ± 4.9	40.0 ± 21.2	24.3 ± 7.0	16.6 ± 1.2	17.6 ± 2.6	13.0 ± 0.8

Two-way ANOVA Test

Significant difference in potassium content between monitoring drains, F_1=20.36 (P<0.01).
Significant difference in potassium content between sampling months, F_2=2.92 (P<0.05).

Seasonal values varied from 14.33 ± 2.31 (Winter) to 21.0 ± 6.43 (Monsoon) in NAJ-1, 11.22 ± 2.97 (Monsoon) to 25.89 ± 15.86 (Winter) in CIV-2, 13.44 ± 4.14 (Monsoon) to 27.22 ± 2.48 (Winter) in POW-3, 11.44 ± 1.71 (Monsoon) to 20.92 ± 2.29 (Summer) in SEN-4, 11.33 ± 3.94 (Monsoon) to 19.0 ± 2.45 (Winter) in BAR-5 and from 7.56 ± 1.2 (Monsoon) to 11.67 ± 1.8 (Winter) in HIN-6. Annual mean values (averages from year 1993 to 1995) of potassium were 17.86, 21.17, 21.97, 17.08, 15.67 and 9.97 in NAJ-1, CIV-2, POW-3, SEN-4, BAR-5 and HIN-6 respectively.

Table RWM-21 shows monthly averages (Mean ± SD) of potassium and the results of ANOVA test between five river monitoring points. The range of potassium content was 1.58 ± 0.59 (Aug.) to 6.07 ± 1.46 (Dec.) at WAZ-1, 3.67 ± 0.47 (Sep) to 8.0 ± 1.63 (May) at ISB-2, 4.0 ± 0.82 (Aug.) to 10.33 ± 1.7 (May) at ITO-3, 5.0 ± 0.82 (July) to 13.33 ± 1.41 (May) at NIZ-4, 3.0 ± 0.82 (Aug.) to 9.33 ± 0.94 (Nov.) at OKH-5 respectively. Analysis of variance showed the existence of significant difference in between the stations (F_1=20.21, P<0.01) as well as between sampling months (F_2=20.21, P<0.05), indicating that potassium content is influenced by seasons, drain water discharge and also river water discharge from Wazirabad barrage.

Seasonal averages ranged from 1.71 ± 0.44 (Monsoon) to 5.29 ± 2.15 (Pre-Monsoon) at WAZ-1, 4.33 ± 0.67 (Monsoon) to 7.67 ± 1.43 (Summer) at ISB-2, 4.44 ± 0.67 (Monsoon) to 9.22 ± 1.13 at ITO-3, 5.22 ± 0.83 (Monsoon) to 12.17 ± 2.11 (Summer) at NIZ-4 and 3.33 ± 0.67 (Monsoon) to 8.17 ± 1.67 (Pre-Monsoon) at OKH-5. The high potassium level during winter might be due to lean discharge of river water from Wazirabad barrage and higher contribution of drain water at Delhi-stretch. Another reason might be due to release of potassium from decomposed and mineralised organic matter during that period. The low potassium content in all stations during monsoon months marked the effect of dilution.

Annual average values of potassium were 3.67 ± 2.25, 6.61 ± 2.75, 7.83 ± 2.75, 9.86 ± 3.33 and 6.47 ± 2.33 at WAZ-1, ISB-2, ITO-3, NIZ-4 and OKH-5 respectively. The maximum concentration of K was observed at Nizamuddin bridge *i.e.* 9.86 (mg/l) and minimum was at Wazirabad barrage *i.e.* 3.67 (un-polluted stretch).

Table RWM-21: Monthly Variation of Potassium (mg/l) in River Water (Year 1993- 1995)

Month	*WAZ-1*	*ISB-2*	*ITO-3*	*NIZ-4*	*OKH-5*
	Mean ± SD	*Mean ± SD*	*Mean ± SD*	*Mean ± SD*	*Mean ± SD*
Jan.	3.37 ± 1.64	6.67 ± 0.47	9.00 ± 0.80	11.00 ± 0.82	6.67 ± 0.47
Feb.	3.83 ± 1.55	7.67 ± 0.47	8.67 ± 1.20	13.33 ± 1.70	7.67 ± 1.25
Mar.	3.80 ± 1.70	8.00 ± 1.41	9.67 ± 3.00	12.33 ± 2.05	8.67 ± 2.05
Apr.	3.17 ± 1.43	7.00 ± 1.63	8.00 ± 0.80	11.67 ± 1.89	7.67 ± 0.47
May	5.87 ± 2.64	8.00 ± 1.63	10.33 ± 1.77	13.00 ± 1.41	6.67 ± 0.47
Jun.	2.27 ± 1.32	7.67 ± 0.47	8.00 ± 3.56	11.67 ± 2.62	5.00 ± 0.00
Jul.	1.75 ± 0.35	5.00 ± 0.00	5.00 ± 0.80	5.00 + 0.82	3.67 ± 0.47
Aug.	1.58 ± 0.59	4.33 ± 0.47	4.00 ± 0.82	5.33 ± 0.47	3.00 ± 0.82
Sep.	1.78 ± 0.31	3.67 ± 0.47	4.33 ± 0.47	5.33 ± 1.25	3.33 ± 0.47
Oct.	5.27 ± 2.53	6.67 ± 0.47	7.33 ± 0.47	8.33 ± 1.70	7.00 ± 1.41
Nov.	5.32 ± 1.69	7.00 ± 0.00	10.00 ± 2.16	10.67 ± 1.70	9.33 ± 0.94
Dec.	6.07 ± 1.46	7.67 ± 0.47	9.67 ± 1.25	10.67 ± 0.47	9.00 ± 0.00

Two-way ANOVA Test

Significant difference in potassium content between river monitoring stations, F_1=20.21 ($P<0.01$).
Significant difference in potassium content between monitoring months, F_2=2.53 ($P<0.05$).

Heavy Metals in Water

Heavy metals occur at very low levels in almost all natural waters due to geo-chemical sources. Some of them (Fe, Cu, Zn, Ni, Co, Mn) are essential nutrients for aquatic life. At higher levels they are toxic. In our country majority of water bodies are invariably polluted by industrial and domestic wastes, which also contain metallic wastes. Therefore, monitoring in water courses become an essential part of water quality management. The measurements and characterisation of heavy metals are necessary to understand their distribution and pathways in the ecosystem and to assess the impact of their discharge into the environment (Ingole and Dhaktode, 1990).

Iron of Drain and River Water

Iron usually exists in natural water both in ferric and ferrous forms. Usually the ferric form is predominant in most of the natural water. Iron exists in the soluble state under reducing conditions. On exposure to air or on addition of oxidants soluble ferrous iron is oxidised to insoluble hydrated ferric oxides rapidly by dissolved oxygen at neutral pH (APHA, 1991). Under alkaline pH conditions of water, iron gets hydrolysed (Stumm and Lee, 1960) and forms insoluble hydroxide (Stumm and Morgan, 1981). The oxides and hydroxides of iron and manganese constitute significant sinks of heavy metals into aquatic system (Jenne, 1976). Ferric iron is only significantly soluble at pH less than 3.5. Iron was found harmful to the fishes even in low concentrations as ferric hydroxide, clogs up gills (Metelev *et al.*, 1971). At pH 6.5 to 7.5, 0.9 mg of iron per litre of water causes the death of fishes. The 1958 WHO international standards for drinking water specify a permissible limit of 0.3 mg/l and an excessive limit of 1.0 mg/l iron.

Monthly average values (Mean ± SD) of iron content in the drains and the result of ANOVA test are given in Table DWM-21. Monthly average values of iron of sampling drains ranged from a minimum of 12.1 ± 2.72 to a maximum of 27.89 ± 9.49 in NAJ-1, 3.34 ± 0.05 to 6.10 ± 0.15 in CIV-2, 3.33 ± 0.05 to 7.42 ± 0.56 in POW-3, 4.35 ± 0.51 to 8.09 ± 0.22 in SEN-4, 3.24 ± 1.06 to 4.07 ± 0.76 in BAR-5. Analysis of variance (Table DWM-21) shows significant difference in 'Fe' content (F_1=29.3, P<0.01) in between sampling drains as well as between sampling months (F_2=2.66, P<0.05).

Table DWM-21: Monthly Variation in Iron (mg/l) of Drain Water (Year 1993-1995)

Month	*NAJ-1* *Mean ± SD*	*CIV-2* *Mean ± SD*	*POW-3* *Mean ± SD*	*SEN-4* *Mean ± SD*	*BAR-5* *Mean ± SD*
Jan.	13.4 ± 2.1	3.4 ± 0.1	3.8 ± 0.5	4.3 ± 0.5	3.4 ± 0.1
Feb.	12.8 ± 0.6	3.3 ± 0.0	3.3 ± 0.0	4.4 ± 0.2	4.0 ± 0.7
Mar.	15.3 ± 2.3	4.4 ± 0.3	3.9 ± 0.0	4.6 ± 0.2	3.2 ± 1.0
Apr.	18.0 ± 1.7	5.2 ± 0.3	5.0 ± 0.0	4.9 ± 0.0	3.3 ± 0.5
May	27.8 ± 9.4	6.1 ± 0.1	7.4 ± 0.5	6.1 ± 0.2	3.4 ± 0.4
Jun.	17.0 ± 2.3	5.9 ± 0.1	8.5 ± 0.4	5.9 ± 0.8	3.8 ± 0.3
Jul.	13.8 ± 0.8	6.0 ± 0.3	8.6 ± 0.3	6.2 ± 0.1	3.8 ± 0.3
Aug.	13.6 ± 0.2	6.0 ± 0.1	8.7 ± 0.9	6.0 ± 0.2	3.8 ± 0.3
Sep.	12.1 ± 2.7	4.1 ± 0.2	9.0 ± 0.3	6.1 ± 0.2	3.9 ± 0.0
Oct.	17.6 ± 0.8	4.1 ± 0.1	8.0 ± 0.1	7.1 ± 0.2	4.0 ± 0.1
Nov.	24.5 ± 2.7	4.9 ± 0.1	7.3 ± 0.2	8.0 ± 0.2	3.7 ± 0.0
Dec.	22.0 ± 1.8	5.2 ± 0.1	7.3 ± 0.4	7.9 ± 0.9	3.8 ± 0.1

Two-way ANOVA Test

Significant difference in iron content between monitoring drains, F_1=29.3 (P<0.01).
Significant difference in iron content between sampling months, F_2=2.06 (P<0.05).

Seasonal values ranged from 13.20 ± 1.83 (Monsoon) to 22.58 ± 9.25 (Summer) in NAJ-1, 3.85 ± 0.69 (Winter) to 5.44 ± 0.70 (Summer) in CIV-2, 4.68 ± 1.6 (Winter) to 8.8 ± 0.61 (Monsoon) in POW-3, 5.19 ± 1.2 (Winter) to 6.15 ± 0.24 (Monsoon) in SEN-4, 3.46 ± 0.69 (Summer) to 3.88 ± 0.17 (Pre-Monsoon) in BAR-5. On basis of annual averages different drain points were arranged in order of NAJ-1 > POW-2 > SEN-4 > CIV-2 > BAR-5.

Table RWM-22 shows monthly average values (Mean ± SD) of iron (mg/l) at five river monitoring stations and the results of two-way ANOVA test. The range of iron was 1.28 ± 0.02 (Jul.) to 2.78 ± 0.08 (Nov.) at WAZ-1, 1.64 ± 0.04 (Aug.) to 3.04 ± 0.19 (Dec.) at ISB-2, 2.34 ± 0.34 (May) to 3.66 ± 0.26 (Nov.) at ITO-3, 2.62 ± 0.37 (May) to 6.2 ± 0.335 (Nov.) at NIZ-4 and 2.98 ± 0.12 (Jun.) to 5.28 ± 0.35 (Nov.) at OKH-5. The data were subjected to two way ANOVA test. The test showed significant difference exist in iron content between different sampling stations (F_1=19.73, P<0.01), as well as between monitoring months (F_2=2.60, P<0.05).

Seasonal averages of iron content of five river monitoring stations varied from 1.45 ± 0.201 (Monsoon) to 2.36 ± 0.427 (Pre-Monsoon) at WAZ-1, 1.82 ± 0.178 mg/l (Monsoon) to 2.52 ± 0.483 (Pre-Monsoon) at ISB-2, 2.6 ± 0.27 (Summer) to 3.65 ± 0.241 (Pre-Monsoon) at ITO-3, 2.82 ± 0.249 (Winter) to 5.61 ± 0.638 (Pre-Monsoon) at NIZ-4, 3.3 ± 0.394 (Summer) to 4.84 ± 0.525 (Pre-Monsoon) at OKH-5. There were two peak seasonal values during post-monsoon and monsoon season which may be attributed to the enrichment of Fe brought from the catchment areas by rain water (Munawar, 1970; Rao, 1971; and Mahan, 1987). The high microbial growth in monsoon months (Pradhan *et al.*, 1991) might also be responsible for the high value of 'Fe' released due to mineralisation of the decomposed body. The lower values during summer and winter season may be attributed to the absence of external source of Fe from catchments areas. Average annual values were 1.99, 2.27, 2.9, 3.63, 3.7 at WAZ-1, ISB-2, ITO-3, NIZ-4 and OKH-5 respectively. The iron content of river water increased suddenly at ISB-2 (ISBT bridge), may be due to mixing of drain water from Najafgarh and Burai drain. The concentration of iron further increased at ITO-3, NIZ-4 and OKH-5 due to further contribution of drain water at Delhi stretch.

Table RWM-22: Monthly Variation of Iron (mg/l) in River Water (Year 1994-1995)

Month	*WAZ-1* *Mean ± SD*	*ISB-2* *Mean ± SD*	*ITO-3* *Mean ± SD*	*NIZ-4* *Mean ± SD*	*OKH-5* *Mean ± SD*
Jan	1.695 ± 0.065	2.050 ± 0.050	2.445 ± 0.055	2.740 ± 0.010	3.200 ± 0.100
Feb	1.700 ± 0.000	2.065 ± 0.085	2.695 ± 0.015	2.735 ± 0.065	3.375 ± 0.115
Mar	1.905 ± 0.075	1.930 ± 0.070	2.755 ± 0.075	2.905 ± 0.045	3.755 ± 0.095
Apr	2.100 ± 0.120	2.295 ± 0.075	2.365 ± 0.065	2.730 ± 0.050	3.135 ± 0.145
May	2.325 ± 0.035	2.480 ± 0.020	2.340 ± 0.340	2.615 ± 0.365	3.375 ± 0.485
Jun	2.640 ± 0.220	2.890 ± 0.090	2.760 ± 0.040	3.040 ± 0.060	2.980 ± 0.120
Jul	1.280 ± 0.020	2.770 ± 0.080	2.920 ± 0.060	3.380 ± 0.020	3.635 ± 0.155
Aug	1.335 ± 0.045	1.640 ± 0.040	3.050 ± 0.070	3.295 ± 0.005	3.620 ± 0.060
Sep	1.725 ± 0.035	2.040 ± 0.060	3.455 ± 0.155	4.030 ± 0.070	3.805 ± 0.025
Oct	1.935 ± 0.045	2.055 ± 0.155	3.645 ± 0.215	5.025 ± 0.135	4.395 ± 0.195
Nov	2.780 ± 0.080	2.980 ± 0.120	3.655 ± 0.265	6.195 ± 0.335	5.280 ± 0.350
Dec	2.470 ± 0.490	3.040 ± 0.190	2.815 ± 0.165	4.860 ± 0.970	3.990 ± 0.310

Two-way ANOVA Test
Significant difference in iron content between river monitoring stations, F_1=19.73 ($P<0.01$).
Significant difference in iron content between monitoring months, F_2=2.66 ($P<0.05$).

Cadmium of Drain and River Water

Cadmium is a bluish white metal. Cadmium is insoluble in water but soluble in acids. Cadmium salts are usually found in wastes from electroplating industries, textile printing, pigment works, lead mines and chemical industries. Cadmium is used as fungicides, insecticides, nematicides, polymerisation catalysts, pigments etc. and also as a contaminant of super phosphate fertilizers (Sittig, 1979). Cadmium has high toxic potential and minute quantities of cadmium are responsible for adverse renal arterial changes in human kidneys. Cadmium tends to concentrate in liver, kidneys, pancreas and thyroid of humans and animals. Once it enters the body it is likely to remain for long time (Nayak and Sawant, 1994; Forstner, 1977; Kudesia, 1980; Forstner and Wittmann, 1981). Cadmium toxicity (50 mg/l) causes vomiting diarrhoea, abdominal pain, loss of bone, deformation, impaired kidneys functioning, impaired reproductive function, hypertension, tumour formation and teratogenic effects (Chatterjee, 1984). The WHO European Standard for cadmium in drinking water is 0.05 mg/l.

Monthly averages (Mean ± SD) of cadmium of five monitoring drains are given in Table DWM-22. Out of five drains monitored, only in Najafgarh drain Cd could be traced. The Cd content varied 0.016

± 0.003 (Jan.) to 0.067 ± 0.002 mg/l (Nov). Seasonal values of cadmium for Najafgarh (DW-22) ranged from a minimum of 0.027 ± 0.012 (Winter) to a maximum of 0.064 ± 0.004 mg/l (Pre-Monsoon). Annual average value of cadmium was 0.038 ± 0.018 mg/l.

Table DWM-22: Monthly Variation in Cadmium (mg/l) of Drain Water (Year 1993-1995)

Month	*NAJ-1*	*CIV-2*	*POW-3*	*SEN-4*	*BAR-5*
	Mean ± SD	*Mean ± SD*	*Mean ± SD*	*Mean ± SD*	*Mean ± SD*
Jan.	0.016 ± 0.003	0.000 ± 0.000	0.000 ± 0.000	0.000 ± 0.000	0.000 ± C.000
Feb.	0.024 ± 0.005	0.000 ± 0.000	0.000 ± 0.000	0.000 ± 0.000	0.000 ± 0.000
Mar.	0.022 ± 0.002	0.000 ± 0.000	0.000 ± 0.000	0.000 ± 0.000	0.000 ± 0.000
Apr.	0.028 ± 0.008	0.000 ± 0.000	0.000 ± 0.000	0.000 ± 0.000	0.000 ± 0.000
May	0.034 ± 0.006	0.000 ± 0.000	0.000 ± 0.000	0.000 ± 0.000	0.000 ± 0.000
Jun.	0.034 ± 0.008	0.000 ± 0.000	0.000 ± 0.000	0.000 ± 0.000	0.000 ± 0.000
Jul.	0.031 ± 0.001	0.000 ± 0.000	0.000 ± 0.000	0.000 ± 0.000	0.000 ± 0.000
Aug.	0.033 ± 0.004	0.000 ± 0.000	0.000 ± 0.000	0.000 ± 0.000	0.000 ± 0.000
Sep.	0.044 ± 0.001	0.000 ± 0.000	0.000 ± 0.000	0.000 ± 0.000	0.000 ± 0.000
Oct.	0.061 ± 0.002	0.000 ± 0.000	0.000 ± 0.000	0.000 ± 0.000	0.000 ± 0.000
Nov.	0.067 ± 0.003	0.000 ± 0.000	0.000 ± 0.000	0.000 ± 0.000	0.000 ± 0.000
Dec.	0.065 ± 0.016	0.000 ± 0.000	0.000 ± 0.000	0.000 ± 0.000	0.000 ± 0.000

Two-way ANOVA Test was not possible, due to inadequate data.

Monthly averages (Mean ± SD) of cadmium mg/l at five river monitoring points have been presented in Table RWM-23. Average values of Cd varied between 0.005 ± 0.0 to 0.017 ± 0.001 at WAZ-1, 0.0 to 0.08 ± 0.01 at ISB-2 and 0.006 ± 0.001 to 0.027 ± 0.001 at OKH-5 (Okhla barrage). The cadmium was not detected at ITO-3 and NIZ-4. Analysis of variance shows significant difference in cadmium content between five monitoring stations (F_1=20.52, P<0.01). However the test showed no significant difference in cadmium level between monitoring months (F_2=0.96).

Seasonal values varied with a minimum of 0.006 ± 0.002 (Monsoon) to a maximum of 0.014 ± 0.004 (Summer) at WAZ-1, 0.0 to 0.008 ± 0.004 (Pre-Monsoon) at ISB-2, and 0.009 ± 0.004 (Monsoon) to 0.002 ± 0.003 (Pre-Monsoon) at OKH-5. The peak values of cadmium were observed during summer due to lean river water flow and high concentration of drain effluents due to high rate of evaporations and low values during monsoon might be due to effect of dilution. Annual average values of cadmium were 0.01, 0.003 and 0.018 at WAZ-1, ISB-2 and OKH-5 respectively. The presence of cadmium at Wazirabad barrage (un-polluted stretch) might be due to plankton and nekton present in the water (Jain and Salman, 1995). The presence of cadmium at ISB-2 was due to drain effluents of Najafgarh drain and Cd at OKH-5 due to Hindon river cut, which carries industrial effluents.

Chromium of Drain and River Water

Chromium salts impart colour to water. The salts of trivalent chromium are physiologically not harmful. It is reported that, the large doses of hexavalent chromium salts lead to corrosive effects in the intestinal tract and to nephritis (Manivasakam, 1989). It is one of the highly toxic metals responsible for various health hazards (Forstner, 1977; Kudesia, 1980; Forstner and Wittmann, 1981). Chromium is hazardous and create complication in the utilization of other nutrients by plants (Khan and Khan,

1983). Chromium compounds are used extensively in industrial processes such as metal pickling, electroplating, aluminium anodising, leather tanning and in the manufacture of paints, dyes, paper, explosives and ceramics. In addition chromates are frequently added to cooling water for corrosion control. The chromium content in natural waters are usually very small. Elevated concentrations can result from mining and industrial processes (Datar and Vashishth, 1990).

Table RWM-23: Monthly Variation of Cadmium (mg/l) in River Water (Year 1994- 1995)

Month	*WAZ-1*	*ISB-2*	*ITO-3*	*NIZ-4*	*OKH-5*
	Mean ± SD	*Mean ± SD*	*Mean ± SD*	*Mean ± SD*	*Mean ± SD*
Jan.	0.005 ± 0.000	Nil	Nil	Nil	0.016 ± 0.001
Feb.	0.008 ± 0.001	Nil	Nil	Nil	0.018 ± 0.001
Mar.	0.009 ± 0.000	Nil	Nil	Nil	0.025 ± 0.001
Apr.	0.012 ± 0.002	Nil	Nil	Nil	0.021 ± 0.001
May	0.017 ± 0.001	Nil	Nil	Nil	0.017 ± 0.001
Jun.	0.018 ± 0.000	Nil	Nil	Nil	0.027 ± 0.001
Jul.	0.005 ± 0.000	0.003 ± 0.003	Nil	Nil	0.007 ± 0.001
Aug.	0.005 ± 0.001	0.003 ± 0.003	Nil	Nil	0.006 ± 0.000
Sep.	0.008 ± 0.001	0.005 ± 0.005	Nil	Nil	0.014 ± 0.002
Oct.	0.009 ± 0.001	0.008 ± 0.008	Nil	Nil	0.019 ± 0.001
Nov.	0.013 ± 0.002	0.007 ± 0.007	Nil	Nil	0.025 ± 0.001
Dec.	0.010 ± 0.002	0.010 ± 0.010	Nil	Nil	0.019 ± 0.001

Two-way ANOVA Test

Significant difference in cadmium content between river monitoring stations, F_1=21.60 ($P<0.01$).
No significant difference in cadmium content between monitoring months, F_2=0.96.

The harmful effects of water borne chromium are digestive track cancer and sometimes lung cancer (Chatterjee,1984). Banat *et al.* (1992) have attributed the chromium accumulation in the Rhine river sediment to industrial and domestic discharges. Chromium also discharged from electroplating units (Mathur *et al.*, 1987) galvanizing and tanning industries in river Rhine (Weijden and Middelburg, 1989). According to WHO International Standards for drinking water, the mandatory limit for hexavalent chromium is 0.05 mg/l.

Monthly averages (Mean ± SD) and the results of analysis of variance of chromium content (mg/l) in drain water are presented in Table DWM-23. Out of five monitoring drains the chromium was not found only in Civil mill drain water. Monthly averages values of chromium varied from a minimum of 0.308 ± 0.012 to a maximum of 0.825 ± 0.039 in NAJ-1, 0.081 ± 0.005 to 0.111 ± 0.025 in POW-3, 0.05 ± 0.004 to 0.01 ± 0.005 in SEN-4, 0.018 ± 0.002 to 0.073 ± 0.005 in BAR-5. Analysis of variance shows significant differences in chromium content between the monitoring stations (F_1=19.7, $P<0.01$). Difference between the monitoring months was not significant (F_2= 0.73).

Seasonal average values ranged from 0.342 ± 0.058 (Winter) to 0.657 ± 0.089 (Pre-Monsoon) in NAJ-1, 0.081 ± 0.009 (Winter) to 0.126 ± 0.044 (Summer) in POW-3, 0.058 ± 0.010 (Monsoon) to 0.098 ± 0.014 (Post-monsoon) in SEN-5. Annual average values of chromium in four drains were 0.56 ± 0.15, 0.102 ± 0.034, 0.077 ± 0.017, 0.043 ± 0.017 in NAJ-1, POW-3, SEN-4 and BAR-5 respectively. The Najafgarh drain carries industrial effluents from Wazirpur industrial area and other industries functioning at its basin. The higher values of chromium observed in Najafgarh drain may be due to industrial water present in it.

Table DWM-23: Monthly Variation in Chromium (mg/l) of Drain Water (Year 1993-1995)

Month	*NAJ-1*	*CIV-2*	*POW-3*	*SEN-4*	*BAR-5*
	Mean ± SD	*Mean ± SD*	*Mean ± SD*	*Mean ± SD*	*Mean ± SD*
Jan.	0.30 ± 0.09	0.00 ± 0.00	0.08 ± 0.00	0.07 ± 0.01	0.04 ± 0.00
Feb.	0.38 ± 0.07	0.00 ± 0.00	0.08 ± 0.01	0.08 ± 0.00	0.04 ± 0.00
Mar.	0.59 ± 0.11	0.00 ± 0.00	0.08 ± 0.02	0.07 ± 0.00	0.05 ± 0.00
Apr.	0.59 ± 0.00	0.00 ± 0.00	0.16 ± 0.00	0.06 ± 0.00	0.04 ± 0.00
May	0.61 ± 0.01	0.00 ± 0.00	0.17 ± 0.01	0.08 ± 0.00	0.04 ± 0.00
Jun.	0.82 ± 0.03	0.00 ± 0.00	0.08 ± 0.01	0.09 ± 0.00	0.05 ± 0.00
Jul.	0.45 ± 0.04	0.00 ± 0.00	0.08 ± 0.01	0.05 ± 0.00	0.01 ± 0.00
Aug.	0.50 ± 0.01	0.00 ± 0.00	0.08 ± 0.00	0.05 ± 0.00	0.01 ± 0.00
Sep.	0.65 ± 0.06	0.00 ± 0.00	0.11 ± 0.02	0.07 ± 0.00	0.03 ± 0.00
Oct.	0.66 ± 0.02	0.00 ± 0.00	0.10 ± 0.01	0.08 ± 0.00	0.03 ± 0.00
Nov.	0.64 ± 0.12	0.00 ± 0.00	0.18 ± 0.00	0.09 ± 0.00	0.06 ± 0.00
Dec.	0.48 ± 0.16	0.00 ± 0.00	0.07 ± 0.00	0.10 ± 0.00	0.07 ± 0.00

Two-way ANOVA Test

Significant difference in chromium content between monitoring drains, F_1=19.7 (P<0.01).
No significant difference in chromium content between sampling months, F_2=0.73.

Table RWM-24 presents for monthly averages (Mean ± SD) and the results of ANOVA test for chromium contents (mg/l) at five river monitoring stations. Though chromium was monitored at five stations (from Wazirabad to Okhla barrage) it was not detected at Wazirabad barrage (unpolluted stretch). The range of chromium content were 0.013 ± 0.001 (Jul.) to 0.030 ± 0.002 (Oct.) at ISB-2, 0.005 ± 0 (Jul.) to 0.025 ± 0.001 (Nov.) at ITO-3, 0.009 ± 0 (Jul.) to 0.037 ± 0.001 (Jan.) at NIZ-4, 0.007 ± 0.001 (Jul.) to 0.037 ± 0.002 (Jun.) at OKH-5. Analysis of variance (Table RWM-24) shows highly significant difference in chromium concentration in between sampling stations (F_1=20.52 P<0.01) and no significant difference in monitoring months (F_2=1.95).

Seasonal averages shows a minimum of 0.018 ± 0.005 (Monsoon) to a maximum of 0.028 ± 0.003 (Pre-Monsoon) at ISB-2, 0.007 ± 0.002 (Monsoon) to 0.019 ± 0.003 (Winter) at ITO-3, 0.011 ± 0.005 (Monsoon) to 0.0321 ± 0.003 (Winter) at NIZ-4, 0.015 ± 0.007 (Monsoon) to 0.032 ± 0.004 (Summer) at OKH-5. The higher concentration of chromium was observed during winter and summer season, may be due to lean river water flow (a little discharge from Wazirabad barrage) and higher percentage contribution of drain water comparison to river water at down stream river monitoring points. The monsoon months showed lowest values, indicating the effect of dilution. Yearly averages of chromium were in the order of 0.026 ± 0.008 > 0.024 ± 0.009 > 0.023 ± 0.006 > 0.015 ± 0.005 at OKH-5, NIZ-4, ISB-2 and ITO-3 respectively.

Copper of Drain and River Water

Copper salts occur in natural surface water only in trace amounts up to about 0.05 mg/l. Copper occurs as native metals and in sulphide ores in nature. It is insoluble in water, but soluble in nitric acid and hot sulphuric acid. Metallic copper is used in all gauges of electric wire, coil and armature windings in electrical industries, made into casting, sheets, rods tubing and wire used in water and gas piping. It forms many important alloys. Aluminium, bronze, copper compounds are used as insecticides and algicides. They are also utilized in electroplating. Copper poisoning leads to toxic effects like ptyalism, nausia, hematuria, convulsions and death (Prasad, 1978). Copper salts are used in water supply systems to control biological growths in reservoirs and distribution pipes and to catalyse the oxidation of manganese. Corrosion of copper containing alloys in pipe fittings may introduce measurable amounts of copper into the water in pipe system. Effluents from electroplating plants constitute one of the important source of heavy metal pollution and normally contain objectionable quantities of copper, lead, zinc, cadmium, chromium, cyanides, oils and grease (Ramaswamy and Somashekar, 1982). However copper is essential to human and the adult daily requirement had been estimated to be 20 mg (APHA, 1991).

Table RWM-24: Monthly Variation of Chromium (mg/l) In River Water (Year 1994-1995)

Month	*WAZ-1*	*ISB-2*	*ITO-3*	*NIZ-4*	*OKH-5*
	Mean ± SD	*Mean ± SD*	*Mean ± SD*	*Mean ± SD*	*Mean ± SD*
Jan.	Nil	0.026 ± 0.002	0.017 ± 0.002	0.037 ± 0.001	0.027 ± 0.001
Feb.	Nil	0.023 ± 0.001	0.022 ± 0.001	0.031 ± 0.001	0.029 ± 0.002
Mar.	Nil	0.026 ± 0.002	0.019 ± 0.001	0.030 ± 0.002	0.028 ± 0.002
Apr.	Nil	0.019 ± 0.001	0.014 ± 0.003	0.025 ± 0.004	0.029 ± 0.001
May	Nil	0.024 ± 0.001	0.013 ± 0.003	0.020 ± 0.004	0.033 ± 0.003
Jun.	Nil	0.024 ± 0.005	0.022 ± 0.002	0.028 ± 0.002	0.037 ± 0.002
Jul.	Nil	0.013 ± 0.001	0.005 ± 0.000	0.009 ± 0.000	0.007 ± 0.001
Aug.	Nil	0.015 ± 0.001	0.006 ± 0.000	0.007 ± 0.001	0.016 ± 0.005
Sep.	Nil	0.025 ± 0.001	0.010 ± 0.002	0.018 ± 0.002	0.021 ± 0.002
Oct.	Nil	0.030 ± 0.002	0.016 ± 0.003	0.021 ± 0.002	0.026 ± 0.000
Nov.	Nil	0.026 ± 0.002	0.025 ± 0.001	0.037 ± 0.001	0.030 ± 0.000
Dec.	Nil	0.025 ± 0.000	0.020 ± 0.000	0.028 ± 0.002	0.034 ± 0.005

Two-way ANOVA Test

Significant difference in chromium content between river monitoring stations, F_1=20.52 (P<0.01).
No significant difference in chromium content between monitoring months. F_2=1.95.

Table DWM-24, shows the monthly averages (Mean ± SD) of copper in drain water and the results of ANOVA test. The averages ranged from 0.176 ± 0.022 to 0.339 ± 0.029 in NAJ-1, 0.020 ± 0 to 0.143 ± 0.007 in CIV-2, 0.078 ± 0.002 to 0.285 ± 0.005 in POW-3, 0.054 ± 0.002 to 0.518 ± 0.002 in SEN-4, 0.017 ± 0.001 to 0.059 ± 0.003 in BAR-5.

Analysis of variance shows significant difference in Cu content between monitoring drains (F_1= 22.8, P<0.01) as well as between sampling months (F_2=2.07, P>0.05).

Seasonal and annual average values of copper (mg/l) varied between 0.193 ± 0.021 (Winter) to 0.318 ± 0.069 (Monsoon) in NAJ-1, 0.034 ± 0.017 (Monsoon) to 0.122 ± 0.033 (Winter) in CIV-2, 0.096 ± 0.025 (Monsoon) to 0.255 ± 0.030 (Summer) in SEN-4, 0.02 ± 0.003 (Monsoon) to 0.054 ± 0.008 (Summer) in BAR-5. Annual average copper values in drain water were 0.275 ± 0.068, 0.084 ± 0.036, 0.179 ± 0.074, 0.116 ± 0.013, 0.039 ± 0.016 at NAJ-1, CIV-2, POW-3, SEN-4 and BAR-5 respectively.

Table DWM-24: Monthly Variation in Copper (mg/l) of Drain Water (Year 1993-1995)

Month	NAJ-1 Mean ± SD	CIV-2 Mean ± SD	POW-3 Mean ± SD	SEN-4 Mean ± SD	BAR-5 Mean ± SD
Jan.	0.176 ± 0.022	0.093 ± 0.002	0.185 ± 0.005	0.067 ± 0.003	0.049 ± 0.007
Feb.	0.198 ± 0.015	0.123 ± 0.038	0.255 ± 0.025	0.076 ± 0.003	0.059 ± 0.003
Mar.	0.228 ± 0.001	0.95 ± 0.005	0.270 ± 0.010	0.86 ± 0.001	0.046 ± 0.002
Apr.	0.246 ± 0.040	0.090 ± 0.000	0.255 ± 0.005	0.518 ± 0.002	0.056 ± 0.003
May	0.339 ± 0.029	0.090 ± 0.020	0.285 ± 0.005	0.099 ± 0.001	0.061 ± 0.007
Jun.	0.326 ± 0.010	0.085 ± 0.005	0.210 ± 0.020	0.109 ± 0.001	0.055 ± 0.007
Jul.	0.326 ± 0.070	0.025 ± 0.005	0.093 ± 0.002	0.054 ± 0.002	0.019 ± 0.001
Aug.	0.336 ± 0.091	0.020 ± 0.000	0.078 ± 0.002	0.074 ± 0.011	0.017 ± 0.001
Sep.	0.293 ± 0.005	0.058 ± 0.002	0.117 ± 0.034	0.080 ± 0.006	0.025 ± 0.001
Oct.	0.321 ± 0.014	0.090 ± 0.005	0.097 ± 0.000	0.080 ± 0.004	0.024 ± 0.06
Nov.	0.301 ± 0.015	0.095 ± 0.005	0.135 ± 0.015	0.077 ± 0.004	0.031 ± 0.006
Dec.	0.217 ± 0.021	0.143 ± 0.007	0.175 ± 0.005	0.074 ± 0.004	0.035 ± 0.005

Two-way ANOVA Test

Significant difference in copper content between monitoring drains, F_1=22.8 (P<0.01).
Significant difference in copper content between sampling months, F_2=2.07 (P<0.05).

Table RWM-25: Monthly Variation of Copper (mg/l) in River Water (Year 1994-1995)

Month	WAZ-1 Mean ± SD	ISB-2 Mean ± SD	ITO-3 Mean ± SD	NIZ-4 Mean ± SD	OKH-5 Mean ± SD
Jan.	Nil	0.028 ± 0.002	0.034 ± 0.002	0.044 ± 0.004	Nil
Feb.	Nil	0.038 ± 0.002	0.031 ± 0.001	0.041 ± 0.002	Nil
Mar.	Nil	0.036 ± 0.003	0.036 ± 0.003	0.037 ± 0.001	Nil
Apr.	Nil	0.040 ± 0.003	0.028 ± 0.002	0.028 ± 0.001	Nil
May	Nil	0.031 ± 0.001	0.032 ± 0.001	0.026 ± 0.002	Nil
Jun.	Nil	0.043 ± 0.003	0.029 ± 0.003	0.030 ± 0.003	Nil
Jul.	Nil	0.018 ± 0.001	0.010 ± 0.000	0.013 ± 0.000	Nil
Aug.	Nil	0.014 ± 0.002	0.010 ± 0.000	0.014 ± 0.000	Nil
Sep.	Nil	0.033 ± 0.003	0.016 ± 0.004	0.026 ± 0.002	Nil
Oct.	Nil	0.036 ± 0.002	0.024 ± 0.005	0.034 ± 0.002	Nil
Nov.	Nil	0.040 ± 0.002	0.021 ± 0.001	0.027 ± 0.001	Nil
Dec.	Nil	0.047 ± 0.005	0.023 ± 0.003	0.029 ± 0.004	Nil

Two-way ANOVA Test

Significant difference in copper content between river monitoring stations, F_1=20.33 (P< 0.01).
No significant difference in copper content between monitoring months, F_2=1.33.

Monthly average values (Mean ± SD) of copper and the results of ANOVA test for copper content in river monitoring stations are presented in Table RWM-25. The minimum/and maximum bimonthly averages (Mean ± SD) of copper were 0.014 ± 0.003 (Aug.) and 0.047 ± 0.005 (Dec.) at ISB-2, 0.01 ± 0 (Jul.) and 0.036 ± 0.002 (Oct.) at ITO-3, and 0.014 ± 0.002 (Aug.) and 0.044 ± 0.004 (Jan.) at NIZ-4 respectively. The copper was not observed at WAZ-1 (Wazirabad barrage) and OKH-5 (Okhla barrage). Analysis of variance test shows significant difference in copper content between different monitoring stations (F_1=20.33, P<0.01). However the difference between sampling months was not significant (F_2=1.33).

Seasonal and annual average values in copper concentration varied from a minimum of 0.021 ± 0.009 (Monsoon) to a maximum of 0.038 ± 0.003 (Pre-Monsoon) at ISB-2, 0.012 ± 0.004 (Monsoon) to 0.031 ± 0.003 (Winter) at ITO-3, 0.017 ± 0.006 (Monsoon) to 0.039 ± 0.006 (Winter) at NIZ-4. The peak concentration of copper observed during winter season might be due to lean river water discharge from Wazirabad barrage and higher percentage contribution of drain water at river monitoring points. The low values during monsoon was the effects of dilution by rain waters. Annual average copper values were in order of 0.033 ± 0.01 > 0.029 ± 0.009 > 0.024 ± 0.009 at ISB-2, NIZ-4 and ITO-3 respectively. Absence of copper at OKH-5 might be due to dilution of copper containing drain effluents.

Zinc of Drain and River Water

Zinc has no known adverse physiological effects upon human beings. In fact it is an essential and beneficial element in human nutrition. From aesthetic considerations however, high concentrations (above 5 mg/l) of zinc of domestic water are undesirable. A bitter astringent taste is produced by zinc. Concentration of zinc above 30 mg/l give water milky appearance and causes a greasy film on boiling. It is now reported that even 5 mg/l can cause a greasy film on boiling. Zinc is commonly found in small quantities on domestic water supplies and industrial waters due to corrosion of galvanized iron and brass in condensing, cooling and distributing systems. Zinc may also enter into a water supply through the discharge of industrial effluent such as galvanising water, zinc plating wastes, viscose rayon wastes and rubber factory waste (Manivasakam, 1989). A relatively higher concentration of zinc may be toxic to aquatic life (Kothandaraman and Viswanathan, 1986).

Table DWM-25 represents for bimonthly average values (Mean ± SD) and the results of ANOVA test for zinc concentration in five monitoring drains. The minimum and maximum monthly average values were 0.320 ± 0.001 and 1.2 ± 0.122 in NAJ-1, 0.560 ± 0.050 and 1.086 ± 0.1 in CIV-2, 0.745 ± 0.185 and 1.675 ± 0.025 in POW-3, 0.425 ± 0.035 and 1.115 ± 0.135 in SEN-4, 0.30 ± 0.02 and 0.745 ± 0.065 in BAR-5 drain respectively. Analysis of variance shows significant difference in zinc concentration between sampling stations (F_1=20.4, P<0.01) as well as between sampling months (F_2=2.66, P<0.05).

Seasonal and annual average values varied between 0.424 ± 0.113 (Summer) to 0.982 ± 0.076 (Pre-Monsoon) in NAJ-1, 0.628 ± 0.132 (Monsoon) to 1.01 ± 0.204 (Pre-Monsoon) in CIV-2, 0.955 ± 0.210 (Monsoon) to 1.273 ± 0.130 (Pre-Monsoon) in POW-3, 0.585 ± 0.239 (Winter) to 0.960 ± 0.097 (Pre-Monsoon) in SEN-4, 0.140 ± 0.134 (Winter) to 0.175 ± 0.044 (Pre-Monsoon) in BAR-5 respectively. Annual averages of zinc content were in order of 1.050 ± 0.337 > 0.836 ± 0.194 > 0.688 ± 0.224 > 0.593 ± 0.138 > 0.558 ± 0.152 in POW-3, CIV-2, SEN-4, NAJ-1 and BAR-5 drain respectively.

Table RWM-26 shows the monthly averages (Mean ± SD) and the results of analysis of variance for zinc content in river water at five monitoring points. The minimum and maximum average values of zinc were 0.125 ± 0.015 (April) and 0.445 ± 0.035 (Nov.) at WAZ-1, 0.295 ± 0.005 (Dec.) and 0.405 ± 0.095 (Jun.) at ISB-2, 0.195 ± 0.005 (Feb.) and 0.470 ± 0.04 (Nov.) at ITO-3, 0.270 ± 0.01 (Jan.) and 0.570 ± 0.05 (Oct.) at NIZ-4 and 0.27 ± 0.01 (Feb.) and 0.620 ± 0.06 (Jul.) at OKH-5. Analysis of variance shows the significant difference in zinc content between different river monitoring stations (F_1=19.81, P<0.01), as well as between the monitoring months (F_2=2.91, P<0.05).

Table DWM-25: Monthly Variation in Zinc (mg/l) of Drain Water (Year 1994-1995)

Month	*NAJ-1*	*CIV-2*	*POW-3*	*SEN-4*	*BAR-5*
	Mean ± SD	*Mean ± SD*	*Mean ± SD*	*Mean ± SD*	*Mean ± SD*
Jan.	0.358 ± 0.032	0.766 ± 0.027	0.745 ± 0.335	0.425 ± 0.105	0.300 ± 0.020
Feb.	0.320 ± 0.001	0.825 ± 0.029	0.790 ± 0.200	0.425 ± 0.035	0.370 ± 0.010
Mar.	0.344 ± 0.106	0.790 ± 0.142	0.745 ± 0.185	0.570 ± 0.010	0.435 ± 0.045
Apr.	0.593 ± 0.033	0.704 ± 0.031	0.820 ± 0.140	0.52 ± 0.040	0.525 ± 0.105
May	0.378 ± 0.011	0.861 ± 0.025	1.040 ± 0.060	0.585 ± 0.085	0.525 ± 0.035
Jun.	0.383 ± 0.014	0.841 ± 0.096	1.370 ± 0.240	0.710 ± 0.070	0.435 ± 0.055
Jul.	0.62 ± 0.032	0.560 ± 0.050	0.795 ± 0.065	0.605 ± 0.025	0.595 ± 0.055
Aug.	0.414 ± 0.054	0.655 ± 0.025	0.890 ± 0.090	0.645 ± 0.035	0.630 ± 0.050
Sep.	0.710 ± 0.050	0.830 ± 0.100	1.180 ± 0.200	0.730 ± 0.050	0.745 ± 0.065
Oct.	0.923 ± 0.045	0.855 ± 0.035	1.230 ± 0.130	0.905 ± 0.075	0.725 ± 0.055
Nov.	1.042 ± 0.048	1.165 ± 0.185	1.315 ± 0.115	1.015 ± 0.085	0.705 ± 0.025
Dec.	1.286 ± 0.122	1.180 ± 0.100	1.675 ± 0.025	1.115 ± 0.135	0.705 ± 0.015

Two-way ANOVA Test

Significant difference in zinc content between monitoring drains, F_1=20.41 (P<0.01).
Significant difference in zinc content between sampling months, F_2=2.66 (P<0.05).

Seasonal and annual averages varied from 0.250 ± 0.033 (Summer) to 0.418 ± 0.038 (Pre-Monsoon) at WAZ-1, 0.283 ± 0.058 (Monsoon) to 0.478 ± 0.064 (Pre-Monsoon) at ISB-2, 0.237 ± 0.041 (Winter) to 0.453 ± 0.046 (Pre-Monsoon) at ITO-3, 0.295 ± 0.026 (Winter) to 0.490 ± 0.094 (Pre-Monsoon) at NIZ-4, 0.30 ± 0.033 (Winter) to 0.583 ± 0.055 (Monsoon) at BAR-5. Peak values of zinc were observed during monsoon and post-monsoon seasons might be due to solubilization of zinc ions present in sediment of river bank by rain water. The lower values of zinc during winter might be due to precipitation of this metal from the water to the sediment due to lean river flow and low temperature (Jain and Salman, 1995). Annual average values (mg/l) were in order of 0.412 > 0.380 > 0.353 > 0.330 > 0.328 at OKH-5, NIZ-4, ISB-2, ITO-3 and WAZ-1 respectively.

Nickel of Drain and River Water

Nickel does not occur in natural water. Metal plating wastes when discharged in to the water courses contribute nickel. Increased use of stainless steel and other nickel containing alloy in water systems also contribute nickel. Nickel compounds are found in many ores and minerals. It is relatively non-toxic elements at lower level, but higher amount (more than 30 mg) may cause changes in muscle, brain, lungs, liver, kidney and can also cause cancer (Ramteke and Moghe, 1986). Nickel is an important non-ferrous metal closely related to iron and cobalt (Carbson, 1956). As an alloying metal it is used in steel, separately or with chrome, or with zinc or simultaneously with zinc and copper or with titanium etc. for making different industrial and strategic products (Smith *et al.*, 1961). Nickel is insoluble in water and soluble in acid. It forms alloys with copper, manganese, zinc, chromium, molybdenum etc. Stainless steel is the most widely used nickel alloy (Sittig, 1979).

Table RWM-26: Monthly Variation of Zinc (mg/l) in River Water Year (1994-1995)

Month	*WAZ-1* *Mean ± SD*	*ISB-2* *Mean ± SD*	*ITO-3* *Mean ± SD*	*NIZ-4* *Mean ± SD*	*OKH-5* *Mean ± SD*
Jan.	0.290 ± 0.010	0.345 ± 0.055	0.245 ± 0.045	0.270 ± 0.010	0.290 ± 0.010
Feb.	0.285 ± 0.015	0.380 ± 0.080	0.195 ± 0.005	0.305 ± 0.025	0.270 ± 0.010
Mar.	0.265 ± 0.025	0.430 ± 0.070	0.255 ± 0.015	0.300 ± 0.010	0.325 ± 0.005
Apr.	0.215 ± 0.015	0.330 ± 0.070	0.305 ± 0.025	0.320 ± 0.040	0.335 ± 0.055
May	0.275 ± 0.015	0.330 ± 0.130	0.295 ± 0.015	0.410 ± 0.050	0.320 ± 0.020
Jun.	0.245 ± 0.035	0.405 ± 0.095	0.385 ± 0.005	0.410 ± 0.020	0.390 ± 0.010
Jul.	0.335 ± 0.015	0.325 ± 0.065	0.390 ± 0.010	0.425 ± 0.035	0.620 ± 0.060
Aug.	0.355 ± 0.005	0.355 ± 0.155	0.345 ± 0.015	0.360 ± 0.020	0.585 ± 0.025
Sep.	0.410 ± 0.020	0.305 ± 0.005	0.370 ± 0.010	0.440 ± 0.020	0.545 ± 0.045
Oct.	0.390 ± 0.010	0.335 ± 0.055	0.435 ± 0.045	0.570 ± 0.050	0.435 ± 0.025
Nov.	0.445 ± 0.035	0.385 ± 0.125	0.470 ± 0.040	0.410 ± 0.050	0.435 ± 0.055
Dec.	0.430 ± 0.070	0.295 ± 0.005	0.270 ± 0.010	0.340 ± 0.050	0.390 ± 0.070

Two-way ANOVA Test

Significant difference in zinc content between river monitoring stations, F_1=19.81 ($P<0.01$).
Significant difference in zinc content between monitoring months, F_2=2.91 ($P<0.05$).

Table DWM-26 shows the monthly averages (Mean ± SD) and the results of analysis of variance for nickel content in monitoring drains. The minimum and maximum average values of nickel (mg/l) were 0.221 ± 0.002 (Jan.) and 0.594 ± 0.326 (Aug.) in NAJ-1, 0.02 ± 0.002 (Jul.) and 0.071 ± 0.002 in CIV-2, 0.055 ± 0.005 (Jul.-Aug.) and 0.180 ± 0.010 (Mar.) in POW-3, 0.043 ± 0.002 (Jul.) and 0.089 ± 0.004 (Dec.) in SEN-4, 0.019 ± 0.001 (Jul.) and 0.06 ± 0.007 (Apr.) in BAR-5. Analysis of variance results shows that, difference in nickel concentration in between monitoring drains (F_1=20.8, $P<0.01$) but no significant difference was observed between sampling months (F_2=1.33).

Annual as well as seasonal average values of nickel varied between 0.253 ± 0.024 (Winter) to 0.440 ± 0.016 (Monsoon) in NAJ-1, 0.029 ± 0.01 (Monsoon) to 0.067 ± 0.005 (Summer) in CIV-2, 0.063 ± 0.012 (Monsoon) to 0.150 ± 0.038 (Summer) in POW-3, 0.055 ± 0.014 (Monsoon) to 0.84 ± 0.005 (Pre-Monsoon) in Sen-4, 0.023 ± 0.005 (Monsoon) to 0.051 ± 0.010 (Summer) in BAR-5 drain. Annual average values were in order of 0.327 ± 0.199 > 0.112 ± 0.044 > 0.072 ± 0.015 > 0.054 ± 0.017 > 0.039 ± 0.013 at NIZ-1, POW-1, SEN-4, CIV-2 and BAR-5 respectively.

Table RWM-27 represents the monthly averages and the results of analysis of variance test for the nickel content at different drain monitoring stations. Monthly averages varied from a minimum of 0.031 ± 0.005 (Sep) to a maximum of 0.061 ± 0.020 (Nov.) at WAZ-1, 0.016 ± 0.002 (Aug.) to 0.071 ± 0.002 (Jan.) at ISB-2, 0.01 ± 0 (Aug.) to 0.057 ± 0.006 (Nov.) at ITO-3, 0.016 ± 0.003 (Aug.) to 0.081 ± 0.002 (Nov.) at NIZ-4, 0.020 ± 0.002 (Aug.) to 0.070 ± 0.001 (Nov.) at OKH-5. The ANOVA test shows significant difference in nickel content between monitoring points (F_1=21.41, $P<0.01$) as well as between monitoring months (F_2=2.96, $P<0.05$).

Seasonal and annual average values are depicted in Fig. RW 27. Seasonal values varied from a minimum of 0.031 ± 0.007 (Monsoon) to a maximum of 0.066 ± 0.015 (Pre-Monsoon) at WAZ-1, 0.026 ± 0.011 (Monsoon) to 0.049 ± 0.009 (Pre-Monsoon) at ITO-3, 0.020 ± 0.005 (Monsoon) to 0.064 ± 0.004 (Winter) at NIZ-4, 0.027 ± 0.008 (Monsoon) to 0.065 ± 0.005 (Pre-Monsoon) at OKH-5. Peak values of nickel were observed during post-monsoon season might be due to lean discharge of river water from Wazirabad barrage and higher quantities of drain water contribution at river monitoring points. Yearly averages (mg/l) of nickel were in order of 0.052 > 0.048 > 0.045 > 0.044 > 0.032 at ISB-2, WAZ-1, NIZ-4, OKH-5 and ITO-3 respectively.

Table DWM-26: Monthly Variation in Nickel (mg/1) of Drain Water (Year 1994-1995)

Month	*NAJ-1*	*CIV-2*	*POW-3*	*SEN-4*	*BAR-5*
	Mean ± SD	*Mean ± SD*	*Mean ± SD*	*Mean ± SD*	*Mean ± SD*
Jan.	0.221 ± 0.002	0.045 ± 0.000	0.105 ± 0.005	0.080 ± 0.004	0.03 ± 0.005
Feb.	0.270 ± 0.013	0.051 ± 0.007	0.155 ± 0.015	0.072 ± 0.001	0.035 ± 0.001
Mar.	0.300 ± 0.004	0.061 ± 0.002	0.180 ± 0.010	0.063 ± 0.018	0.054 ± 0.004
Apr.	0.68 ± 0.002	0.072 ± 0.002	0.155 ± 0.035	0.085 ± 0.001	0.060 ± 0.007
May	0.221 ± 0.002	0.045 ± 0.000	0.105 ± 0.045	0.080 ± 0.008	0.049 ± 0.013
Jun.	0.306 ± 0.037	0.064 ± 0.001	0.135 ± 0.030	0.069 ± 0.009	0.43 ± 0.003
Jul.	0.217 ± 0.006	0.020 ± 0.002	0.055 ± 0.005	0.043 ± 0.002	0.019 ± 0.001
Aug.	0.594 ± 0.326	0.026 ± 0.000	0.055 ± 0.005	0.050 ± 0.006	0.20 ± 0.004
Sep.	0.511 ± 0.478	0.042 ± 0.003	0.078 ± 0.006	0.073 ± 0.005	0.029 ± 0.001
Oct.	0.326 ± 0.030	0.058 ± 0.002	0.082 ± 0.004	0.087 ± 0.001	0.40 ± 0.001
Nov.	0.328 ± 0.018	0.071 ± 0.002	0.097 ± 0.002	0.080 ± 0.004	0.048 ± 0.005
Dec.	0.273 ± 0.005	0.069 ± 0.004	0.122 ± 0.013	0.089 ± 0.004	0.044 ± 0.005

Two-way ANOVA Test

Significant difference in nickel content between monitoring drains. F_1=20.8 ($P<0.01$).

No significant difference in nickel content between sampling months, F_2=1.33.

Table RWM-27: Monthly Variation of Nickel (mg/l) in River Water Year (1994-1995)

Month	*WAZ-1*	*ISB-2*	*ITO-3*	*NIZ-4*	*OKH-5*
	Mean ± SD	*Mean ± SD*	*Mean ± SD*	*Mean ± SD*	*Mean ± SD*
Jan.	0.055 ± 0.025	0.049 ± 0.001	0.025 ± 0.004	0.070 ± 0.016	0.031 ± 0.002
Feb.	0.041 ± 0.002	0.064 ± 0.004	0.028 ± 0.002	0.080 ± 0.022	0.034 ± 0.004
Mar.	0.040 ± 0.011	0.054 ± 0.006	0.027 ± 0.003	0.081 ± 0.002	0.036 ± 0.007
Apr.	0.051 ± 0.010	0.063 ± 0.005	0.037 ± 0.001	0.062 ± 0.009	0.056 ± 0.007
May	0.058 ± 0.009	0.060 ± 0.001	0.038 ± 0.004	0.051 ± 0.002	0.062 ± 0.001
Jun.	0.058 ± 0.010	0.071 ± 0.002	0.049 ± 0.002	0.035 ± 0.010	0.059 ± 0.003
Jul.	0.047 ± 0.026	0.041 ± 0.002	0.015 ± 0.005	0.018 ± 0.019	0.023 ± 0.003
Aug.	0.042 ± 0.014	0.016 ± 0.002	0.010 ± 0.000	0.016 ± 0.003	0.020 ± 0.002
Sep.	0.031 ± 0.005	0.041 ± 0.002	0.025 ± 0.001	0.027 ± 0.006	0.038 ± 0.003
Oct.	0.041 ± 0.019	0.067 ± 0.004	0.041 ± 0.002	0.035 ± 0.009	0.061 ± 0.002
Nov.	0.061 ± 0.020	0.070 ± 0.002	0.057 ± 0.006	0.034 ± 0.008	0.070 ± 0.001
Dec.	0.048 ± 0.005	0.053 ± 0.003	0.031 ± 0.001	0.036 ± 0.002	0.043 ± 0.003

Two-way ANOVA Test

Significant difference in nickel content between river monitoring stations, F_1=21.41 ($P<0.01$).

Significant difference in nickel content between monitoring months, F_2=2.96 ($P<0.05$).

Lead of Drain and River Water

Lead is not found in natural water supply. It may be due to the corrosion reactions and wastes contamination. Pollution due to lead is mainly attributed to the industrial effluents, coal combustion products and petroleum washes from various mechanical workshops garages, because lead is used as antiknocking compounds in petroleum (Katariya, 1994) and also from domestic sewage discharge in the city (Hydn, 1961; Harrison and Laxen, 1984). Automobile exhaust also adds considerable amount of lead particulates to the river water (Pophali *et al.*, 1990) so also the traffic of diesel and petrol vehicles (Agarwal *et al.*, 1978) and lead dust fallout. Lead is a minor element in the earth crust but the significant contribution of lead is from the atmosphere. Lead is classified as highly toxic element in nature (Forstner, 1977; Kudesia, 1980; Forstner and Wiltmann, 1981). Lead in high doses has been recognized as a cumulative general metabolic poison (Suess, 1982).

Table DWM-27 shows the monthly averages (Mean ± SD) of lead content in drain water of Najafgarh. The lead was not found in Civil mill, Sen-nursing home, Power house and Barpulla drains. The maximum concentration of lead was observed in the month of November *i.e.* 0.09 ± 0.004 (mg/l) and minimum in July *i.e.* 0.033 ± 0.005 (mg/l) in Najafgarh drain. Annual and seasonal mean values are depicted in Table DW26. Seasonal values of lead varied from 0.038 ± 0.008 (Monsoon) to 0.075 ± 0.015 (Pre-Monsoon) in Najafgarh drain. Annual mean values (Table DW26) was 0.061 ± 0.02 in drain water of Najafgarh.

The lead was not detected in river water of Yamuna between Wazirabad barrage and Okhla barrage.

Table DWM-27: Monthly Variation in Lead (mg/l) of Drain Water (Year 1994-1995)

Month	*NAJ-1* *Mean ± SD*	*CIV-2* *Mean ± SD*	*POW-3* *Mean ± SD*	*SEN-4* *Mean ± SD*	*BAR-5* *Mean ± SD*
Jan.	0.068 ± 0.005	0.000 ± 0.000	0.000 ± 0.000	0.000 ± 0.000	0.000 ± 0.000
Feb.	0.073 ± 0.005	0.000 ± 0.000	0.000 ± 0.000	0.000 ± 0.000	0.000 ± 0.000
Mar.	0.085 ± 0.001	0.000 ± 0.000	0.000 ± 0.000	0.000 ± 0.000	0.000 ± 0.000
Apr.	0.088 ± 0.008	0.000 ± 0.000	0.000 ± 0.000	0.000 ± 0.000	0.000 ± 0.000
May	0.053 ± 0.015	0.000 ± 0.000	0.000 ± 0.000	0.000 ± 0.000	0.000 ± 0.000
Jun.	0.042 ± 0.004	0.000 ± 0.000	0.000 ± 0.000	0.000 ± 0.000	0.000 ± 0.000
Jul.	0.033 ± 0.005	0.000 ± 0.000	0.000 ± 0.000	0.000 ± 0.000	0.000 ± 0.000
Aug.	0.035 ± 0.007	0.000 ± 0.000	0.000 ± 0.000	0.000 ± 0.000	0.000 ± 0.000
Sep.	0.045 ± 0.005	0.000 ± 0.000	0.000 ± 0.000	0.000 ± 0.000	0.000 ± 0.000
Oct.	0.061 ± 0.004	0.000 ± 0.000	0.000 ± 0.000	0.000 ± 0.000	0.000 ± 0.000
Nov.	0.090 ± 0.004	0.000 ± 0.000	0.000 ± 0.000	0.000 ± 0.000	0.000 ± 0.000
Dec.	0.067 ± 0.002	0.000 ± 0.000	0.000 ± 0.000	0.000 ± 0.000	0.000 ± 0.000

Two-way ANOVA Test was not possible, due to inadequate data.

Pesticides in Drain and River Water

BHC (HCH) of Drain and River Water

The BHC which is HCH (Hexachloro Cyclohexane) is generally included in the class of organochlorine compounds but they are actually isomers prepared from chlorination of benzene and are saturated cylic hydrocarbons. BHC are available in four forms, alpha-BHC, beta-BHC, gama-BHC and delta-BHC depending upon the position of six chlorine atoms. Mehrotra (1985) has given an exhaustive review on the use of HCH and its environmental effects in India. HCH was first introduced in 1949 to control insect pests of the agriculture. Its use in public health started in 1960. The large scale use of HCH has led to its presence in different components of the environment.

In India, HCH residues have been reported from soil (0.025 to 124 ppm), water (0.02 to 6.61 ppm), cereals and animal feed (0.01 to 3.52 ppm), vegetables oil and oil seeds (0.31 to 0.58 ppm), milk and milks products (0.35 to 82 ppm), meat and poultry (0.04 to 1.522 ppm), wildlife as high up to 13.98 and 22.487 ppm in human tissues (Kulshrestha, 1989). Dikshith *et al.*, 1990, reported a total HCH residues of 4.654 (range 1.576 to 7.747) in wells, 6.130 (range 4.320 to 8.718) in hand-pump and 9.941 (range 6.221 to 15.58) in ponds.

Monthly average values (Mean ± SD) and the results of ANOVA test for T-BHC (mg/l) content of drain water are provided in Table DWM 28. Monthly average values varied from 430.00 ± 220.0 to 6650.0 ± 2300 in NAJ-1, 1600 ± 100 to 4390 ± 410 in CIV-2, 365.0 ± 55.0 to 1620 ± 330 in POW-3, 455 ± 5 to 2160 ± 220 in SEN-4 and 400 ± 10 to 2429.0 ± 431.0 in BAR-5. Analysis of variance shows the existence of significant difference in T-BHC content between monitoring drains (F_1=26.31, P<0.01) as well as between monitoring months (F_2=2.86, P<0.05).

Table DWM-28: Monthly Variation in Total BHC (mg/l) of Drain Water (Year 1994-1995)

Month	*NAJ-1*	*CIV-2*	*POW-3*	*SEN-4*	*BAR-5*
	Mean ± SD	*Mean ± SD*	*Mean ± SD*	*Mean ± SD*	*Mean ± SD*
Jan.	6650.00 ± 2300.00	2660.00 ± 20.00	925.00 ± 65.00	1585.00 ± 55.00	1045.00 ± 55.00
Feb.	5272.50 ± 592.50	2620.00 ± 220.00	1070.00 ± 170.00	1230.00 ± 90.00	1490.00 ± 300.00
Mar.	4830.00 ± 600.00	3280.00 ± 300.00	1420.00 ± 460.00	935.00 ± 45.00	1680.00 ± 200.00
Apr.	2740.00 ± 390.00	3109.00 ± 71.00	1110.00 ± 30.00	1090.00 ± 130.00	1725.00 ± 35.00
May	1870.00 ± 30.00	3666.50 ± 313.50	1310.00 ± 320.00	1510.00 ± 270.00	2125.00 ± 35.00
Jun.	935.00 ± 145.00	4390.00 ± 410.00	1620.00 ± 330.00	2025.00 ± 65.00	2429.00 ± 431.00
Jul.	780.00 ± 200.00	1815.00 ± 135.00	435.00 ± 95.00	455.00 ± 5.00	410.00 ± 50.00
Aug.	3550.00 ± 3010.00	1600.00 ± 100.00	365.00 ± 55.00	520.00 ± 10.00	400.00 ± 10.00
Sep.	430.00 ± 220.00	1840.00 ± 50.00	610.00 ± 70.00	1010.00 ± 70.00	810.00 ± 130.00
Oct.	730.00 ± 250.00	2295.00 ± 165.00	834.50 ± 54.50	1925.00 ± 255.00	779.00 ± 219.00
Nov.	1319.00 ± 679.00	2325.00 ± 235.00	905.00 ± 45.00	2160.00 ± 220.00	925.00 ± 235.00
Dec.	2435.00 ± 1415.00	2035.00 ± 95.00	925.00 ± 85.00	2015.00 ± 135.00	820.00 ± 170.00

Two-way ANOVA Test

Significant difference in BHC content between monitoring drains, F_1=26.31 (P<0.01).
Significant difference in BHC content between sampling months, F_2=2.86 (P<0.05).

Seasonal and annual average values of T-BHC (mg/l) in five monitoring drains varied from a minimum of 1024.5 ± 590.34 (Pre-Monsoon) to a maximum of 5104.17 ± 1932.96 (Winter) in NAJ-1, 1751.67 ± 147.81 (Monsoon) to 3611.38 ± 577.26 (Summer) in CIV-2, 470.0 ± 127.54 (Monsoon) to 1365.0 ± 374.13 (Summer) in POW-3, 661.67 ± 251.09 (Monsoon) to 2042.50 ± 265.55 (Pre-Monsoon) in SEN-4, 540.0 ± 207.28 (Monsoon) to 1145.0 ± 311.01 (Winter) in BAR-5. Annual mean values of BHC were 2628.46, 2636.29, 960.8, 1371.7 and 1219.8 in NAJ-1, CIV-2, POW-3, SEN-4 and BAR-5 respectively.

The higher concentration BHC was observed in Najafgarh drain and may be due to contamination of drain water by agricultural run-off.

Monthly average values (Mean ± SD) and the results of analysis of variance for T-BHC content of five river monitoring points are tabulated in Table RWM-28. The bimonthly average values (mg/l) varied from 0 to 203 ± 7.0 at WAZ-1, 149 ± 49.0 to 600.0 ± 160.0 at ISB-2, 62.0 ± 9.0 to 640 ± 20 at ITO-3, 94 ± 4.0 to 760 ± 20.0 at NIZ-4, 119.5 ± 23.5 to 1275.0 ± 95 at OKH-5. Analysis of variance shows significant difference in T-BHC content between monitoring stations (F_1=29.69, P<0.01) and also between monitoring months (F_2=2.78, P<0.05).

Seasonal and annual average values of T-BHC (mg/l) concentration of river varied from 0 (Winter) to 154.5 ± 56.72 (Monsoon) at WAZ-1, 103.67 ± 31.4 (Monsoon) to 550.0 ± 101.98 (Winter) at ISB-2, 72.16 ± 12.84 (Monsoon) to 521.25 ± 115.8 (Summer) at ITO-3, 105.67 ± 11.74 (Monsoon) to 573.75 ± 138.65 (Summer) at NIZ-4, 127.5 ± 21.45 (Pre-Monsoon) to 1052.75 ± 190.06 (Summer) at OKH-5. The maximum content of T-BHC recorded during summer and/or winter season in all down stream stations of Wazirabad barrage might be due to lower water discharge from Wazirabad barrage and higher contribution of drain water at monitoring points. The lower values during monsoon was due to dilution. At Wazirabad barrage the peak values of T-BHC were observed during monsoon, might be due to washing out of agricultural lands of basin areas, by rain water. Annual average values of T-BHC were in order of 530.40 > 386.29 > 361.58 > 308.08 > 51.83 > at OKH-5, ISB-2, NIZ-4, ITO-3 and WAZ-1 respectively.

Table RWM-28: Monthly Variation of Total BHC (mg/l) in River Water (Year 1994- 1995)

Month	*WAZ-1*	*ISB-2*	*ITO-3*	*NIZ-4*	*OKH-5*
	Mean ± SD	*Mean ± SD*	*Mean ± SD*	*Mean ± SD*	*Mean ± SD*
Jan.	Nil	594.0 ± 26.0	378.0 ± 72.0	495.0 ± 35.0	304.5 ± 15.5
Feb.	Nil	581.5 ± 18.5	424.0 ± 26.0	495.0 ± 15.0	675.0 ± 185.0
Mar.	Nil	506.0 ± 4.0	475.0 ± 65.0	625.0 ± 45.0	799.0 ± 31.0
Apr.	Nil	334.0 ± 23.0	600.0 ± 10.0	760.0 ± 20.0	1028.0 ± 68.0
May	Nil	336.5 ± 226.5	640.0 ± 20.0	490.0 ± 70.0	1275.0 ± 95.0
Jun.	45.0 ± 45.0	331.0 ± 171.0	370.0 ± 60.0	420.0 ± 40.0	1109.0 ± 111.0
Jul.	203.0 ± 7.0	149.0 ± 49.0	66.5 ± 1.5	108.0 ± 12.0	323.0 ± 37.0
Aug.	184.5 ± 4.5	248.0 ± 138.0	62.0 ± 9.0	94.0 ± 4.0	277.5 ± 12.5
Sep.	76.0 ± 13.0	300.0 ± 140.0	88.0 ± 5.0	115.0 ± 5.0	172.5 ± 16.5
Oct.	83.5 ± 5.5	439.0 ± 241.0	128.0 ± 18.0	154.0 ± 9.0	135.5 ± 15.5
Nov.	30.0 ± 30.0	478.0 ± 92.0	179.0 ± 11.0	225.0 ± 35.0	119.5 ± 23.5
Dec.	0.0 ± 00.0	600.0 ± 160.0	286.5 ± 23.5	358.0 ± 62.0	147.0 ± 39.5

Two-way ANOVA Test

Significant difference in BHC content between river monitoring stations, F_1=29.69 (P<0.01).

Significant difference in BHC content between monitoring months, F_2=2.78 (P<0.05).

Total-DDT of Drain Water and River Water

The insecticidal property of DDT was discovered by Muller and his workers in 1939. It is an organochlorine insecticide which is a white crystalline solid, tasteless and almost odourless. DDT is a non-systemic stomach and contact insecticide of high persistence on solid surfaces. It is non-phytotoxic and a suspected carcinogen.

During the last decade, many scientific reports have shown the presence of persistent insecticides in man and the environment (Durham, 1965; Kraybill, 1969; USDHEW, 1969). The residues of DDT have been reported in several life forms, food, air and water. Even the mother's milk has been found to be contaminated with these chemicals (IARC, 1974). The man and animals are exposed to these chemicals either directly or indirectly over their entire life time mainly through the food chain. Food is reported to be responsible for more than 90 per cent of the total body intake of these chemicals in general population (Kraybill, 1969; Campbell, 1965). DDT has a strong potential for food chain biomagnification. Concentration of DDT in surface water depends on the soil as well as rains. DDT is highly toxic aquatic invertebrates at concentrations as low as 0.3 (mg/l). Toxic effects include impairment in reproduction and development, cardiovascular modifications and neurological changes.

Table DWM-29 shows monthly averages (Mean ± SD) and the results of analysis of variance of T-DDT (mg/l) content in five monitoring drains. The minimum and maximum values of T-DDT were 202.5 ± 17.5 and 910 ± 24 in NAJ-1, 96.5 ± 1.5 and 385 ± 25 in CIV-2, 0 and 300 ± 20 in POW-3, 30.0 ± 5 and 264.0 ± 96.0 in SEN-4, 103.0 ± 5 and 1230.0 ± 320.0 in BAR-5. Analysis of variance shows significant difference in T-DDT content between monitoring drains (F_1=28.8, P<0.01) as well as between monitoring months (F_2=2.16, P<0.05).

Table DWM-29: Monthly Variation in Total DDT (mg/l) of Drain Water (Year 1994-1995)

Month	*NAJ-1* *Mean ± SD*	*CIV-2* *Mean ± SD*	*POW-3* *Mean ± SD*	*SEN-4* *Mean ± SD*	*BAR-5* *Mean ± SD*
Jan.	320.00 ± 70.00	179.50 ± 3.50	190.00 ± 10.00	91.50 ± 18.50	248.00 ± 38.00
Feb.	485.00 ± 95.00	175.50 ± 14.50	104.00 ± 16.00	67.50 ± 8.50	455.00 ± 25.00
Mar.	490.0 ± 20.00	253.00 ± 43.00	61.50 ± 3.50	93.00 ± 7.00	840.00 ± 80.00
Apr.	700.00 ± 20.00	385.00 ± 25.00	0.00 ± 0.00	85.00 ± 11.00	765.00 ± 95.00
May	910.00 ± 24.00	720.00 ± 260.00	0.00 ± 0.00	75.50 ± 7.50	1130.00 ± 150.00
Jun.	900.00 ± 80.00	100.00 ± 10.00	0.00 ± 0.00	87.00 ± 11.00	1230.00 ± 320.00
Jul.	202.50 ± 17.50	96.50 ± 1.50	0.00 ± 0.00	30.00 ± 5.00	103.00 ± 5.00
Aug.	205.0 ± 25.50	108.00 ± 12.00	0.00 ± 0.00	36.50 ± 6.50	110.00 ± 0.00
Sep.	413.00 ± 117.00	198.00 ± 12.00	59.00 ± 9.00	66.50 ± 11.50	249.00 ± 41.00
Oct.	720.00 ± 360.00	215.00 ± 25.00	139.50 ± 29.50	140.00 ± 20.00	285.00 ± 65.00
Nov.	570.00 ± 60.00	200.00 ± 10.00	215.00 ± 25.00	245.00 ± 35.00	254.00 ± 56.00
Dec.	385.00 ± 125.00	174.00 ± 4.00	300.00 ± 20.00	264.00 ± 96.00	235.00 ± 45.00

Two-way ANOVA Test

Significant difference in DDT content between monitoring drains, F_1=28.8 (P<0.01).
Significant difference in DDT content between sampling months, F_2=2.16 (P<0.05).

Seasonal and annual mean of T-DDT varied from a minimum of 23.56 ± 10.22 (Summer) to a maximum of 23.56 ± 8.30 (Monsoon) in NAJ-1, 39.21 ± 12.78 (Monsoon) to 47.69 ± 12.47 (Winter) in CIV-2, 33.07 ± 4.9 (Monsoon) to 42.49 ± 8.53 (Pre-Monsoon) in POW-3, 43.56 ± 11.63 (Monsoon) to 58.17 ± 8.43 (Winter) in SEN-4, 33.46 ± 9.32 (Summer) to 40.94 ± 5.46 (Pre-Monsoon) in BAR-5. Annual mean values of DDT in NAJ-1, CIV-2, POW-3, SEN-4 and BAR-5 (five drains) were 27.75, 43.10, 39.32, 49.74 and 37.56 mg/l respectively.

Table RWM-29 shows the monthly average values (Mean ± SD) of river water and the results of Analysis of variance. The T-DDT content varied from a minimum of '0' to a maximum of 200 ± 10 (May) at WAZ-1, 0 to 114 ± 8.5 (Jun.) at ISB -2, 0 to 152.0 ± 13.0 (Oct.) at NIZ-4 and 0 to 66.0 ± 13.0 (Oct.) at OKH-5. Two way of analysis variance shows a significant difference in T-DDT content between monitoring stations (F_1=70.96, P<0.01) and also between sampling months (F_2=50.06, P<0.05).

The minimum and maximum values of T-DDT at five river monitoring stations were 52.33 ± 21.28 (Monsoon) and 170.5 ± 73.91 (Pre-Monsoon) at WAZ-1, 11.33 ± 25.34 (Monsoon) and 98.0 ± 23.89 (Summer) in ISB-2, 0 and 73.67 ± 24.67 (Monsoon) at ITO-3, 0 and 137.17 ± 193.25 (Pre-Monsoon) at SEN-4, 0 and 365.25 ± 68.04 (Pre-Monsoon) at OKH-5. Peak values of T-DDT were observed during monsoon and post-monsoon season might be due to washout of agricultural fields by rain water. Annual average values of T-DDT were in order of 80.667 > 64.2 > 55.04 > 33.29 > 12.37 at WAZ-1, ISB-2, NIZ-4, ITO-3 and OKH -5 respectively.

T-DDT was observed in higher concentration at Wazirabad barrage which might be due to contamination of river water by pesticides used in agricultural purpose in the basin area of Yamuna river.

Table RWM-29: Monthly Variation of Total DDT (mg/l) in River Water (Year 1994-1995)

Month	*WAZ-1* Mean ± SD	*ISB-2* Mean ± SD	*ITO-3* Mean ± SD	*NIZ-4* Mean ± SD	*OKH-5* Mean ± SD
Jan.	Nil	62.5 ± 9.2	Nil	Nil	Nil
Feb.	Nil	74.5 ± 7.8	Nil	Nil	Nil
Mar.	Nil	70.0 ± 42.4	Nil	Nil	Nil
Apr.	Nil	95.0 ± 7.1	Nil	Nil	Nil
May	200.0 ± 10.0	113.0 ± 9.9	Nil	Nil	Nil
Jun.	165.0 ± 5.0	114.0 ± 8.5	Nil	Nil	Nil
Jul.	40.0 ± 20.0	Nil	61.5 ± 21.5	115.0 ± 25.0	Nil
Aug.	43.0 ± 15.0	Nil	56.5 ± 3.5	113.0 ± 17.0	Nil
Sep.	74.0 ± 5.0	34.0 ± 8.5	103.5 ± 7.0	183.5 ± 14.5	51.0 ± 13.0
Oct.	103.0 ± 7.0	92.0 ± 6.4	152.0 ± 13.0	195.0 ± 15.0	66.0 ± 13.0
Nov.	238.0 ± 42.0	48.5 ± 21.5	26.5 ± 26.5	54.0 ± 54.0	31.5 ± 31.5
Dec.	105.0 ± 5.0	51.0 ± 4.2	Nil	Nil	Nil

Two-way ANOVA Test

Significant difference in DDT content between river monitoring stations, F_1=70.96 (P<0.01).

Significant difference in DDT content between monitoring months, F_2=50.06 (P<0.05).

Endosulfan of Drain and River Water

Endosulfan is a common member of organochlorinated hydrocarbons. There is ample evidence to show that endosulfan like many other chlorinated hydrocarbons used in agricultural programmes, is lethal to a variety of aquatic and terrestrial organisms even in very small doses. Sediment and water samples from twenty six fish ponds of West Bengal were analysed for the detection of pesticides residue by TLC and at all of these ponds DDT, endosulfan, BHC, methyl parathion residues were detected (Joshi, 1987).

Table DWM-30: Monthly Variation in Total Endosulfan (mg/l) of Drain Water (Year 1994-1995)

Month	*NAJ-1*	*CIV-2*	*POW-3*	*SEN-4*	*BAR-5*
	Mean ± SD	*Mean ± SD*	*Mean ± SD*	*Mean ± SD*	*Mean ± SD*
Jan.	280.50 ± 12.50	244.00 ± 34.00	139.50 ± 139.50	3.50 ± 71.50	335.00 ± 25.00
Feb.	254.00 ± 56.00	303.00 ± 7.00	153.00 ± 153.00	5.00 ± 53.00	650.00 ± 170.00
Mar.	320.00 ± 50.00	349.00 ± 51.00	126.50 ± 126.50	3.50 ± 58.50	620.00 ± 40.00
Apr.	445.00 ± 85.00	370.00 ± 10.00	152.00 ± 152.00	4.00 ± 75.50	790.00 ± 30.00
May	521.50 ± 58.00	460.00 ± 20.00	110.00 ± 110.00	0.00 ± 65.50	1095.00 ± 15.00
Jun.	493.00 ± 97.00	490.00 ± 100.00	105.00 ± 105.00	75.00 ± 76.00	1220.00 ± 240.00
Jul.	137.50 ± 7.50	108.00 ± 12.00	70.50 ± 70.50	2.50 ± 36.50	69.50 ± 30.50
Aug.	125.00 ± 5.00	104.50 ± 4.50	60.50 ± 60.50	4.50 ± 39.00	58.00 ± 10.00
Sep.	364.00 ± 116.0	219.00 ± 21.00	92.50 ± 92.50	3.50 ± 71.00	223.00 ± 37.00
Oct.	474.00 ± 94.00	280.00 ± 40.00	112.00 ± 112.00	16.00 ± 77.50	265.00 ± 35.00
Nov.	370.00 ± 50.00	291.50 ± 18.5	105.00 ± 105.00	5.00 ± 84.00	355.00 ± 5.00
Dec.	233.50 ± 23.50	249.50 ± 3.50	109.50 ± 109.50	18.50 ± 78.00	299.00 ± 9.00

Two-way ANOVA Test
Significant difference in endosulfan content between monitoring drains, F_1=28.8 (P<0.01).
Significant difference in endosulfan content between sampling months, F_2=2.16 (P<0.05).

Monthly averages (Mean ± SD) and the results of analysis of variance for endosulfan content (mg/l) in monitoring drains are shown in Table DWM-30. The average values ranged from 125 ± 5 to 521.5 ± 58.50 in NAJ-1, 104.5 ± 4.5 to 490.0 ± 100 in CIV-2, 60.2 ± 4.5 to 153.0 ± 5 in POW-3, 36.5 ± 6.5 to 84.0 ± 2.0 in SEN-4, 58.0 ± 10.0 to 1220 ± 240.0 in BAR-5. Analysis of variance shows the significant difference in T-Endosulfan content between monitoring drains (F_1=26.8, P<0.01) and also between monitoring months (F_2=2.12, P<0.05).

The minimum and maximum total endosulfan values in five drains were 208.83 ± 128.75 (Monsoon) and 422.0 ± 91.5 (Pre-Monsoon) in NAJ-1, 143.83 ± 55.04 (Monsoon) and 285.75 (Pre-Monsoon) in CIV-2, 74.5 ± 1384 (Monsoon) and 137.17 ± 15.32 (Winter) in POW-3, 48.83 ± 16.2 (Monsoon) and 80.75 ± 3.7 (Pre-Monsoon) in SEN-4, 116.83 ± 80.4 (Monsoon) and 931.25 ± 268.0 (Summer) in BAR-5. Annual mean values of total endosulfan of five drains were in order of 498.29 > 334.83 > 289.04 > 111.3 > 65.5 in BAR-5, NAJ-1, CIV-2, POW-3 and SEN-4.

The minimum and maximum total endosulfan values in five drains were 208.83 ± 128.75 (Monsoon) and 422.0 ± 91.5 (Pre-Monsoon) in NAJ-1, 143.83 ± 55.04 (Monsoon) and 285.75 (Pre-Monsoon) in CIV-2, 74.5 ± 1384 (Monsoon) and 137.17 ± 15.32 (Winter) in POW-3, 48.83 ± 16.2 (Monsoon) and 80.75 ± 3.7 (Pre-Monsoon) in SEN-4, 116.83 ± 80.4 (Monsoon) and 931.25 ± 268.0 (Summer) in BAR-5. Annual mean values of total endosulfan of five drains were in order of 498.29 > 334.83 > 289.04 > 111.3 > 65.5 in BAR-5, NAJ-1, CIV-2, POW-3 and SEN-4.

The monthly average values (Mean ± SD) and the results of analysis of variance of total endosulfan are presented in Table RWM-30. Monthly values varied from a minimum of 0 to a maximum of 43.0 ± 5.6 (Sept.) at WAZ-1, 0 to 100.50 ± 27.6 (May) at ISB-2, 0 to 149.0 ± 19.0 (Nov.) at ITO-3, 0 to 239.0 ± 41.0

(Nov.) at NIZ.- 4 and 0 to 421.0 ± 47.0 (Nov.) at OKH-5. Analysis of variance shows a significant difference in endosulfan content (mg/l) between monitoring stations (F_1=52.83, P<0.01). No significant difference was observed between monitoring months (F_2=1.53).

Table RWM-30: Monthly Variation of Total Endosulfan (mg/l) in River Water Year (1994-1995)

Month	*WAZ-1*	*ISB-2*	*ITO-3*	*NIZ-4*	*OKH-5*
	Mean ± SD	*Mean ± SD*	*Mean ± SD*	*Mean ± SD*	*Mean ± SD*
Jan.	0.00 ± 0.00	52.0 ± 5.7	Nil	Nil	Nil
Feb.	0.00 ± 0.00	69.5 ± 37.5	Nil	Nil	Nil
Mar.	0.00 ± 0.00	47.5 ± 10.6	Nil	Nil	Nil
Apr.	10.0 ± 10.0	80.0 ± 28.3	Nil	Nil	Nil
May	27.5 ± 2.5	100.5 ± 27.6	Nil	Nil	32.5 ± 6.5
Jun.	3.5 ± 17.5	89.5 ± 26.2	Nil	Nil	39.0 ± 6.0
Jul.	23.0 ± 7.0	Nil	Nil	64.5 ± 4.5	26.5 ± 26.5
Aug.	31.0 ± 5.0	Nil	Nil	79.5 ± 6.5	0.00 ± 0.00
Sep.	43.0 ± 0.0	20.0 ± 28.3	62.0 ± 6.00	113.0 ± 7.0	169.5 ± 9.5
Oct	34.5 ± 34.5	77.0 ± 29.7	69.5 ± 6.50	147.5 ± 21.5	305.5 ± 19.5
Nov.	22.5 ± 22.5	61.75 ± 24.7	149.0 ± 19.00	239.0 ± 41.0	421.0 ± 47.0
Dec.	Nil	44.0 ± 21.2	100.0 ± 10.00	204.5 ± 41.5	155.0 ± 155.0

Two-way ANOVA Test

Significant difference in endosulfan content between river monitoring stations, F_1=52.83 (P<0.01).
No significant difference in endosulfan content between monitoring months, F_2=1.53.

The minimum and maximum endosulfan values were 0 (Winter) and 32.33 ± 9.6 (Monsoon) at WAZ-1, 6.67 ± 3.8 (Monsoon) and 53.5 ± 8.56 (Pre-Monsoon) at ISB-2, a (Summer) and 109.25 ± 42.21 (Pre-Monsoon) at ITO-3, 0 (Summer) and 193.25 ± 56.26 (Pre-Monsoon) at NIZ-4, 0 (Winter) and 363.25 ± 68 (Pre-Monsoon) at OKH-5. The maximum values of endosulfan were observed during post-monsoon might be due to contamination of river by washing of agricultural lands. Annual average values at endosulfan were in the order of 95.75 > 70.67 > 39.58 > 31.70 > 18.92 at OKH-5, NIZ-4, ISB-2, ITO-3 and WAZ-1 respectively.

Microbial Assessment of River Water

To assess the quality of water in terms of pathogenic and parasite organisms indicating a health risk for the human population, the use of an indicator system is considered to be indispensable, due to the vast variety of these organisms. The coliform bacteria, especially the thermo-tolerant types are considered to be a convenient indicator for the exposure of a water body to the faeces of warm blooded organisms, including man. Since human water-borne diseases are mainly transferred by exposure to faecal matter, the health risk can be assessed by measuring the presence of these bacteria (D.de Zwart and Trivedy, 1994).

Total Coliforms

The coliforms bacteria are discharged from the human intestine and their presence indicates the possibility of the presence of phathogenic organisms. The coliform group comprises all of the facultative and aerobic gram-negative, non-spore forming rod shaped bacteria. The coliform bacteria include the genera *Escherichia, Citrobacter, Enterobacter* and *Klebsiella*. The *Escherichia coli* are entirely of human origin but their exclusive estimation is difficult and hence the entire coliforms are used as indicator. (Trivedy and Goel, 1984). The membrane filtration (MF) method which gives a direct count of total coliforms was used in river water monitoring.

Monthly average number of total coliforms (Mean ± SD) in river water and the results of analysis of variance are presented in Table RWM-31. Average total coliforms at five river monitoring points varied from a minimum of 613 ± 147 (Dec.) to a maximum of 17267 ± 20318 (Jul.) at WAZ-1, 43633 ± 114398 (May) to 405333 ± 188378 (Mar.) at ISB-2, 169667 ± 107370 (May) to 396000 ± 78254 (Feb.) at ITO-3, 163667 ± 103384 (Jun.) to 377667 ± 69264 (Jan.) at NIZ-4, 223333 ± 33993 (April) to 392667 ± 8994 (Jan.) at OKH-5. Analysis of variance shows a significant difference in total coliforms between river monitoring stations (F_1=20, P<0.01) as well as between monitoring months (F_2=2.07, P<0.05).

Seasonal and annual means of total coliforms varied from 895.0 ± 391.0 (Summer) to 12611.0 ± 15504.6 (Monsoon) at WAZ-1, 178291.7 ± 152604.3 (Summer) to 333666.7 ± 27624 (Winter) at ISB-2, 219500.0 ± 99181.2 (Summer) to 346222.2 ± 74011.1 (Winter) at ITO-3, 235000 ± 64172.9 (Summer) to 353444.4 ± 73798.7 (Winter) at NIZ-4, 279333.3 ± 74402.81 (Monsoon) to 339888.9 ± 56396 (Summer) at OKH-5.

Table RWM-31: Monthly Variation of Total Coliforms (CFU/100 ml) in River Water (Year 1994-1995)

Month	*WAZ-1* *Mean ± SD*	*ISB-2* *Mean ± SD*	*ITO-3* *Mean ± SD*	*NIZ-4* *Mean ± SD*	*OKH-5* *Mean ± SD*
Jan.	1200 ± 668	355000 ± 71447	362667 ± 37959	377667 ± 69264	392667 ± 8994
Feb.	933 ± 189	332667 ± 130956	396000 ± 68254	344333 ± 10209	348667 ± 42468
Mar.	1033 ± 544	405333 ± 188370	318667 ± 29273	345000 ± 40208	316667 ± 41899
Apr.	1200 ± 216	215333 ± 2625	219333 ± 28767	256667 ± 9428	223333 ± 33993
May	733 ± 94	43633 ± 114398	169667 ± 107370	174667 ± 102818	326667 ± 46428
Jun.	613 ± 211	48867 ± 131671	170333 ± 106772	163667 ± 103384	322200 ± 9356
Jul.	17267 ± 20318	255000 ± 63068	346667 ± 18856	312000 ± 55450	255000 ± 49497
Aug.	14200 ± 15416	282667 ± 30070	282333 ± 16049	336667 ± 41899	231667 ± 70985
Sep.	6367 ± 2745	279333 ± 58545	309333 ± 98256	315667 ± 56073	351333 ± 32714
Oct.	2100 ± 1042	306000 ± 40343	319667 ± 109393	320000 ± 112250	366000 ± 80862
Nov.	900 ± 82	275000 ± 55126	325000 ± 21213	313333 ± 47140	313333 ± 91094
Dec.	613 ± 147	323333 ± 43205	303333 ± 78457	275000 ± 109621	255000 ± 39370

Two-way ANOVA Test

Significant difference in number of total coliform between monitoring drains, F_1=20.00 (P<0.01).
Significant difference in number of total coliform between sampling months, F_2=2.06 (P<0.05).

The maximum number total coliforms were detected during monsoon season at Wazirabad which might be due to wash-out of faecal matter and over flowing of sewage drains by rain water. The peak coliforms numbers were observed during summer season at all down stream stations of Wazirabad might be due to lean discharge of river water from Wazirabad barrage and higher concentration of drain effluents discharged into the river Yamuna.

Annual mean of total coliforms observed were in the order of 308527 > 294555 > 293583 > 260180 > 3930 at OKH-5, ITO-3, NIZ-4, ISB-2, WAZ-1 respectively.

According to the "Designated Best-Use Criteria (CPC8-1978)" the limit of number of total coliforms is 500/100 ml for "Class-B" type water. From the monitoring data it was observed that, in all river monitoring points the total coliforms numbers were more than the required limit.

Faecal Coliforms

Monthly average faecal coliforms (Mean ± SD) in river water and the results of analysis of variance are depicted in Table RWM-32. Monthly average faecal coliforms number ranged from 93 ± 90 (May) to 15000 ± 17678 (Jul.) at WAZ-1, 27433 ± 1819 (Jun.) to 280000 ± 16330 (Jan.) at ISB-2, 64967 ± 29973 (Jun.) to 304000 ± 41215 (Jan.) at ITO-3, 6000 ± 22760 (Jan.) to 308333 ± 76630 (Jan.) at NIZ-4, 113333 ± 61283 (Jun.) to 343333 ± 32998 (Jan.) at OKH-5. Analysis of variance shows a significant difference in faecal coliforms between river monitoring stations (F_1=20.94, P<0.01). However, no significance difference was observed between sampling months (F_2=2.06).

Seasonal minimum and maximum values of five river monitoring points were 334 ± 164 (Summer) and 9889 ± 13245 (Monsoon) at WAZ-1, 100233 ± 81010 (Summer) and 270889 ± 20295 (Winter) at ISB-2, 133500 ± 73307 (Summer) and 2890000 ± 76262 (Winter) at NIZ-4, 150167 ± 67133 (Summer) and 291667 ± 85716 (Winter) at OKH-5. Annual mean numbers were in the order of 223417 > 221750 > 213297 > 199494 > 2818 at NIZ-4, OKH-5, ITO-3, ISB-2, WAZ-1 respectively.

Table RWM-32: Monthly Variation of Faecal Coliforms (CFU/100 ml) in River Water (Year 1994-1995)

Month	*WAZ-1*	*ISB-2*	*ITO-3*	*NIZ-4*	*OKH-5*
	Mean ± SD	*Mean ± SD*	*Mean ± SD*	*Mean ± SD*	*Mean ± SD*
Jan.	627 ± 339	280000 ± 16330	304000 ± 41215	308333 ± 76630	343333 ± 32998
Feb.	433 ± 94	274333 ± 19431	299000 ± 61822	272667 ± 37250	220000 ± 142361
Mar.	400 ± 216	217667 ± 12815	266333 ± 41153	226000 ± 25140	150667 ± 93243
Apr.	433 ± 125	127000 ± 34322	159667 ± 16337	175000 ± 7071	166667 ± 9428
May	293 ± 90	28833 ± 850	69267 ± 30279	72000 ± 31284	170000 ± 58878
Jun.	210 ± 83	27433 ± 1819	64967 ± 29973	61000 ± 22760	113333 ± 61283
Jul.	15000 ± 17678	211667 ± 41700	211667 ± 34238	246667 ± 46428	213333 ± 59067
Aug.	10900 ± 12092	248333 ± 13123	241667 ± 20138	280000 ± 40825	209333 ± 65429
Sep.	3767 ± 1725	240333 ± 42664	268333 ± 84097	273333 ± 38586	282667 ± 19345
Oct.	1150 ± 618	251667 ± 45522	213333 ± 23570	273333 ± 65997	326667 ± 67987
Nov.	367 ± 170	238333 ± 39651	176667 ± 54365	236667 ± 52493	256667 ± 87305
Dec.	243 ± 154	248333 ± 10274	284667 ± 58909	256000 ± 86595	208333 ± 44033

Two-way ANOVA Test

Significant difference in number of faecal coliform between monitoring drains, F_1=20.94 (P<0.01).

Significant difference in number of faecal coliform between sampling months, F_2=2.06 (P<0.05).

Water Quality Index

Water Quality Index (WQI) may be defined as a rating reflecting the composite influence on the overall quality of a number of quality characteristics or water quality parameter (Tiwari and Ali,1988).

It reflects the collective influence of various physico-chemical and biological criteria of water (Sahu *et al.*, 1991). Hence WQI may be used as one of the most effective ways to communicate information on water quality trends to the policy makers and general public (Sinha *et al.*,1990a). Although WQI is not absolutely objective, it is more efficient method of monitoring the trends in water quality than those based on existing classifications. Therefore Water Quality Index (WQI) enables reduction of large amount of data to a single reproduction index, that provides distinct managerial and operational advantages (Patnaik and Mohapatra, 1994). Basically and essentially the WQI concept is meant for use in the management of natural streams. A number of attempts have been made to develop Combined Water Quality Index (CWQI) for municipal sewage, but they have met only with limited success. That is the reason why in the present work, all the river monitoring stations of the Yamuna river at Delhi stretch are calculated, season-wise as well as year-wise (from year 1993-95).

Considerable efforts have also been made in the few last decade to develop comprehensive mathematical management of water quality in river basin. The work on water quality modelling for the development of Water Quality Index (WQI) has been undertaken by various investigators. Horton (1965) reported an index number system for rating water quality and Harkins (1974) published a paper with objective study of Water Quality Index.

Due to the manifold uses of water, the quality of water required for each use differs significantly from others. This creates problem for formulation of WQI, because high value of a parameter may be desirable for some purpose whereas it may be indifferent or even detrimental for another. For example, a high value of DO is essential for aquatic life, but it is only of marginal value in drinking water supply and highly undesirable in boiled feed water. This situation poses a number of question like whether DO should be included as a parameter in formulating water quality index and, if included, should it affect the WQI positively or negatively and what weight should be assigned to it? Deninger and Maciunas (1971) suggested that the WQI developed with a specific use orientation does not seem to rate water quality levels in a manner markedly different from the rating made by a general non-specific use of oriented index. Thus water of a certain quality retains that relative water quality rating regardless of use for which it is being considered. Instead of developing a number of indices for various uses of water, it appears to be more meaningful to further develop and refine a sensitive and general water quality index. On the other hand, if one is faced with the task of judging the sensitivity of an index, the knowledge of what use the water is to be put to make the work of determining the relevant parameters much easier.

Water Quality Parameters

The most basic use of water is for drinking and personal hygiene. But the Central Pollution Control Board (1978-79) has classified the river and coastal water bodies into five major classes (*i.e.* A to E) based on Designated- Best-Use criteria (Appendix-I). Based on this classification the desired water quality of the river Yamuna at Delhi stretch is classified as "Class-B" (*i.e.* outdoor bathing, swimming and recreation). So for the present study the use of river water for outdoor bathing, swimming and recreation purpose has been considered.

The following water quality parameters have been selected for calculating the water quality index:

1. pH, 2. Dissolved Oxygen, 3. Biochemical Oxygen Demand, 4. Total Coliforms, 5. Total Dissolved Solids, 6. Total Suspended Solids, 7. Ammonical-Nitrogen.

The first four water quality parameters have been selected from CPCB Designated-Best-Use Standards for out-door bathing (Class-B) and last three parameters from NIEM (Network for Industrial Environmental Management, 1989).

Standard and Weightage

The "Standards" (permissible values of various pollutants) for "Class-B" (outdoor bathing, organised) are shown in the Table W-1.

Table W-I: CPCB Designed-Best-Use Standards for "Class-B" and Unit Weights of Water Quality Parameters

Sl.No.	*Water Quality Parameters*	*Standard*	*Unit weight*
1.	pH	6.5–8.5	0.0040
2.	DO	5 mg/l	0.200
3.	BOD	4 mg/l	0.333
4.	MPN/100 ml	500	0.002
5.	TDS	500 mg/l	0.0002
6.	TSS	100 mg/l	0.01
7.	NH_3-N	1.24 mg/l	0.08

It is obvious that the more harmful a given pollutant is the smaller in its permissible value or standard (Vs) recommended for out-door bathing "Class-B". So the proportional to the recommended standards for the corresponding parameters,

i.e. $Wi = K/Si$ (WQ. 1.1)

where,

Wi: Unit weight for the 'I' th parameters Pi (l= 1.2, ... , N) and

K: Constant of proportionality and

Si: Standard for 'I' th parameter.

For the sake of simplicity we assume that K=1. For pH, we have assumed the same unit weight as that for chloride in both cases.

Calculation of WQI

The WQI, reproduced into a formula for calculating the index involves a series of judgments which ultimately will determine the form and utility of the index. For that out of six different methods mentioned below (Botten *et al.*, 1978) of combined water quality rating curves and associated weightings, one method may be selected by the parameters are to be combined.

1. Unweighted Arithmetic Index
2. Weighted Arithmetic Index
3. Unweighted Solway Index
4. Weighted Solway Index
5. Unweighted Geometric Index and
6. Weighted Geometric Index.

For the present work, weighted arithmetic index has been chosen for calculation of WQIs.

Calculation of water quality index consists of two fundamental steps (Ott, 1978),

1. calculation of the quality ratings (or sub indices) for each the water quality parameters used in the index and
2. aggregation of these sub-indices into the overall index.

The present study involves a very simple method (Tiwari *et al.*, 1986) for calculating the water quality index following the two stages mentioned above.

Let there be N water quality parameters, Pi (I = 1, 2, ..., N) to be taken into account for calculating the WQI. Then the quality rating (or sub index) qi corresponding to the ith parameter Pi is a number reflecting the relative value of this parameter in the polluted water with respect to its "standard" or permissible value. One of the simplest relations to calculate qi is

$$qi = 100\,[(Vo - Vio)/(Vs - Vio)] \qquad \text{(WQ. 1.2)}$$

where,

Vo = The measured value of the 'I'th parameter in the polluted water
Vio = The ideal value of this parameters in pure water
Vs = The standard for the "I" th parameter.

Since in general the ideal Vio = O for the outdoor bathing "Class- B", for most parameters *e.g.* WQ.1.2 assumes the simple form.

$$qi = 100\,(Vo/Vs) \qquad \text{(WQ. 1.3)}$$

Equations WQ. 1.2 and WQ. 1.3 ensure that qi = O, if the 'I'th parameter is totally absent in the polluted water qi = 100, if the amount of this parameters is just equal to its permissible value Si for the drinking water except the following 2 exceptions.

1. For pH the ideal value is 7.0 (for natural water) for "Class-B" (out door bathing purpose) and the permissible value ranges from 6.5 to 8.5 so *e.g.* WQ. 1.2 for this case may be written as

$$qpH = 100\,(V_{pH} – 6.5)/(8.5–6.5) \qquad \text{(WQ. 1.4)}$$

where,

V_{pH} is the observed value of pH and the sign of difference (–) indicates the numerical value of the difference, ignoring its algebric sign.

2. For DO, the ideal value may be taken as 14.6 mg/l (the solubility of oxygen in pure water at 0°C). Since the standard for "Class-B" Water quality is 5.0 mg/l *e.g.* WQ.1.2 reduces to

$$q_{DO} = 100\,[(V_{DO} - 14.6)/(5-14.6)] \qquad \text{(WQ. 1.5)}$$

Where V_{DO} is the observed value of DO.

The overall water quality index (WQI) may be calculated by aggregating the quality ratings (or sub indices) qi linearly. Thus it may be written.

$$WQI = \frac{\left(\sum_{i=1}^{n} q_i w_i\right)}{\left(\sum_{i=1}^{n} w_i\right)} \qquad \text{(WQ. 1.6)}$$

Table W-2: Sample Calculation of Water Quality Index (WQI) for River Yamuna at Wazirabad Barrage Up Stream Water During Winter Season of Year 1995 for "Class-B" Water Quality

Parameters (Pi)	*Observed Value (Vo)*	*Unit (Weight) (Wi)*	*Quality Rating (qi)*	*(q_iW_i)*
pH	8.13	0.004	75.33	0.301
DO	9.13	0.200	56.98	11.396
BOD	3.43	0.333	114.33	38.07
TDS	3.50	0.002	70	0.14
TSS	38.67	0.01	38.67	0.387
NH_3-N	1.02	0.80	82.25	65.8
MPN/100 ml	513	0.002	100.5	0.0201

*All values except pH are in mg/l

$$WQI = \frac{\Sigma q_i W_i}{\Sigma W_i} = \frac{116.114}{1.3492} = 86.06$$

Results and Discussion

A sample calculation of WQI using equations WQ. 1.3 to WQ. 1.6 for all river monitoring stations for (Class-B) bathing, swimming and recreation purpose during winter season of year 1965 is shown in the Table WQ. 3. Similarly WQIs for all seasons as well as annual WQIs from year 1993 to 1995 have been calculated and presented in Table WQ-1 and Fig. WQI-1.

The seasonal WQIs shows a minimum of 61.73 and a maximum of 72.89 in the year 1993, 63.21 and 99.87 in 1994, 82.19 and 105.20 in 1995, at Wazirabad barrage (up stream water). The annual WQIs of this station were 68.16, 74.75 and 93.85 for year 1993, 1994 and 1995 respectively. This implies that the Wazirabad barrage (reference station) satisfies the criteria for "Class-B" water (bathing swimming and recreation purpose) because all values were well below 100, the permissible limit for "Class-B" (except rainy season during year 1995).

The seasonal WQIs (year 1993-95) ranged from a minimum of 143 to a maximum of 700.18 at ISBT bridge, 171.18 to 627.30 at ITO barrage, 171.24 to 539.36 at Nizamuddin bridge and 156.57 to 554.87 at (Okhla barrage).

Maximum values of WQIs were recorded at all the stations during winter (except Wazirabad). The reason may be that during winter the water discharge from Wazirabad barrage was lean. So that the drain effluents discharged to river Yamuna flowed without dilution at Delhi stretch. The lowest values of WQIs were obtained during the rainy season at all the stations (except Wazirabad barrage) due to the effect of dilution of pollutants by rainwater. The annual WQIs of river monitoring station; Wazirabad barrage, ISBT bridge, ITO barrage, Nizamuddin bridge and Okhla barrage were 78.92, 399.8, 384.6, 384.6, 372.4 and 362.2 respectively.

In the present investigation it was observed that the WQIs of all river monitoring stations (except Wazirabad barrage) were well above the permissible limit through out the year. The above finding clearly shows that the river water at Delhi stretch (except Wazirabad) is totally unfit for bathing, swimming and recreation purpose (Class-B).

Water quality models proved to be powerful tools in water resources management as they can incorporate the complexity of the relevant processes in the water body into a utilitarian form for management consideration. Through the use of model both diagnostic and predictive capability is provided, diagnostic in the sense of permitting the identification and isolation of specific factors affecting the water quality and predictive in permitting evaluation of future effects of proposed changes in the water body, its inputs or its uses (Singh and Anandh, 1996).

Thus, the water quality index itself does not give the final answer to the problem of water quality management. It can only represent the standard of quality of water and can point out where the trouble is and what utmost action can be taken for the largest improvement in the quality of a given body of water (Mishra, 1999).

So far as the criteria for "Class-B" (bathing, swimming and recreation) for Yamuna river at Delhi stretch are concerned it is totally unsuitable for bathing, swimming and recreation purpose. The WQIs values were so high that the water quality of Yamuna river from down stream of Wazirabad barrage is not even fit for "Class-D" (propagation of wildlife and fisheries).

Acknowledgements

PKB is grateful to authorities of Central Pollution Control Board, New Delhi and Sambalpur University for providing facilities to undertake the work.

References

Agarwal, V.K., Raj, K.P.S., Panchal, D.I., Balaraman, R., 1978. Heavy metal content in the water and sediments of Sursagar lake of Baroda (Gujarat). *Intern. J. Environment Studies*, 12: 245-250.

Ajmal, A. and Razi-Ud-Din (1988. Studies on the pollution of Hindon river and Kali Nadi (India. In: *Ecology and pollution of Indian rivers* (Ed. Trivedy, R. K.), Ashish Publishing House, New Delhi, pp. 87-112.

Ajmal, M., Razi-Ud-Din and Khan, A., 1985. Physicochemical aspects of pollution in Kali Nadi. *IAWPC Tech. Annual*, 12: 106-114.

Ali, M. and T.N., 1988a. Multiple regression techniques for rapid monitoring of water quality, *J. IWWA.* 2: 237-240.

APHA (1991. *Standard methods for the examination of water and wastewater, 18th Ed.*, APHA-AWWA-WPCF, American Publis. Health Association, Washington, D.C.

Bagde, U.S. and Verma, A.K., 1985. Limnological studies of J.N.U. Lake, New Delhi. Proc. Nat Symp. Pure and Applied Limnology (Ed. Adoni, A. D.. *Bull. Bot. Soc.* Sagar, 32: 16-23.

Banat, K., Forstner, U. and Muller, G., 1992. Schwermetall-annereich-Erungen in den sedimentn wichtiger Flusse in Bereich der Bundesrepublik Deutschland, line Bestandsaufnahme. *Internal rep.* Lab., Sedimentsforsch, Heidelderg.

Baruah, A.K., Sharma, R.N. and Borah, G.C., 1993. Impact of sugar mill and distillery effluents on water quality of river Gelabil, Assam. *Indian J. Environ. Hlth.*, 35: 288-293.

Bhargava, D.S., 1985. Water quality variation and control technology of Yamuna river. *Env. Pollut.*, 37 (Series B): 355-376.

Bhatnagar, G.P. and Sharma, G.P., 1975. Physico-chemical features of sewage polluted lower lake, Bhopal, Environ-agents and their biological effect. *Pro of the Inter. Symp. Int.* 5(4): 212-223.

Bilgrami, K.S. and Datta Munshi, J.S., 1985. *Ecology of river Ganges: Impact of human activities and conversation of aquatic biota* (Patna to Farakka), Final Technical Report submitted to D.O. En, Allied Press, Bhagalpur (India), p. 97.

Billore, D.K., 1981. *Ecological studies on Pichhola lake, Udaipur*. Ph.D. Thesis, Dept. of Botany, School of Basic Sciences and Humanity, University of Udaipur, Udaipur.

Brajnandan, P., Jatly, Y.C. and Singh, A. K., 1985. Periodicity and interrelationship of physico-chemical. factors in ponds. Proc. Nat. Symp. Pure and Appl. Limnology (Ed. Adani, A. D.) *Bull. Bat. Soc.*, Sagar, 32: 1-11.

Campbell, J.E., Richardson, L.A. and Schafer, M.L., 1965. Insecticide residues in the human diet. *Arch. Environ. Hlth.*, 10: 831.

Chatterjee, C.C., 1984. *Human physiology*. Medical Allied Agency, Calcutta.

Cole, G.A., 1979. *A text book of limnology, 2nd Ed.* C.V. Mosby Company, Site Louis, p. 426.

CPCB (1993-94. *Pollution statistics of Delhi*. Issued by Public Relation Cell for mass awareness, Central Pollution Control Board, Delhi.

D.de.Zwart and Trivedi, R.C., 1994. *Manual on integrated water quality evaluation*. Report No. 802023003, Joint Publication of RIVM and Central Pollution Control Board, East Arjun Nagar, Delhi, India.

Datar, M.D. and Vashishtha, R.P., 1990. Investigation of heavy metals in water and silt sediments of Betwa river, *IJEP*, 10: 666-672.

Datar, M.D., Jain, K.C. and Vashishta, R.P., 1990. Distribution of heavy metals in Betwa river water. In: Trivedy, R.K., Ed.) *River Pollution in India*, Ashish Publishing House, New Delhi (India), 179-187.

Datta Munshi, J. and Datta Munshi, J.S., 1995. *Fundamentals of freshwater biology*, Narendra Publishing House, New Delhi, p. 222.

Deniniger, R.A. and Maciunas J.J., 1971. *A water quality index for public water supplies*. Department of Environmental and Industrial Health, University of Michigan, Ann Arber Michigan (U.S.A).

Dikshith, T.S.S., Raizada, R.B., Kumar, S.N., Srivastava, M.K., Kulshrestha, S.K. and Adholia, U.N., 1990. Residues of DDT and HCH in major sources of drinking water in Bhopal, India. *Bull. Environ. Contam. Toxicol.*, 45: 389-393.

Durham, W.F., 1965. *Effects of pesticides on man research in pesticides*. Academic Press Inc., N.Y., p. 63.

Ellis, N.M., Westfalk, B.A. and Ellis, M.D., 1946. Determination of water quality for fish and wildlife service. *U.S. Interior Res. Rept.*, 9: 121-122.

Forstner, U., 1977. *Metal concentrations in metal pollution in freshwater sediments natural background and cultural effects, Interactions between sediments and freshwater*. Goltorman, H.L. (Ed.) Washington/The Hague, Pudoc/Junk, B.V. Pub., pp. 94-103.

Forstner, U. and Wittman, G.T.W., 1981. *Metal pollution in the aquatic environment*, Springer-Verlag, Berlin.

Ghose, N.C. and Sharma, C. B., 1989. *Pollution of Ganga river*. Ashish Publishing House, New Delhi.

Golterman, H.L., 1975. Physicological Limnology. Elsevier Scientific Publ. Co., N.Y. p. 489. *Indian J. Environ. Hlth.*, 21: 131-142.

Harkins, R.D., 1974. An objective water quality index. *J. Water Poll. Cont. Fed.*, 3: 589-591.

Harrison, A.D., 1963. Hydrobiological studies on the Vall river in the Verrenigung area, Part II, The chemistry, bacteriology and invertebrates of the bottom muds. *Hydrobiologia*, 21: 66-89.

Harrison, R.M. and Laxen, D.P.H., 1984. *Lead in atmosphere, in lead pollution: causes and control*. Chapman and Hall Ltd., London, p. 10.

Hegde, G.R., Bengale, U.D. and Bharati, S.G., 1985. Diurnal variation of physico-chemical factors and phytoplankton in a pond and a lake near Dharwad, Karnataka State, India. *Comp. Physiol. Ecol.*, 10: 245-250.

Horton, R.K., 1965. An index-number system for rating water quality. *J. Water Poll. Cont. Fed.*, 3: 300-305.

Hutchinson, G.E., 1957. *A treatise on limnology, Vol. 1: Geography, Physics and Chemistry*. Wiley, N.Y., p. 1015.

Hydn, H.S., 1961. Distribution of uranium and other metals in crude oils U.S., *Goel. Surv. Bull.*, 1100-3: 72-89.

IARC (1974. Monograph on evaluation of carcinogenic risk of chemicals to man some organochlorine pesticide pollution. *Baioteku (Biotech)*, 3: 589.

Imevbore, A.M.A (1978. The chemistry of river Niger in the Kainj reservoir Area. *Arch. Hydrobiol.*, 67: 412-431.

Ingole, S.A. and Dhaktode, S.S., 1990. Pollution studies of Amba river estuary: A comparative study. In: Trivedy, R.K. (Ed.), *River Pollution in India*, Ashish Publishing House, New Delhi (India), pp. 237-253.

Israili, A.W., 1991. Occurrence of heavy metals in Ganga river water and sediments of Western Uttar Pradesh. *Poll. Res.*, 10: 103-109.

Israili, A.W. and Ahmad, M.S., 1993. Chemical characteristics of river Yamuna from Dehradun to Agra. *Indian J. Environ. Hlth.*, 35: 199-204.

Jackson, M.L., 1964. *Chemistry of the soil*. Bear, F.E. (Ed.), 2nd Ed., Reinhold, New York, Chapter 2.

Jain, S. and Salman, S., 1995. Heavy metal concentration in a highly eutrophic lake sediments and overlying water. *Poll. Res.*, 14: 471-476.

Jenne, E.A., 1976. *Trace elements sorption by sediments and soil sites and processes*. Symposium on molybdenum, Marcel Dekker, New York, 2: 425-553.

Joshi, H.C., 1985. Pesticides residue monitoring in the Bhagirath Hoogly stretch of Ganga river system. *Proc. Nat. and Sen. Poll. Control of Env. Manage.*, NEERI Nagpur W.O., 123-124.

Joshi, K.M., 1982. Heavy metal pollution in India status and impact. *Proceeding of the seminar on Status and impact of heavy metal pollution in India*. Centre for Environ. Studies, Anna Univ., Madras.

Juday, C., Birge, E.A., Kemmerer, J.J. and Robinson, R.J., 1928. Phosphorous content of lake waters of Northern Wisconsin. Trans. Wisconsin, *Acad. Sci. Arts and Lett.*, 23: 233-248.

Katariya, H.C., 1994. An evaluation of water quality of Kaliasot river. *IJEP*, 14: 690-694.

Khan, S. and Khan, N.N., 1983. Influence of lead and cadmium on the growth and nutrient concentration of tomoto (*Lycopersicum esculentum*) and egg-plant (*Solanum melongena*). *Plant and Soil*, 74: 387-394.

Kothandaraman, H. and Viswanathan, S., 1986. Coocum, a perspective study, in river Coocum, Let it be a resource, Univ. of Madras, 4: 1-4.10.

Kraybill, H.F., 1969. Significance of pesticide residues in foods relations to total environmental stress. *Canad. Med. Ass. J.* 100: 204.

Krishnamurthy, S.R. and Bharati, S.G.(1995. Evaluation of water pollution in the river Kali near Dandeli (North Kanara district), Karnataka State, India. *Poll. Res.*, 14: 93-98.

Kudesia, V.P., 1989. Physico-chemical properties of Ganga river at Farrukhabad. *Indian J. Environ. Agric.*, 3: 81-82.

Kulshrestha, S.K., 1986-89. *Monitoring and surveillance of residues of organochlorine pesticides and heavy metals in fishes from selected environmental components of Madhya Pradesh and Rajasthan*. Technical Report (19/85/86/RE) Motilal Vigyan Mahavidyalaya, Bhopal.

Kumar, A., 1994. Studies on pollution in the Mayurakshi in south Bihar. *J. Environ. Pollut.*, 2(1): 21-26.

Kumar, A., 1994a. The effect of sewage pollution on the ecology of river Mayurakshi in Bihar, Abs. No. 40. Symp. on Management of ecosystem and sustainable utilization of biological resources at Vellyani (Trivendrum), 7- 9 Sept. 1994.

Kumar, A., 1995. Studies on pollution in the river Mayurakshi in south Bihar. *J. Environ. Pollut.*, 2(3): 137-141.

Kumar, A., 1997. *Limnobiotic studies of River Mayurakshi in Santal Praganas, Bihar*. A Ph.D. thesis submitted to S.K. University, Dumka (Bihar).

Kumar, A. and Sharma, R.K., 1999. Physico-chemical characteristics of river Sengar in district-Etawah (U.P.. *Ind. J. of Environment. Sci.* 3(2): 201-203.

Lal, P.C. and Bhattacharya, K.G., 1990. Trace metal status of the Bharalu river. *IJEP*, 10: 747-751.

Malviya, S., Yadav, M. and Palharya, J.P., 1990. Aspect of water pollution in the river Narmada at Hoshangabad (MP). In: Trivedy, R.K. (Ed.), *River Pollution in India*, Ashish Publishing House, New Delhi (India), pp. 59-70.

Manivaskam, N., 1989. *Physico-chemical examination of water, sewage and industrial effluents*. Pragati Prakashan, Meerut, p. 380.

Mathur, A., Sharma, Y.C., Rupainwar, D.C., Murthi, R.C. and Chandra, (Mrs.) S., 1987. A study of river Ganga at Varanasi with special emphasis on heavy metal pollution. *Poll. Res.*, 6: 37-44.

Mehrotra, K.N., 1985. Use of DDT and BHC and its environmental effects in India, *Proc. Indian North Sci. Acad.*, 51(5): 581-595.

Metelev, V.V., Kanev, A.I. and Dzasokhova, N.G., 1971. *Vodnaya Toksikologya*. Kolos Publishers, Moscow.

Michael, P., 1984. *Ecological methods for field and laboratory Investigations*. Tata MaGraw Hill Publishing Co. Ltd., New Delhi (India).

Mishra, S., 1999. *Quantification of heavy metals in Brahmani river ecosystem at Rourkela and its modelling*. A Ph.D. thesis submitted to Sambalpur University (Orissa).

Mishra, S.R. and Saxena, D.N., 1991. Pollution ecology with reference to physico-chemical characteristics of Morar (Kalpi) river, Gwalior (MP). In: Shastree, N.K. (Ed.) *Current trends in limnology, Vol. 1*, Narendra Publishing House, Delhi (India), pp. 159-184.

Mitra, A.K., 1982. Chemical characteristics of surface water at selected gauging stations in the river Godavari, Krishna and Tungabhadra. *Indian J. Environ. Hlth.*, 24: 165-179.

Mohan, K.S. and Reddy, N.S., 1987. A study of cations in two freshwater lakes of Hyderabad, India. *Poll. Res.*, 6: 5-8.

Moss, B., 1973. The influence of environmental factors on the distribution of freshwater algae: An experimental study II, Role of pH and the carbon dioxide bicarbonate system. *J. Ecol.*, 61: 157-177.

Munawar, M., 1970. Limnological studies on freshwater ponds of Hyderabad, India. II. *Hydrobiologia*, 31: 101-128, 39(3): 307-320.

Murugesan, A.G., Hameed, A.K.M.S.A. and Sukumaran, N., 1994. Water quality profile of the perennial river Tambraparni. *IJEP*, 14: 567-572.

Nayak, M.S. and Sawant, A.D., 1994. Seasonal variation in heavy metal concentration in water of lakes supplying water to Bombay city. *Journal of Indian Water Works Assoc.*, 14: 241-243.

Ohle, W., 1934. Chemisch and physikalishe unterruch ungienhor dentscher seen. *Arch. Hydrobiol*, 26: 386-464.

Patil, P., 1982. *An ecological study of the algal flora of the Lakes of Bhopal*. Ph. D. thesis, Bhopal University, Bhopal: 310.

Phifer, L.D. and Tompson, T.G., 1937. Seasonal variation in the surface water of San, J. Van. Channel-I during the five year period, January, 31st to December, 30th. 1935. *J. Mar. Res.*, 34-53.

Pophali, S., Siddiqui, S. and Khan, L.H., 1990. Sources and distribution of heavy metals in the abiotic components of a polluted urban stream in Bhopal. *IJEP*, 10: 600-603.

Pradhan, S. and Mohanty, R.K., 1991. Occurrence of microbes in Steel Plant effluents. *IJEP*, 11: 201-205.

Prasad, A.S., 1978. *Trace elements and iron in human metabolism*. Plenum Press, New York, USA.

Provasoli, L., 1958. Nutrition and ecology of protozoans. *Rev. Microbiol.*, 12: 278-308.

Rai, H., 1974. Limnological studies on the river Yamuna at Delhi, India. 1. Relation between the chemistry and the state of pollution in the river Yamuna. *Arch. Hydrobiol*, 73(3): 369-393.

Rajesh, K., 1981. *Some aspects of limnological studies in river Jhelum*. Ph. D. Thesis, Kashmir University, J&K, 420.

Ramaswamy, S.N. and Somashekar, R.K., 1982. Ecological studies on algal of electroplating wastes. *Phykos*, 21: 8-12.

Ramteke, D.S. and Moghe, C.A., 1986. *Manual on water and wastewater analysis*. NEERI, Nagpur (India).

Rao, K.L., 1979. *India's water wealth, its assessment, uses and projections*. Orient, Longman, Delhi, p. 210.

Rao, V.S., 1971. An ecological study of three freshwater ponds of Hyderabad, India. *Hydrobiologia*, 38: 213-223.

Raybak, J.J. and Sikorska, V., 1976. Environment. In: Pieczynska, E. (Ed.) *Selected problems of lakes, Littoral ecology*. Univ. of Warsaw.

Sahu, B.K., Panda, R.B., Sinha, B.K. and Nayak, A., 1991. Water quality index of the river Brahmani at Rourkela industrial complex of Orissa. *J. Ecotoxico. Env. Monit.*, 1: 169-175.

Sarwar, S.G. and Rifat, M., 1991. Abiotic features of a freshwater lotic ecosystem of Kashmir (Doodhganga river. *Poll. Res.*, 10: 69-73.

Screenivasan, A. and Sunderaj, R., 1967. Effects of certain wastes on the water quality fisheries of rivers Cauvery and Bhavani. *Environ. Hlth.*, 9: 13-21.

Shah, A.R., 1988. Physico-chemical aspects of pollution in river Jhelum (Kashmir) during 1981-83. In: Trivedy, R.K. (Ed.) *Ecology and pollution of Indian rivers*, Ashish Publishing House, New Delhi (India), pp. 163-207.

Sharma, R.C., Gausain, O.P. and Juyal, C.P., 1990. Ecology of high altitude river Bhilanga of Garhwal Himalaya. In: Trivedy, R.K. (Ed.) *River Pollution in India*, Ashish Publishing House, New Delhi (India), pp. 11-30.

Chapter 59

Limnology of Lentic Freshwater Systems in North Cachar Hills, Assam, India

T. N. Majumdar[1], *A. Gupta*[2] *and M. Daolagupu*[1]

Department of Zoology, Haflong Government College, Haflong – 788 819, Assam

[1]Department of Ecology and Environmental Science, Assam University, Silchar – 788 011, Assam, India

E-mail: agecol@rediffmail.com

ABSTRACT

A study was carried out on the limnology of five lentic systems, at different altitudes in North Cachar Hills district of Assam. The influence of altitude on temperature is evident from the results obtained. Marked seasonal variations were observed in dissolved oxygen and free carbon dioxide content which also showed variations among the different lentic systems. Total alkalinity and conductivity of water were found to be low. Comparatively higher levels of total alkalinity, conductivity, phosphates and nitrates were recorded in the dry season. A strong correlation between primary productivity and phosphates, nitrates and total alkalinity could be observed. Statistical analysis using ANOVA and Duncan's Long range Test revealed significant difference among the different lentic systems in several physico-chemical properties such as dissolved oxygen, total alkalinity, phosphates, nitrates and conductivity of water.

Keywords: *Assam, Lentic system, North Cachar Hills, Physico-chemical properties.*

Introduction

The North Cachar Hills District is one of the two hill districts of Assam. It lies between *c* 25° to 26° North latitude and *c* 92° to 94° East longitude. The district consists of undulating hills, valleys and small patches of plain lands. The total geographical area of North Cachar Hills is approximately 4888 sq. km. On its eastern side are the states of Nagaland and Manipur, while Cachar district of Assam lies towards its south. On the western side is the state of Meghalaya, while Nowgaon and Karbianglong district of Assam are located towards its north (Figure 59.1). The main mountain range of the district is the Borail spreading from east to west in the southern part of the district.

Although extensive work has been done on the limnology of lentic systems in other parts of India (Singh and Ray, 1995; Tiwari, 1999; Sahat *et al.*, 2001; Mariappan and Vasudevan, 2002), few such investigations have been carried out in the lentic bodies of North-East India (Das *et al.*, 2002; Dutta Gupta *et al.*, 2004) with no published account on those in North Cachar Hills district. This article, therefore, attempts to generate base line information on the limnology of five lentic systems at different altitudes in this district. Besides physico-chemical variables, the vegetation near the systems along with aquatic macrophytes were also studied, since they have a pronounced effect on ecosystem processes (Carpenter and Lodge, 1986) and can be used to monitor the quality of water (Dewanji and Matai, 2000).

Materials and Methods

Study Area

For the present investigation five water bodies were selected at three different places in N.C. Hills District. The Fishery Tank, the Haflong Lake and the Upper Lake are situated in Haflong, (altitude

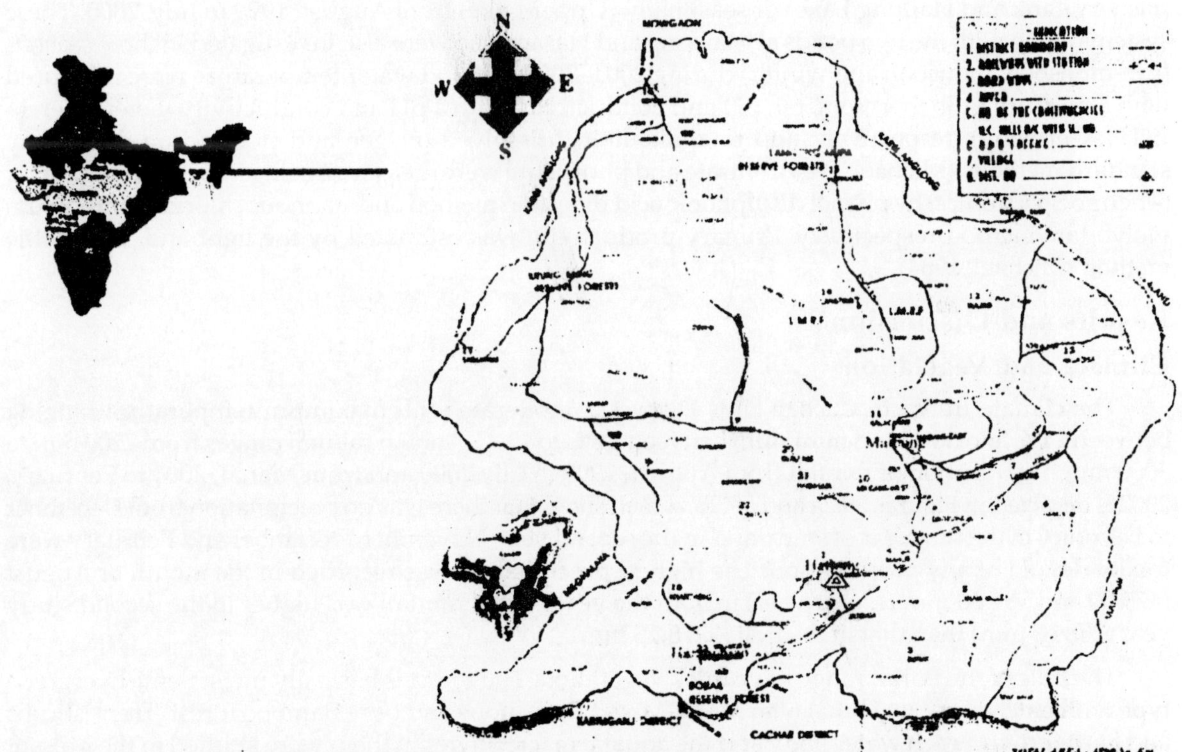

Figure 59.1: Map of N.C. Hills District of Assam, Showing Study Sites

c 935.10 msl.), the headquarter of N.C. Hills District, while a pond each in Maibong (altitude *c* 365,87 msl), a small town in the northern part of the district, and Harangajao (altitude *c* 240.79 msl) in the south, were investigated (Figure 59.1). The Fishery Tank is *c* 250 m long and *c* 50 m wide with an average depth of 2 metres. It is surrounded by dense human habitation, public buildings and roads. The Haflong Lake is an artificial lake *c* 1.0 km long and *c* 135 m wide with an average depth of 2 m. It is situated in a low lying area and receives sewage from a large part of Haflong town. The Upper Lake is located about 100 m above the level of the other two waterbodies in Haflong. It lies between two steep hills and is *c* 250 m long, *c* 30.0 m wide, and *c* 12.0 m deep. Human habitation around the lake is very thinly dispersed. The pond in Maibong, located near the railway station, has a length of *c* 80 m and width and depth of 40 m and 2 m, respectively, and is surrounded by paddy fields and human habitation. The pond in Harangajao is a small waterbody located amidst paddy fields near the Haflong-Silchar Road about 6.0 km. from Harangajao town. It is *c* 45 m long with an average width and depth of 18 m and 2.1 m, respectively.

Climate and Vegetation

Information on the, climate and vegetation of North Cachar Hills District was collected from different secondary sources (Champion and Seth, 1968; Parashar, 1994). Primary data on vegetation in and around the waterbodies were also collected.

Physico-chemical Properties of Water

Physico-chemical properties of water were studied in three lentic system at Haflong either monthly (Fishery Tank and Haflong Lake) or seasonally (Upper Lake) from August, 1999 to July, 2000. These systems along with the two ponds at Maibong and Harangajao were also investigated in three seasons (pre-monsoon, monsoon and winter) during 2001–2002. Air and water temperature were measured with a mercury bulb thermometer, pH and conductivity with a pH and conductivity meter (Merck-WTH, Germany), respectively, and total alkalinity, dissolved oxygen and free carbon dioxide by standard methods (Michael, 1984). Nitrate and phosphate were measured using a spectrophotometer (Thermo Spectronics) by phenol disulphonic acid digestion method and stannous chloride-ammonium molybdate method respectively. Primary productivity was estimated by the light and dark bottle method (Michael, 1984).

Results and Discussion

Climate and Vegetation

The climate in North Cachar Hills District is moderate with maximum temperature ranging between 24°C and 30°C and minimum between 14°C to 19°C. Annual rainfall ranges from 2200 mm to 3000 mm. Distribution of rainfall from August, 1999 to July 2000 and from March, 2001 to February, 2002 is depicted in Figures 59.2a and 59.2b, which show that there was no precipitation from December to February in the first year of study and in the second year, November, December and February were totally devoid of any precipitation. The highest precipitation was recorded in the month of August (475.00 and 569.00 mm respectively) in both the years. Total rainfall was higher in the second study year (2889.3 mm) than that in the first (2278.75 mm).

The hills of the Borail range have thick vegetation. The forest is basically moist Semi-Evergreen type with patches of east Himalayan Moist-Mixed Deciduous and pure Bamboo forest. Trees, shrubs and herbs around each water body and the aquatic macrophytes in them were studied in the present investigation (Table 59.1). Relatively fewer trees species were present in the catchment of Upper Lake

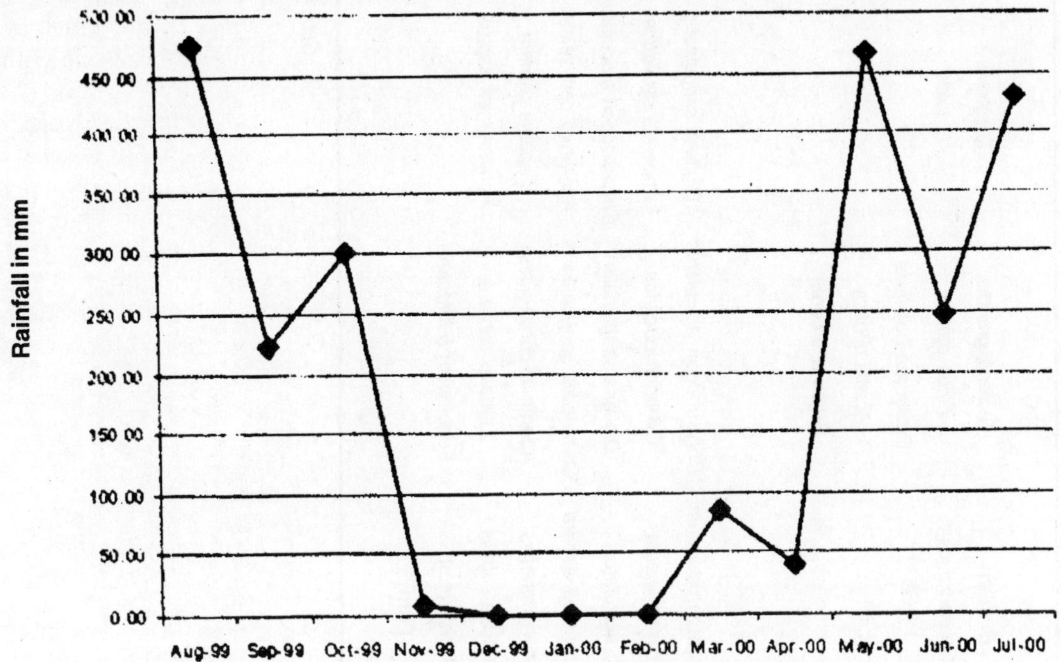

Figure 59.2a: Rainfall in N.C. Hills District from August 1999 to July 2000

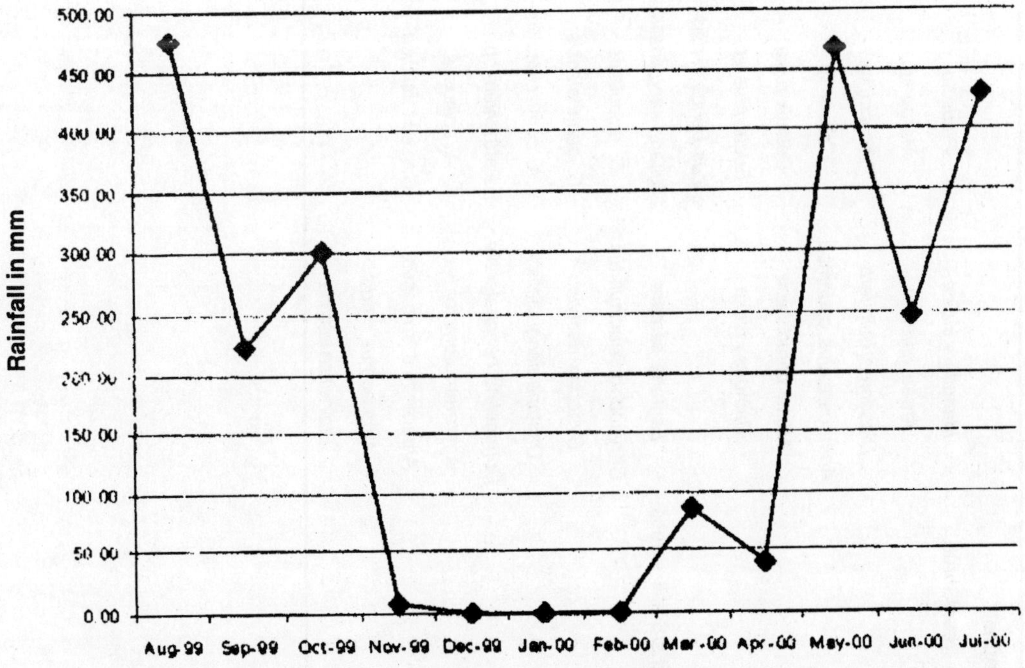

Figure 59.2b: Rainfall in N.C. Hills District from March 2001 to February, 2002

Table 59.1: Vegetation in and Around the Water Bodies of North Cachar Hills, Assam

Types of Vegetation	*Fishery Tank*	*Haflong Lake*	*Upper Lake*	*Maibong Pond*	*Harahgajao Pond*
Marginal					
Trees/Shrub	*Mangifera indica*	*Pinus* sp.	*B. tulda*	*Tectona grandis*	*T. grandis*
	Areea catechu	*Margo* sp.	*M. paradisiaca*	*Ficus carica*	*B. tulda*
	Dioscorea bulbifera	*Acacia* sp.		*Dioscorea bulbifera*	
	Anthocephalus cadamba	*Cassia* sp.		*Acacia nilotica*	
	Musa paradisiaca	*Delonix regia*		*Gmelina arborea*	
	Bambusa tulbla	*Eucalyptus* sp.			
		Ficus benghalensis			
Herbs/Small shrubs	*Lantana camara*	*Digitaria sanguinalis*	*Eupatorium odoratum*	*Chrysopogon aciculatus*	*Digitaria sanguinalis*
	Chrysopogon aciculatus	*C. aciculatus*	*Hedyotis lineata*	*Cynodon dactylon*	*Imperata cylindrica*
	Cynodon dactylon	*C. benghalensis*	*C. aciculatus*	*Digitaria sanguinalis*	*Chrysopogon aciculatus*
	Digitaria sanguinalis	*A. conyzoides*	*Achyranthes aspara*	*Commelina benghalensis*	*Ageratum conyzoides*
	Commelina benghalensis	*Parthenium* sp.	*C. benghatensis*	*Ageratum conyzoides*	*Lantana camara*
	Ageratum conyzoides	*A. spinosus*	*Bambusa subtilis*	*Lantana camara*	*Achyranthes aspara*
	Amaranthus spinosus	*Centella asiatica*	*Blachnum* sp.	*Eupatorium odoratum*	
	Amaranthus viridis		(Fern)		
	Blainvillea latifolia				

Contd....

Table 59.1–Contd...

Types of Vegetation	*Fishery Tank*	*Haflong Lake*	*Upper Lake*	*Maibong Pond*	*Harahgajao Pond*
Aquatic Macrophytes					
Marginal	*Axonopus compressus*	*Axonopus compressus*	*Axonopus compressus*	*Axonopus compressus*	*Axonopus compressus*
	Polygonum perfoliatum	*Sagittaria* sp.	*Alternanthera sessilis*	*Monochoria hastaefolia*	*Sagittaria* sp.
	Polygonum hydropiper	*Cardanthera triflora*	*Amaranthus* sp.	*Sagittaria* sp.	*Amaranthus* sp.
	Ipomea aquatica	*Limnophila heterophylla*	*Polygonum perfoliatum*	*Amaranthus* sp.	*Monochoria hastaefolia*
	Phylanthus nirurii	*Monochoria hastaefolia*	*Polygonum hydropiper*	*Limnophila heterophylla*	*Limnophila heterophylla*
	Amaranthus sp.	*Monochoria vaginalis*			*Ipomea aquatica*
		Amaranthus sp.			
Floating	*Eichhorrnia crassipes*	*Trapa natans*	*Trapa natans*	*Trapa natans*	*Azolla pinnata*
	Azolla pinnata	*Eichhornia crassipes*		*Pistia stratiotes*	*Utricularia* sp.
	Pistia stratiotes	*Pistia stratiotes*		*Azolla pinnata*	*Lemna* sp.
	Utricularia sp.	*Azolla pinnata*		*Lemna* sp.	*Hydrilla* sp.
		Utricularia sp.		*Utricularia* sp.	*Eichhornia crassipes*
		Lemna sp.		*Eichhornia crassipes*	
Submerged	NIL	*Hydrocharis* sp.	*Salvinia* sp.	*Salvinia* sp.	NIL
		Salvinia sp.		*Vallisneria* sp.	
		Vallisneria sp.		*Hydrilla* sp.	
		Hydrilla sp.		*Chara* sp.	
Emergent	NIL	*Nelumbo* sp.	NIL	*Nelumbo* sp.	NIL
		Nymphaea sp.		*Nymphaea* sp.	
		Oryza sp.		*Oryza* sp.	

and Harangajao Pond when compared with those around the other waterbodies studied. Emergent type of aquatic macrophyte was totally absent in Fishery Tank, Upper Lake and Harangajao Pond. Submerged aquatic macrophyte was absent in both Fishery Tank and Harangajao Pond whereas in Upper Lake only one species of submerged (*Salvinia* sp.) and one species of floating macrophyte (*Trapa natans.*) were found covering a very small area of the waterbody. Several species of marginal macrophytes were present in all the lentic systems, studied.

Physico-chemical Properties of Water

The physico-chemical properties of water in differet lentic systems studied are depicted in Tables 59.2 and 59.3. The influence of altitude on temperature is evidenced by the lowest temperature regime in Upper Lake, Haflong, and the highest in Harangajao. It is a well-known fact that no other single factor has so profound influence on aquatic ecosystem than temperature (Welch, 1952). Hydrogen ion concentration (pH) is another important environmental factor influencing the aquatic ecosystem (Welch, 1952). In the present investigation, the Fishery Tank at Haflong showed higher pH values (6.5 to 8.35) than the other systems during the first annual cycle (1999–2000). This probably reflects the effect of periodic liming and manuring of the tank by the Department of Fisheries during that period. In the absence of liming during 2001–2002, the pH regime became much lower (6.5–7.5). Lentic systems located at a lower altitude (< 100 msl) in both Brahmaputra and Barak valleys in Assam exhibited a somewhat higher pH range (Das and Baruah, 2002; Dutta Gupta *et al.*, 2004). In contrast, the pH values recorded in certain lentic systems of Shillong, Meghalaya, which are located at a higher altitude (1495–1900 msl) than those in the present study, were somewhat lower (Gupta *et al.*, 1992). Thus the lentic systems of North Cachar Hills, located at an intermediate altitude, exhibit a pH regime lower than that in the valleys, but higher than those at a higher altitude. Conductivity was found to range from 17.6–78.1 $\mu S\ cm^{-1}$, which is lower than those reported from the other regions of India (Habib *et al.*, 1997). The entry of nutrient-rich sewage into Haflong Lake and liming/ manuring during the first annual cycle in Fishery Tank is reflected in the relatively high conductivity in these two systems. The conductivity in the latter system declined in the second annual cycle when liming was discontinued. The range of dissolved Oxygen is wide in small waterbodies and may vary greatly from one waterbody to the other in the same area (Kaushik *et al.*, 1991). In the present study, dissolved oxygen was found to vary from 1.6 $mg\ l^{-1}$ to 10.4 $mg\ l^{-1}$ in the different study sites. Relatively higher dissolved oxygen values were recorded in Upper lake and Haranbrajao pond (6.2–10 $mg\ l^{-1}$ and 5.2–10.4 $mg\ l^{-1}$ respectively) when compared to those in Fishery tank, Haflong Lake and Maibong Pond (2.7–10 $mg\ l^{-1}$, 1.6–6.4 $mg\ l^{-1}$ and 2.8–6.4 $mg\ l^{-1}$ respectively) as a result of the entry of large amount of organic matter in the form of manure and sewage in the latter. Again, in Haflong Lake lower oxygen values were recorded in the first annual cycle (1.6–6.4 $mg\ l^{-1}$) than that in second annual cycle (4.8–6.4 $mg\ l^{-1}$). This might be due to the accidental drainage of water from this system in the first annual cycle when the water level was much lower. In most of the water bodies higher oxygen values were recorded during winter. Higher values of dissolved oxygen in colder months even in tropical systems were reported by some workers (Kauashik *et al.*, 1989; Prasad, 1990). Higher level of free carbon dioxide (6–13 $mg\ l^{-1}$) could be observed during post monsoon in the first annual cycle (1999–2000) in Fishery Tank, Haflong Lake and Upper Lake. In contrast, in second annual cycle (2001–2002), Fishery Tank, Haflong Lake and Harangajao Pond showed the highest level of free CO_2 (15.6 $mg\ l^{-1}$, 11 $mg\ l^{-1}$, and 8 $mg\ l^{-1}$ respectively) during monsoon whereas it was higher during pre-monsoon in Upper Lake and Maibong Pond (5.2 $mg\ l^{-1}$ and 9 $mg\ l^{-1}$ respectively). The rain absorbs small amount of CO_2 from the atmosphere and delivers it to the water bodies (Unni, 1972). In the second study year rainfall was considerably heavier in the monsoon months. This might be the cause of higher CO_2 during Monsoon.

Table 59.2: Physico-chemical Properties of Water in the Lentic Systems of North Cachar Hills, Assam, in the First Annual Cycle (1999–2000)

Month	Water Body	Air Temp. °C	Water Temp. °C	pH	Conductivity μScm^{-1}	Dissolved Oxygen mgl^{-1}	Free Carbon dioxide mgl^{-1}	Total Alkalinity mgl^{-1}	Phosphates μmgl^{-1}	Nitrates mgl^{-1} μgml^{-1}	Gross Primary Productivity $mgC/m^3/24$ hrs (Range)	Respiration $mgC/m^3/24$ hrs (Range)	Net Primary Productivity $mgC/m^3/24$ hrs (Range)
Aug'99	Fishery Tank	27	28	7.5	73.9	5.8	8	68	26.6	0.94	1576.5±523.53 (975.9–1951.9)	925.9±114.66 (825.8–900.9)	650.6±458.72 (150.1–1051)
	Haflong Lake	27	28	7.5	52	4.2	10	64	15.3	1.2	1764.9 (1576.5–1951.9)	1351.3	412.9 (225.23–600.58)
	Upper Lake	24	26	7.24	20.9	8.6	3	48	5.1	1.8	2402.3 (1726.66–3077.45)	2026.9 (1501.4–2552.4)	375.36 (225.2–525.5)
Sep'99	Fishery Tank	26	28.5	6.75		4.8	12.4	64					
	Haflong Lake	28	29	6.9		3.2	10.5	64					
	Upper Lake												
Oct'99	Fishery Tank	24	25.5	6.5		3.2	13	104					
	Haflong Lake	25.5	26	6.7		3.2	14	68					
Nov'99	Fishery Tank	21	19.5	7.15	78.1	5.6	6	60	127	4.6	1701.63±312.55 (1351.3–1951.9)	925.10±144.66 (825.80–1051.0)	775.73±263.64 (525.5–1051)
	Haflong Lake	20	19	6.8	64.2	4.0	8	64	12.7	4.6	1351.3 (1051–1651.6)	488 (150.14±825.8)	863.33 (150.14–1776.5)
	Upper Lake	19	18	6.5	30.1	6.8	6	30	18.1	5.9	1426.4 (1351.3–1501.4)	825.8 (750.7–900.9)	750.7
Dec'99	Fishery Tank	17	17	6.55		5.6	6	80					
	Haflong Lake	17	16	6.6		3.8	6.8	92					
	Upper Lake												
Jan'00	Fishery Tank	16	14.5	6.78	73.9	3.2	12	92	30	1.6	625.60±188.94 (450.41±825.8)	425.4±263.69 (150.1–675.7)	200.2±86.69 (150.1–300.3)
	Haflong Lake	16	14	6.5	59	3.6	6.8	94	10	2.8	750.7 (450.4–1051)	375.36 (300.3–450.4)	375.36 (150.14–600.58)
	Upper Lake	19	17.5	6.95	24.1	7	2.8	36	12.5	6.0	1126.1	825.8 (750.7–900.9)	300.3 (325.2–375.36)

Contd...

Table 59.2–Contd...

Month	Water Body	Air Temp. °C	Water Temp. °C	pH	Conductivity $\mu S cm^{-1}$	Dissolved Oxygen $mg l^{-1}$	Free Carbon dioxide $mg l^{-1}$	Total Alkalinity $mg l^{-1}$	Phosphates $\mu g l^{-1}$	Nitrates $mg l^{-1}$ $\mu g m l^{-1}$	Gross Primary Productivity $mgC/m^3/24$ hrs (Range)	Respiration $mgC/m^3/24$ hrs (Range)	Net Primary Productivity $mgC/m^3/24$ hrs (Range)
Feb'00	Fishery Tank	20	19	6.95		2.8	13	64					
	Haflong Lake	20	19	7.0		1.6	9	80					
	Upper Lake												
Mar'00	Fishery Tank	23	22.5	7.6		3.6	11.5	60					
	Haflong Lake	22	23	7.1		5.0	8.4	68					
	Upper Lake												
Apr'00	Fishery Tank	24	21	8.2		5.8	6.4	66					
	Haflong Lake	24	20	7.7		6.4	9.5	64					
	Upper Lake												
May'00	Fishery Tank	24.5	25	7.8	64.2	2.7	13	68	17	0.48	1676.63±991.23 (600.6–2552.2)	1151.1±740.63 (300.3–1501.4)	525.53±456.7 (225.2–1051.1)
	Haflong Lake	25	26	7.4	48	4.2	10	66	9.5	0.38	2552.4 (1651.6–3453.3)	1764.8 (1051–2477.4)	825.8 (600.58–975.9)
	Upper Lake	22	21	7.01	22.9	6.2	5.8	40	5.0	2.5	825.8 (450.43–1201.15)	337.8 (75.1–600.6)	487 (375.36–600)
Jun'00	Fishery Tank	25	22	8.2		4.6	12.4	68					
	Haflong Lake	24.5	21	7.3		4.8	11	66					
	Upper Lake												
Jul'00	Fishery Tank	28	29	8.35		4.8	12.4	66					
	Haflong Lake	28.5	29	7.6		4.6	11	68					
	Upper Lake												

Table 59.3: Physico-chemical Properties of Water in the Lentic Systems of North Cachar Hills, Assam, in the Second Annual Cycle (2001–2002)

Season	*Waterbody*	*Air Temp. °C*	*Water Temp. °C*	*pH*	*Conductivity μ Scm⁻¹*	*Dissolved Oxygen (mgl⁻¹)*	*Free Carbon Dicxide (mgl⁻¹)*	*Total Alka-linity (mgl⁻¹)*	*Phosphates μg ml⁻¹*	*Nitrates mgl⁻¹*
Premonsoon (May'01)	Fishery Tank	24	25.2	7.5	23.6	6	8	84	1.08	5.8
	Haflong Lake	23	24	7.6	55.8	5.6	8.4	80	21.2	45.9
	Upper Lake	20	20.5	7.2	19.2	8	5.2	20	–	2.8
	Maibong Pond	26	28	6.52	55.4	2.8	9	56	2.4	2.2
	Harangajao Pond	24	22.5	7	20.2	6.4	5.2	56	1.02	0.76
Monsoon (Aug'01)	Fishery Tank	26	29	6.5	20.2	5.6	15.6	76	0.24	6.7
	Haflong Lake	28	29	7.3	63	4.8	20	68	24.6	23
	Upper Lake	24	26	6.45	18.9	9.6	4.8	20	–	2.05
	Mailbong Pond	22	23	7	59	5.6	7	46	11.68	19.7
	Harangajao Pond	30	32	6.05	18.2	5.2	8	48	1.33	0.88
Winter (Jan'02)	Fishery Tank	20	18	7.15	28.6	6	6	84	4.62	4.85
	Haflong Lake	16	14.5	6.8	74.6	6.4	8.4	94	21.3	45.9
	Upper Lake	18	17	7	20.1	10	3.4	26	–	2.8
	Maibong Pond	20.5	19	6.2	61	6.4	5.6	68	5.35	4.6
	Harangajao Pond	20	18	6.85	17.6	10.4	2.4	20	0.72	0.68

Table 59.4: Comparison Among Physico-chemical Properties of Water in Different Waterbodies Using One Way ANOVA and Duncan's Long Range Test

Period	*Physico-Chemical Parameters*	*F Value*	*df*	*Level of Significance*	*Duncan's LRT*
First Annual Cycle	Dissolved Oxygen	9.8	9,2	$P \leq 0.01$	UL > HL, FT; HL ≅ FT
	Free Carbon Dioxide	0.13	9,2		
	Total alkalinity	5.016	9,2	$P \leq 0.05$	FT, HL < UL
	pH	0.88	9,2		
	Air Temp.	0.09	9,2		
	Water Temp.	0.05	9,2		
	Conductivity of Water	69.63	9,2	$P \leq 0.001$	FT > HL > UL
	Phosphates	0.67	9,2		
	Nitrates	1.34	9,2		
	Gross Primary Productivity	30.07	9,2	$P \leq 0.001$	
	Respiration	0.076	9,2		
	Net Primary Productivity	0.36	9,2		
Second Annual Cycle	Dissolved Oxygen	3.398	10,4		
	Free Carbon Dioxide	2.15	10,4		
	Total Alkalinity	14.36	10,4	$P \leq 0.001$	FT, HL > MP, HL, UL; MP > UL, FT ≅ HL; MP ≅ HP; HP ≅ UL
	Phosphates	41.30	10,4	$P \leq 0.001$	HL > MP, FT, HP, UL; MP > HP, UL
	Nitrates	13.42	10,4	$P < 0.001$	HL > MP, FT, UL, HP
	pH	1.17	10,4		
	Conductivity	64.35	10,4	$P \leq 0.001$	HL > MP > FT, UL, HP
	Air Temp.	0.36	10,4		

Higher total alkalinity could be observed in Fishery Tank (60 mg l^{-1}–104 mg l^{-1}) and Haflong Lake (64 mg l^{-1}–94 mg l^{-1}) followed by Maibong (46–68 mg l^{-1}) and Harangajao Ponds (20–56 mg l^{-1}), while the lowest was in Upper Lake (20–48 mg l^{-1}). Alikunhi (1957) stated that in highly productive waters, alkalinity ought to be more than 100 mg l^{-1}. Alkalinity greater than 100 mg l^{-1} was recorded only once (October 1999 in Fishery Tank) during the present study, although concentrations of 80–94 mg l^{-1} were also recorded in Fishery Tank and Haflong Lake in the post monsoon and winter seasons, while in all the other systems, alkalinity values were consistently much lower. Phosphate levels were higher in Fishery Tank during first annual cycle (26.6–127 µg l^{-1}), which again may be due to manuring and management. Nitrates were found to fluctuate widely between 0.38 mg l^{-1} and 45.9 mg l^{-1}. The least concentration of nitrate was recorded from Harangajao Pond. During the second annual cycle, the range of nitrate content was the highest (23.0–45.9 mg l^{-1}) in Haflong Lake among all the waterbodies studied. This might be due to the sewage disposal into the lake. However, the fluctuations were very narrow during the first annual cycle, when there was a stress in the lake ecosystem due to the accidental draining out of its water. Conductivity, total alkalinity, phosphates and nitrates were relatively high in all the systems during dry seasons which might be due to evaporation of water (Furch and Junk, 1993). In the present study, higher phosphate concentrations were recorded in Haflong Lake, Maibong Pond and Harangajao Pohd during monsoon. Both Maibong and Harangajao Ponds being situated in the vicinity of paddy fields received nutrient enriched surface runoff during monsoon, while the bulk disposal of sewage into Haflong Lake during monsoon might be the reason behind the observed high level of phosphate in these waterbodies during this season. Fluctuations in both gross primary productivity and net primary productivity could be observed in all the waterbodies. The highest GPP was observed in Haflong Lake (2552.4 mgC/m^3/24 hrs) in pre-monsoon, although highest NPP was recorded in the same system during post-monsoon (863.33 mgC/m^3/24 hrs). A strong correlation between primary productivity and phosphates, nitrates and total alkalinity contents of the lentic systems was apparent.

Statistical Analysis

Statistical analysis of data for physico-chemical properties of water in different waterbodies with ANOVA and Duncan's Long Range Test is shown in Table 59.4. The table revealed a significant difference among the lentic systems only in dissolved oxygen, total alkalinity and conductivity during the first annual cycle and in total alkalinity, phosphates, nitrates and conductivity during the second annual cycle. During the first annual cycle, Upper Lake showed significantly higher dissolved oxygen content than Haflong Lake and Fishery Tank, while both Fishery Tank and Haflong Lake showed significantly higher total alkalinity than Upper Lake. Fishery Tank had higher conductivity of water than Hatlong Lake, which in its turn had higher conductivity of water than Upper Lake. During the second annual cycle, significantly higher total alkalinity was observed in both Fishery Tank and Haflong Lake than that in the other systems. Halfong Lake also showed higher concentrations of phosphates, nitrates and conductivity compared to the others.

Acknowledgements

TNM is thankful to Mr. GK. Das, Principal, Haflong Government College, for his help and cooperation. Help rendered by Sri Arun Chandra Das in collection of samples is also acknowledged.

References

Alikunhi, K. H., 1957. Fish culture in India. *Farm Bulletin*, ICAR, New Delhi, 20: 1–6.

Carpenter, S.R. and Lodge, D.M., 1986. Effects of submerged macrophytes on ecosystem processes. *Aquat. Bot.* 26: 341–370.

Champion, H.G. and Seth, S.K., (1968. *Forest Types of India*. Forest Research Institute, Dehradun.

Das, A C., Baruah, B.K., Baruah, D. and Sengupta, S., 2002. Study on wetlands of Guwahati city-1. Water quality of ponds and beets. *Poll. Res.*, 21(4): 511–513.

Dewanji, A. and Matai, S., 2000. Aquatic weeds as indicators of water quality. In: *Environmental Stress: Indication Mitigation and Eco-Conservation*, (Eds.) Yunus, M., Sing, N. and Dekok, L.J. Kluwer Academic Publishers, Boston, p. 251–258.

Dutta Gupta, S., Gupta, S. and. Gupta, A., 2004. Euglenoid blooms in the flood plain wetlands of Barak Valley, Assam, Northeastern India. *J. Environ. Biol.*, 25(3): 369–373.

Furch, K. and Junk, W.J., 1993. Seasonal nutrient dynamics in an Amazonian Lake Arch. *Hydrobiol.*, 128: 277–295.

Gupta, S., Michael, R.G. and Gupta, A., 1992. A preliminary investigation on physico-chemical factors, periphyton and invertebrate communities in a protected water works of Shillong, Meghalaya. In: *Proc. Nat. Acad. Sci.*, India, 62(B)III: 333–337.

Habib, O.A, Tripett, R. and Murphy, K.J., 1997. Seasonal changes in phytoplankton community structure in relation to physico-chemical factors in Lack Lomond, Scotland. *Hydrobiologia*, 350: 63–79.

Kaushik, S., Agarkar, M.S. and Saxena, D.N., 1991. Water quality and periodicity of phytoplantonic algae in Chambal tank, Gwalior, Madhya Pradesh. *Bionature*, 11: 87–94.

Kaushik, S., Sharma, S. and Saksena, M.N., 1989. Physico-chemical factors and the aquatic insect density of a pond receiving cotton mill effluent at Gwalior. *Indian J. Ecol.*, 16(1): 64–67.

Mariappan, P., and Vasudevan, T., 2002. Correlation coefficients of some physico-chemical parameters of drinking water ponds in eastern part of Sivagangai District, Tamil Nadu. *Poll. Res.*, 21(4): 403–407.

Michael, P., 1984. *Ecological Methods for Field and Laboratory Investigations*. Tata McGraw-Hill Publishing Company, Limited, New Delhi.

Parashar, D. (Ed.), 1994. *N.C. Hills: The Enchanting Land and the Bird Mystery of Jatinga*. North Cachar Hills Autonomous District Council, Haflong in Association with Konark Publishers. Pvt. Ltd., New Delhi.

Prasad, D.Y., 1990. Primary productivity and energy flow in Upper Lake, Bhopal (India). *Indian J. Environ. Health*, 32(2): 132–199.

Sahat, T., Manna, N.K., Som Majumdar, S. and Bhattacharjya, I.N., 2001. Primary productivity of the Subhas Sarobar Lake in East Calcutta in relation to some selected physicochemical parameters. *Poll. Res.*, 20(1): 47–52.

Singh, J.P. and Roy, S.P., 1995. Limnobiotic investigation of Kawar Lake, Begusarai, Bihar. *Env. Eco.*, 13(2): 330–335.

Tiwari, D.R., 1999. Physicochemical studies of the Upper Lake Water, Bhopal, Madhya Pradesh, India. *Poll. Res.*, 18(3): 323–326.

Unni, K.S., 1972. An ecological study of the macrophytic vegetation of Doodhadhari lake, Raipur (M.P.). 3. Chemical factors. *Hydrobiologia*, 40: 25–36.

Welch, P.S., 1952. *Limnology*. McGraw-Hill Book Company, New York.

Chapter 60

Studies on Selected Major Elements and Nutrients in Rushikulya Estuary (East Coast of India)

Tapan Rani Mahapatro
Department of Marine Sciences, Berhampur University, Berhampur, Orissa, India

ABSTRACT

Like chlorinity the major elements (Ca, Mg, B and F) in general, exhibited an increasing trend from riverine to estuarine end. Major elements showed higher concentrations observed during summer, whereas lower during monsoon. This indicated the predominance of these elements in seawater. Positive correlations observed between chlorinity and said major elements as well as among the said major elements. Negative correlations observed between the major elements and nutrients. The vertical gradient for chlorinity and major elements also showed increasing trend. The nutrients (PO_4-P, SiO_4-Si, NO_3-N and NH_4-N) showed decreasing trend from riverine to estuarine end. Which indicated their dominance in river water. The nutrients showed negative correlation with chlorinity and major elements. The negative value may be due to the river runoff/terrestrial input to the estuary. The positive relations among nutrients indicated their addition to the estuary from same source. Whereas the negative correlation with chlorinity suggested that the freshwater is the main source of addition of nutrients. The vertical gradient for the nutrients showed decreasing trend.

Keywords: *Chlorinity, Correlation, Major element, Negative, Nutrient Positive.*

Introduction

Estuaries are the environments that play an important role in the transfer of products of continental weathering to the ocean. These are unique water systems as they are transitional regions between fresh river water and saline coastal water. These are the regions of high biological productivity and are often subjected to high degree of pollution. These are also important regions of the coastal ecosystem since their enclosed nature often protects them from extreme winds and waves providing a rich source of nourishment. As a complex system the estuary receives huge quantity of dissolved chemical inputs from a number of sources as runoffs through the course of river (Das *et al.*, 1997). Input of the same from the sea itself cannot be ruled out. The distribution of chemical constituents in the estuarine environment is greatly regulated by physical, chemical and biological processes. The estuarine environment plays a great role in controlling the net fluvial transport and fate of the chemical constituents. Hence studies on the behaviour and distribution of dissolved constituents such as major elements and nutrients are highly important for assessing river inputs into the ocean.

Nutrients are the limiting factors for the growth of plants. Nutrient distribution is principally determined by mixing processes between freshwater of high nutrient and sea water of low nutrient contents. Nutrients are functionally involved in the processes of living organisms. Human activities within river catchments particularly clearing, agriculture and sewage production increase the nutrient content of runoff entering the receiving water body. The incorporation of the nutrients into the cells, tissues and extra cellular structures of living organisms and the excretions and regenerations of these elements in solution will impose additional mechanisms of addition, removal and transport on the geophysical and geo-chemical processes which influence the concentrations of the nutrients and other major elements. The primary agencies for the biological removal of inorganic nitrogen, phosphorous and silicon from solution in estuarine water are unicellular algae of the phytoplankton. Most of the phytoplankton has developed in the brackish layer due to nutrient enrichment (Legovic *et al.*, 1994). The site of these biologically activated mechanisms will depend upon the ecology of the organisms responsible and this will set up quite different patterns of distribution of the elements from those produced by a biological processes. Therefore, biological activity will tend to modify the movements towards geophysical and geo-chemical equilibrium. These effects will also be apparent in the distribution of any element, which is incorporated adventitiously into living organisms in sufficiently large quantities relative to its concentrations in estuarine water (Uthup, 1996). In terms of nutrient limitation of phytoplankton growth, it is generally assumed that growth in the later stages of bloom, in estuaries is either nitrogen or phosphorous limited. This comes from the fact that, in most rivers, phytoplankton growth is phosphorous limited and in the sea, growth is nitrogen limited (Legovic *et al.*, 1994). Hence in estuary the phytoplankton growth would be more in presence of both nitrogen and phosphorous. Nutrients regulate biological processes by limiting or enhancing organic production in almost all types of aquatic environments. The study of nutrients in the dissolved forms would help in understanding the potential availability of life supporting elements in any particular region of the aquatic system (Uthup, 1996). River waters, though relatively less uniform in their composition of major constituents, are still important as principal suppliers to the seas through estuaries. The transport of these constituents, however, depends largely on the geological nature of the drainage area. Concentration levels of major constituents, in general, do not exhibit appreciable changes and they bear a constant ratio with chlorinity in open seas. However, rapid changes in the behaviour of major constituents do occur in estuaries due to changes in physico-chemical parameters such as land runoff, precipitation and evaporation as a result of temporal changes. In addition, the highly productive features of estuaries alter the concentration levels of some bio-geochemically important elements such as fluorine, boron, calcium and magnesium (Barnes and Green, 1970).

It forms shallow and well mixed estuary wherein the influence of tide decreases sharply towards upstream. The estuary remains freshwater dominated during monsoon while the seawater dominated during pre-monsoon and summer, when the influx of freshwater from riverine source is very less/ negligible. The estuary is highly productive. A number of reports have been published on major elements and nutrients from time to time in this estuary (Tripathy *et al.*, 1990; Gouda and Panigrahy, 1992-95; Panigrahy *et al.*, 1999; Mahapatro *et al.*, 2001; Mahapatro and Padhy, 2001a and Mahapatro and Padhy, 2001b). However, no effort has been made so far to trace the path source of major elements and nutrients and to understand the impact of land drainage on the water of Rushikulya estuary. The present communication is an attempt to find out the correlation factor among these parameters and chlorinity.

Materials and Methods

The climate of this region is semi-arid type. Basing on the meteorological condition of the area, five different seasons namely monsoon (July-Sept), post-monsoon (Oct- Nov), winter (Dec-Jan), pre-monsoon (Feb-Apr) and summer (May-June) are experienced. The tides of this estuary are mixed and semi-diurnal type. The sampling was undertaken (Dec 1999–Nov 2000) at seven selected stations (Figure 60.1), which cover the entire estuary, during the seasons mentioned above. The surface water samples were collected using polythene bucket, while bottom water samples were collected 1.5 m below the surface using shallow water sampler. The samples were filtered by 0.45 μm filter paper. Before filling the acid cleaned polythene bottles were rinsed 2/3 times with the respective station water. These filled polythene bottles with leak-proof stoppers were preserved in a refrigerator till analyses, which was completed within a month from the date of collection. Generally, samples preserved in refrigerator gives no change in result (for major elements) even after 1 month of storage. The nutrients were analysed within 24 hours of collection.

Chlorinity determinations were made using argentometric titration procedure (Grasshoff, 1976). Nitrate was determined after its reduction to nitrite followed by photometric method based on the formation of an azodye, which was proposed by Bendschneider and Robinson (1952) and later modified and described in the Manual of Methods in Aquatic Environment Research (1972) was followed for the fabrication and operation of cadmium reductor column. The photometric determination of ortho-phosphate in seawater was based on its formation of phosphomolybdenum complex with acidified molybdate reagent, which on reduction with ascorbic acid results in the formation of a highly coloured molybdenum blue compound. The method adopted in the present investigation was a modified procedure of Murphy and Riley (1962) as given by Grasshoff (1976). Phenol-hypochlorite method (Solorzano, 1969) was adopted for the determination of ammonium in seawater. The determination of dissolved silicon was based on the formation of a yellow silicomolybdic acid complex when an acidified sample was treated with ammonium molybdate solution. This on reduction with ascorbic acid yields intensely coloured molybdenum blue complex. The method of Koroleff (1971) was employed in the present investigation. Calcium and magnesium were estimated by a complexometric titration. In this method (Culkin and Cox, 1966) total concentration of calcium and magnesium was first estimated in an aliquot of sample by titrating with standard EGTA solution at pH 10 using NH_4OH/NH_4Cl buffer and Eriochrome Black-T (EBT) as indicator. At the end point the colour changes from wine red to purple blue. Calcium was then estimated by titrating another (same) aliquot of sample with standard EGTA solution at pH 12 using a buffer (sodium tetra borate, NH_4Cl and NaOH) in presence of Zincon as indicator. At the end point the blue colour of the solution changes to orange-red (Ringbom *et al.*, 1958). The interference of strontium and magnesium in the titration of calcium using EGTA is negligible in riverine and estuarine waters. The amount of magnesium present in the

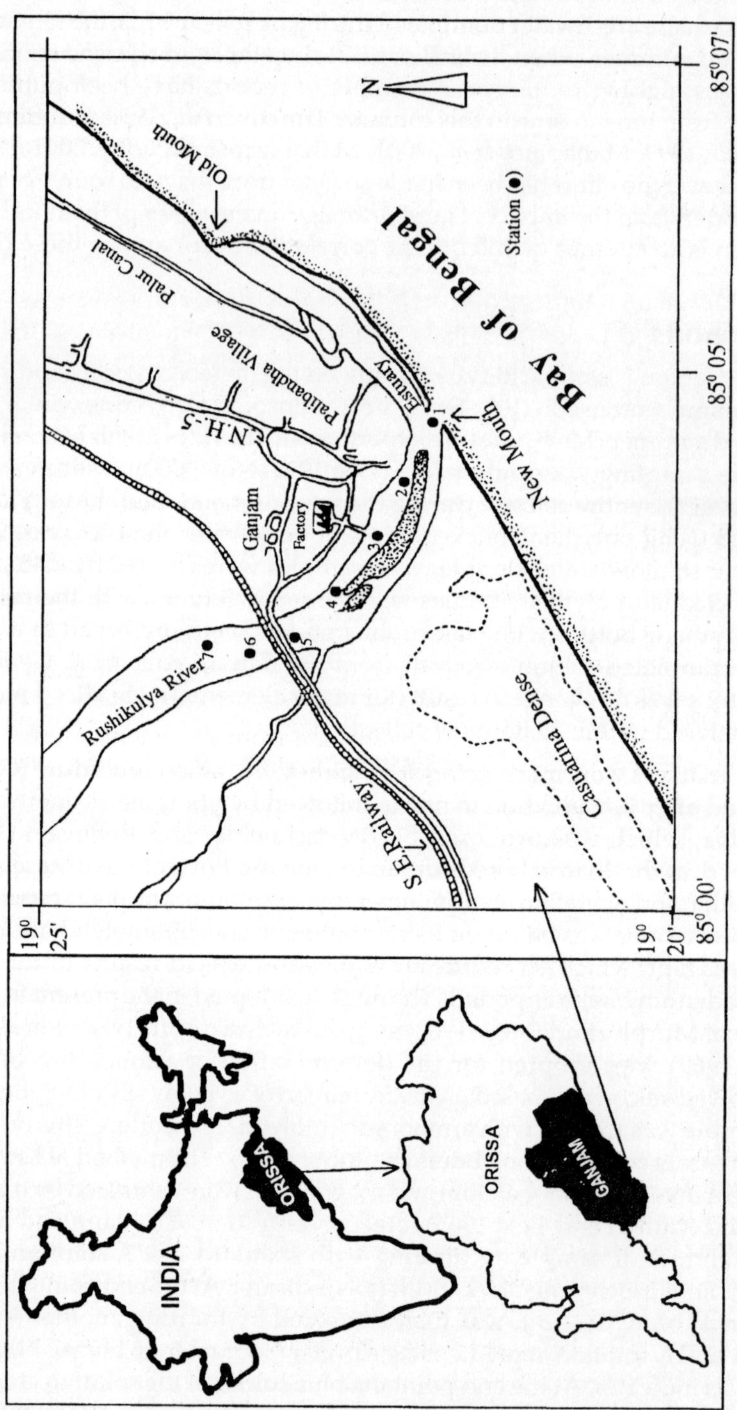

Figure 60.1: Location Map of Rushikulya Estuary Showing Sampling Stations.

aliquot of sample was computed from the difference in the two titres. Fluoride was estimated by alizarine complexone procedure (Greenhalgh and Riley, 1961). Boron was estimated spectrophotometrically by curcumin method (APHA, 1975). The mean, standard error (SE), correlation coefficient and percentage of coefficient of variation (CV) were calculated by MS-Office–2000, Excel Package. Due to very less water at the upstream stations during pre-monsoon and summer, the surface and bottom samples could not be collected.

Results and Discussion

Chlorinity

Chlorinity is considered as the parameter in studying the mixing processes and intrusion of saline waters into estuaries by tidal action. The mixing and diffusion phenomena occurring in estuaries are largely influenced by chlorinity distribution pattern. Wide fluctuations in chlorinity are generally observed in estuaries.

The chlorinity concentrations for surface water ranged from 3.371–8.464, 7.799–12.145, 10.180–14.375, 1.063–6.117 and 2.552–7910 per cent during winter, pre-monsoon, summer, monsoon and post-monsoon respectively. The minimum and maximum surface water chlorinity concentrations were 1.063 and 14.375 per cent (mean 6.674 ± 0.608 per cent) respectively. The percentage of CV for surface water chlorinity was 11.074. The bottom water chlorinity concentrations during winter, pre-monsoon, summer, monsoon and post-monsoon were 3.886–9.438, 8.165–12.444, 10.440–14.486, 1.268–6.803 and 2.895–8.779 per cent respectively. The minimum and maximum bottom water chlorinity concentrations were 1.268 and 14.486 per cent (mean 7.117 ± 0.602 per cent) respectively. The percentage of CV for it was 10.864.

The surface water chlorinity progressively diminished from the estuary mouth at Site 1 and attained the lowest value at Site 7 on the riverine end. The bottom water, chlorinity also exhibited the same trend from estuarine to riverine end but maintained higher concentrations than surface one throughout [Figure 60.2(a)]. The higher concentrations during all seasons at Site 1 may be due to the dominance of saline water at this point. The lower concentrations during all seasons at Site 7 may be due to the dominance of freshwater over saline water. Similar observations were reported elsewhere (Naik, 1978; Nair *et al.*, 1983; Narvekar *et al.*, 1983; Narvekar and Zinge, 1987; Palanichamy and Balasubramanian, 1989; Das *et al.*, 1997). Minimum chlorinity was observed during monsoon. The minimum values during this season may be due to the efflux of more freshwater and land drainage system. Similar chlorinity distribution in this estuary was observed by earlier workers (Mahapatro and Padhy, 2001 b).

Nitrate (NO_3-N)

It is the final oxidant product of nitrogen compounds in saline water and is considered to be the only thermodynamically stable oxidant level of nitrogen in the presence of oxygen in saline water (Sillen, 1961). In the present study surface water nitrate concentration ranged from 2.237–4.428, 1.158–1.976, 0.752–0.910, 2.719–4.492 and 2.589–3.878 µg at l^{-1} during winter, pre-monsoon, summer, monsoon and post-monsoon respectively. The minimum and maximum concentrations for the surface water were 0.752 and 4.492 µg at l^{-1} (mean 2.647 ± 0.208 µg at l^{-1}) respectively. The calculated percentage of CV for the same was 2.032. The bottom water nitrate concentration ranged from 2.153–4.073, 1.151–1.784, 0.752–0.897, 2.719–4.079 and 2.598–3.280 µg at l^{-1} during winter, pre-monsoon, summer, monsoon and post-monsoon respectively. The minimum and maximum concentrations for the bottom water were 0.736 and 4.280 μg at 1^{-1} (mean 2.541 ± 0.199 μg at 1^{-1}) respectively. The calculated percentage of CV for it was 1.186.

In the present study NO_3-N concentrations for both surface and bottom water exhibited a clear increasing trend from estuarine to riverine end (Site 1 to Site 7) during all the seasons. The bottom water NO_3-N concentration was less than the surface one at all the stations [Figure 60.2(b)]. It can be attributed to the outflow of freshwater containing higher concentrations into the sea than the intrusion of seawater at bottom containing low concentration into the estuaries. Highest peak was observed during monsoon at Site 7, whereas lowest peak during summer at Site 1. The highest peak may be due to the cumulative impact of rainfall and river discharge experienced at this station. The distribution of NO_3-N at all stations maintained the order monsoon > post-monsoon > winter > pre-monsoon > summer, overall all these variations are highly significant. Heavy rain and consequent land drainage cause high values during monsoon. Similar studies on concentrations of NO_3-N associated with monsoon rain have been reported from several areas (Sankaranarayanan and Qasim, 1969; Solarzano and Ehrlich, 1975; Mackey and Leatherland, 1976; De Sousa *et al.*, 1981; Nair *et al.*, 1983, 1984 and Chandran and Ramamoorthi, 1984). Further, very low concentrations of NO_3–N observed during summer. This could be attributed to the rapid utilization of NO_3–N by phytoplankton. Phytoplankton production along the East Coast of India is normally high during pre-monsoon and summer (Ganapati and Rao, 1958) as was observed by me and thus phytoplankton assimilation could be taken as a major factor controlling NO_3-N distribution.

Both surface and bottom water NO_3-N were negatively correlated with chlorinity and major elements (B, F, Ca and Mg), whereas positively correlated with other nutrients (NH_4-N, SiO_4-Si and PO_4-P) [Tables 60.1(a) and (b)]. All the correlations were highly significant. The negative correlation indicated that freshwater input is primarily responsible for the nutrient economy in estuarine water and the positive correlation indicated the common source of occurrence of nutrients. Similar reports were putforth for the coastal water of Orissa (Panigrahy *et al.*, 1999; Padmavathi and Satyanarayana, 1999). The negative correlation with chlorinity is also indicated that low saline water is favourable for biotic population and conforms the views of Dehadrai (1970) and Dehadrai and Bhargava (1972). This states that biological productivity was maximum during pre-monsoon and summer. So, the nitrate loss can be attributed entirely to the biological utilization. It is in agreement with the view of De Sousa (1983).

Phosphate (PO_4-P)

The presence of PO_4-P in an estuary can be taken as an index of potential fertility of the ecosystem as a whole (Redfield, 1934). Phosphorous occurs in natural waters solely as PO_4-P. The PO_4-P rich detritus falls and is decomposed by bacteria. The presence of phosphates in large quantities in freshwater indicates pollution through sewage and industrial waste (Bandela *et al.*, 1999).

The surface water PO_4-P concentrations ranged from 0.579–0.763, 0.328–0.386, 0.257–0.297, 0.705–0.836 and 0.672–0.801 μg at 1^{-1} during winter, pre-monsoon, summer, monsoon and post-monsoon respectively. The minimum and maximum concentrations for the same were 0.257 and 0.836 μg at 1^{-1} (mean 0.610 ± 0.036 μg at 1^{-1}) respectively. The calculated percentage of CV for it was 0.039. The bottom water PO_4-P concentrations ranged from 0.633–0.782, 0.334–0.396, 0.261–0.297, 0.690–0.824 and 0.661–0.782 μg at 1^{-1} during winter, pre-monsoon, summer, monsoon and post-monsoon respectively. The minimum and maximum concentrations of the same were 0.261 and 0.824 μg at 1^{-1} (mean 0.589 ± 0.034 μg at 1^{-1}) respectively. The calculated percentage of CV for it was 0.035.

It showed decreasing trend from riverine to estuarine end (Site 1 to Site 7). The bottom water PO_4-P concentrations were higher than surface one at all the stations during the period of study [Figure 60.2(c)], which indicates its depletion at surface due to utilization by phytoplankton and regeneration in the water column at bottom. The PO_4-P concentrations were observed in the order monsoon > post-monsoon > winter > pre-monsoon > summer. Very low concentrations during the pre-monsoon and summer picked up with the advent of the winter. Highest concentration was observed during monsoon at Site 7, whereas lowest concentration was observed during summer at Site 7. Similar report has been reported by Satpathy (1996) in the coastal waters of Kalpakkam. PO_4-P as the seasonal variations have been found to be significant it can be stated that the impact of rainfall and river discharge on PO_4-P changes is prominent. According to Nair *et al.* (1983) rainfall resulting in heavy river discharge leads to transportation of sediments in a big way. Thus an increase in the value of PO_4-P was noticed with the onset of monsoon rains. In addition, anthropogenic inputs of super phosphates applied to the fields as fertilizers, sewage and domestic waste (which contains alkali phosphate used in households as detergent). Similar result was observed by Das *et al.* (1997). The higher concentration at Site 7 during monsoon may be due to freshwater influx. Similar observations were made for the same estuary by several workers (Gouda and Panigrahy, 1995 and Mahapatro *et al.*, 2001)

Both surface and bottom PO_4-P showed highly significant values with the studied elements. With nutrients it showed positive correlation, whereas with major elements and chlorinity it showed negative correlation [Tables 60.1(a) and (b)]. It may be due to desorption of PO_4-P in high chlorinity. Similar reports has been made in Godavari estuary (Padmavati and Satyanarayana, 1999).

Table 60.1(a): Correlation Analysis of Surface Water Chlorinity, Nutrients and Major Elements in Rushikulya Estuary

	Chlorinity	*PO_4-P*	*SiO_4-Si*	*NO_3-N*	*NH_4-N*	*Fluoride*	*Calcium*	*Magnesium*	*Boron*
Chlorinity	1								
PO_4-P	–0.94	1							
SiO_4-Si	–0.98	0.94	1						
NO_3-N	–0.95	0.90	0.97	1					
NH_4-N	–0.95	0.99	0.96	0.94	1				
Fluoride	0.97	–0.97	–0.97	–0.95	–0.98	1			
Calcium	0.97	–0.96	–0.97	–0.93	–0.97	0.97	1		
Magnesium	0.98	–0.97	–0.96	–0.92	–0.98	0.97	0.98	1	
Boron	0.99	–0.97	–0.98	–0.94	–0.97	0.99	0.99	0.99	1

Silicate (SiO_4-Si)

The SiO_4-Si showed significant variations during the study period. During winter, pre-monsoon, summer, monsoon and post-monsoon the surface water concentrations were ranged from 21.352–31.018, 15.761–22.127, 14.589–17.986, 25.129–33.287 and 23.754–30.227 μg at 1^{-1} respectively. The minimum and maximum surface water concentrations were 14.589 and 33.287 μg at 1^{-1} (mean 24.194 ± 0.949 μg at 1^{-1}) respectively. The calculated percentage of CV was 0.016. During winter, pre-monsoon, summer, monsoon and post-monsoon the bottom water concentrations were ranged from 19.341–29.895, 15.747–22.004, 14.581–17.985, 23.683–31.643 and 22.789–28.541 μg at 1^{-1}

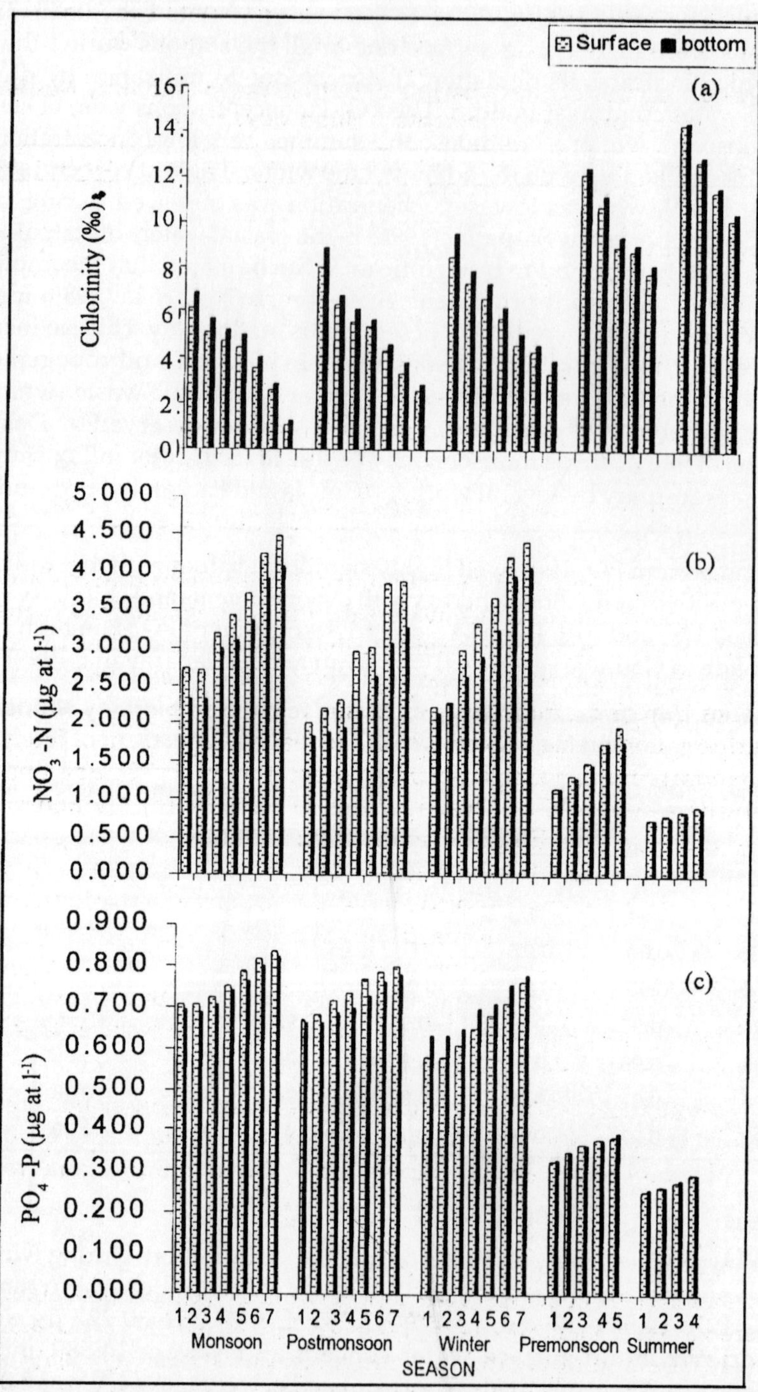

Figure 60.2(a): Seasonal Variations of Chlorinity in Rushikulya Estuary.

Figure 60.2(b): Seasonal Variations of NO_3-N in Rushikulya Estuary.

Figure 60.2(c): Seasonal Variations of PO_4-P in Rushikulya Estuary.

respectively. The minimum and maximum concentrations of it were 14.581 and 31.643 μg at l^{-1} (mean 23.052 ± 0.851 μg at l^{-1}) respectively. The percentage of CV for it was 0.746.

Table 60.1(b): Correlation Analysis of Bottom Water Chlorinity, Nutrients and Major Elements in Rushikulya Estuary

	Chlorinity	*PO_4-P*	*SiO_4-Si*	*NO_3-N*	*NH_4-N*	*Fluoride*	*Calcium*	*Magnesium*	*Boron*
Chlorinity	1								
PO_4-P	–0.92	1							
SiO_4-Si	–0.97	0.90	1						
NO_3-N	–0.92	0.90	0.95	1					
NH_4-N	–0.95	0.99	0.93	0.91	1				
Fluoride	0.97	–0.96	–0.93	–0.90	–0.98	1			
Calcium	0.98	–0.94	–0.95	–0.90	–0.96	0.98	1		
Magnesium	0.98	–0.97	–0.94	–0.90	–0.97	0.98	0.98	1	
Boron	0.99	–0.96	–0.95	–0.92	–0.97	0.99	0.99	0.99	1

The minimum peak was observed during summer at Site 1 and maximum during monsoon at Site 7. The lowest concentrations in the surface waters during summer attained highest concentrations during monsoon at all the stations presumably due to the river water that brings in additional quantities of the nutrients to it. Similar observation was reported by Nair *et al.* (1983) in Ashtamudi estuary. During all the seasons increasing trend from estuarine to riverine end was noticed. It may be to the dominance of riverine water over sea water which carries wide spectrum of finely divided siliceous materials. Similar reports have been putforth by several workers (Gouda and Panigrahy, 1992; Padmavati and Satyanarayana, 1999 and Mahapatro and Padhy, 2001a). The minimum concentrations observed during summer may be due to its involvement in biological process and adsorption onto suspended sediments. Similar reports were published by Anirudhan and Nambisan (1990) in Cochin estuary. The maximum concentrations observed during monsoon may be due to high influx of freshwater during this season. Many of the siliceous materials have been transported to the sea by rivers or by wind (Riley and Chester, 1971). Its concentration during the study period appeared in the order monsoon > post-monsoon > winter > pre-monsoon > summer. The SiO_4-Si exhibited relatively high concentration at surface when compared with bottom during all the seasons [(Figure 60.2(d)]. It may be due to the same reason as mentioned for NO_3-N.

Both surface and bottom SiO_4-Si concentrations were showed highly significant correlation [Tables 60.1(a) and (b)]. Positive correlations were observed among all the nutrients. It showed negative correlations with chlorinity and major elements. It may be due to the same reason as stated for NO_3-N.

Ammonium (NH_4-N)

The surface water NH_4-N concentrations during winter, pre-monsoon, summer, monsoon and post-monsoon were varied from 2.418–3.436, 1.006–1.976, 0.801–0.927, 3.093–4.027 and 2.834–3.618 μg at l^{-1} respectively. The minimum and maximum concentrations were 0.801 and 4.027 μg at l^{-1} (mean 2.546 ± 0.188 μg at l^{-1}) respectively. The percentage of CV for it was 1.060. Similarly, the bottom water NH_4-N concentrations during winter, pre-monsoon, summer, monsoon and post-monsoon were varied from 2.193–3.115, 0.993–1.501, 0.797–0.924, 2.730–3.861 and 2.517–3.367 μg at l^{-1} respectively.

The minimum and maximum concentrations of it were 0.797 and 3.861 μg at l^{-1} (mean 2.336 ± 0.164 μg at l^{-1}) respectively. The percentage of CV for it was 0.812.

The NH_4-N (for both surface and bottom) exhibited an increasing trend from estuarine to riverine end. The surface water NH_4-N concentration showed higher value when compared with bottom one during all the seasons [Figure 60.2(e)]. It may be due to its predominant freshwater origin in the study area. The NH_4-N concentrations during the study period observed in the order monsoon > post-monsoon > winter > pre-monsoon > summer. High concentrations were observed during monsoon may be due to the high influx of freshwater during this season as compared with other seasons. The combined effects of precipitation, river runoff and land drainage during monsoon may be another reason for high concentration during this season. Low concentrations during pre-monsoon and summer may be due to its utilization by phytoplankton. Similar observations were made by several workers (Sarala Devi *et al.*, 1983; Verlencar, 1987; Pradhan, 2000 and Acharya, 2001). Like other nutrients it also showed positive significant correlation. Positive correlations were observed, [Tables 60.1 (a) and (b)] among nutrients, whereas negative correlations were observed with chlorinity and major elements.

Fluoride

It has long been considered as an acute pollutant to natural environment because of the ability of plants and marine organisms to accumulate it (Ares, 1978) and because of its detrimental effects upto aquatic biota, which can be affected by concentrations as low as 1.5 mg/l (Wright and Davison, 1975). According to Carpenter (1969) about 80-90 per cent of fluoride in river water is cyclic derived from sea salts transported through rain. He also suggested that the contribution of dissolved fluoride by weathering and leaching of rocks is very small since most of the fluoride mobilized during weathering is carried in suspension still bound in detrital minerals.

The surface water fluoride concentrations during winter, pre-monsoon, summer, monsoon and post-monsoon varied from 0.35–0.63, 0.74–0.89, 0.83–0.98, 0.25–0.54 and 0.29–0.61 mg/l respectively. The minimum and maximum concentration of the same were 0.25 and 0.98 mg/l (mean 0.592 ± 0.037 mg/l) respectively. The calculated percentage of CV for it was 0.41. The bottom water fluoride concentrations during winter, pre-monsoon, summer, monsoon and post-monsoon varied from 0.49–0.69, 0.78–0.91, 0.83–0.99, 0.32–0.59 and 0.33–0.67 mg/l respectively. The minimum and maximum concentrations of the same were 0.32 and 0.99 mg/l (mean 0.640 ± 0.034 mg/l) respectively. The percentage of CV of it was 0.034.

Maximum fluoride concentration was observed during summer at Site 1, whereas minimum during monsoon at Site 7. It showed a clear increasing trend from riverine to estuarine end. The surface water concentration was less than bottom at every station during all the seasons of study [(Figure 60.2(f)]. Whereas during pre-monsoon and summer the bottom and surface water concentrations showed very less or no difference. It may be due to the partial stratification of bottom water during these seasons. With the onset of monsoon its concentration dropped abruptly and remained low during the entire season. Highest concentrations were noticed in both surface and bottom waters during pre-monsoon and summer showed the regulation of fluoride by highest salt water intrusion. Similar reports have been observed by several workers (De Sousa and Sen Gupta, 1988; Sarma *et al.*, 1993; Padmavati and Satyanarayana, 1999 and Mahapatro and Padhy, 2001b). Its concentration during the study period showed the order summer > pre-monsoon > winter > post-monsoon > monsoon. The post-monsoon and winter showed a recovery period when the fluoride concentration at any point in the estuary showed gradual increase with time reaching its peak during pre-monsoon and summer.

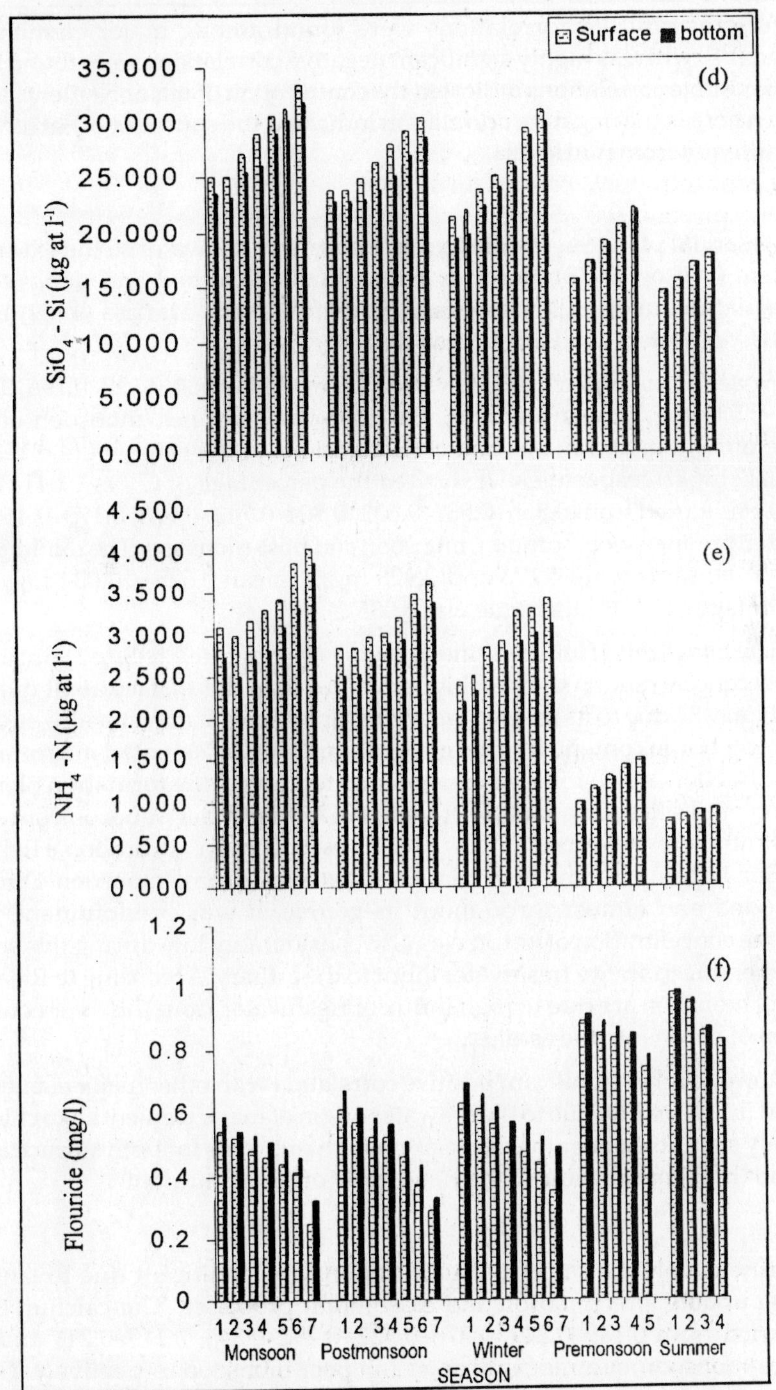

Figure 60.2(d): Seasonal Variations of SiO_4-Si in Rushikulya Estuary.

Figure 60.2(e): Seasonal Variations of NH_4-N in Rushikulya Estuary.

Figure 60.2(f): Seasonal Variations of Fluoride in Rushikulya

Highly significant positive correlations were found among major elements and chlorinity [Tables 60.1(a) and (b)], whereas highly significant negative correlations were found between nutrients and fluoride. The positive correlations indicated the contribution of major elements to the estuary from the same source, whereas the negative correlation indicated the source of occurrence of fluoride and chlorinity were different from nutrients.

Boron

Boron is an essential plant micro-nutrient and occurs in sea water to the extent of 4.5 mg/kg at salinity 35 per cent (Wilson, 1975). Boron is differentially adsorbed and deposited on suspended matter during estuarine mixing (Levinson and Ludwick, 1966). Harriss (1969) has attributed the removal of boron from waters to adsorption on detrital clays.

The surface water boron concentrations varied from 0.313–0.564, 0.627–0.796, 0.741–0.922, 0.122–0.429 and 0.219–0.538 mg/l during winter, pre-monsoon, summer, monsoon and post-monsoon respectively. The minimum and maximum concentrations of the same were 0.122 and 0.922 mg/l (mean 0.498 ± 0.037 mg/l) respectively. It showed the percentage of CV as 0.041. The bottom water boron concentrations varied from 0.356–0.581, 0.632–0.804, 0.742–0.928, 0.178–0.452 and 0.279–0.567 mg/l during winter, pre-monsoon, summer, monsoon and post-monsoon. The minimum and maximum concentrations for the same were 0.178 and 0.928 mg/l (mean 0.525 ± 0.034 mg/l) respectively. It showed the percentage of CV for the same was 0.035.

It showed increasing trend from riverine to estuarine end (Site 7 to Site 1) during all the seasons. The bottom water concentrations showed higher values than the surface at all the stations of study [Figure 60.2(g)]. It may be due to its utilization by phytoplankton. It may not also be the primary factor responsible for lower boron content at surface even though it is an essential micro-nutrient (Narvekar *et al.*, 1981). Some fraction is also removed from seawater during the formation of authigenic silicate and by adsorption onto siliceous oozes (Porrenga, 1967). The order of boron concentrations during study period was summer > pre-monsoon > winter > post-monsoon > monsoon. Highest concentration was observed at Site 1 during summer, whereas lowest at Site 7 during monsoon. During monsoon, the estuary was flooded and almost throughout its course, it was predominant with freshwater. Simultaneously, the concentration of boron was also considerably low during this season and reverse case during summer due to scanty freshwater input to the estuary. According to Rajagopal *et al.* (1981) low values during monsoon are due to high influx of freshwater from the river containing relatively low concentration of boron into the estuary.

The boron showed highly significant positive correlation with other major elements and chlorinity [Tables 60.1(a) and (b)]. It may be due to the seawater origin of major elements like chlorinity. The boron also showed highly significant negative correlation with nutrients for both surface and bottom water [Tables 60.1(a) and (b)]. It may be due to the river water origin of nutrients.

Calcium

In the estuarine waters the calcium concentrations are significant due to land drainage, high rates of biological uptake, precipitation and dissolution processes. The calcium concentration for surface water varied from 0.189–0.249, 0.276–0.322, 0.292–0.343, 0.119–0.211 and 0.146–0.243 g/l during winter, pre-monsoon, summer, monsoon and post-monsoon respectively. The minimum and maximum concentrations for the same were 0.119 and 0.343 g/l (mean 0.226 ± 0.011 g/l) respectively. The coefficient of CV for the same was 0.004. The bottom water calcium concentrations were ranged from 0.195–0.258, 0.278–0.327, 0.292–0.345, 0.137–0.227 and 0.152–0.257 g/l during winter, pre-monsoon, summer, monsoon and post-monsoon respectively. The minimum and maximum concentrations for the same were 0.137 and 0.345 g/l (mean 0.235 ± 0.010 g/l) respectively. The percentage of CV for the same was 0.003.

The calcium concentration showed an increasing trend from (Site 7 to Site 1) riverine, to estuarine end [Figure 60.2(h)] during all seasons. This variation may be due to gradual reduction in the incursion of seawater into the estuary with increasing distance from mouth (Site 1 to Site 7). Site 1 being mouth of the estuary receives more saline water, which leads more calcium concentration than other stations. Similar reports have been putforth by several workers (Sen Gupta and Naik, 1981; Narvekar *et al.*, 1983; Narvekar and Zingde, 1987; Palanichamy and Balasubramanian, 1989). The bottom water calcium concentrations were more than surface one. Culkin and Cox (1966) opinioned that extraction of calcium by living organisms from surface water leading to lower calcium level in surface water and dissolution of dead animals and precipitated calcium carbonate at bottom region could result increased $CaCO_3$ concentration in bottom water. It was also observed that there was very less to no difference between bottom and surface water concentrations during pre-monsoon and summer. It may be due to the saturation of estuarine water during these two seasons. The calcium concentrations for both surface and bottom water were in the order summer > pre-monsoon > winter > post-monsoon > monsoon. For both surface and bottom water higher calcium concentrations were recorded during summer, whereas lower during monsoon. The higher concentration of calcium during summer may be due to scanty freshwater discharge. Another reason for it may be due to the recovery of calcium for gradual attainment of normal estuarine conditions. This is in agreement with the report of Naik (1978) and Palanichamy and Balasubramanian (1989). The lower calcium concentrations observed during monsoon may be due to high influx of freshwater resulted by the heavy rainfall and land drainage. The more freshwater run-off leads to greater dilution, which accounted lowering of calcium concentration than other seasons. Similar observation was also made for Gopalpur creek by Sahu *et al.* (2002).

The calcium showed highly significant correlations with other major elements, chlorinity and nutrients [Tables 60.1(a) and (b)]. Positive correlations were observed among major elements and chlorinity, whereas negative correlations were observed with nutrients. The reason is same as the reasons mentioned for boron.

Magnesium

In estuarine water magnesium contribute significantly to the biological and geo-chemical processes. High concentrations of this element in the estuary, relative to its bio-geochemical reactions tend to make it conservative. The surface water magnesium concentrations for winter, pre-monsoon, summer, monsoon and post-monsoon varied from 0.427–0.602, 0.706–0.871, 0.801–1.021, 0.261–0.464 and 0.328–0.589 g/l respectively. The minimum and maximum concentrations for the same were 0. 261 and 1.021 g/l (mean.0.553 ± 0.037 g/l) respectively. The percentage of CV for the same was 0.042. The bottom water magnesium concentrations for winter, pre-monsoon, summer, monsoon and post-monsoon varied from 0.431–0.631, 0.711–0.871, 0.803–1.025, 0.309–0.498 and 0.334–0.596 g/l respectively. The minimum and maximum concentrations for the same were 0.309 and 1.025 g/l (mean 0.573 ± 0.035 g/l) respectively. The percentage of CV for the same was 0.038.

The distribution pattern of magnesium at seven different stations during five different seasons is given in Figure 60.2(i). The magnesium concentrations were found to be high at Site 1 (mouth), whereas low at Site 7 (riverine end). This variation may be due to gradual increase in the incursion of freshwater into the estuary with increasing distance from mouth (Site 1 to Site 7). As Site 1 is located at the mouth

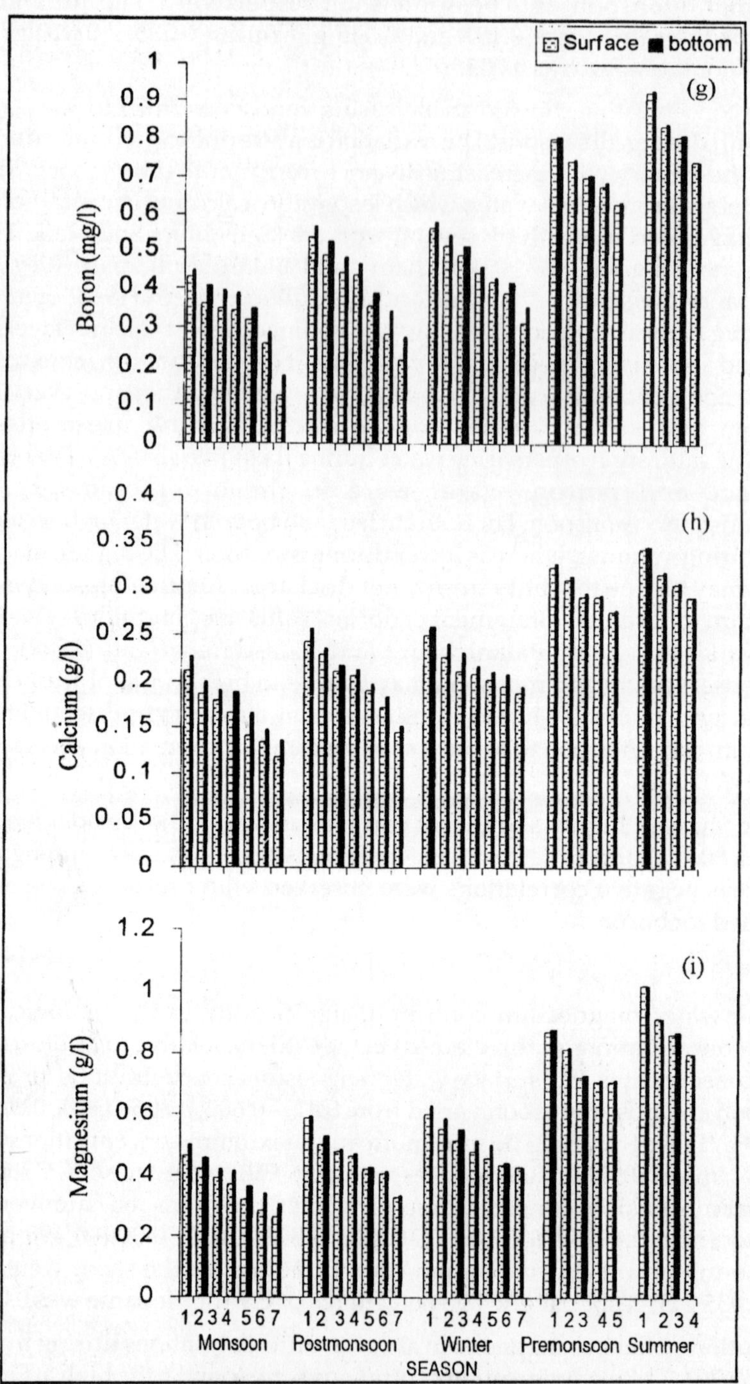

Figure 602(g): Seasonal Variations of Boron in Rushikulya Estuary
Figure 60.2(h): Seasonal Variations of Calcium in Rushikulya Estuary
Figure 60.2(i): Seasonal Variations of Magnesium in Rushikulya Estuary

of the estuary, which is influenced more by neritic waters leading more saline condition, a possible attribute for higher concentration of magnesium at this station. Site 7, which is very much influenced by the freshwater of poor magnesium concentrations, maintains low magnesium level. Other stations are in between, overall there observed an increasing concentration of magnesium from Site 7 to Site 1. Similar reports have been made by several workers elsewhere (Sen Gupta and Naik, 1981; Narvekar *et al.*, 1983; Narvekar and Zingde, 1987; Palanichamy and Balasubramanian, 1989).

The bottom water magnesium concentrations showed higher value as compared to surface one. Bottom to surface water concentration difference was not significant and the minor difference noticed in the present study may be due to some removal mechanisms from surface water. Another reason may be due to the stratification of the estuary by the neritic water at the bottom layer and freshwater dominance at surface layer. Culkin and Cox (1966) are of the same opinion for higher magnesium concentrations in the bottom water.

The magnesium concentrations showed the order of summer > pre-monsoon > winter > post-monsoon > monsoon for both surface and bottom water during the study period. Summer period recorded higher magnesium concentrations, whereas lower during monsoon. The estuary seemed to be influenced mostly by seawater, due to scanty freshwater discharge during most of the periods other than monsoon. This could have resulted in higher concentration of magnesium during summer and pre-monsoon. Much more freshwater had been discharged into the estuary during monsoon period. This could account for the lowering of magnesium concentration during monsoon. The concentration of discussed element was also found to show wider variations during monsoon, whereas more or less stable in nature with higher values observed during summer. During monsoon, the greater dilution would lead to a wide variation, whereas the summer could be recovery period for magnesium along with chlorinity. Palanichamy and Balasubramanian (1989) are of the same opinion for high and low values of magnesium during summer and monsoon respectively.

The magnesium showed highly significant correlations with other major elements, chlorinity and nutrients [Tables 60.1(a) and (b)]. Positive correlations were observed among major elements and chlorinity, whereas negative correlations were observed with nutrients. The reason is same as the reasons mentioned for boron.

Acknowledgement

The author (TRM) is thankful to UGC for the award of JRF and also the Head of the Department for the laboratory facility.

References

Acharya, S.K., 2001. *Studies on water characteristics of Rushikulya estuary (Bay of Bengal*. M.Phil. Thesis, Berhampur University, Berhampur, Orissa, India.

Anirudhan, T.S. and Nambisan, P.N.K., 1990. Distribution of salinity and silicon in Cochin estuary. *Indian J. Mar. Sci.*, 19: 137-139.

APHA (1975. *Standard methods for the examination of water and wastewater*, 14th edn., Washington, D.C.

Ares, J.O., 1978. Fluoride cycling near a coastal emission source. *J. Air Poll. Cont. Asso.*, 28(4): 344-349.

Bandela, N.N., Vaidya, D.P., Lomte, V.S. and Shivanikar, S.V., 1999. The distribution pattern of phosphate and nitrogen forms and their inter-relationships in Barul Dam water. *Poll. Res.*, 18(4): 411-414.

Barnes, R.S.K. and Green, J., 1970. The estuarine environment. In: *Chemical Processes in Estuaries*, John Philips (Ed.), Applied Science Pub. Ltd., London.

Bendschneider, K. and Robinson, R.J., 1952. A new spectrophotometric method for the determination of nitrite in seawater. *J. Mar. Res.*, 11: 87-99.

Carpenter, R., 1969. Factors controlling the marine geochemistry of fluoride. *Geochimca et Cosmochima Acta.*, 33: 1153-1167.

Chandran, R. and Ramamorthi, K., 1984. Hydro-biological studies in the gradient zone of the Vellar estuary. I. Nutrients. *Mahasagar-Bull. Natn. Inst. Oceanogr.*, 17: 133-140.

Culkin, F. and Cox, R.A., 1966. Sodium, potassium, magnesium, calcium and strontium in seawater. *Deep-sea Res.*, 13: 789-804.

Das, J., Das, S.N. and Sahoo, R.K., 1997. Semidiurnal variation of some physico-chemical parameters in the Mahanadi estuary, East Coast of India. *Ind. J. Mar. Sci.*, 26: 323-326.

De Sousa, S.N., 1983. Studies on the behaviour of nutrients in the Mandovi estuary during pre-monsoon. *Estuar. Coast. Shelf Sci.*, 16: 299-308.

De Sousa, S.N. and Sen Gupta R., 1988. Distribution and behaviour of fluoride in Mandovi and Zuari estuaries, central west coast of India. *India J. Mar. Sci.*, 17: 195-201.

De Sousa, S.N., Sen Gupta R., Sujata Sanzgiri and Rajagopal, M.D., 1981. Studies on nutrients of Mandovi and Zuari River Systems. *India J. Mar. Sci.*, 10: 314- 321.

Dehardrai, P.V., 1970. Observations on certain environmental features at the Dona Paula Point in Marmagao Bay, Goa. *Proc. Ind. Acad. Sci.*, 52(B): 56.

Dehardrai, P.V. and Bhargava, R.M.S., 1972. Seasonal organic production in relation to environmental features in Mandovi and Zuari estuaries, Goa. *Ind. J. Mar. Sci.*, 1: 52- 56.

Ganapati, P.N. and Subba Rao, D.V., 1958. Qualitative study of plankton off Lawson's, Bay, Waltair. *Proc. of Ind. Asso. of Sci.*, 48(B): 189-209.

Gouda, R. and Panigrahy, R.C., 1992. Seasonal distribution and behaviour of silicate in the Rushikulya estuary, east-coast of India. *Ind. J. Mar. Sci.*, 21: 111-115.

Gouda, R. and Panigrahy, R.C., 1995. Seasonal distribution and behaviour of nitrate and phosphate in Rushikulya estuary east coast of India. *Ind. J. Mar. Sci.*, 24: 233-235.

Grasshoff, K., 1976. *Methods of sea water analysis.* Verlag chemie, Weinhein, 260 pp.

Greenhalgh, R. and Riley, J.P., 1961. Determination of fluoride in natural waters with particular reference to seawater. *Anal. Chim. Acta.*, 25: 179-186.

Harriss, R.C., 1969. Boron regulation in the oceans. *Nature*, 223: 290-291.

Koroleff, F., 1971. Determination of reactive silicate. *ICES paper.* Cm (c) 43 pp.

Legovic, T., Zutic, V., Grzetic, Z., Cauwet, G., Precali, R. and Vilicic, D., 1994. Eutrophication in the Krka estuary. *Marine Chemistry*, 46: 203-215.

Levinson, A.A. and Ludwick, J.C., 1966. Speculation on the incorporation of boron into orgillaceous sediments. *Geochim. Cosmochim. Acta*, 30: 855-861.

Mackey, D.W. and Leatherland, T.M., 1976. Chemical processes in an estuary receiving major inputs of Industrial and domestic wastes. In: *Estuarine chemistry*, J.D. Burton and P.S. Liss (Eds.), Academic Press, London, pp. 185-218.

Mahapatro, T.R. and Padhy, S.N., 2001a. Seasonal fluctuation of physico-chemical parameters of Rushikulya estuary, Bay of Bengal. *Ind. J. Env. and Ecoplan.*, 5(1): 35-40.

Mahapatro, T.R. and Padhy, S.N., 2001b. Behaviour of fluoride in Rushikulya Estuary, Bay of Bengal. *J. of Env. Res.*, 11(2): 141-144.

Mahapatro, T.R., Padhy, S.N. and Patra, S.P., 2001. Assessment of water quality in Rushikulya Estuary. *J. of Indu. Poll. Cont.*, 17(1): 107-112.

Manual of methods in aquatic environment research (1972. FAO Tech. Paper No. 137, FAO, Rome, 145 pp.

Murphy, J. and Riley, J.P., 1962. A modified single solution method for determination of phosphate in natural waters. *Anal. Chim. Acts*, 27: 31-36.

Naik, S., 1978. Calcium and magnesium concentration in the near shore waters of Goa. *Mahasagar-Bull. Natn. Inst. Oceano.*, 11(3&4): 185-189.

Nair, N.B., Abdul Azis, P.K., Dharmaraj, K., Arunchalam, M., Krishna Kumar, K. and Balasubramanian, N.K., 1983. Ecology of Indian estuaries, Part I: Physico-chemical features of water and sediment nutrients of Ashtamudi estuary. *Ind. J. Mar. Sci.*, 12: 143-150.

Nair, N.B., Abdul Azis, P.K., Krishna Kumar, K., Dharmaraj, K. and Arunachalam, M., 1984. Ecology of Indian estuaries Part VI: Physico-chemical conditions in Kadinamkulam backwater, southwest coast of India. *Ind. J. Mar. Sci.*, 13: 69-74.

Narvekar, P.V. and Zingde, M.D., 1987. Behaviour of boron, calcium and magnesium in Puma and Auranga estuaries (Gujarat), west coast of India. *Ind. J. Mar. Sci.*, 16(1): 46- 50.

Narvekar, P.V., Zingde, M.D. and Kamat Dalal, V.N., 1981. Behaviour of boron, calcium and magnesium in Mindola river estuary (Gujarat. *Indian J. Mar. Sci.*, 10: 90- 92.

Narvekar, P.V., Zingde, M.D. and Kamat Dalal, V.N., 1983. Behaviour of boron, calcium and magnesium in a polluted estuary. *Estuar. Coast. Shelf Sci.*, 16: 9-16.

Padmavathi, D. and Satyanarayana, D., 1999. Distribution of nutrients and major elements in riverine, estuarine and adjoining coastal waters of Godavari, Bay of Bengal. *Ind. J. Mar. Sci.*, 28: 345-354.

Palanichamy, S. and Balasubramanian, T., 1989. Distribution of calcium and magnesium in the Vellar estuary. *Mahasagar*, 22(1): 1-11.

Panigrahy, P.K., Das, J., Das, S.N. and Sahoo, R.K., 1999. Evaluation of the influence of various physico-chemical parameters on coastal water quality around Orissa by factor analysis. *Ind. J. Mar. Sci.*, 28: 360-364.

Porrenga, D.H., 1967. Influence of grinding and heating of layer silicates on boron adsorption. *Geochim Coschim Acta.*, 31: 309-312.

Pradhan, S.K., 2000. *Studies on behaviour of nutrients in the Bahuda Estuary (Bay of Bengal.* M.Phil. Thesis, Berhampur University, Berhampur, Orissa, India.

Rajagopal, M.D., Rajendran, A. and Reddy, C.V.G., 1981. Distribution of dissolved boron in the waters of the Zuari estuary (Goa. *Indian J. Mar. Sci.*, 10: 20-23.

Redfield, A.C., 1934. In: *James Johnstone Memorial volume*, Edited by R.J. Daniel. University Press, Liver Pool, p. 176.

Riley, J.P. and Chester, R., 1971. In: *Introduction to marine chemistry*. Academic Press, London and New York.

Ringbom, A., Pensar, G. and Wanninen, E., 1958. A complexometric titration method for determining calcium in the presence of magnesium. *Anal. Chim. Acta*, 19: 525-531.

Sahu, S.K., Mahapatro, T.R. and Padhy, S.N., 2002. Seasonal variations in calcium and magnesium in the Gopalpur creek, east coast of India. In: *Ecology of polluted waters*, Arvind Kumar (Ed.), APH Publishing Corporation, New Delhi, I: 123-130.

Sankaranarayanan, V.N. and Qasim, S.Z., 1969. Nutrients of the Cochin backwaters in relation to environmental characteristics. *Mar. Biol.*, 2: 236-247.

Sarla Devi, K., Venugopal, P., Remani, K.N., Dominic, Z. and Unnithan, R.V., 1983. Nutrients in some estuaries of Kerala. *Mahasagar-Bull. Natn. Inst. Oceanogr.*, 16: 161-173.

Sarma, V.V., Sudhakar, U. and Varaprasad, S.J.D., 1993. Behaviour of fluoride and dissolved silicon in Gauthami Godavari estuarine environment. *Mahasagar-Bull. Natn. Inst. Oceanogr.*, 26: 105-113.

Satpathy, K.K., 1996. Seasonal distribution of nutrients in the coastal waters of Kalpakkam, east coast of India. *Ind. J. Mar. Sci.*, 25: 221-224.

Sen Gupta, R. and Naik, S., 1981. Studies on calcium, magnesium and sulphate in the Mandovi and Zuari River Systems (Goa. *Ind. J. Mar. Sci.*, 10(1): 24- 34.

Sillen, L.G., 1961. *The physical chemistry of sea water*. Lectures at the International Oceanographic Congress in N.Y., 1959, pp. 549-581.

Solarzano, L. and Ehrlich, B., 1975. Chemical investigations of Loch Etive, Scotland, I. Inorganic nutrients and pigments. *J. Expt. Mar. Bio. and Eco.*, 29: 45-64.

Solorzano, L., 1969. Determination of ammonia in natural waters by phenol-hypochlorite method. *Limnol. Oceanogr.*, 14: 799-801.

Tripathy, S.K., Panigrahy, R.C., Gouda, R. and Panda, D., 1990. Distribution of calcium and magnesium in the Rushikulya estuary (Orissa), east coast of India. *Ind. J. Mar. Sc.*, 19: 212-214.

Uthup, A.T., 1996. *Distribution of nutrients in the sediments of Kayamkulam Estuary*. M.Phil. Thesis, Cochin University of Science and Technology, Cochin, India.

Verlencar, Y.N., 1987. Distribution of nutrients in the coastal and estuarine waters of Goa. *Mahasagar-Bull. Natn. Inst. Oceanogr.*, 20: 205-215.

Wilson, T.R.S., 1975. Salinity and the major elements of sea water. In: *Chemical oceanography*, Vol. I, J.P. Riley and G. Skirrow (Eds.), Academic Press, London, pp. 365-413.

Wright, D.A. and Davison, A.W., 1975. The accumulation of fluoride by marine and intertidal animals. *Env. Poll.*, 8: 1-13.

Index

A

Aerobic biodegradation, 82-96
analytical procedure, 84
basal medium, 83-84
chemicals, 83
experimental procedure, 84
organisms, 83
preparation of inoculum, 84
results and discussion, 84-95
chloride release, 95
effect of meat extract and peptone on 4-CP degradation, 93-95
-- yeast extract on 4-CP degradation, 84-89
-- yeast extract on 2-CP degradation, 89-93

Anabaena doliolum, 375-381
introduction, 375-376
materials and methods, 376-377
results and discussion, 377-379

Analysis of weed flora of agricultural fields, 338-347
introduction, 339
weeds distribution in wet field, 346
weeds distribution in dry field, 346-347

Aqueous and aquated organic solvents, 289-301
introduction, 290-291

Auto-gasification, 117-127
introduction, 118
results and discussion, 119-126
effect of environment, 119
-- heating rate, 119-122
-- particle size, 119
kinetics of auto gasification, 122-123
verification of the model, 123

Awareness of environmental, 1-16
checks, balances and incentives, 14
clean technology and energy conservation, 14
coal mining, 3-5
disposal and utilization of flyash generated, 7-8
environmental awareness review, 15-16
– impact of reform in transmission and distribution, 13
– impacts without reform and due to reform, 13-14
– management and monitoring, 8-11
groundwater utilization and degradation, 6
hydel power generation, 13
introduction, 2
linkage between OERC and the state pollution control board, 15

methodology, 2
plant emissions and acid rain potential, 7
quality of life index in mining areas, 11-12
recommendations, 16
response of people around thermal power plants, 12-13
results, 2
thermal power generation, 2-3
Azodyes, 399-415
actinomycete biodegradation, 406
aerobic treatment, 404-406
algal biodegradation, 406
anaerobic treatment, 407
anaerobic/aerobic sequential step-treatment, 407-408
azo dye degradation by fungi, 406
azo dyes and its intermediates, 400-401
biological treatment, 399-415
factors affecting azo dye biodegradation, 408-410
introduction, 399-400
research need, 410-411
toxicity considerations, 401-402
treatment options, 402-404

B

Biodiversity, 97-103
introduction, 97-98
Bud fly incidence, 228-231

C

Calcium gluconate, 184-194
introduction, 184-185
materials and methods
chemicals, 185
isolation of molds, 185
maintenance of cultures, 185
optimization of fermentation condition for calcium gluconate, 186-187
samples collected, 185
selection and screening, 185-186
results and discussion, 187
effect of temperature, 192-193
identification of selected mold strain, 187-188
impact of different concentrations of dextyrose, 190
– – – nitrogen sources, 191-192
optimization of fermentation condition, 188
potential of mold isolates to produce calcium gluconate, 187
Cenchrus ciliaris, 331-337
Cenchrus ciliaris, 332-337
introduction, 332
Centella asiatica, 221-227
callusing response, 222-223
field transfer, 224-226
introduction, 221-222
material and methods, 222
rooting response, 223-224
shooting response, 223
Chilli powders, 152-155
introduction, 152-153
Cirrinus mrigala, 211-215
introduction, 211-212
Class of exponential family of distributions, 145-151
introduction, 145-147
binomial distribution, 148-149
geometric distribution, 150-151
particular cases, 147-148
Coastal belt of Kerela, 216-220
Comparison of energy interaction, 134-144
Composition of vermicast, 104-110
Ctenopharyngodon idella, 128-133
Culex quinquefasciatus, 357-359
Cyanobacterial diversity, 416-450
observations and discussion, 448-449
systematic enumeration, 417

D

Defoliation studies in maize, 451-467
calorific energy distribution in maize plant, 452
contribution of various plant parts in grain formation, 452
culm leaves, 452
cultivar and seasonal response to defoliation, 457-458
defoliation and dry matter accumulation, 455-456
– – phenology, 458-459
– stem reserve, 457
– as a simulator of biotic and abiotic stresses, 459-460
– relative to absorption and distribution of nutrients, 456-457
effect of intensity and timing of defoliation, 454-455
forage yield and quality, 460
grain yield and yield characters, 460-461
husk leaves, 453

introduction, 451-452
tassel, 453-454
Diabetic rats, 174-183
introduction, 175-176
Dynamics of bud fly incidence, 228-231
introduction, 229
results and discussion, 229
degree of bud infestation as affected by varietal reaction, 231
dynamics of the pest incidence, 229-230
relationship of bud fly incidence what weather factors, 230-231

E

Ecofriendly management of solid wastes, 232-240
characteristics and management of pulp industrial solid wastes, 233
composing of industrial solid wastes, 235
decomposition of pulp industrial sludges, 235-236
effect of compost application on crop growth and productivity, 237-238
impact of land application of compost on soil fertility and productivity, 236-237
– – pulp industrial solid wastes on soil properties, 233-234
– – pulp industrial solid wastes on crop productivity, 234
nutrient content of pulp industrial solid wastes, 233
Effect of weed control treatments, 156-159
Environmental concepts and environmental problems, 17-23
environmental management, 20
exposure to environmental problems, 21-23
favourite subjects, 19
general awareness, 19
environmental problems, 19-20
introduction, 17-18
material and methods, 18-19
objectives of the study, 18
results and discussion, 19
sources of knowledge, 20-21
Environmental problems, 17-23
Eucalyptus globules, 199-206
agrochemical features of test crops, 200
introduction, 200
Euglenophyceae of Hatnur dam, 352-356
Evaluation of antimicrobial efficacy, 160-170
experimental section, 162
materials and methods, 161-162
poisoned food technique, 163
results and discussion, 163-169

F

Freshwater lakes of Kolar Taluk, 382-390
materials and methods, 383
results and discussion, 384
bicarbonate, 386
calcium, 386-387
carbonate, 386
chloride, 388
conductivity, 384-385
fluoride, 388-389
hardness, 385-386
iron, 387-388
magnesium, 387
nitrates, 388
pH, 384
potassium, 387
sodium, 387
sulphates, 388
TDS, 386

G

Groundwater quality of the Bharatpur area, 46-52
chloride, 50
colour, 47
dissolved oxygen, 48
electrical conductivity, 51
fluoride, 51
introduction, 47
magnesium hardness, 50
nitrate, 50-51
odour, 47
pH, 48
results and discussion, 47
sulphate, 50
temperature, 47-48
total alkalinity, 48-50
– dissolved solids, 51
– hardness, 50
Gulf of mannar resources, 326-330
discussion, 327-329
introduction, 326-327
materials and methods, 327
results, 327

H

Hatnur dam, 352-356

I

Impact of health education, 364-368

K

Kadinamkulam lake, 62
 introduction, 62-63
Kerela, 216-220
Kolar Taluk, 382-390
Kulachel area of Kanyakumari district, 171-173
 introduction, 171

L

Laisoipat lake, 270-277
Lake of Udaipur city of Rajasthan, 256-263
Larvicidal and ovicidal activity, 111-116
 bioassay, 113
 introduction, 112
 mosquito culture, 112
 plant collection and extraction, 112
 results and discussion, 113-116
Lentic freshwater systems, 546-558
Limnology of lentic freshwater systems, 546-558
 introduction, 546-547
 materials and methods, 547-548
 results and discussion, 548
  climate and vegetation, 548-552
  physico-chemical properties of water, 552-557
 statistical analysis, 557

M

Maize, 451-467
Management of sucking insect pests, 246-251
 materials and methods, 247-248
 results and discussion, 248-249
Mangroves, 302-305
 conservation and management, 305
 degradation and destruction of mangroves, 304-305
 distribution, 303
 ecological importance of mangroves, 303-304
 introduction, 302-303
 values of mangroves, 304
Market potential for two wheelers, 39-45
 cyclical variations, 42-43
 database, 40
 irregular variations, 41-42
 model description, 40
 secular trend and growth rate, 43-44
 time series approach to sales forecasting, 41
 variability in sales with respect to time series components, 44
Molecular interactions, 369-374
 introduction, 370
Mulches on soil properties, 391-398
 crop productivity, 395-397
 design and layout of experiment, 392
 introduction, 391-392
 results and discussion, 392
 soil moisture retention, 392-393
 soil properties, 393-395
 study site, 392

O

Ovicidal activity, 111-116

P

Pennisetum glaucum, 264-269
Pennisetum glaucum: introduction, 265
Pericallia ricini, 195-198
Pericallia ricini: introduction, 195-196
Post-harvest soil fertility status, 311-319
 introduction, 312
 nitrogen use efficiency, 315
  experimental, 47
 quality of the Bharatpur area
  calcium hardness, 50
  carbonate and non-carbonate hardness, 50

R

Ratio of weeds, 207-210
 introduction, 208
 materials and methods, 208
 results and discussion, 208-210
River Yamuna, 472

S

Shigometric studies, 360-363
 introduction, 360-361
Silvopastoral systems, 53-61
 crown width vs. pasture production, 57, 59-60
 data measurement, 55-56
 introduction, 54-55
 materials and methods, 55-57
 pasture yield, 57-58
 site characteristics, 55
 tree growth and production models, 57
 trunk diameter growth model, 56
Solanum trilobatum, 306-310
Sorghum-wheat sequence, 278-288
Spodoptera litura, 348-351
 dry pod yield, 350
 foilage damage, 349
 introduction, 348-349

Status of insect pests, 252-255
  incidence of natural enemies of pest fauna, 255
  introduction, 253
  pest incidence and their succession, 253-255
Structure and optical properties, 320-325
  introduction, 320-321
Studies on selected major elements and nutrients, 559-576
  ammonium, 567-568
  boron, 570
  calcium, 570-571
  chlorinity, 563
  fluoride, 568-570
  introduction, 560-561
  magnesium, 571-573
  materials and methods, 561-563
  nitrate, 563-564
  phosphate, 564-565
  silicate, 565-567

T

Temple town in South India, 70-81
  biochemical oxygen demand, 76
  chemical oxygen demand, 76
  chlorides, 75-76
  DO, 75
  elesctrical conductivity, 73-74
  pH, 73
  temperature, 73
  total coliforms, 76-80
Tiruchirappalli, 241-245
  materials and methods, 242
  results, 242-243

V

Vehicular pollution, 24-38
  approaches and action plan, 25
  CNG supply network, 26-29
  constraints, 34
  conversion of vehicle fleet to CNG mode, 29-30
  effect of abetment and control measures on ambient air, 35
  emission warranty, 33
  future plans of implementing agencies, 35
  implementing agencies, 25-26
  improvement in fuel quality, 32
  -- vehicle technology, 31-32
  inspection and maintenance center, 34
  other measures, 34
  phasing out of old vehicles, 31
  premixed 2T oil dispensers, 33
  prevention of fuel adulteration, 32-33
  provision of new ISBTs, 33-34
Vermicast, 104-110

W

Wastewater quality of major drains, 468-545
  alkalinity, 501-503
  ammonical nitrogen, 495-497
  biochemical oxygen demand, 490-492
  cadmium, 517-518
  calcium, 507-509
  chemical oxygen demand, 488-490
  chloride, 503-505
  chromium, 518-520
  climate, 470
  copper, 520-523
  dissolved oxygen, 485-486
  endosulfan of drain and river water, 531-533
  environmental status of Delhi, 469-470
  heavy metals in water, 515
  importance of river Yamuna, 472
  industries and industrial wastewater, 471-472
  introduction, 468-469
  magnesium, 509-511
  methodology, 472
    selection of sampling points, 472
    water and wastewater sampling, 472-473
  microbial assessment of river water, 533
  nickel, 524-526
  oxygen absorbance value, 486-488
  pesticides, 527-529
  physico-chemical characteristic, 476-478
  population, 470
  potassium of drain and river water, 513-515
  results and discussion, 539-540
  sewerage system, 470
  sodium, 511-513
  total DDT, 529-531
  – dissolved solids, 483-485
  – Kjeldhal nitrogen, 492-493
  – suspended solids, 481-483
  water quality index, 535-536
  -- parameters, 536-537
  water supply, 470
  zinc, 523-524